Algorithmic and Computational Robotics: New Directions

Algorithmic and Computational Robotics: New Directions

The Fourth Workshop on the Algorithmic Foundations of Robotics

edited by

Bruce Randall Donald
Dartmouth College

Kevin M. Lynch
Northwestern University

Daniela Rus
Dartmouth College

CRC Press is an imprint of the
Taylor & Francis Group, an **informa** business
AN A K PETERS BOOK

2000
The Workshop on the Algorithmic Foundations of Robotics

First published 2001 by A K Peters, Ltd.

Published 2018 by CRC Press
Taylor & Francis Group
6000 Broken Sound Parkway NW, Suite 300
Boca Raton, FL 33487-2742

First issued in paperback 2019

ISBN 13: 978-0-367-44726-7 (pbk)
ISBN 13: 978-1-56881-125-3 (hbk)

Visit the Taylor & Francis Web site at
http://www.taylorandfrancis.com

and the CRC Press Web site at
http://www.crcpress.com

Library of Congress Cataloging-in-Publication Data

Workshop on the Algorithmic Foundations of Robotics (4th : 2000 : Dartmouth College)
Algorithmic and computational robotics : new directions : the fourth Workshop on the Algorithmic Foundations of Robotics / edited by Bruce Donald, Kevin Lynch, Daniela Rus.
p. cm.
Workshop held March 16-18, 2000, Dartmouth College.
Includes bibliographical references and index.
ISBN 1-56881-125-X (alk. paper)
1. Robotics--Congresses. 2. Algorithms--Congresses. I. Donald, Bruce R. II. Lynch, Kevin (Kevin M.) III. Rus, Daniela. IV. Title.

TJ210.3 .W664 2000
629.8'92--dc21 00-049355

Cover illustration: Mechanical drawing of Leonardo da Vinci.
From Leonardo da Vinci, Cod. Atl., fol. 357 r-a.Reynal and Company, New York, p.498. Copyright in Italy by the Istituto Geografico De Agostini – Novara – 1956.

Contents

Foreword

Robot algorithms are abstractions of computational processes that control or reason about motion and perception in the physical world. The computation may be implemented in software, hard-wired electronics, biomolecular mechanisms, or purely mechanical devices. Because actions in the physical world are subject to physical laws and geometric constraints, the design and analysis of robot algorithms raises a unique combination of questions in control theory, computational and differential geometry, and computer science. Algorithms serve as a unifying theme in the multi-disciplinary field of robotics.

The Fourth International Workshop on the Algorithmic Foundations of Robotics (WAFR) brought together a group of about sixty researchers to discuss recent trends and important future directions of research on the algorithmic foundations of robotics. Held at Dartmouth College on March 16-18, 2000, the workshop was chaired by Bruce Randall Donald (Dartmouth), Kevin Lynch (Northwestern), and Daniela Rus (Dartmouth). The workshop consisted of six invited talks and twenty-four contributed presentations in a single track. Each paper was rigorously refereed by the program chairs plus at least three members of the program committee. The program committee consisted of the conference chairs plus Pankaj Agarwal (Duke), Srinivas Akella (Rensselaer Polytechnic Institute), Nancy Amato (Texas A&M), Antonio Bicchi (University of Pisa), Kamal Kant Gupta (Simon Fraser), Leslie Kaelbling (MIT), Makoto Kaneko (University of Hiroshima), David Kriegman (UIUC), Steve LaValle (Iowa State), Dinesh Manocha (UNC), Jim Ostrowski (University of Pennsylvania), John Reif (Duke), and Elisha Sacks (Purdue).

Topics reported at WAFR 2000 included geometric algorithms, minimalist and underactuated robotics, robot controllability, manufacturing and assembly, holonomic and nonholonomic motion planning, manipulation planning, sensor-based planning, task-level planning, grasping, navigation, biomimetics, medical robotics, self-assembly, modular and reconfigurable robots, and distributed manipulation. This book contains the proceedings of WAFR 2000. A number of these papers are clearly destined to be landmarks in the field. For example, the paper "Meso-Scale Self-Assembly" by Gracias, Choi, Weck, and Whitesides was voted number three in the "Top Five Must Read Papers Recommended by the Carnegie Mellon School of Computer Science Faculty" for the year 2000. It is worth noting that numbers one and two were each written more than 40 years ago, and "Meso-Scale Self-Assembly" was voted ahead of number four, Judge Jackson's findings of fact against Microsoft. This demonstrates the value of interdisciplinary interaction at workshops such as WAFR.

We are very grateful to Dartmouth College and Sandia National Laboratories for their generous financial support of WAFR. We would like to thank all participants for their contributions. We thank Dartmouth Project Assistant David Bellows for his work in organizing WAFR and in typesetting the book and Fred Henle for his advice and assistance with LaTeX and image enhancement.

Bruce Randall Donald

Kevin M. Lynch

Daniela Rus

June, 2000

Participants

Pankaj Agarwal	Duke University
Srinivas Akella	Rensselaer Polytechnic Institute
Nancy Amato	Texas A&M
Boris Aronov	Polytechnic University
Devin Balkcom	Carnegie Mellon University
Matthew Berkemeier	Utah State University
Antonio Bicchi	University of Pisa
Amy Briggs	Middlebury College
Herve Bronimann	Polytechnic University
Joel Burdick	California Institute of Technology
Zack Butler	Carnegie Mellon University
Ming C. Lin	University of North Carolina Chapel Hill
Greg Chirikjian	Johns Hopkins University
Anne Collins	Duke University
Bruce Randall Donald	Dartmouth College
Michael Erdmann	Carnegie Mellon University
Robert Ghrist	Georgia Institute of Tech.
Ken Goldberg	University of California Berkeley
W. Eric L. Grimson	MIT
Kamal Gupta	Simon Fraser University
John H. Reif	Duke University
Dan Halperin	Tel Aviv University
Li Han	Texas A&M
John Hollerbach	University of Utah
David Hsu	Stanford University
Wesley Huang	Rensselaer Polytechnic Institute
Seth Hutchinson	University of Illinois
Leonidas J. Guibas	Stanford University
Xuerong Ji	University of North Carolina Charlotte
Satoshi Kagami	University of Tokyo
Fumio Kanehiro	University of Tokyo
Makoto Kaneko	Hiroshima University
Lydia Kavraki	Rice University
Eric Klavins	University of Michigan
Daniel Koditschek	University of Michigan
James Kuffner	University of Tokyo
Steven LaValle	Iowa State University

Florent Lamiraux	LAAS-CNRS
Jean-Claude Latombe	Stanford University
Jean-Paul Laumond	LAAS-CNRS
Kevin M. Lynch	Northwestern University
Dinesh Manocha	University of North Carolina Chapel Hill
Matt Mason	Carnegie Mellon University
William Messner	Carnegie Mellon University
Mark Moll	Carnegie Mellon University
An Nguyen	Stanford University
Jim Ostrowski	University of Pennsylvania
Elon Rimon	Technion University Israel
Alfred Rizzi	Carnegie Mellon University
Daniela Rus	Dartmouth College
Elisha Sacks	Purdue University
Daniel Scharstein	Middlebury College
Edward Scheinerman	Johns Hopkins University
A. Frank van der Stappen	Utrecht University
Ileana Streinu	Smith College
Robert Sun	Duke University
George Whitesides	Harvard University
Jing Xiao	University of North Carolina Charlotte
Mark Yim	Xerox PARC
Tao (Mike) Zhang	University of California Berkeley
Li Zhang	Stanford University
Yan Zhuang	University of California Berkeley

Meso-Scale Self-Assembly

David H. Gracias, *Harvard University, Cambridge, MA*
Insung Choi, *Harvard University, Cambridge, MA*
Marcus Weck, *Harvard University, Cambridge, MA*
George M. Whitesides, *Harvard University, Cambridge, MA*

This paper describes the application of self-assembly to the formation of structured aggregates (both static and dynamic) of millimeter sized objects. We argue that MEso-scale Self-Assembly (MESA) provides a framework for fabricating two- and three- dimensional structures (with objects ranging in size between 10 nm and 10 mm), and illustrates broadly important principles underlying the behavior of complex, natural systems.

1 Introduction

Self-Assembly is the spontaneous organization of molecules or objects, under steady state or equilibrium conditions, into stable aggregates, by non-covalent forces; these aggregates are not necessarily at the global minimum in energy [14, 15]. Meso-Scale Self-Assembly (MESA) refers to self assembly carried out with objects ranging in size from 10 nm to 10 mm using forces having lateral extension with similar size [8, 6, 1, 9, 12, 17, 13, 7, 4, 11]. MESA bridges the gap between molecular self-assembly (e.g. crystallization [16]; and protein-ligand recognition [10]) and the conventional fabrication of macroscopic machines. This paper briefly outlines the most important concepts governing self-assembly at these length scales. It also describes the application of MESA in (a) fabricating artificial crystals and three dimensional structures [9, 12, 17], (b) mimicking molecular self assembly [4, 11, 3, 5], and (c) designing dynamical systems that show tailored complexity [2].

2 Forces

We begin by describing two static systems; in these systems, the capillary interactions between individual units are responsible for MESA. (i) One system uses MESA to form 2D aggregates. It employs selective patterning of the faces of the object into hydrophilic and hydrophobic sets. These objects self assemble, while floating at the interface between a hydrophobic liquid (perfluorodecalin, PFD) and a hydrophilic one (water) [6, 1]. In this system, PFD wets the hydrophobic faces and forms "positive" menisci (menisci that extend above the plane of the interface); water wets the hydrophilic faces and forms "negative" menisci (menisci that extend below the plane). Menisci of the same shape interact attractively with each other, while a mismatch in the shape of the menisci results in weak or repulsive interactions between the objects. Since capillary forces have decay lengths on the order of millimeters to nanometers, they can be used to assemble objects across this range of sizes. (ii) The second system uses MESA to form three-dimensional (3D) aggregates. In this system liquid films (e.g. of low melting-point alloys [9], or polymeric adhesives [12]) coat the faces of the assembling units. These films have a high interfacial free energy in contact with water. When two faces come into contact, the films coalesce; this coalescence minimizes the exposed area of the interface and decreases the free energy of the system.

We also describe a system based on dynamical self-assembly [2] that incorporates a combination of attractive and repulsive forces. In this system, millimeter-sized magnetic disks are placed at a liquid-air interface, and subjected to the magnetic field produced by a rotating permanent magnet. The attractive forces occur between the magnetic units and the rotating magnet. The repulsive forces are due to hydrodynamic interactions associated with vortices in the fluid in the vicinity of the individual spinning magnets.

3 Two-Dimensional MESA

We assembled two-dimensional (2D) aggregates using pieces made of PDMS, an organic polymer: polydimethylsiloxane; (density = 1.05 g/mL) at the PFD (density = 1.91 g/mL) /water (density = 1.00 g/mL)

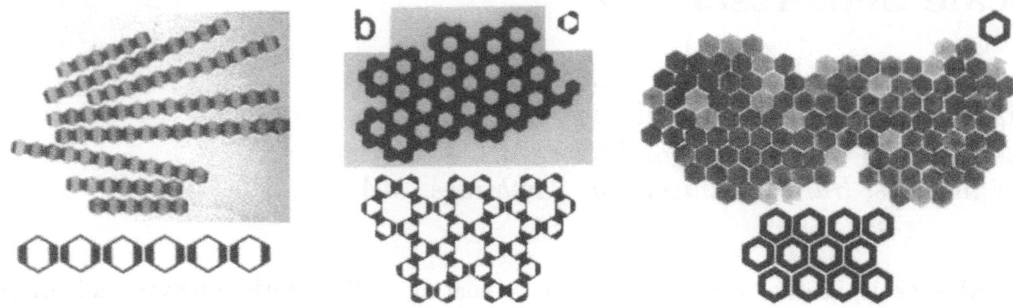

Figure 1: *Photographs of self-assembled arrays of hydrophobically patterned hexagons, floating in the plane of the PFD / water interface, and agitated by an orbital shaker. All the hexagons have centrosymmetric patterns of hydrophobic faces; the dark faces are hydrophobic, while the light ones are hydrophilic. (A) Two faces (opposite) of individual hexagons are hydrophobic; this pattern results in a linear array on assembly. (B) Alternate faces of individual hexagons are hydrophobic; this pattern generates a porous 2D array. (C) All the faces of individual hexagons are hydrophobic; this pattern generates a 2D, close-packed array.*

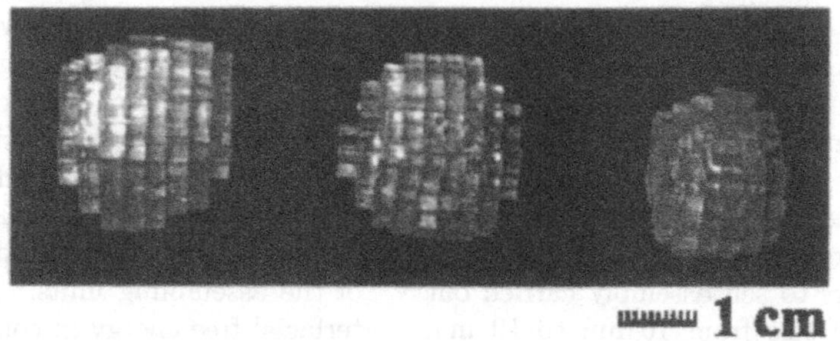

Figure 2: *3D assemblies containing 300 cubes. All the faces of the individual cubes were coated with a hydrophobic polymer; assembly was carried out in water.*

interface [6, 1, 7, 4, 11]. The PDMS objects assembled upon agitation using an orbital shaker. We chose this interface for four reasons: (a) PDMS does not swell in contact with either liquid; (b) capillary forces at the interface are strong (interfacial energy=0.05 J/m^2); (c) PDMS has a density intermediate between that of water and PFD, and the pieces therefore float at the interface; (d) a thin film of PFD remains between the faces of the objects when they assemble, and acts as a lubricant; this film allows the objects to move laterally relative to one another, and minimizes the free energy of the system. Figure 1 shows photographs of aggregates formed with different hexagonal units at the PFD/water interface.

4 Three-Dimensional MESA

MESA also provides an alternate strategy for the fabrication of 3D microstructures. We assembled large, regular arrays (crystals) of millimeter sized polymeric polyhedra. The faces of these polyhedra were coated with either a photcurable adhesive [12] or molten solder [9]. When these liquid-coated polyhedra were suspended in water and agitated, collisions between them allowed contact and coalescence of the liquid films; this coalescence minimized the interfacial area between the hydrophobic liquid and water, and provided the thermodynamic driving force for the self-assembly. In experiments using pieces coated with photocurable adhesive, after self-assembly was complete, the structures were exposed to UV radiation; this exposure cured the lubricant and gave the aggregates sufficient mechanical strength that they could be manipulated and characterized. Figure 2 shows assembled 3D structures that contain 300 cubes; these structures formed from suspensions of 1000 cubes, using a photocurable adhesive. Figure 3 shows 3D porous structures formed by the as-

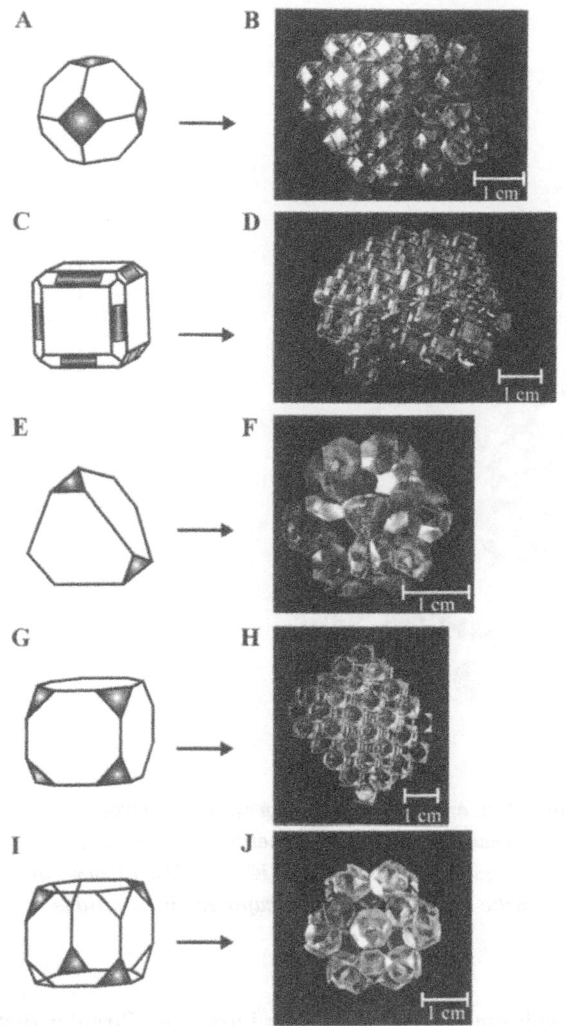

Figure 3: ***3D self-assembly of porous structures using a high surface tension, low-melting point solder. Polyhedra on the left, (dark regions represent areas covered with solder) assembled into the structures shown on the right.***

sembly of polyhedra whose faces were covered with a solder having high interfacial tension with water.

5 Molecule-Mimetic Self-Assembly: Hierarchical and Templated MESA

One of the most interesting applications of MESA is as a vehicle with which to abstract and illustrate important concepts in self-assembly modeled on those exhibited by molecules and to use them to join mesoscale objects. Examples of processes that have been abstracted from organic and biological chemistry and used in MESA include hierarchical [4] and templated [3] self-assembly, and DNA recognition [5]. We describe results of these experiments carried out at the PFD / water interface through the interaction of hydrophilic and hydrophobic menisci as described earlier.

The optical photographs in Figure 4 illustrate the translation of the molecular concepts of hierarchical and templated self-assembly into the mesoscopic world. The results shown in Figure 4A demonstrate hierarchical self-assembly. Two interactions of different strengths were responsible for the self-assembly. Initially, a strong capillary interaction between hydrophobic faces formed structures containing four components; a weaker interaction between the concave hydrophobic faces subsequently caused aggregation into larger structures. Figure 4B illustrates templated MESA. These objects interacted among themselves through concave hydrophobic regions; the self-assembly was directed using round, hydrophilic objects (the templates). On addition of the templating structures, the system formed the inclusion structures shown in Figure 4B; inclusion of these templates strengthened and extended the hexagonal lattice in these structures. We observed that the formation of the cyclic structures began with the pre-orientation of the trefoil objects around the circular template, followed by the self-assembly of these trefoil objects.

6 Biomimetic MESA: Models of DNA Recognition

Figure 5A, B, and C are photographs illustrating a model of sequence-specific assembly: this model is based on sequence specific molecular recognition in DNA and RNA. In particular it is based on a long strand, a specific sequence of which can be recognized by a short probe with a complementary sequence. These strands are based on individual objects containing hydrophobic lock and key functionalities that are connected via a PDMS thread. The two strands associated by pairing at the complementary sequences, with probabilities that clearly illustrate the principle, but are much lower than corresponding molecular systems.

This process is the first, primitive step toward modeling the formation of double-stranded DNA by the

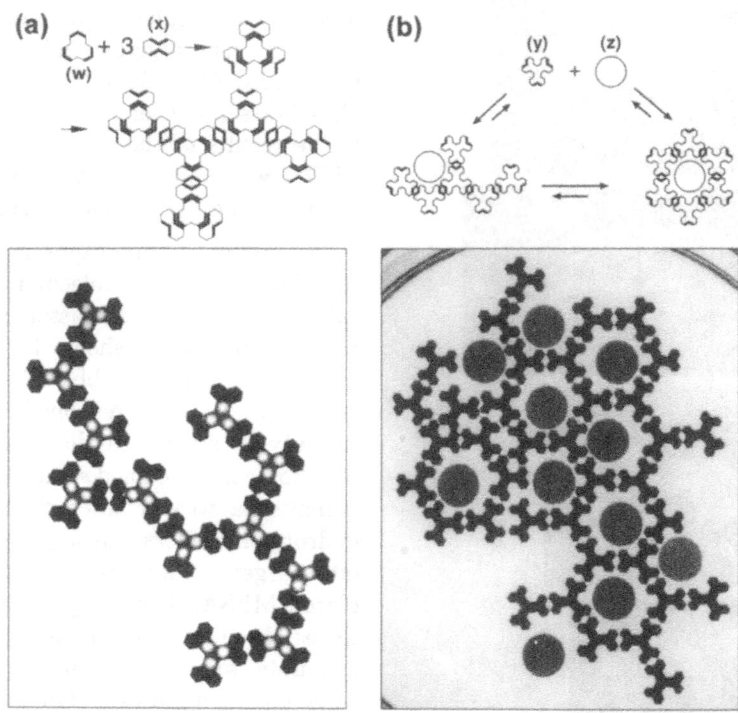

Figure 4: *(A) Hierarchical MESA in 2D. Initially four pieces (one (w) and three (x)) aggregate via strong capillary interactions at the hydrophobically patterned faces (dark regions). A weaker capillary interaction between two concave hydrophobic faces enables these small aggregates to grow into larger structures. (B) Templated MESA in 2D. Eleven circular pieces (z), act as templates around which 37 pieces (y), assemble into cyclic hexamers through capillary interactions at the hydrophobic patterned faces (dark regions).*

association of short sequences of single-stranded DNA. Although this model has obvious limitations, it contains several features analogous to molecular systems, including cooperativity, directionality, and sequence specificity.

7 Dynamic Self-Assembly

We have also examined dynamic self-assembly systems, that is, systems that develop order only when dissipating energy. These systems are particularly interesting for their possible relevance to issues of complexity and emergence and ultimately, perhaps to understanding very complex systems, such as the cell. Here we describe the formation of dynamic patterns of unexpected regularity and complexity by millimeter-sized magnetic disks subjected to a rotating magnetic field and spinning at the liquid-air interface. Circular disks made up of PDMS doped with magnetite (1 mm inner diameter, 400 μm thick) were placed in a dish filled with a liquid (typically ethylene glycol/water or glycerin/water with kinematic viscosities 1-50 cp). A bar magnet (5.6 cm x 4 cm x 1 cm) was then rotated below the dish (2.4 cm below the air-liquid interface), at an angular velocity of 200-1200 r.p.m. Figure 6 is a schematic diagram of the experimental set-up. Figure 7 shows the dynamic patterns formed with different numbers of disks. A variety of patterns were observed, ranging from unnucleated (n<5) to nucleated (n>5) to bistable patterns (n=10, 12). The wealth of phenomena displayed by this system makes it a good one to study complexity: the interactions involved are simple enough to be studied in detail but the patterns are far from obvious.

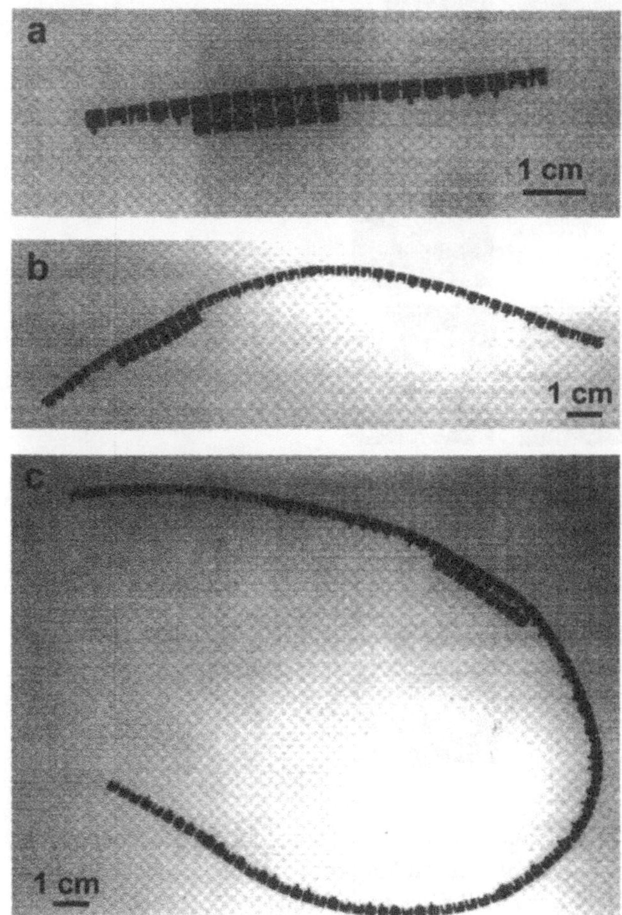

Figure 5: *Examples of a two-dimensional model for sequence-specific self-assembly. In all three cases, a long strand and a short probe were placed at the water/PFD interface and swirled at a frequency of 60 r.p.m. on an orbital shaker. These objects interacted through hydrophobic lock and key structures. Self-assembly between the long strand and the probe at a pre-determined recognition site took place with high probability, in less than two hours. Figures show the self-assembled structures based on sequence-specific recognition between (a) a 22-membered single strand and a seven-membered probe, (b) a larger 50-membered single strand and an eight-membered probe, and (c) the largest single strand examined (100-membered) and a nine-membered probe.*

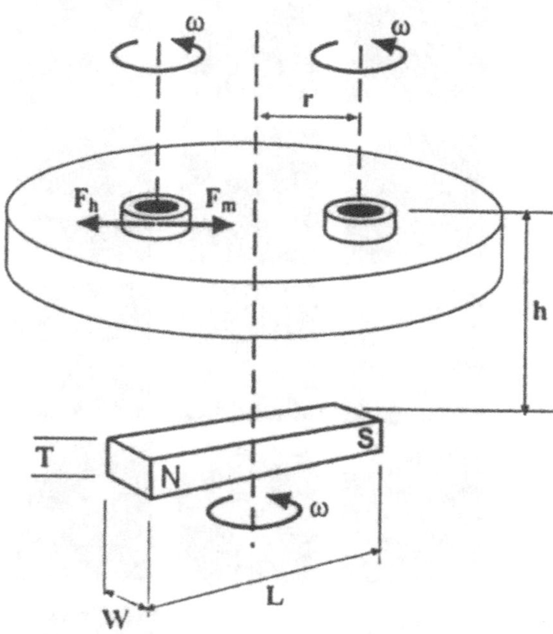

Figure 6: *The scheme of the experimental setup for the dynamic self-assembly. A bar magnet rotates below a dish filled with liquid. Magnetically doped disks are placed on the interface, and are fully immersed in the liquid except for their top surface. The disks spin around their axes. A magnetic force (F_m) attracts the disks towards the center of the dish, and hydrodynamic force (F_h) pushes them apart from each other.*

Acknowledgments

Financial support was provided in part by the National Science Foundation (grant CHE-9801358 and ECS-9729405), the National Institutes of Health (grant GM 30367), and DARPA. M.W. thanks the German Academic Exchange Service (DAAD) for a postdoctoral fellowship.

References

[1] A. Terfort, N. Bowden, and G. M. Whitesides. Three-dimensional self-assembly of millimeter-scale components. *Nature*, 386:162–164, 1997.

[2] B. A. Grzybowski, H. A. Stone and G. M. Whitesides. Dynamic self-assembly of magnetized, millimeter-sized objects rotating at the liquid-air interface. *Submitted.*

[3] I. S. Choi, M. Weck, B. Xu, N. L. Jeon and G. M. Whitesides. Mesoscopic templated self-assembly at the fluid-fluid interface. *Langmuir.* (In Press).

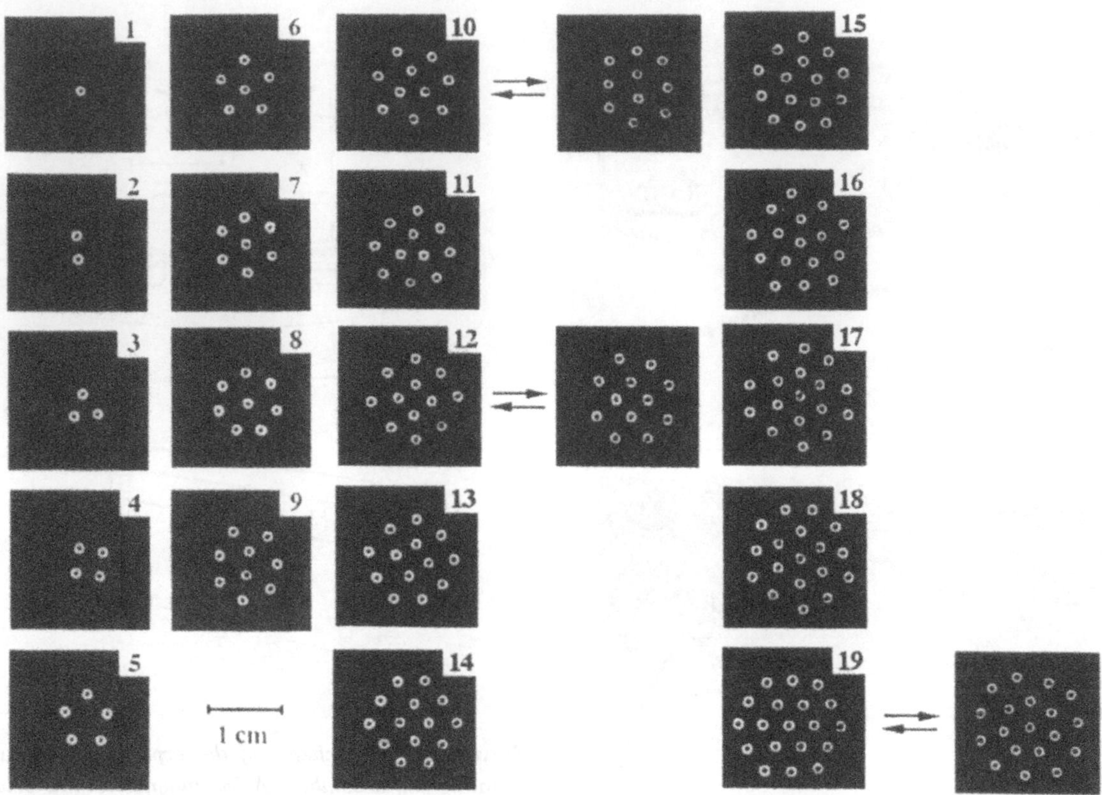

Figure 7: *The dynamic patterns formed by various numbers (n) of disks rotating at the ethylene glycol-air interface, 27 mm above the plane of the external magnet. The disks are composed of polyethylene shell (white) of outer diameter 1.27 mm, filled with PDMS doped with 25 % w/w of magnetite (black core). All disks spin around their centers at ω=700 r.p.m, and the entire aggregate slowly (<2 r.p.m.) precesses around its center. For n<5, the aggregates do not have a "nucleus"-all disks are precessing on the rim of a circle. For n>5, nucleated structures appear. For n=10 and n=12, the patterns are bistable in the sense that the two observed patterns interconvert irregularly with time. For n=19, the hexagonal pattern (left) appears only above ω 800 r.p.m., but can be "annealed" down to 700 r.p.m. by slowly decreasing the spinning rate. Without annealing, a less symmetric pattern exists at ω=700 r.p.m.*

[4] I.S. Choi, N. Bowden and G. M. Whitesides. Macroscopic, hierarchical, two-dimensional self-assembly. *Angewandte Chemie International Edition*, 38:3078–3081, 1999.

[5] M. Weck, I. S. Choi, N. L. Jeon and G. M. Whitesides. Assembly of mesoscopic analogs of nucleic acids. *Journal of the American Chemical Society.* (In Press).

[6] N. Bowden, A. Terfort, J. Carbeck and G. M. Whitesides. Self-assembly of mesoscale objects into ordered two-dimensional arrays. *Science*, 276:233–235, 1997.

[7] N. Bowden, I. S. Choi, B. A. Grzybowski and G. M. Whitesides. Mesoscale self-assembly of hexagonal plates using lateral capillary forces: Synthesis using the "capillary bond". *Journal of the American Chemical Society*, 121:5373–5391, 1999.

[8] P. A. Kralchevsky and K. Nagayama. Capillary forces between colloidal particles. *Langmuir*, 10:23–36, 1994.

[9] T. L. Breen, J. Tien, S. R. J. Oliver, T. Hadzic and G. M. Whitesides. Design and self-assembly of open, regular, 3d mesostructures. *Science*, 284:948–951, 1999.

[10] M. M. C. A. L. Lehninger, D. L. Nelson. . *Principles of Biochemistry 2nd Ed.* Worth Publishers, 1993.

[11] I. S. Choi, N. Bowden and G. M. Whitesides. Shape-selective recognition and self-assembly of mm-scale components. *Journal of the American Chemical Society*, 121:1754–1755, 1999.

[12] T. L. B. J. Tien and G. M. Whitesides. Crystallization of millimeter-scale objects with use of capillary forces. *Journal of the American Chemical Society* , 120:12670–12671, 1998.

[13] N. B. Y. X. L. Isaacs, D. N. Chin and G. M. Whitesides. *Self-assembling Systems on Scales from Nanometers to Millimeters: Design and Discovery*. D. N. Reinhoudt, John Wiley and Sons, New York, 1999.

[14] J.-M. Lehn. *Supramolecular Chemistry: Concepts and Perspectives: A Personal Account.* VCH: Weiheim, 1995.

[15] M. V. M. M. Mrksich and G. M. Whitesides. *Using Self-Assembled Monolayers to Study the Interactions of Man-Made Materials with Proteins; 2nd ed.* Landes Co.: Austin, TX, 1997.

[16] J. W. Mullin. *Crystallization*. London, Butterworths, 1961.

[17] W. T. S. Huck, J. Tien and G. M. Whitesides. Three-dimensional mesoscale self-assembly. *Journal of the American Chemical Society*, 120:8267–8268, 1998.

Robust Geometric Computing in Motion

Dan Halperin, *Tel Aviv University, Israel*

Transforming a geometric algorithm into an effective computer program is a difficult task. This transformation is made particularly hard by the basic assumptions of most theoretical geometric algorithms concerning complexity measures and (more crucially) the handling of robustness issues. The paper starts with a discussion of the gap between the theory and practice of geometric algorithms, together with a brief review of existing solutions to some of the problems that this dichotomy brings about.

We then turn to an overview of the CGAL *project and library. The* CGAL *project is a joint effort by a number of research groups in Europe and Israel to produce a robust software library of geometric algorithms and data structures. The library is now available for use with significant functionality. We describe the main goals and results of the project.*

The central part of the paper is devoted to arrangements (i.e., space subdivisions induced by geometric objects) and motion planning. We concentrate on the maps and arrangements part of the CGAL *library. Then we describe two packages developed on top of* CGAL *for constructing robust geometric primitives for motion algorithms.*

1 Introduction

For over two decades, research in Computational Geometry has yielded numerous efficient algorithms and data structures for solving problems arising in a diversity of areas from statistics and chemistry to GIS and robotics [42],[54],[68]. Whereas the original focus of this field was almost exclusively theoretical, these days a lot of attention is given to practical solutions and their software implementation. Transforming a theoretical geometric algorithm into an effective running program is a hard task. The difficulties in this process are rooted in the assumptions of the theoretical study concerning complexity measures, the arithmetic model and the treatment of degenerate input.

In this paper we discuss the gap between the theory and practice of geometric algorithms. We then describe efforts to settle this gap and facilitate the successful implementation of geometric algorithms in general and of algorithms for geometric arrangements and motion planning in particular.

Most of the algorithms developed in Computational Geometry aim for efficient worst-case asymptotic running time and storage. The computational model used is the so-called *real* RAM model that allows for infinite precision arithmetic operations [64]. Every operation on a (small) constant number of simple geometric objects (such as line segments, circles, planes) takes 'unit' time. In addition the input to the algorithm is assumed to be in *general position*, namely degenerate configurations are excluded. For example, no three of a set of input points may lie on a single line. We collectively refer to the issues of arithmetic precision and the treatment of degenerate input as *robustness issues*.

Each of the assumptions we have just mentioned needs to be revised in practice. Some of the problems are not unique to geometric algorithms, such as the fact that the standard asymptotic measures of algorithm performance may hide prohibitively large constants. This pitfall has been observed many times in geometry, for example in the context of range searching [37], vertical decomposition [50], and construction of Minkowski sums [4]; the latter example will be described in more detail in the sequel.

Sometimes the worst-case resource bounds are deterringly high since they cover the treatment of even the most pathological non-realistic input instances. This has led to the study of realistic input models (a.k.a. "fatness"). Making additional assumptions on the input (based on the specific nature of a problem at hand) and tailoring the algorithms accordingly can result in much more efficient algorithms that are often simpler [5],[27],[48],[76],[77].

The unit-cost model for operations on a constant number of simple geometric objects is also questionable, especially when we aim for robust computation. Already in two-dimensions and for collections of simple algebraic curves (e.g., circular arcs), algorithms need to determine the sign of a polynomial of a rather high degree [14]. Carrying out one such operation robustly may be costly. Redesigning algorithms so that they use lower-degree predicates was recently proposed as a means to enhance robustness [13],[14].

Indeed, robustness issues seem to be the most critical in the passage from theory to practice in geometric algorithms. Ignoring these issues can result in unreliable or incorrect programs. This has led to an intensive study in recent years. We briefly mention the main approaches to handling robustness issues next and refer the reader to recent surveys on the topic [69],[79] for further information.

Let us illustrate the problem of arithmetic precision in geometric computing. Consider the two polygons on the left-hand side of Figure 1: does the small polygon fit inside the cavity in the larger polygon? We find the answer by computing the Minkowski sum of the larger polygon and a copy of the smaller polygon rotated by 180° degrees (the details are deferred to Section 4.1). Suppose the answer is yes as the figure implies. To give a correct answer we must be able to identify a singular point on the boundary of the Minkowski sum (in the middle of the sum on the right-hand side of Figure 1) which is the intersection point of many line segments. The standardly available computer arithmetic and number types could not in general give such answers.

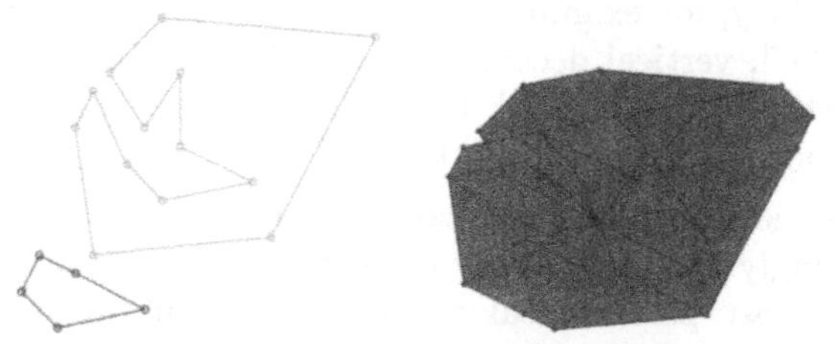

Figure 1: *The small polygon can fit into the cavity in the larger polygon (left-hand side) as indicated by a singular point in the middle of the Minkowski sum of the larger polygon with a rotated copy of the small one (right-hand side).*

A solution is given by using *exact arithmetic* (see, e.g., [17],[18],[20],[25],[80]). We can use rational numbers and maintain the numerator and denominator as unlimited length integers (supported by several software packages). Precise predicates are also supported when they involve square roots [30] or even taking the k-th root [19],[59].

Exact arithmetic predicates can be rather time consuming. There are several adaptive evaluation schemes that save computing time by resorting to expensive computations only when they cannot determine the correct answer by simpler means like standard computer arithmetic. One form of adaptive evaluation is a *floating-point filter* which makes use of the hardware-supported floating-point arithmetic at the initial stage. For the use of such techniques in exact geometric computing see [29],[40],[60],[74]. The packages described in Sections 3 and 4 make extensive use of LEDA's filters [60] to speed up their running time; see for example the experimental results reported in [39] for a comparison of exact arithmetic with or without filters.

If we are not using exact arithmetic, precision problems become especially noticeable at or near degeneracies. Burnikel et al. propose to handle degeneracies directly [21] for certain two-dimensional problems. In three- and higher-dimensional problems, however, directly handling degeneracies is an extremely tedious and error-prone task. Symbolic perturbation was suggested as a general means to overcome degeneracies [32],[34],[78] provided that exact arithmetic is available. If one insists on using only fixed-precision arithmetic then several methods that approximate the geometric objects were proposed to guarantee robustness and/or remove degeneracies; see, for example, [43],[50],[53],[61],[75].

Because of all the difficulties in implementing geometric algorithms (as discussed above) several computational-geometry groups have decided to put up a carefully designed and implemented library with an emphasis on robustness and generality [63]. This resulted in CGAL: The Computational Geometry Algorithms Library.

The CGAL project[1] officially started in September 1996. The actual work on the library's kernel started

[1]The CGAL project and its successor project GALIA were both funded by the European Union. Throughout the paper we refer to both projects as the CGAL project.

earlier [35] and drew on the experience from still earlier projects including: the geometric part of LEDA [59], C++gal [2], PlaGeo and SpaGeo[3], and the XYZ-workbench [70]. In the next section we give an overview of the CGAL project and library, the main design goals that guided the development of the library and a brief description of its various parts.

CGAL is definitely not the only software development effort of computational geometry algorithms. Implementations of geometric algorithms have been reported in the literature and were made available for various problems; see [7] and [8]. These include libraries and workbench efforts. Schirra [69] comments that as far as libraries/workbenches are concerned, two have paid special attention to robustness issues: The XYZ-workbench and LEDA.

This paper concentrates on *arrangements* and *motion planning*. Arrangements are space subdivisions induced by geometric objects and they are closely related to the study of collision-free path planning for objects moving among obstacles. Each of these topics, as well as the connection between them, has been the subject of extensive research. We give more background about these areas and point to bibliography in the relevant sections. In Section 3 we focus on the maps and arrangements part of the library which has been developed at Tel Aviv University. We explain the connection between arrangements and motion planning in Section 4 and describe two packages developed on top of CGAL that provide robust primitives for motion algorithms. Concluding remarks and directions for further research are given in Section 5.

2 The CGAL Project and Library

The Computational Geometry Algorithms Library CGAL is a software library of robust geometric algorithms and data structures.

The participating groups in the CGAL project are from ETH Zurich (Switzerland), The Free University Berlin (Germany), Halle University (Germany), INRIA Sophia-Antipolis (France), Max Planck Institute Saarbrucken (Germany), RISC Linz (Austria), Tel Aviv University (Israel) and Utrecht University (The Netherlands).

Among the major goals in the design of the library were robustness, generality, and efficiency [16],[35],[36]. In general **robustness** is achieved in CGAL through exact computation of geometric predicates, which often require the use of special number types. The user is free to choose different number types such as the machine float or double, but then the algorithms are not always guaranteed to produce the correct output. There is a clear indication in CGAL under which conditions the algorithms are certified to work correctly.

Number types could be easily imported from other packages such as LEDA or *The GNU Multiple Precision Arithmetic Library* [44] and smoothly used with CGAL. The ability of the user to specify different number types for the same algorithm is one aspect of the **generality** of the library. The project has adapted [16] the so-called *generic programming* paradigm that makes heavy use of the C++ template mechanism [9]. In addition to number types users can choose a representation type for point coordinates: Cartesian or homogeneous. Major flexibility and generality is achieved through the use of 'traits classes'. A traits class brings together all the data types and operations needed by a certain algorithm and is passed as an additional parameter to the algorithm. In Section 3.2 we illustrates how traits classes enhance the generality of the code for representing two-dimensional subdivisions.

For a detailed discussion of the above and additional goals in the design of CGAL see [16],[35],[36].

Notice that there is no global strategy for handling degenerate input in CGAL. If an algorithm is claimed to perform correctly as long as exact predicates are provided, this means in particular that it could handle degenerate input. While this is often doable and even desirable for two-dimensional problems [21] it may be unnecessary or extremely difficult for higher-dimensional problems. In Section 4.2 we propose an efficient method for removing degeneracies from three-dimensional arrangements.

CGAL consists of three parts. The first is the kernel, which consists of primitive constant-size geometric objects and predicates on them (e.g., points and orientation test for points). The second part, the basic library, consists of a large collection of fundamental algorithms (e.g., constructing convex hulls) and data

[2]F. Avnaim, C++gal: A C++ library for geometric algorithms, INRIA Sophia-Antipolis, 1994.

[3]G.-J. Giezeman, PlaGeo, a library for planar geometry, and SpaGeo, a library for spatial geometry, Utrecht University, 1994.

structures (e.g., kd-trees) parametrized by trait classes. The maps and arrangements package that we describe in the next section is part of the basic library. The third part incorporates non-geometric support facilities such as I/O support for debugging and for interfacing CGAL to various visualization tools. The full functionality of the library can be found in the manuals of the various parts [24]. Additional documentation includes an installation guide and a "getting started" manual guiding the novice how to use the library through a series of program examples.

CGAL can be used as a stand-alone library. However, there are close connections between CGAL and LEDA—the library of efficient data structures and algorithms. There are provisions in CGAL for easy use of LEDA's special number types, graphical window and graphical output to a postscript file. For a succinct listing of the principal differences between CGAL and the geometric part of LEDA see [60, Section 9.11].

3 Maps and Arrangements

Given a collection S of geometric objects (such as lines, planes, or spheres) the *arrangement* $\mathcal{A}(S)$ is the *subdivision* of the space where these objects reside into cells as induced by the objects in S. Figure 2 illustrates a planar arrangement of segments which consists of vertices, edges and faces: a *vertex* is either a segment endpoint or the intersection point of two (or more) segments, an *edge* is a maximal portion of a segment not containing any vertex, and a *face* is a maximal region of the plane not containing any vertex or edge. We assume below some familiarity with arrangements; for more information about arrangements see, for example, [3],[31],[46],[73].

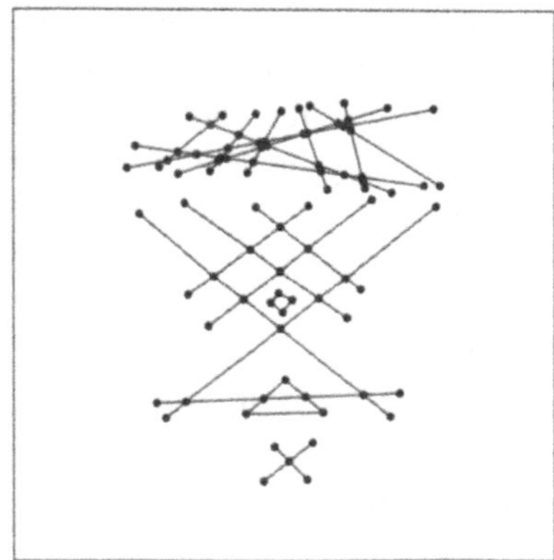

Figure 2: *An arrangement of segments.*

Arrangements are a useful tool in the study of geometric problems in robotics and other areas. We use them to partition a space into cells such that certain invariants are maintained in each cell. The objects in S that define the arrangement are the loci of critical points in space where the invariants change. Arrangements lead to efficient solutions in, among other areas, robot motion planning (as we will explain in more detail in Section 4.1), assembly planning, and visibility problems [46],[73].

Motivated by these applications the Tel Aviv site of the CGAL project has been responsible for the development of a package to support the construction and manipulation of arrangements of general curves. Our two-dimensional package is described in this section; many technical details that we omit here can be found in the CGAL user manual [24, Part II, Chapters 8 through 10] and in [39] and [51]. In Section 4.2 we describe implementations for three-dimensional arrangements of triangles and of polyhedral surfaces built on top of CGAL.

3.1 From Maps to Arrangements

Subdivisions could be used in CGAL in one of four levels, where each higher level is built on top of the previous one: (i) topological maps (which are not necessarily planar), (ii) planar maps, (iii) planar maps with intersections, and (iv) arrangements. We will restrict our description here to planar maps and arrangements.

A planar map in CGAL is defined as the subdivision of the plane into cells by a collection of pairwise interior-disjoint x-monotone curves. The choice to restrict the curves to be x-monotone at this level makes it easy to apply algorithmic methods such as the *vertical decomposition* [28] to otherwise general curves. The planar map level is the natural level to represent a simple plane subdivision such as a geographic map or a room with polygonal obstacles.

Planar maps in CGAL support traversal over faces, edges and vertices of the map, traversal over the edges of a face and around a vertex and efficient *point location* (namely identifying the feature of a map where a query point is located). They are dynamic and support insertion of new edges into or removal of edges out of an

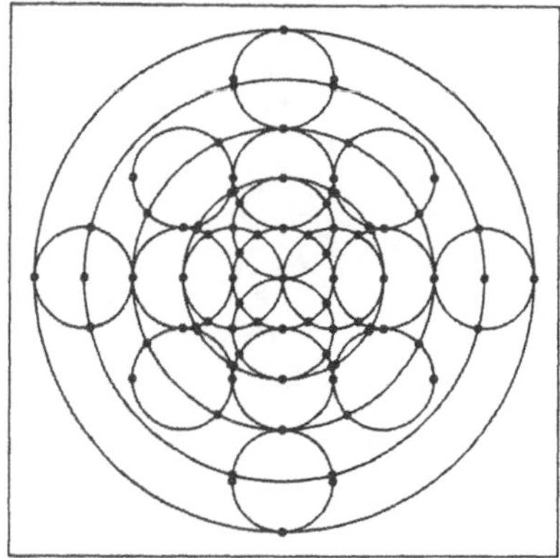

Figure 3: *An arrangement of circles [52].*

existing map. The representation is based on the *Doubly Connected Edge List* (DCEL) [28, Chapter 2]. This representation belongs to a family of edge-based data structures in which each edge is represented as a pair of opposite *half edges*. The representation supports inner components (holes) inside the faces.

The curves at the arrangement level are the most general: they may intersect and are not necessarily x-monotone. The arrangement level employs the planar map level and preprocesses the input curves by decomposing each one of them into x-monotone subcurves, and further subdivides them at their intersection points with the other curves. A data structure which we call the *hierarchy tree* maintains the curve-history of each of the edges in the final planar map (namely the original curves and sub-curves in which this edge is contained). In addition to the planar-map functionality, the arrangement level supports overlapping curves and the traversal of all the curves containing a query edge.

3.2 The Separation of Combinatorics and Numerics

As mentioned in Section 2, algorithms in CGAL are passed a so-called traits class that contains the needed data types and operations. For planar maps, the traits class is an abstract interface of predicates and functions that wraps the access of an algorithm to the geometric (rather than combinatorial) inner representation. We have formulated the requirements from the planar map's traits so they make as few assumptions on the curves as possible. Packing those predicates and functions under one traits class gives flexibility in choosing the geometric representation of the objects (Homogeneous, Cartesian) and flexibility in choosing the geometric kernel (LEDA, CGAL, or a user-defined kernel). The traits class enables the users to employ the maps and arrangements package with different kinds of curves that they define (as required by their applications). The only restriction is that they obey the predefined interface.

The planar map traits class defines the two basic geometric objects of the map: the point and the x-monotone curve. In addition several types of predicates are required: For example, a predicate that returns whether a point is above, below, or on a given curve, and a predicate that compares the y-coordinate of two curves at a given x-coordinate.[4]

We supply ready-made traits for line segments, circular arcs, and polylines while using various number types. We mainly rely on LEDA's special number types for efficient exact arithmetic.

3.3 Point Location

Point location is a basic service required from a subdivision data structure: Given a query point p report the face of the subdivision containing p. Planar maps (and the levels above them) provide this service. We provide three algorithms for point location as well as a mechanism for the users to supply their own point-location algorithm.

The algorithms we have implemented are: (i) a naive algorithm, which goes over all the edges in the map to find the location of the query point; (ii) an efficient algorithm (the default one) which is based on the randomized incremental construction of a search structure through the vertical decomposition of the map [15],[62],[72], and (iii) a "walk-along-a-line" (walk, for short) algorithm that is an improvement over the naive approach: it finds the point's location by walking along a vertical line from "infinity" towards the query point. We remind the reader that our point location implementation handles general finite planar maps. The subdivision is not necessarily *monotone* (each face's boundary is a union of x-monotone chains) nor connected. In addition the input may be x-degenerate.

We have observed an interesting behavior of the point-location algorithms: During the construction of

[4]The full list of requirements can be found in [24].

a map (or an arrangement), using the walk algorithm is faster than using the presumably fast vertical-decomposition based search structure. The latter answers queries in logarithmic time whereas the walk algorithm may require up to linear time per query. The explanation to this phenomenon is the significant overhead in constructing the search structure (which does not express itself as strongly in the asymptotic bounds). See [39] for a comparison of the performance of our point location algorithms.

Besides the naive algorithm, implementing the point-location algorithms for possibly degenerate arrangements of arbitrary curves is non-trivial. Already the walk algorithm raises difficulties; Mehlhorn and Näher describe these difficulties for their walk point-location in (straight edge) Delaunay triangulations [60]. The naive point location was used in our implementation to verify the correctness of the other algorithms through the use of a so-called "double checker": It is a debug point-location strategy[5] that to the planar map package looks as a regular strategy but for every query it runs two different algorithms, compares their results and notifies the developer in case of a mismatch.

We have also developed an adaptive point location scheme for arrangements of parametric curves [52]. The idea is to bound a curve inside a polygon (which serves as a polygonal approximation of the curve) and perform operations on the resulting arrangement of polygons. If we are unable to tell which original curves lie on the boundary of a face containing a query point, we refine the polygonal approximation until we can answer the query; see Figure 4 for an illustration on a collection of Bézier curves. This scheme makes use of the arrangement polygonal-line traits.

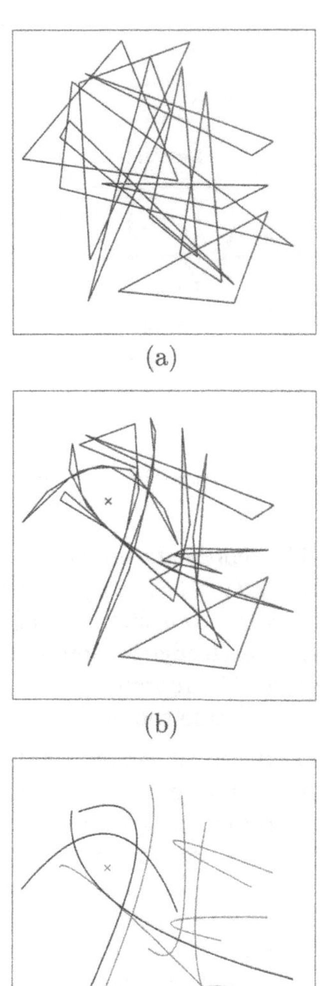

Figure 4: *An arrangement of polygons each bounding a Bézier curve before (a) and after (b) the first point location; (c) displays the underlying arrangement of Bézier curves and in bold line the curves that bound the face containing the query point.*

3.4 On-line Zone Construction

We conclude this section with another problem on planar arrangements of lines for which we implemented and compared four different algorithms: identifying the faces of an arrangement that are intersected by an additional curve γ (these faces are called the *zone* of γ); the curve is given to the algorithm on-line piece by piece. This work is motivated by the efficient exploration of the area bisectors of a polygon [11] related to using MEMS arrays for part orienting [12]. The experiments led to useful observations regarding the use of number types in arrangement computation and about the benefits of caching the results of geometric computation, both constructions and predicates, to save exact computation time (for example, once computed we retain the intersection point of two lines for later use by the algorithm instead of recomputing it). An experimental comparison between the algorithms is reported in [6].

[5]The use of "strategy" here refers to the *strategy pattern* [41].

4 Robust Motion Primitives

Arrangements are useful in solving motion planning problems. In this section we explain this connection and describe two tools that are based on arrangements for producing robust primitives for motion algorithms.

The first tool constructs exact polygonal Minkowski sums. Minkowski sums describe configuration space obstacles for translational motion planning and related problems. Following the typical CGAL approach robustness of the tool is achieved through the use of exact arithmetic. Moreover, we directly handle degeneracies; the solution depicted in Figure 1 exemplifies the handling of a degenerate configuration by our package.

The second tool computes *approximate* three-dimensional swept volumes, which describe the space occupied by a polyhedron as it is moved along a trajectory in space. Swept volumes have many applications including the verification that the motion of an object along a certain trajectory can be carried out without collision with the environment objects. In applications where very high precision is not required, we propose a method which perturbs the surfaces that determine the swept volume so as to remove all possible degeneracies and which allows for computing robustly when using the standard floating-point arithmetic.

4.1 Exact Polygonal Minkowski Sums

Given two sets P and Q in $\mathbb{R}^2$, their *Minkowski sum* (or vector sum), denoted by $P \oplus Q$, is the set $\{p+q \mid p \in P, q \in Q\}$. Minkowski sums are used in a wide range of applications, including robot motion planning [58], assembly planning [47], and computer-aided design and manufacturing (CAD/CAM) [33].

Consider, for example, a planar setting with an obstacle P and a robot Q that moves by translation. We can choose a reference point r rigidly attached to Q and suppose that Q is placed such that the reference point coincides with the origin. If we let Q' denote a copy of Q rotated by 180° degrees, then $P \oplus Q'$ is the locus of placements of the point r where $P \cap Q \neq \emptyset$. In the study of motion planning the space of all possible placements of the reference point is called the *configuration space* of Q and this sum is called a *configuration space obstacle* because Q collides with P when translated along a path π exactly when the point r, moved along π, intersects $P \oplus Q'$.

We distinguish three types of points in the configuration space according to the placements of the robot that they represent: *free placements,* where the robot does not intersect any obstacle, *forbidden placements,* where the robot intersects the interior of an obstacle, and *semi-free placements,* where the robot is in contact with the boundary of an obstacle, but does not intersect the interior of any obstacle. The collection of all the points in the configuration space that represent semi-free robot placements partitions the configuration space into *free regions* and *forbidden regions.* In a motion-planning problem with two degrees of freedom, for instance, the points representing semi-free placements of the robot lie on several curves in a two-dimensional configuration space. We call these curves *constraint curves.* Note that each such curve is induced by the contact of a robot boundary feature and an obstacle boundary feature. We are therefore interested in studying the partitioning of 2D space by a collection of constraint curves and this is where motion planning and the study of arrangements meet. This discussion extends to motion planning with more degrees of freedom and higher dimensional arrangements and has been the subject of an intensive study; see e.g., [22],[45],[49],[71],[73].

There has been much work on obtaining sharp bounds on the size of the Minkowski sum of two sets in two and three dimensions, and on developing fast algorithms for computing Minkowski sums. If P is a polygonal set with m vertices and Q is another polygonal set with n vertices, then $P \oplus Q$ is a portion of the arrangement of $O(mn)$ constraint segments, where each segment is the Minkowski sum of a vertex of P and an edge of Q, or vice versa. Therefore the size of $P \oplus Q$ is $O(m^2n^2)$ and it can be computed within that time; this bound is tight in the worst case [55]; see Figure 5. The sum has lower worst-case complexity when one of the polygons or both are convex.

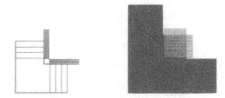

Figure 5: *P and Q are polygons with horizontal and vertical teeth with m and n vertices respectively. The complexity of $P \oplus Q$ is $\Theta(m^2n^2)$.*

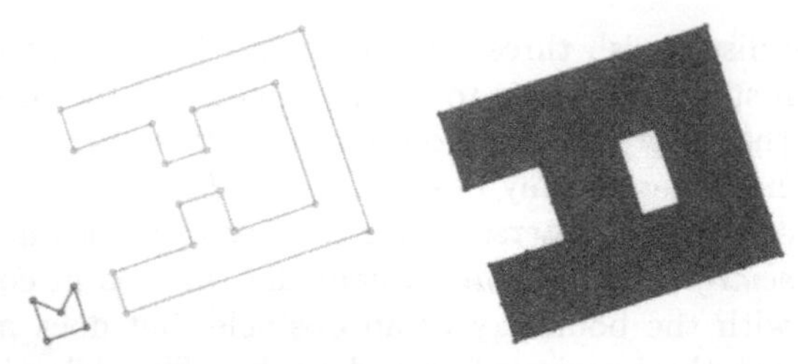

Figure 6: *Tight passage: the desired target placement for the small polygon is inside the inner room defined by the larger polygon (left-hand side). In the configuration space (right-hand side) the only possible path to achieve this target passes through the line segment emanating from the hole in the Minkowski sum [38].*

We devised and implemented three algorithms for computing the Minkowski sum of two polygonal sets [4], [38]. Our main goal was to produce a *robust* and exact implementation. This goal was achieved by employing the CGAL planar map package (Section 3) while using exact number types. We are currently using our software to solve translational motion planning problems in the plane. We are able to compute collision-free paths even in environments cluttered with obstacles, where the robot could only reach a destination placement by moving through tight passages, practically moving in contact with the obstacle boundaries. See Figure 6 for an illustration. This is in contrast with most existing motion planning software for which tight or narrow passages constitute a significant hurdle. We remark that our tool can only handle translational motion; steps towards robust handling of polygon rotation are discussed in [23].

The input to our algorithms consists of two *polygonal sets* P and Q (each being an arbitrary collection of simple polygons), with a total of m and n vertices respectively. Our algorithms consist of the following three steps: **(1)** Decompose the polygons of P into convex subpolygons $P_1, P_2, \ldots, P_s$ and the polygons of Q into convex subpolygons $Q_1, Q_2, \ldots, Q_t$, **(2)** For each $i \in [1..s]$ and for each $j \in [1..t]$ compute the Minkowski *subsum* $P_i \oplus Q_j$, and **(3)** Construct the union of all the subsum polygons $P_i \oplus Q_j$ computed in Step 2; the output is represented as a planar map. The three algorithms differ in the way they compute the union (Step 3). The ability to discover the semi-free placement of Figure 1 or the tight passage of Figure 6 requires the program to identify the simultaneous intersection of many edges or the overlap of edges. For further details on the algorithms and a comparison of their performance, see [38].

As the bound $\Theta(m^2n^2)$ indicates, even for input polygons with a moderate number of vertices the output can be huge. This, together with our choice to use exact arithmetic, necessitates speeding up the algorithm in other ways. We discovered that the choice of the decomposition in Step 1 can have a dramatic effect on the running time of the Minkowski-sum algorithms. (Note that in the asymptotic complexity measures the choice of decomposition is meaningless: every reasonable decomposition, including an arbitrary triangulation, produces subpolygons of total complexity $O(k)$ where k is the complexity of the polygon.) This led us to investigate what constitute good decompositions for efficient construction of Minkowski sums [4].

We studied and experimented with various well-known decompositions as well as with several new decomposition schemes. Among our findings are that in general: (i) triangulations are too costly (although they can be produced quickly, they considerably slow down the Minkowski-sum computation), (ii) what constitutes a good decomposition for one of the input polygons depends on the other input polygon—consequently, we developed a procedure for simultaneously decomposing the two polygons such that a "mixed" objective function is minimized, (iii) there are optimal decomposition algorithms that significantly expedite the Minkowski-sum computation, but the decomposition itself is expensive to compute — in such cases simple heuristics that approximate the optimal decomposition perform very well. The decomposition methods, the experiments, and the conclusions drawn from them are reported in [4].

4.2 Degeneracy-Free Approximate Swept Volumes

A *swept volume* is the geometric space occupied by an object moving along a prescribed trajectory in a given time interval. The motion can be translational and rotational. We call the moving object a *generator* and its motion a *sweep*. See Figure 7 for an illustration.

Swept volumes play an important role in many geometric applications, such as geometric modeling, robot workspace computation, numerical control cutter path generation, and assembly design. There is rich liter-

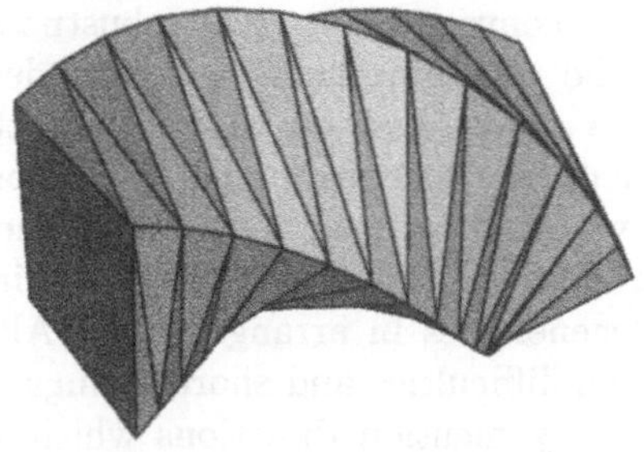

Figure 7: *The volume swept by a square moving along a helical trajectory.*

ature on swept volumes which is beyond the scope of this paper to review; see, e.g., [1] for a bibliography.

Consider the following question: Given an assembly of parts, a specific part P in the assembly, and a potential path π for removing P out of the assembly—can P be moved along π without colliding with the other parts of the assembly? We can answer this question by computing the swept volume of P along π and check it for collision with the other parts. We were specifically motivated by problems in large assemblies and manual removal of parts for maintenance and repair.

We chose to implement an algorithm proposed by Abrams and Allen [2] which handles polyhedral bodies moving along an arbitrary trajectory in a motion that can be translational and rotational, and outputs a polyhedral approximation of the appropriate swept volume; we will refer to this algorithm as ASW (approximate swept volume). The algorithm generates a set $\mathcal{P}$ of polyhedral surfaces such that the boundary of the desired swept volume is the boundary of the outer cell of the arrangements $\mathcal{A}(\mathcal{P})$. (ASW ignores voids.) Our overall plan was: (i) use ASW to produce the set $\mathcal{P}$ and then (ii) use an implementation of the space sweep algorithm [26], adapted from triangles to polyhedral surfaces, to compute the outer cell of $\mathcal{A}(\mathcal{P})$.

However, the algorithm described in [26] assumes general position (its implementation already has to handle a large number of cases even when assuming general position). Abrams and Allen also report that they could not find robust software for arrangements that could lead to a robust implementation of their solution [2]. Unlike two-dimensional arrangements where directly handling degeneracies is a satisfying solution (as in Section 4.1 above), full treatment of degeneracies in three-dimensional arrangements is an arduous task. Moreover, because of the nature of the problem at hand and the approximate solution we have chosen for it, the recognition of all the degeneracies in itself is irrelevant. Identification of all the degeneracies is only meaningful as a way to extract a consistent topology of the outer cell boundary—but we will obtain the same goal in a much simpler way. We apply a "controlled" perturbation to the surfaces in $\mathcal{P}$ so that all degeneracies are removed [66],[67]. One of our main goals here is to allow to manipulate the swept volume robustly with standard floating-point arithmetic. We describe our scheme next.

Since we aim to use standard floating-point arithmetic, we are unable to tell for sure whether a degeneracy exists. We can only tell that a *potential degeneracy* exists. To define such a potential degeneracy formally, we use a *resolution parameter*, $\varepsilon > 0$, which is a small positive real number. Two polyhedral features (e.g., a vertex and a non-incident facet) are assumed too close (and therefore potentially degenerate) whenever they are less than ε away from each other. We assume that ε is given as an input parameter according to the machine precision, the type of the arithmetic operations and the depth of the *expression tree* [69]. In our algorithm this tree's depth is a small constant, thus ε is a constant that can be determined and bounded independently of the input size.

Next we define a perturbation radius δ, which is proven to be sufficiently small (δ depends on ε and the input polyhedral surfaces). Any polyhedral surface in $\mathcal{P}$ or any feature thereof will be perturbed by at most δ. Specifically, each vertex of any surface in $\mathcal{P}$ is moved by a Euclidean distance of at most δ from its original placement.

Our perturbation scheme is "controlled" in two ways. First, by determining the size of δ we set a tradeoff that controls the magnitude of the perturbation versus the efficiency of the computation of the perturbation— the larger the δ the faster the computation. Second, unlike various popular heuristic perturbation schemes (e.g., "heuristic epsilons" [69]), our perturbation guarantees that the resulting collection of polyhedral surfaces is degeneracy free by carrying out a controlled incremental insertion process where we do not proceed to the next iteration before the arrangement induced by the subcollection of surfaces or subsurfaces of $\mathcal{P}$ we inserted so far is degeneracy free. Successful completion of the process is guaranteed for δ values above a threshold which is determined by the analysis of the procedure.

The thesis [65] contains a detailed report on all the degeneracies arising in arrangements of polyhedral surfaces and the swept volume computation, gives bounds on the perturbation radius as a function of ε and the input polyhedral surfaces, and describes the implementation of the scheme together with experimental results.

Our implementation of the swept volume algorithm uses CGAL in various ways (geometric primitives, predicates, affine transformations, assertions, and preconditions testing). Most notably it uses the *polyhedral surface* package developed by Kettner [57].

This work is an extension of an earlier work on arrangements of spheres as they arise in molecular modeling [50]. In both cases (molecular modeling and swept volumes) we show by experiments [50], [67] that also on highly degenerate input (as depicted in Figure 8) the controlled perturbation scheme works efficiently even when δ is fairly small.

5 Conclusions

We described advancement in *robust* implementation of geometric algorithms. After reviewing the CGAL project and library we concentrated on the maps and arrangements part of CGAL and showed how this and other parts of CGAL are useful for devising robust primitives for motion planning.

Regarding arrangements, the next goal would be to develop packages for three- and higher-dimensional arrangements comparable in their robustness and generality to the two-dimensional package described in Section 3. A major obstacle to achieving this goal is the treatment of degeneracies (as far as precision is concerned we could use exact arithmetic wherever possible). We mention next several possible directions to confront degeneracies in arrangements. Although we point out the difficulties and shortcomings of each direction we only mention directions which we believe are feasible, perhaps in restricted form (that is, only in three dimensions or only for arrangements of certain simple types of objects).

Figure 8: *Swept volume with many intersections (and potential degeneracies) in the middle [66].*

- Certifying a correct result only for input in general position. This is feasible (in the sense that asserting general position is easier than computing a degenerate arrangement) but impractical since quite often the input of physical-world problems is not in general position.
- Directly handling all degeneracies (as we do in the two-dimensional case). This seems to be a tremendous task already for arrangements of triangles in three dimensions. What is missing is a systematic and concise way to express and compute the topology of the arrangement at the neighborhood of degeneracies.
- Applying symbolic perturbation (see the Introduction). In this solution the difficulty seems to be in the interpretation of the output (postprocessing) [69].
- Tightly approximating the arrangement, possibly using only fixed precision arithmetic (as in Section 4.2). There are many conceivable ways to achieve this goal, one of which is extending the scheme that we propose above, controlled perturbation, to other objects and to higher dimensions. Implementing the scheme is not difficult. However, giving a theoretical guarantee for its viability for arrangements of complex objects is a difficult task [65].

For motion planning, supporting only two- and (currently partially) three-dimensional arrangements means we could solve robustly and accurately problems with a limited number of degrees of freedom.[6]

[6]The swept volume application described in Section 4.2

Progress in the implementation of algorithms for higher-dimensional arrangements would result in robust algorithms for systems with more degrees of freedom. Notably, probabilistic roadmap (PRM) techniques already offer a solution for motion planning problems with many degrees of freedom [10],[56]. However, densely cluttered environments and narrow passages in the workspace are difficult for PRM techniques; in fact tight passages are impossible to identify without exact predicates. We anticipate that hybridizing the two approaches, namely PRMs and exact (or tightly approximate) arrangements, will lead to stronger motion planning algorithms.

Finally, note that CGAL offers much additional functionality that could be useful for algorithms in robotics and motion planning, including convex hulls, Voronoi diagrams, triangulations of point sets, various optimization algorithms (e.g., smallest enclosing sphere), and multi-dimensional search structures.

Acknowledgements

The work described in this paper was done by or has been made possible by the efforts of many individuals—the participants of the CGAL and GALIA projects. The full list of participating sites is given at the beginning of Section 2.

The major contributions at the Tel Aviv site which are described in Sections 3 and 4 have been made by Eyal Flato, Iddo Hanniel, Oren Nechushtan, and Sigal Raab. Others who contributed at Tel Aviv are: Sariel Har-Peled, Chaim Linhart, Michal Ozery, and more recently Eti Ezra, Shai Hirsch, and Eli Packer. The author also thanks Stefan Schirra and Lutz Kettner for their comments on a draft of this paper.

Work reported in this paper has been supported by ESPRIT IV LTR Projects No. 21957 (CGAL) and No. 28155 (GALIA), by the USA-Israel Binational Science Foundation, by The Israel Science Foundation founded by the Israel Academy of Sciences and Humanities (Center for Geometric Computing and its Applications), by a Franco-Israeli research grant "factory of the future" (monitored by AFIRST/France and The Israeli Ministry of Science), and by the Hermann Minkowski – Minerva Center for Geometry at Tel Aviv University.

approximately solves a problem with six degrees of freedom. However for solving motion planning problems *exactly* through the usage of arrangements, we need k-dimensional arrangements for problems with k degrees of freedom.

References

[1] K. Abdel-Malek. Swept volumes: Bibliography. www.icaen.uiowa.edu/~amalek/sweep/bibliog.htm.

[2] S. Abrams and P. Allen. Swept volumes and their use in viewpoint computation in robot work-cells. In *Proc. IEEE Intl. Sympos. on Assembly and Task Planning*, pages 188–193, 1995.

[3] P. Agarwal and M. Sharir. Arrangements. In J.-R. Sack and J. Urrutia, editors, *Handbook of Computational Geometry*, pages 49–119. Elsevier Science Publishers B.V. North-Holland, Amsterdam, 1999.

[4] P. K. Agarwal, E. Flato, and D. Halperin. Polygon decomposition for efficient construction of Minkowski sums. Manuscript. www.math.tau.ac.il/~flato/TriminkWeb, Tel Aviv University, 1999.

[5] P. K. Agarwal, M. Katz, and M. Sharir. Computing depth orders for fat objects and related problems. *Comput. Geom. Theory Appl.*, 5:187–206, 1995.

[6] Y. Aharoni, D. Halperin, I. Hanniel, S. Har-Peled, and C. Linhart. On-line zone construction in arrangements of lines in the plane. In *Proc. of the 3rd Workshop of Algorithm Engineering*, volume 1668 of *Lecture Notes Comput. Sci.*, pages 139–153. Springer-Verlag, 1999.

[7] N. Amenta. Directory of computational geometry software. www.geom.umn.edu/software/cglist/.

[8] N. Amenta. Computational geometry software. In J. E. Goodman and J. O'Rourke, editors, *Handbook of Discrete and Computational Geometry*, chapter 52, pages 951–960. CRC Press LLC, Boca Raton, FL, 1997.

[9] M. Austern. *Generic Programming and the STL — Using and Extending the C++ Standard Template Library*. Addison-Wesley, 1999.

[10] J. Barraquand, L. E. Kavraki, J. C. Latombe, T.-Y. Li, R. Motwani, and P. Raghavan. A random sampling scheme for robot path planning. *Internat. J. Robot. Res.*, 16(6):759–774, 1997.

[11] K.-F. Böhringer, B. Donald, and D. Halperin. On the area bisectors of a polygon. *Discrete Comput. Geom.*, 22:269–285, 1999.

[12] K.-F. Böhringer, B. R. Donald, and N. C. MacDonald. Upper and lower bounds for programmable vector fields with applications to MEMS and vibratory plate part feeders. In J.-P. Laumond and M. H. Overmars,

editors, *Algorithms for Robotic Motion and Manipulation*, pages 255–276. A. K. Peters, Wellesley, MA, 1996.

[13] J.-D. Boissonnat and F. P. Preparata. Robust plane sweep for intersecting segments. Report TR 3270, INRIA, Sophia Antipolis, Sept. 1997.

[14] J.-D. Boissonnat and J. Snoeyink. Efficient algorithms for line and curve segments intersection using restricted prediactes. In *Proc. 15th Annu. ACM Sympos. Comput. Geom.*, 1999.

[15] J.-D. Boissonnat and M. Yvinec. *Algorithmic Geometry*. Cambridge University Press, UK, 1998.

[16] H. Brönniman, L. Kettner, S. Schirra, and R. Veltkamp. Applications of the generic programming paradigm in the design of CGAL. Technical Report MPI-I-98-1-030, Max-Planck-Insitut für Informatik, 66123 Saarbrücken, Germany, 1998.

[17] H. Brönnimann, I. Emiris, V. Pan, and S. Pion. Computing exact geometric predicates using modular arithmetic with single precision. In *Proc. 13th Annu. ACM Sympos. Comput. Geom.*, pages 174–182, 1997.

[18] H. Brönnimann and M. Yvinec. Efficient exact evaluation of signs of determinants. In *Proc. 13th Annu. ACM Sympos. Comput. Geom.*, pages 166–173, 1997.

[19] C. Burnikel, R. Fleischer, K. Mehlhorn, and S. Schirra. Efficient exact geometric computation made easy. In *Proc. 15th Annu. ACM Sympos. Comput. Geom.*, pages 341–350, 1999.

[20] C. Burnikel, J. Könnemann, K. Mehlhorn, S. Näher, S. Schirra, and C. Uhrig. Exact geometric computation in LEDA. In *Proc. 11th Annu. ACM Sympos. Comput. Geom.*, pages C18–C19, 1995.

[21] C. Burnikel, K. Mehlhorn, and S. Schirra. On degeneracy in geometric computations. In *Proc. 5th ACM-SIAM Sympos. Discrete Algorithms*, pages 16–23, 1994.

[22] J. Canny. *The Complexity of Robot Motion Planning*. ACM - MIT Press Doctoral Dissertation Award Series. MIT Press, Cambridge, MA, 1987.

[23] J. Canny, B. R. Donald, and E. K. Ressler. A rational rotation method for robust geometric algorithms. In *Proc. 8th Annu. ACM Sympos. Comput. Geom.*, pages 251–260, 1992.

[24] *The CGAL User Manual, Version 2.1*, 2000. www.cs.uu.nl/CGAL.

[25] K. L. Clarkson. Safe and effective determinant evaluation. In *Proc. 33rd Annu. IEEE Sympos. Found. Comput. Sci.*, pages 387–395, Oct. 1992.

[26] M. de Berg, L. J. Guibas, and D. Halperin. Vertical decompositions for triangles in 3-space. *Discrete Comput. Geom.*, 15:35–61, 1996.

[27] M. de Berg, M. J. Katz, A. F. van der Stappen, and J. Vleugels. Realistic input models for geometric algorithms. In *Proc. 13th Annu. ACM Sympos. Comput. Geom.*, pages 294–303, 1997.

[28] M. de Berg, M. van Kreveld, M. Overmars, and O. Schwarzkopf. *Computational Geometry: Algorithms and Applications*. Springer-Verlag, Heidelberg, Germany, 1997.

[29] O. Devillers and F. P. Preparata. A probabilistic analysis of the power of arithmetic filters. *Discrete amd Computational Geometry*, 20:523–547, 1998.

[30] T. Dubé, K. Ouchi, and C.-K. Yap. Tutorial for the `real/expr` package. 1996.

[31] H. Edelsbrunner. *Algorithms in Combinatorial Geometry*, volume 10 of *EATCS Monographs on Theoretical Computer Science*. Springer-Verlag, Heidelberg, West Germany, 1987.

[32] H. Edelsbrunner and E. P. Mücke. Simulation of simplicity: A technique to cope with degenerate cases in geometric algorithms. *ACM Trans. Graph.*, 9(1):66–104, 1990.

[33] G. Elber and M.-S. Kim, editors. *Special Issue of Computer Aided Design: Offsets, Sweeps and Minkowski Sums*, volume 31. 1999.

[34] I. Z. Emiris, J. F. Canny, and R. Seidel. Efficient perturbations for handling geometric degeneracies. *Algorithmica*, 19(1–2):219–242, Sept. 1997.

[35] A. Fabri, G. Giezeman, L. Kettner, S. Schirra, and S. Schönherr. The CGAL kernel: A basis for geometric computation. In M. C. Lin and D. Manocha, editors, *Proc. 1st ACM Workshop on Appl. Comput. Geom.*, volume 1148 of *Lecture Notes Comput. Sci.*, pages 191–202. Springer-Verlag, 1996.

[36] A. Fabri, G. Giezeman, L. Kettner, S. Schirra, and S. Schönherr. On the design of CGAL, the Computational Geometry Algorithms Library. Technical Report MPI-I-98-1-007, Max-Planck-Institut Inform., 1998. To appear in *Software — Practice and Experience*.

[37] P.-O. Fjällström, J. Petersson, L. Nilsson, and Z. Zhong. Evaluation of range searching methods for contact searching in mechanical engineering. *Internat. J. Comput. Geom. Appl.*, 8:67–83, 1998.

[38] E. Flato and D. Halperin. Robust and efficient construction of planar Minkowski sums. In *Abstracts 16th European Workshop Comput. Geom.*, pages 85–88, Eilat, 2000.

[39] E. Flato, D. Halperin, I. Hanniel, and O. Nechushtan. The design and implementation of planar maps in CGAL. In *Proc. of the 3rd Workshop of Algorithm Engineering*, volume 1668 of *Lecture Notes Comput. Sci.*, pages 154–168. Springer-Verlag, 1999.

[40] S. Fortune and C. J. Van Wyk. Static analysis yields efficient exact integer arithmetic for computational geometry. *ACM Trans. Graph.*, 15(3):223–248, July 1996.

[41] E. Gamma, R. Helm, R. Johnson, and J. Vlissides. *Design Patterns – Elements of Reusable Object-Oriented Software.* Addison-Wesley, 1995.

[42] J. E. Goodman and J. O'Rourke, editors. *Handbook of Discrete and Computational Geometry.* CRC Press LLC, Boca Raton, FL, 1997.

[43] M. Goodrich, L. J. Guibas, J. Hershberger, and P. Tanenbaum. Snap rounding line segments efficiently in two and three dimensions. In *Proc. 13th Annu. ACM Sympos. Comput. Geom.*, pages 284–293, 1997.

[44] T. Granlund. *GNU MP, The GNU Multiple Precision Arithmetic Library, version 2.0.2*, June 1996.

[45] D. Halperin. Robot motion planning and the single cell problem in arrangements. *Journal of Intelligent and Robotic Systems*, 11:45–65, 1994.

[46] D. Halperin. Arrangements. In J. E. Goodman and J. O'Rourke, editors, *Handbook of Discrete and Computational Geometry*, chapter 21, pages 389–412. CRC Press LLC, Boca Raton, FL, 1997.

[47] D. Halperin, J.-C. Latombe, and R. H. Wilson. A general framework for assembly planning: The motion space approach. In *Proc. 14th ACM Symp. on Comput. Geom.*, pages 9–18, 1998.

[48] D. Halperin and M. H. Overmars. Spheres, molecules and hidden surface removal. *Comput. Geom. Theory Appl.*, 11:83–102, 1998.

[49] D. Halperin and M. Sharir. Arrangements and their applications in robotics: Recent developments. In K. Goldberg, D. Halperin, J.-C. Latombe, and R. Wilson, editors, *Proc. Workshop Algorithmic Found. Robot.*, pages 495–511. A. K. Peters, Wellesley, MA, 1995.

[50] D. Halperin and C. R. Shelton. A perturbation scheme for spherical arrangements with application to molecular modeling. *Comput. Geom. Theory Appl.*, 10:273–287, 1998.

[51] I. Hanniel. The design and implementation of planar arrangements of curves in CGAL. M.Sc. thesis, Dept. Comput. Sci., Tel Aviv University, Tel Aviv, Israel, 2000. In preparation.

[52] I. Hanniel and D. Halperin. Two-dimensional arrangements in CGAL and adaptive point location for parametric curves. Manuscript. Tel Aviv University, 2000.

[53] J. Hobby. Practical segment intersection with finite precision output. *Comput. Geom. Theory Appl.*, 13:199–214, 1999.

[54] W. D. Jones. Computational geometry, a community bibliography. Overview at sal.cs.uiuc.edu/~jeffe/compgeom/compgeom.html.

[55] A. Kaul, M. A. O'Connor, and V. Srinivasan. Computing Minkowski sums of regular polygons. In *Proc. 3rd Canad. Conf. Comput. Geom.*, pages 74–77, Aug. 1991.

[56] L. Kavraki, P. Svestka, J. Latombe, and M. Overmars. Probabilistic roadmaps for fast path planning in high dimensional configuration spaces. *IEEE Tr. on Rob. and Autom.*, 12:566–580, 1996.

[57] L. Kettner. Designing a data structure for polyhedral surfaces. In *Proc. 14th Annu. ACM Sympos. Comput. Geom.*, pages 146–154, 1998.

[58] J.-C. Latombe. *Robot Motion Planning.* Kluwer Academic Publishers, Boston, 1991.

[59] K. Mehlhorn, S. Näher, M. Seel, and C. Uhrig. *The LEDA User Manual, Version 4.1.* Max-Planck-Insitut für Informatik, 66123 Saarbrücken, Germany, 1999.

[60] K. Melhorn and S. Näher. *The LEDA Platform of Combinatorial and Geometric Computing.* Cambridge University Press, 1999.

[61] V. Milenkovic. Verifiable implementations of geometric algorithms using finite precision arithmetic. In D. Kapur and J. L. Mundy, editors, *Geometric Reasoning.* North-Holland, Amsterdam, Netherlands, 1988.

[62] K. Mulmuley. *Computational Geometry: An Introduction Through Randomized Algorithms.* Prentice Hall, Englewood Cliffs, NJ, 1994.

[63] M. H. Overmars. Designing the Computational Geometry Algorithms Library CGAL. In *Proc. 1st ACM Workshop on Appl. Comput. Geom.*, volume 1148 of *Lecture Notes Comput. Sci.*, pages 113–119. Springer-Verlag, May 1996.

[64] F. P. Preparata and M. I. Shamos. *Computational Geometry: An Introduction.* Springer-Verlag, New York, NY, 1985.

[65] S. Raab. Controlled perturbation for arrangements of polyhedral surfaces with application to swept volumes. M.Sc. thesis, Dept. Comput. Sci., Bar Ilan University, Ramat Gan, Israel, 1999.

[66] S. Raab. Controlled perturbation of arrangements of polyhedral surfaces with application to swept volumes. In *Proc. 15th Annu. ACM Sympos. Comput. Geom.*, 1999.

[67] S. Raab and D. Halperin. Controlled perturbation of arrangements of polyhedral surfaces with application to swept volumes. Full version, manuscript, Tel Aviv University, 2000.

[68] J.-R. Sack and J. Urrutia, editors. *Handbook of Computational Geometry*. North-Holland, 1999.

[69] S. Schirra. Robustness and precision issues in geometric computation. In J.-R. Sack and J. Urrutia, editors, *Handbook of Computational Geometry*, pages 597–632. Elsevier Science Publishers B.V. North-Holland, Amsterdam, 1999.

[70] P. Schorn. *Robust algorithms in a program library for geometric computation*. Ph.D. thesis, ETH Zürich, Switzerland, 1991. Report 9519.

[71] J. T. Schwartz and M. Sharir. On the "piano movers" problem II: General techniques for computing topological properties of real algebraic manifolds. *Adv. Appl. Math.*, 4:298–351, 1983.

[72] R. Seidel. A simple and fast incremental randomized algorithm for computing trapezoidal decompositions and for triangulating polygons. *Comput. Geom. Theory Appl.*, 1(1):51–64, 1991.

[73] M. Sharir and P. Agarwal. *Davenport-Schinzel Sequences and Their Geometric Applications*. Cambridge University Press, 1995.

[74] J. Shewchuk. Adaptive robust floating-point arithmetic and fast robust geometric predicates. *Discrete Comput. Geom.*, 18:305–363, 1997.

[75] K. Sugihara. On finite-precision representations of geometric objects. *J. Comput. Syst. Sci.*, 39:236–247, 1989.

[76] A. F. van der Stappen. *Motion Planning amidst Fat Obstacles*. Ph.D. dissertation, Dept. Comput. Sci., Utrecht Univ., Utrecht, Netherlands, 1994.

[77] A. F. van der Stappen, D. Halperin, and M. H. Overmars. The complexity of the free space for a robot moving amidst fat obstacles. *Comput. Geom. Theory Appl.*, 3:353–373, 1993.

[78] C. K. Yap. A geometric consistency theorem for a symbolic perturbation scheme. *J. Comput. Syst. Sci.*, 40(1):2–18, 1990.

[79] C. K. Yap. Robust geometric computation. In J. E. Goodman and J. O'Rourke, editors, *Handbook of Discrete and Computational Geometry*, chapter 35, pages 653–668. CRC Press LLC, Boca Raton, FL, 1997.

[80] C. K. Yap. Towards exact geometric computation. *Comput. Geom. Theory Appl.*, 7(1):3–23, 1997.

Controlled Module Density Helps Reconfiguration Planning

An Nguyen, *Stanford University, Stanford, CA*
Leonidas J. Guibas, *Stanford University, Stanford, CA*
Mark Yim, *Xerox Palo Alto Research Center, Palo Alto, CA*

In modular reconfigurable systems, individual modules are capable of limited motion due to blocking and connectivity constraints, yet the entire system has a large number of degrees of freedom. The combination of these two facts makes motion planning for such systems exceptionally challenging. In this paper we present two results that shed some light on this problem. First we show that, for a robotic system consisting of hexagonal 2D modules, the absence of a single excluded configuration is sufficient to guarantee the feasibility of the motion planning problem (for any two connected configurations with the same number of modules). We also provide an analysis of the number of steps in which the reconfiguration can be accomplished. Second, we argue that skeletal metamodules, which are scaffolding-like structures in 2D built out of normal modules, offer an interesting alternative. General shapes can be built out of these metamodules and, unlike the case for shapes built directly out of modules, a metamodule can collapse and pass through the interior of its neighboring metamodules, thus eliminating all blocking constraints. This tunneling capability makes the motion planning problem easier and allows faster reconfiguration as well, by providing a higher bandwidth conduit, the interior of the shape, through which the modules can flow. The conclusion of our work is that it is worthwhile to study subclasses of shapes that (1) approximate closely arbitrary shapes, while also (2) simplifying significantly the motion planning problem.

1 Introduction

A modular reconfigurable robot, sometimes known as a *metamorphic robot* consists of many identical modules, each with limited ability to connect, disconnect, and move around its neighbors. Originally proposed by Yim [12], Chirikjian [2], and Murata [7], these robots have since been further studied by these and several other authors [13, 14, 1, 4, 6, 5, 8, 9, 10, 11, 15]. A modular reconfigurable robot offers many advantages over conventional robots [7]. Such a robot has high fault tolerance, since each module can be replaced by any other module in the robot; the homogeneity of the modules allows low cost mass production; finally, the robot can assume a wide variety of conformations or shapes, and thus can be easily adapted to performing numerous tasks.

A reconfigurable robot changes its shape by moving one or more of its modules locally. The motion of each individual module in the configuration must satisfy certain constraints, as illustrated in Figure 1. In general, there are two types of constraints: the connectivity constraint requires all modules to stay connected at all times to a fixed base (for power distribution, inter-module communication, etc), and the blocking constraint requires a module not to collide with other modules during its motion. A module may also need to support itself on its neighboring modules in order to move.

Though each module can have very limited motion, or none at all, the large number of modules present means that the system has many degrees of freedom — in general, roughly proportional to the number of modules present. The homogeneity of the modules further complicates the reconfiguration problem, as it is not clear which module in the original configuration should go to which location in the final configuration. Thus motion planning for modular reconfigurable robots is a challenging combinatorial task; in fact, under several scenarios, the local blocking constraints can make the task infeasible.

In this paper we propose to address this problem by focusing on selected subsets of the allowed module configurations. By imposing certain constraints that limit how locally packed the modules can be, we demonstrate that the reconfiguration problem can become always feasible, or at least significantly easier. We give two

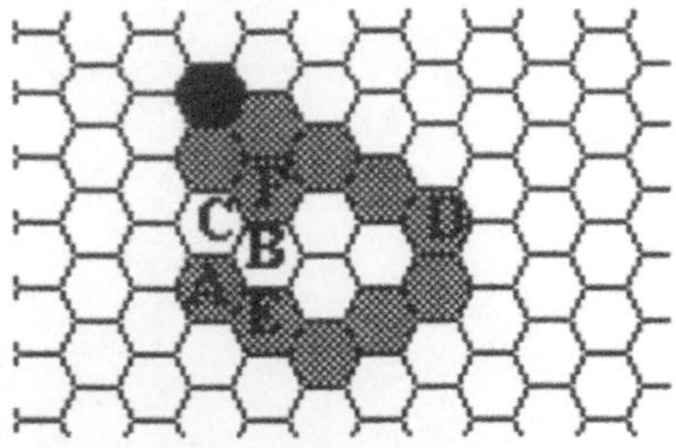

Figure 1: *This figure shows a reconfigurable robot composed of rigid hexagonal modules. Each module can move to a nearby empty space by rotating itself around a corner it shares with some other supporting module. The allowable motions of a module are restricted. The base, always shown in black in this and all subsequent figures, can never move. The module at D cannot move anywhere since such motion would leave the configuration disconnected. The module at A cannot move to B around supporting module E because such motion is blocked by F. The module at A cannot move to C either, because there is no module supporting the move.*

results, both for modular reconfigurable robots whose modules are rigid and correspond to a regular tiling of space. First, we show that for robots composed of 2D rigid hexagonal modules on a planar lattice, the reconfiguration problem is always solvable if both the initial and final configurations do not contain the excluded pattern (occupied, empty, occupied) parallel to any of the three axis of the lattice. We also analyze the complexity of our morphing algorithm in this case. Second, we propose the use of a *metamodule* structure for a 2D hexagonal lattice. A metamodule is effectively a scaled copy of the original module, built out of the original modules but with free space in its interior. Structures built entirely out of these metamodules are easier to reconfigure, as a metamodule has the ability to fold up and tunnel through any of its neighbors, thus removing the troublesome blocking constraints for the original modules.

In both cases there is still a very rich space of allowed robot configurations. Our intuition is that the shapes the robot needs to assume so as to perform various functions can be constructed or, at least, well-approximated by configurations satisfying the stated packing constraints. Thus, by restricting ourselves to such less dense structures we gain simplicity in motion planning without any significant sacrifice in functionality.

2 Related Work

Here we survey briefly earlier related work on reconfigurable robots. Murata et al. built a "self-assembling machine" [7]. This machine is a robot constructed out of 2D fractum modules. The reconfiguration of such a robot is done with random local motions to improve the "fitness" of a configuration with respect to the final configuration. Due to the randomness of the planning, the convergence is slow, and not practical for large problems [7]. Pamecha et al. [9] built a robot from regular hexagonal modules. The hexagonal module is designed so that it has no blocking constraints. They, too, solved the reconfiguration problem using simulated annealing to improve the matching distance to the final configuration. Upper and lower bounds on the optimal number of moves required were also given [10, 3]. Murata and Kotay et al. independently built 3D modular robots using 3D units [8] and 3D molecules [6] and discussed motion planning for their robots. Rus and Vona [11] built a system out of cubic compressible modules called *Crystalline Atoms*. They gave a *melt and grow* algorithm to do the motion planning between any two configurations. Another way to build 3D modular robots was proposed by Yim et al. [14], using a 3D rhombic dodecahedron as the key element. Zhang et al. [15] gave several heuristic algorithms for distributed parallel motion planning of such an robot.

The extant literature has not addressed the issue of the feasibility of the motion planning problem. Several of the proposed robots including Murata's fractum modules [8] and Yim's rhombic dodecahedron [14] (with 3, 5, or 7-sided constraints), admit of infeasible reconfiguration problems. Though in some cases the feasibility can be restored by building flexible modules to overcome the neighbor blocking constraints, this adds mechanical and control complexity, and therefore cost to the system.

3 The Reconfiguration Problem

A fundamental problem for a robot of this type is to plan how to accomplish its reconfiguration. Given initial and final configurations, we would like to compute the motions of the individual modules in the configuration that transform the initial configuration into the final one. In theory, reconfiguration can be done by searching the graph of all possible configurations, with edges between two configurations representing the reachability from one to other in a single step of some module in the configuration. This searching approach is not practical since the number of possible configurations grows exponentially with the number of modules in the configuration. The approach is thus intractable, even for robots with modest numbers of modules.

Note that the optimal number of steps to transform one configuration to another is a metric defined on the space of all configurations. We refer to this number as the distance in the configuration space. Any optimal planner can be used to compute this distance, and conversely, any way to compute this distance function yields a way to do the motion planning. The computation of the optimal metric function would probably be as hard as the motion planning itself. Current approaches for this problem define heuristic approximations to this function that can be computed quickly and potentially decrease the distance to the final configuration.

4 Difficulty with Motion Planning

It is unfortunate that in the presence of motion constraints, two configurations that are very close in most intuitive distance notions may not be at all close to each other in the configuration space. For example, consider two solid disk-like shapes packed with modules except for a single-module hole near their centers, but in a slightly different location in the two disks, see Figure 2. These shapes are approximately the same by virtually all geometric and topological measures. Yet, the number of motion steps required to transform one disk to another is large, since any motion plan must unpack the disk before being able to move some module into the empty hole.

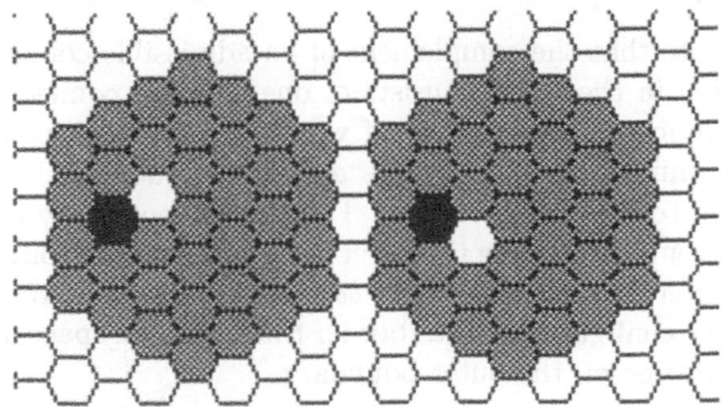

Figure 2: *Configurations that are approximately the same in geometric and topological sense may be far apart in the configuration space.*

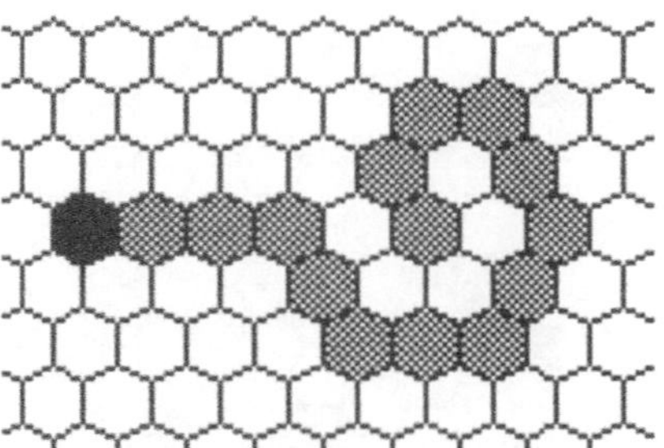

Figure 3: *Here is an example of an immobile configuration. The black module at one end of the chain is the fixed base, and the module at the other end cannot move because of blocking constraint. All other modules cannot move because of connectivity constraint.*

Before we discuss further difficulties, note that a robot may be *immobile*, with none of its modules able to move without violating their motion constraints. Such immobile configurations are typically the result of the blocking constraints (in practice the connectivity constraint always exists, but it cannot cause immobilization by itself.) An example of immobilization for a robot composed of rigid hexagons is shown in Figure 3. Robots made of fracta by Murata et al. or rhombic dodecahedron by Yim et al. with 7-sided blocking constraints are known to have immobile configurations. Robots made of rhombic dodecahedra with 3- and 5-sided constraints can also be immobile. The construction of such immobile configurations is similar to that of the configuration shown in Figure 3.

If the initial or final configuration is immobile, it will be impossible to find a motion plan between the two configurations. In general, the existence of an immobile configuration is likely to make the reconfiguration problem harder, even when it is feasible. There exist immobile configurations where a slight change in a configuration can make it mobile again — thus the distance in the configuration space between two configurations can be very sensitive to local perturbations. It would be hard for a heuristic measure to capture this delicate distinction. Another problem with immobile configurations is that there may exist an *immobile branch* even in a mobile configuration. Any motion planning involving immobile branches would require construction or destruction of the branches,

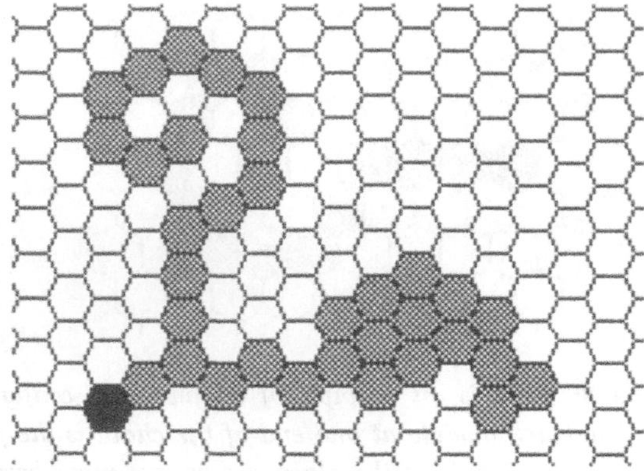

Figure 4: *An example of an immobile branch in a mobile configuration.*

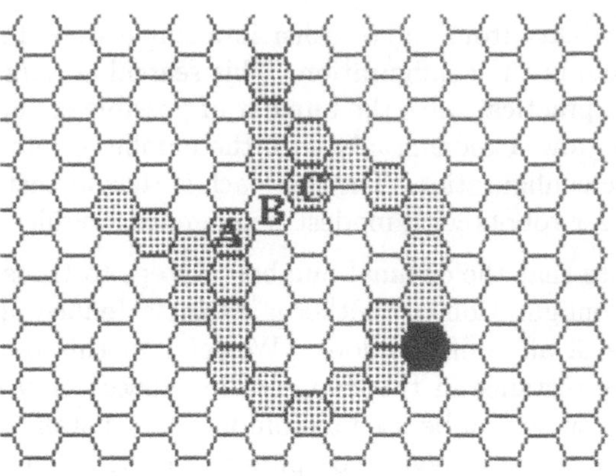

Figure 5: *The above configuration is inadmissible because the empty grid B is between two occupied grids A and C. If a module is added to the configuration at B, the configuration becomes admissible. The configuration also becomes admissible if the module at A or C is removed.*

and it would be hard to define a heuristic that captures that.

Given the difficulty for solving the motion planning problem in general, it is desirable to restrict ourselves to a simpler problem. As already mentioned, we propose to restrict ourselves to a subset of all possible configurations, large enough to adequately approximate an arbitrary configuration, yet restricted enough to make the motion planning more tractable.

5 Reconfiguration for Planar Hexagonal Modules

We consider the case of a reconfigurable robot composed of rigid hexagonal modules in the plane, as illustrated in Figure 1. We assume that all configurations of interest must be connected, and that a particular module, called the base, is immobile. The base can be used for transmitting power and communication into the system. We will refer to a possible location for a module as a grid cell, or simply a *grid*. The only motion allowed for a module is the rigid rotation around a vertex it shares with some other module in the configuration, subject to the condition that such rotation is not blocked by any module nor leaves the configuration disconnected.

We call a robot configuration *admissible* if it is connected and there is no empty grid cell directly in between two occupied grid cells in the configuration; see Figure 5. Admissible configurations have a very nice property, as shown in the following proposition.

Proposition 1 *Any admissible configuration can be transformed into a straight line configuration.*

As an immediate corollary of proposition 1:

Corollary 2 *Any admissible configuration can be transformed into any other admissible configuration with the same number of modules.*

Proof of proposition 1:

We construct a transformation from an admissible configuration to a straight configuration in two phases. First, we transform the admissible configuration to an admissible chain configuration (to be defined below), and then convert that chain to a straight line configuration.

Note that the complement of an admissible configuration in the plane consists of one or more connected components, exactly one of which is infinite. We call this infinite component the *outer free space* (with respect to the configuration). The module boundary line segments separating the outer free space and the configuration are called the *outer boundary*, and the modules in the configuration that border the outer free space are said to be on the outer boundary.

For ease of discussion, we assume that the hexagon lattice is oriented as in the previous figures. We use lattice coordinates with the x-axis in the minus 30 degree direction, and the y-axis in the vertical direction.

In the first phase, we use a greedy algorithm. Let us choose an extreme module in the configuration, say choose among all modules with maximal x coordinate the one with maximal y coordinate. Clearly, this module is on the outer boundary of the configuration. We grow a "tail" by repeatedly locating available modules that can be moved towards the extreme module we just selected, moving them there and then appending them to the end of this growing tail. This process ends when there is no module available for extending the tail.

The search for a module to add to the tail is done greedily. Starting from the extreme position, we search among all modules on the outer boundary of the configuration, say in the counter-clockwise direction, for the first module that can move clockwise without violating any motion constraints. We then let that module roll into the free space, along the outer boundary toward the extreme position, and then along the existing tail until it reaches the tip. The feasibility of such a motion is guaranteed with the help of the following lemma:

Lemma 3 *A module moving along the outer boundary of an admissible configuration in counter-clockwise (clockwise) direction can only be blocked at a position at which its blocker module is capable of moving in the that same direction.*

Proof of lemma 3:

We only consider the case when the direction is counter-clockwise. The clockwise case is similar. See Figure 6.

Say that in the original configuration, before the moving module moves to B, grid A is occupied, and grids B and C are empty. The admissibility of the configuration implies that grids D and F are empty. Thus the only reason that a moving module cannot move forward from B to C is that E is occupied. The admissibility again dictates that H and I are empty. There are now two possibilities:

- If K is occupied, then L is empty, and thus the module at A can roll counter-clockwise over K to I.
- If K is empty, then there must be a module at J so that A is not disconnected from the rest of the configuration. Grid M must then be empty, and the module at A can roll counter-clockwise over module at J to K.

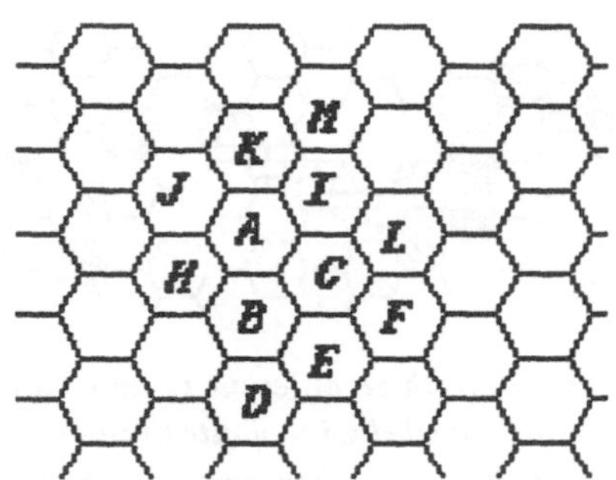

Figure 6: *This figure is used in the proof of lemma 3. A is a grid on the outer boundary of a configuration, and B and C are on the outer free space. If a module, moving in the counter-clockwise direction along the outer boundary of the configuration, can reach B but cannot reach C, then the lemma asserts that the module at A can move counter-clockwise to either K, or I. Other grid labels in the figure are used in the proof of the lemma.*

In both situations, the module at A can move counter-clockwise, and thus lemma 3 is proved. ◇

Since we choose the first module that can move counter-clockwise, the lemma guarantees that such a module can move counter-clockwise without ever being blocked toward the extreme module. The module can then move to the tip of the constructed tail. We need to make sure, however, that the configuration is admissible at the end of each move.

Note that the removal of a module from an admissible configuration always leaves it admissible as long as it is still connected. We only need to ensure that when we add the module to the tip of the tail, the configuration is still admissible. Since the tail points right up from a module at an extreme position, most modules in the tail are too far to affect the admissibility of the rest of the configuration. We need to be cautious only when building the first few modules in the tail.

Let the coordinates of the extreme module be (x_0, y_0). It is easy to verify that the following scheme for building the tail will keep the configuration admissible, see Figure 7:

- if $(x_0 - 1, y_0 + 1)$ is empty, the tail can be build at $(x_0 + 1, y_0 + 1), (x_0 + 2, y_0 + 2), (x_0 + 3, y_0 + 3), \ldots$;
- if $(x_0 - 1, y_0 + 1)$ is occupied, and $(x_0 - 1, y_0)$ is also occupied, the tail can be build at $(x_0, y_0 + 1), (x_0 + 1, y_0 + 1), (x_0 + 2, y_0 + 2), (x_0 + 3, y_0 + 3), \ldots$;

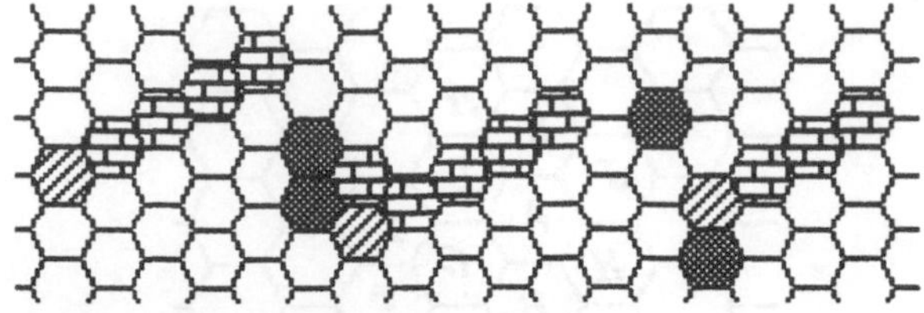

Figure 7: *There are three different types of tails. In each sub-figure, the grid shaded with slanted lines is the extreme module, gray grid(s) are other occupied grids, and grids shaded with bricks are grids in the tail to be constructed.*

- if $(x_0 - 1, y_0 + 1)$ is occupied, but $(x_0 - 1, y_0)$ is empty, then $(x_0 - 1, y_0 - 1)$ is also empty, and in order for (x_0, y_0) to be connected, $(x_0, y_0 - 1)$ is occupied. The tail can be build at $(x_0 + 1, y_0), (x_0 + 2, y_0 + 1), (x_0 + 3, y_0 + 2), \ldots$.

The first phase eventually ends. At that time, no piece in the configuration is free to move except the module at the end of the constructed tail.

Lemma 4 *The configuration at the end of phase 1 is a chain with one end at the end of the newly constructed tail and with the other end at the fixed base.*

Before we prove lemma 4, let us prove:

Lemma 5 *Any admissible configuration has a module on the outer boundary that can move without any motion constraint.*

Proof of lemma 5:

We call a module, say X, in a configuration a *connector* if the removal of X from the configuration makes the configuration disconnected. For a connector X, we define the *branch* at X as the subconfiguration consisting of X and the component that becomes disconnected from the base upon the removal of X.

Consider an extreme module in the configuration that is different from the fixed base. This module is on the outer boundary, and its motion (clockwise or counter-clockwise) into the outer free space doesn't violate any blocking constraint. If this module is not a connector, it is the desired module.

If it is a connector, consider the branch at this connector as a configuration, with the connector as the fixed base. By an inductive argument, we can assume that there is a module capable of moving without constraints. This module is on the outer boundary of the new configuration and thus is on the outer boundary of the original configuration. ◇

Proof of lemma 4:

In this proof, we use the term *loop* for a circular list of distinct modules, each being a neighbor to the modules located right before and after it in the circular list. See Figure 8.

Figure 8: *The black loop represents the largest loop in the configuration. A is the loop connector for the fixed base. B is the loop connector for the the tail. C is an extreme module on the loop that has no motion constraints for its motion to the outer free space. D is an extreme module on the loop, but is a connector. A and B happen to be extreme modules of the loop.*

Consider the configuration obtained after phase 1. Assume, for the sake of contradiction, that this configuration is not a chain, so there is a loop in the configuration. Among all the loops, choose the one enclosing the largest number of grids. Call this loop $\mathcal{L}$. From any module, say X, outside of loop $\mathcal{L}$, there is one or more paths connecting the module to the loop, all of which must share a common module Y contacting the loop (otherwise, the loop can be augmented to enclose more grids inside.) We call Y the *loop connector* for X. The removal of Y would make X disconnected from $\mathcal{L}$, and thus make the configuration disconnected. Thus, a loop connector is a connector as previously defined.

Consider an extreme module M in loop $\mathcal{L}$. Without loss of generality, we can assume that M is neither the loop connector for the tip of the tail constructed, nor the loop connector for the fixed base (if the loop connectors for those modules are defined). We can make this assumption, since a loop has more than two extreme modules, and thus we can always choose an extreme module different from the above two connectors. Note that with our choice of $\mathcal{L}$, module M must be on the outer boundary of the configuration.

There are two possible cases:

- If M has no neighboring module outside of the loop then, due to its extremality, M can move into the outer free space without any motion constraints.
- If M has as a neighbor module N outside of the loop, then M is a loop connector for N. Consider the branch at M as a configuration with fixed base M. By lemma 5, there is a module on the boundary of this new configuration that can move into the outer free space of the new configuration. It is clear that this module can move into the outer free space of the original configuration.

Thus, in both cases, we have a contradiction: there is a module capable of moving to the tail built at the end of phase 1. This proves that the assumption we made is false, and thus lemma 4 holds. In other words, the greedy algorithm in phase 1 transforms any admissible configuration into an admissible chain. ◇

In phase 2, we show that it is possible to transform an admissible chain configuration into a straight line configuration. The chain configuration may be winding around the fixed base, and a simple way to make it straight is to unwind it. Starting with the extreme module we used in phase 1, we repeatedly choose a different extreme module on the chain, then build a tail at this location. The tail will have more and more modules after each iteration, since the previous extreme module and all the modules in the previous tail are in the new tail. Eventually, the fixed base becomes one of the extreme modules, and the tail contains all modules other than the fixed base.

If the original configuration has n modules, the first phase would take $\mathcal{O}(n^2)$ steps since each module moves at most $\mathcal{O}(n)$ steps. In the second phase, the cost for each tail relocation is $\mathcal{O}(n^2)$. The number of tail relocation can be bounded by the following lemma:

Lemma 6 *The number of tail relocation steps in phase 2 is* $\mathcal{O}(n^{1/2})$

Proof: At any moment, consider the configuration excluding the tail. Assume that this configuration has m, $(m \leq n)$ modules, then its diameter d (measured in terms of grid modules) is at least $\Theta(m^{1/2})$. Let P and Q be two modules with distance d. It is easy to see that in two tail relocations, by relocating the tail to P, then to Q (or to Q, then to P), the tail has at least d more modules; the remaining configuration has $\Theta(m^{1/2})$ less modules. Straightforward analysis shows that the number of tail relocations must be at most $\mathcal{O}(n^{1/2})$. ◇

From lemma 6, the number of steps in the second phase is $\mathcal{O}(n^{5/2})$. This worse case can be achieved for a winding chain of n modules around the fixed base.

This completes our proof of proposition 1. ◇

6 Robots Composed of Metamodules

In this section, we explore a different approach to reconfiguration planning. Instead of restricting ourselves to admissible configurations, we consider arbitrary configurations that can be built out of a different building block, a *skeletal metamodule* or in short a *metamodule.* A skeletal metamodule is a collection of modules in a scaffolding-like structure, designed with the purpose of relaxing the basic module blocking constraints. The metamodules themselves pack in a lattice structure; the ones we consider have an overall shape that is a scaled-up copy of the basic module shape.

Figure 9 shows a robot of rigid hexagons, built from skeletal metamodules. The metamodule in Figure 9 still has a hexagonal shape, but it is an enlarged copy of the original module shape and has an empty interior.

A metamodule can decompose itself into modules, which can move to another nearby location, then reassemble themselves back into a metamodule. This can happen on the surface of the structure, just like with regular modules, but now with no blocking constraints. More interestingly, in a configuration of metamodules that are sufficiently large, a metamodule can move to a neighboring empty position *through the interior* of a supporting metamodule neighboring both the metamodule and the nearby empty location. This can be achieved by moving out of the way certain modules of the supporting metamodule, thus creating a gate to

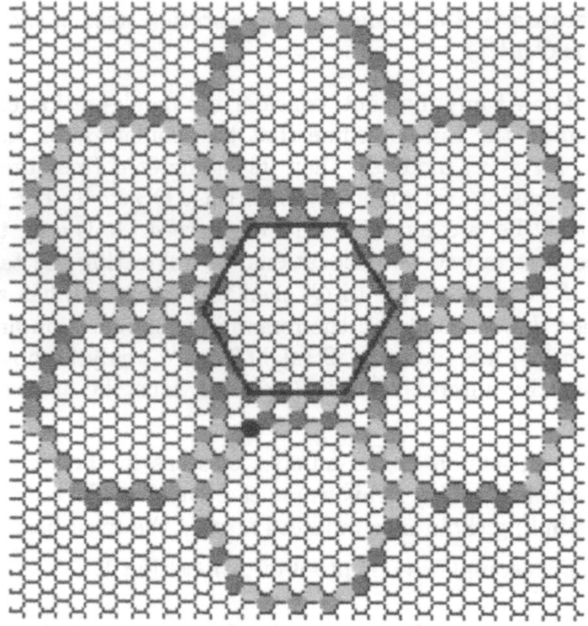

Figure 9: *This figure illustrates a configuration of several hexagonal metamodules. Each metamodule consists of 36 hexagonal modules, yet collectively behaves like a single hexagonal module. Unlike a regular hexagonal module, a metamodule have no blocking constraints, and can move by tunneling through other metamodules.*

its interior, and then allowing modules of the moving metamodule to go inside. The supporting metamodule can then close the gate and open another gate toward the empty location, letting the modules inside it move out and to assemble back to a metamodule again; see Figure 10. This neighbor to neighbor motion is again without any blocking constraint. In the same way a metamodule can actually tunnel through the interior of the structure and then reassemble itself when it reaches the boundary on the other side. Thus the metamodule can be thought of as a module of a new metamorphic robot, but now this module has no blocking constraints and can pass through the interior of other modules.

As the above example shows, metamodules can be built out of regular reconfigurable robot modules and then they themselves can be treated as building blocks for a reconfigurable robot. Any sufficiently large shape built out of basic modules can be well approximated by metamodules, or an exact scaled version of the original shape can be constructed directly out of metamodules (if scale is not important) by using a metamodule in-

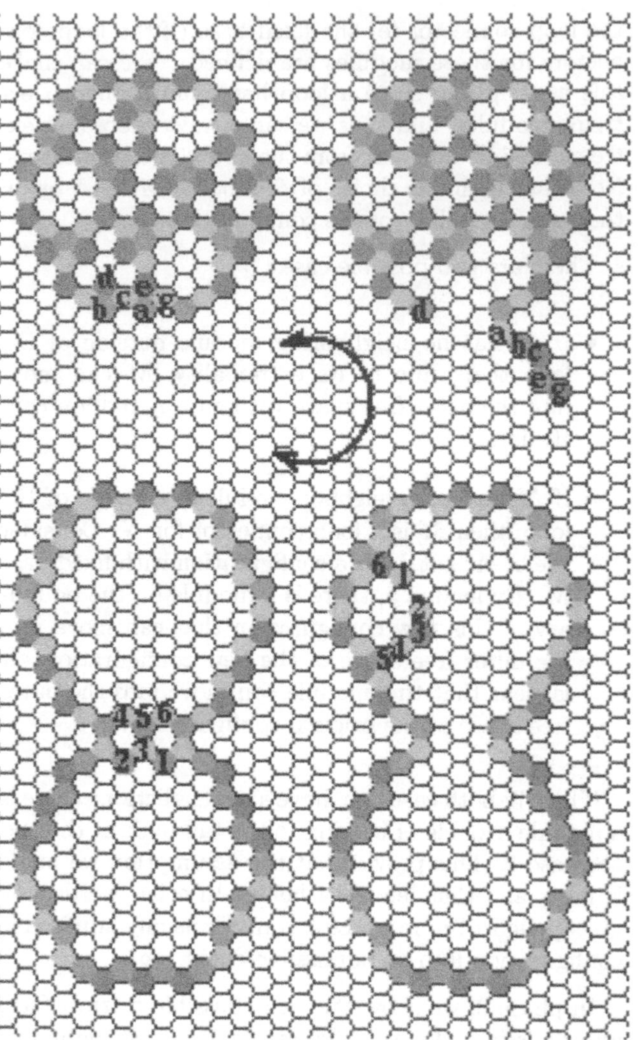

Figure 10: *This figure illustrates the process in which a metamodule, collapsed inside another metamodule, comes out to form a metamodule again. On the top we see a metamodule with another collapsed metamodule inside it. The grids labeled by letters form the gate that opens to let the collapsed metamodule out. On the bottom we see the second metamodule almost fully assembled. Its final modules are in the grids labeled by numbers — after they come out the transformation is complete.*

stead of a module as the basic unit. Thus metamodules configurations can be used in lieu of basic module configurations, with little loss in modeling power.

Metamodules have several advantages over standard modules. If we have a heuristic algorithm to do motion planning for a reconfigurable robot of hexagon modules, that algorithm can be used directly for a robot of hexagonal metamodules. Furthermore, heuristic measures for the distance between two configurations will most likely result in better quality plans in the metamodule case, as the blocking constraints have been eliminated. For example, the heuristics proposed by Chirikjian and Pamecha to do motion planning [10], and the bound on the number of steps [3] are valid for metamodule configurations.

Metamodules also have the additional tunneling capability. Besides moving on the boundary of the configuration, like regular modules, metamodules have the flexibility to go through the interior of other modules and thus they can make shortcuts through the configuration. If we let many modules move at the same time, the interior is now available to provide a much higher bandwidth channel for the motion than is possible on the surface alone. This tunneling capability of metamodules can be also used to create compressible modules, such as those proposed by Rus and Vona [11].

6.1 Restricted Configurations of Metamodules

In this section, we propose a new way to do parallel motion planning for a restricted class of metamodule configurations. This combines the idea of metamodules and that of restricted configurations. We allow modules to move concurrently to speed up reconfiguration. The approach is motivated by the work of Rus and Vona [11].

Suppose that we have two layers of metamodules as shown in Figure 11. If we keep the metamodules on the bottom layer fixed, we can collapse all metamodules on the top layer, move the modules in parallel to

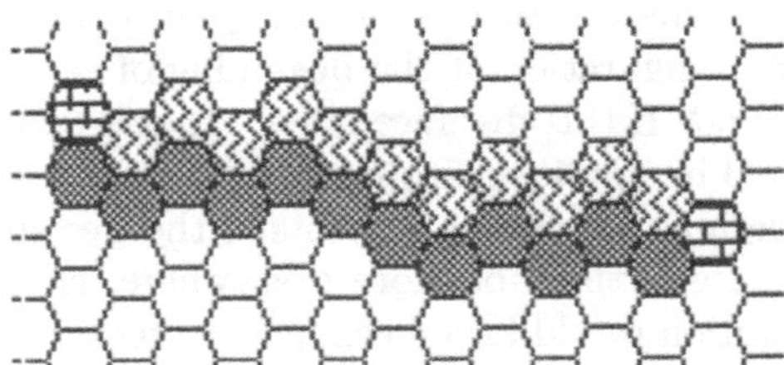

Figure 11: *The top layer of metamodules can move to the right over the bottom layers in one parallel time step. This is equivalent to moving the metamodule on the top layer from the extreme left position to the extreme right position in one parallel step.*

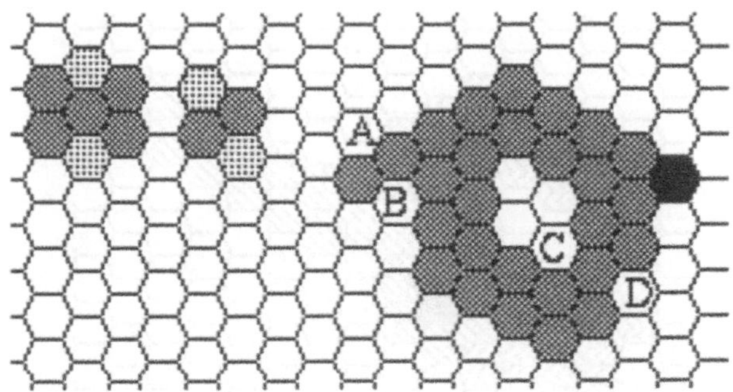

Figure 12: *The two local combinations of modules on the left are prohibited in a* fat *configuration. The dark and light gray grids represent occupied and unoccupied grids respectively. The configuration on the right is not* fat, *because the fat condition is violated in two different places. If two modules are added to the configuration at B and C, the configuration will become* fat.

the right, then reconstruct the metamodules on top. This can be thought of as a shifting of the top layer over the bottom layer by one module, in one time step. The overall effect of the operation is to move the leftmost metamodule of the top layer all the way to the right in one time step, independent of the number of metamodules on the layers.

We will use the above operation as a basic step for our motion planning. It is preferable that the configuration we deal with is not too skinny. We restrict ourselves to a new class of *fat configurations* in which two local combinations of metamodules, as explained in Figure 12, are prohibited. This restriction essentially requires that a *fat* configuration has locally, width two or more; again, it should be clear that this restriction has little affect on the class of shapes that can be approximated. We will use the term *nonfat* for configurations that are not *fat*.

Fat configurations have a nice property, as shown in the following proposition:

Proposition 7 *Any* fat *configuration of n metamodules can be transformed into a straight line* fat *configuration in $\mathcal{O}(n)$ parallel steps.*

As an immediate corollary of proposition 7:

Corollary 8 *Any* fat *configuration of n metamodules can be transformed into any other* fat *configuration with the same number of modules in $\mathcal{O}(n)$ parallel steps.*

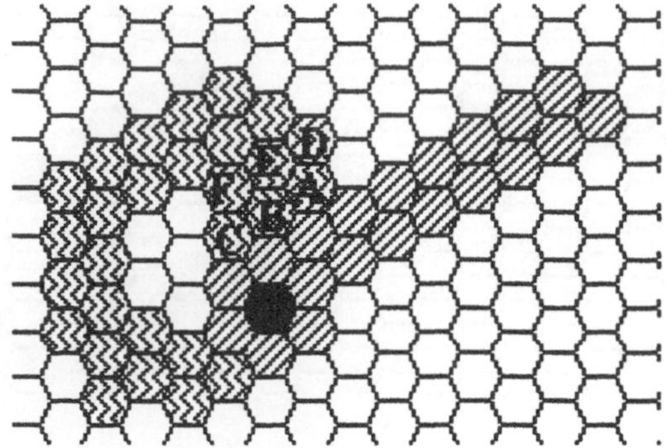

Figure 13: *The metamodules in a* fat *configuration can move to generate the rounded straight line to the right. The algorithm applied on this configuration first moves A, then moves both B and C (and breaks the only loop in the configuration), then moves D, F, and E to the tail.*

Proof of proposition 7:

Without loss of generality, it suffices to show that we can transform a *fat* configuration to a *fat* straight line configuration with a rounded end (shown shaded with slanted lines in Figure 13.) Furthermore, we can assume that the six metamodules surrounding the base metamodule are already in the original configuration. This is so we can preprocess the original configuration and postprocess the final configuration to the desired form in $\mathcal{O}(n)$ time.

We again will use a greedy algorithm. Starting from the tip of the tail constructed so far, we search in the counter-clockwise direction along the boundary of the configuration to find the first metamodule not already in the tail satisfying either of the following two conditions:

- The removal of the metamodule leaves the configuration connected and *fat*. In this case, we use the parallel shearing operation to shift an entire layer of the configuration on the boundary, with the net effect of moving the selected metamodule in $\mathcal{O}(1)$ steps to the tip of the tail. If the layer on the boundary includes metamodules in both the upper and lower layers of the tail, we do the shearing in two steps, the first shearing uses only one of the two layers of the tail, and the second shearing uses the remaining layer.
- The removal of the metamodule leaves the configuration connected but *nonfat*, and further removal of one of its neighboring metamodule leaves the configuration still connected and *fat* again. In this case, we move both modules to the tip of the tail in two consecutive time steps.

Note that the fixed base should not be in any layer being shifted, and this is guaranteed since there are modules surrounding the fixed base at all times. Also, although the configuration at the end of each step may not be *fat*, this only happens at the tip of the tail (when the tail first touches some other part of the configuration.) For the analysis, it is safe to assume that the configuration is always *fat*, since we can pretend that the tip of the tail is not yet there when we search for the next metamodule.

Before further analysis, let us give some intuition behind the two cases above. In the first case, the operation takes away a metamodule on the boundary, making the configuration thinner. Since the configuration is *fat*, the topology of the configuration does not change. In the second case, the configuration has some thin cross section in a loop. The removal of a module in that thin section makes the configuration *nonfat*, yet the additional metamodule removal eliminates the skinny cross section. The loop is disconnected, and the topology of the configuration changes, see Figure 13.

We claim that the above greedy algorithm moves all metamodules in the configuration to the tail. Let us assume for the sake of contradiction that there are some metamodules not in the tail when the greedy algorithm fails to find any metamodules to move. Consider the free space, the complement of the configuration. If there are two or more connected components of the free space, consider the smallest distance between the component contacting the tail and all remaining components. This distance is not one, since we assume that the configuration at the beginning of each search is always *fat*. If this distance is two, the second condition would be applicable to remove the two metamodules, merging the two components of the free space. If the distance is three or more everywhere, clearly the first condition would have been used to remove further metamodules. Thus the free space is connected. In order for the first condition not to be applicable, the metamodules not in the tail are on one or more thin chains originating from the tail. This is a contradiction, since the metamodules at the tip of these chains would have been removed by the first condition.

The $\mathcal{O}(n)$ number of parallel steps follows trivially, since the remaining configuration has one less metamodule after every $\mathcal{O}(1)$ time steps. ◇

6.2 Reconfiguration by Tunneling

If the metamodules are large enough to allow tunneling, then another method can be used for reconfiguring an arbitrary configuration into a line, in $\mathcal{O}(n)$ steps. The intuition here is that a metamodule at a boundary of a configuration can follow a tunneling path through a configuration emerging at a growing tail that is forming a line (and consequently between two arbitrary configurations). Every metamodule in that path can immediately follow this leader metamodule and emerge at the end of the tail. The appropriate metaphor here is of a turning a rubber glove inside out by pushing in at the fingers and having them come out the other end, one at a time, appended to each other in a straight line, see Figure 14.

The first step is to impose a tree structure on the configuration. This is done by labeling the metamodules with increasing numbers in a breadth first traversal sequence, starting with the metamodule where the tail will grow as number 0. The breadth first numbering can be achieved with simple incremental message passing.

The following algorithm is purely local, in that every metamodule runs the same program (a small finite state machine) and no communication is required between metamodules except for knowledge of their neighboring metamodules and their numbers (in the labeling).

There are four states that every metamodule can be in:

- S_1 — the metamodule is part of the original structure (not tunneling nor part of the tail)
- S_2 — the metamodule is tunneling towards the 0 metamodule.
- S_3 — the tunneling metamodule has reached the 0 metamodule and is now tunneling in the tail.
- S_4 — the metamodule has grown the tail and so has stopped tunneling.

There are three rules for transitioning between states:

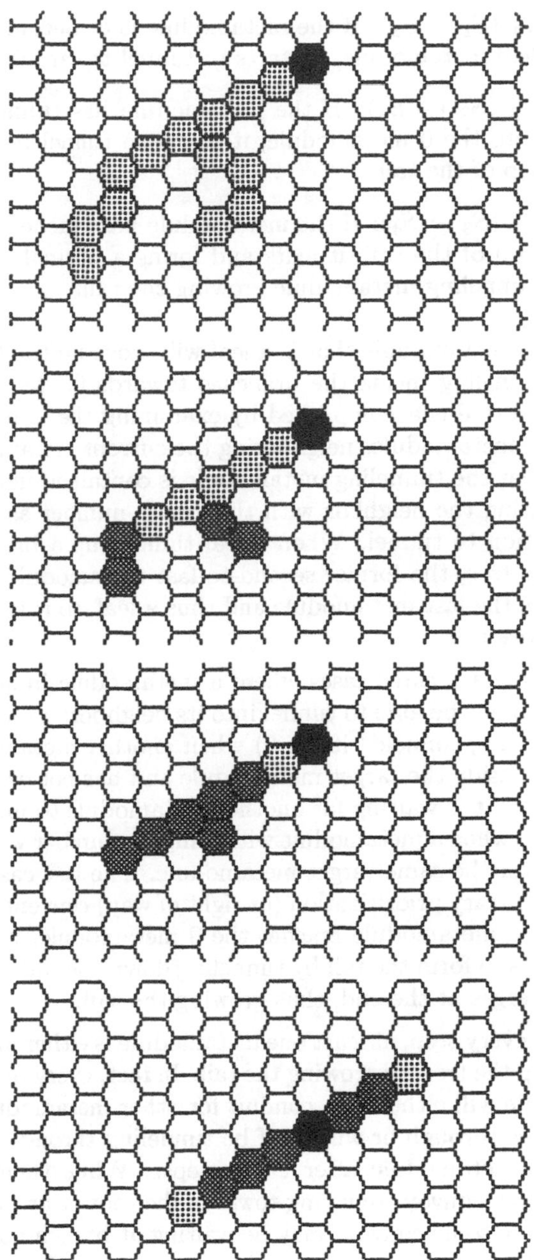

Figure 14: *Several snapshots of the tunneling process as it transforms the top configuration to a straight line. Metamodules tunnel toward the base, then extend out to form a straight line. In these figures, light gray grids contain regular metamodules, and dark gray grids contain metamodules with another metamodule tunneling inside them.*

- T_1 ($S_1 \mapsto S_2$): If the metamodule is or becomes a leaf metamodule, it starts to tunnel toward 0.
- T_2 ($S_2 \mapsto S_3$): If the metamodule has tunneled into the 0 metamodule, it starts to tunnel to the tip of the tail.
- T_3 ($S_3 \mapsto S_4$): If the metamodule has reached the end of the tail, it exits and forms a normal non-tunneling metamodule growing the tail.

Any metamodule that is a leaf will move up the tree by tunneling inside the branches towards 0. Moving up the tree is accomplished by examining the number of the metamodules neighboring the current metamodule that the tunneling metamodule is contained in and targeting the neighbor with the lowest number as the direction to tunnel. When a leaf tunnels up a branch of the tree, the former second-to-last metamodule becomes the last metamodule and thus a leaf, so it starts to move up.

There are three cases where a metamodule in state S_2 will not be able to tunnel into its neighbor: 1) when it is being tunneled itself, 2) when another metamodule is inside the target metamodule and has not exited because it is waiting for another metamodule to move, and 3) when a metamodule with a higher number wants to enter the same target metamodule. The last case is an arbitrary prioritization (or right of way) convention. Once a metamodule reaches the 0 metamodule, it attempts to form the tail by tunneling down the tail until it emerges at the end, thus growing the tail.

At every step, at least one metamodule is either moving up the tree, or growing the tail. In fact, every metamodule will either be a conduit for other metamodules to pass through or will itself be tunneling through another metamodule after $\mathcal{O}(n)$ steps. While there is always a convoy tunneling towards the root or growing the tail, some convoys may be waiting at branch nodes for other convoys to finish their tunneling. Clearly a metamodule only moves $\mathcal{O}(n)$ steps before it reaches its eventual goal position in the tail, as its motion follows a path in the tree and through the tail. What is more interesting is that a metamodule M also never waits a total of more than $\mathcal{O}(n)$ steps before reaching its destination. We can see this as follows. After the initial $\mathcal{O}(n)$ steps, metamodule M waits only when it is being tunneled through, or when it is part of a tunneling convoy that is waiting at a branch node up the tree for another convoy to pass through. In both cases another module, the *blocker* of M from a different subtree, is moving though the branch node for every time step that our module M has to wait. But once a metamodule has been the blocker of M, it can never block M again at a branch node, as it has now become an ancestor of M in the tree. Thus no node needs more than $\mathcal{O}(n)$ moving or waiting steps to reach its final destination.

Since we move from the leaves of the branch inward, connectivity is never broken. Furthermore this algorithm operates in a very simple and purely local fashion. All of this assumes that only one metamodule can be "tunneled" inside another metamodule at one time. Metamodules with larger interiors would have more "bandwidth" which would mean less traffic problems and fewer steps.

7 Conclusion

In this paper we have explored certain restrictions on the allowed structure of reconfigurable robots with the aim of making the motion planning problem easier, without unduly confining the set of shapes the robot can assume. Though all our results are in 2D, we believe that appropriate extensions to 3D are possible, and in some cases straightforward. For example, we can build 3D rhombic dodecahedron metamodules that consist of an edge skeleton of a scaled copy of the basic RD module. Because the parallelogram facets of these metamodules are actually solid only along their edges, other collapsed metamodules can enter, tunnel through, and exit a given metamodule easily, without any of the gate opening or closing operations we need in 2D. In 3D as well, the traffic management of convoys of modules tunneling through the robot interior seems easier than in 2D. Since the tunneling reconfiguration algorithm imposes a tree structure which is independent of the dimensionality of the physical modules, it will work for this 3D RD case as well.

Acknowledgments

Leonidas Guibas and An Nguyen were supported in part by National Science Foundation grants CCR–9623851 and IRI–9619625, as well as by US Army MURI grant DAAH04–96-1-0007. An Nguyen was also supported by a Stanford Graduate Fellowship.

References

[1] Casal, A., Yim, M., "Self-Reconfiguration Planning for a Class of Modular Robots," SPIE Symposium on Intelligent Systems and Advanced Manufacturing, Sept. 1999.

[2] Chirikjian, G., "Kinematics of a Metamorphic Robotic System," in *Prc. 1994 IEEE Int. Conf. Robotics and Automation, San Diego, CA.*

[3] Chirikjian, G., Pamecha, A., "Bounds for Self-Reconfiguration of Metamorphic Robots," *JHU Technical Report, RMS-9-95-1.*

[4] Chirikjian, G. et al., "Evaluating Efficiency of Self-Reconfiguration in a Class of Modular Robots," in *Journal of Robotic Systems, June 1996.*

[5] Kotay, K., et al., "The Self-reconfiguring Robotic Molecule: Design and Control Algorithms," Algorithmic Foundations of Robotics, 1998.

[6] Kotay, K. et al., "The Self-reconfiguring Robotic Molecule," in *Proceedings of the 1998 IEEE International Conference on Robotics & Automation, May 1998.*

[7] Murata, S. et al., "Self-Assembling Machine", in *Proceedings of the 1994 IEEE International Conference on Robotics & Automation, May 1994.*

[8] Murata, S. et al., "A 3-D Self-Reconfigurable Structure", in *Proceedings of the 1998 IEEE International Conference on Robotics & Automation, May 1998.*

[9] Pamecha, A. et al., "Design and Implementation of Metamorphic Robots" in *Proceedings of the 1996 ASME Design Engineering Technical Conference and Computers in Engineering Conference, Aug 1996.*

[10] Pamecha, A., Chirikjian, G., "Useful Metrics for Modular Robot Motion Planning," in *IEEE Transactions on Robotics and Automation, Vol. 13, No 4, August 1997.*

[11] Rus, D., Vona, M., "Self-reconfiguration Planning with Compressible Unit Modules," in *Proceedings of the 1999 IEEE International Conference on Robotics & Automation, May 1999.*

[12] Yim, M., "A Reconfigurable Modular Robot with Many Degrees of Locomotion," in *Proc. 1993 JSME Int. Conf. Advanced Mechatronics.*

[13] Yim, M., "Locomotion With a Unit-Modular Reconfigurable Robot", Stanford University PhD Thesis, 1994.

[14] Yim, M. et al., "Rhombic Dodecahedron Shape for Self-Assembling Robots," Xerox PARC, SPL TechReport P9710777, 1997.

[15] Zhang, Y., et al., "Distributed Control for 3D Shape Metamorphosis," submitted *Autonomous Robots Journal, special issue on self-reconfigurable robots*, 1999.

Positioning Symmetric and Non-Symmetric Parts using Radial and Constant Force Fields

Florent Lamiraux, *LAAS-CNRS, Toulouse, France*
Lydia E. Kavraki, *Rice University, Houston, TX*

Part positioning is an important task in manufacturing. New approaches have been proposed to perform this task using force fields implemented on an active surface. A part placed on such a surface is subjected to a resultant force and torque and moves toward a stable equilibrium configuration. Such force fields can be implemented using different technologies. In this paper, we study the combination of a unit radial field with a small constant force field. In prior work we proved that such a combination can uniquely position a class of non-symmetric parts. In this paper, we propose a more complete modeling of this combination which allows us to devise a method to determine all the equilibrium configurations of a part in the above force fields. This method works for both symmetric and non-symmetric parts. Beyond the method, this paper reports a comprehensive study of the action of radial and constant potential fields over parts with an original characterization of local minima of the lifted potential field.

1 Introduction

During manufacturing, parts typically stored in boxes have to be manipulated and oriented before assembly. This task is critical in manufacturing since it strongly affects the productivity of the assembly line. Orientation has been traditionally performed by vibratory bowl feeders. However, these devices are designed for a given part and need to be modified if the shape of the part changes.

Recent work has investigated alternative ways of orienting parts, and emphasis has been placed on simple programmable devices [1, 6, 9, 15, 14]. Part positioning without sensing has become very popular over the past few years since it is easy to implement and the methods proposed can be very robust [1, 2, 3, 6, 7, 8, 9, 16]. Our work brings new contributions in this domain.

One of the pioneering papers in this area, [9], proposed to orient a polygonal part by a sequence of squeezes performed by two parallel jaw grippers. Given a polygonal part, the paper described an algorithm to compute the best sequence of squeezes that uniquely orients the convex hull of the part no matter what the initial configuration is. In [16], the parts are on a conveyor belt and rotate by contact with passive fences.

Another approach is based on force fields implemented in an horizontal plane [6, 13]. A part lying on the field is subjected to a resultant force and torque that move the part toward a stable equilibrium configuration if one exists. A series of papers (see [4] for detailed references) established the fundamentals for part manipulation using force fields. Current technology permits the implementation of certain force fields in the microscale with MEMS actuators [6] or air jets [2] and in the macroscale using mechanical devices [14]. Vibrating plates can also be used to produce certain force fields in the plane [3].

In work which is summarized in [4, 6] the properties of force fields that are suitable for sensorless manipulation were analyzed and manipulation strategies were proposed. Several fields were investigated including the squeeze, radial, and inertial fields and combinations thereof. Notably, the algorithm of [9] still applies when the jaw grippers are replaced by squeeze force fields. Later work [10] introduced the elliptic potential field which gave rise to two stable equilibria for non symmetric parts.

Complexity and Uniqueness of Positioning

There are two main issues when positioning[1] parts using force fields. The first one is the number of basic operations, (*i.e.*, number of force fields used) and the second one is the number of distinct positions that the

[1]Orienting a part consists in specifying the orientation of the part. Positioning a part consists in specifying the position and orientation.

strategy yields. Both issues have been addressed previously and work has been done both on finding shorter sequences of operations [6] and on finding a unique operation that positions a part. In [10], a quadratic potential field is used to position parts. Any part with distinct principal moments of inertia has exactly two stable equilibrium configurations in these fields. The other parts have undefined orientation. Addressing the above issues at the same time would consist in finding a single field that uniquely positions a part. It was conjectured in [6] that the combination of a unit radial field with a constant force field would fulfill this task. In [5], we proved that this conjecture is true under the condition that when the part is subjected to the unit radial field only, the center of mass of the part in equilibrium is not at the origin of the field. Thus, for almost any non-symmetric part, there exists a force field that uniquely positions this part.

Contributions and Organization of the Paper

Symmetric parts like a rectangle or a regular polygon for instance do not fulfill the above condition: When placed in a unit radial field, at equilibrium, the center of mass of the part will coincide with the origin of the unit radial field, due to symmetry. The goal of this paper is to determine all the stable equilibrium configurations of a general part, symmetric or not, subjected to the combination of a unit radial field with a small constant field. There are two main contributions in this paper. The first one is a characterization of the stable equilibrium configurations of a part using a general modeling based on the curve of translational equilibria for fixed orientations and its partial derivatives. This modeling is a generalization of results previously published [5, 12]. The second contribution is a method to compute the above mentioned partial derivatives and an algorithm that determines all the equilibrium configurations of a given part.

This paper is organized as follows. In Section 2, we give some notation and definitions. In Section 3, we first define the unit radial field that we then combine with a constant field. We define the equilibrium curve as the locus of the translational equilibria for fixed orientation. After pointing out the smoothness of this curve, we state the central result of the paper that expresses the stable equilibrium configurations of a part w.r.t. the equilibrium curve and its partial derivatives. Section 4 first describes the practical computations of these partial derivatives. An algorithm to determine all the stable equilibrium configurations of a part is then given. Finally, an example of positioning a symmetric part is presented. We conclude in Section 5.

2 Notation and Definitions

This section puts forward the notation and definitions used in the paper. We first define the action of a force field over a part and the notion of a stable equilibrium configuration in this context. Then we outline some properties of force fields deriving from a potential function. Most of these definitions have been introduced in [4].

If E is a set, we denote by $\text{int}(E)$ and ∂E the interior and boundary of E. Throughout the paper, $\mathbf{r}$ denotes a point of the plane of coordinates (ξ, η) in a frame attached to the plane. We consider a part in the plane occupying surface S with center of mass G at $(0,0)$ in a reference configuration $\mathbf{q}_0$. We suppose that S is a compact set and that its boundary ∂S is a zero-measure subset of the plane. The configuration space of the part is $\mathcal{C} = \mathbf{R}^2 \times \mathbf{S}^1$, where $\mathbf{S}^1$ is the unit circle. A configuration $\mathbf{q} = (x, y, \theta) \in \mathcal{C}$ corresponds to a rigid-body transformation $\varphi_{\mathbf{q}}$ in the plane transforming $\mathbf{r} = (\xi, \eta)$ into

$$\varphi_{\mathbf{q}}(\mathbf{r}) = \begin{pmatrix} x + \xi\cos\theta - \eta\sin\theta \\ y + \xi\sin\theta + \eta\cos\theta \end{pmatrix}.$$

We denote by $S_{\mathbf{q}}$ the subset occupied by the part in configuration $\mathbf{q}$.

When the part is placed on a force field $\mathbf{f}(\mathbf{r})$, the resultant force and torque it is subjected to are the following:

$$\mathbf{F}(\mathbf{q}) = \int_{S_{\mathbf{q}}} \mathbf{f}(\mathbf{r})d\mathbf{r} \qquad (1)$$

$$M(\mathbf{q}) = \int_{S_{\mathbf{q}}} (\mathbf{r} - \mathbf{r}_G) \times \mathbf{f}(\mathbf{r})d\mathbf{r} \qquad (2)$$

where $\mathbf{r}_G = (x, y)$ is the position of the center of mass G, when the part is at configuration $\mathbf{q}$.

We say that $\mathbf{q}$ is an *equilibrium configuration* if the resultant force and torque vanish at $\mathbf{q}$. Moreover, an equilibrium configuration $\mathbf{q}$ is said to be *stable* if, subjected to a small perturbation, the part stays in the neighborhood of $\mathbf{q}$.

Potential Fields The class of force fields deriving from a potential function are very helpful in mechanics and electrostatics since the equilibrium configurations of a point-particle subjected to such a field are exactly the local minima of the potential field. This property extends naturally for "reasonable" force fields from the plane to the configuration space via the notion of *lifted potential field* [4].

Definition 1 (Lifted Potential Field)
Let $\mathbf{f}(\mathbf{r})$ be a force field in the plane deriving from a potential function u, *i.e.*, $\mathbf{f}(\mathbf{r}) = -\nabla u(\mathbf{r})$. The function

$$U(\mathbf{q}) = \int_{S_\mathbf{q}} u(\mathbf{r})d\mathbf{r} = \int_S u(\varphi_\mathbf{q}(\mathbf{r}))d\mathbf{r}$$

over $\mathcal{C}$ is called the *lifted potential field* induced by u.

The lifted potential field $U(\mathbf{q})$ is thus defined by integrating the value of the plane potential field over the surface $S_\mathbf{q}$ occupied by the part in configuration $\mathbf{q}$. In [6] some properties of the lifted potential field are established. The most interesting among these properties is that the partial derivatives of the lifted potential field w.r.t. x, y and θ are exactly the opposite of respectively the coordinates of the resultant force and the resultant torque.

$$\begin{aligned} \frac{\partial U}{\partial x}(\mathbf{q}) &= -F_x(\mathbf{q}) \\ \frac{\partial U}{\partial y}(\mathbf{q}) &= -F_y(\mathbf{q}) \\ \frac{\partial U}{\partial \theta}(\mathbf{q}) &= -M(\mathbf{q}). \end{aligned}$$

The main consequence of these equalities is that stable equilibrium configurations are equivalent to local minima of the lifted potential. For this reason, we focus our attention in the following sections on the local minima of the lifted potential field induced by radial and constant fields.

3 Radial and Constant Fields

In this section we study the properties of combinations of a unit radial force field with a small constant force field. Both fields derive from a potential. We give a method to determine the stable equilibrium configurations (*i.e.*, local minima) of a part subjected to such a combination in the general case. See also [5], where some of the material below was originally developed.

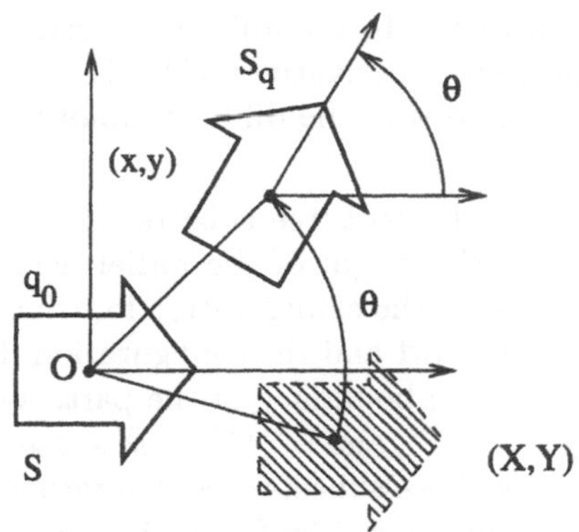

Figure 1: *Parameterization of $\mathcal{C}$ with the system of coordinates $\mathbf{q} = (X, Y, \theta)$. $\varphi_\mathbf{q}$ corresponds to a translation of (X, Y) followed by a rotation of angle θ about the origin.*

3.1 Unit Radial Field

We call unit radial field the force field $\mathbf{f}(\mathbf{r}) = -\frac{\mathbf{r}}{\|\mathbf{r}\|}$. This field has constant magnitude, is oriented toward the origin of the frame attached to the plane and derives from the potential function:

$$v(\mathbf{r}) = \|\mathbf{r}\| = \sqrt{\xi^2 + \eta^2}. \quad (3)$$

v is clearly symmetric by rotation about the origin. Thus if $\mathbf{q}'$ is obtained from $\mathbf{q}$ by a rotation about the origin, the values of the lifted potential field at $\mathbf{q}$ and $\mathbf{q}'$ are the same. To take fully advantage of this property, we are going to use a new system of coordinates (X, Y, θ) for the configuration of the part, illustrated in Figure 1 and defined by:

$$\begin{aligned} X &= x\cos\theta + y\sin\theta \\ Y &= -x\sin\theta + y\cos\theta. \end{aligned}$$

Expressed in this system of coordinates, the lifted potential field V corresponding to v depends only on X and Y and can be written as follows:

$$V(X, Y, \theta) = \int_S v(X + \xi, Y + \eta)d\xi d\eta. \quad (4)$$

Because of the independence on θ, we will consider V as a function of (X, Y) only. Notice that when $\theta = 0$, $X = x$ and $Y = y$. $V(X, Y) = V(x, y, 0)$ can thus be considered as the lifted potential field of the part in translation. This interpretation can be helpful to understand some of the forthcoming developments.

Smoothness of V In the following sections, we will use a lot the partial derivatives of V. For this reason, we state in this section the differentiability properties of this function.

$\mathcal{C}$ is partitioned into three subsets. The configurations for which the origin of the radial field is in the interior of the part, the configuration for which the origin is outside the part and the configurations for which the origin is on the boundary of the part. We denote respectively $\mathcal{C}^{in}$, $\mathcal{C}^{out}$, and $\mathcal{C}^{bound}$ these subsets. To decide in which of these subsets a configuration (X, Y, θ) lies, we need only to consider (X, Y) in the new system of coordinates. (see Figure 3.1 for intuition):

$$\begin{array}{lcl} \mathbf{q} \in \mathcal{C}^{in} & \Leftrightarrow & (-X, -Y) \in \text{int}(S) \\ \mathbf{q} \in \mathcal{C}^{out} & \Leftrightarrow & (-X, -Y) \notin S \\ \mathbf{q} \in \mathcal{C}^{bound} & \Leftrightarrow & (-X, -Y) \in \partial S. \end{array}$$

As far as differentiability is concerned, V has the following properties.

Proposition 2 *(i) V is of class C^∞ over $\mathcal{C}^{in}$ and over $\mathcal{C}^{out}$,*

(ii) V is of class C^2 over $\mathcal{C}$ and

$$\frac{\partial V}{\partial X}(X,Y) = \int_S \frac{\partial v}{\partial \xi}(X+\xi, Y+\eta)d\xi d\eta \quad (5)$$

$$\frac{\partial V}{\partial Y}(X,Y) = \int_S \frac{\partial v}{\partial \eta}(X+\xi, Y+\eta)d\xi d\eta \quad (6)$$

$$\frac{\partial^2 V}{\partial X^2}(X,Y) = \int_S \frac{\partial^2 v}{\partial \xi^2}(X+\xi, Y+\eta)d\xi d\eta \quad (7)$$

$$\frac{\partial^2 V}{\partial Y^2}(X,Y) = \int_S \frac{\partial^2 v}{\partial \eta^2}(X+\xi, Y+\eta)d\xi d\eta \quad (8)$$

$$\frac{\partial^2 V}{\partial X \partial Y}(X,Y) = \int_S \frac{\partial^2 v}{\partial \xi \partial \eta}(X+\xi, Y+\eta)d\xi d\eta. \quad (9)$$

The classical results of integration theory, regarding differentiating inside the integral cannot be applied as such to the unit radial field because of the non differentiability at the origin. The proof of this proposition is based on a family v_h of approximations of v, C^∞ everywhere and equal to the radial field outside the disc of radius h centered at the origin. We do not give here the proof of this Proposition. This proof can be found in [11].

It is interesting to notice that the lifted potential field V is C^∞ although v is not even differentiable. The singularity of v at the origin affects the smoothness of V only where the boundary of the part passes above this singularity.

Minimum of V and Pivot Point Due to symmetry, the unit radial field obviously cannot uniquely position a part. However, for a fixed value of θ, any part in translation has a unique stable equilibrium. This property is a consequence of Proposition 3 below. This proposition was also given in [5]. It is repeated here for completeness. Notice that it provides a different proof for the existence and uniqueness of the pivot point defined below than the one given in [4].

Proposition 3 *V verifies the following properties:*

(i) The Hessian of V Hess $V(X,Y)$ is positive definite everywhere in $\mathbf{R}^2$,

(ii) V has a unique local minimum over $\mathbf{R}^2$.

Proof: As Hess V is a 2 by 2 matrix, (i) is equivalent to:

$$\forall (X,Y) \in \mathbf{R}^2, \quad \begin{array}{c} \text{tr Hess } V(X,Y) > 0 \\ \text{det Hess } V(X,Y) > 0 \end{array}$$

where tr and det are respectively the trace and determinant operators. According to (7-9) and (3), the second order partial derivatives of V are:

$$\frac{\partial^2 V}{\partial X^2}(X,Y) = \int_S \frac{(Y+\eta)^2}{((X+\xi)^2+(Y+\eta)^2)^{3/2}} d\xi d\eta$$

$$\frac{\partial^2 V}{\partial Y^2}(X,Y) = \int_S \frac{(X+\xi)^2}{((X+\xi)^2+(Y+\eta)^2)^{3/2}} d\xi d\eta$$

$$\frac{\partial V}{\partial X \partial Y}(X,Y) = \int_S \frac{-(X+\xi)(Y+\eta)}{((X+\xi)^2+(Y+\eta)^2)^{3/2}} d\xi d\eta.$$

It is straightforward from these expressions that tr Hess $V = \frac{\partial^2 V}{\partial X^2} + \frac{\partial^2 V}{\partial Y^2}$ is positive everywhere. The determinant of the Hessian of V:

$$\text{det Hess } V(X,Y) = \left(\frac{\partial^2 V}{\partial X^2}\frac{\partial^2 V}{\partial Y^2} - (\frac{\partial^2 V}{\partial X \partial Y})^2 \right)(X,Y)$$

is the sum of two terms, each of which is a product of two integrals over S. Replacing these products by integrals over the Cartesian product $S^2 = S \times S$, we get

$$\left(\int_S f(\xi,\eta)d\xi d\eta \right) \left(\int_S g(\xi,\eta)d\xi d\eta \right) = \int_{S^2} f(\xi_1,\eta_1)g(\xi_2,\eta_2)d\xi_1 d\eta_1 d\xi_2 d\eta_2.$$

If we condense the notation as follows, $X_i = (X + \xi_i)$, $Y_i = (Y + \eta_i)$ for $i = 1, 2$, and omit $d\xi_1 d\eta_1 d\xi_2 d\eta_2$, we have:

$$\begin{aligned}\det \text{Hess } V(X,Y) &= \int_{S^2} \frac{X_1^2 Y_2^2}{(X_1^2+Y_1^2)^{3/2}(X_2^2+Y_2^2)^{3/2}} \\ &- \int_{S^2} \frac{X_1 Y_1 X_2 Y_2}{(X_1^2+Y_1^2)^{3/2}(X_2^2+Y_2^2)^{3/2}} \\ &= \int_{S^2} \frac{Y_1^2 X_2^2 - X_1 Y_1 X_2 Y_2}{(X_1^2+Y_1^2)^{3/2}(X_2^2+Y_2^2)^{3/2}}.\end{aligned}$$

In the first integral, (X_1, Y_1) and (X_2, Y_2) have a symmetric role and can be switched so that $X_1^2 Y_2^2$ can be replaced by $\frac{1}{2}(X_1^2 Y_2^2 + X_2^2 Y_1^2)$ and:

$$\begin{aligned}\det \text{Hess } V(X,Y) &= \frac{1}{2}\int_{S^2} \frac{X_1^2 Y_2^2 + X_2^2 Y_1^2 - 2X_1 Y_1 X_2 Y_2}{(X_1^2+Y_1^2)^{3/2}(X_2^2+Y_2^2)^{3/2}} \\ &= \frac{1}{2}\int_{S^2} \frac{(X_1 Y_2 - X_2 Y_1)^2}{(X_1^2+Y_1^2)^{3/2}(X_2^2+Y_2^2)^{3/2}} \\ &> 0.\end{aligned}$$

Thus Hess V is positive definite everywhere. This ensures us that if V has a local minimum, it is unique. Moreover, as $v(\mathbf{r})$ tends toward infinity when $\|\mathbf{r}\|$ tends toward infinity, $V(X, Y)$ also tends toward infinity as (X, Y) diverges. This property implies the existence of a local minimum of V. □

We denote by (X_0, Y_0) the unique minimum of V. The set of equilibrium configurations of the part under the radial field is the following $\{(X_0, Y_0, \theta), \theta \in \mathbf{S}^1\}$. Let us express this curve in the standard system of coordinates:

$$x = X_0 \cos\theta - Y_0 \sin\theta \quad (10)$$

$$y = X_0 \sin\theta + Y_0 \cos\theta. \quad (11)$$

The stable equilibrium configurations are obtained by rotation of the part about the origin of the radial field. The point of the part situated at the origin in these configurations is called the *pivot point* and denoted P [6]. In the stable equilibrium configuration corresponding to $\theta = 0$ (remember that in this case, $X = x$, $Y = y$.), the center of mass is translated to (X_0, Y_0). Thus in configuration $\mathbf{q}_0$, the pivot point is at $(-X_0, -Y_0)$ (see Figure 2).

3.2 Radial-Constant Field

The previous section established the existence and uniqueness of the pivot point for a part subjected to the unit radial field. In this section, we perturb the radial field by adding a small constant field in order to break the symmetry. We are going to show that for each fixed orientation θ of the part, the corresponding lifted potential field has a unique minimum in (X, Y). When now θ varies, the curve of these minima is C^1 and C^∞ if the pivot point is not on the boundary of the part. We call this curve the *equilibrium curve*. Then we will give a characterization of the local minima of the lifted potential using the equilibrium curve.

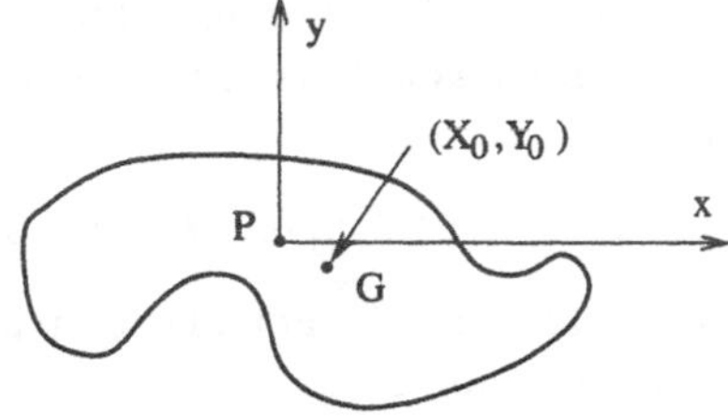

Figure 2: *In a unit radial field, at the equilibrium configuration corresponding to $\theta = 0$, the center of mass is at (X_0, Y_0).*

We consider now the following potential function in the plane (Figure 3):

$$u(\mathbf{r}) = v(\mathbf{r}) + \delta\eta$$

where v is the unit radial field and δ is a positive constant. The second term $\delta\eta$ corresponds to the constant

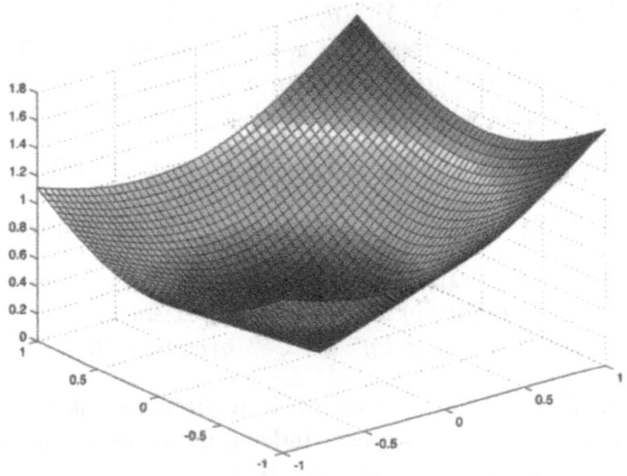

Figure 3: *Potential function of the combination of a unit radial force field with a constant force field.*

force field $-\delta(0,1)$. The lifted potential field for a given value of δ is expressed as follows in the (X, Y, θ) system of coordinates:

$$U_\delta(X,Y,\theta) = V(X,Y) + \delta|S|\,(\sin\theta\, X + \cos\theta\, Y) \quad (12)$$

where $|S|$ is the area of S. For clarity purposes, we define the following functions

$$U_{\theta,\delta}(X,Y) = U(X,Y,\theta,\delta) = U_\delta(X,Y,\theta)$$

where the variables we put in the subscript are considered constant.

Minimum of $U_{\delta,\theta}$ and Equilibrium Curve The second term of the right hand side of Expression (12) is linear in (X, Y). For this reason, $U_{\theta,\delta}$ has the same second order partial derivatives as V and Hess $U_{\theta,\delta} =$ Hess V is positive definite (Proposition 3). For any fixed value of θ and $\delta < 1$, $u(\mathbf{r})$ tends toward infinity with $\|\mathbf{r}\|$. Thus $U_{\theta,\delta}(X,Y)$ tends toward infinity with (X, Y) and $U_{\theta,\delta}$ has a unique local minimum. We denote by $(X^*(\theta,\delta), Y^*(\theta,\delta))$ this local minimum. We can express it in the standard system of coordinates by:

$$\begin{aligned} x^*(\theta,\delta) &= \cos\theta\, X^*(\theta,\delta) - \sin\theta\, Y^*(\theta,\delta) \quad (13)\\ y^*(\theta,\delta) &= \sin\theta\, X^*(\theta,\delta) + \cos\theta\, Y^*(\theta,\delta). \quad (14) \end{aligned}$$

For each value of $\delta < 1$, these local minima define a curve of parameter θ that we call equilibrium curve. We are going to show now that this curve is of class C^1. It is of class C^∞ for small values of δ if the pivot point is not on the boundary of the part.

Notice that $(X^*, Y^*)(\theta, 0) = (X_0, Y_0)$ is the minimum of V and therefore is independent of θ. The smoothness of the equilibrium curve will prove useful in the sequel where its partial derivatives are used.

Proposition 4 *The equilibrium curve is smooth.*

(i) X^*, Y^*, x^* and y^* are continuously differentiable.

(ii) if $(-X_0, -Y_0) \notin \partial S$ (i.e., the pivot point is not on the boundary of the part), there exists δ_0 such that X^*, Y^*, x^*, and y^* are C^∞ over $\mathbf{S}^1 \times [0, \delta_0]$.

Proof: (i) This proposition is a direct result of the implicit function theorem. Indeed, if we define the following function from $\mathbf{R}^4$ into $\mathbf{R}^2$:

$$F : (X,Y,\theta,\delta) \to \begin{pmatrix} \frac{\partial U}{\partial X}(X,Y,\theta,\delta) \\ \frac{\partial U}{\partial Y}(X,Y,\theta,\delta). \end{pmatrix}$$

From (12), as V is of class C^2 (Proposition 2), F is of class C^1. By definition, the equilibrium curve minimizes the lifted potential field for fixed θ and δ and therefore fits the following implicit representation:

$$F(X^*, Y^*, \theta, \delta) = 0.$$

The differential of the partial function $F_{\theta,\delta}$ of the variables (X, Y) is exactly the Hessian of V. From Proposition 3, this differential is invertible everywhere. According to the implicit function theorem, these conditions imply that X^* and Y^* can be expressed as C^1 functions of (θ, δ). As the equilibrium curve is unique, these C^1 functions are necessarily the formerly defined $X^*(\theta,\delta)$ and $Y^*(\theta,\delta)$.

(ii) If the pivot point is not on the boundary of the part, $-(X^*(\theta,0), Y^*(\theta,0)) \notin \partial S$. By the continuity of X^* and Y^*, there exists a δ_0 such that for any $\theta \in \mathbf{S}^1$ and $0 \le \delta \le \delta_0$, $-(X^*(\theta,\delta), Y^*(\theta,\delta)) \notin \partial S$. In other words, if we follow the equilibrium curve for a small δ, the origin of the field remains completely inside or completely outside the part and $(X^*, Y^*, \theta, \delta)$ remains in a domain where F is smooth (from Proposition 2). Therefore, according to the implicit function theorem, X^* and Y^* are also smooth. Relations (13) and (14) imply that x^* and y^* have the same differentiability properties as X^* and Y^*. □

From now on, we will assume that the pivot point is not on the boundary of the part, so that the partial derivatives of the equilibrium curves are all defined for small δ.

Properties of the equilibrium curve We now point out a property of the equilibrium curves that will constitute the basis of our method to determine the local minima of U. For a fixed value of δ the local minima of U are obviously on the equilibrium curve associated to δ. We are going to show that these local minima are the points where (x^*, y^*) crosses the y axis from $x < 0$ to $x > 0$. For that, we define:

$$U^*_\delta(\theta) = U_\delta(X^*(\theta,\delta), Y^*(\theta,\delta), \theta)$$

to be the minimum value of the lifted potential field for given θ and δ. The variation of $U^*_\delta(\theta)$ along the equilibrium curve is given by the following proposition.

Proposition 5 *For any $\theta \in \mathbf{S}^1$,*

$$\frac{dU^*_\delta}{d\theta}(\theta) = \delta|S|\, x^*(\theta, \delta).$$

Proof: For clarity, we omit δ in the notation of this proof. By definition $U^*_\delta(\theta) = U_\delta(X^*(\theta), Y^*(\theta), \theta)$. Differentiating this expression w.r.t. to θ leads to:

$$\begin{aligned}\frac{dU^*_\delta}{d\theta}(\theta) &= \frac{\partial U_\delta}{\partial X}(X^*(\theta), Y^*(\theta), \theta)\frac{dX^*}{d\theta}(\theta) + \\ &\quad \frac{\partial U_\delta}{\partial Y}(X^*(\theta), Y^*(\theta), \theta)\frac{dY^*}{d\theta}(\theta) + \\ &\quad \frac{\partial U_\delta}{\partial \theta}(X^*(\theta), Y^*(\theta), \theta) \\ &= \frac{\partial U_\delta}{\partial \theta}(X^*(\theta), Y^*(\theta), \theta) \\ &= \delta|S|\,(\cos\theta\, X^*(\theta) - \sin\theta\, Y^*(\theta)) \\ &= \delta|S|\, x^*(\theta)\end{aligned}$$

using expression (12) and the fact that the partial derivatives of U_δ w.r.t. X and Y vanish at (X^*, Y^*). □

This proposition leads directly to the following property.

Proposition 6 *For any fixed value of $\delta < 1$, the two following properties are equivalent:*

(i) *(X, Y, θ) is a local minimum of U_δ,*

(ii) *$X = X^*(\theta, \delta)$, $Y = Y^*(\theta, \delta)$ and the equilibrium curve crosses the y-axis from left to right when θ increases: $x^*(\theta, \delta) = 0$ and $\frac{\partial x^*}{\partial \theta}(\theta, \delta) > 0$.*

Figure 4 shows the value of the lifted potential along the equilibrium curve for a given part.

We are going now to devise a method for computing these equilibrium configurations for small values of δ and for any kind of part, symmetric or not.

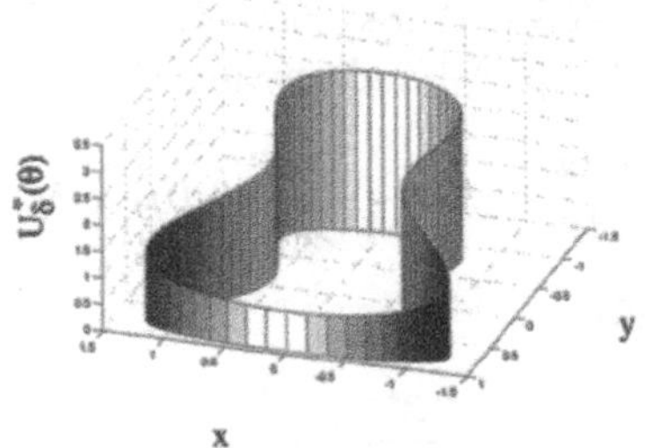

Figure 4: *Along the equilibrium curve, the variation of the value of the lifted potential is proportional to x^*. Thus the stable equilibrium configurations of the part are those where the equilibrium curve crosses the y-axis from left to right.*

3.3 Computation of the Stable Equilibrium Configurations of a Part for Small δ

In [5], we proved that for a part with distinct pivot point and center of mass $(X_0, Y_0) \neq (0,0)$, the stable equilibrium configuration is unique. The proof is based on the expression (10) of the equilibrium curve of parameter $\delta = 0$ and on the continuity of $x^*(\theta, \delta)$ and its partial derivatives. More precisely, $x^*(\theta, 0)$ vanishes for two values of θ: θ_1 and $\theta_2 = \theta_1 + \pi$. $x^*(\theta, 0)$ is increasing at one of these two values (say θ_2) and decreasing at the other one:

$$\begin{array}{ll} x^*(\theta_1, 0) = 0 & \frac{\partial x^*}{\partial \theta}(\theta_1, 0) < 0 \\ x^*(\theta_2, 0) = 0 & \frac{\partial x^*}{\partial \theta}(\theta_2, 0) > 0. \end{array}$$

By continuity of x^* and $\frac{\partial x^*}{\partial \theta}$, we proved that for small values of δ, $x^*(\theta, \delta)$ still vanishes only twice, around θ_1 and θ_2 and therefore the stable equilibrium configuration is unique.

When the pivot point and the center of mass are the same (that is when $(X_0, Y_0) = (0,0)$), this reasoning does not apply since from (10) and (13), $x^*(\theta, 0) = 0$ for all θ. The following theorem however, states that the criterion applied previously to $x^*(\theta, 0)$ can be applied to the first non-uniformly zero partial derivative of x^* w.r.t. δ, evaluated for $\delta = 0$: $\frac{\partial^n x^*}{\partial \delta^n}(\theta, 0)$. More precisely, if $\frac{\partial^n x^*}{\partial \delta^n}(\theta, 0)$ has only simple roots, *i.e.*, vanishes with non zero slope then for small δ, $x^*(\theta, \delta)$ also has only simple roots close to the roots of $\frac{\partial^n x^*}{\partial \delta^n}(\theta, 0)$ (Figure 5). Among those roots, some represent stable equilibrium configurations, the other ones are unstable.

Theorem 7 *For a given part, if there exists an integer $n \geq 0$ such that:*

(i) *for any k such that $0 \leq k \leq n-1$, $\frac{\partial^k x^*}{\partial \delta^k}(\theta, 0) = 0$ uniformly over $\mathbf{S}^1$,*

(ii) *$\frac{\partial^n x^*}{\partial \delta^n}(\theta, 0)$ vanishes only at a finite number of points $(\theta_1, ..., \theta_{2m})$, and*

(iii) *for any $1 \leq l \leq m$, $\frac{\partial^{n+1} x^*}{\partial \theta \partial \delta^n}(\theta_{2l}, 0) > 0$, $\frac{\partial^{n+1} x^*}{\partial \theta \partial \delta^n}(\theta_{2l-1}, 0) < 0$,*

then for small values of δ, the part has exactly m stable equilibrium configurations. These configurations converge toward (X_0, Y_0, θ_{2l}) in the (X, Y, θ) system of coordinates, $1 \leq l \leq m$, when δ tends toward 0.

Proof: Let us first notice that $\frac{\partial^{n+1} x^*}{\partial \theta \partial \delta^n}(\theta_p, 0)$ represents the slope at θ_p of $\frac{\partial^n x^*}{\partial \delta^n}(\theta, 0)$ seen as a function of θ.

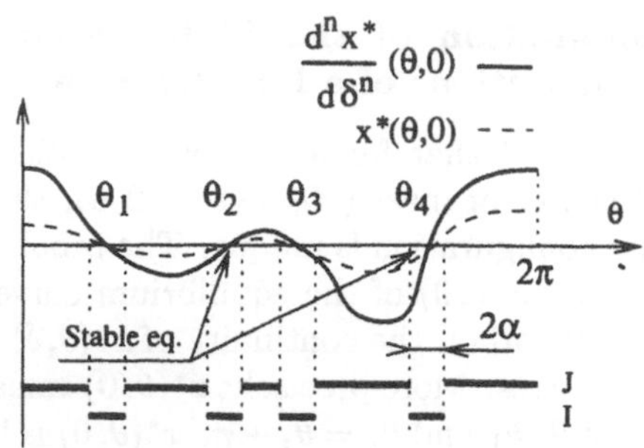

Figure 5: *The first non-uniformly zero partial derivative $\frac{\partial^n x^*}{\partial \delta^n}(\theta,0)$ of x^* w.r.t. δ provides all the information about the local minima of U_δ for small values of δ. The values of θ where this function vanishes with positive slope (θ_2 and θ_4 on this example) are close (by continuity) to the values where $x^*(\theta,\delta)$ vanishes with positive slope for small δ. According to Proposition 6, these values of θ are stable equilibrium configurations.*

Therefore condition (iii) simply means that $\frac{\partial^n x^*}{\partial \delta^n}(\theta,0)$ changes sign at each θ_p for $1 \leq p \leq 2m$.

All the partial derivatives of x^* are continuous. Thus, from condition (iii), there exist two positive numbers α and δ_1 such that for any positive real number $\delta < \delta_1$ and any integer l between 1 and m,

$$\begin{array}{ll} \forall\theta \in (\theta_{2l}-\alpha, \theta_{2l}+\alpha), & \frac{\partial^{n+1} x^*}{\partial\theta\partial\delta^n}(\theta,\delta) > 0 \\ \forall\theta \in (\theta_{2l-1}-\alpha, \theta_{2l-1}+\alpha), & \frac{\partial^{n+1} x^*}{\partial\theta\partial\delta^n}(\theta,\delta) < 0. \end{array} \tag{15}$$

Let us denote by I the union of the intervals of θ defined above:

$$I = \bigcup_{1\leq p\leq 2m} (\theta_p - \alpha, \theta_p + \alpha)$$

and by $J = \mathbf{S}^1 \setminus I$ the complement of I as shown on Figure 5.

The proof consists of two parts:

1. We first prove that over each interval constituting I, for small fixed δ, the slope $\frac{\partial x^*}{\partial \theta}$ of x^* keeps a constant sign and therefore x^* vanishes only once over each of these intervals.

2. Then, we show that, for small δ, x^* does not vanish over J.

1. Differentiating the equation defined in (i) w.r.t. θ, we get that, for any $k \leq n-1$,

$$\frac{\partial^{k+1} x^*}{\partial\theta\partial\delta^k}(\theta,0) = 0$$

uniformly over $\mathbf{S}^1$. If we take a fixed θ in the interval $(\theta_{2l}-\alpha, \theta_{2l}+\alpha)$, and we consider $\frac{\partial x^*}{\partial \theta}(\theta,\delta)$ as a function of δ that we temporarily denote by $f(\delta)$, the above equality can be rewritten:

$$\frac{\partial^k f}{\partial \delta^k}(0) = 0 \quad \text{for} \quad 0 \leq k \leq n-1. \tag{16}$$

Moreover, (15) implies that $\forall\delta \in [0,\delta_1]$,

$$\frac{\partial^n f}{\partial \delta^n}(\delta) > 0. \tag{17}$$

Therefore, using Taylor-Lagrange formula, for any $\delta \in [0,\delta_1]$, there exists β, $0 \leq \beta \leq 1$ such that:

$$\begin{aligned} f(\delta) &= \sum_{k=0}^{n-1} \frac{\partial^k f}{\partial \delta^k}(0)\frac{\delta^k}{k!} + \frac{\partial^n f}{\partial \delta^n}(\beta\delta)\frac{\delta^n}{n!} \qquad (18) \\ &= \frac{\partial^n f}{\partial \delta^n}(\beta\delta)\frac{\delta^n}{n!} > 0. \qquad (19) \end{aligned}$$

This establishes that for any $\delta < \delta_1$ and any $\theta \in (\theta_{2l}-\alpha, \theta_{2l}+\alpha)$, $1 \leq l \leq m$, $\frac{\partial x^*}{\partial \theta}(\theta,\delta) > 0$. Thus for a fixed $\delta \in [0,\delta_1]$, the function $x^*(\theta,\delta)$ of θ is increasing over $(\theta_{2l}-\alpha, \theta_{2l}+\alpha)$ and cannot vanish more than once over this interval. Using the same reasoning, we can establish that $x^*(\theta,\delta)$ is decreasing over the intervals $(\theta_{2l-1}-\alpha, \theta_{2l-1}+\alpha)$ and cannot vanish more than once in each of them either.

2. From condition (ii), when θ remains in J, $\frac{\partial^n x^*}{\partial \delta^n}(\theta,0)$ does not vanish. As J is compact, $\left|\frac{\partial^n x^*}{\partial \delta^n}(\theta,0)\right|$ admits a positive lower bound over J, that we denote by M:

$$M = \min\left\{\left|\frac{\partial^n x^*}{\partial \delta^n}(\theta,0)\right|, \theta \in J\right\} > 0. \tag{20}$$

From the uniform continuity of $\frac{\partial^n x^*}{\partial \delta^n}(\theta,\delta)$ over the compact set $J \times [0,\delta_1]$, there exists δ_2, $0 < \delta_2 \leq \delta_1$ such that for any $\delta \in [0,\delta_2]$ and any $\theta \in J$,

$$\left|\frac{\partial^n x^*}{\partial \delta^n}(\theta,\delta) - \frac{\partial^n x^*}{\partial \delta^n}(\theta,0)\right| < M$$

and using (20),

$$\left|\frac{\partial^n x^*}{\partial \delta^n}(\theta,\delta)\right| > 0 \tag{21}$$

We can apply again Taylor-Lagrange relation. To keep the same notation, $f(\delta)$ denotes now $x^*(\theta,\delta)$ considered as a function of δ for a fixed value of $\theta \in J$.

From condition (i) we get that condition (16) is satisfied and from (21) that $\frac{\partial^n f}{\partial \delta^n}(\delta)$ does not vanish over $[0, \delta_2]$ and thus keeps a constant sign over this interval. Condition (17) (or its counterpart $\frac{\partial^n f}{\partial \delta^n}(\delta) < 0$) is thus satisfied over $[0, \delta_2]$. We can reuse (18) and (19) (replacing $>$ by $<$ if $\frac{\partial^n f}{\partial \delta^n}(\delta) < 0$) to conclude that $x^*(\theta, \delta)$ does not vanish over $[0, \delta_2]$ and has the same sign as $\frac{\partial^n x^*}{\partial \delta^n}(\theta, 0)$.

If we now consider a fixed $\delta < \delta_2$ and if we recall that $\frac{\partial^n x^*}{\partial \delta^n}(\theta, 0)$ changes sign between two subintervals constituting J, we can conclude that $x^*(\theta, \delta)$ vanishes *exactly* once over each $(\theta_l - \alpha, \theta_l + \alpha)$, $1 \le l \le 2m$ at a value of θ that we denote by $\tilde{\theta}_l(\delta)$. The stable equilibrium configurations of the part are expressed in the (X, Y, θ) system of coordinates by $(X^*(\tilde{\theta}_{2l}(\delta), \delta), Y^*(\tilde{\theta}_{2l}(\delta), \delta), \tilde{\theta}_{2l})$. Since α can be chosen as small as desired by making δ tend toward 0, $\lim_{\delta \to 0} \tilde{\theta}_{2l}(\delta) = \theta_{2l}$. X^*, Y^* being continuous, the stable equilibrium configurations converge toward (X_0, Y_0, θ_{2l}) as δ tends toward 0. □

Let us point out that in most cases, the first non-uniformly zero partial derivative $\frac{\partial^n x^*}{\partial \delta^n}(\theta, 0)$ has simple roots (we will see later that these partial derivatives are trigonometric polynomials in θ) and Theorem 7 can be used. This theorem extends results from [12]. In the next section we use Theorem 7 to compute all the equilibrium configurations of a part under the radial-constant force field.

4 An Algorithm to Compute all Equilibrium Configurations of a Part

This section is organized as follows. First we explain how to iteratively compute expressions of the partial derivatives $\frac{\partial^k x^*}{\partial \delta^k}(\theta, 0)$ w.r.t. the partial derivatives of the lifted radial potential field V evaluated at $(0,0)$: $\frac{\partial^n V}{\partial X^k \partial Y^{n-k}}(0,0)$. These expressions enable us to compute the equilibrium configurations of a part using Theorem 7. We describe all the necessary operations in the form of an algorithm. Then we make some comments about the computation of $\frac{\partial^n V}{\partial X^k \partial Y^{n-k}}(0,0)$. Finally, an example illustrates the algorithm.

4.1 Expressions of $\frac{\partial^k x^*}{\partial \delta^k}(\theta, 0)$

Differentiating Equation (13) yields:

$$\frac{\partial^k x^*}{\partial \delta^k}(\theta, \delta) = \cos\theta \, \frac{\partial^k X^*}{\partial \delta^k}(\theta, \delta) - \sin\theta \, \frac{\partial^k Y^*}{\partial \delta^k}(\theta, \delta).$$

To compute an expression of $\frac{\partial^k x^*}{\partial \delta^k}(\theta, 0)$ we need thus to compute expressions of $\frac{\partial^k X^*}{\partial \delta^k}(\theta, 0)$ and $\frac{\partial^k Y^*}{\partial \delta^k}(\theta, 0)$.

By definition, (X^*, Y^*) minimizes $U_{\theta,\delta}$. Thus the partial derivatives $\frac{\partial U_{\theta,\delta}}{\partial X}$ and $\frac{\partial U_{\theta,\delta}}{\partial Y}$ vanish at (X^*, Y^*). Using expression (12), this statement is equivalent to:

$$\begin{aligned}
\frac{\partial V}{\partial X}(X^*(\theta,\delta), Y^*(\theta,\delta)) + \delta|S|\sin\theta &= 0 \\
\frac{\partial V}{\partial Y}(X^*(\theta,\delta), Y^*(\theta,\delta)) + \delta|S|\cos\theta &= 0.
\end{aligned}$$

By successively differentiating this system w.r.t. δ and by evaluating the resulting system at $\delta = 0$ we will get expressions of $\frac{\partial^k X^*}{\partial \delta^k}(\theta, 0)$ and $\frac{\partial^k Y^*}{\partial \delta^k}(\theta, 0)$. We show below the first step of this process by computing an expression of $\frac{\partial x^*}{\partial \delta}(\theta, 0)$. Then we will address the general case with higher order $\frac{\partial^k x^*}{\partial \delta^k}(\theta, 0)$.

Distinct Pivot Point and Center of Mass If $(X_0, Y_0) \ne (0,0)$, we express this vector in polar coordinates: $(X_0, Y_0) = (\rho\cos\psi, \rho\sin\psi)$, $\rho > 0$. Then, from (10),

$$x^*(\theta, 0) = \rho\cos(\theta + \psi).$$

$x^*(\theta, 0)$ vanishes with positive slope for $\theta = \frac{3\pi}{2} - \psi$ and Theorem 7 (with $n = 0$) permits to conclude that when δ tends toward 0, the stable equilibrium configuration is unique and tends toward $(X_0, Y_0, \frac{3\pi}{2} - \psi)$ in the new system of coordinates, that is $(0, -\rho, \frac{3\pi}{2} - \psi)$ in the standard system of coordinates.

Same Pivot Point and Center of Mass We assume now that $(X_0, Y_0) = (0,0)$. Let us differentiate once the above system w.r.t. δ. We get:

$$\begin{aligned}
\frac{\partial^2 V}{\partial X^2}\frac{\partial X^*}{\partial \delta} + \frac{\partial^2 V}{\partial X \partial Y}\frac{\partial Y^*}{\partial \delta} + |S|\sin\theta &= 0 \quad (22) \\
\frac{\partial^2 V}{\partial X \partial Y}\frac{\partial X^*}{\partial \delta} + \frac{\partial^2 V}{\partial Y^2}\frac{\partial Y^*}{\partial \delta} + |S|\cos\theta &= 0 \quad (23)
\end{aligned}$$

where the partial derivatives of V are evaluated at $(X^*(\theta,\delta), Y^*(\theta,\delta))$ and $\frac{\partial X^*}{\partial \delta}$, $\frac{\partial Y^*}{\partial \delta}$ are evaluated at (θ, δ). If we take $\delta = 0$ in this system, we get:

$$\text{Hess } V(0,0) \begin{pmatrix} \frac{\partial X^*}{\partial \delta}(\theta, 0) \\ \frac{\partial Y^*}{\partial \delta}(\theta, 0) \end{pmatrix} = -|S| \begin{pmatrix} \sin\theta \\ \cos\theta \end{pmatrix}. \quad (24)$$

According to Proposition 3, the above Hessian is invertible, so that expressions of $\frac{\partial X^*}{\partial \delta}(\theta, 0)$ and $\frac{\partial Y^*}{\partial \delta}(\theta, 0)$

and thus of $\frac{\partial x^*}{\partial \delta}(\theta, 0)$ are obtained by inversion of system (24). Let us notice that these expressions are trigonometric polynomials of θ.

To get higher order derivatives of X^* and Y^*, we need to differentiate several times (24). Let us point out a property of this system.

Proposition 8 *Differentiating $k-1$ times system (22-23) ($k \geq 2$) yields a system of the form:*

$$\text{Hess}\, V(X^*(\theta,\delta), Y^*(\theta,\delta)) \begin{pmatrix} \frac{\partial^k X^*}{\partial \delta^k}(\theta,\delta) \\ \frac{\partial^k Y^*}{\partial \delta^k}(\theta,\delta) \end{pmatrix} = E_k \tag{25}$$

where E_k is a polynomial expression of variables $\frac{\partial^l X^*}{\partial \delta^l}(\theta,\delta)$, $\frac{\partial^l Y^*}{\partial \delta^l}(\theta,\delta)$ for $1 \leq l \leq k-1$. The coefficients of this polynomial are functions of the $\frac{\partial^i V}{\partial X^j \partial Y^{i-j}}(X^*(\theta,\delta), Y^*(\theta,\delta))$, for $2 \leq i \leq k+1$ and $0 \leq j \leq i$.

Proof: This proposition can be easily proved by induction. For $k = 2$, if we differentiate once (22-23), we find that the proposition is satisfied with:

$$E_2 = -\begin{pmatrix} V_{3,0}X_1^{*2} + 2V_{2,1}X_1^*Y_1^* + V_{1,2}Y_1^{*2} \\ V_{2,1}X_1^{*2} + 2V_{1,2}X_1^*Y_1^* + V_{0,3}Y_1^{*2} \end{pmatrix}$$

where $V_{i,j} = \frac{\partial^{i+j} V}{\partial X^i \partial Y^j}(X^*(\theta,\delta), Y^*(\theta,\delta))$, $X_1^* = \frac{\partial X^*}{\partial \delta}(\theta,\delta)$, $Y_1^* = \frac{\partial Y^*}{\partial \delta}(\theta,\delta)$.

Now, we assume that the proposition is satisfied at order k, that is E_k in (25) is of the correct form. Let us differentiate again this equality w.r.t. δ. The left-hand side becomes:

$$\text{Hess}\, V(X^*(\theta,\delta), Y^*(\theta,\delta)) \begin{pmatrix} \frac{\partial^{k+1} X^*}{\partial \delta^{k+1}}(\theta,\delta) \\ \frac{\partial^{k+1} Y^*}{\partial \delta^{k+1}}(\theta,\delta) \end{pmatrix} + \begin{pmatrix} V_{3,0}X_1^* + V_{2,1}Y_1^* & V_{2,1}X_1^* + V_{1,2}Y_1^* \\ V_{2,1}X_1^* + V_{1,2}Y_1^* & V_{1,2}X_1^* + V_{0,3}Y_1^* \end{pmatrix} \begin{pmatrix} X_k^* \\ Y_k^* \end{pmatrix}.$$

Note that the first term of this sum is exactly the left hand side of (25) at order $k+1$. The second term can be included in E_{k+1}: it is of the correct form.

It remains to check that if E_k is of the form described in the proposition, so is E_{k+1}. This test is straightforward. □

The expressions involved in the above proposition are rather complex. That is why we only give their form instead of writing them here. However, it is important to notice the structure of the successive relations expressed by (25). If we take $\delta = 0$ in these equations, we get linear systems in $(\frac{\partial^k X^*}{\partial \delta^k}(\theta,0), \frac{\partial^k Y^*}{\partial \delta^k}(\theta,0))$. These systems are invertible since Hess V is positive definite. By inverting these systems iteratively, we get successive expressions of $\frac{\partial^k X^*}{\partial \delta^k}(\theta,0)$ and $\frac{\partial^k Y^*}{\partial \delta^k}(\theta,0)$ and thus of $\frac{\partial^k x^*}{\partial \delta^k}(\theta,0)$ w.r.t. the partial derivatives of V at $(0,0)$. It can be verified that these expressions are trigonometric polynomials in θ. This iterative procedure is the core of the algorithm presented below.

```
minima ← ∅
//Pivot point and center of mass distinct.
If (∂V/∂X(0,0) ≠ 0) or (∂V/∂Y(0,0) ≠ 0)
        (X0, Y0) ← minimize(V(X,Y))
        (ρ, ψ) ← polar-coord(X0, Y0)
        minima ← {(0, −ρ, 3π/2 − ψ)}
        exit;
endif;
//Same pivot point and center of mass.
x[0] ← 0;
expr[1] ← expression (22-23);
n ← 0;
while (x[n] ≡ 0) do
        n ← n + 1;
        system ← evaluate(expr[n],δ = 0);
        (X[n], Y[n])←
                solve(system,(∂^n X*/∂δ^n(θ,0), ∂^n Y*/∂δ^n(θ,0)));
        x[n] ← cos θ X[n] − sin θ Y[n];
        expr[n + 1] ← differentiate(expr[n],δ);
od;
xθ ← differentiate(x[n], θ);
(θ1, ..., θm) ← solve(x[n] = 0, θ);
for each i between 1 and m do
        If evaluate(xθi, θ = θi) > 0
                minima ← minima ∪ {(0, 0, θi)}
        endif
od;
```

Table 1: *Algorithm that computes the local minima of the lifted radial-constant field U_δ for small δ.*

4.2 The Algorithm

Using the above developments, the roots of the first non-uniformly zero $\frac{\partial^k x^*}{\partial \delta^k}(\theta,0)$ can be computed by the algorithm presented in Table 1. This algorithm needs the partial derivatives of V evaluated at $(0,0)$ and a function, `minimize`, that computes the unique minimum of V.

The Partial Derivatives of V. If the part is polygonal, we can express exactly the partial derivatives of V and thus evaluate them for any (X, Y). In this case, `minimize` can numerically minimize V using a gradient method and return an approximation of (X_0, Y_0).

If the part is not a polygon, the partial derivatives of V can be computed numerically. We do not give details about these numerical computations. Some discussion can be found in [11].

4.3 Example

Let us compute the stable equilibrium configurations of the part represented in Figure 6. For this polygonal part, symmetric by rotation of angle $\frac{\pi}{2}$, we have computed using Maple the expressions of $\frac{\partial V}{\partial X}$ and $\frac{\partial V}{\partial Y}$. These expressions are very long, we cannot report them here. Differentiating symbolically these expressions, we obtained higher order derivatives of V that we evaluated at $(0,0)$:

$$\begin{aligned}
&\frac{\partial^2 V}{\partial X^2}(0,0) = \frac{\partial^2 V}{\partial Y^2}(0,0) = \\
&\frac{8\sqrt{10}}{5}(\text{Argsinh } 3 - \text{Argsinh } \frac{1}{2}) \\
&\frac{\partial^2 V}{\partial X \partial Y}(0,0) = 0 \\
&\frac{\partial^3 V}{\partial X^3}(0,0) = \frac{\partial^3 V}{\partial X^2 \partial Y}(0,0) = 0 \\
&\frac{\partial^3 V}{\partial X \partial Y^2}(0,0) = \frac{\partial^3 V}{\partial Y^3}(0,0) = 0 \\
&\frac{\partial^4 V}{\partial X^4}(0,0) = \frac{\partial^4 V}{\partial Y^4}(0,0) = \frac{\sqrt{2}}{2} - 3 \\
&\frac{\partial^4 V}{\partial X^3 \partial Y}(0,0) = \frac{\partial^4 V}{\partial X \partial Y^3}(0,0) = 0 \\
&\frac{\partial^4 V}{\partial X^2 \partial Y^2}(0,0) = \frac{\sqrt{2}}{2}.
\end{aligned}$$

We apply now the successive steps of the algorithm described in Table 1 to the part.

To simplify the notation, let us write $\lambda = \frac{\partial^2 V}{\partial X^2}(0,0) = \frac{\partial^2 V}{\partial Y^2}(0,0)$. Substituting $\delta = 0$ in (22-23) yields:

$$\begin{aligned}
\lambda \frac{\partial X^*}{\partial \delta}(\theta, 0) + |S| \sin\theta &= 0 \\
\lambda \frac{\partial Y^*}{\partial \delta}(\theta, 0) + |S| \cos\theta &= 0.
\end{aligned}$$

Inverting this system yields:

$$\frac{\partial X^*}{\partial \delta}(\theta, 0) = -\frac{|S|}{\lambda} \sin\theta \qquad \frac{\partial Y^*}{\partial \delta}(\theta, 0) = -\frac{|S|}{\lambda} \cos\theta.$$

Substituting these equalities in (13) gives as expected:

$$x^*(\theta, 0) = 0.$$

Now, differentiating again (22-23) and substituting $\delta = 0$, we get:

$$\begin{aligned}
\lambda \frac{\partial^2 X^*}{\partial \delta^2} + V_{3,0} X_1^{*2} + 2V_{2,1} X_1^* Y_1^* + V_{1,2} Y_1^{*2} &= 0 \\
\lambda \frac{\partial^2 Y^*}{\partial \delta^2} + V_{2,1} X_1^{*2} + 2V_{1,2} X_1^* Y_1^* + V_{0,3} Y_1^{*2} &= 0
\end{aligned}$$

where $V_{i,j} = \frac{\partial^{i+j} V}{\partial X^i \partial Y^j}(0,0)$, $X_1^* = \frac{\partial X^*}{\partial \delta}(\theta, 0)$, $Y_1^* = \frac{\partial Y^*}{\partial \delta}(\theta, 0)$ and $\frac{\partial^2 X^*}{\partial \delta^2}$ and $\frac{\partial^2 Y^*}{\partial \delta^2}$ are evaluated at $(\theta, 0)$. From this system, we extract expressions of the second-order partial derivatives of X^* and Y^*:

$$\frac{\partial^2 X^*}{\partial \delta^2}(\theta, 0) = 0 \qquad \frac{\partial^2 Y^*}{\partial \delta^2}(\theta, 0) = 0.$$

Thus:

$$\frac{\partial^2 x^*}{\partial \delta^2}(\theta, 0) = 0.$$

To get the third-order partial derivatives of X^* and Y^*, we need to differentiate twice (22-23) w.r.t. δ and take $\delta = 0$. This yields two equations of the form:

$$\begin{aligned}
\lambda \frac{\partial^3 X^*}{\partial \delta^3}(\theta, 0) + \ldots &= 0 \\
\lambda \frac{\partial^3 Y^*}{\partial \delta^3}(\theta, 0) + \ldots &= 0
\end{aligned}$$

where ... stands for already evaluated expressions. From this system, we get:

$$\begin{aligned}
\frac{\partial^3 X^*}{\partial \delta^3}(\theta, 0) &= 2048 \frac{\sin\theta((2\sqrt{2}+6)\cos^2\theta + \sqrt{2} - 6)}{\lambda^4} \\
\frac{\partial^3 Y^*}{\partial \delta^3}(\theta, 0) &= -2048 \frac{\cos\theta((2\sqrt{2}+6)\cos^2\theta - 3\sqrt{2})}{\lambda^4}
\end{aligned}$$

and finally:

$$\frac{\partial^3 x^*}{\partial \delta^3}(\theta, 0) = \frac{1024\sqrt{2} + 3072}{\lambda^4} \sin 4\theta.$$

We recall that $\lambda = \frac{8\sqrt{10}}{5}(\text{Argsinh } 3 - \text{Argsinh } \frac{1}{2}) > 0$.

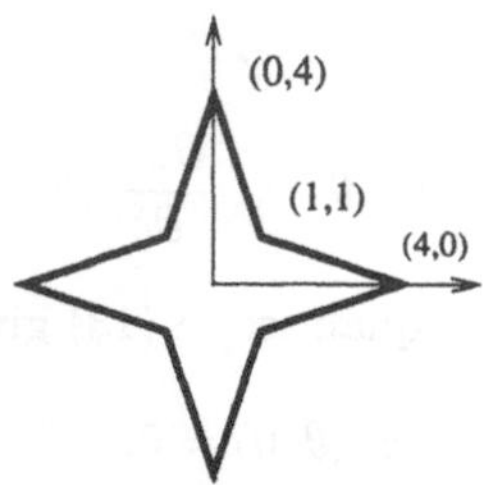

Figure 6: *Polygonal part symmetric by rotation of $\frac{\pi}{2}$.*

x^* satisfies the hypotheses of Theorem 7 for $n = 3$. $\frac{\partial^3 x^*}{\partial \delta^3}(\theta, 0)$ vanishes 8 times, 4 times with negative slope and 4 times with positive slope (for $\theta = i\frac{\pi}{2}$, $i = 0, 1, 2, 3$). We conclude that for small values of δ, the part has 4 stable equilibrium positions converging toward $(0, 0, i\frac{\pi}{2})$ when δ tends toward 0. The part is thus positioned up to part symmetry by the force field.

5 Conclusion

We have proposed a method to compute all stable equilibrium configurations of a part subjected to the combination of a unit radial force field with a small constant field. This method is general and can be applied to any part, symmetric or not. In addition, this paper reports a comprehensive study of the action of radial and small constant potential fields over parts. It proposes an interesting characterization of local minima of the lifted potential field using the equilibrium curve and its partial derivatives for $\delta = 0$. Some questions remain open however. All these results are asymptotic for small δ. The general case with any δ between 0 and 1 is far more difficult since we do not have any expression of the equilibrium curve in general and the stable equilibrium configurations are the roots of the equilibrium curve. For this case, numerical methods will probably be necessary. The rate of convergence under the fields described in this paper, as well as under previously proposed fields, is an open question.

Acknowledgments

This work was performed while Florent Lamiraux was with the Department of Computer Science at Rice University. The authors are grateful to Karl Böhringer and Bruce Donald for many discussions. Work on this paper by Lydia Kavraki and Florent Lamiraux was partially supported by NSF IRI-970228 and NSF CISE SA1728-21122N.

References

[1] S. Akella, W. Huang, K. Lynch, and M. Mason. Planar manipulation on a conveyor with a one joint robot. In G. Giralt and G. Hirzinger, editors, *International Symposium on Robotics Research*, pages 265–276. Springer, London, 1996.

[2] D. Biegelsen, W. Jackson, A. Berlin, and P. Cheung. Air jet arrays for precision positional control of flexible media. In *Proc. Int. Conf. on Micromechatronics for Information and Precision Equipment*, pages 631–634, Tokyo, Japan, 1997.

[3] K. Böhringer, V. Bhatt, and K. Goldberg. Sensorless manipulation using transverse vibrations of a plate. In *Proc. IEEE Int. Conf. on Rob. and Autom.*, pages 1989–1996, 1995.

[4] K. Böhringer, B. Donald, and N. MacDonald. Programmable vector fields for distributed manipulation, with application to mems actuator arrays and vibratory part feeders. *International Journal on Robotics Research*, 18:168–200, Feb. 1999.

[5] K.-F. Böhringer, B. R. Donald, L. E. Kavraki, and F. Lamiraux. Part orientation with one or two stable equilibria using programmable vector fields. to appear in the IEEE Transactions on Robotics and Automation.

[6] K.-F. Böhringer, B. R. Donald, and N. C. MacDonald. Upper and lower bounds for programmable vector fields with applications to MEMS and vibratory plate parts feeders. In J. Laumond and M. Overmars, editors, *Algorithms for Robotic Motion and Manipulation*, pages 255–276, Natick, MA, 1997. AK Peters.

[7] J. Canny and K. Y. Goldberg. Risc for industrial robotics: recent results and open problems. In *Proc. IEEE Int. Conf. on Rob. and Autom.*, pages 1951–1958, 1994.

[8] M. Erdmann and M. Mason. An exploration of sensorless manipulation. *IEEE Tr. on Rob. and Autom.*, 4(4):369–379, 1988.

[9] K. Y. Goldberg. Orienting polygonal parts without sensors. *Algorithmica*, 10:201–225, 1993.

[10] L. E. Kavraki. Part orientation with programmable vector fields: Two stable equilibria for most parts. In *Proc. IEEE Int. Conf. on Robotics and Automation*, pages 2446–2451, 1997.

[11] F. Lamiraux and L. Kavraki. Part positioning using radial and constant fields: Modeling and computation of all equilibrium configurations. in preparation.

[12] F. Lamiraux and L. E. Kavraki. Positioning symmetric parts using a combination of a unit-radial and a constant force fields. to appear in the IEEE International conference on Robotics and Automation 2000, 2000.

[13] W. Liu and P. Will. Part manipulation on an intelligent motion surface. In *International Conference on Intelligent Robots and Systems*, pages 399–404, Pittsburg, PA, Aug. 1995. IEEE/RSJ.

[14] J. Luntz, W. Messner, and H. Choset. Velocity field design for parcel manipulation on the virtual vehicule, a discrete distributed actuator array. In P. Agarwal, L. E. Kavraki, and M. Mason, editors, *Robotics: The Algorithmic Perspective*, pages 35–47. AK Peters, Natick, MA, 1998.

[15] D. S. Reznik and J. Canny. Universal part manipulation in the plane with a single horizontally-vibrating plate. In P. Agarwal, L. E. Kavraki, and M. Mason, editors, *Robotics: The Algorithmic Perspective*, pages 35–47. AK Peters, Natick, MA, 1998.

[16] J. Wiegley, K. Goldberg, M. Peshkin, and M. Brokowski. A complete algorithm for designing passive fences to orient parts. In *Proc. Int. Conf. on Rob. and Autom.*, pages 1133–1139, 1996.

Complete Distributed Coverage of Rectilinear Environments

Zack J. Butler, *Carnegie Mellon University, Pittsburgh, PA*
Alfred A. Rizzi, *Carnegie Mellon University, Pittsburgh, PA*
Ralph L. Hollis, *Carnegie Mellon University, Pittsburgh, PA*

Complete coverage of an unknown environment is a valuable skill for a variety of robot tasks such as floor cleaning and mine detection. Additionally, for a team of robots, the ability to cooperatively perform such a task can significantly improve their efficiency. This paper presents a complete algorithm DC_R (distributed coverage of rectilinear environments) which gives robots this ability. DC_R is applicable to teams of square robots operating in finite rectilinear environments and executes independently on each robot in the team, directing the individual robots so as to cooperatively cover their shared environment relying only on intrinsic contact sensing to detect boundaries. DC_R exploits the structure of this environment along with reliable position sensing to become the first algorithm capable of generating cooperative coverage without the use of either a central controller or knowledge of the robots' initial positions. We present a completeness proof of DC_R, which shows that the team of robots will always completely cover their environment. DC_R has also been implemented successfully in simulation, and future extensions are presented which will enable instantiation on a real-world system.

1 Introduction

The *coverage* problem, that of planning a path for a sensor, effector, or robot to reach every point in an environment, is one that appears in a number of domains. The problem of *sensor-based* coverage, that of planning such a path from sensor data in the absence of *a priori* information about the environment, is limited to robotics, but also applies to a number of different tasks. What is common to all these problems, whether a spray painting task on a known surface or a mine detection task with little or no initial information, is a need for assurance of complete coverage. For known areas, a path can be correctly generated off-line [1], but in the sensor-based case, the usual solution is instead to use a strict geometric algorithm about which completeness can be proven for any environment of a given class. Tasks which may be accomplished by multiple robots introduce additional complexity, since each point in the environment need only be reached by one of several robots, and in order to cooperate, the robots must know (or discover) each other's location. However, using multiple robots gives the potential for increased efficiency in terms of total time required.

While a number of sensor-based coverage algorithms have been proposed, in most, the algorithm begins by assuming the environment to be simply shaped (e.g. simply connected, monotone, convex, etc.). To cover its environment, the robot may then execute a simple coverage path until it discovers evidence that contradicts the initial assumption, at which point one of several strategies is used to ensure coverage on all sides of the newly discovered obstacle. An algorithm presented by Lumelsky et al. in [2] and extended in [3] produces coverage of $\mathcal{C}^2$ environments for robots with finite non-zero sensing radius by recursively building a subroutine stack to ensure all areas of the environment are covered. This algorithm does not explicitly build a map, in contrast to sensor-based coverage work by Acar [4] based on a planned coverage strategy outlined in [5]. In [4], a cellular decomposition of the environment is constructed and used to form a graph which in turn is used to plan coverage — when a specific cell has been covered, the robot uses the structure of the graph to plan a path to an unexplored area, and when the graph has no unexplored edges, coverage is complete. The cellular decomposition approach of [5] also inspired the algorithm presented in [6], which in turn is the basis for the current work. In [6], we presented an algorithm for coverage of rectilinear environments by a single robot using only intrinsic contact sensing. This algorithm also explicitly leveraged the degeneracies of the environment (degenerate in the $\mathcal{C}^2$ sense) by decomposing the free space into a set of rectangular cells.

In contrast to the more commonly studied coverage tasks mentioned above, the inspiration for the current work comes from a manufacturing environment. The *minifactory*, an automated assembly system under development in the Microdynamic Systems Laboratory[1], has been built within a framework that provides for rapid design, programming, and deployment [7].

A minifactory includes several types of independent robots, but this work concentrates on the *couriers*, small tethered robots that operate on a set of tileable *platens* which form the factory floor. The couriers have micron-level position sensing but only intrinsic contact sensing to detect the boundaries of their environment. In addition, each is equipped with an upward-pointing optical sensor to locate LED beacons placed on overhead robots as calibration targets. One of the tasks for the couriers is to collectively explore the as-built factory from unknown initial positions to generate a complete factory map. This task has led to the investigation of coverage algorithms for teams of robots, with the restrictive environment providing a simplified domain to consider. The algorithm developed here therefore applies to teams of square robots with intrinsic contact sensing operating in a shared, connected rectilinear environment with finite boundary and area. In addition, the robots in the team will not know their relative initial positions or orientations, however, due to the structure of the environment, their orientation will be one of four distinct values (i.e. with axes aligned with the environment boundaries) and cannot change.

Like the work described here, previous work in distributed robotics has presented the use of a common algorithm executed by each robot in a team (without a central controller) to achieve a specific task [8, 9, 10]. For example, in the work of Donald et al. [8], several distributed algorithms were presented to perform a cooperative manipulation task. There, however, the goal was to recast a simple provable algorithm in such a way that explicit communication was unnecessary, but could rather be implicit in the task mechanics. In our work, the environment is static, and so this reduction is not available, and the underlying algorithm (single-robot coverage) is much more complex. Other work on decentralized control of cooperative mobile robots has generally focused on the creation of a certain group behavior (as simple as foraging [9] or as complex as playing soccer [10]) without proof of the correctness or completeness of the individual or group algorithms.

On the other hand, research into algorithms for complete coverage of an environment by cooperating robots has so far used a central controller deploying robots from known locations, which is not satisfactory for the minifactory problem. For example, Gage's work [11] uses random walks by a large team with a common home position to generate probabilistically complete coverage. A fairly abstract algorithm presented by Rao et al. [12] uses a small team of point-sized robots with infinite range sensing to build a visibility graph of a polygonal environment. In contrast, work by Rekleitis et al. [13] uses cooperating robots with mutual remote sensing abilities, but with explicit cooperation to reduce mapping errors rather than to increase efficiency.

2 DC_R: Overview

DC_R is an algorithm developed for square robots that use only intrinsic contact sensing, and produces complete coverage of any finite rectilinear environment while using cooperation between robots to produce coverage more efficiently. It consists of three distinct components, shown schematically in Figure 1. The first, CC_{RM}, covers the environment by incrementally building a cellular decomposition $\mathbf{C}$ ($\mathbf{C} = \{C_0, \ldots, C_n\}$). It uses only $\mathbf{C}$ and the robot's current position $p = (p_x, p_y)$ (i.e. time-based history is not used) to direct the robot to continue coverage. CC_{RM} is based on the work in [6], and performs coverage without taking into account any other robots in its team. However, with the other components of DC_R properly designed, it performs coverage equally capably in a cooperative

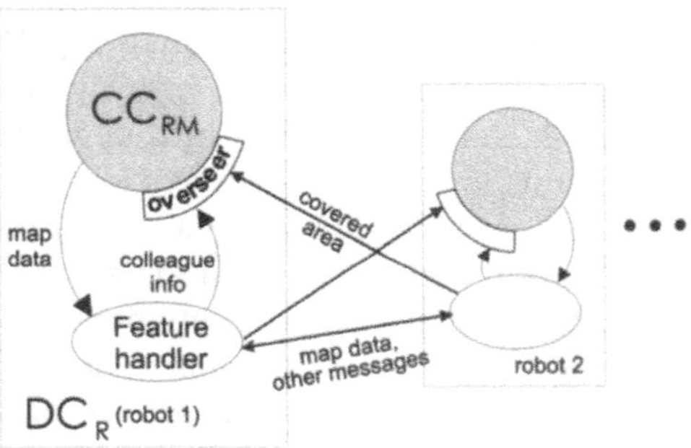

Figure 1: *A schematic rendition of the algorithm DC_R, a copy of which is run independently by each robot performing coverage.*

[1]More information at http://www.cs.cmu.edu/~msl

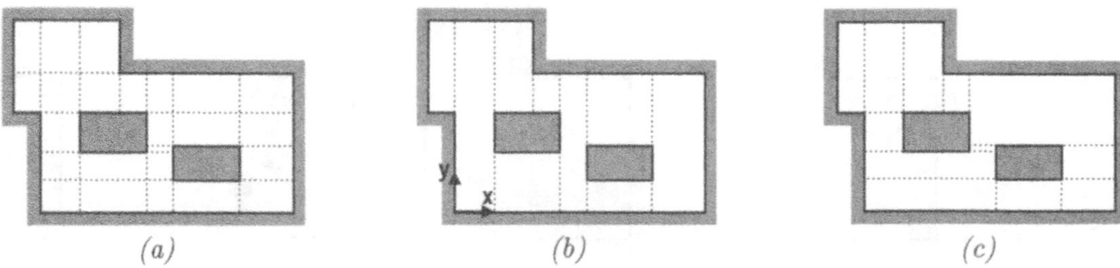

Figure 2: *Examples of (a) the sweep-invariant decomposition, (b) an oriented rectilinear decomposition, and (c) a possible generic rectilinear decomposition of a rectilinear environment.*

setting. The second part of DC_R, the *feature handler*, derives pre-specified types of features from **C** and communicates with other robots in the team to determine the relative position of the various robots. Finally, the *overseer* takes incoming data from colleagues and alters **C**. This is done without the explicit "knowledge" of CC_{RM}, but because CC_{RM} uses no state other than **C** and p, the overseer can alter **C** to incorporate the new data, and as long as this is done such that **C** remains in a state admissible to CC_{RM} (as described in Sec. 2.4), coverage continues.

2.1 Cell Decompositions under DC_R

Before discussing the makeup of DC_R, it may be instructive to describe the type of cell decompositions that will be constructed. This discussion will focus only on the geometry of the cells in the decomposition, but in each type of decomposition, every cell also includes "connections" (geometric references) to each of its neighbors that allow for the straightforward creation of an adjacency graph of the decomposition. One way to uniquely decompose a known rectilinear environment is the *sweep-invariant* decomposition (SID), as shown in Figure 2a. In a SID, each obstacle or boundary edge is extended until a perpendicular wall is reached, with all extended edges defining cell boundaries — as a result, a cell's edge may be determined by an arbitrarily distant boundary component. Thus, it is not possible to incrementally construct the SID without splitting cells that have already been completed, since determining a cell's extent could require knowledge of features that are arbitrarily distant from the cell itself. This difficulty of incremental construction makes sensor-based coverage based on the SID impractical. A different, nearly unique type of decomposition is the *oriented rectilinear* decomposition, shown in Figure 2b. This decomposition is produced by extending all vertical edges (the x values of these edges are called *interesting points*, analogous to critical points of a sweep through a $\mathcal{C}^2$ environment) to form cell boundaries, and can be incrementally constructed and easily covered with the use of seed-sowing paths as described below. This decomposition is produced by DC_R when performing coverage without colleagues.

Multiple robots performing cooperative coverage in a shared environment will not necessarily have the same orientation, and therefore will not necessarily build the same oriented rectilinear decomposition. This means that neither of the above decompositions can be built in a cooperative setting. Instead, the cell decompositions that are constructed under DC_R fall into a class referred to here as *generic rectilinear* decompositions (GRDs), an example of which can be seen in Figure 2c. A GRD consists of cells that are rectangular and supersets of cells of the SID. In addition, in a *valid* GRD, no two cells have overlapping area ($\forall i, j\colon C_i \cap C_j = \emptyset$) and all cells contain connections to all neighbors across their common edges. It should also be noted that during the performance of DC_R, a valid GRD may consist of multiple disconnected components. This is because robots only share completed cells, but may meet in an area that neither has completed. Also, for a given environment, different initial positions of robots may lead to different GRDs, since the nature and timing of the cooperation may change.

2.2 CC_{RM} Description

To perform cooperative coverage, each robot must first be able to perform coverage by itself. This is the job of CC_{RM}, a sensor-based coverage algorithm for a rectangular robot with only intrinsic contact sensing operating in rectilinear environments. It operates in cycles,

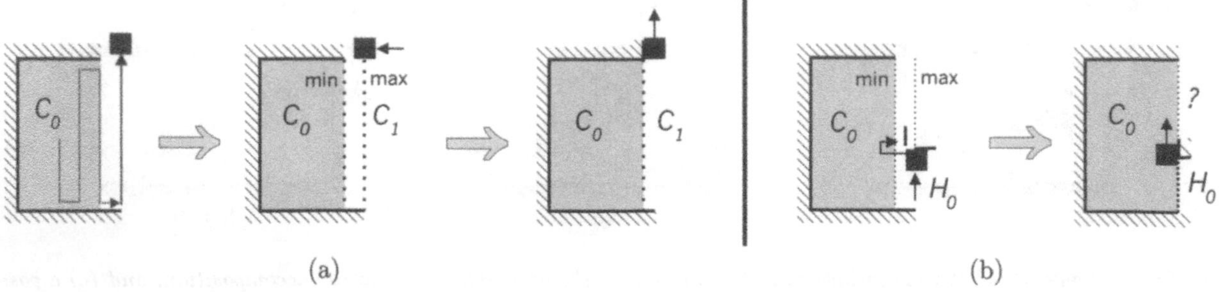

Figure 3: *Some examples of how CC_{RM} discovers and localizes interesting points.*

with the length of each cycle being a single straight-line trajectory of the robot. At the beginning of each cycle, it examines the structure of **C** around the current position, and uses an ordered list of rules to determine the next trajectory (both direction and distance) required to continue coverage. The rules are structured so as to produce complete coverage. Once a trajectory is submitted to the underlying robot, the robot moves until one of three types of *coverage events* occurs. A coverage event occurs when the maximal distance of the trajectory is achieved, a collision takes place, or contact that is to be maintained with a wall is lost. CC_{RM} then updates **C** given the type of event, chooses a new trajectory, and the cycle repeats.

Under CC_{RM}, each cell C_i is described by its minimum known and maximum potential extents, C_{i_n} and C_{i_x} respectively. Associated with each edge of the cell is a list of *intervals*, each of which is a connected line segment describing the cell's neighbor across that edge. Each interval therefore points to a wall, another cell or a *placeholder*, a line segment denoting the entrance to unexplored area.

To cover each cell, CC_{RM} generates a *seed-sowing* path as shown in the leftmost portion of Figure 3a, in which the robot travels along paths parallel to its y axis and as far apart as the width of the robot. These continue until an interesting point is detected, such as in the middle portion of Figure 3a. When this occurs, CC_{RM} updates **C** based on the type of coverage event. In the case of Figure 3a, a new cell (C_1) is added around p with uncertainty in the boundary between it and the previous cell. A rule then fires based on this uncertainty to localize the interesting point before coverage continues in the new cell. Another case of detection and localization of an interesting point is shown in Figure 3b, which uses similar rules, although in this case a placeholder H_0 is built rather than a new cell.

Since the first applicable rule determines the next trajectory, rules for interesting points are tested first, followed by the seed-sowing rule, as follows:

1. If p is in a cell, call that cell C_c and continue at rule 2. Otherwise, if p_x is within w (the width of the robot) of a complete cell, go into that cell.
2. If C_c has a side edge with finite uncertainty, move to localize that edge.
3. If C_c has a side edge at a known position that is not completely explored, investigate the closest unexplored point in it.
4. If C_c has unknown floor or ceiling, go in $-y$ or $+y$ respectively.
5. If C_c is not covered from its left edge to its right edge (note that if either edge is unknown, this will always be true), continue seed-sowing.

If none of these rules apply to C_c, then it must be *complete*. A cell is defined as complete when each of its edges are at known location and are spanned by a set of intervals, and its interior has been completely covered with seed-sowing strips. If C_c is complete, the following rules are used, in order:

6. If there is an incomplete cell in **C**, plan a path to it and follow the first step of that path.
7. If C_c has a placeholder neighbor, build a new cell from it and enter the new cell.
8. Choose a placeholder from **C**, plan a path to the cell it adjoins and take the first step of the path.

Path planning is done by implicitly creating a graph from the adjacency relationships of the cells in **C**, and searching that graph for a path to the destination. The search is done depth-first from the destination back to p, and the destination is chosen deterministically, so that even though the plan is regenerated after each trajectory, the planned path will remain the same as the plan is executed. Traversal of each cell is done simply with straight line motions. Finally, if none of these rules fire, there must be no placeholders or incomplete cells, and so **C** must be completely covered and its boundary explored, meaning coverage is complete.

Once cooperation is achieved, a cell may have another (complete) cell above or below it, as seen in Figure 2c. To allow CC_{RM} to continue seed-sowing in such cells, a construct called an *exploration boundary* has been developed. Exploration boundaries are virtual walls placed at the floor and ceiling of each cell obtained from a colleague, and have the property of allowing the robot to pass though them only when the robot is in a complete cell. This allows CC_{RM} to perform seed-sowing in a cell that does not have walls at its floor or ceiling, as shown in Figure 4a, but does not impede path planning once the cell is complete, as shown in Figure 4b. It should also be noted that in any case, a completed cell will always have an *attached* floor and ceiling, where an attached cell edge is one that is adjacent to a wall or another complete cell (not a placeholder) at every point.

A second addition made to CC_{RM} that only has impact during cooperation is that a robot can "claim" a placeholder as it builds a cell from it. The robot passes this claim on to the overseers of its colleagues, and so the other robots in the team will not travel to that area to perform coverage. This helps minimize the double covering of area, but also implies trustworthiness of the robots. However, the claiming of area does not effect completeness of DC_R, so it can be implemented only when desired.

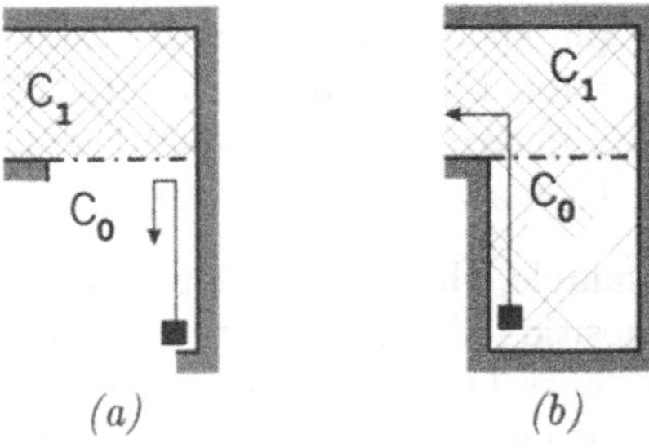

Figure 4: *The effects of an exploration boundary (dash-dot line) when the robot is in (a) an incomplete cell and (b) a complete cell.*

2.3 Feature Handler Description

The feature handler is quite independent from the other two components in its behavior. Its task is to use the data in **C** to generate colleague relationships between robots, however, it does not alter **C**. Therefore, the specific types of features and algorithms used by the feature handler can change from one system to the next without affecting the rest of DC_R. For example, in the current implementation, the feature handlers look for distances between unlabeled landmarks (beacons) that are common to two robots' maps. However, other feature types such as wall lengths or beacon labels (if present) could also be used, depending on the particular system. The generic behavior of a feature handler is to inform all other robots in the team of the values for all instances of a specified feature type in **C**. When two robots discover matching features, their feature handlers symmetrically compute the relative transforms between their local coordinate systems using appropriate geometric algorithms. These transforms are then used by the overseer when incorporating data obtained from colleagues.

The other mandatory job of the feature handler is to transfer data to all colleagues at the correct times. When a colleague relationship is formed, all complete cells must be given to the colleague, and when a cell in **C** first becomes complete, that cell must be reported to all colleagues. This ensures that the information available to each robot is consistent, which in turn maintains the attached edge property of cells described above.

2.4 Overseer Description

The overseer has the task of incorporating all data from colleagues into **C**, a job complicated by the requirement that **C** must remain admissible to CC_{RM}. The addition of an incoming cell C_{new} to **C** is done in two stages. In the first stage, zero or more new cells are added to **C** to account for the area of C_{new}. Then, for each added cell, its intervals are assigned to walls, existing cells or newly created placeholders. An example of the action of the overseer is shown in Figure 5.

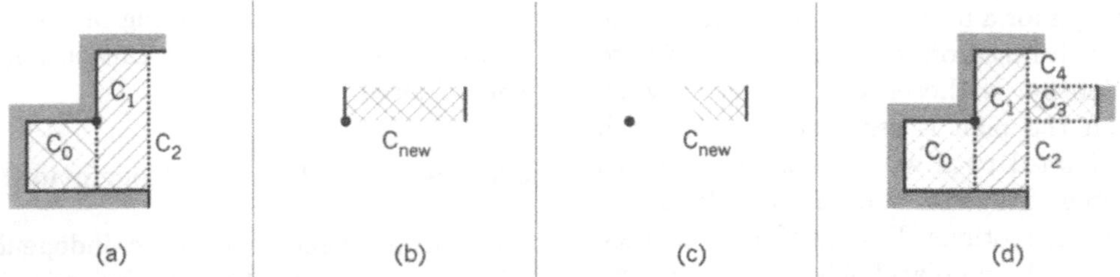

Figure 5: *An example of adding new area by the overseer, in which the initial cell decomposition is depicted in Figure 5a and the incoming cell C_{new} in Figure 5b. The dot in each section of the figure represents a common real-world point.*

The cell C_{new} arrives described in the coordinate system of the sending robot, and so it is first transformed into the local coordinate system using the transform provided by the feature handler. Also at this time, all intervals in C_{new} that do not point to walls are modified to point to "unidentified free space" rather than a specific cell or placeholder, since any such neighbor information in C_{new} is meaningful only to the robot that sent it.

To determine the area of cells to be added to $\mathbf{C}$, $\mathbf{C_{com}}$ is defined as the set of all complete cells in $\mathbf{C}$, and $\mathbf{C_{inc}} = \mathbf{C} - \mathbf{C_{com}}$. For the example in Figure 5, $\mathbf{C_{com}} = \{C_0, C_1\}$ and $\mathbf{C_{inc}} = \{C_2\}$. C_{new} is then intersected with each cell in $\mathbf{C_{com}}$ as follows:

- $\forall C_i \in \mathbf{C_{com}}$:
  - If $C_i \cap C_{new} = \emptyset$, do nothing.
  - If C_{new} is wider (larger in x) than C_i:
    - If $C_{new,right} > C_{i,right}$, make a copy of C_{new} called C_x, set $C_{x,left} = C_{i,right}$, and call the overseer with C_x.
    - Similarly (note no else here) for $C_{new,left} < C_{i,left}$.
    - If C_{new} is also taller than C_i, make a copy of C_{new} called C_x, set $C_{x,left} = C_{i,left}$ and $C_{x,right} = C_{i,right}$, then call the overseer with C_x.
  - Else if C_{new} is taller than C_i, perform similar tests on top and bottom (no third test).

If C_{new} (or its descendants) survive this process (such as the area shown in Figure 5c, added to $\mathbf{C}$ as C_3 in Figure 5d), it will consist only of area new to $\mathbf{C}$, i.e. $C_{new} \cap \mathbf{C_{com}} = \emptyset$. In addition, whether or not the area has been divided, each cell will still have at least two attached edges (either walls or other complete cells, possibly also provided by the sender of C_{new}). Each new cell is then intersected with every cell $C_i \in \mathbf{C_{inc}}$. This intersection process is designed to retain the complete C_{new} and eliminate any overlap with incomplete cells (shrinking or deleting the incomplete cells as necessary). Since an incomplete cell must be attached on its floor and ceiling, C_{new} cannot be taller than any $C_i \in \mathbf{C_{inc}}$. The intersection is therefore performed as follows (note that it is not recursive, since each C_{new} is now a fixed size):

- $\forall C_i \in \mathbf{C_{inc}}$:
  - If $C_{i_x} \cap C_{new} = \emptyset$, do nothing.
  - If $C_{i_n} \cap C_{new} = \emptyset$, reduce C_{i_x} so that it does not overlap C_{new}, skip to next C_i.
  - If $C_{i_n} \cup C_{new} = C_{new}$, replace C_i with C_{new}, skip to next C_i.
  - If C_{new} is the same height as C_i, there must be a partial overlap in the x direction, so reduce C_{i_n} (on either its left or right as appropriate) to abut C_{new}.
  - Otherwise, there must be partial overlap in the y direction:
    - If $C_{new,ceil} < C_{i,ceil}$, replace C_i with a cell C_x, set $C_{x,floor} = C_{new,ceil}$ and keep only placeholders attached to C_x, and create an interval in C_x to point to C_{new}.
    - Similarly (again no else) for $C_{new,floor} > C_{i,floor}$.

In the example, this intersection process results in the decomposition shown in Figure 5d. Finally, each unassigned interval i in C_{new} is given the correct neighbor(s). This is done by determining which cell's maximum extent (if any) is across from the two ends of i (these cells are denoted C_{top} and C_{bot}). i is then assigned as follows:

- If $C_{top} = C_{bot} = \emptyset$, build a new placeholder H_n equal in size to i and set i's neighbor to H_n.
- If $C_{top} = C_{bot} \neq \emptyset$:
  - If C_{top} is complete, set i's neighbor to C_{top}. Also, find the interval in C_{top} that corresponds to i and connect it to C_{new}.
  - If C_{top} is incomplete and i is horizontal, split i, connect it to C_{top} for $i \cap C_{top_n}$ and build a placeholder for $i \cap C_{top_x}$.
  - If C_{top} is incomplete and i is vertical, connect i to C_{top} over the y extent of C_{top_n} as long as C_{top_n} is within one robot width of i, build a placeholder otherwise. If $C_{top,n}$ adjoins C_{new}, find the corresponding interval in C_{top} and connect it to C_{new}.
- If $C_{top} \neq C_{bot}$, this should only occur if i is horizontal and C_{top} and C_{bot} are each either an incomplete cell or $\emptyset$. In this case, connect i where adjacent to C_{top_n}, C_{bot_n} to those cells, and build placeholders for the remainder of i.

3 Proof

To prove that DC_R leads to complete coverage of any finite rectilinear environment by any number of robots (in the absence of interrobot collisions), it is first necessary to show the completeness of CC_{RM}, since DC_R run by a single robot is exactly CC_{RM}. We then show that any cooperation regardless of its timing and nature does not interfere with the progress of CC_{RM}. These statements, combined with the reactive nature of CC_{RM} (and therefore the decoupled nature of coverage and cooperation under DC_R), imply that DC_R is complete.

Proposition 1 *CC_{RM} continues coverage to completion in a finite rectilinear environment in the absence of cooperation that alters the robot's current cell.*

Completeness of CC_{RM} is shown through the construction and analysis of a finite state machine (FSM) which represents all possible behavior of CC_{RM}. To construct the FSM, an equivalence relation is defined over all possible cell decompositions $\mathbf{C}$ and robot positions p such that any two $(\mathbf{C}, p)$ pairs that cause the same rule to be applied in the same way are considered equivalent. The resulting equivalence classes define the states of the FSM. All possible motions of the robot under CC_{RM} (without cooperation) are then enumerated,

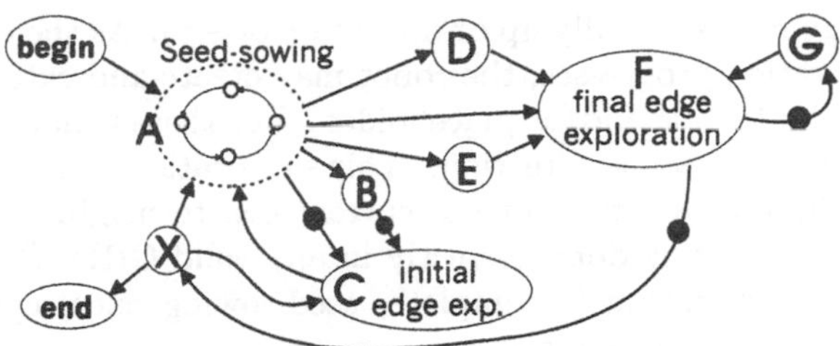

Figure 6: *A simplified version of the FSM representation of CC_{RM}. Contents of the nodes are described in the context of the proof. Gray dots represent the completion of a cell.*

starting from the initial condition of an empty map. Each motion has from one to three (uncontrollable) outcomes, each of which is a different type of coverage event and is represented by a transition in the FSM. Completeness is then demonstrated by showing that there are no terminal states other than complete coverage, and no cycles that do not result in a monotonically increasing measure of progress toward coverage. In addition, since states of the FSM are uniquely determined by the robot's current cell C_c, the FSM is valid for any valid GRD that includes C_c (no matter how it was constructed). A simplified version of the FSM, in which some nodes consist of several states and transitions, is shown in Figure 6. Note that all loops in Figure 6 contain a cell completion event, which indicates progress toward complete coverage. While the details of the FSM are beyond the scope of this presentation, a brief description of each node may give some insight to the structure of the FSM as well as CC_{RM} itself.

Node A contains the states that describe seed-sowing. In the absence of another incomplete cell overlapping the current cell, seed-sowing continues, cycling through the four states in Node A, until an interesting point is discovered. This discovery can be made in five different ways, each leading to one of five successor states of Node A.

Node B contains a single state, the situation shown in the middle of Figure 3a. From this state, a single motion completes the previous cell, and the robot is then in a cell with one known and partially explored edge. All states for which the current cell has this structure are contained in Node C, in which the robot explores this first side of the cell. Node C contains two cycles which correspond to the exploration of that edge monotonically up or down ($+y$ or $-y$). As the exploration progresses, the robot may create and extend wall intervals and/or placeholders, but there is never a cause to reverse direction. This exploration may also include attachment of the current cell to neighboring cells, which is done correctly in any valid GRD. Once this exploration is complete, seed-sowing once again begins in this new cell.

Nodes D and E each represent a single state which will always lead to Node F. Node D represents a state in which a corner is discovered by losing contact while moving along the floor or ceiling of a cell during seed-sowing. Node E represents the case shown in Figure 3b, in which a placeholder is built and the final edge of a cell localized. In both of these cases, the cell's edge is then at a known location. Node F in turn applies to all cells where all but one edge are known and explored. It is similar to Node C, in that it contains several states that describe the exploration of an edge, and the robot moves monotonically along the edge. There are some internal differences between the nodes, and when the exploration is complete in Node F, so is the cell, with one exception. This exception is for the first cell (C_0), for which exploration of its first known edge is also described by Node F, leading to Node X. This can only happen once, however, so this cycle cannot be repeatedly traversed without completing a cell.

Finally, Node X describes the state where the robot's current cell is complete, as well as when the first edge of C_0 has been explored. In the latter case, since C_0 is not yet complete, CC_{RM} returns to Node A to perform seed-sowing toward the other side of C_0. Otherwise, there are three possibilities. If there is an incomplete cell in $\mathbf{C}$, the robot will enter it and restart seed-sowing in Node A. Otherwise, if a placeholder exists, CC_{RM} will build a cell from the placeholder and begin to explore the cell's near edge as described by Node C. Finally, if and only if Node X is reached and no incomplete cells or placeholders exist, coverage is complete and DC_R terminates.

Proposition 2 *The action of the overseer always leaves* $(\mathbf{C}, p)$ *in the domain of* CC_{RM}.

This proposition has both global and local (to p) implications. Globally, the overseer must always produce a valid GRD, since this is assumed in Proposition 1. In addition, the local construction of $\mathbf{C}$ must be such that $\mathbf{C}$ and p form a state which is represented in the FSM described above. Global correctness is demonstrated by showing that the area of any cells added to $\mathbf{C}$ is correct and that all mutual connections between cells are constructed correctly. Local correctness is then shown via an enumeration of the possible effects on the robot's current cell by the overseer.

Proof that added area is correct relies on the fact that all complete cells in a GRD are supersets of SID cells (including the cell obtained by the overseer C_{new}). Therefore, $\forall C_i \in \mathbf{C_{com}}, (C_{new} - C_i)$ is also a superset of SID cells. To confirm that the area added by the overseer from C_{new} is in fact $(C_{new} - C_i)$, C_{new} is written as $C_l \cup C_m \cup C_r$, where C_l is the area to the left of C_i, C_r the area to the right of C_i, and C_m the remainder of C_{new}. C_l and C_r will be fed back to the overseer if non-null, at which point they will remain unchanged relative to C_i. If C_m is larger than C_i, it will also be given to the overseer, but will be subject to the vertical intersection test, which in turn sends $C_m - C_i$ to the overseer. Otherwise, $C_m \subset C_i$, or equivalently $C_m - C_i = \emptyset$. In either case, the total area added based on C_{new} and C_i is $(C_l \cup C_r \cup [C_m - C_i]) = (C_{new} - C_i)$. For incomplete cells $\{C_i : C_i \in \mathbf{C_{inc}}, C_i \cap C_{new} \neq \emptyset\}$, it must then be shown that after C_{new} is added, all known edges of C_i lie on edges of the SID. Since in all cases, edges of C_i that are moved will be coincident with edges of C_{new}, which is a valid GRD cell, this condition is also satisfied.

To show that each added cell is correctly connected to its neighbors, it must be shown that determining the neighbors at each end of an interval (as is done by the overseer) is sufficient to determine its overall disposition. This is in turn true if no interval is adjacent to more than two cells, and no cell lies adjacent to an interval without reaching one end of it. Proofs of these statements rely on the property that each complete cell in a GRD will always have two attached opposing edges.

To show that no cell can lie in the middle of an interval i, assume that such a cell exists (call it C_x). By definition, C_x must not have a neighbor on the side that attaches to i, otherwise that neighbor would be present instead of C_{new}. C_x must therefore have neighbors on its sides perpendicular to i. However, if these neighbors are walls, i will end at the edges of C_x, which contradicts the original assumption. Otherwise, the two neighbors must themselves have neighbors in the directions perpendicular to i. Eventually this chain of

Cell relation	description	$p \in C_{new}$	$p \notin C_{new}$
$C_{new} \cap C_{c_x} = \emptyset$	no overlap	—	no effect†
$C_{new} \cap C_{c_n} = \emptyset, C_{new} \cap C_{c_x} \neq \emptyset$	overlap maxsize only	case 1 in text	Continue in C_c
$C_{new} \cap C_{c_n} = C_{c_n}$	cell subsumed	in Node X	case 2 in text
$C_{new} \cap C_{c_n} \subset_y C_{c_n}$	top/bottom replaced	in Node X	Continue in small C_c
	middle replaced	as above, but see case 3 in text	

†It is possible for the intervals on an edge of C_c to be modified by the addition of an adjacent cell, but this will not change the state of CC_{RM} except perhaps to complete C_c.

Table 1: *Effects of the overseer on the robot's current cell* C_c.

cells must adjoin a wall, at which point i's neighbor will be a cell, not null as assumed.

Proving that an interval i will not adjoin more than two cells is also shown by contradiction. First, assume i is a vertical interval in C_{new} adjacent to two existing cells. The two cells (C_a and C_b, complete or incomplete) must therefore be unattached on the side facing C_{new}. C_a and C_b must therefore have attached floors and ceilings (including their mutual edge), and both must therefore have been created by a robot with the same sweep direction. But this is not possible, as one would have to be created first (even if both came from different robots), and would therefore have to extend to a true wall, not just to the other cell. Two cells like this C_a and C_b therefore cannot exist, and so a vertical interval cannot have more than one neighbor. For a horizontal interval, the argument works exactly the same way for complete cells. However, in this case it is possible to have at most two incomplete cells adjacent to i if and only if they are C_0 and C_1.

Finally, to show that the action of the overseer does not leave the robot in a position from which coverage cannot continue, the possible actions of the overseer in the neighborhood of p must be investigated. If the robot is in a complete cell at the time of cooperation, this cell will not be changed, and so the state of CC_{RM} is likewise unchanged. Otherwise, the robot's current cell C_c is incomplete, and so the possible intersections of C_{new} and C_c must be enumerated. However, C_{new} must be no taller than C_{c_n}, since any known neighbors above and below C_c must be walls or complete cells. The enumeration of all cases of intersection of C_c and C_{new} is therefore as shown in Table 1, with the resultant state of CC_{RM} also dependent on whether p is within C_{new}.

Three of the results from this intersection are nontrivial, and are presented here in more detail. Case 1 is shown in Figure 7a, and is handled by connecting the interval in C_{new} to C_c even though the minimum extents of the two cells do not adjoin. This connection is prescribed in Section 2.4. This puts CC_{RM} in Node X, but allows the robot to reenter C_c, as it is required to do, since C_c is still incomplete. CC_{RM} then resumes seed-sowing, since C_c still does not extend to C_{new}. Case 2, shown in Figure 7b, is handled by rule 1 of CC_{RM} (this is in fact the only case in which this rule produces motion). Now in Node X, a move in the x direction will always succeed, since C_{new} must be as tall as C_c, and if there was a wall between p and C_c, it would have been discovered already. The robot then continues from this complete cell. Finally, Case 3 in Table 1 is one in which multiple incomplete cells are created, such as in Figure 5d, which could potentially cause problems for seed-sowing. However, these new cells will always share a known edge, and coverage can only continue away from that edge. Also, the potential failure of seed-sowing under CC_{RM} can only occur when moving through the known but unexplored edge of an incomplete cell. Therefore, the robot will always complete one of the new cells (even with successors) and return to the other without triggering a failure of CC_{RM}.

Proposition 3 *Propositions 1 and 2 are sufficient to prove completeness of* DC_R.

Proposition 2 ensures that regardless of the input to the overseer, CC_{RM} will always find itself in the FSM described in Proposition 1, from which it will continue to perform coverage, making monotonic progress. Also, since the overseer can only increase the area spanned by $\mathbf{C_{com}}$ (or leave it unchanged), the act of cooperation also describes monotonic progress toward complete coverage.

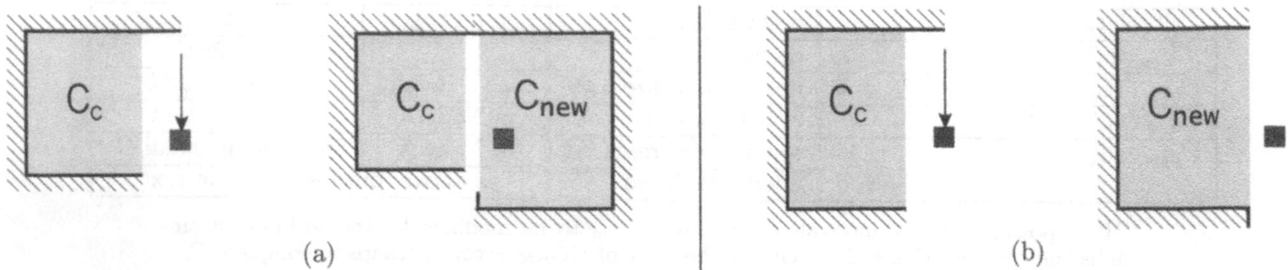

Figure 7: *(a) A case where the robot can't get to an incomplete cell and (b) a case where the robot is left outside* **C**.

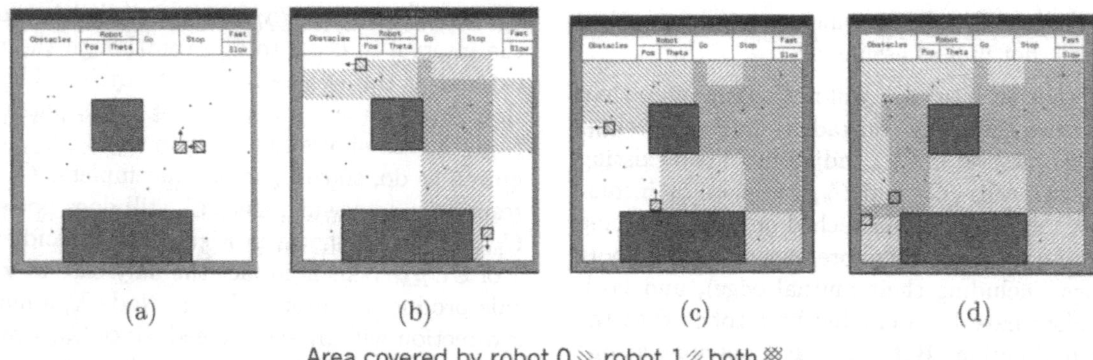

Figure 8: *Screenshots of a two-robot coverage run in progress: (a) Initial positions (b) Just after colleague relationship is formed (c) Each robot exploring a different region (d) Coverage is complete.*

4 Implementation / Example

DC_R has been implemented in simulation, and a series of screen shots of a single run are shown in Figure 8. The simulation gives an overall view of the environment as coverage progresses, and also displays a view (not shown here) of the cell decomposition internal to each robot. In the simple example shown here, the two robots R_0 and R_1 performed coverage independently until a time just before Figure 8b was taken, at which point their feature handlers decided that sufficient information was present in their maps to determine their relative position. Note that at this point, R_0 is no longer performing seed-sowing over the full width of the environment. By the time Figure 8c was taken, each robot was working on a different area of the environment, after which they ended up in the same area. R_1 then claimed the area at the lower left before R_0 could, and so R_0 simply waited in a "safe" place for R_1 to finish. Finally, when coverage is complete in Figure 8d, note that only about half of the area has been visited by both robots. While clearly not optimal with respect to time or total distance for the pair, it does show that each robot spends less time covering than it would without cooperation.

In addition to the simulation, CC_{RM} has been successfully implemented in the minifactory environment on a single courier, and the relationship of the algorithm to the underlying robot control will remain the same in DC_R. However, some extensions to DC_R will ease its transition into a real-world robot system. For example, the simulation currently incorporates small non-cumulative position error, which is allowed for in most cases, but not in a rigorous (or even completely correct) manner. Further analysis of the effects of this and other types of uncertainty will make DC_R applicable to a wider variety of robot systems. Also, while the simulation can effect collisions between robots, DC_R currently uses simple methods to attempt to avoid colleagues and make progress. These methods often succeed, but are prone to failure in complex environments, and more intelligent strategies must be developed, preferably ones that allow the completeness proof to be retained with minimal modification. Our eventual goal is the implementation of DC_R on the minifactory hardware to verify its utility in a real-world system.

5 Conclusion

In this paper, an algorithm DC_R has been presented with which a team of independent robots can cooperatively cover their shared environment. It comprises a reactive coverage algorithm CC_{RM} which operates without explicit knowledge of cooperation and two additional components (the feature handler and the overseer) which maintain cooperative relationships with other robots to increase efficiency of the team. This decoupling of coverage from cooperation enabled the completeness proof of DC_R presented in this paper, demonstrating that any team of square robots with intrinsic contact sensing can successfully cover a finite rectilinear environment efficiently.

Acknowledgments

The authors would like to thank Howie Choset and Ercan Acar for helpful discussions about coverage. This research was funded in part by NSF grant DMI-9527190. Zack Butler was supported in part by an NSF Graduate Research Fellowship.

References

[1] Y. S. Suh and K. Lee, "NC milling tool path generation for arbitrary pockets defined by sculptured surfaces," *Computer Aided Design*, vol. 22, no. 5, pp. 273–284, 1990.

[2] V. Lumelsky, S. Mukhopadhyay, and K. Sun, "Dynamic path planning in sensor-based terrain acquisition," *IEEE Trans. on Robotics and Automation*, vol. 6, no. 4, pp. 462–472, 1990.

[3] S. Hert, S. Tiwari, and V. Lumelsky, "A terrain covering algorithm for an AUV," *Autonomous Robots*, vol. 3, pp. 91–119, 1996.

[4] E. Acar and H. Choset, "Critical point sensing in unknown environments for mapping," in *IEEE Int'l Conf. on Robotics and Automation*, 2000.

[5] H. Choset and P. Pignon, "Coverage path planning: The boustrophedon decomposition," in *Intl. Conf. on Field and Service Robotics*, 1997.

[6] Z. J. Butler, A. A. Rizzi, and R. L. Hollis, "Contact sensor-based coverage of rectilinear environments," in *Proc. of IEEE Int'l Symposium on Intelligent Control*, Sept. 1999.

[7] A. A. Rizzi, J. Gowdy, and R. L. Hollis, "Agile assembly architecture: An agent-based approach to modular precision assembly systems," in *Proc. of IEEE Int'l. Conf. on Robotics and Automation*, pp. 1511–1516, April 1997.

[8] B. R. Donald, J. Jennings, and D. Rus, "Information invariants for distributed manipulation," *International Journal of Robotics Research*, vol. 16, no. 5, pp. 673–702, 1997.

[9] A. Drogoul and J. Ferber, "From Tom Thumb to the dockers: Some experiments with foraging robots," in *From Animals to Animats II*, pp. 451–460, MIT Press, 1993.

[10] P. Stone and M. Veloso, "Task decomposition, dynamic role assignment and low-bandwidth communication for real-time strategic teamwork," *Artificial Intelligence*, vol. 110, pp. 241–273, June 1999.

[11] D. W. Gage, "Randomized search strategies with imperfect sensors," in *Mobile Robots VIII*, pp. 270–279, 1993.

[12] N. Rao, V. Protopopescu, and N. Manickam, "Cooperative terrain model acquisition by a team of two or three point-robots," in *Proc. of IEEE Int'l. Conf. on Robotics and Automation*, pp. 1427–1433, April 1996.

[13] I. Rekleitis, G. Dudek, and E. Milios, "Multi-robot exploration of an unknown environment, efficiently reducing the odometry error," in *Int'l Joint Conf. in Artificial Intelligence*, (Nagoya, Japan), pp. 1340–1345, August 1997.

Closed-Loop Distributed Manipulation Using Discrete Actuator Arrays

Jonathan E. Luntz, *University of Michigan, Ann Arbor, MI*
William Messner, *Carnegie Mellon University, Pittsburgh, PA*
Howie Choset, *Carnegie Mellon University, Pittsburgh, PA*

Distributed manipulation systems induce motions on objects through the application of many external forces. An actuator array performs distributed manipulation using a planar array of many small stationary actuators (which we call cells) each of which applies a force in the plane to larger objects which rest atop many cells at once. An actuator array can transport and orient flat objects in the plane. The authors have developed a table-top scale actuator array consisting of many motorized wheels. In such a macroscopic array, a fairly small number of cells support the object. The work of the authors builds upon and extends the work of researchers in microelectromechanical (MEMS) actuator arrays by explicitly modeling the discreteness in the system, including the set of supports, distribution of weight, and generation of traction forces and by using the resulting model to directly design classes of actuation fields (sets of wheel speeds). Using a constant (open-loop) wheel velocity field, discreteness causes undesirable behavior such as unstable rotational equilibria, suggesting the use of object feedback. Discrete distributed control algorithms are derived by inverting the dynamics of manipulation (the relationship between wheels speeds and forces on the object) to come up with wheel velocity fields which effect the desired forces and moments on the object. These algorithms reduce the many-input-three-output control problem to a three-input-three-output control problem. Using these algorithms, the authors demonstrate the stability of closed-loop discrete distributed manipulation.

1 Introduction

An actuator array performs distributed manipulation where many small stationary elements (which we call cells) cooperate to manipulate larger objects. Objects lie on a regular array as they are transported and oriented. Many cells support the object simultaneously, and as it moves, the set of supporting cells changes. While supporting the object, each cell provides a traction force on it, and the combined action of all the cells supporting the object determines the motion of the object. Current applications of actuator arrays range from micromechanical systems transporting pieces of silicon wafer to macroscopic arrays of motorized wheels transporting cardboard boxes.

Actuator arrays represent a recent development in robotic manipulation. Typical work addresses the task of bringing an object to a particular position and orientation using open-loop modes of operation. The action of the cells is pre-programmed and constant, establishing a force field in which the object moves [1, 2]. These analyses considered microelectromechanical (MEMS) scale applications, where the large number of actuators justifies the assumption that the forces from the discrete actuators are a continuous force field applied over the area of the object. Under this assumption, researchers used potential field theory to predict motions and stable poses of objects.

In this work, we examine a macroscopic actuator array, the Modular Distributed Manipulator System (MDMS). Each cell of the MDMS is made up of a pair of motorized roller wheels which together can apply a directable traction force to objects such as cardboard boxes. The MDMS uses distributed control and therefore is fully programmable. An 18 cell prototype MDMS has been built and tested. On the MDMS, as few as four or six cells support an object such that continuous approximations fail and we must account for discreteness when analyzing the MDMS.

Discreteness causes imprecision in open-loop manipulation, particularly in the orientation of objects. For example, under certain modeling assumptions appropriate to the MDMS, the net moment acting on an object is not a direct function of the object's orientation, and orientation can only be done to cell resolution [5]. In addition, some objects having stable rotational equilibria on a continuous array have unstable rotational equilibria on a discrete array [4]. Therefore, closed-loop object control is necessary to precisely orient objects. In a closed-loop mode of operation, the action of the cells updates based on the object's motion. Feedback may be obtained from local sensing at each cell or from some global sensor, such as a vision system. Other work has been done in closed-loop distributed manipulation [6] using vibrating plates to generate continuous force fields acting on small (point) objects rather than pointwise forces acting on large objects.

This paper focuses on the derivation of closed-loop manipulation strategies for the MDMS. Section 2 reviews the discrete dynamics of manipulation on the MDMS from the previous WAFR conference. Section 3 derives closed-loop policies by inverting the dynamics. Section 4 analyzes the dynamics and stability of closed-loop manipulation. Concluding remarks are made in Section 5.

2 Dynamics of Manipulation

As the MDMS manipulates an object, the object passes from one set of supporting cells to the next. To understand the motion of the object, we examine the planar dynamics of an object while it rests on a single arbitrary set of cells. To compute the traction forces on the object from each wheel, we first compute the supporting (normal) forces and then apply a viscous-type friction law.

Consider an object of weight W, whose center of mass is located at $\vec{X}_{cm} = [\ x_{cm} \quad y_{cm}\ ]^{\mathrm{T}}$ resting on n cells. The object is supported by vertical normal forces $\vec{N} = [\ N_1 \quad \ldots \quad N_n\ ]$ whose determination requires consideration of equilibrium of the object in both the vertical (z) direction and in rotation about the x and y axes. Equilibrium provides three equations, and since we have n supports, the system is statically indeterminant, and we must consider flexibility in the system.

We assume each support is a linear spring. Physically, this flexibility is either a flexible suspension under each wheel as shown in Figure 1 or, as in the prototype, flexibility in the surface of the bottom of the object. We assume the bottom of the object is flat such that the spring deflections (and hence the normal forces) distribute linearly under the object. The 3 equilibrium

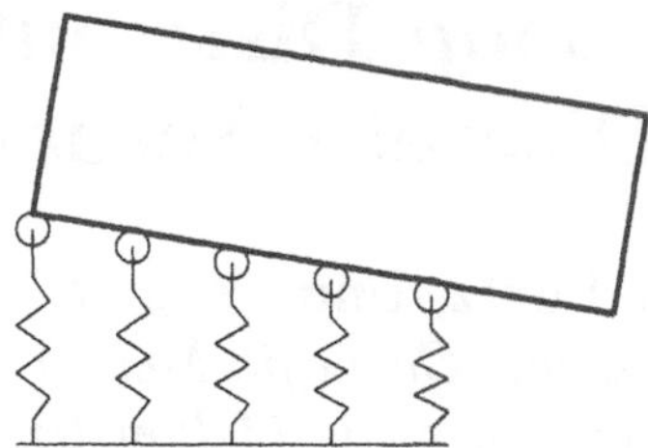

Figure 1: *Flexible supports act as springs to distribute the load.*

equations combined with n instances of this compatibility constraint form a set of $n+3$ equations and $n+3$ unknowns from which we can solve for $\vec{N}$ as a function of object position [5].

$$\vec{N}^{\mathrm{T}} = W\mathbf{B}^{\mathrm{T}}\left(\mathbf{B}\mathbf{B}^{\mathrm{T}}\right)^{-1}\left(\begin{bmatrix}1\\0\\0\end{bmatrix} + \begin{bmatrix}0 & 0\\1 & 0\\0 & 1\end{bmatrix}\vec{X}_{cm}\right) \tag{1}$$

where the matrix $\mathbf{B}$ contains the positions of the cells currently supporting the object.

$$\mathbf{B} = \begin{bmatrix}1 & \ldots & 1\\ x_1 & \ldots & x_n\\ y_1 & \ldots & y_n\end{bmatrix} \tag{2}$$

The horizontal forces on the object are derived from the normal forces through the use of a viscous-type friction law (see Figure 2). The horizontal force from each cell $\vec{f}_i$ is proportional to a coefficient of friction μ, that cell's normal force N_i, and the vector difference between the velocity of the wheel and the velocity of the object at the point of the cell.

$$\vec{f}_i = \mu\left(\vec{V}_i - \dot{\vec{X}}_{cm} + \omega\begin{bmatrix}0 & 1\\-1 & 0\end{bmatrix}\left(\vec{X}_i - \vec{X}_{cm}\right)\right)N_i \tag{3}$$

where ω is the rotation speed of the object. This traction force from each cell is summed over all the cells.

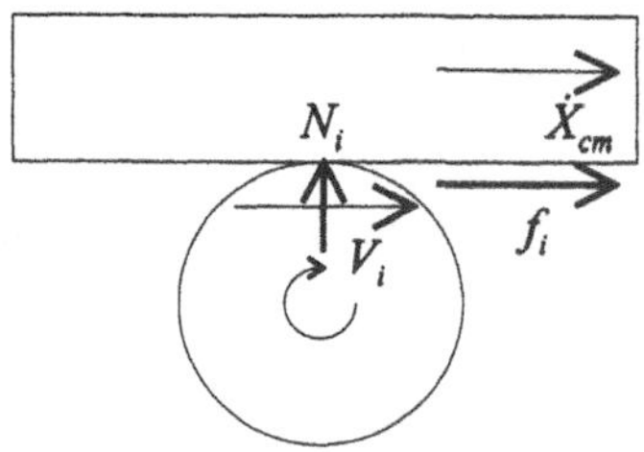

Figure 2: *Interaction between wheel and object.*

Define a wheel velocity matrix $\mathbf{V}$ as

$$\mathbf{V} = \begin{bmatrix} \vec{V}_x \\ \vec{V}_y \end{bmatrix} = \begin{bmatrix} V_{1_x} \cdots V_{n_x} \\ V_{1_y} \cdots V_{n_y} \end{bmatrix} = \begin{bmatrix} \vec{V}_1 \cdots \vec{V}_n \end{bmatrix} \tag{4}$$

Summing vectorially, the net horizontal force is

$$\vec{f} = \mu \mathbf{V}\, \vec{N}^{\mathrm{T}} - \mu \dot{\vec{X}}_{cm} W \tag{5}$$

Observe that the net horizontal force is not a function of the object's rotation speed - the terms multiplying ω are identically zero [5]. Furthermore, the second term in this equation is a constant linear damping term. Substituting $\vec{N}$ from Equation 1 yields

$$\vec{f} = \underbrace{\mu W \mathbf{V}\mathbf{B}^{\mathrm{T}} \left(\mathbf{B}\mathbf{B}^{\mathrm{T}}\right)^{-1} \begin{bmatrix} 0 & 0 \\ 1 & 0 \\ 0 & 1 \end{bmatrix}}_{\mathbf{k_s}} \vec{X}_{cm} \underbrace{+ \mu W \mathbf{V}\mathbf{B}^{\mathrm{T}} \left(\mathbf{B}\mathbf{B}^{\mathrm{T}}\right)^{-1} \begin{bmatrix} 1 \\ 0 \\ 0 \end{bmatrix}}_{\vec{f}_o} - \mu \dot{\vec{X}}_{cm} W \tag{6}$$

The 2×2 spring constant matrix $\mathbf{k_s}$, and the offset force vector $\vec{f}_o$ are constant while the object rests on a particular set of supports.

We compute the net torque similarly [5] resulting in the following expression for the net torque acting on the object as a function of position and rotational speed (and set of supports).

$$\tau = \underbrace{\mu W \vec{R} \mathbf{B}^{\mathrm{T}} (\mathbf{B}\mathbf{B}^{\mathrm{T}})^{-1} \begin{bmatrix} 1 \\ 0 \\ 0 \end{bmatrix}}_{\tau_o} + \underbrace{\mu W \vec{R} \mathbf{B}^{\mathrm{T}} (\mathbf{B}\mathbf{B}^{\mathrm{T}})^{-1} \begin{bmatrix} 0 & 0 \\ 1 & 0 \\ 0 & 1 \end{bmatrix}}_{\vec{k}_{s_\tau}} \vec{X}_{cm} + \vec{X}_{cm} \times \left(\vec{f}_o + \mathbf{k_s} \vec{X}_{cm}\right) - \mu\omega\left(\vec{\mathcal{X}} \vec{N}^{\mathrm{T}} - W \vec{X}_{cm}^{T} \vec{X}_{cm}\right) \tag{7}$$

where $R_i = \vec{X}_i \times \vec{V}_i$ and $\mathcal{X}_i = \vec{X}_i^{\mathrm{T}} \vec{X}_i = (x_i^2 + y_i^2)$. The vector $\vec{k}_{s_\tau}$ relates torque to position, and τ_o is a scalar constant torque. The net applied torque and damping are both effectively position dependent. The term multiplying ω is always negative, and hence dissipative (although not constant). Note that while an object rests on a particular set of supports, torque on the object is not a function of orientation. This is important for determining stable orientations.

3 Position and Orientation Feedback

Our strategy to implement position and orientation feedback is to apply velocity fields which reduce the array to a single "virtual" actuator capable of generating a desired force and torque on the object. Since force and torque change as the object moves, we must continuously recompute and adjust the wheel speeds. We assume sensing of the object's position and orientation is done externally, for example, by a vision system. It is desirable to distribute the computation to reduce communication such that each cell need only be aware of the desired net force and torque and its location relative to the center of the object, and not the state of each cell.

3.1 Applying Both Force and Torque

We compute velocity fields by inverting the relationship between wheel speeds and net force and torque. As a first attempt, we specify both a force and torque to compute the velocity field. Without loss of generality, we set the origin of the system to lie at the center of the object such that the cells move rather than the object. Also, we ignore the damping terms in Equations 6 and 7. For the purpose of feedback control, since the damping is mainly a function of the object's motion, we treat damping as a property of the manipulated object rather than the actuators. The expressions for the action of our virtual actuator (i.e. the net force and torque on the object) reduce to:

$$\vec{f} = \vec{f}_o = \mu W \mathbf{V}\mathbf{B}^{\mathrm{T}} \left(\mathbf{B}\mathbf{B}^{\mathrm{T}}\right)^{-1} \begin{bmatrix} 0 & 0 \\ 1 & 0 \\ 0 & 1 \end{bmatrix} \quad \text{and}$$
$$\tau = \tau_o = \mu W \vec{R} \mathbf{B}^{\mathrm{T}} \left(\mathbf{B}\mathbf{B}^{\mathrm{T}}\right)^{-1} \begin{bmatrix} 1 \\ 0 \\ 0 \end{bmatrix} \tag{8}$$

We can rewrite these equations in terms of the stacked velocity vector formed by stacking the transposes of the x and y component rows of $\mathbf{V}$.

$$\begin{bmatrix} f_{o_x} \\ f_{o_y} \\ \tau_o \end{bmatrix} = \mu W \cdot \underbrace{\left[\begin{array}{ccc|ccc|ccc} 1&0&0&0&0&0&0&0&0 \\ \hline 0&0&0&1&0&0&0&0&0 \\ \hline 0&0&0&0&0&0&1&0&0 \end{array}\right] \left[\begin{array}{c|c|c} \left(\mathbf{B}\mathbf{B}^{\mathrm{T}}\right)^{-1}\mathbf{B} & 0_{3\times n} & 0_{3\times n} \\ \hline 0_{3\times n} & \left(\mathbf{B}\mathbf{B}^{\mathrm{T}}\right)^{-1}\mathbf{B} & 0_{3\times n} \\ \hline 0_{3\times n} & 0_{3\times n} & \left(\mathbf{B}\mathbf{B}^{\mathrm{T}}\right)^{-1}\mathbf{B} \end{array}\right] \left[\begin{array}{c|c} I_n & 0_n \\ \hline 0_n & I_n \\ \hline -\mathbf{D_y} & \mathbf{D_x} \end{array}\right]}_{\mathbf{Q}} \cdot$$

$$\left[\begin{array}{c}\vec{V}_x^{\mathrm{T}} \\ \hline \vec{V}_y^{\mathrm{T}}\end{array}\right] \tag{9}$$

where $\mathbf{D_x}$ and $\mathbf{D_y}$ are diagonal matrices containing the x and y components of the cell positions. We pseudo-invert this underconstrained set of equations to solve for the stacked velocity vector.

$$\left[\begin{array}{c}\vec{V}_x^{\mathrm{T}} \\ \hline \vec{V}_y^{\mathrm{T}}\end{array}\right] = \frac{1}{\mu W}\mathbf{Q}^{\mathrm{T}}\left(\mathbf{Q}\mathbf{Q}^{\mathrm{T}}\right)^{-1}\begin{bmatrix} f_{o_x} \\ f_{o_y} \\ \tau_o \end{bmatrix} \tag{10}$$

Unfortunately, it is not possible to algebraically reduce $\mathbf{Q}^{\mathrm{T}}\left(\mathbf{Q}\mathbf{Q}^{\mathrm{T}}\right)^{-1}$ to a more useful form. Each cell's wheel speeds depend on the positions of all cells currently supporting the object. Therefore, while a field computed in such a manner will provide the desired force and torque, its computation cannot be distributed, and a centralized controller must give speed commands to each cell. This is not practical for a large system because of bandwidth limitations.

3.2 Superposition of Fields

Because traction force from each cell varies linearly with wheel speeds, net forces and torque add with the superposition of velocity fields. Therefore, we adopt the strategy of superimposing two fields: a field to apply a force without a torque and a field to apply a torque without a force, where each field operates under distributed control.

To compute a field to apply a force with no torque, we invert the relationship between wheel speeds and net force. By using the Penrose pseudo-inverse, we obtain the velocity field which provides the desired net force with the minimum sum of squares of wheel speeds. This minimizes extraneous motions, and hopefully (with no guarantee) the resulting field generates no net torque.

The expression for the net force in terms of the stacked velocity vector is

$$\begin{bmatrix} f_{o_x} \\ f_{o_y} \end{bmatrix} = \mu W \left[\begin{array}{ccc|ccc} 1 & 0 & 0 & 0 & 0 & 0 \\ \hline 0 & 0 & 0 & 1 & 0 & 0 \end{array}\right] \left[\begin{array}{c|c} \left(\mathbf{B}\mathbf{B}^{\mathrm{T}}\right)^{-1}\mathbf{B} & 0_{3\times n} \\ \hline 0_{3\times n} & \left(\mathbf{B}\mathbf{B}^{\mathrm{T}}\right)^{-1}\mathbf{B} \end{array}\right] \left[\begin{array}{c}\vec{V}_x \\ \hline \vec{V}_y\end{array}\right] \tag{11}$$

We pseudo-invert this underconstrained set of equations to solve for the stacked velocity vector.

$$\left[\begin{array}{c}\vec{V}_x^{\mathrm{T}} \\ \hline \vec{V}_y^{\mathrm{T}}\end{array}\right] = \frac{1}{\mu W}\left[\begin{array}{c|c} \mathbf{B}^{\mathrm{T}} & 0_n \\ \hline 0_n & \mathbf{B}^{\mathrm{T}} \end{array}\right] \left[\begin{array}{c|c} 1 & 0 \\ 0 & 0 \\ 0 & 0 \\ \hline 0 & 1 \\ 0 & 0 \\ 0 & 0 \end{array}\right] \begin{bmatrix} f_{o_x} \\ f_{o_y} \end{bmatrix} = \frac{1}{\mu W}\left[\begin{array}{c|c} 1 & 0 \\ \vdots & \vdots \\ 1 & 0 \\ \hline 0 & 1 \\ \vdots & \vdots \\ 0 & 1 \end{array}\right] \begin{bmatrix} f_{o_x} \\ f_{o_y} \end{bmatrix} \tag{12}$$

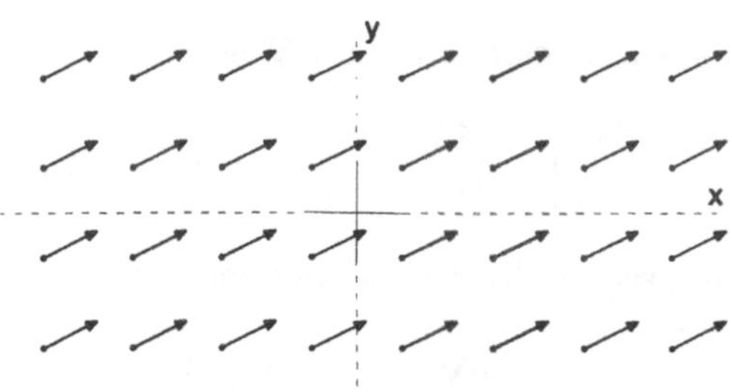

Figure 3: *The uniform field applies an arbitrary net force on an object with no net torque.*

Figure 3 shows the resulting uniform field. Our methodology ensures that the net force is independent of the set of supports such that we can apply the same net force without knowing which cells the object rests on. The velocities of the cells are decoupled, and are only functions of the net force to be applied. This field easily operates under distributed control, where a centralized controller need only broadcast the desired net force to all the cells.

This uniform field can be shown to apply no torque. While it may seem intuitive that this be true, it is not trivial. If we specified each cell's force rather than wheel speeds, there would be a net torque when the object centroid and the centroid of the supporting cells did not coincide. Since the traction force from each cell depends on the supporting force, however, the distribution of weight ensures that all moments balance. The algebraic proof is omitted.

The same methodology used to compute the uniform field applies to the application of an arbitrary torque. The expression for the net torque in terms of the stacked velocity vector is

$$\tau_o = \mu W \begin{bmatrix} 1 & 0 & 0 \end{bmatrix}\left(\mathbf{B}\mathbf{B}^{\mathrm{T}}\right)^{-1}\mathbf{B}\left[\begin{array}{c|c} -\mathbf{D_y} & \mathbf{D_x} \end{array}\right]\left[\begin{array}{c}\vec{V}_x^{\mathrm{T}} \\ \hline \vec{V}_y^{\mathrm{T}}\end{array}\right] \tag{13}$$

We pseudo-invert this underconstrained set of equations to solve for the stacked velocity vector.

$$\left[\begin{array}{c}\vec{V}_x^{\mathrm{T}} \\ \hline \vec{V}_y^{\mathrm{T}}\end{array}\right] = \frac{1}{\mu W}\left[\begin{array}{c} -\mathbf{D_y} \\ \hline \mathbf{D_x} \end{array}\right]\left(\left[\begin{array}{c|c} -\mathbf{D_y} & \mathbf{D_x} \end{array}\right]\left[\begin{array}{c} -\mathbf{D_y} \\ \hline \mathbf{D_x} \end{array}\right]\right)^{-1}\mathbf{B}^{\mathrm{T}}\begin{bmatrix} 1 \\ 0 \\ 0 \end{bmatrix}\tau_o \tag{14}$$

This expression can be algebraically reduced since $\mathbf{D_x}$ and $\mathbf{D_y}$ are diagonal matrices. After some algebra, the

Figure 4: *The computed rotational field applies an arbitrary net torque on an object.*

wheel velocities decouple and the velocity of each wheel becomes

$$V_{x_i} = \frac{1}{\mu W}\frac{-y_i}{x_i^2+y_i^2}\tau_o \quad \text{and} \qquad V_{y_i} = \frac{1}{\mu W}\frac{x_i}{x_i^2+y_i^2}\tau_o \tag{15}$$

Figure 4 shows the resulting computed rotational field. The vector formed by each cell's two wheel speeds is perpendicular to its position relative to the center of the object with a magnitude inversely proportional to distance. Again, our methodology ensures that the net torque is independent of the set of supports. This field also operates under distributed control, where a centralized controller need only broadcast the desired torque and current object location to all the cells.

This rotational field, however, has two problems. First, since the wheels speeds are inversely proportional to their distance from the object center, some wheel speed commands may become arbitrarily large, and the wheel speeds will saturate. In practice, only one cell will be close enough to the object center to saturate, and its contribution will be limited.

The second problem is that a net force is applied by the rotational field. We obtain the expression for the net force by substituting Equations 15 into Equation 11. Considering the x component of Equation 11 (the y component is handled similarly), the expression for the net force is

$$\begin{aligned} f_{o_x} &= \mu W \, [1\,0\,0] \left(\mathbf{B}\mathbf{B}^{\mathrm{T}}\right)^{-1} \mathbf{B} \begin{bmatrix} \frac{1}{\mu W}\frac{-y_1}{x_1^2+y_1^2} \\ \vdots \\ \frac{1}{\mu W}\frac{-y_n}{x_n^2+y_n^2} \end{bmatrix} \\ &= [1\,0\,0] \begin{bmatrix} n & \sum x_i & \sum y_i \\ \sum x_i & \sum x_i^2 & \sum x_i y_i \\ \sum y_i & \sum x_i y_i & \sum y_i^2 \end{bmatrix} \begin{bmatrix} \sum \frac{-y_i}{x_i^2+y_i^2} \\ \sum \frac{-x_i y_i}{x_i^2+y_i^2} \\ \sum \frac{-y_i^2}{x_i^2+y_i^2} \end{bmatrix} \end{aligned} \tag{16}$$

This expression, in general is not equal to zero and generates an extra force which was not intended by this field. This disturbance force ($\vec{f_d}$) is dependent on the application of torque. While it can be shown that there are many positions and orientations at which this disturbance force vanishes [3] and its effect may be small, we desire an alternate method for generating a torque without generating a disturbance force.

3.3 Known Torque Direction Method

To derive a field to generate a torque with no force, we exploit the distribution of weight among the cells to compute wheel velocities in such a way that no net force is applied. Rotational equilibrium of the object about the x and y axes ensures that the normal forces are distributed such that

$$0 = \begin{bmatrix} x_1 & \cdots & x_n \\ y_1 & \cdots & y_n \end{bmatrix} \begin{bmatrix} N_1 \\ \vdots \\ N_n \end{bmatrix} \tag{17}$$

We can represent the net force on the object as

$$\vec{f} = \mu \begin{bmatrix} V_{1_x} & \cdots & V_{n_x} \\ V_{1_y} & \cdots & V_{n_y} \end{bmatrix} \begin{bmatrix} N_1 \\ \vdots \\ N_n \end{bmatrix} \tag{18}$$

These two equations are of the same form with the position matrix replaced by the velocity matrix. If we set the wheel velocity matrix to match the position matrix, the net force will be zero. In addition, we can swap the two rows of the velocity matrix and negate the top row while maintaining zero force to generate the following velocity field.

$$V_{i_x} = -\alpha y_i \qquad \text{and} \qquad V_{i_y} = \alpha x_i \tag{19}$$

where α is a constant with which we can scale the field. This is a kinematic rotational field since the wheel speeds vary linearly with position as they do in rigid body rotation as shown in Figure 5. This field is decoupled and operates under distributed control. The net torque this field applies is

$$\tau_l = \sum_i \tau_i = \mu\alpha \sum_i N_i \left(y_i^2 + x_i^2\right) \tag{20}$$

This torque is not constant; it changes with both object position and set of supports. However, because each cell contributes positively to the summation of

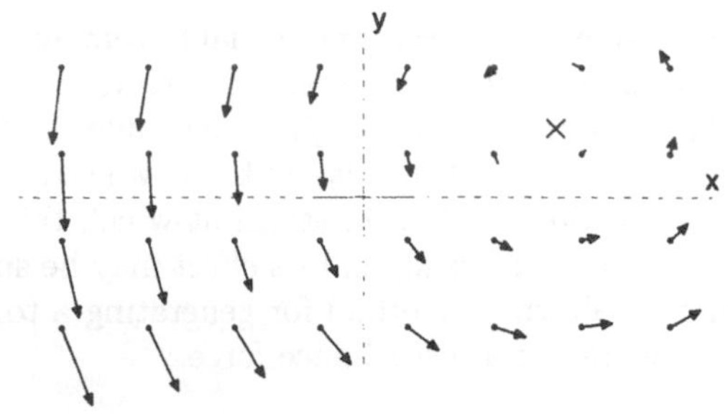

Figure 5: *The kinematic rotational field applies a torque of known direction without a net force.*

torques, the *direction* of the torque is known. Therefore, we eliminate the disturbance force generated by the computed rotational field at the expense of uncertainty in the applied torque magnitude. This field is still useful for feedback by setting a constant α and multiplying Equation 20 by the desired torque.

4 Closed-Loop Dynamics and Stability

Given these three superimposable manipulation fields, we can now implement separate proportional control feedback loops for position and orientation as shown in Figure 6. The two loops are completely independent except for the disturbance force generated by the orientation loop in the case of the computed rotational field.

Because the force applied by the uniform field is precise, and the damping term from Equation 6 is constant and linear, the position loop has simple linear second order time invariant dynamics. (The other terms in Equation 6 are identically zero because the origin of the field follows the object.) The closed-loop transfer

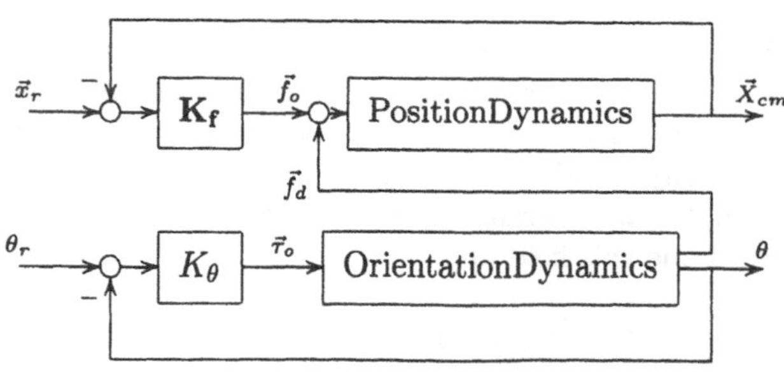

Figure 6: *Feedback control structure. Superimposable fundamental fields allow separate position (top) and orientation (bottom) loops.*

function from position reference input x_r to the object's position x_{cm} under proportional gain K_{f_x} is

$$\frac{X_{cm}(s)}{X_r(s)} = \frac{K_{f_x}}{ms^2 + \mu W s + K_{f_x}} \tag{21}$$

(Actually, a pair of transfer functions exist, one for x and one for y.) The closed-loop transfer function from the disturbance input, f_{d_x} (from the computed rotational field), to object position is

$$\frac{X_{cm}(s)}{F_{d_x}(s)} = \frac{1}{ms^2 + \mu W s + K_{f_x}} \tag{22}$$

In the case of the kinematic rotational field, the position loop is independent of the orientation loop and is already stable. In the case of the computed rotational field, the position loop (which is a stable linear time invariant system with a disturbance) is asymptotically stable as long as the disturbance asymptotically approaches zero. Assuming the orientation loop is itself asymptotically stable, the orientation error (and therefore the torque command) asymptotically approaches zero. The disturbance force asymptotically approaches zero since it varies in proportion to the torque command. Therefore, the entire manipulation dynamics are asymptotically stable as long as the orientation loop is itself asymptotically stable.

4.1 Rotational Stability Under The Computed Rotational Field

The closed-loop rotational dynamics under the computed rotational field have a linear form, with mass, spring, and damping coefficients, but the damping is a function of the set of supports and the object's translational position. The computed rotational field is such that the torque applied is precise (ignoring saturation) and with a proportional gain of K_τ is equal to $K_\tau(\theta_r - \theta)$. The rotational damping coefficient from Equation 7 is simply $\mu \vec{\mathcal{X}} \vec{N}^{\mathrm{T}}$, or equivalently, $\mu \sum N_i \left(x_i^2 + y_i^2\right)$, which refer to as $b(\vec{X}_{cm}(t), \theta)$. The other terms (and the second portion of the damping coefficient) in Equation 7 are identically zero since the origin of the field follows the object. The rotational dynamics of the object under the computed rotational field therefore are

$$J\ddot{\theta} + b(\vec{X}_{cm}(t), \theta)\dot{\theta} + K_\tau \theta = K_\tau \theta_r \tag{23}$$

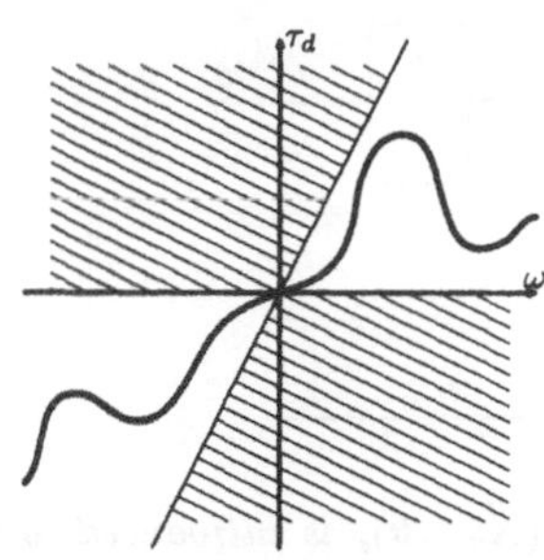

Figure 7: *Rotational damping is a sector nonlinearity where rotation speed (ω) maps to damping torque (τ_d). In this example, the nonlinearity lies in the sector* $(0, 2)$.

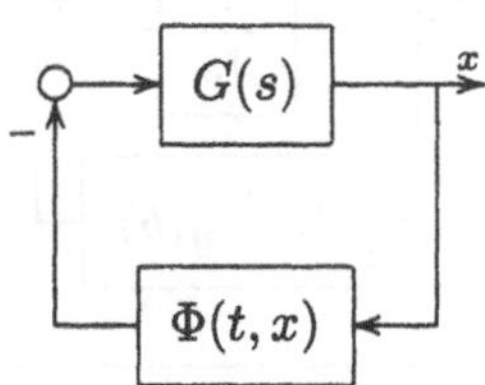

Figure 8: *Standard form of a system to be analyzed by the circle criterion*

The damping coefficient $b(\vec{X}_{cm}(t), \theta)$ is not a function of the wheel speeds, but it is a function of the position of the object and the set of supports (and implicitly the orientation). The coefficient changes discontinuously, but is bounded both above and below. This nonlinearity is deterministic, but is time-varying since it depends on the object position (which is independent of orientation).

While it is difficult to come up with good upper and lower bounds on the damping coefficient, we can base extremely conservative bounds on the facts that the object is of finite size (contained in a circle of radius r_{max}) and that all of the supporting forces are positive. These limits ensure that the damping force is a sector nonlinearity (lying in a sector (k_1, k_2) such that $k_1 < b(\vec{X}_{cm}(t), \theta) < k_2$). Figure 7 shows an example of such a sector nonlinearity.

Figure 9 shows the nonlinear orientation feedback loop for the computed rotational field. State-dependent damping makes the orientation loop truly nonlinear, although it is *instantaneously* linear such that we may perform some block diagram manipulations such as passing gains through the nonlinearity and combining the nonlinearity in parallel with itself.

We can show that this loop is stable using the circle criterion for the sector nonlinearity [7]. A sector

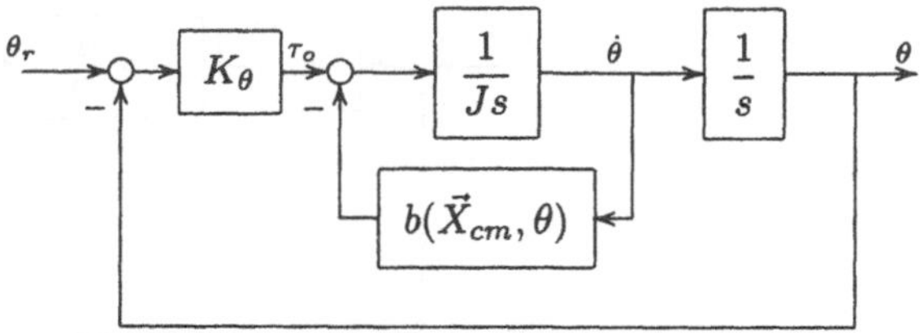

Figure 9: *The orientation loop for the computed rotational field is a second order system with (nonlinear) variable damping coefficient* $b(\vec{X}_{cm}(t), \theta)$.

nonlinearity is a function $\Phi(t, x)$ such that $k_1 x^2 \leq x\Phi(t, x) \leq k_2 x^2$ for all $t \geq 0$. The damping sector nonlinearity is a variable gain, $k_1 \leq b(\vec{X}_{cm}(t), \theta) \leq k_2$ such that $\Phi(t, \theta) = \phi(\vec{X}_{cm}(t), \theta)\theta$. Define the disk $\mathbf{D}(k_1, k_2)$ as the circle whose center lies on the real axis and crosses the real axis at $-\frac{1}{k_1}$ and $-\frac{1}{k_2}$ as shown by the circle in Figure 13. The circle criterion states that a system of the form shown in Figure 8 with a linear system $G(s)$ with ρ unstable poles connected in feedback by a sector nonlinearity, $\Phi(t, x)$, will be globally asymptotically stable if the Nyquist plot of the forward-path linear system encircles the disk $\mathbf{D}(k_1, k_2)$ exactly ρ times counter-clockwise.

Under the computed rotational field, the orientation loop readily transforms into the standard form of a system to be analyzed by the circle criterion. For zero reference, the block diagram in Figure 9 is equivalent to the form in Figure 10. It can be shown that the Nyquist plot of the forward path always lies in the right half plane and cannot enter or encircle the disk $D(k_1, k_2)$. Since there are no unstable poles in this forward path, by the circle criterion, the closed-loop orientation dynamics are globally asymptotically sta-

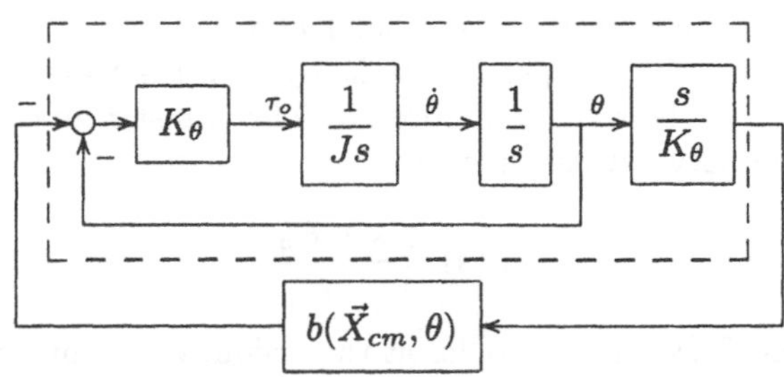

Figure 10: *The rearranged orientation feedback loop under the computed rotational field has linear forward path and the sector nonlinearity* $b(\vec{X}_{cm}, \theta)$ *in feedback.*

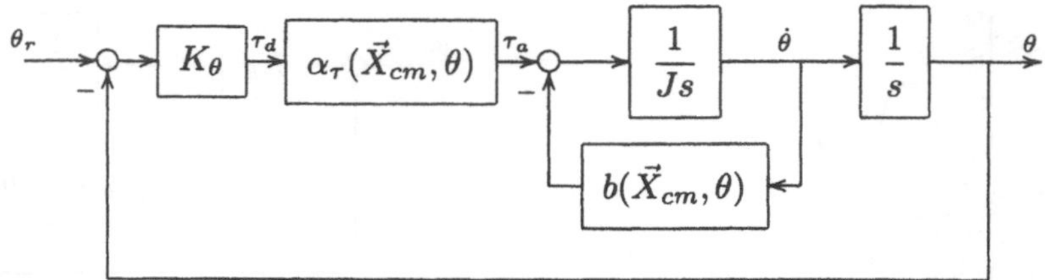

Figure 11: *Using the kinematic rotation field, a second sector nonlinearity, $\alpha_\tau(\vec{X}_{cm}, \theta)$, is introduced to the orientation feedback loop.*

ble under the computed rotational field for all values of proportional gain. Since this loop is asymptotically stable, the entire dynamics will be stable.

4.2 Rotational Stability under the Kinematic Rotational Field

Figure 11 shows the orientation loop using the kinematic rotation field. Because the applied torque (Equation 20), is not precise, an additional nonlinearity is introduced from the desired torque to the applied torque. The applied torque varies from the desired torque by a state dependent gain which we call $\alpha_\tau(\vec{X}_{cm}, \theta)$. This coefficient has the same form as $b(\vec{X}_{cm}, \theta)$ within the multiplicative constant α. Since there are now two sector nonlinearities, it is more difficult to manipulate the block diagram to the proper form to apply the circle criterion.

Figure 12 shows the rearranged version of this loop. To rearrange the loop, assume the input is zero and combine the gain K_θ with the integrator block. Since the two nonlinearities differ only by a multiplicative constant, α, factor out α and define $K'_\theta = \alpha K_\theta$. The two nonlinearities (one with the integrator block), now called $\phi(\vec{X}_{cm}, \theta)$, are in parallel. Since $\phi(\vec{X}_{cm}, \theta)$ is a pointwise-linear state-dependent time-varying gain, combine the two parallel paths to form the block diagram at the right of Figure 12. The transfer function of the forward path is now

$$G'(s) = \frac{s + K'_\theta}{Js^2} \tag{24}$$

Since $G'(s)$ has two poles at the origin, we circumvent the origin by passing the Nyquist contour to the right which treats these two poles as unstable. (Alternatively, one could shift the two poles slightly away from the origin by applying appropriate block diagram loop

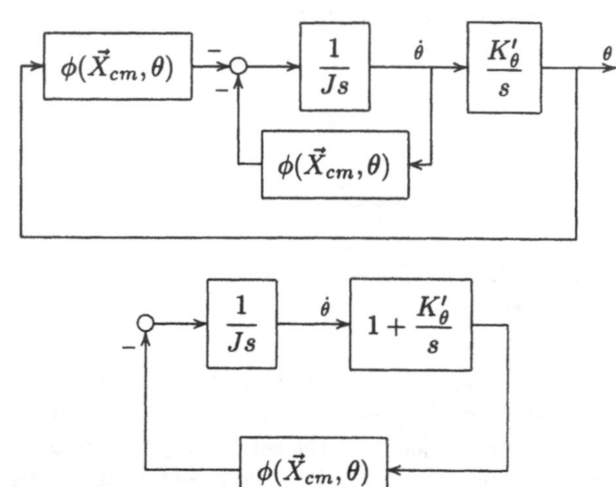

Figure 12: *The block diagram for the orientation feedback loop with the kinematic rotational field can be rearranged to apply the circle criterion. The two nonlinearities are in parallel (left) and since they have the same form, can be combined (right).*

transformations.) Therefore, to satisfy the circle criterion, the Nyquist plot must encircle the disk $\mathbf{D}(k_1, k_2)$ twice.

Figure 13 shows the Nyquist plot of $G'(s)$. The solid portion of the Nyquist plot follows the function $y = \pm\sqrt{\frac{-x}{K'_\theta}}$, where x and y refer to the real and imaginary components of $G(s)$. The dashed portion of the Nyquist plot loops around counterclockwise at infinity. This figure shows that given values of k_1 and k_2 there is a range of gains K'_θ for which the disk $\mathbf{D}(k_1, k_2)$ is encircled twice. The square root curve shifts inward as the gain is increased, and after a certain gain, the Nyquist plot enters the disk and we can no longer guarantee the closed-loop system to be stable.

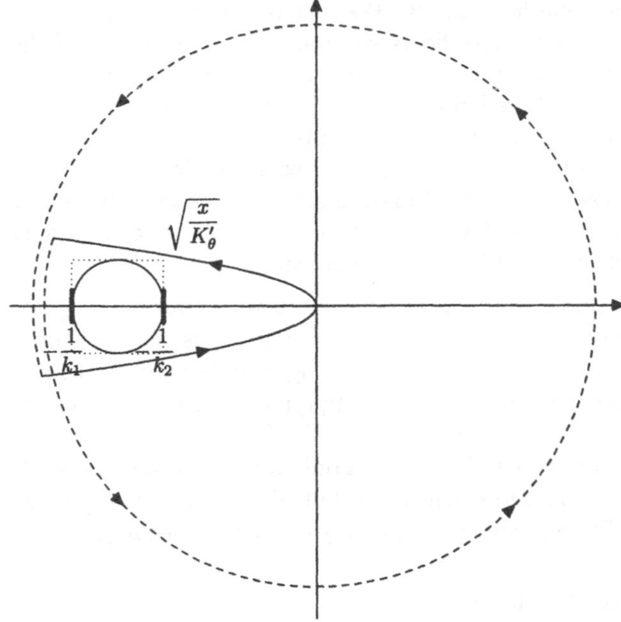

Figure 13: *Application of the circle criterion to the orientation loop using the kinematic rotational field. The circle represents the disk* $\mathbf{D}(k_1, k_2)$ *where* k_1 *and* k_2 *provide the bounds on the sector nonlinearity. The solid and dashed path shows the Nyquist plot in the complex plane of the linear forward path. The dashed portion of the path represents an encirclement at infinity. The solid portion of the path is of the form of a square root. Note that* $\mathbf{D}(k_1, k_2)$ *is encircled twice.*

Rather than solve the cubic equation to determine the range of gains for which the Nyquist plot does not enter the disk, we can obtain a sufficient condition for asymptotic stability by guaranteeing that the Nyquist plot will not enter the square circumscribing the disk (shown by the dotted square in Figure 13). The upper right corner of the square is located at

$$-\frac{1}{k_2} + \frac{\frac{1}{k_1} - \frac{1}{k_2}}{2} j = -\frac{1}{k_2} + \frac{k_2 - k_1}{2k_1 k_2} j \qquad (25)$$

This point lies inside the square root portions of the Nyquist plot if the following condition holds.

$$\sqrt{\frac{\frac{1}{k_2}}{K'_\theta}} > \frac{k_2 - k_1}{2k_1 k_2} \quad \Longrightarrow \quad K'_\theta < \frac{4{k_1}^2 k_2}{(k_2 - k_1)^2} \qquad (26)$$

Therefore, determining bounds k_1 and k_2 on $\phi(\vec{X}_{cm}, \theta)$, allows the selection of a gain to guarantee global asymptotic stability of the rotational closed-loop system under the kinematic rotation field. It is interesting to examine limiting cases for these bounds. In the limit where $k_1 = k_2$, the denominator in Equation 26 goes to zero, and the system will be stable at any gain. This is expected since this is simply the linear case. In the limit where $k_2 \gg k_1$, the bound becomes $K'_\theta < \frac{4{k_1}^2}{k_2}$ which may be a very small number. Therefore, it is important to have good bounds on the nonlinearity to allow a reasonable range of gains.

We do not currently have good bounds on $\phi(\vec{X}_{cm}, \theta) = \mu \vec{\mathcal{X}} \vec{N}^{\mathrm{T}}$, although we know finite bounds exist. The bounds on this nonlinearity are functions of the size and shape of an object. For a given object, we could obtain exact bounds by computationally moving the object over its entire configuration space (which repeats on a regular array).

Alternatively, we could obtain better bounds with a more general, geometric analysis for a class of objects (namely rectangular objects). This would require an analysis of the discrete sampling of the object's shape over its range of motion, which is quite difficult in itself, and not within the scope of this work. Better bounds, however, would allow for the use of higher gains while guaranteeing stability. Also, the circle criterion itself only provides a sufficient condition for stability. We believe that the closed-loop system under the kinematic rotation field is stable at all gains.

4.3 Closed-Loop Simulation

Figure 14 shows simulations of the closed-loop system under both rotational fields. Under the computed rotational field (left) the object reaches equilibrium exactly at both the desired position and orientation. Since this is approximately a set of mass-damper systems, by adjusting the gains, we can adjust the overshoot and settling time of each component of the closed-loop response. The curvature of the path of the center of the object (which otherwise would have been straight) demonstrates the effect of the disturbance force. Under the kinematic rotational field (right) the object moves in a straight line to the equilibrium position, and the rotational motion is greatly unaffected by the torque uncertainty.

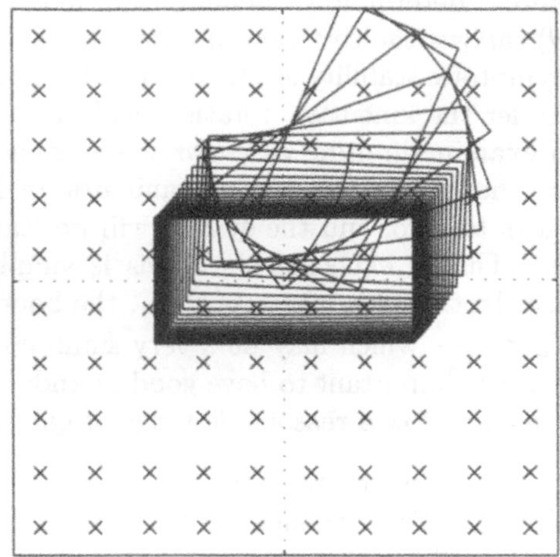

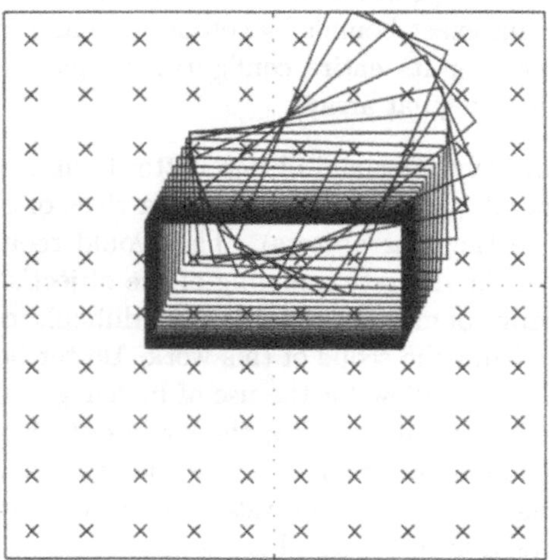

Figure 14: *Simulation of the closed-loop system under the computed rotational field (top) and the kinematic rotational field (bottom). The object is shown as the moving rectangle, with a curve tracing the motion of it's center. Each cell is marked with an 'X'.*

5 Conclusions

In this paper we have derived closed-loop control strategies for manipulating objects. While more complicated to implement than open-loop control, these strategies address precision limitations apparent in the open-loop operation of discrete actuator arrays. We derived a class of fields which reduce the array of actuators to a single virtual actuator for which simple feedback methods apply. Because of the design methodology used, these fields eliminate the dependence of the dynamics on the set of supports. Because of the use of distributed control, limitations are present in these strategies in the form of coupling from torque to force, or in the form of torque uncertainty. We proved that these limitations do not affect closed-loop stability nor do they hinder precision of equilibria, but they may affect closed-loop performance.

Our future efforts will be toward performance guarantees for closed-loop manipulation using distributed control obtained by generating good bounds on the discrete nonlinearities and applying robust control methods. We will also extend the feedback methods presented here to include not only manipulation to a single position and orientation, but also path and trajectory following with multiple objects simultaneously.

References

[1] K. Böhringer, B. Donald, R. Mihailovich, and N. MacDonald. A theory of manipulation and control for microfabricated actuator arrays. In *Proceedings, IEEE International Conference on Robotics and Automation*, 1994.

[2] L. Kavraki. Part orientation with programmable vector fields: Two stable equilibria for most parts. In *Proceedings, IEEE International Conference on Robotics and Automation*, 1997.

[3] J. Luntz, W. Messner, and H. Choset. Discreteness issues of manipulation using actuator arrays. In *Workshop on Distributed Manipulation, IEEE International Conference on Robotics and Automation*, 1999.

[4] J. Luntz, W. Messner, and H. Choset. Open loop orientability of objects on actuator arrays. In *Proceedings, IEEE International Conference on Robotics and Automation.*, 1999.

[5] J. Luntz, W. Messner, and H. Choset. Velocity field design for parcel manipulation on the modular distributed manipulator system. In *Proceedings, IEEE International Conference on Robotics and Automation.*, 1999.

[6] D. Reznik, E. Moshkoich, and J. Canny. Building a universal planar manipulator. In *Proceedings, Workshop on Distributed Manipulation at the International Conference on Robotics and Automation*, 1999.

[7] M. Vidyasagar. *Nonlinear Systems Analysis.* Prentice-Hall Inc., Englewood Cliffs, NJ, 1978.

Kinematic Tolerance Analysis with Configuration Spaces

Leo Joskowicz, *The Hebrew University, Jerusalem, Israel*
Elisha Sacks, *Purdue University, West Lafayette, IN*

This paper is a survey of our research on kinematic tolerance analysis of mechanical systems with parametric part tolerances. We present a general algorithm for planar systems and illustrate it with a design case study. The algorithm constructs a variation model for the system, derives worst-case bounds on the variation, and helps designers find unexpected failure modes, such as jamming and blocking. The variation model is a generalization of the configuration space representation of nominal part contacts. The algorithm handles general planar systems of curved parts with contact changes, including open and closed kinematic chains. It constructs a variation model for each interacting pair of parts then derives the overall system variation at a given configuration by composing the pairwise variation models via sensitivity analysis and linear programming. We demonstrate the algorithm on a gear selector mechanism in an automotive transmission with 100 functional parameters. The analysis, which takes less than a minute on a workstation, indicates that the critical kinematic variation occurs in third gear and identifies the parameters that cause the variation.

1 Introduction

This paper describes our research in kinematic tolerance analysis of mechanical systems. The task is to estimate the worst-case or average error in critical system parameters due to manufacturing variation. This analysis plays a key role in improving design quality and in reducing development time. In current practice, tolerance analysis is an imperfect, difficult, and time consuming activity. To keep the analysis affordable and on time, only those aspects of a design that are presumed to be critical are considered. These typically include safety items, selected clearances, and areas where part interferences are expected. Unproven assumptions and simplifications are often made to speed up the analysis. These limitations cause significant risks and uncertainties despite large analysis efforts.

The major steps in tolerance analysis are tolerance specification, variation modeling, and sensitivity analysis. Tolerance specification defines the allowable variation in the shapes and configurations of the parts of a system. The most common are parametric and geometric tolerance specifications [16, 25]. Variation modeling produces mathematical models that map tolerance specifications to system variations. Sensitivity analysis estimates the worst-case and statistical variations of critical properties in the model for given part variations. Designers iterate through these steps to synthesize systems that work reliably and that optimize other design criteria, such as cost and durability.

The tolerancing objectives are to produce designs that can be assembled and that function correctly despite manufacturing variation. In assembly tolerancing, general part variations must be modeled, so geometric tolerance specifications are the norm. Statistical sensitivity analysis is appropriate because guaranteed assembly is more expensive than discarding a few defective products. Most algorithms perform tolerance analysis on the final assembled configuration [2], although recent research explores toleranced assembly sequencing [12] and fixturing [4]. In functional tolerancing, the relevant part variations occur in functional features whose descriptions are parametric. Parametric tolerances, which are simpler than geometric tolerances, are best suited to capture these variations. Worst-case analysis is most appropriate because functional failures that occur after product delivery can be unacceptable.

Our research addresses functional kinematic tolerance analysis. Kinematic tolerancing is the most important form of functional tolerancing because kinematic function, which is described by motion constraints due to part contacts, largely determines mechanical system function. The task is to compute the variation in the part motions due to variations in the tolerance parameters. Variation modeling derives the functional relationship between the tolerance parameters and the system kinematic function. Sensitivity analysis determines the variation of this function over the allowable parameter values.

We illustrate kinematic tolerance analysis on an intermittent gear mechanism (Figure 1). The mechanism consists of a constant-breath cam, a follower with two pawls, and a gear with inner teeth. The cam and the gear are mounted on a fixed frame and rotate around their centers; the follower is free. Rotating the cam causes the follower to rotate in step and to reciprocate along its length (the horizontal axis in the figure). The follower engages a gear tooth with one pawl (snapshot a), rotates the gear 57 degrees, disengages, rotates independently for 5 degrees while the gear dwells (snapshot b), then engages the gear with its opposite pawl and repeats the cycle (snapshot c). The mechanical function is conversion of rotary motion into alternate rotation and dwell. The dwell time is determined by the gear tooth spacing. Quantitative tolerance analysis bounds the variation in the gear rotation, which is the critical parameter in precision indexing. Qualitative analysis detects failure modes, such as jamming, when one pawl cannot disengage because the other prematurely touches the gear.

Creating a variation model is the limiting factor in kinematic tolerance analysis. In most cases, the analyst has to formulate and solve systems of algebraic equations to obtain the relationship between the tolerance parameters and the kinematic function. The analysis grows much harder when the system topology changes, that is, when different parts interact at various stages of the work cycle, such as in the example above. Contact changes represent qualitative changes in the system function. They occur in the nominal function of higher pairs, such as gears, cams, clutches, and ratchets. Part variation produces unintended contact changes in systems whose nominal designs prescribe permanent contacts, such as joint play in linkages. The analysis has to determine which contacts occur at each stage of the work cycle, to derive the resulting kinematic functions, and to identify potential failure modes due to unintended contact changes, such as play, under-cutting, interference, and jamming. Once the variation model is obtained, sensitivity analy-

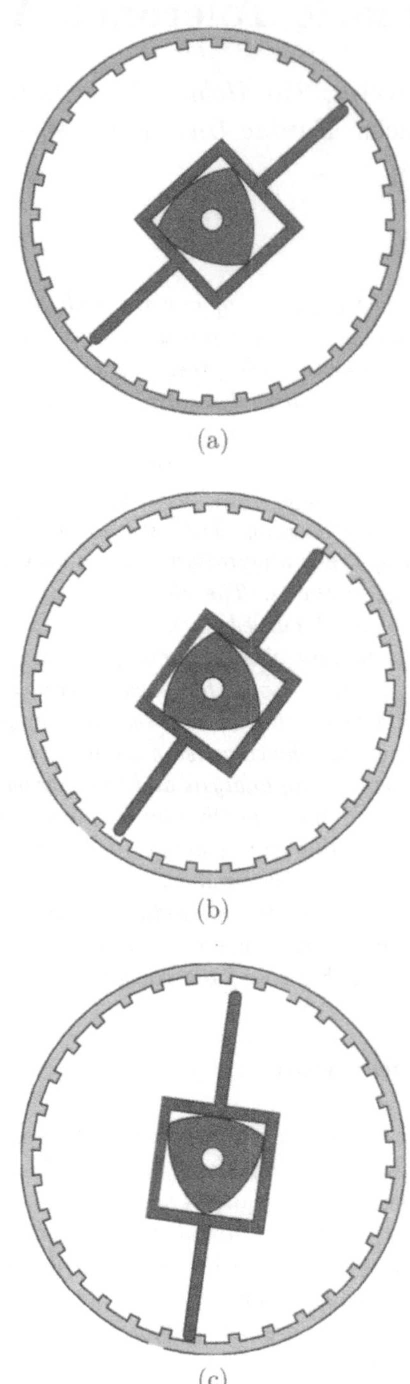

Figure 1: *Intermittent gear mechanism: (a) upper follower pawl engaged, (b) follower disengaged, (c) lower pawl engaged.*

sis can be performed by linearization, statistical analysis, or Monte Carlo simulation to quantify the variation in each mode [2].

We have developed a general kinematic tolerance analysis algorithm that addresses these issues [20]. The algorithm constructs a variation model for the system, derives worst-case bounds on the variation, and helps designers find failure modes. The variation model is a generalization of our configuration space representation of the nominal part contacts. The algorithm handles general planar systems of curved parts with contact changes, including open and closed kinematic chains. It analyzes systems with 50 to 100 parameters in under a minute, which permits interactive tolerancing of detailed functional models.

In this paper, we survey our research on kinematic tolerance analysis [9, 19, 20, 21]. The diverse prior results are presented in an integrated manner with realistic examples. We describe the configuration space representation, explain its role in tolerance analysis, and outline the analysis algorithm for general planar systems. We demonstrate the algorithm on an industrial application: Tolerancing a gear selector mechanism with 100 functional parameters. We conclude with a discussion of future work and of applications in path planning with geometric uncertainty.

2 Previous Work

Previous work on kinematic tolerance analysis of mechanical systems falls into three increasingly general categories: static (small displacement) analysis, kinematic (large displacement) analysis of fixed contact systems, and kinematic analysis of systems with contact changes. Static analysis of fixed contacts, also referred to as tolerance chain or stack-up analysis, is the most common. It consists of identifying a critical dimensional parameter (a gap, clearance, or play), building a tolerance chain based on part configurations and contacts, and determining the parameter variability range using vectors, torsors, or matrix transforms [5, 26]. Recent research explores static analysis with contact changes [1, 4, 8]. Configurations where unexpected failures occur can easily be missed because the software leaves their detection to the user. Kinematic analysis of fixed contact mechanical systems, such as linkages, has been thoroughly studied in mechanical engineering [6]. It consists of defining kinematic relations between parts and studying their kinematic variation [3]. Most commercial computer-aided tolerancing systems include this capability for planar and spatial mechanisms [24]. These methods are impractical for systems with many contact changes, such as a chain drive, and can miss failure modes due to unforeseen contact changes. Our method overcomes these limitations by automating variation modeling and analysis for general planar systems.

3 Configuration Space

We model nominal kinematic function within the configuration space representation of rigid body interaction. Configuration space is a general representation for systems of rigid parts that is widely used in robot motion planning [11, 15]. We construct a configuration space for each pair of interacting parts in the mechanical system. The configuration space is a manifold with one coordinate per part degree of freedom. Interactions of pairs of fixed-axes planar parts are modeled with two-dimensional spaces [18], whereas interactions between general planar pairs are modeled with three-dimensional spaces [17]. In both cases, points specify the relative configuration (position and orientation) of one part with respect to the other. We perform contact analysis by computing a configuration space for each pair of parts.

Configuration space partitions into three disjoint sets that characterize part interaction: Blocked space where the parts overlap, free space where they do not touch, and contact space where they touch without overlap. Blocked space represents unrealizable configurations, free space represents independent part motions, and contact space represents motion constraints due to part contacts. The spaces have useful topological properties. Free and blocked space are open sets whose common boundary is contact space. Contact space is a closed set comprised of algebraic patches that represent contacts between pairs of part features. Patch boundary curves represent simultaneous contacts between two pairs of part features.

We illustrate these concepts on the gear/follower pair of the intermittent gear mechanism (Figure 2). We compute the configuration space of the gear relative to the follower. The gear frame is at the center of the outer boundary circle and its x axis is horizontal in part a. The follower frame is at the center of the inner

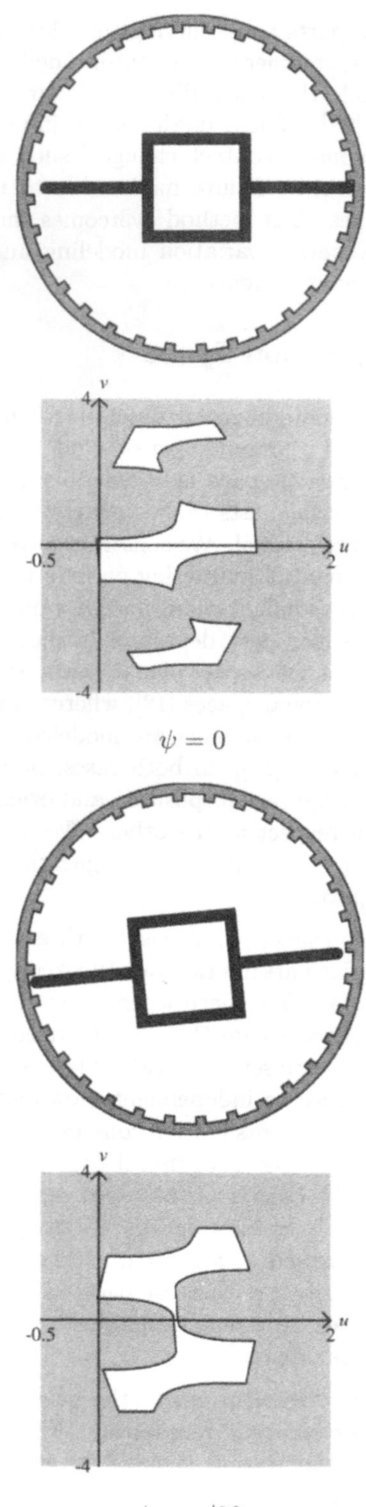

$\psi = 0$

$\psi = \pi/32$

Figure 2: *Gear/follower configuration space slices.*

square profile and its x axis is parallel to the pawls. The configuration space coordinates are the position (u, v) and the orientation ψ (in radians) of the gear frame in the follower frame.

We first examine two configuration space slices where the gear translates at a fixed orientation of $\psi = 0$ and $\psi = \pi/32$. The distance between these orientations is maximal because the configuration space is ψ-periodic with period $\pi/16$. Free space is white, blocked space is grey, and contact space is black. At $\psi = 0$, the free space consists of three connected components (the configuration shown in the figure is in the middle one). All part motions stay in the component containing their starting configuration. The complex shape of the contact space encodes the way that the gear can slide along the follower pawls. At $\psi = \pi/32$, the free space consists of a single component. The contact curves have changed because different gear teeth can touch the follower pawls at this orientation. The narrow channel between the upper and lower regions is traversed when the gear disengages one pawl and engages the other.

The configuration space of a pair is a complete representation of the part contacts. Contacts between pairs of features correspond to contact patches (curve segments in two dimensions and surface patches in three). The patch geometry encodes the motion constraint and the patch boundary encodes the contact change conditions. Part motions correspond to paths in configuration space. A path is legal if it lies in free and contact space, but illegal if it intersects blocked space. Contacts occur at configurations where the path crosses from free to contact space, break where it crosses from contact to free space, and change where it crosses between neighboring contact patches.

Figure 3 shows a detail of the three-dimensional configuration space where the gear translates and rotates. The labels a, b, and c mark the configurations displayed in Figure 1. The configuration follows the motion path (shown in grey) up (increasing u) along the highlighted patch, into the page (decreasing v), then left (increasing ψ). It then leaves the patch, crosses free space (a black region), and enters another patch. (A more detailed description of the indexing mechanism, including a faulty variant, appears elsewhere [22]).

The configuration space representation generalizes from pairs of parts to systems with more than two parts. A system of n planar parts has a $3n$-dimensional

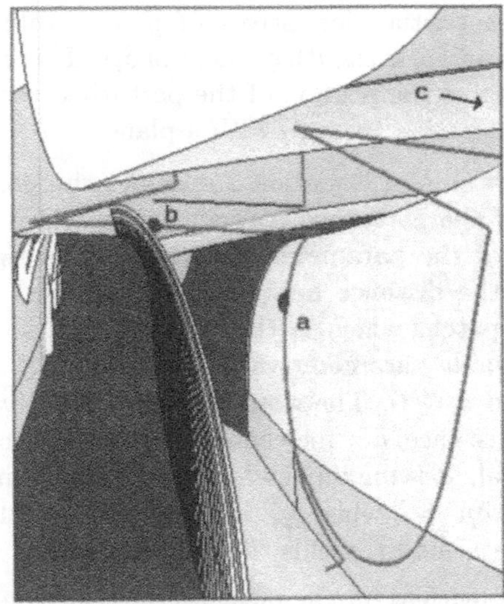

Figure 3: *Detail of the gear/follower configuration space. The horizontal axis is the u translation, the axis into the page the v translation, and the vertical axis the ψ rotation.*

configuration space whose points specify the n part configurations. A system configuration is free when no parts touch, is blocked when two parts overlap, and is in contact when two parts touch and no parts overlap. Computing the complete high-dimensional mechanism configuration space is impractical. Instead, we construct the relevant system contact patches from the pairwise spaces.

We have developed a configuration space computation program for planar pairs whose part boundaries consist of line segments and circular arcs [17, 18]. These features suffice for most engineering applications with the exception of involute gears and precision cams, which are best handled by specialized methods [7, 14]. The program computes an exact representation of contact space: A graph whose nodes represent contact patches and whose arcs represent patch adjacencies. Each node contains a contact function that evaluates to zero on the patch, is positive in nearby free configurations, and is negative in nearby blocked configurations. Each graph arc contains a parametric representation of the boundary curve between its incident patches.

4 Kinematic Variation

We model kinematic variation by generalizing configuration spaces to toleranced parts. The contact patches of a pair are parameterized by the touching features, which depend on the tolerance parameters. As the parameters vary around their nominal values, the contact patches vary in a band around the nominal contact space, which we call the contact zone. The contact zone defines the kinematic variation in each contact configuration: Every pair that satisfies the part tolerances generates a contact space that lies in the contact zone. Kinematic variations do not occur in free configurations because the parts do not interact.

We illustrate contact zones on the gear/follower pair. The gear is parameterized by the inner radius $r_i = 15.7mm$, the outer radius $r_o = 16.4mm$, and the ratio $\mu = 0.3$ between the angular width of a tooth and the angular spacing between teeth. The follower is parameterized by the cam radius $r_c = 5mm$ (which determines the size of the inner square), the pawl thickness $w = 0.5mm$, and the length $l = 9.8mm$ of the arms. The worst-case parameter variations are $\pm 0.01mm$. The full configuration space has a three-dimensional contact zone that is hard to visualize. Instead, we examine configuration space slices, which have planar contact zones (Figure 4). The contact zone is bounded by the curves that surround the nominal contact curves. Its width varies with the sensitivity of the nominal contact configuration to the tolerance parameters.

Each contact patch generates a region in the contact zone that represents the kinematic variation in the corresponding feature contact. The region boundaries encode the worst-case kinematic variation over the allowable parameter variations. They are smooth functions of the tolerance parameters and of the part configurations in each region. They are typically discontinuous at patch boundaries because the adjacent patches depend on different parameters, as can be seen in the gaps between adjacent contact zone boundary curves in Figure 4. The variation at boundary configurations is the maximum over the neighboring patch variations. The contact zone regions represent the quantitative kinematic variation, while the relations among regions represent qualitative variations, such as possible jamming, under-cutting, and interference. For example, the narrow channel in the center of free space can close, which prevents the follower from switching between pawl contacts, because the contact zones of the channel sides overlap.

The contact zone is obtained from the parametric part models and the nominal contact patches. Each

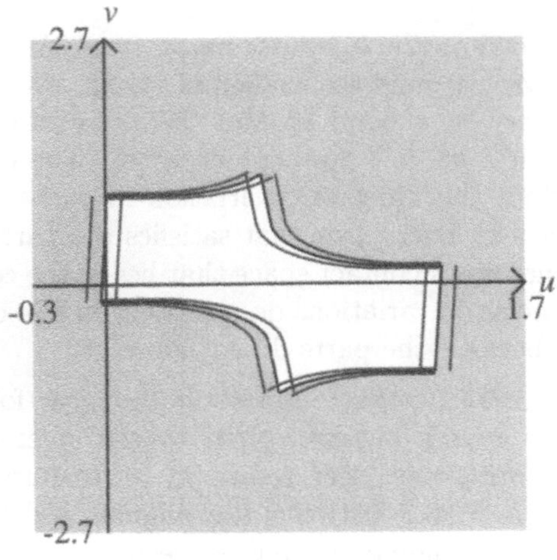

$\psi = 0$

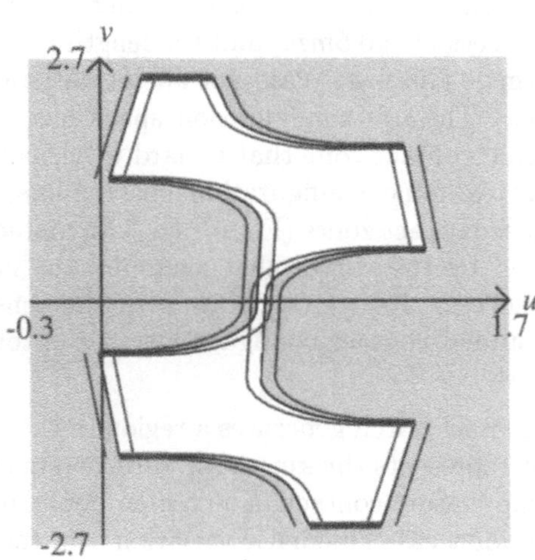

$\psi = \pi/32$

Figure 4: *Contact zones for gear/follower configuration space slices.*

patch satisfies a contact equation $g(u, v, \psi) = 0$, which we rewrite as $g(u, v, \psi, \mathbf{p}) = 0$ to make explicit the dependence on the vector $\mathbf{p}$ of tolerance parameters. A parameter perturbation of $\delta\mathbf{p}$ leads to a perturbed patch that satisfies

$$g(u + \delta u, v + \delta v, \psi + \delta\psi, \mathbf{p} + \delta\mathbf{p}) = 0. \quad (1)$$

Following the standard tolerancing approximation which considers only the first-order effects on kinematic variation, we obtain the linear expression:

$$\frac{\partial g}{\partial u}\delta u + \frac{\partial g}{\partial v}\delta v + \frac{\partial g}{\partial \psi}\delta\psi = -\sum_i \frac{\partial g}{\partial p_i}\delta p_i \quad (2)$$

where the partial derivatives of g are evaluated at (u, v, ψ) and δp_i is the ith element of $\delta\mathbf{p}$. This equation approximates the portion of the perturbed patch near the configuration (u, v, ψ) with a plane.

The left side of Equation 2 specifies the normal direction of the perturbed contact patch, which is independent of the parameter variations. The right side specifies the distance between the perturbed and the nominal patch, which is the kinematic variation, for any allowable parameter variation $l_i \leq \delta p_i \leq u_i$ with $l_i \leq 0$ and $u_i \geq 0$. The worst-case kinematic variation (largest distance) occurs when the right side is maximal or minimal. It is maximal when every term is maximal, i.e., when $\delta p_i = l_i$ when $g_{pi} > 0$ and $\delta p_i = u_i$ otherwise. Switching u_i and l_i yields the minimal value.

The derivatives $\partial g/\partial p_i$ measure the sensitivity of the contact configuration to the tolerance parameters. Designers can make small changes in the kinematic function by changing the parameters in accordance with the sensitivities. The following table shows the sensitivities of the gear/follower pair to the design parameters in the $\psi = 0$ slice.

parameter	maximum	minimum	average
r_c	0.87	0.00	0.58
w	3.00	1.00	2.33
l	1.00	0.00	0.67
r_i	1.00	0.00	0.39
r_o	1.00	0.00	0.31
μ	1.54	0.05	0.68

The values are from 84 contact configurations. Linear interpolation between these configurations approximates the entire contact space to three significant digits, hence the sample sensitivities are representative of the entire space. The average pawl thickness sensitivity is four times that of the other parameters, which makes it a good candidate for small design changes.

The contact zone model generalizes from pairs to systems. The contact space is a semi-algebraic set in configuration space: a collection of points, curves, surfaces, and higher dimensional components. As the tolerance parameters vary around their nominal values, the components vary in a band around the nominal contact space, which is a higher-dimensional analog of the three-dimensional contact zone of a pair. It is impractical to compute the full contact zone, so we sample it at critical configurations. These configurations can be specified by a designer or can by derived by simulating the system function and sampling periodically. Several simulations may be needed to sample all the operating modes of the system.

We compute the system variation at a sample configuration by determining which pairs of parts are in contact, obtaining the corresponding parameterized contact equations from the pairwise configuration spaces, and solving a linear optimization problem. The variables are the part coordinate variations $(\delta x_i, \delta y_i, \delta\theta_i)$ and the tolerance parameters p_i. The constraints come from the tolerances and from the contact patches. The tolerances provide two constraints per parameter $l_i \leq p_i \leq u_i$. We collect the contact equations into a vector equation:

$$\mathbf{g}(\mathbf{x}, \mathbf{p}) = 0 \tag{3}$$

with $\mathbf{x}$ part coordinates and $\mathbf{p}$ tolerance parameters. We linearize the contact equations around the current configuration and the nominal parameter values to get:

$$D_{\mathbf{x}}\mathbf{g}\delta\mathbf{x} + D_{\mathbf{p}}\mathbf{g}\delta\mathbf{p} = 0 \tag{4}$$

with $D_{\mathbf{x}}\mathbf{g}$ the Jacobian matrix with respect to $\mathbf{x}$ and $D_{\mathbf{p}}\mathbf{f}$ the Jacobian matrix with respect to $\mathbf{p}$. This equation is the system analog of Equation 2. It approximates the portion of the perturbed configuration space near $\mathbf{x}$ with a hyper-plane. Additional constraints model driving motions, part play, and dynamical effects. The objective functions are the maxima and minima of the coordinate variations. We solve one linear program for each function to obtain the system variation. The details and examples appear elsewhere [20].

5 Case Study: The Gear Selector Mechanism

We have applied the kinematic tolerance analysis program to automotive power train design where the shortcomings of the current tolerance analysis practice are acute [23]. The orders to purchase the transfer lines, machines, and tools are placed when the tolerance analysis begins. When the tolerance analysis reveals a problem, modifications to the design often require very expensive changes to the manufacturing process. Engineers must conduct a statistical analysis to determine if the probability of trouble is acceptable, decrease the part tolerances, which increases costs, or accept higher warranty costs and customer dissatisfaction. During production, downstream consequences

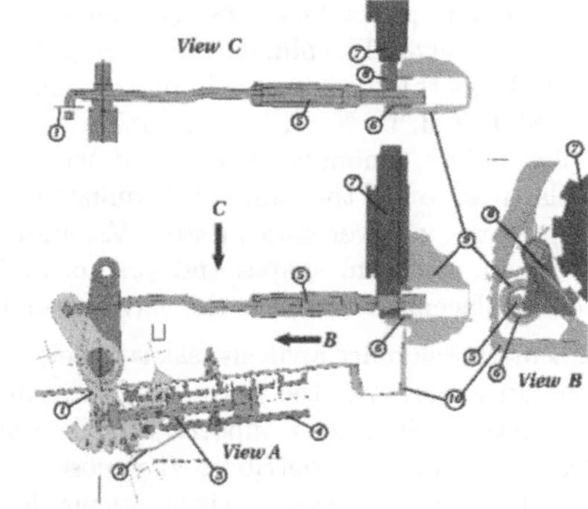

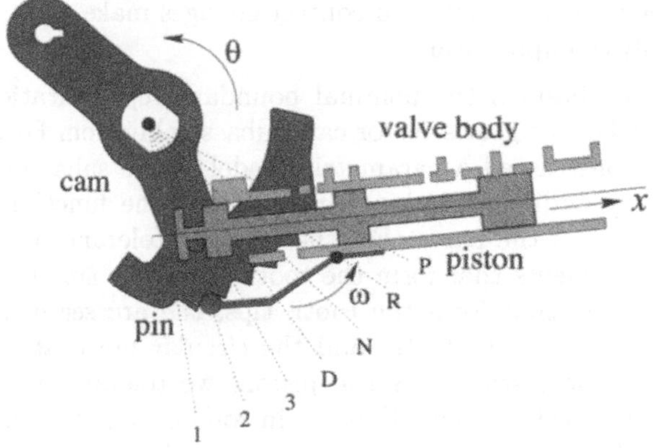

Figure 5: *CAD drawing of gear selector mechanism (top) and cam/pin/piston assembly (bottom).*

of yet-to-be discovered tolerance problems can stop production. Many tolerance problems reflect incorrect simplifying assumptions where engineers ignored an important feature or where the geometric complexity of the parts, their motions, and their interactions produced unexpected effects.

We summarize the analysis of the cam/piston assembly in a gearshift mechanism. Figure 5 shows the four main parts of the cam/piston assembly: the rooster cam, the pin, the piston, and the valves body. There are three moving parts, each with one degree of freedom, that form two 2D configuration spaces. The cam rotates around an axis at its center. Its angular position is controlled by the shift stick (not shown). Its side pin is mounted on the piston's left end and causes it to slide back and forth inside the valves body, which is fixed. The different piston positions open and close the conducts on the valve. The pin, which is spring-loaded, temporarily locks the rooster cam in one of seven settings labeled 1, 2, 3, D, N, R, and P. Each of the cam settings determines a nominal opening of the valves. The angular position of the cam is determined by the pin that pushes the lower cam profile. Variations in the pin, piston, and cam shapes and positions affect the piston displacement and thus the valve opening.

The kinematic tolerance analysis task is to determine the maximum variation of the piston displacement for each cam setting. It is also important to determine which feature variations contribute the most to the piston variation: the cam axial position, its profile, the pin radius, or others. The many part features, complex kinematic relations, and contact changes make manual analysis impractical.

We obtained the nominal boundary representation model of the gear selector cam subassembly from Ford. We constructed a parametric model of the subassembly by adding variation parameters to the functional features of the parts. For the cam, we toleranced the line segments that form the tooth sides, the small arc segments that form the tooth tips, the arc segments that connect the teeth, and the circular pin that engages the piston. For the piston, we toleranced the two vertical segments that are in contact with the cam pin. For the pin, we toleranced the single, circular feature. Line segments were toleranced by varying the coordinates of the two endpoints; arc segments were toleranced by varying the radius and the center coordinates. To account for uncertainties in the position of the rotation axes, we also toleranced the centers of rotation of the cam and the pin. Since we chose the piston as the reference part of the assembly, there was no need to tolerance the orientation of its translation axis. The model has in total 86 tolerance parameters for the cam, 8 for the piston, 5 for the pin, and 99 overall. We assigned every parameter an independent tolerance of ± 0.1 mm. Constructing the parametric models and inputting them to the program took one person two hours.

To determine the kinematic variations, we computed the nominal configuration spaces and the contact zones of the cam/piston and cam/pin pairs, as shown in Figures 6 and 7. The computation took about 20 seconds, using a Lisp program running on an Indigo 2 Workstation. In the cam/piston zone, the piston offset, x, has a worst-case variation of between 0.41mm and 0.45 mm over the functional range of the cam angle, θ. In the cam/pin zone, the pin orientation, ω, has a worst-case variation of between 0.013 radians and 0.018 radians. We computed the kinematic variation of the system at the configuration $\theta = 0.371$ radians, $\omega = -0.0008$ radians, $x = 260.8$ mm where the gear selector is in third gear (Figure 5). The worst-case variation of x is 0.9 mm—roughly half from each pair. The main factors in the cam/piston variation are the cam center horizontal position (25%), vertical position (25%), tooth base x position (25%), and the x coordinates of the piston vertical segments (10%). The cam/pin varia-

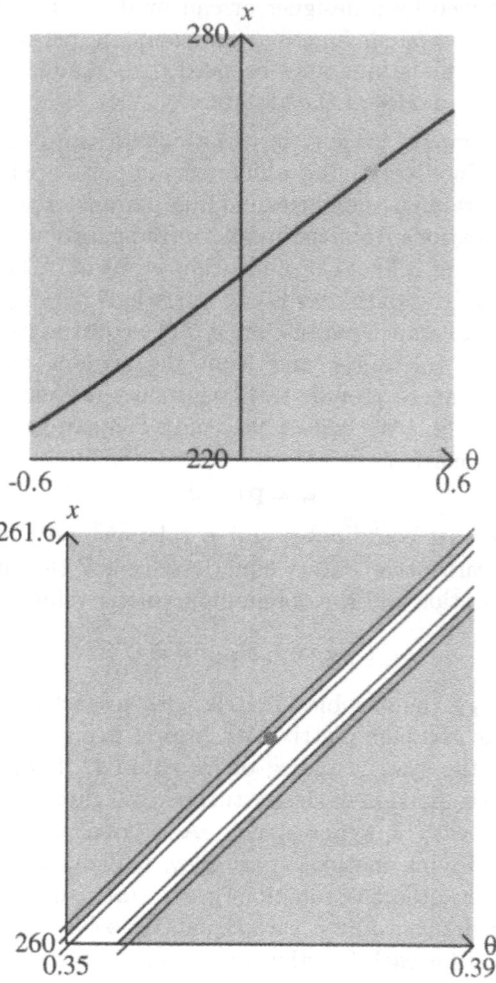

Figure 6: *Comb/piston configuration space and contact zone detail.*

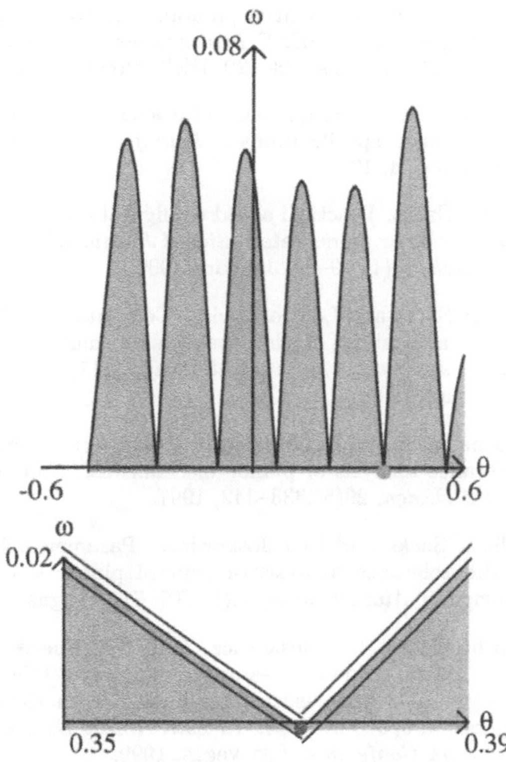

Figure 7: *Comb/pin configuration space and contact zone detail.*

tion is evenly distributed among the parameters of the touching features and the part centers of rotation.

6 Conclusion

We described a kinematic tolerance analysis algorithm for mechanical systems with parametric part tolerances and presented a design case study. The algorithm constructs a variation model for the system, derives worst-case bounds on the variation, and helps designers find unexpected failure modes, such as jamming and blocking. We have applied our program to detailed parametric models of a variety mechanisms including a Geneva cam pair, a 35mm camera shutter, a movie camera film advance, and a micro-mechanical gear discriminator. In all cases, the program produced useful and interesting qualitative and quantitative results.

We plan to extend the scope of our algorithm to spatial tolerance analysis of planar systems. Planar pairs must be analyzed as spatial pairs to study axis misalignment due to manufacturing variation, assembly error, or wear. We have developed a configuration space computation algorithm for spatial pairs whose parts move along fixed axes [10]. The contact zone model applies to these spaces. To implement it, we need to program the partial derivatives of the contact functions, $g(u, v, \mathbf{p})$, for every type of contact, e.g. plane/cylinder, cylinder/sphere, and cylinder/circle. We are also studying tolerance synthesis within the configuration space representation and are extending the algorithm to variational part models produced by modern CAD tools, such as ProEngineer and Catia.

Kinematic tolerance analysis has potential relevance to robot path planning with uncertain geometry. The analysis provides a detailed understanding of how small variations in geometry effect nominal contact relations. This knowledge could be useful in compliant motion and in manipulation where the robot plans based on the domain geometry. The maximum material estimate represents a single point in the contact zone, hence provide less information than the full zone. In coarse path planning, the extra information is probably unnecessary because the robot wants to avoid obstacles robustly, not to interact with them.

Acknowledgments

Ralf Schultheiss and Uwe Hinze collaborated in the case study. Sacks is supported by NSF grants CCR-9617600 and CCR-9505745 and by the Purdue Center for Computational Image Analysis and Scientific Visualization. Joskowicz is supported by a grant from the Authority for Research and Development, The Hebrew University of Jerusalem, Israel. Both are supported by a Ford University Research Grant, the Ford ADAPT200 project, and by grant 98/536 from the Israeli Academy of Science.

References

[1] Eric Ballot and Pierre Bourdet. A computation method for the consequences of geometric errors in mechanisms. In *Proc. of the 5th CIRP Int. Seminar on Computer-Aided Tolerancing*, Toronto, 1997.

[2] K. W. Chase and A. R. Parkinson. A survey of research in the application of tolerance analysis to the design of mechanical assemblies. *Research in Engineering Design*, 3(1):23–37, 1991.

[3] Kenneth Chase, Spencer Magleby, and Charles Glancy. A comprehensive system for computer-aided tolerance analysis of 2d and 3d mechanical assemblies. In *Proc. of the 5th CIRP Int. Seminar on Computer-Aided Tolerancing*, Toronto, 1997.

[4] Jingliang Chen, Ken Goldberg, Mark Overmars, Dan Halperin, Karl Bohringer, and Yan Zhuang. Shape tolerance in feeding and fixturing. In P.K. Agarwal, L. E. Kavraki, and M. T. Mason, editors, *Robotics, The Algorithmic Perspective: 3rd Workshop on Algorithmic Foundations of Robotics (WAFR)*. A. K. Peters, 1998.

[5] André Clemént, Alain Rivière, Phillipe Serré, and Catherine Valade. The ttrs: 13 constraints for dimensioning and tolerancing. In *Proc. of the 5th CIRP Int. Seminar on Computer-Aided Tolerancing*, Toronto, 1997.

[6] G. Erdman, Arthur. *Modern Kinematics: developments in the last forty years.* John Wiley and Sons, 1993.

[7] Max Gonzales-Palacios and Jorge Angeles. *Cam Synthesis.* Kluwer Academic Publishers, Dordrecht, Boston, London, 1993.

[8] Matsamoto Inui and Masashiro Miura. Configuration space based analysis of position uncertainties of parts in an assembly. In *Proc. of the 4th CIRP Int. Seminar on Computer-Aided Tolerancing*, 1995.

[9] Leo Joskowicz, Elisha Sacks, and Vijay Srinivasan. Kinematic tolerance analysis. *Computer-Aided Design*, 29(2):147–157, 1997. reprinted in [13].

[10] Ku-Jim Kim, Elisha Sacks, and Leo Joskowicz. Kinematic analysis of spatial fixed-axes pairs using configuration spaces. Technical Report CSD-TR 99-036, Purdue University, 1999.

[11] Jean-Claude Latombe. *Robot Motion Planning.* Kluwer Academic Publishers, Boston, 1991.

[12] Jean-Claude Latombe and Randall Wilson. Assembly sequencing with toleranced parts. In *Third ACM Symposium on Solid Modeling and Applications*, 1995.

[13] Jean-Paul Laumond and Mark Overmars, editors. *Algorithms for Robotic Motion and Manipulation.* A. K. Peters, Boston, MA, 1997.

[14] Faydor L. Litvin. *Gear Geometry and Applied Theory.* Prentice Hall, New Jersey, 1994.

[15] T. Lozano-Pérez. Spatial planning: A configuration space approach. In *IEEE Transactions on Computers*, volume C-32, pages 108–120. IEEE Press, 1983.

[16] Aristides A. G. Requicha. Mathematical definition of tolerance specifications. *Manufacturing Review*, 6(4):269–274, 1993.

[17] Elisha Sacks. Practical sliced configuration spaces for curved planar pairs. *International Journal of Robotics Research*, 18(1):59–63, January 1999.

[18] Elisha Sacks and Leo Joskowicz. Computational kinematic analysis of higher pairs with multiple contacts. *Journal of Mechanical Design*, 117(2(A)):269–277, June 1995.

[19] Elisha Sacks and Leo Joskowicz. Parametric kinematic tolerance analysis of planar mechanisms. *Computer-Aided Design*, 29(5):333–342, 1997.

[20] Elisha Sacks and Leo Joskowicz. Parametric kinematic tolerance analysis of general planar systems. *Computer-Aided Design*, 30(9):707–714, August 1998.

[21] Elisha Sacks, Leo Joskowicz, Ralf Schultheiss, and Uwe Hinze. Computer-assisted kinematic tolerance analysis of a gear selector mechanism with the configuration space method. In *25th ASME Design Automation Conference*, Las Vegas, 1999.

[22] Elisha Sacks, Charles Pisula, and Leo Joskowicz. Visualizing three-dimensional configuration spaces for mechanical design. *Computer Graphics and Applications*, 19(5):50–53, 1999.

[23] Ralf Schultheiss and Uwe Hinze. Detect the unexpected - how to find and avoid unexpected tolerance problems in mechanisms. In *Global consistency of tolerances, Proc. of the 6th CIRP Int. Seminar on Computer-Aided Tolerancing, F. van Houten and H. Kals eds., Kluwer*, 1999.

[24] O.W. Solomons, F. van Houten, and H. Kals. Current status of cat systems. In *Proc. of the 5th CIRP Int. Seminar on Computer-Aided Tolerancing*, Toronto, 1997.

[25] Herbert Voelcker. A current perspective on tolerancing and metrology. *Manufacturing Review*, 6(4):258–268, 1993.

[26] Daniel Whitney, Olivier Gilbert, and Marek Jastrzebski. Representation of geometric variations using matrix transforms for statistical tolerance analysis. *Research in Engineering Design*, 6(4):191–210, 1994.

Deformable Free Space Tilings for Kinetic Collision Detection

Pankaj K. Agarwal, *Duke University, Durham, NC*
Julien Basch, *Stanford University, Stanford, CA*
Leonidas J. Guibas, *Stanford University, Stanford, CA*
John Hershberger, *Mentor Graphics Corp., Wilsonville, OR*
Li Zhang, *Stanford University, Stanford, CA*

We present kinetic data structures for detecting collisions between a set of polygons that are moving continuously in the plane. Unlike classical collision-detection methods that rely on bounding volume hierarchies, our method is based on deformable tilings of the free space surrounding the polygons. The basic shape of our tiles is that of a pseudo-triangle, *a shape sufficiently flexible to allow extensive deformation, yet structured enough to make detection of self-collisions easy. We show different schemes for maintaining pseudo-triangulations using the framework of kinetic data structures, and analyze their performance. Specifically, we first describe an algorithm for maintaining a pseudo-triangulation of a point set, and show that this pseudo-triangulation changes only quadratically many times if points move along algebraic arcs of constant degree. We then describe an algorithm for maintaining a pseudo-triangulation of a set of convex polygons. Finally, we extend our algorithm to the general case of maintaining a pseudo-triangulation of a set of moving simple polygons. These methods can be extended to situations where the polygons deform as well as move.*

1 Introduction

Collision detection between moving objects is a fundamental problem in computational simulations of the physical world. Because of its universality, it has been studied by several different communities, including robotics, computer graphics, computer-aided design, and computational geometry. Several methods have been developed for the case of rigid bodies moving freely in two and three dimensions. Though a physical simulation involves several other computational tasks, such as motion dynamics integration, graphics rendering, and collision response, collision detection remains one of the bottlenecks in such a system. A commonly used approach to expedite the collision detection between complex shapes is based on hierarchies of simple bounding volumes surrounding each of the objects. For a given placement of two non-intersecting objects, their respective hierarchies are refined only to the coarsest level at which the primitive shapes in the two hierarchies can be shown to be pairwise disjoint.

Motion in the physical world is in general continuous over time, and many systems attempt to speed up collision checking by exploiting this temporal coherence, instead of repeating a full collision check *ab initio* at each time step [22]. Swept volumes in space or space-time have been used towards this goal [6, 17, 19]. Though fixed time-sampling is customary for motion integration, collisions tend to be rather irregularly spaced over time. If we know precisely the motion laws of the objects, then it makes sense to try to predict exactly when collisions will happen, instead of hoping to locate them with time sampling. There have been a few theoretical papers in computational geometry along these lines [8, 13, 23], but their results are not so useful in practice because they use complex data structures and are only applicable for limited types of motion.

In this paper, we consider the situation where we have multiple moving or deforming polygonal objects in the plane. Such a setting can arise in various simulation and motion planning tasks. For example, in several manufacturing processes, the mechanical parts are modeled as planar polygons [1, 5, 4] for parts feeding and orienting tasks. In motion planning for multiple mobile robots amidst possibly moving obstacles, the robots and obstacles are usually projected to the floor, and the problem is simplified to a two-dimensional problem to improve efficiency. More importantly, we hope that extensions of the major techniques developed in this paper will be able to handle three dimensional objects.

Recently, Basch et al. [2] and Erickson et al. [9] presented algorithms for detecting collision between two polygons using the *kinetic data structures* framework, which was originally introduced by Basch et al. [3, 12]. These kinetic algorithms avoid many of the problems that arise in fixed time-sampling methods. A kinetic data structure, or KDS for short, is built on the idea of maintaining a discrete attribute of objects in motion by animating a proof of its correctness through time. The proof consists of a set of elementary conditions, called *certificates*, based on the kinds of tests performed by ordinary geometric algorithms — CCW (counterclockwise) tests in our case. Those certificates that can fail as a result of the motion of the polygons are placed in an event queue, ordered according to their earliest failure time. When a certificate fails, the proof needs to be updated. Unless a collision has occurred, we perform this update and continue the simulation. In contrast to fixed time step methods, for which the fastest moving object gates the time step for the entire system, a kinetic method is based on *events* (the certificate failures) that have a natural significance in terms of the problem being addressed (collision detection in this case).

Unlike the previous hierarchy-based algorithms, the algorithm by Basch et al. [2] maintains a decomposition of the common exterior of the two moving polygons (similar approach was used by Mount [20] for the static problem of intersection detection.). The cells of this decomposition deform continuously as the objects move. As long as all the cells in the decomposition remain disjoint, the decomposition itself acts as a KDS proof of non-collision between the objects. At certain times, cells become invalid because they self-intersect and the decomposition has to be modified. Unfortunately, in their approach an extension to collision detection between many polygons is very expensive — such a decomposition has to be built separately for every pair of polygons.

In this paper we present an algorithm for detecting collision between arbitrary sets of simple polygons as they move and/or deform in the plane. As in [2], we maintain a decomposition of the common exterior of polygons, the free space, into deformable tiles. Ideally, we would like to maintain a decomposition so that the self-intersection of a cell of the decomposition is easy to detect, the decomposition is simple to update when a self-intersection occurs, and the decomposition conforms to the motion of the polygons so that self-intersections do not happen too many times. An obvious choice for the decomposition is a triangulation of the free space. Such a triangulation was used by Held et al. [16] to track a moving object flying among fixed obstacles. Although a triangulation satisfies the first two criteria, it contains too many cells and therefore has to be updated frequently when all the defining shapes are in motion. We will therefore use a *pseudo-triangulation* as the decomposition, which has considerably fewer cells compared to a triangulation. Pseudo-triangles can flex as the objects move, and therefore the combinatorial structure of the tiling needs fewer updates. At the same time, the cells in a pseudo-triangulation have sufficiently simple shapes so that their self-intersections are easy to detect and the triangulation is easy to update. Pseudo-triangulations have been used in the past, but primarily for various visibility problems [7, 21]. An additional benefit of our structure is that it can gracefully adapt to object shapes that are themselves flexible. As our polygons move they can also deform and change shape. This additional flexibility impacts our data structure primarily on the number of events it has to process — but its basic nature and certification remains unchanged.

In Section 2, we describe our model for motion, define pseudo-triangulations, and describe the certificates needed to maintain a pseudo-triangulation. In Section 3, we first describe how to maintain a pseudo-triangulation for a set $\mathcal{P}$ of n moving points in the plane, and show that it can be maintained in an output-sensitive manner. For low-degree algebraic motion, the pseudo-triangulation changes about $O(n^2)$ times. We can further refine the pseudo-triangulation to maintain a triangulation of $\mathcal{P}$, which changes about $O(n^{7/3})$ times if each point in $\mathcal{P}$ is moving with fixed velocity. To our knowledge, this is the first triangulation (without Steiner points) that changes a sub-cubic number of times, even for linear motions.

In Section 4, we describe a scheme for maintaining the pseudo-triangulation of a set $\mathcal{P}$ of k disjoint convex polygons with a total of n vertices. We show that the greedy vertical pseudo-triangulation proposed by Pocchiola and Vegter [21] can be maintained efficiently. A nice feature of this (or any minimal) pseudo-triangulation is that the number of cells is only $O(k)$, which is typically much smaller than n, the total complexity of the polygons. On the other hand, the size of any triangulation has to be at least $\Omega(n)$. We will show that we need a kinetic data structure of size $O(k)$ to maintain this pseudo-triangulation. But unlike the set of points case, we do not have sharp bounds on the number of events.

Finally, in Section 5, we combine our algorithm for the convex case with the one in [2] to construct a pseudo-triangulation for a set of pairwise-disjoint simple polygons moving in the plane. It can easily be updated when a certificate fails. The number of certificates needed to maintain the correctness of the pseudo-triangulation is within an $O(\log n)$ factor of the size of a *minimum link subdivision* separating the polygons [24]. By this property, our separation proof automatically adapts to the complexity of the relative placement of the polygons, and its size will vary between $O(k)$ (when the polygons are far from each other) and $O(n)$ (when they are closely intertwined). A compact separation proof is important when objects are allowed to change their *motion plan* unpredictably since the number of certificates represents the cost of recomputing the event times. Recently, Kirkpatrick et al. [18] were able to suggest a method that kinetically maintains a decomposition of the free space whose size is within a constant factor of the size of a minimum link subdivision. Unlike our method, in which the tiling is determined only by the current positions of the polygons, in their method, the tiling depends on the history of the motions; this makes it hard to analyze the efficiency of their method.

Traditionally, collision detection has been divided into two phases: the *broad phase*, in which one uses a simple-minded algorithm (typically a bounding box check) to determine which pairs of objects might collide and thus need further testing, and the *narrow phase*, in which these candipdate pairs get a more detailed collision test using a sophisticated algorithm. Note that our approach based on deformable free space tilings completely obviates this distinction. The tiling effectively 'hides' features of objects that are far away and treats the objects as equivalent to their convex hulls. As the objects get closer and more intertwined, their features get progressively revealed and participate in collision checks. Furthermore our framework can be extended in a straightforward way to deformable objects, a setting that no previous collision detection method had successfully addressed. Though this paper describes how to implement the deformable tiling idea only in 2D, we are hopeful that 3D extensions will also be possible.

2 Models of Motion and Pseudo-Triangulations

In this section we discuss our models of motion, define pseudo-triangulations, and discuss how we maintain them as objects move continuously.

2.1 Models of Motion

There are two reasons for defining motion models. First, we need to be able to compute the certificate failure times. Second, we wish to be able to give bounds on the maximum number of events that can happen. We describe the model for which we bound the number of events first.

A rigid polygon P is described at rest by a reference point o, by an orthonormal basis (x, y), and by the coordinates (p_x, p_y) of each vertex p of P in this orthonormal frame centered at o. An *algebraic motion* of P is given by a trajectory $o(t)$ of the reference point and by an orthonormal basis $(x(t), y(t))$, such that the coordinates of o, x, y are algebraic functions of time of bounded degree. The position of v at time t, denoted by $v(t)$, is $o(t) + p_x x(t) + p_y y(t)$. If P is a point, then o is the same as P and the vectors x, y are not needed. We will use $P(t)$ to denote the polygon P at time t. The *degree of motion* of $\mathcal{P}$ is the maximum degree of a polynomial defining the coordinates of o_i, x_i, y_i, for $1 \leq i \leq k$. If the degree of motion is 1, we say that the motion is *linear*. We call a motion *translational* if the coordinates of vectors x_i, y_i, for all $1 \leq i \leq k$, are constant.

Although the above model differs from that used in practice in describing rotational motion, it is flexible enough to approximate arbitrary motions to any desired accuracy, for a limited time. Note, however, that an algebraic rotation has necessarily non-uniform angular velocity, and can cover only a constant number of full turns. If we parameterize the basis $(x(t), y(t))$ appropriately, we can represent algebraically some special cases, such as when all the objects rotate with angular velocities that are rational multiples of a common angular velocity. Rotations with general angular velocities can either be approximated using power series, or the model can be extended to handle such rotations, as long as the total number of full turns is bounded. What is crucial for our analyses is that our kinetic certificates can fail only a bounded number of times each during the course of the motions (this is what is called a *pseudo-algebraic motion*).

For a deformable polygon, we assume that its deformation is described by the motion of its vertices. An *algebraic deformation* is a deformation where the motion of each vertex can be described by an algebraic function of bounded degree. Further, we also assume that basic structural properties of shapes are preserved during the motion — *deformable convex polygons* always have convex shapes during the entire deformation, and the shapes of *deformable simple polygons* are always simple. In the latter case, we need to be able to detect self-collisions in order to insure simplicity. As we shall see, our methods can be used to detect self-collisions as well.

For a real collision detection system or simulation, the equations of motion will typically be given by partial differential equations or bounds on acceleration. The kinetic setting applies to such situations as long as we can compute or give lower bounds on the failure time of the certificates we are concerned with. In the remaining of this paper, we assume that such a computation takes $O(1)$ time.

2.2 Pseudo-Triangulation

A *pseudo-triangle* in the plane is a simple polygon whose boundary consists of three concave chains, called *side chains*, that join at their endpoints. The three endpoints of a pseudo-triangle are called *corners*. The edges incident upon the corners are called *corner edges* (Figure 1 (i)). For a vertex p on the boundary of a pseudo-triangle, we denote p_L, p_R be the predecessor and the successor of p along $\partial\Delta$ in the counterclockwise direction.

Let S be a polygon with holes; some of the holes may be degenerate, i.e., they can be points or segments. A *pseudo-triangulation* $\mathcal{T}(S)$ of S is a planar subdivision of the closure of S, so that each face of $\mathcal{T}(S)$ is a pseudo-triangle and so that the interior of each face lies in the interior of S; see Figure 1 (ii) for an example. In other words, $\mathcal{T}(S)$ is a collection of pseudo-triangles with pairwise-disjoint interiors, each lying inside S, that cover S. The vertices of $\mathcal{T}(S)$ are

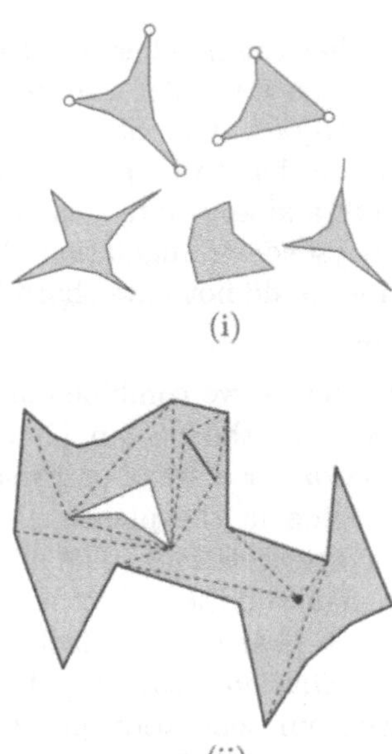

Figure 1: *(i) The examples of some pseudo-triangles and non-pseudo-triangles. The first two figures are pseudo-triangles, but the others are not. (ii) An example of pseudo triangulation of a polygon with holes.*

the same as the vertices of S, and each edge of $\mathcal{T}(S)$ is either an edge of ∂S or a segment whose interior lies inside S; edges of the latter type are called *diagonals*.

Let $\mathcal{P} = \{P_1, \ldots, P_n\}$ be a set of k simple polygons with a total of n vertices. We allow the polygons to be degenerate, in the sense that each P_i might be a point or a line segment, but we assume that all polygons are bounded. We also assume that the objects are in general position. Let $\overline{\mathcal{P}}$ and $\mathcal{P}^\circ$ denote the convex hull and interior of $\mathcal{P}$, respectively. We define the *free space* $\mathcal{F}(\mathcal{P})$ of $\mathcal{P}$ to be $\overline{\mathcal{P}} \setminus \mathcal{P}^\circ$. The boundary of $\mathcal{F}(\mathcal{P})$, denoted by $\partial\mathcal{F}(\mathcal{P})$, consists of polygon boundaries and convex hull edges.

Since a polygon (even with degenerate holes) can always be triangulated and a triangulated planar subdivision is obviously a pseudo-triangulation, a pseudo-triangulation of $\mathcal{F}(\mathcal{P})$ always exists. A pseudo-triangulation $\mathcal{T}$ is called *minimal* if the union of any two faces in $\mathcal{T}$ is not a pseudo-triangle. If $\mathcal{T}$ is not minimal, then there exists a diagonal d in $\mathcal{T}$ so that $\mathcal{T}$ remains a pseudo-triangulation after the removal of d from it. Although minimal pseudo-triangulations are not unique, we can prove the following result on the size of minimal pseudo-triangulations of $\mathcal{F}(\mathcal{P})$.

Lemma 1 *Let $\mathcal{P}$ be a set of k pairwise-disjoint simple polygons. If m of these polygons are points, and there are a total of r reflex vertices in $\mathcal{P}$, then a pseudo-triangulation of $\mathcal{F}(\mathcal{P})$ is minimal if and only if it contains $2k - m + r - 2$ pseudo-triangles.*

2.3 Maintaining a Pseudo-Triangulation

We are interested in maintaining $\mathcal{T}(\mathcal{P})$, a pseudo-triangulation of $\mathcal{F}(\mathcal{P})$, as $\mathcal{F}(\mathcal{P})$ deforms continuously. Specifically, we want to maintain $\mathcal{T}(\mathcal{P})$ as a kinetic data structure. As mentioned in the introduction, this is accomplished by maintaining a proof of the correctness of $\mathcal{T}(\mathcal{P})$, which consists of a small set of elementary conditions called *certificates*. As long as the certificates remain valid, the current combinatorial structure of $\mathcal{T}(\mathcal{P})$ is valid. It is only when a certificate fails that $\mathcal{T}(\mathcal{P})$ needs to be updated. A certificate failure is called an *event*. All the events are placed in an event queue according to their failure time. The structure is then maintained by processing those events one by one. The efficiency of such a data structure depends on the size of the proof, the number of events, and the number of certificates that need to be updated at each event.

In our set-up, we certify that each face in $\mathcal{T}(\mathcal{P})$ is a pseudo-triangle and that the faces cover $\mathcal{F}(\mathcal{P})$. In view of the definition of $\mathcal{T}(\mathcal{P})$, we need to certify the following two conditions for each pseudo-triangle Δ in $\mathcal{T}(\mathcal{P})$ (see Figure 2(i)):

1. Each side chain C is concave, i.e., for each interior vertex p of C, the angle $\angle p_L p p_R > \pi$ (or the signed area of $\triangle p_L p p_R$ is negative). We call these certificates *reflex* certificates.
2. The side chains of a pseudo-triangle Δ join only at their endpoints, i.e., for each corner vertex q of Δ, the angle $\angle q_L q q_R > 0$ (or the signed area of $\triangle q_L q q_R$ is positive). We call such certificates *corner* certificates.

When the reflex certificate of p fails (Figure 2(ii)), we connect the vertices p_L, p_R, adjacent to p, by an edge to maintain the pseudo-triangularity of Δ. To maintain the pseudo-triangulation, we may also have to update other pseudo-triangles, depending on the edges adjacent to p. If both edges adjacent to p are boundary edges, then we add a triangle pp_Lp_R to $\mathcal{T}(\mathcal{P})$. If only one edge, say pp_L, is a diagonal edge, then pp_L is adjacent to another pseudo-triangle Δ_1. In this case, we delete the edge pp_L and merge the triangle pp_Lp_R into the pseudo-triangle Δ_1. If both pp_L, pp_R are diagonal edges, we then delete one of pp_L or pp_R: which edge is deleted depends on the specific pseudo-triangulation we are maintaining; we will explain the choice for each pseudo-triangulation later in the paper.

When the corner certificate of q fails (Figure 2(iii)), we collapse the corner. Suppose when the concave certificate corresponding to q fails, q_L lies on the edge qq_R. We then delete the edge qq_R and add the edge q_Lq_R to maintain the pseudo-triangularity of Δ. This process also effectively adds the point q_L to the side chain of the other pseudo-triangle incident to the edge qq_R. Since there is only one way to update the pseudo-triangulation when a corner certificate fails, we will not describe updates for this case in the following sections.

In addition, we also have to maintain the boundary $\partial\mathcal{F}(\mathcal{P})$. The edges of $\partial\mathcal{F}(\mathcal{P}) \setminus \partial\overline{\mathcal{P}}$ are automatically maintained by the algorithm. By regarding $\partial\overline{\mathcal{P}}$ as a concave chain with respect to the exterior of $\overline{\mathcal{P}}$, it can also be maintained using reflex certificates for each vertex on the hull. We omit the details here.

Since the pseudo-triangulation is not a canonical structure (there may be many pseudo-triangulations of $\mathcal{F}(\mathcal{P})$), it is not clear which pseudo-triangulation we are maintaining. To avoid this ambiguity, we first describe a static algorithm that constructs a "specific" pseudo-triangulation of $\mathcal{F}(\mathcal{P})$. We then assume that, at any time t, the kinetic data structure maintains the pseudo-triangulation that the static algorithm would have constructed on $\mathcal{P}$ at time t. To maintain this invariant, we need additional certificates. When these certificates fail, we usually update the structure by the *flipping* operation: For any diagonal edge e, consider the two pseudo-triangles Δ_1, Δ_2 that contain e. The union of Δ_1, Δ_2 forms a pseudo-quadrangle $\diamond$—a cell whose boundary consists of four concave chains. Besides e, there is exactly one other diagonal edge inside $\diamond$, which is called the *shadow edge* of e. The quadrangle $\diamond$ can be decomposed into two pseudo-triangles in two ways, one by adding e and the other by adding e's shadow edge. The flipping operation replaces e by e'.

3 Pseudo-Triangulation for Points

In this section, we assume $\mathcal{P}$ to be a set of n points. we define the *incremental pseudo-triangulation* for $\mathcal{P}$,

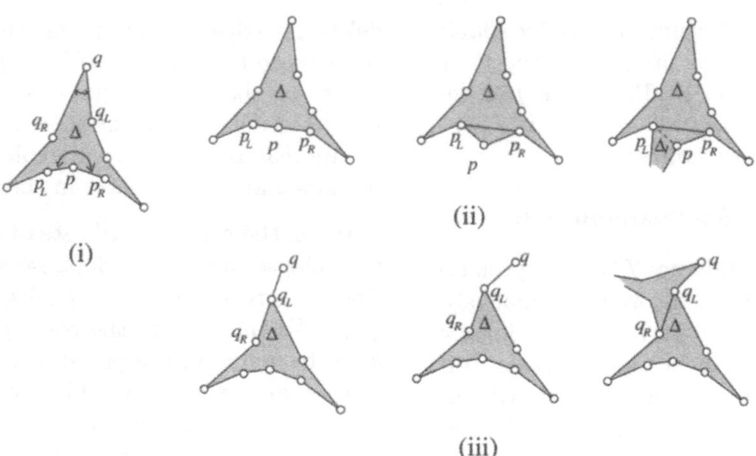

Figure 2: *The certificates to prove pseudo-triangularity.*

and show how to maintain it efficiently. We also show how it can be refined to maintain a triangulation of $\mathcal{P}$.

3.1 Incremental Pseudo-Triangulation

A pseudo-triangulation for points can be built in an incremental manner as follows. We first sort the points in increasing order of their x-coordinates. Suppose that $p_1, p_2, \ldots, p_n$ is the sorted sequence, and denote by $\mathcal{P}_i$ the first i points. We insert the points from left to right. Upon inserting p_i, we draw the two tangent segments from p_i to the convex hull $\overline{\mathcal{P}}_{i-1}$. Denote by $u(p_k)$ and $d(p_k)$ the left endpoints of the upper and lower tangent segments to $\mathcal{P}_{k-1}$ from p_k, respectively (Figure 3). The concave chain on $\overline{\mathcal{P}}_{k-1}$ between $u(p_k)$ and $d(p_k)$ and line segments $p_k u(p_k)$, $p_k d(p_k)$ form a pseudo-triangle $\Delta(p_k)$ — its boundary consists of a concave chain and two single edges. Let $C(p_k)$ denote the concave chain on $\Delta(p_k)$. The three corners of $\Delta(p_k)$ are the points p_k, $d(p_k)$, and $u(p_k)$. After processing every point in $\mathcal{P}$, we obtain a pseudo-triangulation, which is called the *incremental pseudo-triangulation (IPT)* of $\mathcal{P}$. This process resembles to the classical incremental construction of convex hull [10, 11].

3.2 Maintaining the IPT

As the points move, $IPT(\mathcal{P})$ changes continuously, but its combinatorial structure changes only at discrete times.

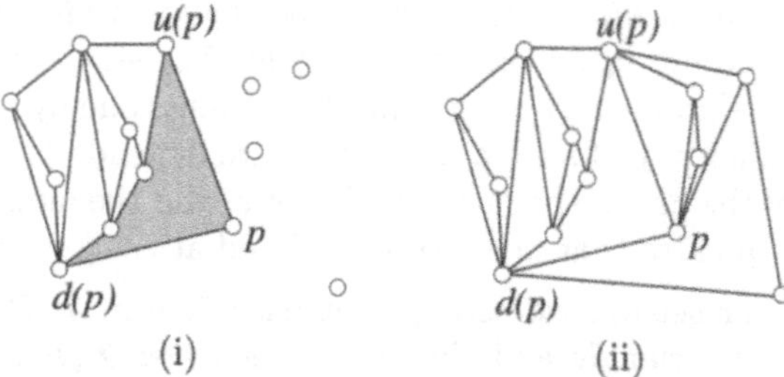

Figure 3: *The incremental pseudo-triangulation of a point set. In (i), the sweep line has just passed p; (ii) shows the final pseudo-triangulation.*

Lemma 2 *As long the x-ordering of the points in $\mathcal{P}$ does not change, the reflex and corner certificates certify the incremental pseudo-triangulation.*

In view of this lemma, as long as the x-ordering does not change, we have to update $IPT(\mathcal{P})$ only when a reflex or corner certificate fails. When a corner event fails, the structure can be updated as described in Section 2.3. When a reflex certificate fails, we have two choices of which edge to delete — such choice is not difficult, and we omit the details in this version of the paper.

Next, we consider the case in which two points exchange in x-order (we call this an x-event). Although such an event does not affect the validity of the pseudo-triangulation, it does cause a change to $IPT(\mathcal{P})$ since the incremental ordering of the points changes. Suppose that p passes q from the left at time t and p is

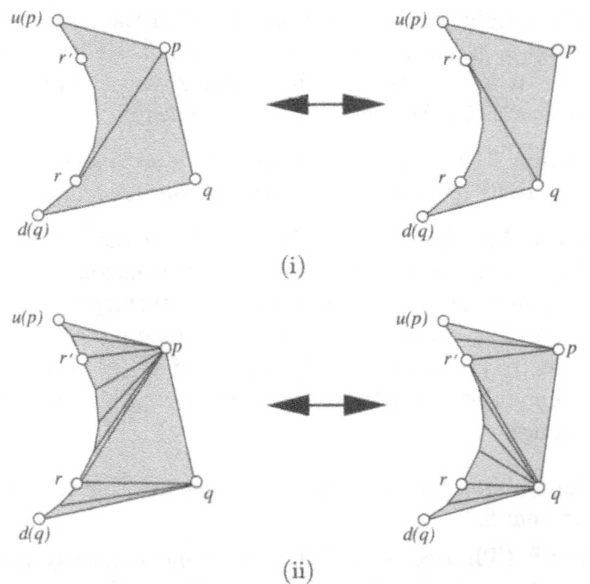

Figure 4: *(i) Two points exchange in x-order. (ii) Fan triangulation of IPT($\mathcal{P}$).*

above q (Figure 4). Let r be $d(p)$ just before time t, and let r' be $u(q)$ immediately after time t. Then the structure is updated by switching the edge rp with the edge $r'q$ — a flipping operation defined in Section 2.3. The point r' can be found by computing a tangent segment from q to the concave chain between $u(p)$ and $d(q)$, which can be done in $O(\log n)$ time.

Notice that in this description, each event changes $IPT(\mathcal{P})$. Thus, the method is output-sensitive. We have shown that

Theorem 3 *The IPT of n points can be maintained in an output-sensitive manner. Each update of the structure takes $O(\log n)$ time.*

In the following, we shall bound the number of changes to $IPT(\mathcal{P})$ for points in constant degree algebraic motion.

3.3 Combinatorial Changes to $IPT(\mathcal{P})$

We now obtain an upper bound on the number of combinatorial changes to $IPT(\mathcal{P})$, i.e., the number of events, under the assumption that the degree of motion of $\mathcal{P}$ is fixed. We define a function $\phi_p^q(t)$ for two distinct points $p, q \in \mathcal{P}$: if at time t, q is to the left of p, then $\phi_p^q(t)$ is the slope of the line that passes through p and q; otherwise, $\phi_p^q(t)$ is undefined. Since the motion has constant degree, the x-order of a pair of points can switch only a constant number of times. Thus, each $\phi_p^q(t)$ consists of a constant number of arcs, all of which are portions of a fixed-degree polynomial. Consider the family of functions

$$\Phi_p = \{\phi_p^q \,|\, q \in \mathcal{P}, q \neq p\}.$$

By our assumptions, any two arcs in Φ_p intersect a constant number of times, say, $s-2$ times. For a point q to be $d(p)$ at time t, $\phi_p^q(t)$ must have the largest value among all the points in $\mathcal{P}$. This is to say, the number of changes to $d(p)$ is the same as the combinatorial complexity of the upper envelope of Φ_p, which is bounded by $\lambda_s(n)$. ($\lambda_s(n)$ is the maximum length of an (n, s) Davenport-Schinzel sequence and is roughly linear.) Similarly, we can bound the number of changes to $u(p)$ by $\lambda_s(n)$. Summing over all the points in $\mathcal{P}$, we thus have:

Theorem 4 *When the points of $\mathcal{P}$ move algebraically with constant degree, $IPT(\mathcal{P})$ changes $O(n\lambda_s(n))$ times, where s is a constant that depends on the degree of the motion.*

3.4 Fan Triangulation of $\mathcal{P}$

It is easy to obtain a triangulation of $\mathcal{P}$ from $IPT(\mathcal{P})$. We connect each point p to every interior point on $C(p)$, thereby creating a fan inside each $\Delta(p)$. We call such a triangulation the *fan triangulation* of $\mathcal{P}$. (Figure 4 (ii)). Although $IPT(\mathcal{P})$ changes only nearly quadratically many times, we are not able to prove a similar bound on the number of changes in the fan triangulation. It is easy to verify that a reflex or corner event causes only $O(1)$ changes to the fan triangulation. The problem comes from the x-ordering events, which we process by switching a fan of p to a fan of q, or vice versa (Figure 4 (ii)). The cost of such switching is proportional to the length of the chain between r and r', where r and r' are the same as defined in Section 3.2. In the worst case, it might be $\Theta(n)$. This gives us a naïve bound of $O(n^3)$ on the number of changes to the fan triangulation.

For linear motion, we can obtain a better upper bound on the number of changes to the fan triangulation by a global argument. To our knowledge, this is the first triangulation for which a sub-cubic bound can be proved, even for linear motions.

Theorem 5 *For points in linear motion, the number of changes to the fan triangulation is bounded by* $O(n^{4/3}\lambda_s(n))$, *where s is a constant.*

In the following, we prove Theorem 5. For each point p, denote by $H_d(p)$ the half-space below the line passing through p and r. Consider a point $v \in \mathcal{P}$ on the concave chain C between r and r' at the time t when an x-event happens. It can be shown that at least one of $u(v)$ and $d(v)$, say $d(v)$, is adjacent to v in C at time t.

Since v lies between r and r', the points p, q must lie below the line that passes through v and $d(v)$, i.e., $p, q \in H_d(v)$. Furthermore, there is no point in $\mathcal{P} \cap H_d(v)$ to the left of p and q. Otherwise, it would contradict that v appears on the concave chain between $u(p)$ and $d(q)$. We define the function $\xi_v^p(t)$ to be the x-coordinate of p if p is in $H_d(v)$ and to the right of v at time t and undefined otherwise. Set $\Xi_v = \{\xi_v^p \mid p \in \mathcal{P}\}$. The point v is on the concave chain between r and r' when p, q exchange their x-ordering if and only if $\xi_v^p(t)$ and $\xi_v^q(t)$ are both defined and they are the smallest amongst all the ξ_v's at time t, i.e., if the point $(t, \xi_v^p(t))(= (t, \xi_v^q(t)))$ is a vertex on the lower envelope of Ξ_v. Denote by δ_v the complexity of the lower envelope of Ξ_v. The number of changes to the fan triangulation in this case is thus bounded by $\sum_{v\in\mathcal{P}} \delta_v$.

Unfortunately, we cannot use a Davenport-Schinzel sequence argument to prove a near-linear bound on δ_v because the graphs of functions $\xi_v^p(t)$ may consist of $\Omega(n)$ arcs. In fact, there are examples in which the graphs of functions in Ξ_v consist of $\Omega(n^2)$ arcs in total. We therefore need a more refined analysis to bound $\sum_v \delta_v$.

When a point q enters or leaves $H_d(v)$, it may or may not appear on the lower envelope of Ξ_v. For a point $v \in \mathcal{P}$, denote by γ_v the number of arc endpoints that appear on the lower envelope of Ξ_v.

Lemma 6 $\sum_{v\in\mathcal{P}} \gamma_v = O(n\lambda_s(n))$.

Proof: When an event of this type happens, $IPT(P)$ changes. Thus, in total, it is bounded by $O(n\lambda_s(n))$ by Theorem 4. □

Consider a point that travels along the edges of the arrangement $\mathcal{A}(\Xi)$ of a set Ξ of t-monotone algebraic arcs. The point moves t-monotonically along an edge in $\mathcal{A}(\Xi)$. When it comes to an intersection between two arcs, it either continues to travel on the same arc or make a right turn if it is possible. Call the trajectory of such a point a *pseudo-concave chain*. Clearly, in a line arrangement, this definition gives us exactly a concave chain. We can prove the following.

Lemma 7 *The lower envelope of the arrangement of* Ξ_v *consists of* γ_v *disjoint (pseudo) concave chains.*

Proof: Imagine that a point travels on the lower envelope of Ξ_v. It needs to stop only if it encounters an endpoint of an arc. Such an endpoint corresponds to a discontinuity of the function ξ_v^p for some p, i.e., corresponds to p coming in or going out of $H_d(v)$. Further, the endpoint is on the lower envelope. This happens γ_v times. □

Combining the Lemma 6 and 7, we are able to prove Theorem 5.

Proof: (Theorem 5) If the points move linearly with a constant velocity, then the x-coordinate of a point is a linear function of time. Thus, the lower envelope of Ξ_v consists of γ_v disjoint concave chains in an arrangement of n lines. According to the results in [14, 15], the total complexity of these chains is bounded by:

$$\delta_v = O(\max(n\gamma_v^{1/3}, n^{2/3}\gamma_v^{2/3}))\,.$$

Therefore, the total complexity is bounded by:

$$\sum_{v\in\mathcal{P}} \delta_v = \sum_v O(\max(n\gamma_v^{1/3}, n^{2/3}\gamma_v^{2/3}))\,.$$

By Lemma 6, $\sum_v \gamma_v = O(n\lambda_s(n))$. Thus, $\sum_v \delta_v$ is maximized when $\gamma_v = \Theta(\lambda_s(n))$, for all v's. This gives us the bound of $O(n^{4/3}\lambda_s(n))$. □

We do not know whether it is possible to extend this result to higher degree algebraic motions or whether this bound is tight.

4 Pseudo-Triangulation for Convex Polygons

We now consider the case in which $\mathcal{P}$ is a set of k convex polygons with a total of n vertices. We describe a different pseudo-triangulation for $\mathcal{P}$, called the *greedy vertical pseudo-triangulation*, which is based on notions developed by Pocchiola and Vegter [21]. They introduced the *greedy pseudo-triangulation* as a tool to compute the visibility complex of a set of convex polygons. We adapt their algorithm for our collision detection application.

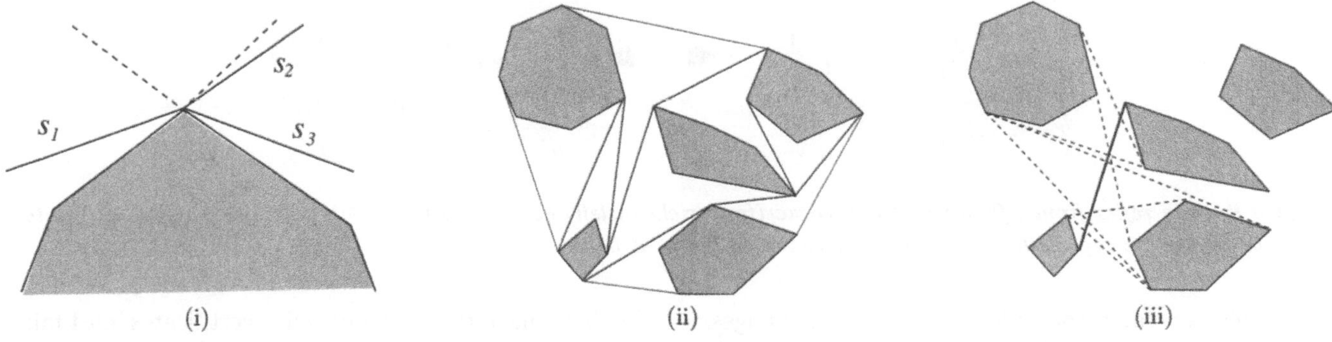

Figure 5: *(i) Intersecting and non-intersecting tangents: s_1, s_2 are intersecting, but s_1, s_3 and s_2, s_3 are non-intersecting tangents. (ii) Greedy vertical pseudo-triangulation. (iii) Left-to-right property. The solid segment is an edge in GVPT . The dotted ones are the free tangents it crosses.*

4.1 Greedy Pseudo-Triangulation for Convex Polygons

There are exactly four bi-tangent segments between any two disjoint convex polygons. A bi-tangent is called *free* if it does not intersect the interior of any object in $\mathcal{P}$. Let s and s' be two free bi-tangents with a common endpoint v, which is a vertex of a polygon $P \in \mathcal{P}$. There are three disjoint wedges formed by the polygon edges incident upon v and between s and s'. We say that s and s' *cross at* v if each of these wedges spans an angle less than π; see Figure 5.[1] We say that two free bi-tangents *intersect* if either they share an interior point or they have a common endpoint and they cross at that endpoint. For any linear ordering $\prec$ on the free tangents, we can build a corresponding greedy triangulation $\mathcal{T}_{\prec}(\mathcal{P})$ as follows. We first sort the tangents by $\prec$ ordering. We then scan through this list and maintain a set S of tangents that we have added to the pseudo-triangulation so far. At each step, we pick the next tangent segment s in the sorted sequence and check whether it intersects any tangents in S. If s does not intersect any segment of S, we add it to S; otherwise, we discard s. After processing all the segments, we obtain a set of non-intersecting free tangents. It is shown in [21] that $|S| = 3k - 3$, and that together with the object boundaries, S forms a pseudo-triangulation of $\mathcal{F}(\mathcal{P})$ consisting of $2k - 2$ pseudo-triangles. By Lemma 1, this procedure constructs a minimal pseudo-triangulation.

For any line segment s, define $\theta(s)$ to be the minimum angle by which we have to rotate s in counterclockwise direction so that it becomes vertical. We order all the free tangents in increasing order of $\theta(\cdot)$. The greedy pseudo-triangulation created using this ordering is called the *greedy vertical pseudo-triangulation (GVPT)* and is denoted by $\mathcal{G}(\mathcal{P})$ (Figure 5 (ii)). It is shown in [21] that $\mathcal{G}(\mathcal{P})$ can be constructed in $O(k \log n)$ time, provided that each polygon is represented as an array storing its vertices in a clockwise (or counterclockwise) order.

The following local rule determines whether a free bi-tangent is in $\mathcal{G}(\mathcal{P})$.

Lemma 8 (Left-to-right property) *A bi-tangent segment s is in $\mathcal{G}(\mathcal{P})$ if and only if $\theta(s) < \theta(s')$ for each free bi-tangent s' intersected by s (Figure 5 (iii)).*

Recall that if we replace an edge in a pseudo-triangulation with its shadow edge, we still obtain a pseudo-triangulation. It turns out that in $\mathcal{G}(\mathcal{P})$ the shadow edge of an edge s is minimal in the ordering $\prec$ among all free bi-tangents crossing s. This implies the following property:

Lemma 9 (Local property) *Suppose that s is an edge in $\mathcal{T}_{\prec}(\mathcal{P})$. Denote by $\prec'$ the same ordering as $\prec$ except assigning s as the maximum element of $\prec'$. Then $\mathcal{T}_{\prec'}(\mathcal{P})$ can be obtained from $\mathcal{T}_{\prec}(\mathcal{P})$ by replacing s with its shadow edge.*

[1]Intuitively, if we smooth the polygon P by taking the Minkowski sum of P with a disk of a sufficiently small radius, say δ, and we translate s and s' by at most δ so that they become tangents to the resulting polygons at their endpoints, then s and s' cross at v if and only if the translated copies of s and s' intersect.

Figure 6: *The reflex(from left to right) and corner(from right to left) events and the updates. When a reflex certificate fails, we choose the edge with smaller slope, as shown in the right figure.*

By this lemma, if the ordering of an edge changes, we just perform a local flip operation to fix the greedy pseudo-triangulation.

4.2 Greedy Pseudo-Triangulation Maintenance

We now describe how to maintain $\mathcal{G}(\mathcal{P})$ as a KDS so that it can be updated efficiently as the polygons in $\mathcal{P}$ move or deform. As described in Section 2.3, we maintain corner and reflex certificates for each pseudo-triangle in $\mathcal{G}(\mathcal{P})$. In addition, for each diagonal edge s of $\mathcal{G}(\mathcal{P})$, we maintain a *diagonal certificate* to certify that $\theta(s) < \theta(s')$, where s' is the shadow edge of s in $\mathcal{G}(\mathcal{P})$. This adds the requirement to maintain the shadow edges although such edges do not appear in $\mathcal{G}(\mathcal{P})$. The shadow edges can also be maintained using corner and reflex certificates. We will show that these certificates are sufficient to maintain $\mathcal{G}(\mathcal{P})$.

As in Section 3, we can prove the following analogue to Lemma 2.

Lemma 10 *If no diagonal certificate fails, then $\mathcal{G}(\mathcal{P})$ changes only when one of the corner or reflex certificates fails.*

Proof: The greedy pseudo triangulation can change only when a tangent stops or starts to be free, when the slope ordering of two intersecting tangents changes, or when two intersecting free tangents stop or start to intersect. Without any diagonal certificate failure, all these cases are reduced to when three objects are collinear. (three objects are *collinear* if there is a line tangent to all of them.) Suppose that the collinear objects are P_1, P_2, P_3 and the line ℓ is tangent to them in that order. Consider the line segments s_1,s_2,s_3 on ℓ that connect P_1P_2, P_2P_3, and P_3P_1, respectively. By the left to right property, for such a collinearity to change $\mathcal{G}(\mathcal{P})$, it must be the case that exactly two of these edges are in $\mathcal{G}(\mathcal{P})$ right before the event happens. Further, one of the corner or reflex certificates must fail when such an event happens. □

When a corner or reflex certificate fails, we can update $\mathcal{G}(\mathcal{P})$ as described in Section 2.3, except for one subtle issue in the case of reflex certificates. When an interior vertex p of a side chain $\ldots p_L p p_R \ldots$ ceases to be a reflex vertex, we add the edge $p_L p_R$ and we have to delete one of the edges $p_L p$ and $p p_R$. Instead of deleting one of them arbitrarily, we delete $p_L p$ if $\theta(p_L p) > \theta(p p_R)$; see Figure 6.

Next, we consider diagonal certificates. If the certificate of a diagonal edge s fails, then either s or its shadow edge (see Section 2 for the definition) is vertical. When such an event happens, the slope ordering of an edge jumps from the minimum to maximum, or vice versa. By the local property of greedy pseudo-triangulations, we can simply perform a flipping operation to s to maintain $\mathcal{G}(\mathcal{P})$.

Since $\mathcal{G}(\mathcal{P})$ has $O(k)$ diagonals and pseudo-triangles, the number of corner and diagonal certificates is $O(k)$. We assume that the polygons remain convex at all times, therefore it suffices to maintain reflex certificates for a vertex only if it is adjacent to a diagonal edge, i.e., a free bi-tangent. The number of such vertices is also $O(k)$. Hence, we obtain the following:

Theorem 11 *$\mathcal{G}(\mathcal{P})$ can be maintained by using a structure with $O(k)$ certificates. Each event can be processed in time $O(\log n)$.*

This data structure works even if the polygons deform continuously over time, as long as they remain convex at all times, though the number of events might be affected. The rigid motion can give us better bounds as stated in Theorem 13.

4.3 Combinatorial changes to $\mathcal{G}(\mathcal{P})$

Next, we bound the number of changes to $\mathcal{G}(\mathcal{P})$ if the degree of motion of $\mathcal{P}$ is fixed. It can be shown that a corner or reflex certificate fails when three polygons of $\mathcal{P}$ become collinear, i.e., a line is tangent to three polygons. Hence, a combinatorial change in $\mathcal{G}(\mathcal{P})$ happens only when three polygons of $\mathcal{P}$ are collinear or when a bi-tangent becomes vertical. The number of such events can be bounded by the following lemma.

Lemma 12 *Suppose that P_1, P_2, P_3 are convex polygons with n_1, n_2, n_3 vertices, respectively, and the degree of their motion is constant. They can become collinear $O(n_1 + n_2 + n_3)$ and $O(n_1n_2 + n_1n_3 + n_2n_3)$ times for translational and rigid motions, respectively. The bi-tangent between P_1, P_2 can become vertical $O(1)$ and $O(n_1 + n_2)$ times for translational and rigid motions, respectively.*

Applying this lemma to all triplets of $\mathcal{P}$, we obtain the following weak upper bounds on the number of changes to $\mathcal{G}(\mathcal{P})$.

Theorem 13 *Let $\mathcal{P}$ be a set of k polygons with a total of n vertices. $\mathcal{G}(\mathcal{P})$ changes $O(kn^2)$ times if the degree of motion of $\mathcal{P}$ is fixed. If the motion is translational, the number of changes is only $O(k^2n)$. If the polygons may deform, the number of changes is bounded by $O(n^3)$.*

5 Pseudo-Triangulation for Simple Polygons

In this section, we consider the case in which $\mathcal{P}$ is a set of k pairwise-disjoint simple polygons with a total of n vertices. We describe a method for maintaining a pseudo-triangulation of $\mathcal{P}$ by combining our algorithm for convex polygons with the algorithm by Basch et al. [2] for two simple polygons.

5.1 The Mixed Pseudo-Triangulation

For each $P \in \mathcal{P}$, define the *relative geodesic cycle* $C(P)$ of P to be the shortest cycle in $\mathcal{F}(\mathcal{P})$ with the same homotopy type as ∂P. The region enclosed by $C(P)$ is called the *relative convex hull* and is denoted by $\overline{P}$. We decompose $\mathcal{F}(\mathcal{P})$ into two parts $\mathcal{F}_1(\mathcal{P}) = \mathcal{F}(\mathcal{P}) \cap (\bigcup \overline{P})$ and $\mathcal{F}_2(\mathcal{P}) = \mathrm{cl}(\mathcal{F}(\mathcal{P}) \setminus \bigcup \mathrm{int}(\overline{P}))$. We compute pseudo-triangulations $\mathcal{T}(\mathcal{F}_1(\mathcal{P}))$ and $\mathcal{T}(\mathcal{F}_2(\mathcal{P}))$ separately. These two triangulations together give the *mixed pseudo-triangulation* of $\mathcal{P}$.

First, we consider $\mathcal{F}_1(\mathcal{P})$. Since the interiors of all the relative convex hulls are pairwise disjoint, we can

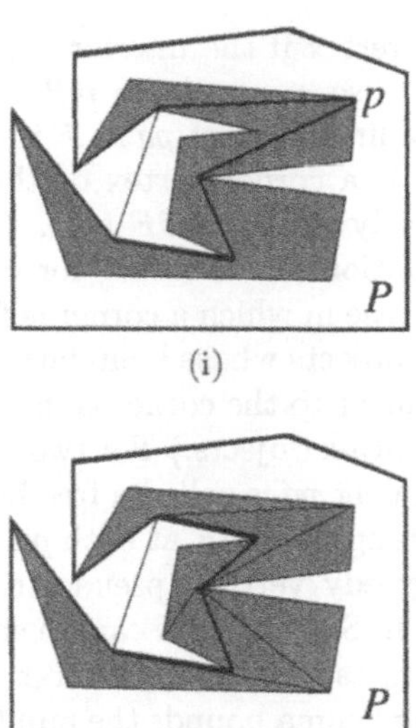

Figure 7: *(i) The relative geodesic cycle of P pinned at the vertex p. (ii) The external relative geodesic triangualtion. The thickened cycle is the geodesice cycle corresponding to the root node of T(P).*

compute a pseudo-triangulation of each of them separately. We triangulate each $\overline{P}$ as in [2]. Following [2], we define the *pinned relative geodesic cycle* of P, with respect to a pinning point set B (a subset of the vertices of P), to be the shortest cycle in $\mathcal{F}(\mathcal{P})$ that passes through the vertices in B and has the same homotopy type as ∂P. The *pinned relative convex hull* of P with respect to B is the region enclosed by the pinned relative geodesic cycle (Figure 7(i)). If B contains all the vertices of P, the pinned relative convex hull of P is P itself. Consider a balanced binary tree $T(P)$ with the vertices of P as leaves, in the order they appear on ∂P; each internal node stores the vertex from its left child. We can then build $O(\log |P|)$ pinned relative geodesic cycles, where the pinning set of each cycle is the set of vertices at a given level of $T(P)$. By overlaying these cycles, we obtain a pseudo-triangulation of $\overline{P} \setminus P$ (Figure 7(ii)). This pseudo-triangulation is called the *external relative geodesic triangulation.* Refer to [2] for the maintenance and bounds on the number of changes of this structure.

Now, we consider $\mathcal{F}_2(\mathcal{P})$. In the following, we generalize the greedy pseudo-triangulation, defined in Section 4, to a connected polygonal subdivision. Consider a connected polygonal region $\mathcal{F}$. A vertex v of $\mathcal{F}$ is called a *corner* vertex if the interior angle at v is less than 180°; otherwise, v is called a *reflex* vertex. For a point p on ∂F, a line segment pq in $\mathcal{F}$ is called *tangent* to ∂F at p if p is a corner vertex or the line passing through p, q locally supports ∂F at p. (The intuition behind the definition of "tangent" for corner vertices comes from the case in which a corner is formed by two separate convex objects whose boundaries touch. Then a "tangent" segment to the corner is indeed tangent to one of the two convex objects.) For two vertices p, q on ∂F, the line segment pq is called a free bi-tangent if pq lies in $\mathcal{F}$ and is tangent to ∂F at both p and q. We now construct the greedy vertical pseudo-triangulation of $\mathcal{F}$, as described in Section 4. It can be shown that the algorithm constructs a minimal pseudo-triangulation of $\mathcal{F}$. The following lemma bounds the number of pseudo-triangles in such a pseudo-triangulation.

Lemma 14 *If ∂F consists of k connected components and m corner vertices, the number of pseudo-triangles in the greedy pseudo triangulation of $\mathcal{F}$ is $O(k+m)$.*

The relative geodesic triangulation $\mathcal{T}(\mathcal{F}_1(\mathcal{P}))$ and greedy pseudo-triangulation $\mathcal{T}(\mathcal{F}_2(\mathcal{P}))$ together form a pseudo-triangulation of $\mathcal{F}(\mathcal{P})$, which we call the *mixed pseudo-triangulation* (Figure 8). Next, we bound the size of the mixed pseudo-triangulation.

We call a diagonal edge of the mixed pseudo-triangulation a *bridge* edge if it connects two different polygons of $\mathcal{P}$. As we will see later, bridge edges play an important role in maintaining the pseudo-triangulation.

Lemma 15 *$\partial\mathcal{F}_2(\mathcal{P})$ consists of $O(k)$ connected components and $O(k)$ bridge edges and corner vertices.*

Proof: Clearly, the number of connected components of $\mathcal{F}_2(\mathcal{P})$ is bounded by $O(k)$, as $\mathcal{P}$ contains k disjoint objects. Consider a connected component $\mathcal{F}$ of $\mathcal{F}_2(\mathcal{P})$ and a corner vertex p on $\partial\mathcal{F}$. Suppose that the two adjacent vertices to p on $\partial\mathcal{F}$ are q_1, q_2. One of q_1 and q_2 must be on the same object as p, and the other must be on a different object. This implies that the number of corner vertices is bounded by the number of bridge edges on $\partial\mathcal{F}$. The number of bridge edges on $\overline{\mathcal{P}}$ is clearly bounded by $O(k)$. For those edges not on the convex hull, consider the planar graph G where a node in G corresponds to an object in $\mathcal{P}$ and an edge between two nodes corresponds to a bridge edge on $\partial\mathcal{F}_2(\mathcal{P})$ that connects the corresponding two objects.

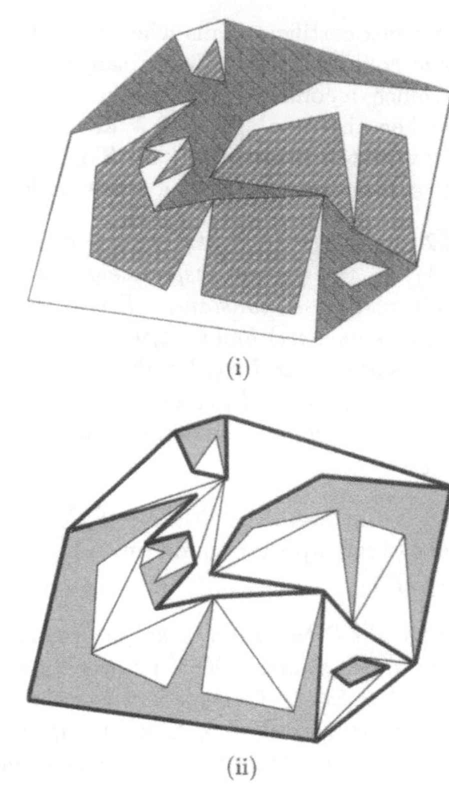

Figure 8: *The mixed pseudo-triangulation. (i) $\mathcal{F}_1(\mathcal{P})$ and $\mathcal{F}_2(\mathcal{P})$ are shaded differently. (ii) Mixed pseudo-triangulation; thick edges denote the boundary between $\mathcal{F}_1(\mathcal{P})$ and $\mathcal{F}_2(\mathcal{P})$.*

Then, the number of bridge edges is bounded by the number of edges in the graph G. The graph G is planar, but it may contain duplicate edges. By bounding the number of faces with only two edges on the boundary, we can still bound the number of edges of G, thus the number of bridge edges, by $O(k)$. □

Lemmas 14 and 15 imply the following.

Corollary 16 *The greedy triangulation $\mathcal{T}(\mathcal{F}_2(\mathcal{P}))$ of $\mathcal{F}_2(\mathcal{P})$ has $O(k)$ pseudo-triangles.*

Next, we bound the size of $\mathcal{T}(\mathcal{F}_1(\mathcal{P}))$. We will focus on the bridge edges in $\mathcal{F}_1(\mathcal{P})$. Basch et al. [2] have shown that any line segment lying inside $\mathcal{F}(\mathcal{P})$ crosses $O(\log n)$ bridge edges, and thus the number of bridge edges is $O(\tau \log n)$, where τ is the size of the min-link separator between two simple polygons. Here, we generalize their result to our setup by arguing that any free line segment can cross at most two relative convex hulls.

Lemma 17 *Any free line segment crosses* $O(\log n)$ *bridge edges of* $\mathcal{T}(\mathcal{F}_1(\mathcal{P}))$.

Proof: Consider a free line segment pq. Suppose that pq crosses m relative convex hulls. As in [2], pq can cross $O(m \log n)$ bridge edges. We claim that pq intersects $\partial\overline{P}$, the boundary of the relative convex hull of a polygon P, in at most one point. Indeed, if pq intersects $\partial\overline{P}$ at two points, then we can shortcut $\partial\overline{P}$ between these two points, thereby contradicting the assumption that $\partial\overline{P}$ is a geodesic. This implies that pq can intersect at most two relative convex hulls, which completes the proof of the lemma. □

A *minimum link subdivision* is a polygonal subdivision of the plane, using as few (line segment) edges as possible, such that each $P \in \mathcal{P}$ is contained in its own face of the subdivision. A *minimum link separator* for a polygon $P \in \mathcal{P}$ is a simple polygon homotopic to ∂P with as few edges as possible.

Consider any Jordan cycle C in $\mathcal{F}(\mathcal{P})$ that is homotopic to ∂P. Clearly, C must intersect all the bridge edges incident to P because C separates P from other objects in $\mathcal{P}$. Suppose that C consists of τ line segments. Since each line segment on C crosses at most $O(\log n)$ bridge edges connected to P and each bridge edge is crossed at least once, we can conclude that the number of bridge edges incident to P is bounded by $O(\tau \log n)$. Therefore, we have the following result on the number of bridge edges.

Lemma 18 *The number of bridge edges in* $\mathcal{T}(\mathcal{F}_1(\mathcal{P}))$ *is bounded by* $O(\kappa \log n)$, *where* κ *is the number of edges in a minimum link subdivision of* $\mathcal{P}$. *For each polygon* P *in* $\mathcal{P}$, *the number of bridge edges connected to* P *is bounded by* $O(\tau(P) \log n)$, *where* $\tau(P)$ *is the number of edges in a minimum link separator of* P.

Thus, we obtain the following bounds on the size of the mixed pseudo-triangulation by combining Corollary 16 and Lemma 18.

Theorem 19 *The mixed pseudo-triangulation has size* $O(n)$, *and among all the diagonal edges, only* $O(\kappa \log n)$ *edges are bridge edges, where* κ *is the number of edges in a minimum link subdivision of* $\mathcal{P}$.

5.2 Mixed Pseudo-Triangulation Maintenance

As shown in [2], the external relative geodesic triangulation can be maintained by corner and reflex certificates only, and there are local rules to decide which edge to select when a reflex certificate fails. Therefore, the mixed triangulation can be easily maintained, although it comprises two completely different structures. As argued before, the number of certificates in the certification structure of $\mathcal{T}(\mathcal{P})$ is proportional to the number of bridge edges in $\mathcal{T}(\mathcal{P})$. According to Lemmas 15 and 18, we may bound the certificates for certifying $\mathcal{T}(\mathcal{P})$ as follows.

Theorem 20 *The number of certificates in* $\mathcal{T}(\mathcal{P})$ *is bounded by* $O(\kappa \log n)$, *where* κ *is the number of edges in the minimum link subdivision of* $\mathcal{P}$. *For each polygon* P *in* $\mathcal{P}$, *the number of certificates involving* P *is bounded by* $O(\tau(P) \log n + k)$, *where* $\tau(P)$ *is the number of edges in the minimum link separator of* P.

A trivial bound on the number of changes to the mixed pseudo-triangulation for polygons in rigid motion is $O(n^3)$, which is the number of times when three vertices become collinear.

6 Conclusions

We have shown how to efficiently maintain free space pseudo-triangulations for points, as well as convex and simple polygons moving around in the plane. In addition to collision detection, the pseudo-triangulation structure provides a simple way for walking around the free space and performing operations such as ray shooting and other visibility queries. Unlike any previous collision detection method, our deformable tiling can be adapted to work for deformable objects, thus enabling new applications where simulation of flexible shapes is important. An extension of our ideas to 3D presents the next obvious challenge.

Acknowledgments

Pankaj Agarwal was supported in part by National Science Foundation research grant CCR–93–01259, by Army Research Office MURI grant DAAH04–96–1–0013, by a Sloan fellowship, by a National Science Foundation NYI award and matching funds from Xerox Corp, and by a grant from the U.S.-Israeli Binational Science Foundation. Julien Basch, Leonidas

Guibas, and Li Zhang were supported in part by National Science Foundation grant CCR–9623851 and by US Army MURI grant DAAH04-96-1-0007. We wish to acknowledge the contribution of Otfried Cheong and Mark de Berg to preliminary work related to part of this paper.

References

[1] E. M. Arkin, M. Held, and J. S. B. Mitchell. Manufacturing: An application domain for computational geometry. In *First Regional Symposium on Manufacturing Science and Technology*, pages 29–36, October 12-13 1995.

[2] J. Basch, J. Erickson, L. J. Guibas, J. Hershberger, and L. Zhang. Kinetic collision detection for two simple polygons. In *Proc. 9th ACM-SIAM Sympos. Discrete Algorithms*, pages 102–111, 1999.

[3] J. Basch, L. J. Guibas, and J. Hershberger. Data structures for mobile data. In *Proc. 8th ACM-SIAM Sympos. Discrete Algorithms*, pages 747–756, 1997.

[4] R.-P. Berretty, K. Goldberg, M. Overmars, and F. V. der Stappen. On fence design and the complexity of push plans for orienting parts. In *Proc. 13th Annu. ACM Sympos. Comput. Geom.*, pages 21–29, 1997.

[5] K.-F. Böhringer, B. R. Donald, and N. C. MacDonald. Upper and lower bounds for programmable vector fields with applications to MEMS and vibratory plate part feeders. In J.-P. Laumond and M. H. Overmars, editors, *Algorithms for Robotic Motion and Manipulation*, pages 255–276. A. K. Peters, Wellesley, MA, 1996.

[6] S. Cameron. Collision detection by four-dimensional intersection testing. In *Proc. IEEE Internat. Conf. Robot. Autom.*, pages 291–302, 1990.

[7] B. Chazelle, H. Edelsbrunner, M. Grigni, L. J. Guibas, J. Hershberger, M. Sharir, and J. Snoeyink. Ray shooting in polygons using geodesic triangulations. *Algorithmica*, 12:54–68, 1994.

[8] D. Eppstein and J. Erickson. Raising roofs, crashing cycles, and playing pool: Applications of a data structure for finding pairwise interactions. In *Proc. 14th Annu. ACM Sympos. Comput. Geom.*, pages 58–67, 1998.

[9] J. Erickson, L. J. Guibas, J. Stofi, and L. Zhang. Separation-sensitive kinetic collision detection for convex objects. In *Proc. 9th ACM-SIAM Sympos. Discrete Algorithms*, pages 327–336, 1999.

[10] R. L. Graham. An efficient algorithm for determining the convex hull of a finite planar set. *Inform. Process. Lett.*, 1:132–133, 1972.

[11] R. L. Graham and F. F. Yao. Finding the convex hull of a simple polygon. *J. Algorithms*, 4:324–331, 1983.

[12] L. J. Guibas. Kinetic data structures: A state of the art report. In *Proc. 3rd Workshop on Algorithmic Foundations of Robotics*, pages 191–209, 1998.

[13] P. Gupta, R. Janardan, and M. Smid. Fast algorithms for collision and proximity problems involving moving geometric objects. *Comput. Geom. Theory Appl.*, 6:371–391, 1996.

[14] D. Halperin and M. Sharir. On disjoint concave chains in arrangements of (pseudo) lines. *Inform. Process. Lett.*, 40(4):189–192, 1991.

[15] D. Halperin and M. Sharir. Corrigendum: On disjoint concave chains in arrangements of (pseudo) lines. *Inform. Process. Lett.*, 51:53–56, 1994.

[16] M. Held, J. T. Klosowski, and J. S. B. Mitchell. Evaluation of collision detection methods for virtual reality fly-throughs. In *Proc. 7th Canad. Conf. Comput. Geom.*, pages 205–210, 1995.

[17] P. M. Hubbard. Collision detection for interactive graphics applications. *IEEE Trans. Visualization and Computer Graphics*, 1(3):218–230, Sept. 1995.

[18] D. Kirkpatrick, J. Snoeyink, and B. Speckmann. Kinetic collision detection for simple polygons. In *To appear in 16th Sympos. Comput. Geom.*, 2000.

[19] B. Mirtich. *Impulse-based Dynamic Simulation of Rigid Body Systems*. Ph.D. thesis, Dept. Elec. Engin. Comput. Sci., Univ. California, Berkeley, CA, 1996.

[20] D. M. Mount. Intersection detection and separators for simple polygons. In *Proc. 8th Annu. ACM Sympos. Comput. Geom.*, pages 303–311, 1992.

[21] M. Pocchiola and G. Vegter. Topologically sweeping visibility complexes via pseudo-triangulations. *Discrete Comput. Geom.*, 16:419–453, Dec. 1996.

[22] M. K. Ponamgi, D. Manocha, and M. C. Lin. Incremental algorithms for collision detection between general solid models. In *Proc. ACM SIGGRAPH Sympos. Solid Modeling*, pages 293–304, 1995.

[23] E. Schömer and C. Thiel. Efficient collision detection for moving polyhedra. In *Proc. 11th Annu. ACM Sympos. Comput. Geom.*, pages 51–60, 1995.

[24] S. Suri. *Minimum link paths in polygons and related problems*. Ph.D. thesis, Dept. Comput. Sci., Johns Hopkins Univ., Baltimore, MD, 1987.

Real-time Global Deformations

Yan Zhuang, *University of California, Berkeley, CA*
John Canny, *University of California, Berkeley, CA*

Real-time simulation and animation of 3D global deformations is the bottleneck of many applications, such as a surgical simulator. In this paper, we present a system that simulates physically realistic large deformations of soft objects in real-time. We achieve the physical realism by modeling the global deformation using geometrically nonlinear finite element methods. We obtain real-time dynamic simulation of models of reasonable complexity by preprocessing the LU-factorization of a small number of large matrices. To reduce the time and space required for such a preprocess, and the time of the back-substitution, we apply modified nested dissection to reorder the finite element mesh. We also introduce an efficient method that handles deformable object collisions with almost no extra cost.

1 Introduction

Physically realistic modeling and manipulation of deformable objects has been the bottleneck of many applications, such as human tissue modeling, character animation, surgical simulation, etc. Among the potential applications, a virtual surgical training system is the most demanding for the real-time performance because of the requirement of real-time interaction with virtual human tissue.

In this paper, we address the bottleneck problem of real-time simulation of physically realistic large *global deformations* of 3D objects. In particular we apply the finite element method (FEM) to model such deformation. By *global deformation*, we mean deformations, such as large twisting or bending of an object, which involve the entire body, in contrast to poking and squeezing, which involve a relatively small region of the deformable object.

First, we point out that the application of linear elastic finite element methods to simulating large global deformation leads to unacceptable distortions (Figure 1 and 2). To avoid such distortions, we simulate global deformations using nonlinear finite element methods (Section 3).

Secondly, we achieve real-time simulation by restricting time steps, which are not constants, to a small set of values. Finite element methods usually require solving a large sparse linear system at each time step. The restriction of time steps leads to a small number of possible matrices to invert. This enables us to pre-compute all inverse matrices, represented by its LU-factorization. With the precomputed LU-factorization, only back-substitution is needed at each time step of the simulation. In Section 5, we discuss our modified nested dissection algorithm, which reduces both the time for LU-factorization and that of back-substitution.

Finally, we propose an efficient collision handling method for FEM in Section 6. Simulating deformable object collisions using a penalty method [21] requires tiny time steps to generate visually satisfactory animations. A general impulse collision [1] is considered more efficient and accurate but still requires more computational power than collision-free dynamics. In Section 6 we present an extremely simple and efficient collision time integration scheme, which makes the time integration of collision dynamics as cheap as that of collision-free dynamics.

2 Related Work

Our work of modeling and simulating a deformable object falls into the realm of physically based modeling. Witkin et al. [25] summarizes the methods and principles of physically based modeling, which has emerged as an important new approach to computer animation and computer graphics modeling.

In general, there are two different approaches to modeling deformable objects: The mass spring model

Figure 1: *The left image shows a beam at its initial configuration with a fixed left end and a free right end. The middle image shows the* distorted *deformation under gravity, using linear strain. The right image shows the* undistorted *deformation, under the same gravitational force, using quadratic strain (equation (5) and (6)).*

and the finite element model. Gibson and Mirtich [9] gives a comprehensive review of this subject.

The mass spring model has had good success in creating visually satisfactory animations. Waters [23] uses a spring model to create a realistic 3D facial expression. Provot et al. [18] describes a 2D model for animating cloth, using double cross springs. Promayon et al. [17] presents a mass-spring model of 3D deformable objects and develops some control techniques.

Despite the success in some animation applications, the mass spring models do not model the underlying physics accurately, which makes it unsuitable for simulations that require more accuracy. The structure of the mass spring is often application dependent and hard to interpret. The animation results often vary dramatically with different spring structures. The distribution of the mass to nodes is somewhat (if not completely) arbitrary. Despite its inaccuracy, it does not have visual distortion and it is computationally cheap to integrate over time because the system is, by its very nature, a set of independent algebraic equations, which requires no matrix inversions to solve.

As an alternative, finite element methods (*FEM*) model the continuum much more accurately and their underlying mathematics are well studied and developed. Another similar method is the finite difference method, which is less accurate but simpler and appropriate for some applications. Indeed a linear finite difference method over a uniform mesh is just a special case of FEM. Its accuracy and mathematical rigorousness make FEM a better choice for applications such as surgical simulations.

Terzopoulos et al. [21, 20, 22] applies both finite difference and finite element methods in modeling elastically deformable objects. Celniker et al. [15] applies FEM to generate primitives that build continuous deformable shapes designed to support a new free-form modeling paradigm. Pieper et al. [16] applies FEM to computer-aided plastic surgery. Chen [3] animates human muscle using a 20 node hexahedral FEM mesh. Keeve et al. [11] develops a static anatomy-based facial tissue model for surgical simulation using the FEM. Most recently, Cotin et al. [5] presents real-time elastic deformation of soft tissues for surgery simulation, which only simulates static deformations.

James and Pai [10] model real-time static local deformations using the boundary element method (BEM). BEM has the advantage of solving a smaller system because it only deals with degrees of freedom on the surface of the model. However, the resulting system is dense. It is difficult to apply boundary element method to model non-homogeneous material.

Our work differs from the previous work by either one or all of the following: (1) We simulate large global deformations instead of small local deformations; (2) we simulate the dynamic behavior of soft objects rather than the static deformation.

3 Nonlinear Elasticity with FEM

By *global deformations*, we mean deformations that are large and involve the entire body, such as high amplitude bending and twisting (Figure 1 and 2). These types of deformation often occur to soft objects, such as tissue in surgical simulations.

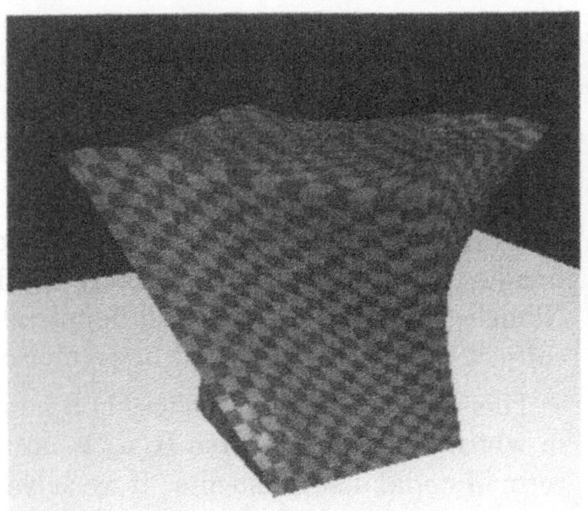

Figure 2: *The bottom of the object is fixed and its top is twisted. The top in the left image is distorted (grown bigger) because it is simulated using linear elasticity. The right image shows that the same distortion does not occur with nonlinear elasticity.*

The theory of elasticity is a fundamental discipline in studying continuum material. It consists of a consistent set of differential equations that uniquely describe the state of stress, strain and displacement of each point within an elastic deformable body. It consists of equilibrium equations relating the stresses; kinematics equations relating the strains and displacements, constitutive equations relating the stresses and strains; and boundary conditions relating to the physical domain. The theory was first developed by Louis-Marie-Henri Navier, Dimon-Denis Poisson, and George Green in the first half of the 19th century [24].

Synthesizing those equations allows us to establish a relationship between the deformation of the object and the exerted forces. However an analytic expression of such relationship is impossible, except for a small number of simple problems. *Finite element methods (FEM)* are one way to solve such a set of differential equations. From now on, we will discuss elasticity within the context of finite element methods.

When the geometry of the deformable object is complicated, it is impossible to obtain an analytic solution of an elastic deformation. FEM solves this problem by subdividing the object into small sub-domains with simple shapes (tetrahedra, hexahedra, etc.), called finite elements. The sub-division (mesh) does not only approximate the original geometry, but also leads to a discrete representation of the deformation.

In particular, we apply a *displacement based* finite element method to simulate such deformation. Namely displacements at vertices of the mesh, called nodes, will be calculated. The values at other points within the element are interpolated by continuous functions, usually low order polynomials, using the nodal values. The global equations (the relationship between all the nodal values) are obtained by assembling elementwise equations by imposing inter-element continuity of the solution and balancing of inter-element forces.[1] This essentially requires solving the following system of differential equations:

$$\mathbf{M}\ddot{\mathbf{u}} + \mathbf{D}\dot{\mathbf{u}} + \mathbf{R}(\mathbf{u}) = \mathbf{F} \tag{1}$$

where $\mathbf{u}$ is the $3n$-dimensional nodal displacement vector; $\dot{\mathbf{u}}$ and $\ddot{\mathbf{u}}$, the respective velocity and acceleration vectors; $\mathbf{F}$, the external force vector; $\mathbf{M}$, the $3n \times 3n$ mass matrix; $\mathbf{D}$, the damping matrix; and $\mathbf{R}(\mathbf{u})$, the internal force vectors due to deformation. n is the number of nodes in the FEM model [29].

To our best knowledge the published research ([16, 3, 11, 5]) assumes small deformations in their virtual environment. The most simulated deformations are those caused by squeezing and poking at a relatively small surface region. The small deformation assumption leads to the often used linear elasticity model, which is based on the following linear strain approximations:

[1] Detailed discussions of FEM can be found in [19, 28].

$$\epsilon_x = \frac{\partial u}{\partial x} \tag{2}$$

$$\gamma_{xy} = \frac{\partial u}{\partial y} + \frac{\partial v}{\partial x} \tag{3}$$

where x, y and z are the independent variables of the cartesian frame, and u, v and w are the corresponding displacement variables at the given point. Other terms of the strain at point (x, y, z), ϵ_y, ϵ_z, γ_{yz} and γ_{zx}, are defined similarly.

This linear strain makes the internal force vector linear with respect to nodal displacement vector. Namely it simplifies Equation (1) to the following *linear* system:

$$\mathbf{M}\ddot{\mathbf{u}} + \mathbf{D}\dot{\mathbf{u}} + \mathbf{K}\mathbf{u} = \mathbf{F} \tag{4}$$

This allows a preprocessing step that computes the constant stiffness matrix $\mathbf{K}$ and its LU factorization. This preprocessing step has been the key factor to real-time performance in previous works, such as [5], which animates deformations using a sequence of static equilibria.

The problem with this linear strain approximation is that it does not model finite rotation correctly. As a result, it introduces distortions when large global deformations occur (Figure 1 and 2), because global deformations usually involve finite rotation of part of the object relative to the rest of it.

To further illustrate this distortion, let us subject an undeformed object to a rigid body rotation. Apparently, the rotation should not introduce any deformation to the object. Namely the strain at any point within the object should be zero. However equations (2) and (3) give a nonzero strain. This "artificial" strain leads to distortion, because the body has to deform in a certain way to balance the stress caused by such an "artificial strain".

To avoid the distortion, as shown in Figures (1) and (2), we model the deformation using the exact strain, which is quadratic as following:

$$\epsilon_x = \frac{\partial u}{\partial x} + \frac{1}{2}\left[\left(\frac{\partial u}{\partial x}\right)^2 + \left(\frac{\partial v}{\partial x}\right)^2 + \left(\frac{\partial w}{\partial x}\right)^2\right] \tag{5}$$

$$\gamma_{xy} = \frac{\partial u}{\partial y} + \frac{\partial v}{\partial x} + \left[\frac{\partial u}{\partial x}\frac{\partial u}{\partial y} + \frac{\partial v}{\partial x}\frac{\partial v}{\partial y} + \frac{\partial w}{\partial x}\frac{\partial w}{\partial y}\right] \tag{6}$$

The other 4 terms of the strain are defined similarly. It is easy to verify that the above nonlinear strain handles arbitrary large rigid body motions correctly. Namely no artificial strain will be introduced when we subject the object to a rigid body motion.

This quadratic strain makes (1) a nonlinear system, in which the internal force $\mathbf{R}(\mathbf{u})$ is no longer a linear term of nodal displacements. If we solve this nonlinear system using an implicit integration scheme such as [2], real time simulation is impossible for reasonably large meshes.

We observe that a soft material such as live tissue has small stiffness in all directions (not necessarily isotropic). This makes explicit time integration schemes appropriate because we can take large time steps. We apply the explicit Newmark scheme to equation (1), which leads to the following equations:

$$\mathbf{u}_{n+1} = \mathbf{u}_n + \dot{\mathbf{u}}_n \Delta t_n + \frac{1}{2}\ddot{\mathbf{u}}_n \Delta t_n^2 \tag{7}$$

$$(\mathbf{M}+\frac{1}{2}\Delta t_n \mathbf{D})\ddot{\mathbf{u}}_{n+1} = \mathbf{F}_{n+1} - \mathbf{R}(\mathbf{u}_{n+1}) - \mathbf{D}(\dot{\mathbf{u}}_n + \frac{1}{2}\ddot{\mathbf{u}}_n \Delta t_n) \tag{8}$$

$$\dot{\mathbf{u}}_{n+1} = \dot{\mathbf{u}}_n + \frac{1}{2}(\ddot{\mathbf{u}}_n + \ddot{\mathbf{u}}_{n+1})\Delta t_n \tag{9}$$

Newmark scheme converts a nonlinear system to 3 linear systems (7), (8) and (9). The order of updating is (7), (8) and then (9).

4 Preprocessing the Inverse Matrices

The bottleneck of Newmark scheme is solving Equation (8). It requires inverting a large sparse matrix $\mathbf{M} + \frac{1}{2}\Delta t_n \mathbf{D}$. This matrix is not a constant matrix because the time step Δt_n may vary over time. Inverting a different large sparse matrix makes real time performance impossible.

To achieve real-time performance, Zhuang and Canny [27] approximated this matrix with its row-lumped diagonal matrix. This is equivalent to diagonalizing both the mass matrix $\mathbf{M}$ and the damping

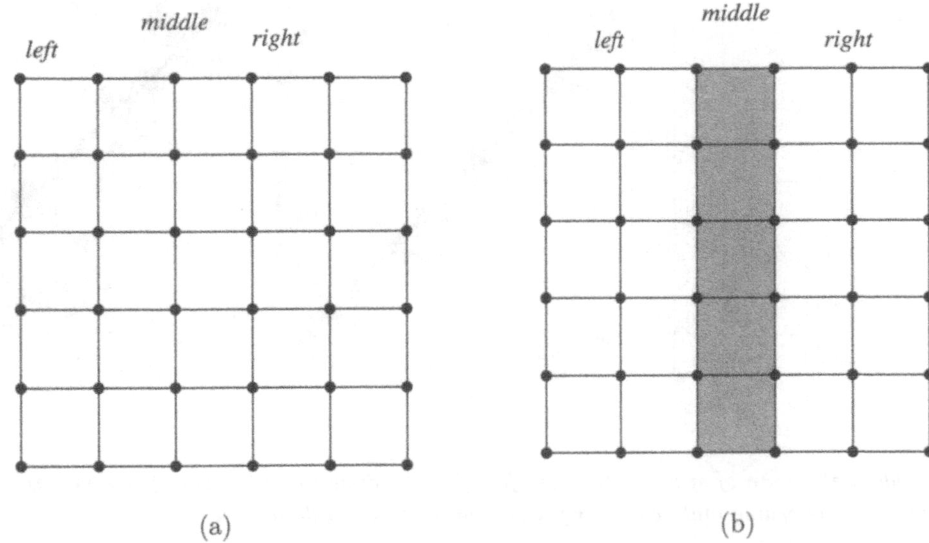

Figure 3: *(a) The dissector in the regular nested dissection algorithm is a layer of nodes. (b) The modified nested dissection algorithm uses a layers of elements (shaded), cut by a plane (or a line in 2D), as the dissector.*

matrix $\mathbf{D}$. The diagonalization of $\mathbf{M}$ is acceptable because it still preserves the global inertia property of the object, although it does not preserve the local moment of inertia. The diagonalization of the damping matrix may lose important viscous elasticity property of the material. For simulations that require more physical realism, diagonalization of matrix $\mathbf{D}$ is not appropriate.

In this section, we propose a different treatment by preprocessing. The matrices $\mathbf{M}$ and $\mathbf{D}$ are contants. The only variable is Δt_n. The time step Δt_n depends on the stability requirement of the system and the collision handling requirement. Instead of approximating these two matrices as [27] does, we restrict the time step Δt_n to a small set of values. Let T be the largest time step allowed, we define the restricted set of allowed time steps as $\{T/2^i | i = 0, 1, \ldots, m\}$. We choose the value m such that $T/2^m < T_{min}$, where T_{min} is the minimum time step in the worst case.

By restricting time steps to such a small set of values, we only have $m + 1$ possible matrices needed for the entire simulation. We can therefore pre-compute the $m + 1$ inverse matrices before the simulation begins.

Instead of precomputing the inverse of matrix $(\mathbf{M} + \frac{1}{2}\Delta t_n \mathbf{D})$, we precompute its LU-factorization. Given the LU-factorization, solving Equation (8) only requires back-substitution. The time for back-substitution is determined by the number of nonzeros in the LU-factorization of $(\mathbf{M} + \frac{1}{2}\Delta t_n \mathbf{D})$. In Section 5, we discuss how to reduce number of nonzeros in the LU-factorization.

5 Nested Dissection

A typical finite element simulation has to solve a large sparse linear system of large number of nonzero entries. For example, a $10 \times 10 \times 10$ linear hexahedral mesh for 3D linear elasticity gives a sparse matrix of 3993×3993 with 242435 non-zeros, which is about 1.5% of the size of a dense matrix with the same dimensions. To solve such a system efficiently, we have to avoid operating on zeros as much as we can. However how efficiently we can do so largely depends on the sparsity of the matrix: *the position of the nonzero entries.*

Given a finite element model of a physical problem, the values of non-zeros of $(\mathbf{M} + \frac{1}{2}\Delta t_n \mathbf{D})$ are determined by the underlying model, while the positions of those non-zeros are determined by the indices of the variables. For convenience, let us denote matrix $(\mathbf{M} + \frac{1}{2}\Delta t_n \mathbf{D})$ by $\mathcal{A}$. The entry (i, j) of $\mathcal{A}$ is nonzero if and only if the variable x_i and x_j are related. Given such a sparse matrix $\mathcal{A} = \mathcal{L}\mathcal{U}$, the LU-factorization takes $O(\sum_j d_j)$ space and $O(\sum_j d_j^2)$ time, where d_j is

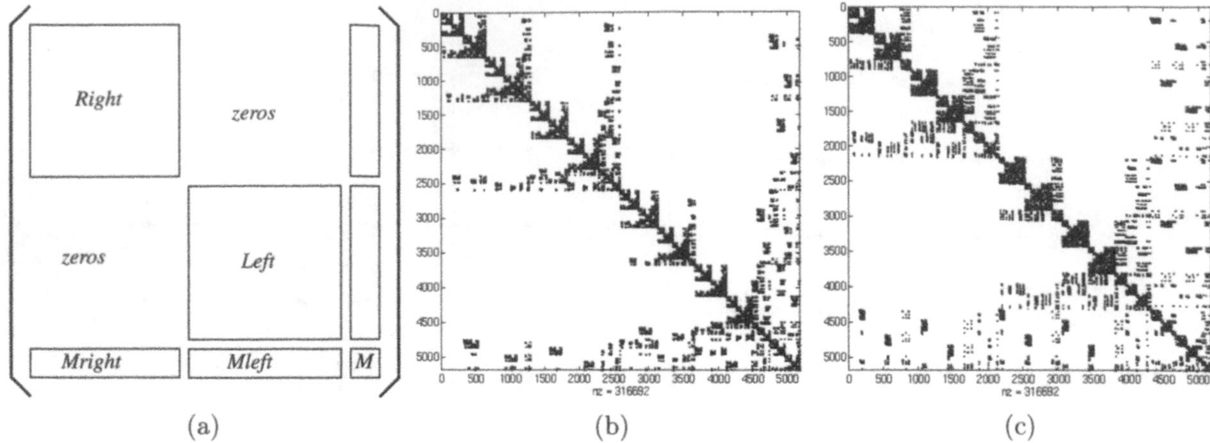

Figure 4: *(a) The block structure of sparse matrix $\mathcal{A}$ after the first dissection. (b) The sparse matrix structure generated by nested dissection. (c) The sparse matrix ordering using non-gird nested dissection.*

the number of non-zeros in each column vector of $\mathcal{L}$ [8]. If we assume that no exact numerical cancellations can occur, $\mathcal{L}$ will have non-zeros below the diagonal everywhere that $\mathcal{A}$ does. We define *fills* to be the below-diagonal entries in which $\mathcal{L}$ is nonzero and the corresponding entry of $\mathcal{A}$ is zero.[2]

Different ordering of the row and column vectors of the matrix $\mathcal{A}$ has no effect on the underlining physical problem that we are solving. However it dramatically change the number of fills. In order to reduce the space and running time for LU-factorization and the time of the corresponding back-substitution, we would like to minimize the number of fills. Unfortunately finding the order that gives the smallest fills is an NP-complete problem [26].

For sparse matrix that arises from *regular* finite element mesh, George[6] proposed a heuristic called *nested dissection* for ordering the variables of the system such that it gives a small number of fills.

Unfortunately a FEM mesh is often unstructured. In this section, we propose a modified nested dissection that works on any *unstructured* finite element mesh.

5.1 Modified Nested Dissection

For the simplicity of the presentation, let us consider a 2-dimensional finite element mesh, where each node has one degree of freedom.[3] A mesh of n nodes leads to a sparse matrix $\mathcal{A}$ of size $n \times n$. An entry (i, j) is nonzero if and only if the node i and j are in the same element.

For the mesh generated on regular grid (3(a)) The regular nested dissection [6] algorithm recursively divide the unordered nodes into 3 groups: *left, middle* and *right.* The group *middle* is just a set of nodes that completely separate *left* and *right.* This algorithm orders the nodes such that its sparse matrix has fractal sparsity as shown in Figure 4(b). After the first step of recursion, we immediately get 2 blocks of zeros as shown in Figure 4(a), because *left* and *right* are *not* directly related.

Unfortunately regular nested dissection algorithm requires a finite element mesh defined on regular grid. For an unstructured mesh, it would be difficult to find *middle* to dissect the mesh. In computer graphics, most meshes are unstructured. This motivated us to extend regular nested dissection to unstructured finite element mesh.

Instead of separating the set of unordered nodes using a layer of nodes, we "cut" the mesh by a axis-aligned plane. It is easy to compute the set of elements cut by this plane. We let *middle* be the set of unordered nodes in the elements cut by this plane and continue recursively as the regular nested dissection.

[2] $\mathcal{A}$ is symmetric.

[3] This can be easily generalized to multiple degrees of freedom and 3-dimensional meshes.

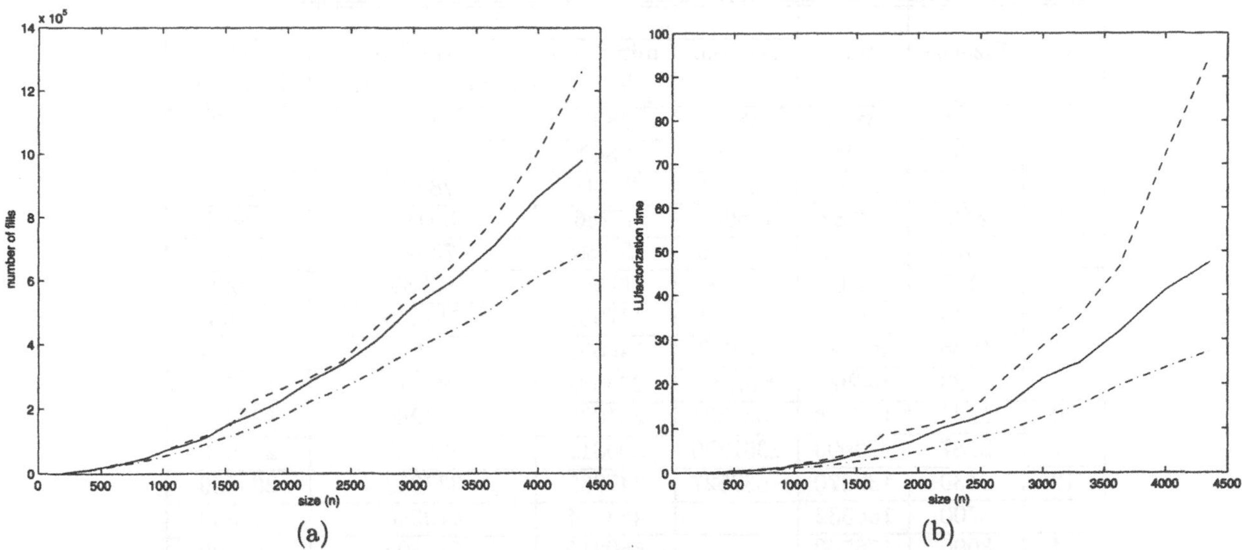

Figure 5: *The result for minimum degree ordering is plotted in dashed line, non-grid nested dissection in solid line, and the standard nested dissection in dash-dot line. (a) Number of fills. (b) Running time.*

This modified nested dissection can be applied to any unstructured finite element meshes, including tetrahedral meshes. Figure 3(b) shows one step of such a dissection. It also leads to the block structure as shown in Figure 4(a), except that the size of the M-block is bigger. At the end, we still get an ordering that gives a sparse matrix with fractal sparsity (Figure 4(c)). Due to the larger size of matrix M, the two "wings" ($M - left$ and $M - right$) are wider at each level.

The pseudo code of the modified nested dissection is as following:

```
Modified-nested-dissection(ℰ, top, perm)
  if ℰ.length = 0 then
     return.
  else if ℰ.length = 1 then
     for each node j in ℰ[0] do
        if perm[j] != −1 do
           perm[j] = top
           top − −
        endif
  else
     Find-dissector(ℰ).
     middle = all the elements cut by the dissector.
     left = elements to the "left" of the dissector.
     right = elements to the "right" of the dissector.
     Modified-nested-dissection(middle, top, perm).
     Modified-nested-dissection(left, top, perm).
     Modified-nested-dissection(right, top, perm).
  endif
```

Before calling this function the first time, we initialize each entry of the permutation array *perm* to -1 (order unassigned), and we compute the centroid of each element. The function *Find-dissector* simply computes the 3 medians along x, y, and z direction and compare the number of elements cut by the axis-aligned planes thru the medians and return as the dissector the plane with the minimum cut.

5.2 Running Time

It takes $O(n)$ time to find the median given a list of n numbers [4]. The depth of the recursion is apparently $O(\log n)$. Thus the total running time is $O(n \log n)$.

5.3 Numerical Experiments

In order to measure the performance of the modified nested dissection algorithm, we compare its fills and LU-factorization time with that of regular nested dissection and that of *minimum-degree* algorithm [12, 7]. All the matrices are derived from a 3-dimension finite element mesh of different size. The comparison of number of fills is listed in Table 1 and that of the LU-factorization time is listed in Table 2.

Minimum-degree ordering is an alternative ordering proposed to handle general matrix. It is an greedy algorithm that does the ordering directly on the connec-

Test	Size(n)	nnz	Fills			
			random	minimum degree	modified nested-dissection	nested dissection
1	192	6348	3636	1836	2034	1980
2	375	15285	25722	9342	10269	8586
3	648	30060	89199	31041	28872	26145
4	882	42384	196641	47286	50481	42462
5	1029	52131	254718	74880	73845	57537
6	1176	60360	332766	96381	88785	71685
7	1344	69888	487701	117900	112365	91971
8	1536	82956	657612	149652	153936	116514
9	1728	94266	813186	225801	182817	141741
10	1944	107118	1036521	261072	224280	175743
11	2187	123993	1361250	300312	286929	224973
12	2430	138870	1687527	347985	337383	264609
13	2700	155532		452574	410346	317898
14	3000	176700		550215	519804	385200
15	3300	195630		644067	595296	442053
16	3630	216588		787410	706131	514395
17	3993	242535		1100817	861705	610515
18	4356	266004		1262097	979875	684378

Table 1: *Number of fills for different ordering.*

tivity graph defined by the matrix. Our numerical experiments show that our modified nested dissection algorithm has an apparent advantage over the minimum-degree ordering, in both space and running time. Our modified nested dissection algorithm produces an order that has less fills, than the minimum-degree ordering, in 17 of 18 tests, while it has a better LU-factorization time in all 18 test. The result is plotted in Figure 5. This suggests that the geometry of the mesh gives better heuristic than its connectivity graph.

Also it is worth noting that the modified nested dissection ordering requires significantly less time for LU-factorization while only having slightly less fills than the minimum-degree ordering. This shows that the modified nested dissection leads to better sparsity: Non-zeros are more optimally positioned in the matrix.

6 Collision Handling

For collisions involving deformable objects, the collision time can be assumed finite (unlike the instantaneous collision of rigid bodies). This allows a larger time step for numerical integration.

The popular penalty methods [21, 20, 22] model the collision by adding an artificial spring of large stiffness at the point of collision. This stiff spring requires tiny integration time steps to stably simulate a collision. Various experiments show that the ratio between a collision free integration time step and that of a penalty collision is on the order of hundreds if not more.

This tempts us to develop new collision-handling methods that avoid adding extra artificial stiffness into the system. We will illustrate our collision-handling method, using a special case: collision between a rigid body and a single node of the FEM mesh of the deformable body (Figure 6). Later in this section, we will show that it is straightforward to extend this method to handle general collisions of deformable objects.

Consider the collision between a moving deformable body and a moving rigid body (Figure 6). To simplify the discussion, we use the moving frame attached to the moving rigid body instead of the fixed world frame. Namely all quantities are relative to the moving rigid body. Assume that at time t_n, the node p on the deformable object, with *relative* velocity $\hat{\mathbf{v}}(p)_n$, is colliding with the rigid surface of outward normal $\hat{\mathbf{n}}$.

Test	Size(n)	nnz	LU factorization time (seconds)			
			random	minimum degree	modified nested-dissection	nested dissection
1	192	6348	0.07	0.05	0.04	0.04
2	375	15285	0.55	0.18	0.17	0.17
3	648	30060	3.38	0.70	0.58	0.53
4	882	42384	10.66	1.07	1.11	0.87
5	1029	52131	15.31	1.98	1.77	1.23
6	1176	60360	22.75	2.67	2.15	1.54
7	1344	69888	40.08	3.54	2.93	2.02
8	1536	82956	61.19	4.65	4.44	2.80
9	1728	94266	86.39	8.86	5.39	3.56
10	1944	107118	123.27	9.88	7.00	4.52
11	2187	123993	192.59	11.31	9.99	6.05
12	2430	138870	274.96	14.15	11.87	7.45
13	2700	155532		21.49	14.91	9.39
14	3000	176700		28.49	21.20	12.51
15	3300	195630		35.26	24.78	15.20
16	3630	216588		46.60	31.96	19.76
17	3993	242535		87.84	41.00	23.56
18	4356	266004		94.44	47.41	27.31

Table 2: *Time measured on HP9000/715.*

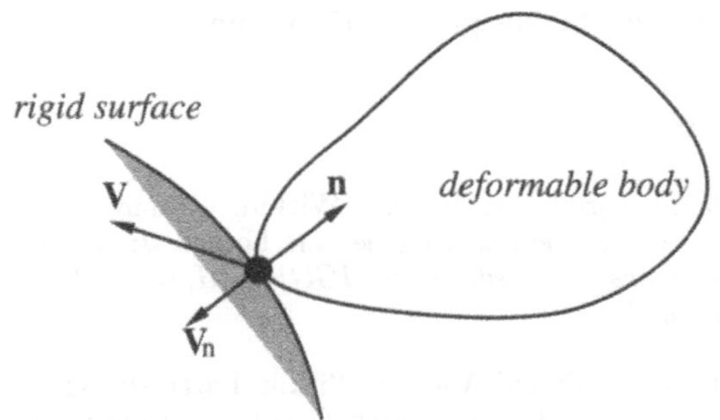

Figure 6: *A flexible body collides with a rigid body.*

The non-penetration constraint requires that the normal component of the relative velocity of point p drops to zero at the moment of collision in the moving frame. Unlike a rigid body collision, the flexible body will maintain contact with the rigid body for a nonzero period of time. We enforce the non-penetration constraint at node p by setting the normal component of $\hat{\mathbf{v}}(p)_{n+1}$ to zero as following:

$$\hat{\mathbf{v}}(p)_{n+1} = \hat{\mathbf{v}}(p)_n + (\hat{\mathbf{v}}(p)_n \cdot \hat{\mathbf{n}})\hat{\mathbf{n}} \qquad (10)$$

By Equation (9), we get:

$$\hat{\mathbf{a}}(p)_{n+1} = \frac{2\hat{\mathbf{v}}(p)_{n+1}}{\Delta t_n} - \frac{2\hat{\mathbf{v}}(p)_n}{\Delta t_n} - \hat{\mathbf{a}}(p)_n \qquad (11)$$

If we choose $\Delta t_{n+1} = \Delta t_n$, by Equation (7), we have:

$$\hat{\mathbf{u}}_{n+2} \cdot \hat{\mathbf{n}} = \hat{\mathbf{u}}_n \cdot \hat{\mathbf{n}} \qquad (12)$$

This shows that the non-penetration constraint is enforced after two time steps, because there is no relative motion of the deformable body normal to the surface of the rigid body.

This collision-handling integration scheme can be considered a special case of impulse [1]. For rigid body collisions, an impulse requires extremely small time steps for numerical integration because the rigid body collision is considered to occur instantaneously. However, for deformable body collisions, the collision time is finite. By delaying the non-penetration constraint by two time steps, we are able to integrate the impulse using large time steps.

This collision integration scheme can be generalized to collisions between deformable bodies and collisions that involve multiple point contacts. Multiple point collisions are modeled as a set of simultaneous single point collisions.

Unlike a general impulse [1, 13, 14], we do not have to distinguish the case that the colliding deformable objects quickly bounce away from each other and that one sticks to or slides on the surface of the other. The bouncing collision, the sticking and sliding contacts, are handled by exactly the same collision integration constraint. This collision constrain adds little extra cost to the dynamic simulation.

7 Conclusions and Future Work

We presented a simulation system that simulates global 3D deformations in real-time. Due to the distortion associated with linear strain, we simulate the global deformation using geometrically nonlinear finite element methods. The nonlinear FEM formulation is derived from the application of the nonlinear exact strain.

It is in general too expensive to solve such a nonlinear FEM system in real time. In order to achieve real-time performance, we pre-compute the LU-factorization of a small number of large sparse matrices. Such preprocessing is possible because we restrict the time steps to a small set of values. Our experiments show that usually we only need no more than 3 different values for time steps.

To reduce the time and space for LU-factorization and the time of back-substitution, we apply nested dissection to reorder the vertices in the finite element mesh. Such a reordering does not change the physical model that we are simulating. But it dramatically reduce the number of nonzeros in the LU-factorization.

We modified the regular nested dissection algorithm so that it works on unstructured finite element mesh. The modified nested dissection ordering takes 30% to 50% more time for the LU-factorization than the regular nested dissection, however it is more general than the regular nested dissection: *it is able to handle any unstructured finite element mesh*.

Our current implementation simply uses a cutting plane and separate the mesh using the elements intersected by the cutting plane. It seems that there is a better algorithm using sweeping-line (plane). By using a sweeping-line (plane) approach, we may be able to find a "thin" layer of nodes (rather than a layer of elements) to separate the mesh. This way, we would be able to generalize the regular nested dissection algorithm to unstructured mesh without giving away any performance. We are currently studying this approach.

We also introduced an efficient collision constraint. This constraint enables us to simulate the collisions with little extra computation, compared to a collision free simulation step. Our experiments show that this collision constraint handles collision much more efficient than penalty method.

Currently our system is able to produce real-time graphics for a mesh of several hundred vertices. We are interfacing our system to haptic devices, such as a Phantom, so that users can interact with deformable objects in real-time.

Acknowledgments

Supported by a Multi-Disciplinary Research Initiative grant for 3D Visualization, sponsored by BMDO with support from ONR. We thank Panayiotis Papadopoulos for sharing with us his FEM expertise and Jonathan Shewchuk for his insight in 3D meshing.

References

[1] David Baraff and Andrew Witkin. Dynamic simulation of non-penetrating flexible bodies. In *Computer Graphics: Proceedings of SIGGRAPH*, pages 303–308. ACM, 1992.

[2] David Baraff and Andrew Witkin. Large steps in cloth simulation. In *Computer Graphics: Proceedings of SIGGRAPH*, pages 303–308. ACM, 1998.

[3] David Chen. *Pump It Up: Computer Animation of a Biomechanically Basded Model of Muscle Using the Finite Element Method.* PhD thesis, MIT, 1992.

[4] Thomas H. Cormen, Charles E. Leiserson, and Ronald L. Rivest. *Introduction to Algorithms*. The MIT Press, 10 edition, 1993.

[5] Stéphane Cotin, Hervé Delingette, and Nicholas Ayache. Real-time elastic deformations of soft tissues for surgery simulation. *IEEE Transcation on Visualization and Computer Graphics*, 5(1):62–73, January-March 1999.

[6] Alan George. Nested dissection of a regular finite element mesh. *SIAM Jounal of Numerical Analysis*, 10(2), 1973.

[7] Alan George and Joseph W. H. Liu. A fast implementation of the minimum degree algorithm using quotient graphs. *ACM Transaction on Mathematical Software*, 6(3), September 1980.

[8] Alan George and Joseph W. H. Liu. *Computer Solution of Large Sparse Positive Definite Systems*. Prentice-Hall, Inc., 1981.

[9] Sarah F. Gibson and Brian Mirtich. A servey of deformable models in computer graphics. Technical Report TR-97-19, Mitsubishi Electric Research Laboratories, Cambridge, MA, November 1997.

[10] Doug L. James and Dinesh K. Pai. Artdefo: Accurate real time deformable objects. *Computer Graphics: Proceedings of Siggraph*, pages 65–72, August 1999.

[11] E. Keeve, S. Girod, P. Pfeifle, and B. Girod. Anatomy-based facial tissue modeling using the finite element method. *IEEE Visualization*, 1996.

[12] Joseph W. H. Liu. Modification of minimum-degree algorithm by multiple elemination. *ACM Transaction on Mathematical Software*, 11(2), June 1985.

[13] Brian Mirtich and John Canny. Impulse-based dynamic simulation. In K. Goldberg, D. Halperin, J.C. Latombe, and R. Wilson, editors, *The Algorithm Foundations of Robotics*. A. K. Peters, Boston, MA, 1995. Proceedings from the workshop held in February, 1994.

[14] Brian Mirtich and John Canny. Impulse-based simulation of rigid bodies. In *Symposium on Interactive 3D Graphics*, New York, 1995. ACM Press.

[15] G. Celniker nad G. Gossard. Deformable curve and surface finite elements for free form shage design. *Computer Graphics*, 25(4), 1991.

[16] S. Peiper, J Rosen, and D. Zeltzer. Interactive graphics for plastic surgery: A task-level analysis and implementation. In *Symposium on Interactive 3D Graphics*, 1992.

[17] E. Promayon, P. Baconnier, and C. Puech. Physically-based deformations constrained in displacements and volume. In *EUROGRAPHICS*, 1996.

[18] X. Provot. Deformation constrains in a mass-spring model to describe rigid cloth behavior. *Computer Interface*, 1995.

[19] J. N. Reddy. *An Introduction to the Finite Element Method*. McGraw-Hill, Inc., 2nd edition, 1993.

[20] D. Terzopoulos and K. Fleischer. Modeling inelastic deformation: Viscoelasticity, plasticity, fracture. *Computer Graphics*, 22, August 1988.

[21] D. Terzopoulos, J. Platt, A. Barr, and K. Fleischer. Elastically deformable models. *Computer Graphics*, 21, July 1987.

[22] D. Terzopoulos and K. Waters. Physically-based facial modeling, analysis and animation. *Journal of Visualization and Computer Animation*, 1990.

[23] K. Waters. A muscle model for animating three-dimensional facial expression. *Computer Graphics*, 21(4), July 1987.

[24] H. M. Westergaard. *Theory of Elasticity and Plasticity*. Dover Publications, Inc., 1964.

[25] A. Witkin and et al. . An introduction to physically based modeling. Course Notes, 1993.

[26] M Yannakakis. Computing the minimum fill-in is np-complete. *SIAM Journal of Algrbraic Discrete Methods*, 2, 1981.

[27] Yan Zhuang and John Canny. Real-time simulation of physically realistic global deformations. *IEEE Visualization: Late Breaking Hot Topics*, October 1999.

[28] O. C. Zienkiewicz and R. L. Taylor. *The Finite Element Method: Basic Formulation and Linear Problems*, volume 1. McGraw-Hill Book Company, 4th edition, 1989. linear finite element method, linear elasticity.

[29] O. C. Zienkiewicz and R. L. Taylor. *The Finite Element Method: Solid and Fluid Mechanics Dynamics and Non-Linearity*, volume 2. McGraw-Hill Book Company, 4th edition, 1989.

Motion Planning for Kinematic Stratified Systems with application to Quasi–Static Legged Locomotion and Finger Gaiting

Bill Goodwine, University of Notre Dame, Notre Dame, IN
Joel W. Burdick, California Institute of Technology, Pasadena, CA

We present a general motion planning algorithm for robotic systems with a "stratified" configuration space. Such systems include quasi-static legged robots and kinematic models of object manipulation by finger repositioning. Our method extends a nonlinear motion planning algorithm for smooth systems to the stratified case, where the relevant dynamics are not smooth. The method does not depend upon the number of legs or fingers, nor is it based on foot placement or finger placement concepts. Examples demonstrate the method.

1 Introduction

This paper considers the motion planning problem for systems whose governing equations of motion impose a "stratified" structure on the system's configuration space (see Section 3). Stratification naturally arises in the context of legged locomotion and object manipulation via finger repositioning. These operations are characterized in part by the system making and breaking contact with its environment. The configuration spaces (or c-space) of these systems are "stratified" into subsets that correspond to different contact states. The governing dynamical equations depend upon the contact state, and the dynamical equations are discontinuous during the making and breaking of contact.

The goal of our motion planning scheme is to determine the control inputs (*e.g.*, mechanism joint variable trajectories) which will steer the walking robot from a starting to a desired final configuration, or to manipulate the grasped object from an initial to a final configuration via a combination of finger repositioning and finger motions. The planner must simultaneously plan the mechanism's motion during a single contact state, as well as determine when to change contact states. This paper presents a general motion planning methodology for this class of systems, which includes all quasi-static legged locomotors and many kinematic models of multi-fingered hand manipulation. The method is independent of the number of legs (or fingers) and many other aspects of a robot's morphology. In the legged locomotion context, it is distinct from previous planning methods in that it is not based on foot placement concepts, and therefore the computationally burdensome calculation of foot placement can be avoided. Instead, our approach focuses on control inputs.

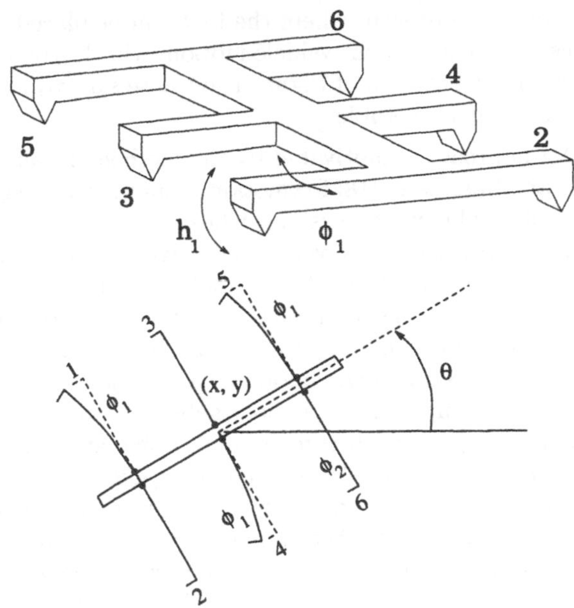

Figure 1: *(a) Schematic of simple hexapod robot; (b) Definition of kinematic variables.*

As a concrete example of when such a planner is needed, consider the hexapod in Figure 1 (this model will be explored in Section 5). Each leg has only two degrees of freedom—the robot can only lift its legs up and down and move them forward and backward. Conventional hexapods are designed with three independent degrees of freedom per leg. The limited control authority in this design may be desirable in practical situations because it decreases the robot's mechanical complexity. This leg geometry can also probably be implemented at very small size scales using MEMS technology. However, such decreased kinematic complexity comes at the cost of requiring a more sophisticated control and motion planning theory. Note that for this robot, it is not immediately clear if it can move "sideways."

The issue of this mechanism's ability to move sideways is the *controllability problem.* In Refs. [9, 8] we present controllability tests for stratified systems. We assume in this paper that a given system is controllable in the stratified sense. Otherwise, it is not possible to track an arbitrary trajectory. Given the assumption of controllability, this paper addresses how to plan the robot's leg (or finger) movements so that it can approximately follow a given trajectory. A conventional "foot-placement" approach, where the foot can be placed as necessary to implement vehicle motion, will clearly not work for the hexapod of Figure 1, because sideways leg placement is impossible.

Our approach is motivated by the method of Lafferriere and Sussmann [18] for motion planning for a class of nonlinear kinematic systems whose equations of motion are smooth. However, since legged robots (and grasping hands) intermittently make and break contact, their equations of motion are not smooth. Hence, the method of Ref. [18] cannot be directly applied. Section 3 introduces the notion of a *stratified* c-space, which is decomposed into various subspaces (or strata) depending upon which combination of feet are in contact with the ground. We extend the approach of Ref. [18] by using the stratified c-space structure in a novel way. *It is likely that other methods for steering smooth systems (such as Ref. [22]) can be similarly extended by adopting our framework.* A main contribution of this work is the introduction of a geometric framework that supports the extension of prior nonholonomic motion planning techniques to this class of systems.

Our approach is general and thus works independently of the number of legs (fingers). It may be true that for a given quasi-static legged robot, one could develop a specific motion planner that would perform as well, or possibly better, than the technique described in this paper. The key advantage of this approach is its generality. It is particularly well suited to the task of quickly designing a planner during the preliminary stages of legged robot system design. While the techniques outlined in this paper are equally applicable for locomotion and hand manipulation, the bulk of the paper will focus on locomotion, with the application to hand manipulation briefly sketched at the end of the paper. An interesting observation, which is not explored in this paper, is that our technique is equally applicable to both locomotion and manipulation.

There is a *vast* literature on the analysis and control of legged robotic locomotion. Prior efforts have typically focused either on a particular morphology (*e.g.* biped [17], quadruped [19, 2], or hexaped [32]) or a particular locomotion assumption (*e.g.* quasi-static [32] or hopping [25]). Less effort has been devoted to uncovering principles that span all morphologies and assumptions. Some general results do exist. For example, the bifurcation analysis in Ref. [5], many optimal control results such as those in Ref. [3] and the fundamental conservation of momentum and energy results that underlie Raibert's hopping results [25] have general applicability. However, none of these methods directly use the inherent geometry of stratified configuration spaces to formulate results which span morphologies and assumptions. Our work makes a novel connection with recent advances in nonlinear geometric control theory. We believe that this connection is a useful and necessary step towards establishing a solid basis for locomotion engineering.

In contrast to robotic legged locomotion, many results in robotic grasping and manipulation are formulated in a manner that is independent of the morphology of the gripper [24]. Vast efforts have been directed toward the *analysis* of grasp stability and force closure [26, 27, 31], motion planning assuming continuous contact [20, 33, 13] and haptic interfaces and other sensing [6, 29, 28]. *Finger gaiting,* where fingers make and break contact with the object has been less extensively considered. Finger gaiting has been implemented in certain instances [23, 15, 7] and also partially considered theoretically [14, 4, 10]. Perhaps the approach which most closely mirrors that of the subject of this proposal is in [24] where notions of controllability and observability from "standard" control theory are applied to grasping (however, these results are limited to the linear case and do not allow for fingers to intermittently contact the object).

Section 2 briefly reviews standard ideas, and summarizes the motion planning method of Ref. [18]. Section

3 introduces our notion of a stratified c-space. Section 4 presents our algorithm in the context of quasi-static legged locomotion, while Section 5 applies this algorithm to the system of Figure 1. Section 6 sketches the application of these ideas to multi-fingered hand manipulation, and presents an example.

2 Background

We assume the reader is familiar with the basic notation and formalism of differential geometry and nonlinear control theory, as in Ref. [16]. The following definitions and classical theorems are reviewed so that the starting point of our development will be clear.

The equations of motion for smooth kinematic nonholonomic systems take the form of a driftless nonlinear affine control system evolving on a configuration manifold, M:

$$\dot{x} = g_1(x)u_1 + \cdots + g_m(x)u_m \quad x \in M. \tag{1}$$

Since we restrict our analysis to quasi-static locomotion and kinematic models of multi-fingered manipulation, the governing equations of motion will piecewise take the form of Equation (1) on each strata. Recall that the Lie bracket between two control vector fields, $g_1(x)$ and $g_2(x)$, is computed as:

$$[g_1(x), g_2(x)] = \frac{\partial g_2(x)}{\partial x} g_1(x) - \frac{\partial g_1(x)}{\partial x} g_2(x)$$

and can be interpreted as the leading order term that results from the sequence of flows

$$\phi_\epsilon^{-g_2} \circ \phi_\epsilon^{-g_1} \circ \phi_\epsilon^{g_2} \circ \phi_\epsilon^{g_1}(x) = \epsilon^2[g_1, g_2](x) + \mathcal{O}(\epsilon^3), \tag{2}$$

where $\phi_\epsilon^{g_1}(x_0)$ represents the solution of the differential equation $\dot{x} = g_1(x)$ at time ϵ starting from x_0.

Campbell-Hausdorff Formula. The flow along the vector field g_i can be considered by its *formal exponential* of g_i, denoted by:

$$\phi_t^{g_i}(x) := e^{tg_i}(x) = (I + tg_i + \frac{t^2}{2} g_i^2 + \cdots) \tag{3}$$

where terms of the form g_i^k are partial differential operators. In order to use Equation (3), composition must be from left to right, as opposed to right to left for flows, *e.g.*, $\phi_{t_2}^{g_2} \circ \phi_{t_1}^{g_1} = e^{g_1 t_1} e^{g_2 t_2}$, where both sides of this equation mean "flow along g_1 for time t_1 and then flow along g_2 for time t_2." The relationship between the flow along vector fields sequentially is given by the Campbell–Baker–Hausdorff formula [34].

Theorem 1 *Given two smooth vector fields g_1, g_2 the composition of their exponentials is given by:*

$$e^{g_1} e^{g_2} = e^{g_1 + g_2 + \frac{1}{2}[g_1, g_2] + \frac{1}{12}([g_1,[g_1,g_2]] - [g_2,[g_1,g_2]]) \cdots} \tag{4}$$

where the remaining terms may be found by equating terms in the (non-commutative) formal power series on the right- and left-hand sides.

2.1 Trajectory Generation for Smooth Systems

This section reviews the motion planning method of Ref. [18] for smooth kinematic systems described by a single equation having the form of Equation (1).

A nonholonomic control system often does not have enough controls to directly drive the system along a given trajectory, *i.e.*, the number m in Equation (1) is less than the c-space dimension. In the method of Ref. [18], this deficit is managed by using an "extended system," where "fictitious controls," corresponding to higher order Lie bracket motions, are added. If enough Lie brackets are added to the system to span all possible motion directions (which is possible if the system is locally controllable), then the motion planning problem becomes trivial for the extended system.

The *extended system* is constructed by adding Lie bracket directions to the original system from Equation (1),

$$\dot{x} = b_1 v^1 + \cdots b_m v^m + b_{m+1} v^{m+1} + \cdots + b_s v^s \tag{5}$$

where $b_i = g_i$ for $i = 1, \ldots, m$, and the $b_{m+1}, \ldots, b_s$ correspond to higher order Lie brackets of the g_i, chosen so that $\dim(\text{span}\{b_1, \ldots, b_s\}) = \dim(T_x M)$. The v^i's are called *fictitious inputs* since they may not correspond with any actual system inputs. The higher order Lie brackets must belong to the Philip Hall basis [30, 21] for the Lie algebra. The control inputs v^i which steer the extended system can be found as follows. To go from a point p to a point q, define a curve $\gamma(t)$ connecting p and q (a straight line would work, but is not necessary). After determining $\gamma(t)$, simply solve:

$$\dot{\gamma}(t) = g_1(\gamma(t))v^1 + \cdots + g_s(\gamma(t))v^s \tag{6}$$

for the fictitious controls v_i. This will involve inverting a square matrix or determining a pseudo-inverse, depending on whether or not there are more b_i's than the dimension of the configuration space.

To find the actual controls, first determine the Philip Hall basis for the Lie algebra generated by $g_1, \ldots, g_m$, and denote by $B_1, B_2, \ldots, B_s$ a collection of basis elements such that when they are evaluated as vector fields, they form a basis space of vector fields. All flows of Equation (1) can be represented in the form:

$$S(t) = e^{h_s(t)B_s} e^{h_{s-1}(t)B_{s-1}} \cdots e^{h_2(t)B_2} e^{h_1(t)B_1} \quad (7)$$

for some functions $h_1, h_2, \ldots, h_s$, called the (backward) Philip Hall coordinates. Furthermore, $S(t)$ satisfies the formal differential equation:

$$\dot{S}(t) = S(t)(B_1 v_1 + \cdots + B_s v_s); \quad S(0) = 1. \quad (8)$$

If we define the *adjoint mapping*:

$$\mathrm{Ad}_{e^{-h_i B_i}} B_j = e^{-h_i B_i} B_j e^{h_i B_i},$$

then it is straight–forward to show that:

$$\mathrm{Ad}_{e^{-h_i B_i} \cdots e^{-h_{j-1} B_{j-1}}} B_j \dot{h}_j = \left(\sum_{k=1}^{s} p_{j,k}(h) B_k \right) \dot{h}_j, \quad (9)$$

for some polynomials $p_{j,k}(h)$. (For a complete derivation, see Ref. [21]). Equating coefficients of Equation (8) with the derivative of Equation (7), and using Equation (9), yields differential equations having the form

$$\dot{h} = A(h)v \qquad h(0) = 0. \quad (10)$$

These equations specify the evolution of the backward Philip Hall coordinates in response to the fictitious inputs, which were found via Equation (6).

Next one must determine the actual inputs from the Philip Hall coordinates. It is easier to determine the real inputs using the forward rather than backward Philip Hall coordinates. The transformation from the backward to forward coordinates is an algebraic one (see [18]). For systems which are nilpotent[1] of order two, or which can be well approximated as nilpotent of order two, the transformation between forward and backward Philip Hall coordinates can be avoided. In these cases, the actual controls can be obtained from the fictitious controls by use of Lie-bracket-like motions where necessary. In practice, this will often be the case, since physical systems that require Lie bracket motions of order greater than two are inconvenient to control since many motions are needed to effect even a small motion in a higher-order Lie bracket direction. For this reason, and for purposes of the clarity of presentation, we limit our attention to second order brackets. However, there is no theoretical limitation on the order of brackets.

If the system is nilpotent, this method exactly steers the system to the desired final state. Else, the system is steered to a point that is, at worst, half the distance to the desired state [18]. The algorithm can be iterated to generate arbitrary precision. This iterated method also includes the notion of a "critical" step length. Ref. [18] estimates the critical step length bound, and shows via simulations that the actual critical length is typically larger than the estimated bound.

3 Stratified Configuration Spaces

The method reviewed in Section 2.1 can not be directly used for legged or multi-fingered robots because their governing equations of motion are not smooth. To adapt this method (and similar nonholonomic motion planning methods) to these systems, we use the notion of a *stratified* configuration space. While the stratified concept is equally applicable to locomotion and multi-fingered manipulation, the language of locomotion is used below for simplicity.

Let S_0 denote a robot's configuration manifold, which describes the robot's position and orientation as well as all of the mechanism's joint variables. The robot's possible configurations will be subjected to constraints if one or more of its feet (fingers) are in contact with the ground (object). The set of configurations corresponding to one contact is generically a codimension one submanifold of S_0. Let $S_i \subset S_0$ denote the codimension one submanifold of S_0 that corresponds to all configurations where only the ith foot contacts the terrain. That the $\{S_i\}$ are submanifolds can be demonstrated by noting that set of points corresponding to ground contact can be described by the preimage of a function describing the foot's height. We generally assume that S_i, is, at least locally, defined by a level set of a function $\Phi_i(x) : S_0 \to \mathbf{R}$.

When both the ith and j^{th} feet are on the ground, the corresponding set of states is a codimension 2 submanifold of S_0 that is formed by the intersection of the two single contact submanifolds. Denote the intersection of

[1] A system of the form Equation (1) is said to be *nilpotent of order k* if all the Lie brackets between control vector fields of order greater than k are 0.

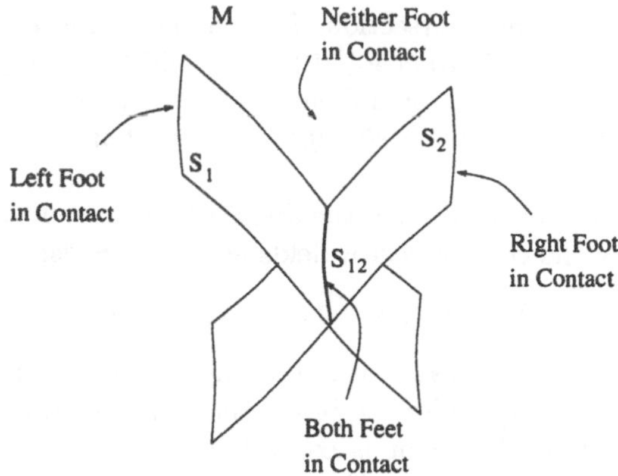

Figure 2: *Abstract depiction of the stratified structure of a biped robot c-space.*

S_i and S_j, by $S_{ij} = S_i \cap S_j$. The structure of the configuration manifold for a biped is abstractly illustrated in Figure 2. For systems with larger numbers of legs (fingers), further intersections, corresponding to more complicated contact states, can be similarly defined in a recursive fashion: $S_{ijk} = S_i \cap S_j \cap S_k = S_i \cap S_{jk}$, etc. We denote an arbitrary intersection set (or "stratum") by $S_I = S_{i_1 i_2 \cdots i_n}$, $I = \{i_1 i_2 \cdots i_n\}$. We assume that S_I is a regular submanifold of S_0. This is generically true for rigid body mechanisms. If the strata $S_{i_1}, S_{i_2}, \ldots, S_{i_k}$ are locally described by the functions $\Phi_{i_1}, \Phi_{i_2}, \ldots, \Phi_{i_k}$, then S_I will be a submanifold of S_0 if the functions $\Phi_{i_1}, \Phi_{i_2}, \ldots, \Phi_{i_k}$ are functionally independent. If the functions Φ_I correspond to foot heights, this functional independence will be satisfied.

We say that the robot c-space is *stratified* [2]. Classically, a *regularly stratified* set $\mathcal{X}$ is a set $\mathcal{X} \subset \mathbb{R}^m$ decomposed into a finite union of disjoint smooth manifolds, called *strata*, satisfying the Whitney condition. The dimension of the strata varies between zero, which are isolated point manifolds, and m, which are open subsets of $\mathbb{R}^m$. The Whitney condition requires that the tangent spaces of two neighboring strata "meet nicely," and for our purposes this condition is generically satisfied (see Ref. [12] for details).

In the classical definition of a stratification [12], stratum $\mathcal{X}_i$ consists of the submanifold S_i with all lower dimensional strata (that arise from intersections of S_i with other submanifolds) removed. By abuse of notation, we will refer to the submanifolds S_i, S_{ij}, S_{ijk}, etc, as strata. We will term the highest codimension stratum containing the point x as the *bottom stratum*, and any other submanifolds containing x as *higher strata*. When making comparisons among different strata, we will refer to higher codimension (*i.e.* lower dimensional) strata as *lower strata*, and lower codimension (*i.e.* higher dimensional) strata as *higher strata*.

Whenever an additional foot contacts the ground, the robot is subjected to additional constraints. For "point-like" feet, this may be a holonomic constraint; whereas some contacts are better characterized by nonholonomic constraints. Regardless of the constraint type, the system's equations of motion will change in a non-smooth manner. Otherwise, the system's equations of motion are smooth, though generally different in each strata. Hence, the discontinuities are localized to regions of transition between strata. Furthermore, we assume that on each stratum, S_i, our control system may be subjected to constraints *in addition to* those present on S_0. On any given stratum, the system is subjected to at least all the constraints present on all the higher strata whose intersection defines that stratum. Thus, when the system transitions from S_0 to S_i, if the system is going to evolve on the stratum S_i for some finite time, the system must not only satisfy all the constraints that are present on the stratum, but also the constraint $\mathbf{d}\Phi_i(x)\dot{x} = 0$.

The equations of motion at $x \in S_I$ are written as

$$\dot{x} = g_{I,1}(x)u^{I,1} + \cdots g_{I,n_I}(x)u^{I,n_I}, \qquad (11)$$

where n_I depends upon the codimension of S_I and the nature of the additional constraints imposed on the system in S_I. We assume that the vector fields in the equations of motion for any given stratum are well defined at all points in that stratum, including points contained in any substrata of that stratum. For example, the vector fields $g_{0,i}(x)$ are well defined for $x \in S_i$. Note, however, that they do *not* represent the system's equations of motion in the substrata, but, nonetheless, are still well defined as vector fields.

Figure 3 illustrates, via a graph-like structure, a four-level stratification, which corresponds to a four-legged walker. A node corresponds to a stratum, and the presence of an edge connecting nodes indicates that it is possible to move between the strata that are connected by the edge. The ability to move between two

[2] Note that the terms "stratification" and "strata" are also used in other contexts to describe the topology of orbit spaces of Lie group actions, and are a slight generalization of the notion of a foliation [1].

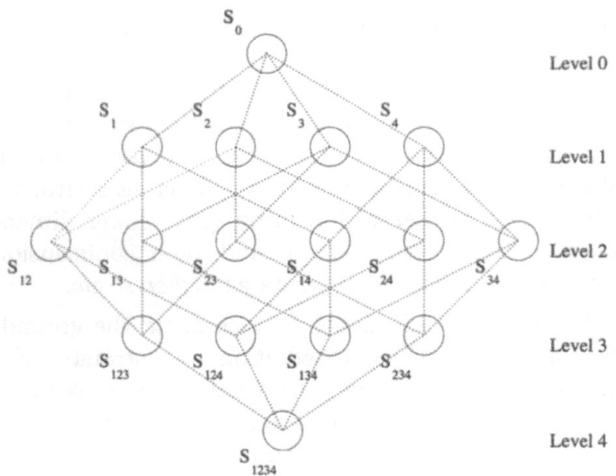

Figure 3: *Four Level Stratification.*

strata depends upon the mechanics of a given problem, and will generally be obvious from the characteristics of a given problem. Whether or not edges between nodes are permissible is considered in more detail in Ref. [8].

We specify a *gait* as an ordered sequence of strata:

$$\mathcal{G} = \{S_{I_1}, S_{I_2}, \ldots, S_{I_n}, S_{I_{n+1}} = S_{I_1}\}, \tag{12}$$

where n is the number of different contact states in the gait. In this ordered sequence, the first and last element are identical, indicating that the gait is a closed loop in the strata graph. For the gait to be meaningful, the system must be able to switch from stratum S_{I_i} to $S_{I_{i+1}}$ for each i. We further assume that the specified gait or gaits satisfy the gait controllability conditions of Ref. [8] so that arbitrary trajectories can be tracked.

In summary, we assume that the only discontinuities present in the equations of motion are due to transitions on and off of the strata. We also make a similar assumption regarding the control vector fields restricted to any stratum, *i.e.*, the control vector fields restricted to any stratum are smooth away from points contained in intersections with other strata. When a configuration manifold is consistent with the above description, we will refer to it as a *stratified configuration space*.

Definition 2 Let S_0 be a manifold, and n functions $\Phi_i : S_0 \mapsto \mathbb{R}$, $i = 1, \ldots, n$ be such that the level sets $S_i = \Phi_i^{-1}(0) \subset S_0$ are regular submanifolds of S_0, for each i, and the intersection of any number of level sets, $S_{i_1 i_2 \cdots i_m} = \Phi_{i_1}^{-1}(0) \cap \Phi_{i_2}^{-1}(0) \cap \cdots \cap \Phi_{i_m}^{-1}(0)$, $m \leq n$, is also a regular submanifold of S_0. Then S_0 and the functions Φ_n define a *stratified configuration space.*

For a given strata, S_I, the *distribution* defined by the span of the control vector fields active on S_I is:

$$\Delta_{S_I} = \text{span}\ \{g_{S_{I,1}}, \ldots, g_{S_{I,n_I}}\}.$$

The *involutive closure* of Δ_{S_I}, denoted by $\overline{\Delta}_{S_I}$, is the closure of Δ_{S_I} under Lie bracketing. A basic assumption is that the robot is controllable. The controllability of a given gait, Equation (12), can be determined as follows. Let $\mathcal{D}_1 = \overline{\Delta}_{I_1}$. If $S_{I_{i-1}} \subset S_{I_i}$, then $\mathcal{D}_i = \mathcal{D}_{i-1} + \overline{\Delta}_{I_i}$. Else, if $S_{I_i} \subset S_{I_{i-1}}$, then $\mathcal{D}_i = (\mathcal{D}_{i-1} \cap TS_{I_i}) + \overline{\Delta}_{I_i}$ In Ref. [8] it s shown that if $dim(\mathcal{D}_n) = dim(T_{x_0} S_B)$ the system is *gait controllable* from x_0 (i.e., the system can reach open nbhd of x_0 in the bottom strata). For a more rigorous discussion and summary of stratified system controllability, see Refs. [9, 8].

4 Legged Trajectory Generation

This section extends the procedure outlined in Section 2 to kinematic systems having a stratified c-space. We focus on quasi-static legged locomotion in this section. However, all of these ideas can be readily extended to the finger gaiting problem. Section 6 sketches the extension to hand manipulation.

Assume that the robot starts at a configuration p and seeks to reach a final configuration q. By a configuration, we mean the position and orientation of the body, as well as the states of the legs. We assume that both p and q lie in the same bottom stratum, denoted by S_B. This corresponds to the legged robot starting and stopping with the same set of feet in contact with the ground. Eliminating this requirement is a simple extension of the algorithm described below.

The switching behavior associated with stratified systems can not be accounted for in the methods of Section 2.1. However, the method of Section 2.1 can be extended to legged and fingered robotic systems via the notion of a *stratified extended system* on S_B.

4.1 The Stratified Extended System

On each strata, only one set of governing equations is in effect. Generally, the equations of motion in the

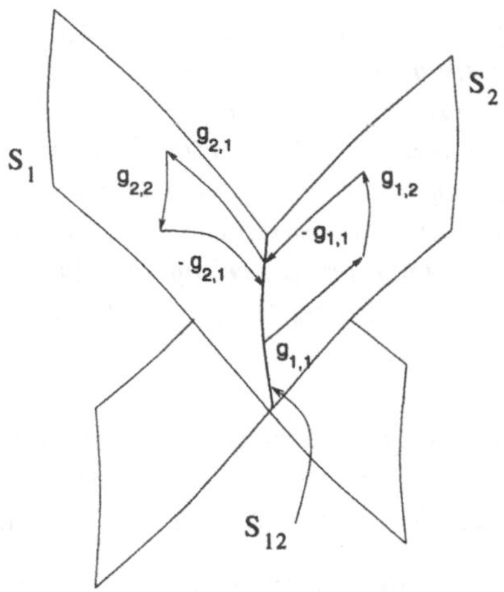

Figure 4: *Sequence of Flows.*

bottom strata will be different than those in higher strata. Furthermore, it will be typically true that the goal q can not be reached by remaining in S_B. Hence, some switching amongst the strata will be necessary. However, since the bottom strata is defined by the intersection of higher strata, the equations of motion in the higher strata are valid at points arbitrarily close to the bottom strata. As shown below, it is possible to consider the vector fields associated with each stratum in one common space. In this case, that common space will be the bottom stratum. This concept will be encapsulated below in the definition of a "stratified extended system." We first introduce some examples to show how we can consider vector fields defined on different strata in a common space. Additional examples that deal with more subtle issues can be found in Ref. [11].

Example 1 *Consider the conceptual biped configuration space as shown in Figure 2. Assume that on stratum S_{12}, the vector field $g_{1,1}$ moves the system off of S_{12} and onto S_1, and correspondingly, $g_{2,1}$ moves the system off of S_{12} onto S_2. Also, we consider the vector fields $g_{1,2}$ and $g_{2,2}$, defined on S_1 and S_2 respectively. Consider the following sequence of flows, starting from the point $x_0 \in S_{12}$*

$$
\begin{aligned}
x_f \;=\; & \underbrace{\phi^{t_6}_{-g_{2,1}}}_{S_{12}\leftarrow S_2} \circ \underbrace{\phi^{t_5}_{g_{2,2}}}_{on\ S_2} \circ \underbrace{\phi^{t_4}_{g_{2,1}}}_{S_2\leftarrow S_{12}} \circ \underbrace{\phi^{t_3}_{-g_{1,1}}}_{S_{12}\leftarrow S_1} \circ \\
& \underbrace{\phi^{t_2}_{g_{1,2}}}_{on\ S_1} \circ \underbrace{\phi^{t_1}_{g_{1,1}}}_{S_1\leftarrow S_{12}} (x_0).
\end{aligned}
\tag{13}
$$

The notation "$S_{12} \leftarrow S_1$" means that the flow takes the system from S_1 to S_{12} and "on S_1" means that the flow lies entirely in S_1. This sequence of flows is illustrated in Figure 4. In this sequence, the system first moved off of the bottom stratum into S_1, flowed along the vector field $g_{1,2}$, flowed back onto the bottom stratum, off of the bottom stratum onto S_2, along vector field $g_{2,2}$ and back to the bottom stratum.

Notice that from the Campbell-Baker-Hausdorff formula (Equation (4)), if the Lie bracket between two vector fields is zero, then their flows commute. Thus, if

$$[g_{1,1}, g_{1,2}] = 0 \quad and \quad [g_{2,1}, g_{2,2}] = 0, \tag{14}$$

we can reorder the sequence of flows in Equation (13) by interchanging the flow along $g_{1,1}$ and $g_{1,2}$ and the flows along $g_{2,1}$ and $g_{2,2}$ as follows

$$
\begin{aligned}
x_f &= \underbrace{\phi^{t_5}_{g_{2,2}} \circ \phi^{t_6}_{-g_{2,1}}}_{interchanged} \circ \phi^{t_4}_{g_{2,1}} \circ \underbrace{\phi^{t_2}_{g_{1,2}} \circ \phi^{t_3}_{-g_{1,1}}}_{interchanged} \circ \phi^{t_1}_{g_{1,1}}(x_0) \\
&= \underbrace{\phi^{t_4}_{g_{2,2}} \circ \phi^{t_2}_{g_{1,2}}(x_0)}_{on\ S_{12}},
\end{aligned}
\tag{15}
$$

if $t_1 = t_3$ and $t_4 = t_6$.

Note that $g_{1,2}$ and $g_{2,2}$ are vector fields in the equations of motion for strata S_1 and S_2, respectively, but not on stratum S_{12}. However, the sequence of flows in Equation (13) occurs on different strata, where the flows are governed by vectors fields associated with each stratum. This flow yields the same net result as the net flow in Equation (15), where the vector fields are evaluated on the bottom stratum, even though they are not part of the equations of motion there. Furthermore, we note that if the vector fields $g_{1,2}$ and $g_{2,2}$ are tangent to the substratum S_{12}, then the resulting flow given in Equation (15) will remain in S_{12}. In fact, it is implicitly required in the above argument that at least $g_{1,2}$ is tangent to S_{12}.

If the bottom stratum, S_B, is described by the level set of a function, Φ_B, and if a vector field, $g_{1,2}$ is not tangent to S_B, then, $\langle \mathbf{d}\Phi_B, g_{1,2}\rangle = f_1 \neq 0$. Also, since the vector field $g_{1,1}$ moves the foot out of contact, we similarly have $\langle \mathbf{d}\Phi_B, g_{1,1}\rangle = f_2 \neq 0$. Then, the vector field, $\tilde{g}_{1,2} = g_{1,2} - \frac{f_1}{f_2} g_{1,1}$, is tangent to S_B because

$$\langle \mathbf{d}\Phi_B, \tilde{g}_{1,2}\rangle = \langle \mathbf{d}\Phi_B, g_{1,2}\rangle - \frac{f_1}{f_2}\langle \mathbf{d}\Phi_B, g_{1,1}\rangle = 0. \tag{16}$$

Henceforth, we will just assume that the vector field on the higher stratum is tangent to the lower stratum, and note that if it is not tangent, we can modify it to be so in the above manner.

The above example shows how one can effectively determine the influence of a control that is defined in a higher stratum on the net evolution of the system in the lower stratum. The following example shows how motions that are analogous to Lie Bracket motions can be realized by controls on *different* higher strata.

Example 2 *Consider the sequence of flows*

$$x_f = \phi^{t_{12}}_{-g_{2,1}} \circ \phi^{t_{11}}_{-g_{2,2}} \circ \phi^{t_{10}}_{g_{2,1}} \circ \phi^{t_9}_{-g_{1,1}} \circ \phi^{t_8}_{-g_{1,2}} \circ \phi^{t_7}_{g_{1,1}}$$
$$\circ \phi^{t_6}_{-g_{2,1}} \circ \phi^{t_5}_{g_{2,2}} \circ \phi^{t_4}_{g_{2,1}} \circ \phi^{t_3}_{-g_{1,1}} \circ \phi^{t_2}_{g_{1,2}} \circ \phi^{t_1}_{g_{1,1}}(x_0)$$

The first six flows in this example are the same as in Example 1. Following the first six flows are six more wherein the flows that are entirely on S_1, i.e., the flow along $g_{1,2}$, and entirely on S_2, i.e., the flow along $g_{2,2}$, are in the negative direction. If the Lie brackets are zero as in Equation (14), and $t_i = t_{i+2}$, $i = 1, 4, 7, 10$ these flows can be rearranged as

$$x_f = \phi^{t_{11}}_{-g_{2,2}} \circ \phi^{t_8}_{-g_{1,2}} \circ \phi^{t_5}_{g_{2,2}} \circ \phi^{t_2}_{g_{1,2}}(x_0).$$

Now, if $t_2 = t_5 = t_8 = t_{11}$,

$$\begin{aligned} x_f &= \phi^{t_{11}}_{-g_{2,2}} \circ \phi^{t_8}_{-g_{1,2}} \circ \phi^{t_5}_{g_{2,2}} \circ \phi^{t_2}_{g_{1,2}}(x_0) \\ &= \phi^{t^2}_{[g_{1,2},g_{2,2}]} + \mathcal{O}(t^3)(x_0), \end{aligned}$$

where $t = t_2 = t_5 = t_8 = t_{11} \ll 1$. Thus, this sequence provides a net flow in S_{12} in the direction of the Lie bracket between vector fields which are in the equations of motion on different strata, S_1 and S_2.

In Examples 1 and 2, it was required that certain Lie brackets be zero. While one could simply check that these conditions are met in a given situation, the following assumption will guarantee this condition.

Assumption 3 *If it is necessary to lift a foot from the ground during a gait cycle, we assume that the robot can directly control, (via a single control, or a combination of control inputs), the height of that foot relative to the ground. Furthermore, for each stratum comprising the given gait, we assume that the system's equations of motion are independent of the foot height. I.e., the robot's motion is independent of whether a particular foot is very close to the ground, or very far from the ground, but may be dependent upon whether or not a foot is in or out of contact with the ground. When this is so, the Lie bracket of the vector field controlling foot height with any other vector field is zero, and the decoupling requirement is satisfied. Additionally, the tangency requirements for canceling the flows associated with raising and lowering the foot will automatically be satisfied.*

This is arguably a strict assumption. However, for kinematic, legged robots this assumption will almost always be satisfied (see Section 5 for an example).

Examples 1 and 2 show that in given a stratified system, the vector fields on any stratum (other than vector fields corresponding to lifting or replacing feet) can be considered as part of the equations of motion in the bottom stratum if either certain Lie bracket and tangency conditions are met, or if Assumption 3 is satisfied. If the vector fields are not tangent to the bottom stratum, they are modified as in Example 1.

We have shown above that it is possible to consider vector fields in higher strata as part of the equations of motion for the system on the bottom stratum. Based on this observation, we introduce the following.

Definition 4 The *extended stratified system* on the bottom strata, S_B, is the driftless affine system comprised of the vector fields on the bottom strata, chosen vector fields from the higher strata, and Lie brackets of vector fields from S_B and higher strata. I.e., it is a system taking the form:

$$\begin{aligned} \dot{x} &= b_1(x)v_1 + \cdots b_m(x)v_m \\ &+ \underbrace{b_{m+1}v_{m+1} \cdots + b_nv_n}_{\text{from higher strata}} \\ &+ \underbrace{b_{n+1}v_{n+1} + \cdots + b_pv_p}_{\text{any Lie brackets}}, \end{aligned} \tag{17}$$

where the $\{b_1, \ldots, b_p\}$ span T_xS_0, the inputs $v_1, \ldots, v_n$ are real, and the inputs $v_{n+1}, \ldots, v_p$ are fictitious.

With this definition, we have effectively increased the class of vector fields that we may employ when using the motion planning algorithm presented in Section 2.

4.2 The Motion Planning Algorithm

For motion planning, the method of Section 2 could be used in conjunction with the stratified extended system. The basic idea is to use the stratified extended system to plan the motion in the bottom stratum in order to obtain the fictitious inputs. We can determine the actual inputs by the method in Section 2 with the modification that whenever the system must flow along a vector field in a higher stratum, it switches to that stratum by lifting the appropriate feet, flowing along the vector field, and then replacing the appropriate feet, as in Example 1.

Specifically, the algorithm to generate trajectories that move the system from initial configuration p to final configuration q is as follows.

1. Construct the *extended stratified system*, Equation (17), on the bottom strata, S_B.
2. Find a nominal trajectory, $\gamma(t)$, that connects p and q. Given $\gamma(t)$, solve
$$\dot{\gamma}(t) = b_1(x)v_1 + \cdots + b_p(x)v_p,$$
for the fictitious inputs, v_i. As discussed in Section 4.3, it may be necessary to decompose the entire trajectory from the initial point to final point into smaller subtrajectories.
3. Solve the stratified extended system for the fictitious control inputs. I.e., solve for the backward Philip Hall coordinates by solving the differential equations (from Equation (10)).
4. For each path segment in each strata, compute the actual controls that steer the system along $\gamma(t)$. This solution might require the transformation of the backward Philip Hall coordinates to forward Philip Hall coordinates.
5. Flow along each first order vector field, and approximate higher order vector fields as illustrated in Example 1. In general, it will be necessary to switch strata between some of these flows.

Before we illustrate this method in Section 5, we consider the issues of gait efficiency and stability. With regard to gait efficiency, note that the straight–forward application of the method of Section 2 may result in an inordinate amount of strata switches. That is because the sequence of flows in Equation (7) are arranged by *order*, and, from a gait efficiency point of view, it is desirable to have them arranged by strata. It is possible to regroup this sequence of flows by strata if the Lie bracket between any vector fields (considered restricted to the bottom stratum) from different strata are zero. If this is true, Examples 1 and 2 show that it is possible to reorder the flows to obtain the same net result. Flows corresponding to the same stratum could be grouped together. In physical terms, this grouping will reduce the number of times that a particular foot must make and break contact with the ground. The example in Section 5 does not satisfy this assumption.

4.3 Gait Stability

There is not an inherent mechanism in the straight–forward application of the method of Section 2 to guarantee the stability of the gait. Recall that the method is based on the selection of a trajectory for the extended system, $\gamma(t)$, from which the fictitious inputs are determined. It is important to note that the actually realized trajectory will generally *not* be $\gamma(t)$. Thus, merely picking an initial trajectory $\gamma(t)$ which is always stable is not sufficient. One also must guarantee that the method's inherent deviations from the initial trajectory lie within the stability bounds.

Stability considerations can be incorporated into the method as follows. Assume that there is a means for determining the stability of the system, such as a scalar–valued function of the configuration, $\Psi(x)$. For convenience, assume that when $\Psi(x) < 0$, the system is unstable, when $\Psi(x) > 0$, the system is stable, and when $\Psi(x) = 0$, the system is on the stability boundary. In our analysis, the initial trajectory, $\gamma(t)$, must be selected such $\Psi(\gamma(t)) > 0$.

The overall approach is to, when necessary, take steps that are "small enough" to ensure that the system remains stable. Since the flow sequences are composed of small motions and a norm is necessary to measure the length of a flow, we will either consider the system locally in $\mathbb{R}^n$ or equip the configuration manifold with a metric. Given a desired step along the trajectory, $\gamma(t)$, $t \in [0,1]$, let $\mathcal{R} = \min\{\|x - c\|, \quad c \in \Psi^{-1}(0)\}$, *i.e.* the distance from the step's starting point to the closest point on the stability boundary. We want to ensure that the system's trajectory does not intersect the set $\Psi^{-1}(0)$. Let x_s and x_f denote the starting and final trajectory points. Without loss of generality, let $\gamma(t) = x + t(x_f - x_s)$ be a desired straight line path between the starting and end points. Also, let $\Delta = \|x_f - x_s\|$. Recall that the fictitious inputs, v^i were determined by solving an equation of the form $\dot{\gamma}(t) = g_1(\gamma(t))v^1 + \cdots + g_s(\gamma(t))v^s$ for the v^i. We have that $\|v^i\| < C\|\dot{\gamma}(t)\| = C\Delta$, for some constant C. By the method of construction of the real inputs from the fictitious inputs, we have that $\|u^i\| < C\Delta^{1/k}$, where k is the degree of nilpotency of the system, or the degree of the nilpotent approximation.

Pick a ball, $\mathcal{B}$, of radius $\mathcal{R}$, and let K be the maximum norm of all the (first order) vector fields, g_i for all points in $\mathcal{B}$. Recall that the real inputs, u^i were given by a sequence of inputs which approximate the flow of the extended system. Denote this sequence by u^i_j, where the superscript indexes the input, and the subscript indexes its position in the sequence. The maximum distance that the system can possibly flow from the starting point, x_s, is given by the sum of the distances of the individual flows. Let $x_m = \max_{t\in[0,1]}\{\|x(t) - x_s\|\}$ denote the point in the flow that is maximally distant from the starting point (this is not necessarily the final point, x_f). To guarantee stability, we must show that $\|x_m - x\| < \mathcal{R}$. However, this distance, $\|x_m - x\|$ is necessarily bounded by the sum of the norms of each individual flow associated with one real control input, u^i_j, *i.e.*,

$$\|x_m - x\| \leq \sum_{i,j} \| \int_0^1 g_i u^i_j dt \|.$$

However, $\|u^i_j\| \leq C\Delta^{1/k}$ and $\|g_i(x)\| \leq K \ \forall x \in \mathcal{B}$. Thus,

$$\|x_m - x\| \leq \sum_{i,j} KC\Delta^{1/k}, \tag{18}$$

and since $\Delta = \|x_f - x\|$, by choosing the desired final point close enough to the starting point, the trajectory will not intersect the stability boundary.

Note that since Δ is raised to the power of $1/k$, if k is large, it may be necessary to make Δ exceedingly small in order to ensure stability. However, the bound expressed in Equation (18) is very conservative since it sums the length of a bound on each individual flow in the series. In actuality, because the largest flows correspond to the Lie brackets of order k, simply summing their component lengths will give a conservative bound. Given these two observations, an appropriate step length may often be best determined experimentally.

These very same observations also apply to obstacle avoidance. If the robot traverses an environment with obstacles, we assume that the nominal trajectory is designed by an holonomic or rigid body planner in such a manner as to avoid obstacles. Ensuring that the actual trajectory also avoids the obstacles, requires that the nominal trajectory be analogously broken into sufficiently small steps to ensure that the actual trajectory remains sufficiently close to it.

5 Example

We illustrate our approach by generating control inputs that will steer the hexapod of Figure 1 to walk over flat terrain (see Section 6 for an example involving manipulation of a curved object, which is analogous to locomotion over uneven terrain). The key difficulty in this example is the fact that the legs are kinematically insufficient, making sideways motion difficult. Assume that the robot walks with a tripod gait [3], alternating movements of legs 1–4–5 with movements of legs 2–3–6. With the tripod gait, this robot has four control inputs. The inputs u_1 and u_2 respectively control the forward and backward angular leg displacements of legs 1–4–5 and legs 2–3–6, while inputs u_3 and u_4 respectively control the height of legs 1–4–5 and 2–3–6.

The equations of motion can be written as follows:

$$\begin{aligned} \dot{x} &= \cos\theta\,(\alpha(h_1)u_1 + \beta(h_2)u_2) \\ \dot{y} &= \sin\theta\,(\alpha(h_1)u_1 + \beta(h_2)u_2) \\ \dot{\theta} &= l\alpha(h_1)u_1 - l\beta(h_2)u_2 \end{aligned}$$

$$\begin{aligned} \dot{\phi}_1 &= u_1; \qquad \dot{\phi}_2 = u_2 \\ \dot{h}_1 &= u_3 \qquad \dot{h}_2 = u_4. \end{aligned}$$

where (x, y, θ) represents the body's configuration, ϕ_i is the front to back angular deflection of the legs, l is the leg length, and h_i is the height of the legs off the ground. The functions $\alpha(h_1)$ and $\beta(h_2)$ are defined by:

$$\alpha(h_1) = \begin{cases} 1 & \text{if } h_1 = 0 \\ 0 & \text{if } h_1 > 0 \end{cases} \qquad \beta(h_2) = \begin{cases} 1 & \text{if } h_2 = 0 \\ 0 & \text{if } h_2 > 0 \end{cases}.$$

[3]Ref. [9] shows that the hexapod is small time locally gait controllable when a tripod gait is used.

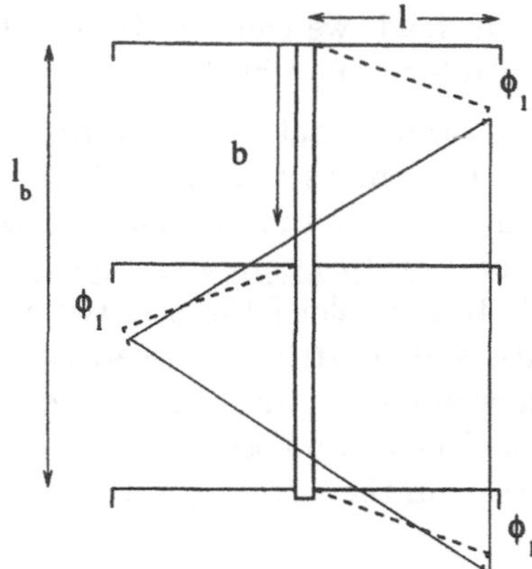

Figure 5: *Stability Margin for Hexapod Tripod Gait.*

Since the robot walks in a tripod gait, stability is ensured if the robot's center of mass remains above the triangle defined by the tripod of feet which are in contact with the ground. For the motion of legs 1–4–5, the robot's center of mass must be at least $b = \frac{l_b}{4} + l \sin\phi_1$ from the front of the robot to ensure stability, where l_b denotes the body's length. See Figure 5. Alternatively, if the center of mass is located a distance b from the front of the robot, then stability is ensured during the motion if both of these constraints are satisfied

$$\begin{aligned} \phi_1 &< \sin^{-1}\left((b - \frac{l_b}{4})/l\right) \\ \phi_2 &> -\sin^{-1}\left((\frac{3l_b}{4} - b)/l\right). \end{aligned}$$

Denote the stratum when all the feet are in contact ($\alpha = \beta = 1$) by S_{12}, the stratum when tripod one is in contact ($\alpha = 1, \beta = 0$), by S_1, the stratum when tripod two is in contact ($\alpha = 0, \beta = 1$), by S_2 and the stratum when no legs are in contact ($\alpha = \beta = 0$), by S_0. Note that this system satisfies the requirements of Assumption 3 since, regardless of the values of α and β, the vector fields moving the foot out of contact with the ground are of the form $\left\{\frac{\partial}{\partial h_i}\right\}$, and the equations of motion are independent of the foot heights, h_i.

The equations of motion in the bottom strata, S_{12} (where all the feet maintain ground contact), are:

$$\begin{pmatrix} \dot{x} \\ \dot{y} \\ \dot{\theta} \\ \dot{\phi}_1 \\ \dot{\phi}_2 \end{pmatrix} = \begin{pmatrix} \cos\theta & \cos\theta \\ \sin\theta & \sin\theta \\ l & -l \\ 1 & 0 \\ 0 & 1 \end{pmatrix} \begin{pmatrix} u_1 \\ u_2 \end{pmatrix} \qquad (19)$$

where (x, y, θ) represents the planar position of a reference frame attached to the robot's center. ϕ_1 is the angle of legs 1–4–5 and ϕ_2 is the angle of legs 2–3–6. The variables u_3 and u_4 are both 0 since the legs maintain ground contact. Let $g_{12,1}$ and $g_{12,2}$ represent the first and second columns in Equation (19).

If legs 1–4–5 are in contact with the ground, but legs 2–3–6 are not in contact, the equations of motion are:

$$\begin{pmatrix} \dot{x} \\ \dot{y} \\ \dot{\theta} \\ \dot{\phi}_1 \\ \dot{\phi}_2 \\ \dot{h}_2 \end{pmatrix} = \begin{pmatrix} \cos\theta & 0 & 0 \\ \sin\theta & 0 & 0 \\ l & 0 & 0 \\ 1 & 0 & 0 \\ 0 & 1 & 0 \\ 0 & 0 & 1 \end{pmatrix} \begin{pmatrix} u_1 \\ u_2 \\ u_4 \end{pmatrix} \qquad (20)$$

where h_i is the height of the corresponding set of legs and u_3 is constrained to be 0. Label columns one, two and three in Equation (20) $g_{1,1}, g_{1,2}$ and $g_{1,3}$, respectively. If legs 2–3–6 are in ground contact and legs 1–4–5 are not, the equations of motion are:

$$\begin{pmatrix} \dot{x} \\ \dot{y} \\ \dot{\theta} \\ \dot{\phi}_1 \\ \dot{\phi}_2 \\ \dot{h}_1 \end{pmatrix} = \begin{pmatrix} 0 & \cos\theta & 0 \\ 0 & \sin\theta & 0 \\ 0 & -l & 0 \\ 1 & 0 & 0 \\ 0 & 1 & 0 \\ 0 & 0 & 1 \end{pmatrix} \begin{pmatrix} u_1 \\ u_2 \\ u_3 \end{pmatrix} \qquad (21)$$

where u_4 is constrained to be 0. The columns in Equation (21) are denoted $g_{2,1}$, $g_{2,2}$, and $g_{2,3}$.

For motion planning purposes, we must select enough vector fields to span the tangent space of the bottom stratum, S_{12}. A simple calculation shows that the set of vector fields,

$$\{g_{12,1}, g_{12,2}, g_{1,2}, g_{2,1}, [g_{12,1}, g_{12,2}]\}$$

spans $T_x S_{12}$ for all $x \in S_{12}$. Note that $[g_{12,1}, g_{12,2}] = (-2l\sin\theta, 2l\cos\theta, 0, 0, 0)^T$. This Lie algebra is *not* nilpotent, and thus the extended system will only be a nilpotent approximation.

The *stratified extended system* is constructed from the extended system that uses the vector fields from all strata.

$$\dot{x} = g_{12,1}v^1 + g_{12,2}v^2 + g_{1,2}v^3 + g_{2,1}v^4 + [g_{12,1}, g_{12,2}]v^5 \qquad (22)$$

or, in greater detail,

$$\begin{pmatrix} \dot{x} \\ \dot{y} \\ \dot{\theta} \\ \dot{\phi}_1 \\ \dot{\phi}_2 \end{pmatrix} = \begin{pmatrix} \cos\theta & \cos\theta & 0 & 0 & -2l\sin\theta \\ \sin\theta & \sin\theta & 0 & 0 & 2l\cos\theta \\ l & -l & 0 & 0 & 0 \\ 1 & 0 & 0 & 1 & 0 \\ 0 & 1 & 1 & 0 & 0 \end{pmatrix} \begin{pmatrix} v^1 \\ v^2 \\ v^3 \\ v^4 \\ v^5 \end{pmatrix}.$$

Let the starting and ending configurations be:

$$\begin{aligned} p &= (x, y, \theta, \phi_1, \phi_2, h_1, h_2) = (0,0,0,0,0,0,0) \\ q &= (x, y, \theta, \phi_1, \phi_2, h_1, h_2) = (1,1,0,0,0,0,0) \end{aligned}.$$

A path that connects these points is $\gamma(t) = (t, t, 0, 0, 0, 0, 0)$. Equating $\dot{\gamma}(t)$ with with the stratified extended system and solving for the fictitious controls yields:

$$\begin{pmatrix} v^1 \\ v^2 \\ v^3 \\ v^4 \\ v^5 \end{pmatrix} = \frac{1}{2l} \begin{pmatrix} l(\cos\theta + \sin\theta) \\ l(\cos\theta + \sin\theta) \\ -l(\cos\theta + \sin\theta) \\ -l(\cos\theta + \sin\theta) \\ (\cos\theta - \sin\theta) \end{pmatrix},$$

or, since $\theta(t) = 0$, and if we let $l = 1$,

$$\begin{pmatrix} v^1 \\ v^2 \\ v^3 \\ v^4 \\ v^5 \end{pmatrix} = \frac{1}{2} \begin{pmatrix} 1 \\ 1 \\ -1 \\ -1 \\ 1 \end{pmatrix}.$$

For a system which is nilpotent of order 2, we have from Equation (9) (where the g's from Equation (22) are substituted for the B's in Equation (9) in the order that they appear in Equation (22):

$$\begin{aligned} \dot{h}_1 &= v^1, \quad \dot{h}_2 = v^2, \\ \dot{h}_3 &= v^3, \quad \dot{h}_4 = v^4, \\ \dot{h}_5 &= v^5 + h_1 v^2 \end{aligned}$$

which yields:

$$\begin{aligned} h_1(1) &= \frac{1}{2} \quad h_2(1) = \frac{1}{2} \\ h_3(1) &= -\frac{1}{2} \quad h_4(1) = -\frac{1}{2} \\ h_5(1) &= \frac{3}{4}. \end{aligned}$$

Since the nilpotent approximation is of order two, there is no need to transform to forward Philip Hall coordinates. Instead, we can directly construct a sequence of controls to move in the desired direction.

Let $\circ$ denote concatenation of control inputs. For example, $u_1 \circ u_2$ denotes that $u_1 = 1$ for time $h_1(1)$ followed by $u_2 = 1$ for time $h_2(1)$. Considering the vector fields on S_{12}, ($g_{12,1}, g_{12,2}$ and $[g_{12,1}, g_{12,2}]$), the system needs to flow along the first two vector fields for $\frac{1}{2}$ seconds, and construct a piece-wise approximation to the flow along the third Lie bracket vector field for $\frac{3}{4}$ seconds. The control sequence to approximately move the system in the direction of the flow of the Lie bracket is:

$$u_1 \circ u_2 \circ -u_1 \circ -u_2 \tag{23}$$

where each of the individual control inputs is equal to one for $\sqrt{\frac{3}{4}}$ seconds (recall Equation (2)). To flow along $g_{11,1}$, $u_1 = 1$ for $\frac{1}{2}$ seconds. Similarly to flow along $g_{12,1}$, $u_2 = 1$ for $\frac{1}{2}$ seconds.

On the higher strata, to flow along $g_{1,1}$, $u_1 = -1$ for $\frac{1}{2}$ seconds and to flow along $g_{2,1}$, $u_1 = -1$ for $\frac{1}{2}$ seconds. In order to execute these flows, the robot must switch from the bottom stratum to the higher strata when executing a control input associated with a fictitious input for a higher strata.

Thus, the total control sequence is:

$$\begin{aligned} &\sqrt{\tfrac{3}{4}}(u_1 \circ u_2 \circ -u_1 \circ -u_2) \\ &\circ \tfrac{1}{2}u_2 \circ \tfrac{1}{2}u_1 \circ \epsilon u_4 \circ (-\tfrac{1}{2}u_2) \circ (-\epsilon u_4) \\ &\circ \epsilon u_3 \circ -(\tfrac{1}{2}u_1) \circ (-\epsilon u_3). \end{aligned}$$

The first four terms in the sequence approximate the Lie bracket motion on the bottom stratum. The $\sqrt{\frac{3}{4}}$ term denotes the length of time each control input is "on." The next two terms are the contribution of the u_1 and u_2 terms individually on the bottom stratum. The next term represents a small flow associated with removing legs 2-3-6 out of contact with the ground, and the following term corresponds to legs 2-3-6 moving back to their initial position. Since the legs are not in contact with the ground, this motion does not cause the body of the robot to move. The next input corresponds to legs 2-3-6 moving back into contact with the ground. The next three inputs correspond to legs 1-4-5 performing an analogous motion.

Figure 6 shows the path of the robot's center as it follows a straight line trajectory, which is broken into four equal segments. Due to the nilpotent approximation,

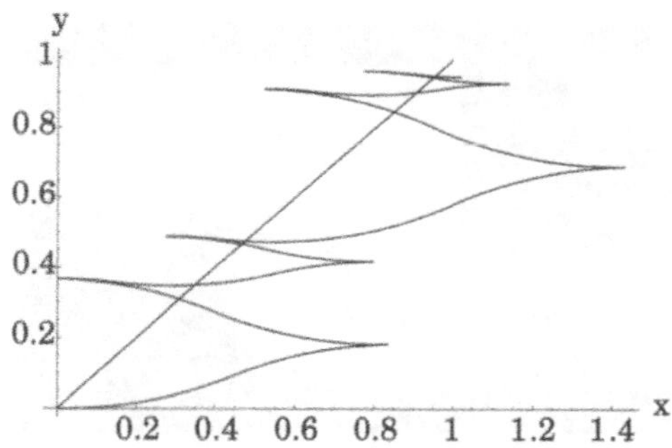

Figure 6: *Straight Trajectory.*

there is some small final error. Better accuracy can be obtained by use of a higher order nilpotent approximation or a second iteration of the algorithm from the robot's ending position. Note that the main body axis is oriented along the x-axis in this example. Since the legs can not move immediately sideways, the robot's motion must include "parallel-parking-like" behavior to follow this line.

There is no inherent limitation in the method which requires the trajectory to be broken down into subsegments, however, there are two reasons to do so. First, since the method is based upon decomposing a desired trajectory into flows along the Philip Hall basis vector fields, the final trajectory is only related to the desired trajectory in that the end points are the same (or approximately the same for nilpotent approximations). Breaking the path into segments leads to better overall tracking. Second, robot stability requirements may also demand smaller steps.

The approach is general enough that approximate tracking of arbitrary trajectories is possible. Figure 7 shows the hexapod following an ellipse while maintaining a constant angular orientation. Figure 8 shows the results when a smaller step size is used. In the first simulation, the elliptical trajectory is broken into 30 segments. In the second, it is divided into 60 segments. In this example, part of the trajectory tracking error is due to the nilpotent approximation, but another contribution to the error is the simplicity of the model. Some directions are more "difficult" for the system to execute than others due to the kinematic limitations of the leg design. Because this mechanism can not execute "crab-like" gaits, its tracking error during sideways motions increases, as this direction corresponds to a Lie bracket direction.

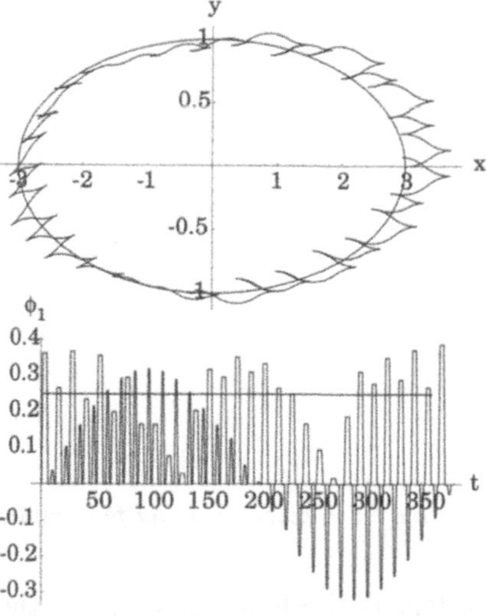

Figure 7: *Elliptical Trajectory.*

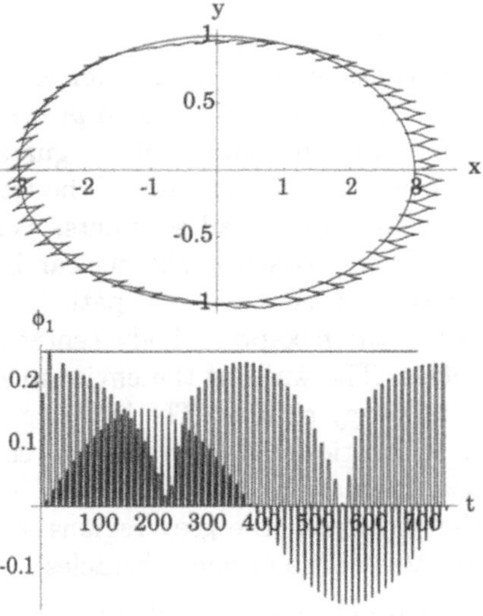

Figure 8: *Elliptical Trajectory with Smaller Steps.*

Also plotted in these figures is the stability criterion. Let the body length be 2 units of length and let the center of mass be located a distance of 0.75 units from the front of the robot. Then, the stability criterion is $\phi_1 < 0.25$ [rad] and $\phi_2 > -.85$ [rad]. In Figures 7 and 8 the stability limits for ϕ_1 are indicated by the straight horizontal lines. In the first case, where the robot takes bigger steps, the stability condition is violated. However, in the second case it is not.

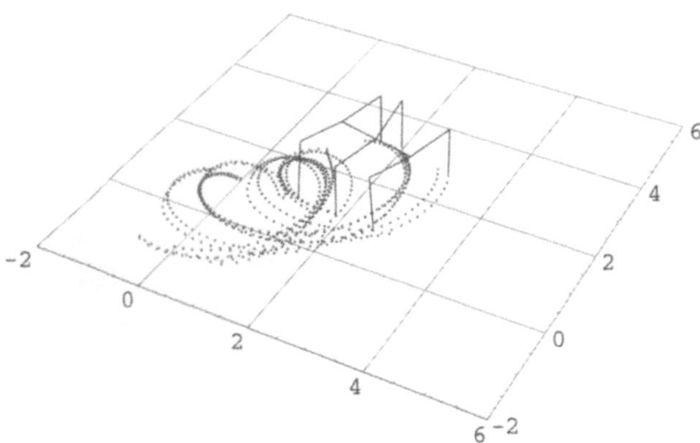

Figure 9: *Hexapod leaving footprints.*

Figure 9 depicts the footprints left by the hexapod as it follows a straight line diagonal path while simultaneously rotating at a constant rate. The complex pattern of the footfalls suggests that any technique that is based on foot placement would be very difficult to apply to this system.

Finally, we consider obstacle avoidance. While the nominal initial trajectory $\gamma(t)$ must *a priori* avoid any obstacles, this constraint alone will not guarantee that the actual motion avoids obstacles. If the trajectory is divided into sufficiently small segments, as suggested in Section 4.3, then obstacle avoidance can be realized. Figure 10 shows a desired nominal path (indicated by a black line) of the hexapod's body center through a set of obstacles. The walls of the environment are indicated by dark grey regions. The lighter grey regions correspond to locations of the vehicle's center where some vehicle orientations may cause the hexapod to intersect the walls (i.e., the grey regions are the projected silhouettes of the c-space obstacles).

To make the problem more challenging, we also specify that the robot rotates at a uniform rate as it follows

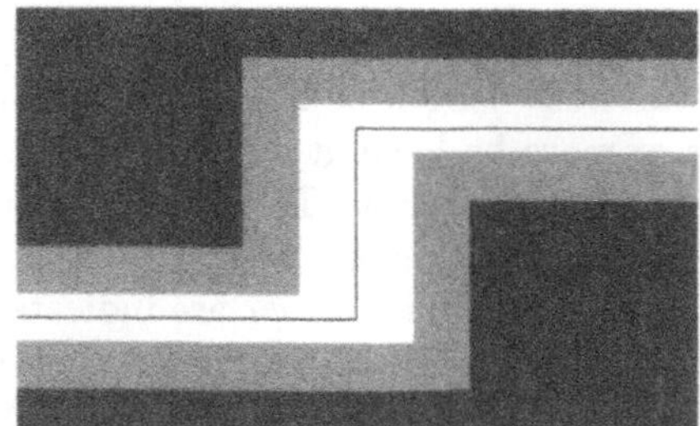

Figure 10: *Nominal obstacle avoidance trajectory.*

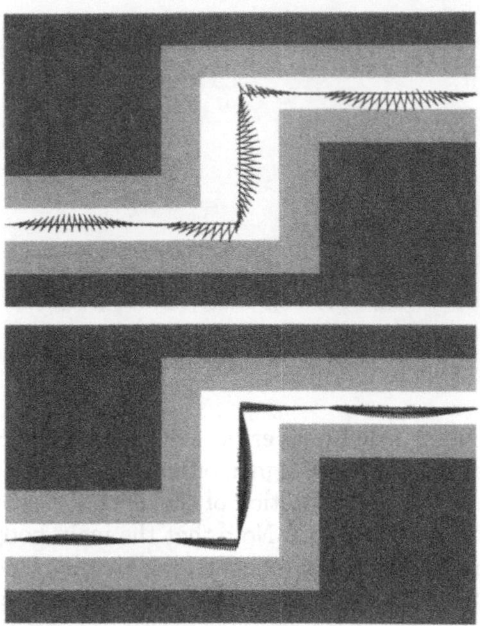

Figure 11: *(a) Obstacles not avoided; (b)Obstacles avoided.*

the nominal trajectory. A real–world scenario where this might be desirable is a patrol robot that must constantly scan in all directions. Figure 11(a) shows the path of robot's center of mass when the trajectory is not finely divided enough to satisfy the criteria of Section 4.3 (it is subdivided into 100 subtrajectories). Since the path of the center of mass intersects the lighter grey regions during portions of its motion, the robot would realistically bump into the walls in this example. However, if the nominal trajectory is sufficiently subdivided (into 300 subtrajectories in this case) to satisfy the requirement of Section 4.3, the robot avoids the walls, as illustrated in Figure 11(b).

6 Multi-fingered Hand Manipulation

The methodology described above can be almost immediately applied to object manipulation via finger gaiting in a multi-fingered hand. The application of this approach leads to a manipulation planning strategy that is independent of the geometry of the grasped object and independent of the manipulating hand's morphology. The method is also independent of the type of contact between the finger and object (*e.g.*, "point contact with friction," "soft finger," etc.) and indepen-

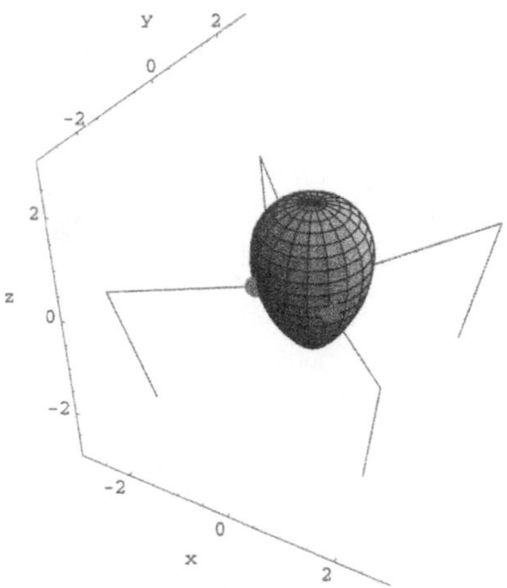

Figure 12: *Four fingers manipulating an object.*

dent of the morphology of the manipulating "fingers" (*i.e.*, independent of the number of joints, etc.).

Consider the "egg–shaped" object in Figure 12 whose surface is parameterized by

$$c(u,v) = \begin{pmatrix} \left(1+\frac{u}{\pi}\right)\cos u \cos v \\ \left(1+\frac{u}{\pi}\right)\cos u \sin v \\ \frac{3}{2}\sin u \end{pmatrix}, \quad \begin{matrix} u \in (\frac{-\pi}{2}, \frac{\pi}{2}) \\ v \in (-\pi, \pi) \end{matrix}.$$

This object is to be manipulated by four, three DOF fingers whose kinematic model is shown in Figure 13. A "point contact with friction" model is assumed.

The stratified c-space will consist of a total of 16 different strata, corresponding to all the possible combinations of finger contacts. However, as will be clear shortly, the system is manipulable if it is restricted to only 5 strata: When all four fingers are in contact plus each of the four cases where only one of the fingers is out of contact. Denote these strata as $S_{1234}, S_{123}, S_{124}, S_{134}$, and S_{234} where the subscripts denote which fingers are in contact with the object.

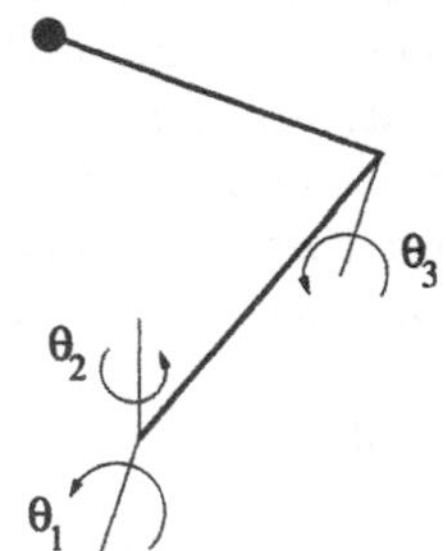

Figure 13: *Finger kinematics.*

Since the nominal trajectory stays away from the fingers' kinematic singularities, the finger tip velocities can be considered as system inputs. This input choice will simplify the computations and make the equations of motion satisfy Equation (14). One can not generally choose the inputs in this way, (for example, when the the finger tips are in rolling contact with the object); however, the more general cases still fits within the framework of the stratified motion planning method outlined in Section 3.

The equations of motion for such a grasped system are straight–forward, though possibly tedious, to derive (see [21] for details). The equations of motion on the bottom stratum are of the form

$$\dot{x} = g_1(x)u_1 + \cdots + g_6(x)u_6,$$

and on the higher strata are of the form

$$\dot{x} = g_1(x)u_1 + \cdots + g_6(x)u_6 \;+\; g_7(x)u_7 + \cdots g_9(x)u_9,$$

where the first 6 inputs are associated with the finger tip velocities for the three fingers contacting the object, and inputs $7--9$ are the three degrees of freedom for the finger that is not in contact with the object. Note that $g_7(x)$ through $g_9(x)$ will take the form $(0, \cdots, 1, \cdots, 0)$ since they are the unconstrained finger tip velocities of the finger which is not contacting the object, and thus they will satisfy Equation 14. Therefore, they may be incorporated into the equations of motion for the bottom stratified extended system.

Incorporating these unconstrained finger tip velocity vector fields for each of the four higher strata gives a stratified extended system of the form:

$$\begin{aligned} \dot{x} &= \underbrace{g_1(x) + \cdots + g_6(x)u_6}_{\text{on } S_{1234}} \\ &+ \underbrace{g_7(x)u_7 + g_8(x)u_8 + g_9(x)u_9}_{\text{from } S_{123}} \\ &\vdots \\ &+ \underbrace{g_{16}(x)u_{16} + g_{17}(x)u_{17} + g_{18}(x)u_{18},}_{\text{from } S_{234}} \end{aligned}$$

where all the vector fields except those on the first line correspond to free finger tip motion. Tedious calculations show that $\{g_1, \ldots, g_{18}\}$ spans the tangent space to the c-space, so the system is stratified manipulable. Since no Lie brackets are necessary to make the system stratified manipulable, this system is already in extended form, and the actual control inputs are the same as the "fictitious" inputs presented in Sections 2 and 3.

Assume that the initial and final configurations are identical (as illustrated in Figure 12), and that the desired motion is a pure rotation of 2π about the axis $\omega = (\frac{1}{\sqrt{3}}, \frac{1}{\sqrt{3}}, \frac{1}{\sqrt{3}})$. Using exponential coordinates, then, the object's nominal configuration as a function of time is given by Rodrigues' formula:

$$\gamma(t) = e^{\hat{\omega}2\pi t} = I + \hat{\omega}\sin 2\pi t + \hat{\omega}^2(1 - \cos 2\pi t), \quad t \in [0, 1].$$

For the object's initial and final configuration in Figure 12, each finger is oriented at an angle of $\pi/4$ relative to the x- and y-axes. As the object rotates, each finger's nominal configuration is such that it contacts the object along that same axis. This can be determined by equating the forward kinematics for each finger with the point on the object's surface that intersects the respective $\pi/4$ radial from the origin, and then, using the kinematics of each finger, determine the desired joint configurations. For this particular example, this trajectory is difficult to compute analytically, but is simple to do numerically for each step of the system's motion. The desired trajectory is decomposed into 10 subsegments, and a sequence of six "snapshots" from the manipulation is shown in Figure 14.

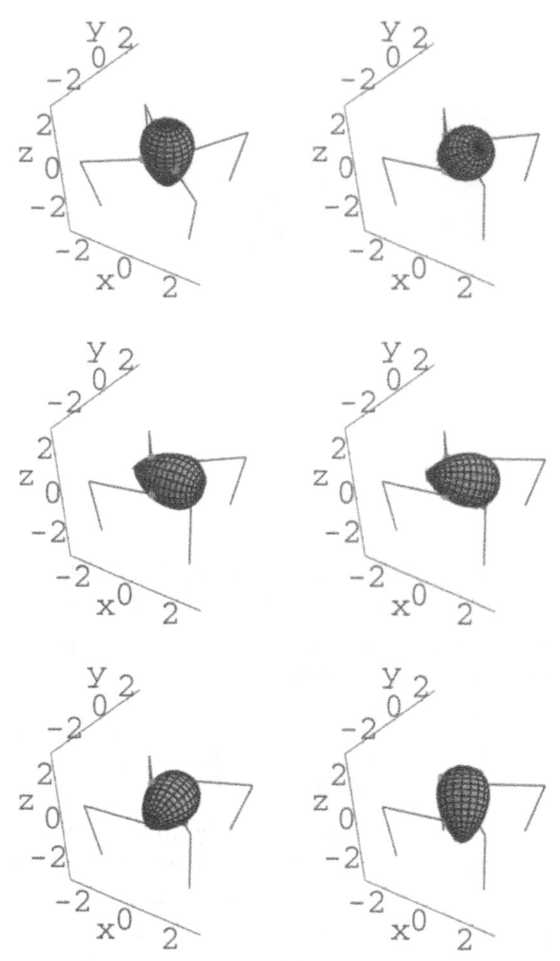

Figure 14: *Snapshots of computer simulation of object manipulation via finger repositioning.*

7 Conclusions

Our method provides a general means to generate rajectories for many types of legged robotic and multifingered systems. The simulations indicate that the approach is rather simple to apply. The method is independent of the number of legs (fingers) and is not based on foot (finger) placement principles. For a given legged robot mechanism, a specifically tuned legplacement-based algorithm may lead to motions which use fewer steps or results in less tracking error. However, for the purposes of initial design and evaluation of a legged mechanism, our approach affords the robot design engineer an automated way to implement a realistic trajectory generation scheme for a quasi-static robot of nearly arbitrary morphology. More importantly, we believe that our approach provides an evolutionary path for future research and generalizations.

Since many interesting robotic systems (such as bipeds) are not kinematic, an algorithm for solving the trajectory generation problem for such systems is necessary. However, since the state of the art for solving the trajectory generation problem for smooth systems with drift is still in its infancy, it may be difficult to make headway along these lines until more complete results for the smooth case become known.

Acknowledgments

This work was partially supported by a grant from the Office of Naval Research.

References

[1] R. Abraham, J. E. Marsden, and T. Ratiu. *Manifolds, Tensor Analysis, and Applications*. Springer–Verlag, second edition, 1988.

[2] Matt Berkemeier. Modeling the dynamics of quadrupedal running. *Int. J. Robotics Research*, 16(9):971–985, 1998.

[3] C.H. Chen, K. Mirza, and D.E. Orin. Force control of planar power grasp in the digits system. In *4th Int. Symp. on Robotics and Manufacturing*, pages 189–194, 1992.

[4] I-M Chen and J.W. Burdick. A qualitative test for n–finger force–closure grasps on planar objects with applications to manipulation and finger gaits. In *Proc. IEEE Int. Conf. on Robotics and Automation*, pages 814–820, 1993.

[5] J. J. Collins and Ian Stewart. Hexapodal gaits and coupled nonlinear oscillator models. *Biological Cybernetics*, 68:287–298, 1993.

[6] Ronald S. Fearing and T.O. Binford. Using a cylindrical tactile sensor for determining curvature. *IEEE Trans. Robotics and Automation*, 7(6):806–817, 1991.

[7] R.S. Fearing. Implementing a force strategy for object reorientation. In *Proc. IEEE Int. Conf. on Robotics and Automation*, pages 96–102, 1986.

[8] B. Goodwine and J.W. Burdick. Controllability of kinematic systems on stratified configuration spaces. IEEE Trans. Automatic Control (to appear), 1999.

[9] Bill Goodwine and Joel Burdick. Gait controllability for legged robots. In *Proc IEEE Int. Conf on Robotics and Automation*, 1998.

[10] Bill Goodwine and Joel Burdick. Stratified motion planning with application to robotic finger gaiting. Proc. IFAC World Congress, 1999.

[11] J. William Goodwine. *Control of Stratified Systems with Robotic Applications*. PhD thesis, California Institute of Technology, 1998.

[12] Goresky and Macpherson. *Stratified Morse Theory*. Springer–Verlag, New York, 1980.

[13] L. Han, Y.S. Guan, Z.X. Li, Q. Shi, and J.C. Trinkle. Dextrous manipulation with rolling contacts. In *Proc. IEEE Int. Conf. on Robotics and Automation*, pages 992–997, Albuquerque, NM, 1997.

[14] J. Hong and G. Lafferriere. Fine manipulation with multifinger hands. In *IEEE Int. Conf. on Robotics and Automation*, pages 1568–1573, 1990.

[15] Maw Kae Hor. *Control and Task Planning of the Four Finger Manipulator*. PhD thesis, NYU, 1987.

[16] Alberto Isidori. *Nonlinear Control Systems*. Springer–Verlag, second edition, 1989.

[17] S. Kajita and K. Tani. Study of dynamic biped locomotion on rugged terrain. In *IEEE Int. Conf. on Robotics and Automation*, pages 1405–1411, Sacramento, CA, 1991.

[18] G. Lafferriere and Hector J. Sussmann. A differential geometric approach to motion planning. In X. Li and J. F. Canny, editors, *Nonholonomic Motion Planning*, pages 235–270. Kluwer, 1993.

[19] J.K. Lee and S.M. Song. Path planning and gait of walking machine in an obstacle-strewn environment. *J. Robotics Systems*, 8:801–827, 1991.

[20] D. J. Montana. The kinematics of contact and grasp. *Int. J. of Robotics Research*, 7(3):17–25, 1988.

[21] Richard M. Murray, Zexiang Li, and S. Shankar Sastry. *A Mathematical Introduction to Robotic Manipulation*. CRC Press, Inc., 1994.

[22] R.M. Murray and S.S. Sastry. Nonholonomic motion planning: Steering using sinusoids. *IEEE Trans. Automatic Control*, 38:700–716, 1993.

[23] T. Okada. Object handling system for manual industry. *IEEE Transactions on Systems, Man and Cybernetics*, 9(2):79–89, 1979.

[24] Domenico Prattichizzo and Antonio Bicchi. Dynamic analysis of mobility and graspability of general manipulation systems. *IEEE Transactions on Robotics and Automation*, 14(2):241–258, April 1998.

[25] M.H. Raibert. *Legged Robots that Balance*. MIT Press, 1986.

[26] E. Rimon and J.W. Burdick. Configuration space analysis of bodies in contact – i. *Mechanism and Machine Theory*, 30(6):897–912, August 1995.

[27] E. Rimon and J.W. Burdick. Configuration space analysis of bodies in contact – ii. *Mechanism and Machine Theory*, 30(6):913–928, August 1995.

[28] K. Salisbury, D. Brock, T. Massie, N. Swarup, and C. Zilles. Haptic rendering: Programming touch interation with virtual objects. In *Proc. Symp. on Interactive 3D Graphics*, pages 123–130, 1995.

[29] K. Salisbury and C. Tarr. Haptic rendering of surfaces defined by implicit functions. In *Proc. ASME Int. Mechanical Engineering Congress and Exposition*, pages 61–67, 1997.

[30] Jean-Pierre Serre. *Lie Algebras and Lie Groups.* Springer–Verlag, 1992.

[31] K.B. Shimoga. Robot grasp synthesis algorithms: A survey. *The International Journal of Robotics Research*, 15(3):230–266, 1996.

[32] S.M. Song and K.J. Waldron. *Machines that walk: the Adaptive Suspension Vehicle.* MIT Press, 1989.

[33] J.C. Trinkle and R.P. Paul. Planning for dexterous manipulation with sliding contacts. *Int. J. of Robotics Research*, 9(3):24–48, 1990.

[34] V.S. Varadarajan. *Lie Groups, Lie Algebras, and Their Representations.* Springer-Verlag, 1984.

Manipulation of Pose Distributions

Mark Moll, *Carnegie Mellon University, Pittsburgh, PA*
Michael A. Erdmann, *Carnegie Mellon University, Pittsburgh, PA*

For assembly tasks parts often have to be oriented before they can be put in an assembly. The results presented in this paper are a component of the automated design of parts orienting devices. The focus is on orienting parts with minimal sensing and manipulation. We present a new approach to parts orienting through the manipulation of pose distributions. Through dynamic simulation we can determine the pose distribution for an object being dropped from an arbitrary height on an arbitrary surface. By varying the drop height and the shape of the support surface we can find the initial conditions that will result in a pose distribution with minimal entropy. We are trying to uniquely orient a part with high probability just by varying the initial conditions. We will derive a condition on the pose and velocity of an object in contact with a sloped surface that will allow us to quickly determine the final resting configuration of the object. This condition can then be used to quickly compute the pose distribution. We also present simulation and experimental results that show how dynamic simulation can be used to find optimal shapes and drop heights for a given part.

1 Introduction

In our research we are trying to develop strategies to orient three-dimensional parts with minimal sensing and manipulation. That is, we would like to bring a part from an unknown position and orientation to a known orientation (but possibly unknown position) with minimal means. In general, it is not possible to orient a part completely without sensors, but it is sufficient if a particular orienting strategy can bring a part into one particular orientation with high probability. The sensing is then reduced to a binary decision: a sensor only has to detect whether the part is in the right orientation or not. If not, the part is fed back to the parts orienting device. Assuming the orienting strategy succeeds with high probability, on average it takes just a few tries to orient the part. An alternative view of this type of manipulation is to consider it as manipulation of the pose distribution. The goal then is to find the pose distribution with minimal entropy, thereby maximally reducing uncertainty.

1.1 Example

In this paper we will discuss the use of dynamic simulation for the design of support surfaces that reduce the uncertainty of a part's resting configuration. As the support surface is changed, the probability distribution function (pdf) of resting configurations will change as well. The pdf will also vary with the initial drop position above the surface. The following figure and paragraph illustrate the basic idea:

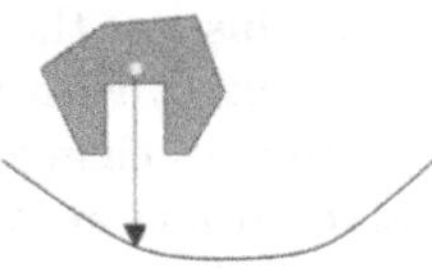

Figure 1: *A part with an initially unknown orientation is dropped on a surface.*

A part with an initially unknown orientation is released from a certain height and relative horizontal position with respect to the bowl. The only forces acting on the part are gravity and friction. We assume the bowl doesn't move. We can compute the final resting configuration for all possible initial orientations. This will give us the pdf of stable poses. The goal is to find the drop height, relative position and bowl shape that will maximally reduce uncertainty. In this paper we assume for simplicity that the initially unknown orientation is uniformly random, but our approach also works for different prior distributions.

	Stable Poses						
							Entropy
quasistatic approximation	0.20	0.13	0.16	0.21	0.14	0.16	1.78
dynamic, flat surface, drop height is $h = 0$	0.18	0.16	0.14	0.34	0.05	0.13	1.66
dynamic, bowl shape is $y = 0.24x^2$, $h = 0.28$, initial hor. pos. $x_0 = -0.41$	0.24	0.03	0.03	0.50	0.08	0.15	1.35

Table 1: *Probability distribution function of stable poses for two surfaces. The initial velocity is zero and the initial rotation is uniformly random.*

Table 1 shows three different pose distributions. Each stable pose corresponds to a set of contact points (marked by the black dots in the table). For an arbitrarily curved support surface the stable poses do not necessarily correspond to edges of the convex hull of the part. We therefore define a stable pose as a set of contact points. This means that any two poses with the same set of contact points are considered to be the same as far as the pose distribution is concerned. In our example the support surface is a parabola $y = ax^2$ with parameter a. Other parameters are the drop height, h, and the initial horizontal position of the drop location, x_0. We limit the surface to parabolas for illustrative purposes only; in general we would use a larger class of possible shapes (see Section 4.1).

The first row in the table shows the pdf assuming quasistatic dynamics. In this case the surface is flat and the part is released in contact with the surface. The second row shows how the pdf changes if we model the dynamics. The initial conditions are the same as for the quasistatic case, yet the pdf is significantly different. The third row shows the pdf for the optimized values for a, h and x_0.

The objective function over which we optimize is the entropy of the pose distribution. If $p_1, \ldots, p_n$ are the probabilities of the n stable poses, then the entropy is $-\sum_{i=1}^{n} p_i \log p_i$. This function has two properties that make it a good objective function: it reaches its global minimum whenever one of the p_i is 1, and its maximum for a uniform distribution. By searching the parameter space we can find the a, h and x_0 that minimize the entropy. In the third row of the table the pose distribution is shown with minimal entropy[1]. The table makes it clear that even with a very simple surface we can reduce the uncertainty greatly by taking advantage of the dynamics.

[1]This is a local minimum found with simulated annealing and might not be the global minimum.

1.2 Outline

In Section 3 we will explain the notion of capture regions and introduce an extension and relaxation of this notion in the form of so-called quasi-capture regions. These quasi-capture regions allow for fast computation of pose distributions. In Section 4 we will present our simulation and experimental results. Finally, in Section 5 we will discuss the results presented in this paper. But first we will give an overview of related work in the next section.

2 Related Work

2.1 Parts Feeding and Orienting

One of the most comprehensive works on the design of parts feeding and assembly design is [12], which describes vibratory bowls as well as non-vibratory parts feeders in detail. The APOS parts feeding system is described by Hitakawa [26]. Berretty et al. [6] present an algorithm for designing a particular class of gates in vibratory bowls. Berkowitz and Canny [4, 5] use dynamic simulation to design a sequence of gates for a vibratory bowl. The dynamics are simulated with Mirtich's impulse-based dynamic simulator, *Impulse* [35]. Christiansen et al. [16] use genetic algorithms to design a near-optimal sequence of gates for a given part. Optimality is defined in terms of throughput. Here, the behavior of each gate is assumed to be known. So, in a sense [16] is complementary to [5]: the latter focuses on modeling the behavior of gates, the former finds an optimal sequence of gates given their behavior. Akella et al. [1] introduced a technique for orienting planar parts on a conveyor belt with a one degree-of-freedom

(DOF) manipulator. Lynch [32] extended this idea to 3D parts on a conveyor belt with a two DOF manipulator. Wiegley et al. [43] presented a complete algorithm for designing passive fences to orient parts. Here, the initial orientation is unknown.

Goldberg [22] showed that it is possible to orient polygonal parts with a frictionless parallel-jaw gripper without sensors. Marigo et al. [34] showed how to orient and position a polyhedral part by rolling it between the two hands of a parallel-jaw gripper. Grossman and Blasgen [25] developed a manipulator with a tactile sensor to orient parts in a tray. Erdmann and Mason [18] developed a tray-tilting *sensorless* manipulator that can orient planar parts in the presence of friction. If it isn't possible to bring a part into a unique orientation, the planner would try to minimize the number of final orientations. In [19] it is shown how (with some simplifying assumptions) three-dimensional parts can be oriented using a tray-tilting manipulator. Zumel [45] used a variation of the tray tilting idea to orient planar parts with a pair of moveable palms.

In recent years a lot of work has been done on programmable force fields to orient parts [9, 10, 28]. The idea is that a kind of force "field" (implemented using, e.g., MEMS actuator arrays) can be used to push the part in a certain orientation. Kavraki [28] presented a vector field that induced two stable configurations for most parts. Böhringer et al. [9, 10] used Goldberg's algorithm [22] to define a sequence of "squeeze fields" to orient a part. They also gave an example how programmable vector fields can be used to simultaneously sort different parts and orient them.

2.2 Stable Poses

To compute the stable poses of an object quasistatic dynamics is often assumed. Furthermore, usually it is assumed that the part is in contact with a flat surface and is initially at rest. Boothroyd et al. [11] were among the first to analyze this problem. An $O(n^2)$ algorithm for n-sided polyhedrons, based on Boothroyd et al.'s ideas, was implemented by Wiegley et al. [44]. Goldberg et al. [21] improve this method by approximating some of the dynamic effects. Kriegman [30] introduced the notion of a *capture region*: a region in configuration space such that any initial configuration in that region will converge to one final configuration. Note that his work doesn't assume quasistatic dynamics; as long as the part is initially at rest and in contact, and the dynamics in the system are dissipative, the capture regions will be correct. The capture regions will in general not cover the entire configuration space.

2.3 Collision and Contact Analysis

For rigid body collisions several models have been proposed. Many of these models are either too restrictive (e.g., Routh's model [39] constrains the collision impulse too much) or allow physically impossible collisions (e.g., Whittaker's model [42] can predict arbitrarily high increases of system kinetic energy). Recently, Chatterjee and Ruina [15] proposed a new collision rule, which avoids many of these problems. Chatterjee introduced a new collision parameter (besides the coefficients of friction and restitution): the coefficient of *tangential* restitution. With this extra parameter a large part of allowable collision impulses can be accounted for, and at the same time this collision rule restricts the predicted collision impulse to the allowable part of impulse space. This is the collision rule we will use (see [36] for details).

Instead of having algebraic laws, one could also try to model object interactions during impact. This approach is taken, for instance, by Bhatt and Koechling [7, 8], who modeled impacts as a flow problem. While this might lead to more accurate predictions, it is obviously computationally more expensive. Also, in order to get a good approximation of the pdf of resting configurations, this level of accuracy might not be required. On the other hand, it is also possible to combine the effects of multiple collisions that happen almost instantaneously. Goyal et al. [23, 24] studied these "clattering" motions and derived the equations of motion.

Given a collision model and the equations of motion, one can simulate the motion of a part. In cases where there are a large number of collisions or with contact modes that change frequently one can simulate the dynamics using so-called impulse-based simulation [35]. However, there are limits to what systems one can simulate. Under certain conditions the dynamics become chaotic [13, 20, 29]. We are mostly interested in systems that are *not* chaotic, but where the dynamics can not be modeled with a quasistatic approximation. In Section 4.1 a number of "chaos plots" are shown that are very similar to the one in [29].

2.4 Shape Design

The shape of an object and its environment imposes constraints on the possible motions of an object. Caine [14] presents a method to visualize these motion constraints, which can be useful in the design phase of both part and manipulator. In [31] the mechanics of entrapment are analyzed. That is, Krishnasamy discusses conditions for a part to "get trapped" and "stay trapped" in an extrusion in the context of the APOS parts feeder. Sanderson [40] presents a method to characterize the uncertainty in position and orientation of a part in an assembly system. This method takes into account the shape of both part and assembly system. In [33] the optimal manipulator shape and motion are determined for a particular part. The problem here was not to orient the part, but to perform a certain juggler's skill (the "butterfly"). With a suitable parametrization of the shape and motion of the manipulator, a solution was found for a disk-shaped part that satisfied their motion constraints. Although the analysis focuses just on the juggling task, it shows that one can simulate and optimize dynamic manipulation tasks using a suitable parametrization of manipulator (or surface) shape and motion.

3 Analytic Results

3.1 Quasi-Capture Intuition

In our efforts to analyze pose distributions in a dynamic environment, we have been working on a generalization of so-called "capture regions" [30] that we have termed *quasi-capture regions*. Specifically, for a part in contact with a sloped surface, we would like to determine whether it is captured, i.e., whether the part will converge to the closest stable pose. For simplicity, let the surface be a tilted plane.

Definition 1 Let a *pose* be defined as a point in configuration space such that the part is contact with the surface.

We assume that friction is sufficiently high so that a part cannot slide for an infinite amount of time. In general capture depends on the whole surface and everything that happens after the current state, but the friction assumption and our definition of pose allow us to define quasi-capture (in Section 3.2) of the part in terms of local state. The closest stable pose can be defined as follows:

Definition 2 We define a *stable pose* to be a pose such that there is force balance when only gravity and contact forces are acting on the part. The *closest stable pose* is the stable pose found by following the gradient of the potential energy function (using, e.g., gradient descent) from the current pose.

We can now define quasi-capture regions:

Definition 3 A *quasi-capture region* is the largest possible region in configuration phase space such that (a) all configurations in this region have the same closest stable pose and (b) no configuration in a quasi-capture region has enough (kinetic and potential) energy to leave this region either with a rolling motion or one collision-free motion.

Ideally these quasi-capture regions would induce a partition of configuration phase space, so that for each point in phase space we would immediately know what its final resting configuration is. Of course, this is not the case in general, since with a sufficiently large velocity an object can reach any stable pose. But if we restrict the velocity to be small to begin with, then we are able to quickly determine the pose distribution. It has been our experience that without the use of quasi-capture regions a lot of computation time is spent on the final part/surface interactions (e.g., clattering motions) before the part reaches a stable pose. In other words, *with our analytic results it is possible to avoid computing a potentially large number of collisions.*

In our analysis we have focused on the two dimensional case. To illustrate the notion of capture, we will start with another example. Consider a rod of length l with center of mass at distance R from each vertex. One can visualize this rod as a disk with radius R and uniform mass, but with contact points only at the ends of the rod (see Figure 2).

Note that the endpoints of the rod are numbered. We will refer to these endpoints later. Let the "side" of the rod where the center of mass is above the rod be the *high energy side*, and the other side be the *low energy side*. We can then define that the rod is "on" the high energy side if and only if the center of mass is

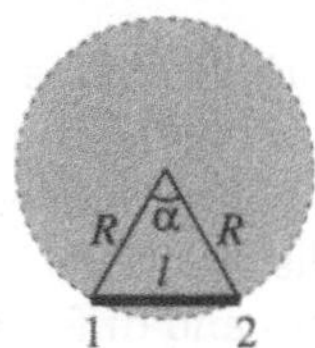

Figure 2: *A rod with an off-center center of mass.*

between and above the endpoints of the rod. Suppose the rod is in contact with a flat, horizontal surface. For the rod to make a transition from one side to the other, it will have to rotate, either by rolling or by bouncing. At some point during the transition the center of mass will pass over one of the endpoints of the rod. Its potential energy at that point will always be greater than or equal to the potential energy at the start of the transition. Hence, to make that transition the rod has to have a minimum amount of kinetic energy. This can be written more formally as

$$\tfrac{1}{2}m\|\mathbf{v}\|^2 > -mg\Delta h. \tag{1}$$

Figure 3 illustrates this requirement.

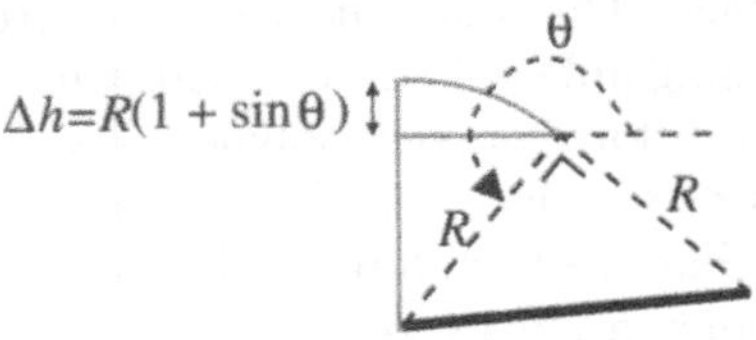

Figure 3: *Capture condition for a rod.*

For a polygonal object in contact with a surface with constant slope we will derive in Section 3.2 a lower bound on the norm of the velocity such that for all velocities below that bound the part will be quasi-captured. As we vary the position of the center of mass with respect to the rod endpoints, the slope of the support surface and the drop height, the bound for the capture velocity will change. This bound will also depend on the relative orientation of the contact point with respect to the center of mass.

For a sloped surface the capture condition is not as simple as for the horizontal surface. By bouncing and rolling down the slope, the rod can increase its kinetic energy. We have derived an upper bound on how far the rod can bounce. This gives an upper bound on the increase in kinetic energy. So the quasi-capture condition can now be stated as: the current kinetic energy plus the maximum gain in kinetic energy has to be less than the energy required to rotate to the other side. To guarantee that the rod is indeed captured, we have to make sure that the maximum gain in kinetic energy is less than the decrease in kinetic energy due to a collision. There are some additional complicating factors. For instance, a change in orientation can increase the kinetic energy, but to rotate to the other side the rod has to rotate back, undoing the gain in kinetic energy.

3.2 Quasi-Capture Velocity

What we will prove is a sufficient condition on the pose and velocity of the rod such that it is quasi-captured. The condition will be of the following form: if the current kinetic energy plus the maximal increase in kinetic energy is less than some bound, the rod is quasi-captured. This bound depends on the current orientation, the current velocity, the slope of the surface and the geometry of the rod. Because of the way we have set up our generalized coordinates (see [36] for details), the kinetic energy is $\frac{1}{2}m\|\mathbf{v}\|^2$. In other words, the mass is just a constant scalar. Without loss of generality we can assume $m = 2$. That way the kinetic energy is simply $\|\mathbf{v}\|^2$. We will write v for $\|\mathbf{v}\|$.

Theorem 4 *The rod with a velocity vector of length v and in contact with the surface is in a quasi-capture region if the following condition holds:*

$$v^2 + \frac{2v\cos\xi\sin\phi}{\cos^2\phi}\left(v\sin(\xi+\phi)+\sqrt{v^2\sin^2(\xi+\phi)-2gd_n\cos\phi}\right) - 2g\left(\frac{d_n}{\cos\phi}+R\varepsilon\right) \le -2gR\left(1+\cos\left(\tfrac{\alpha}{2}+\phi\right)\right),$$

where ξ is the direction of the velocity vector that will result in the largest increase of kinetic energy, θ is the relative orientation of the contact point, $d_n = R(\cos\frac{\alpha}{2} - \sin(\theta+\phi))$ and $\varepsilon = \cos(\frac{\alpha}{2}+\phi) - \frac{\cos(\alpha/2)}{\cos\phi} + \max\left(\tan\phi,\ 2\sin\frac{\alpha}{2}\sin\phi\right)$.

Proof: See appendix A. □

Note that for $\phi = 0$ this bound reduces to $v^2 \le -2gR(1+\sin\theta)$. In other words, this bound is as tight as possible when the surface is horizontal.

One can compute ξ numerically, but the appendix also gives a good analytic approximation. The theorem above gives a sufficient condition on the velocity and pose of the rod such that it cannot rotate to the other side during one bounce. But suppose there is a

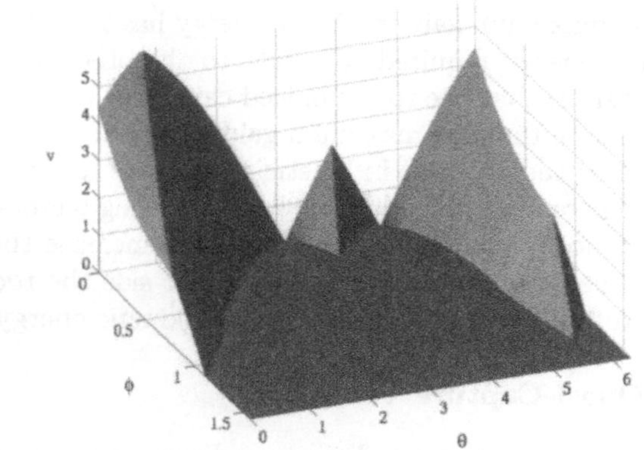

Figure 4: *Quasi-capture velocity as a function of the slope of the surface and the orientation of the rod*

sequence of bounces, each of them increasing the kinetic energy. It is possible that the rod satisfies the quasi-capture condition, but is still able to rotate to the other side in more than one bounce. Thus, the theorem by itself is not enough to guarantee that the rod will converge to its closest stable orientation. In the analysis above we have ignored the dissipation of kinetic energy during collisions. If in the case the capture condition is true the dissipation of kinetic energy is larger than the increase due to the bounce, the rod will indeed be captured after an arbitrary number of bounces. To make sure this is the case the coefficients of restitution can not be too large.

In Figure 4 the quasi-capture velocity is plotted as a function of the slope of the surface and the orientation of the rod. The slope ranges from 0 to $\frac{\pi}{2}$ and the orientation ranges from 0 to 2π. Note that the orientation of the rod is not the same as the relative orientation of the contact point. However, for each combination of ϕ and θ the relative orientation of the contact point can be easily computed. The other relevant parameter values for this plot are: $R = 1m$, $g = -9.81m/s^2$ and $\alpha = \frac{\pi}{2}$. The little bump in the middle corresponds to the rod being captured on the high-energy side. The bigger bumps on the left and right correspond to being captured on the low-energy side.

4 Simulation and Experimental Results

4.1 Dynamic Simulation

To numerically compute the pose distribution of parts, we have written two dynamic simulators. One is based on David Baraff's Coriolis simulator [2, 3], which can simulate the motions of polyhedral rigid bodies. Coriolis takes care of the physical modelling. Our simulator then computes pose distributions for different (parametrized) support surfaces and different initial conditions.

Our simulator uses simulated annealing to optimize over the surface parameters and drop location with respect to the surface. The objective function is to minimize the entropy of the pose distribution. Initially the sampling of orientations of the object is rather coarse, so that the resulting pose distribution is not very accurate. But as the simulator is searching, the simulated annealing algorithm is restarted with an increased sample size and the best current solution as initial guess. This way we can quickly determine the potentially most interesting parameter values and refine them later. Our implementation is based on the one given in [37, pp. 444–455].

Surfaces are parametrized using wavelets [41, 17]. Wavelet transforms are similar to the fast Fourier transform, but unlike the fast Fourier transform basis functions (sines and cosines) wavelet basis functions are localized in space. This localization gives us greater flexibility in modeling different surfaces compared with the fast Fourier transform or, say, polynomials. There are many classes of wavelet basis functions. We are using the Daubechies wavelet filters [17] and in particular the implementation as given in [37, pp. 591–606]. To reduce an arbitrary surface to a small number of coefficients we first discretize the function describing the surface. We then perform a wavelet transform and keep the largest components (in magnitude) in the transform to represent the surface. When we minimize the entropy, we optimize over these components. We can either keep the smaller components of our initial wavelet transform around or set them to zero.

Development of a second simulator was started, because Coriolis had some limitations. In particular, the collision model could not be changed and we wanted to experiment with Chatterjee's collision model [15]. The second simulator also allowed us to optimize for our specific dynamics model. In our model there is only one moving object, and the only forces acting on it are gravity and friction. Currently, the simulator only handles two-dimensional objects, but in the future it might be extended to handle three dimensions as well. It uses the analytic results from the previous section to stop simulating the motion of the part once it is captured.

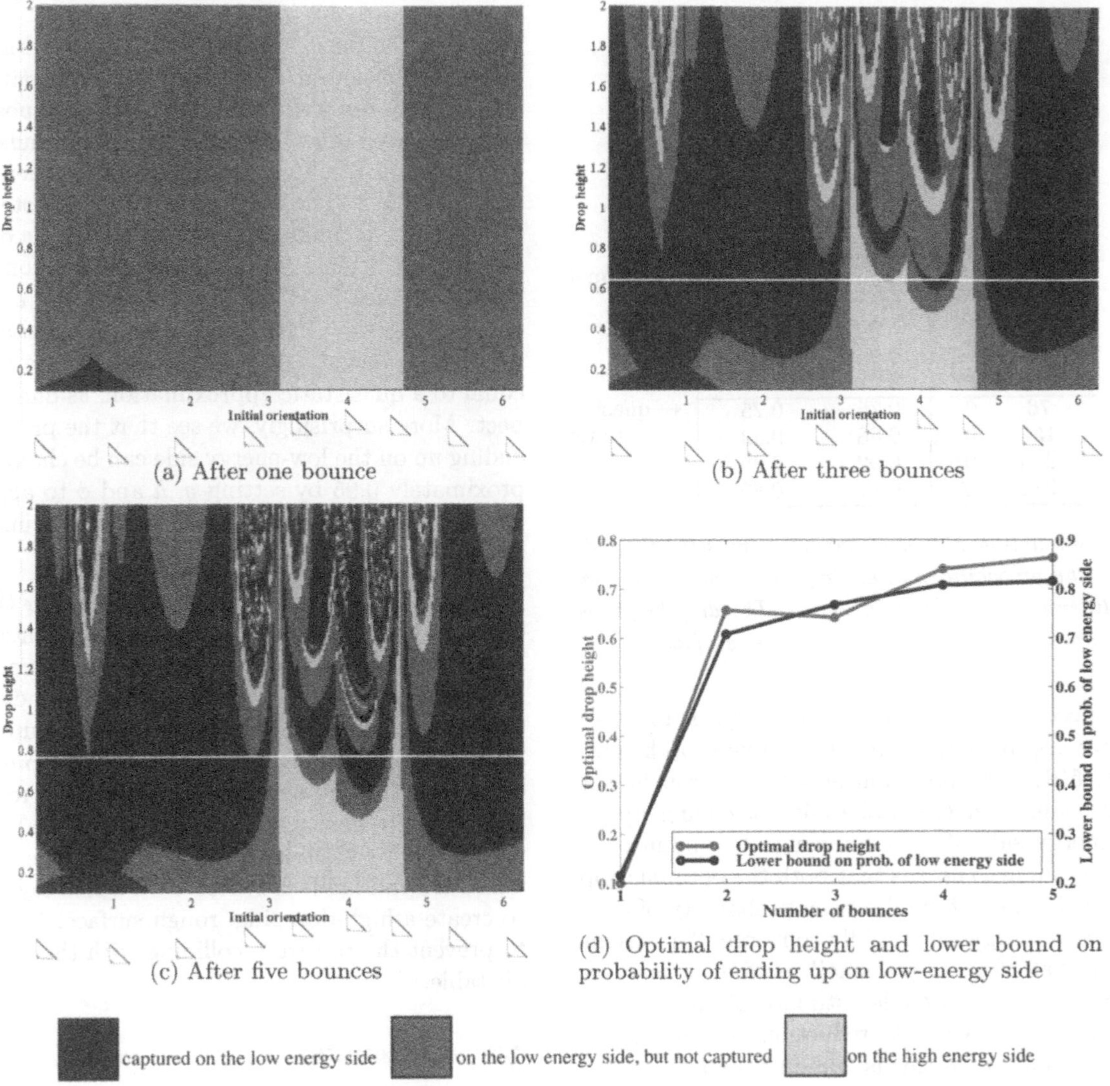

Figure 5: ***Quasi-capture regions for the low energy side of the rod on a 7° slope.*** *The rod parameters are* $\alpha = \pi/2$, $R = 0.05m$, *gravity,* g, *equals* $-1.20m/s^2$, *the coefficient of friction,* μ, *is 5 and the restitution parameters are:* $e = 0.1$, $e_t = -0.2$.

Using the simulator we can compute the quasi-capture regions for the rod. Figures 5(a)–(c) show the quasi-capture regions for the low energy side after one, three and five bounces. The dark areas correspond to initial orientations and initial velocities that result in the rod being quasi-captured. The zero orientation is defined as the orientation where endpoint 1 is to the right of the center of mass. The triangles below the X-axes show the pose of the rod corresponding to the orientation at that point of the X-axes.

Let the optimal drop height be defined as the drop height that maximizes the probability of ending up on the low energy side. Then dropping the rod with uniformly random initial orientation from the optimal drop height will reduce uncertainty about its orientation maximally (unless there exists a drop height that will result in an even higher probability for the high energy side). In Figures 5(a)–(c) the drop height that results in the maximum probability of ending up on the low energy side is marked by a horizontal line. Af-

g (m/s^2)	h (mm)	ϕ	Prob. low energy side Sim.	Exp.	
-0.68	58	20°	0.85	0.94	
-0.68	122	20°	0.90	0.94	
* -0.68	186	20°	0.91	0.93	
-0.68	246	20°	0.93	0.96	
-1.5	58	20°	0.85	0.93	
* -1.5	122	20°	0.90	0.92	
-1.5	186	20°	0.91	0.97	
-1.5	246	20°	0.93	0.97	← optimal
-2.6	58	20°	0.85	0.94	
-2.6	122	20°	0.90	0.93	
-2.6	186	20°	0.91	0.93	
* -2.6	246	20°	0.91	0.94	
-2.6	76	0°	0.75	0.75	← quasi-static
-2.6	156	0°	0.88	0.83	
* -2.6	220	0°	0.92	0.85	
-2.6	284	0°	0.87	0.89	

Table 2: *Simulation and experimental results for the rod. Shown are the probabilities of ending up on the low-energy side for different values for g, h and ϕ. The drop height is measured from the center of the disk to the surface.*

ter each successive bounce this drop height is likely to be a better approximation of the optimal drop height. In Figure 5(d) the approximate optimal drop height and lower bound on the probability of ending up on the low-energy side after one through five bounces is shown. One thing to note is that both the optimal drop height and the lower bound on the probability of ending up on low-energy side rapidly converge. This seems to suggest that after only a small number of bounces we could make a reasonable estimate of the optimal drop height and uncertainty reduction. Further study is needed to find out if this is true in general.

4.2 2D Results

To verify the simulations we also performed some experiments. Our experimental setup was as follows. We used an air table to effectively create a two-dimensional world. By varying the slope of the air table we can vary the gravity. At the bottom of the slope is the surface on which the object will be dropped. The angle ϕ of the surface in the plane defined by the air table can, of course, be varied.

The rod of the previous section has been implemented as a plastic disk with two metal pins sticking out from the top at an equal distance from the center of the disk. When released from the top of the air table the disk can slide under the surface and will only collide at the pins. Experimentally we determined the pose distribution of the rod for different values for g, h and ϕ by determining the final stable pose for 72 equally spaced initial orientations. Our simulation and experimental results of some tests have been summarized in table 2. The rows marked with an asterisk have been used to estimate the moment of inertia of the rod and the coefficients of friction and restitution. The estimated values for these parameters are: $e = 0.404$, $e_t = -0.136$, $\rho = 0.0376$ and $\mu = 4.71$. Note that for a low drop height and a horizontal surface the pdf is equal to a quasistatic approximation, as one would expect. More surprisingly, we see that the probability of ending up on the low-energy side can be changed to approximately 0.95 by setting g, h and ϕ to appropriate values. In other words, we can reduce the uncertainty almost completely.

One can identify several error sources for the differences between the simulation and experimental results. First, there are measurement errors in the experiments: in some cases slight changes in the initial conditions will change the side on which the rod will end up. Second, since the simulations are run with finite precision, it is possible that numerical errors affect the results. Finally, the physical model is not perfect. In particular, the rigid body assumption is just false. The surface on which the rod lands is coated with a thin layer of foam to create a high-damping, rough surface. This is done to prevent the rod from colliding with the sides of the air table.

4.3 3D Results

We have not generalized our analytic results to three dimensions yet, but we can still use our optimization technique to find a good surface and drop height for a given object. For the dynamic simulation we rely now on Baraff's Coriolis simulator. Figure 6(a) shows an orange insulator cap[2] at rest on flat, horizontal surface. The contact points are marked by the little spheres. In Figure 6(b) the bowl resulting from the simulated annealing search process is shown. The initial shape is a paraboloid: $f(x,y) = (x^2 + y^2)/20$. This shape is reduced to a triangulation of a 8×8 regular grid. The part is always released on the left-hand side of the bowl.

[2]This object has been used before as an example in [21, 30, 38].

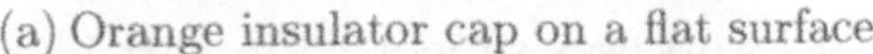

(a) Orange insulator cap on a flat surface

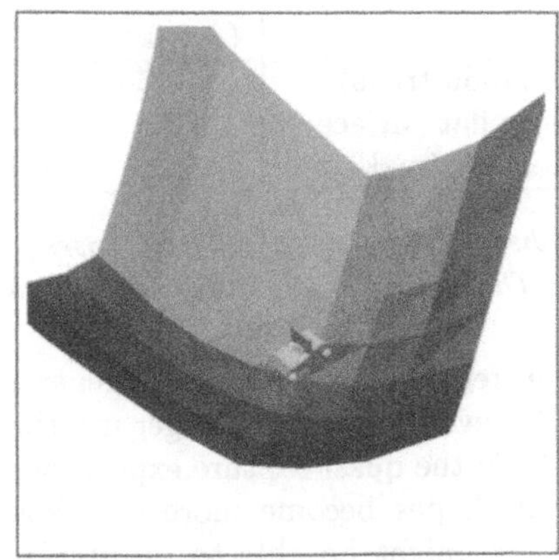

(b) ... and on an optimized bowl

Figure 6: ***Result of optimizing a surface for the orange insulator cap.***

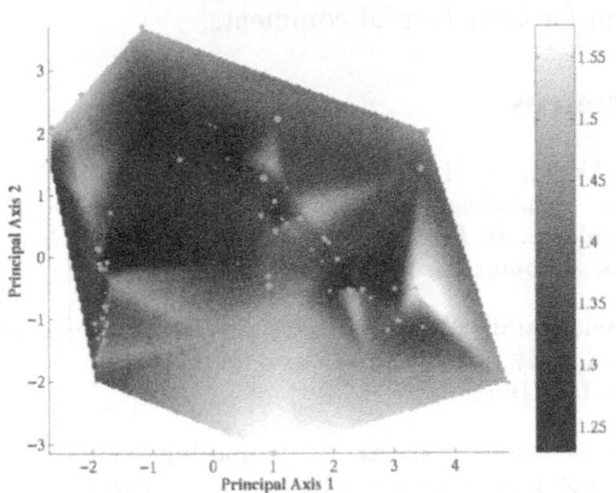

Figure 7: ***Entropy as a function of the two principal axes of the searched five-dimensional parameter space.***

We optimized over the four largest wavelet coefficients of the initial shape and the drop height. The search for the optimal bowl and drop height is visualized in Figure 7. The five-dimensional parameter space is projected onto a two-dimensional space using Principal Component Analysis [27]. Each point corresponds to a bowl shape evaluation, i.e., for each point a pose distribution is computed. The size of each point is proportional to the sample size used to determine the pose distribution. Computing a pose distribution by taking 600 samples takes about 40 minutes on a 500 MHz Pentium III. The surface in Figure 7 is a cubic interpolation between the points. The dark areas correspond to areas of low entropy. Notice that most of the points are in or near a dark area.

Table 3 compares the simulation results with experimental data from [21]. The format is the same as in table 1, except that the stable poses are now written as vectors. These vectors are the outward pointing normals (w.r.t. the center of mass) of the planes passing through the contact points. That way, a face with many vertices in contact with the surface will always be represented by the same vector, no matter which subset of the vertices is actually in contact. In the experimental setup of [21] the part was dropped from one conveyer belt onto another. The initial drop height was 12.0 cm. In the experiments the part had an initial horizontal velocity of 5.0 cm/s. The second row corresponds to computing the pose distribution when the part is dropped from 12.0 cm (but with initial velocity set to 0). The third row corresponds to a local minimum returned by the simulated annealing algorithm.

5 Discussion

We have shown a sufficient condition on the position and velocity of the simplest possible "interesting" shape (i.e., the rod) that guarantees convergence to the nearest stable pose under some assumptions. This condition gives rise to regions in configuration phase-space, where each point within such a region will converge to the same stable pose. We have coined the term quasi-capture regions for these regions, since they are very similar to Kriegman's notion of capture regions. In simulations quasi-capture always seems to imply capture, but further research is needed to prove this claim.

	Stable Poses						Entropy
	$(-1,0,0)$	$(0,-1,0)$	$(0,1,0)$	$(.8,0,.6)$	$(.7,0,-.7)$	$(0,0,-1)$	
Experimental, flat (1036 trials)	0.271	0.460		0.197	0.050	0.022	1.58
Dynamic simulation, flat surface	0.355	0.207	0.221	0.185	0.019	0.014	1.48
Dynamic simulation, optimal bowl	0.622	0.125	0.154	0.096	0.003	0.000	1.09

Table 3: *Probability distribution function of stable poses for two surfaces. The initial velocity is zero and the initial rotation is uniformly random. The experimental data is taken from [21]. There,* $(0,-1,0)$ *and* $(0,1,0)$ *are counted as one pose.*

The quasi-capture regions also apply to general polygonal shapes. However, we can no longer use the symmetry of the rod. So the quasi-capture expressions for general polygonal shapes become more complex. On the other hand, we might be able to orient planar parts by using a setup similar to the one described in Section 4 and attaching two pins to the top of the part. Generalizing the quasi-capture regions to three dimensions is non-trivial and is an interesting direction for future research.

The simulation and experimental results show that the simulator is not 100% accurate, but that it is a useful tool for determining the most promising initial conditions for uncertainty reduction. In other words, the optimum predicted by the simulator will probably be near-optimal in the experiments. We can then experimentally search for the true optimum.

Another area where quasi-capture regions may be applied is in computer animation. Before a part comes to rest, there are many interactions between the part and the support surface. It turns out that these interactions are computationally very expensive. With our capture regions we can eliminate the last "clattering" motions of the part, since we can predict what the final pose will be. For applications where fast animation is more important than physical accuracy, a pre-computed motion can be substituted for the actual motion.

With future research we hope to improve the constraints on the quasi-capture velocity by taking into account more information, such as the direction of the velocity vector. If improving the quasi-capture bounds is impossible, it might be possible to get better approximations for pose distributions. As noted in Section 4.1 it is possible to get a good estimate of the maximal uncertainty reduction after only a small number of bounces of the rod. So another interesting line of research would be to find out how accurate these approximations are in general. We are also planning to do more experiments to verify our current and future analytic results.

Acknowledgments

This work was supported in part by the National Science Foundation under grant IRI-9503648. The authors are grateful to Al Rizzi, Matt Mason and Devin Balkcom for their helpful comments.

References

[1] S. Akella, W. H. Huang, K. M. Lynch, and M. T. Mason. Sensorless parts orienting with a one-joint manipulator. In *Proc. 1997 IEEE Intl. Conf. on Robotics and Automation*, 1997.

[2] David Baraff. Coping with friction for non-penetrating rigid body simulation. *Computer Graphics*, 25(4):31–40, July 1991.

[3] David Baraff. Issues in computing contact forces for non-penetrating rigid bodies. *Algorithmica*, pages 292–352, October 1993.

[4] Dina R. Berkowitz and John Canny. Designing parts feeders using dynamic simulation. In *Proc. 1996 IEEE Intl. Conf. on Robotics and Automation*, pages 1127–1132, April 1996.

[5] Dina R. Berkowitz and John Canny. A comparison of real and simulated designs for vibratory parts feeding. In *Proc. 1997 IEEE Intl. Conf. on Robotics and Automation*, pages 2377–2382, Albuquerque, New Mexico, 1997.

[6] Robert-Paul Berretty, Ken Goldberg, Lawrence Cheung, Mark H. Overmars, Gordon Smith, and A. Frank van der Stappen. Trap design for vibratory bowl feeders. In *Proc. 1999 IEEE Intl. Conf. on Robotics and Automation*, pages 2558–2563, Detroit, MI, May 1999.

[7] Vivek Bhatt and Jeff Koechling. Partitioning the parameter space according to different behavior during 3d impacts. *ASME Journal of Applied Mechanics*, 62 (3):740–746, September 1995.

[8] Vivek Bhatt and Jeff Koechling. Three dimensional frictional rigid body impact. *ASME Journal of Applied Mechanics*, 62(4):893–898, December 1995.

[9] Karl Friedrich Böhringer, Vivek Bhatt, Bruce R. Donald, and Kenneth Y. Goldberg. Algorithms for sensorless manipulation using a vibrating surface. *Algorithmica*, 1997. Accepted for publication.

[10] Karl Friedrich Böhringer, Bruce Randall Donald, and Noel C. MacDonald. Programmable vector fields for distributed manipulation, with applications to MEMS actuator arrays and vibratory parts feeders. *Intl. J. of Robotics Research*, 18(2):168–200, February 1999.

[11] C. Boothroyd, A. H. Redford, C. Poli, and L. E. Murch. Statistical distributions of natural resting aspects of parts for automatic handling. *Manufacturing Engineering Transactions, Society of Manufacturing Automation*, 1:93–105, 1972.

[12] Geoffrey Boothroyd, Corrado Poli, and Laurence E. Murch. *Automatic Assembly.* Marcel Dekker, Inc., New York; Basel, 1982.

[13] M. Bühler and D. E. Koditschek. From stable to chaotic juggling: Theory, simulation, and experiments. In *Proc. 1990 IEEE Intl. Conf. on Robotics and Automation*, pages 1976–1981, 1990.

[14] Michael E. Caine. *The Design of Shape from Motion Constraints.* PhD thesis, MIT Artificial Intelligence Laboratory, Cambridge, MA, September 1993. Technical Report 1425.

[15] Anindya Chatterjee and Andy L. Ruina. A new algebraic rigid body collision law based on impulse space considerations. *ASME Journal of Applied Mechanics*, 1998. Accepted for publication.

[16] Alan D. Christiansen, Andrea Dunham Edwards, and Carlos A. Coello Coello. Automated design of part feeders using a genetic algorithm. In *Proc. 1996 IEEE Intl. Conf. on Robotics and Automation*, volume 1, pages 846–851, 1996.

[17] Ingrid Daubechies, editor. *Different Perspectives on Wavelets*, volume 47 of *Proceedings of Symposia in Applied Mathematics*, 1993. AMS.

[18] Michael A. Erdmann and Matthew T. Mason. An exploration of sensorless manipulation. *IEEE J. of Robotics and Automation*, 4(4):369–379, August 1988.

[19] Michael A. Erdmann, Matthew T. Mason, and George Vaněček, Jr. Mechanical parts orienting: The case of a polyhedron on a table. *Algorithmica*, 10:226–247, 1993.

[20] Rasmus Feldberg, Maciej Szymkat, Carsten Knudsen, and Erik Mosekilde. Iterated-map approach to die tossing. *Physical Review A*, 42(8):4493–4502, October 1990.

[21] Ken Goldberg, Brian Mirtich, Yan Zhuang, John Craig, Brian Carlisle, and John Canny. Part pose statistics: Estimators and experiments. *IEEE Trans. on Robotics and Automation*, 15(5), October 1999.

[22] Kenneth Y. Goldberg. Orienting polygonal parts without sensors. *Algorithmica*, 10(3):201–225, August 1993.

[23] S. Goyal, J. M. Papadopoulos, and P. A. Sullivan. The dynamics of clattering I: Equation of motion and examples. *J. of Dynamic Systems, Measurement, and Control*, 120:83–93, March 1998.

[24] S. Goyal, J. M. Papadopoulos, and P. A. Sullivan. The dynamics of clattering II: Global results and shock protection. *J. of Dynamic Systems, Measurement, and Control*, 120:94–102, March 1998.

[25] David D. Grossman and Michael W. Blasgen. Orienting mechanical parts by computer-controlled manipulator. *IEEE Trans. on Systems, Man, and Cybernetics*, 5:561–565, September 1975.

[26] Hajime Hitakawa. Advanced parts orientation system has wide application. *Assembly Automation*, 8(3):147–150, 1988.

[27] I. T. Jolliffe. *Principal Components Analysis.* Springer-Verlag, New York, 1986.

[28] Lydia E. Kavraki. Part orientation with programmable vector fields: Two stable equilibria for most parts. In *Proc. 1997 IEEE Intl. Conf. on Robotics and Automation*, pages 2446–2451, Albuquerque, New Mexico, 1997.

[29] Zhang Kechen. Uniform distribution of initial states: The physical basis of probability. *Physical Review A*, 41(4):1893–1900, February 1990.

[30] David J. Kriegman. Let them fall where they may: Capture regions of curved objects and polyhedra. *Intl. J. of Robotics Research*, 16(4):448–472, August 1997.

[31] Jayaraman Krishnasamy. *Mechanics of Entrapment with Applications to Design of Industrial Part Feeders.* PhD thesis, Dept. of Mechanical Engineering, MIT, May 1996.

[32] Kevin M. Lynch. Toppling manipulation. In *Proc. 1999 IEEE Intl. Conf. on Robotics and Automation*, pages 2551–2557, Detroit, MI, May 1999.

[33] Kevin M. Lynch, Naoji Shiroma, Hirohiko Arai, and Kazuo Tanie. The roles of shape and motion in dynamic manipulation: The butterfly example. In *Proc. 1998 IEEE Intl. Conf. on Robotics and Automation*, 1998.

[34] Alessia Marigo, Yacine Chitour, and Antonio Bicchi. Manipulation of polyhedral parts by rolling. In *Proc. 1997 IEEE Intl. Conf. on Robotics and Automation*, pages 2992–2997, 1997.

[35] Brian Mirtich and John Canny. Impulse-based simulation of rigid bodies. In *Proc. 1995 Symposium on Interactive 3D Graphics*, April 1995.

[36] Mark Moll and Michael A. Erdmann. Manipulation of pose distributions. Technical Report CMU-CS-00-111, Dept. of Computer Science, Carnegie Mellon University, 2000. http://www.cs.cmu.edu/ mmoll/publications.

[37] William H. Press, Saul A. Teukolsky, William T. Vetterling, and Brian P. Flannery. *Numerical Recipes in C: The art of Scientific Computing*. Cambridge University Press, second edition, 1992.

[38] Anil Rao, David Kriegman, and Ken Goldberg. Complete algorithms for reorienting polyhedral parts using a pivoting gripper. In *Proc. 1995 IEEE Intl. Conf. on Robotics and Automation*, May 1995.

[39] Edward John Routh. *Dynamics of a System of Rigid Bodies*. MacMillan and Co., London, sixth edition, 1897.

[40] Arthur C. Sanderson. Parts entropy methods for robotic assembly system design. In *Proc. 1984 IEEE Intl. Conf. on Robotics and Automation*, pages 600–608, 1984.

[41] G. Strang. Wavelets and dilation equations: A brief introduction. *SIAM Review*, 31:613–627, 1989.

[42] E. T. Whittaker. *A Treatise on the Analytical Dynamics of Particles and Rigid Bodies*. Dover, New York, fourth edition, 1944.

[43] Jeff Wiegley, Ken Goldberg, Mike Peshkin, and Mike Brokowski. A complete algorithm for designing passive fences to orient parts. In *Proc. 1996 IEEE Intl. Conf. on Robotics and Automation*, apr 1996.

[44] Jeff Wiegley, Anil Rao, and Ken Goldberg. Computing a statistical distribution of stable poses for a polyhedron. In *30th Annual Allerton Conf. on Communications, Control and Computing*, 1992.

[45] Nina B. Zumel. *A Nonprehensile Method for Reliable Parts Orienting*. PhD thesis, Robotics Institute, Carnegie Mellon University, Pittsburgh, PA, January 1997.

Appendix A: Proof of Theorem 4

Definition 5 Let a bounce be defined as the flight path between two impacts.

The closest distance between the rod and the slope during one bounce can be described by

$$d(t) = \tfrac{1}{2}g(\cos\phi)t^2 + (v_y\cos\phi + v_x\sin\phi)t - d_{\theta(t)}, \quad (2)$$

where v_x and v_y are the translational components of the velocity and $d_{\theta(t)}$ is a component that depends on the orientation. Let the rod be in contact at $t = 0$. Then $d(0) = 0$ (and therefore $d_{\theta(0)} = 0$). Let $\hat{t}$ be the smallest positive solution to $d(t) = 0$. The change in height is then $\Delta h = \frac{1}{2}g\hat{t}^2 + v_y\hat{t}$, so that the change in v^2 is $\Delta v^2 = 2g\Delta h = g^2\hat{t}^2 + 2v_yg\hat{t}$. To find the maximum Δv^2 for all velocity vectors of length v we can parametrize the translational velocity as $v_x = v\cos\xi$ and $v_y = v\sin\xi$, and maximize over ξ. This ignores the rotational component of the velocity, but the following lemma shows that for a certain value of $d_{\theta(\hat{t})}$ the resulting solution for Δv^2 is an upper bound for the true maximal increase of v^2.

Definition 6 Let the *ideal orientation* be defined as the orientation where the rod is parallel to the surface and the center of mass is below the rod.

Lemma 7 *We can always increase the rod's kinetic energy after a bounce by allowing it to rotate around the center of mass "for free" (i.e., without using energy) to the ideal orientation (ignoring penetrations of the surface) and then letting it continue to fall while maintaining this orientation. However, if the rod is already in the ideal orientation after the bounce, its kinetic energy cannot be increased.*

Proof: One can easily verify that rotating around the center of mass to the ideal orientation of the bounce maximizes distance between the rod and the surface. This distance will always be greater than or equal to 0. If we allow the rod to continue to fall until it hits the surface, its kinetic energy will increase. □

From this lemma it follows that by assuming the rod rotates "for free" to the ideal orientation the increase in kinetic energy due to one bounce is an upper bound on the true increase of kinetic energy. With this lemma

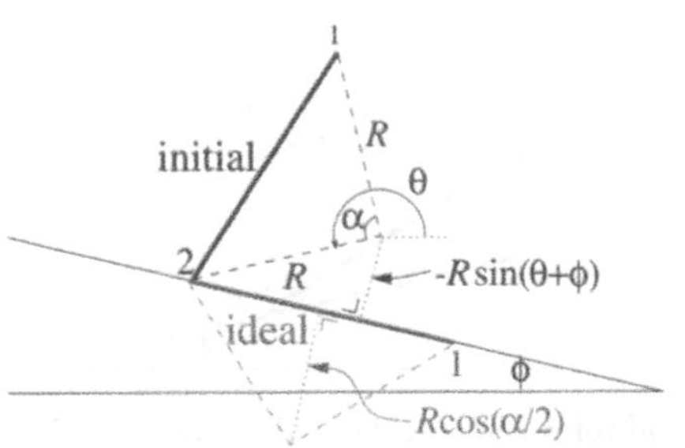

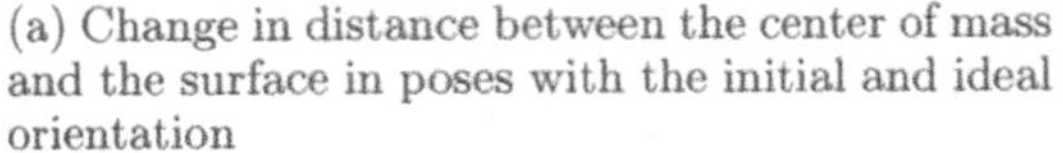
(a) Change in distance between the center of mass and the surface in poses with the initial and ideal orientation

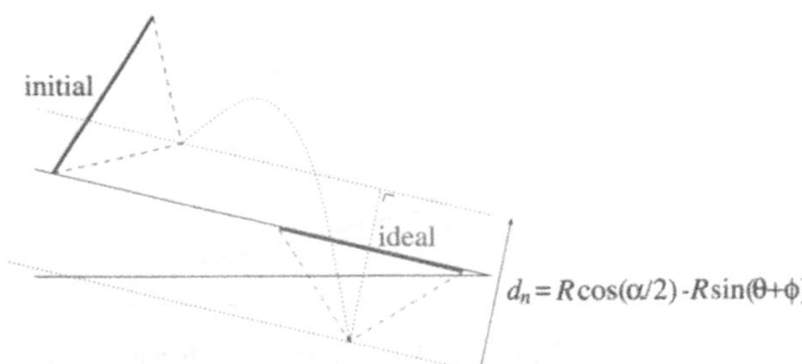

(b) Trajectory of the center of mass during a bounce

Figure 8: *Increase in kinetic energy when rotating to the ideal orientation*

computing the next contact point is a lot easier. Let θ be the relative orientation of the contact point at $t=0$. $\theta=0$ corresponds to the contact point being to the right of the center of mass. The signed distance from the center of mass to the surface at $t=0$ is then $-R\sin(\theta+\phi)$, as shown in Figure 8(a). One can easily verify that in the ideal orientation the relative orientation of endpoint 1 is $\frac{\pi}{2}-\frac{\alpha}{2}-\phi$. Let $\hat{\theta}$ be equal to this relative orientation. In the pose where the rod is in contact with the surface and has the ideal orientation the signed distance from the center of mass to the surface is $-R\sin(\hat{\theta}+\phi)=-R\cos\frac{\alpha}{2}$. So in total the center of mass travels at most a distance $R(\cos\frac{\alpha}{2}-\sin(\theta+\phi))$ in the direction normal to the surface during one bounce. Let d_n be equal to this distance. To solve for the time of impact we can treat the rod as a point mass centered at the center of mass and replace $d_{\theta(\hat{t})}$ in Equation 2 with $-d_n$. Equation 2 is then simply a paraboloid in t. The distance function now measures the distance between the center of mass and the dotted line parallel to the surface shown in Figure 8(b). This approach is not limited to the case where our new orientation is the ideal orientation. Suppose an oracle would tell us that the new orientation is $\tilde{\theta}$. Then we can solve for the time of impact by substituting $R(\sin(\tilde{\theta}+\phi)-\sin(\theta+\phi))$ for $d_{\theta(\hat{t})}$ in Equation 2.

The following lemma gives a bound on the velocity needed to *roll* to the other side.

Lemma 8 *If the rod is in rolling contact, then to be able to roll to the other side the following condition has to hold:* $v^2 > -2gR(1+\sin\theta+(\mathrm{sign}(\cos\theta)-1)\sin\frac{\alpha}{2}\sin\phi)$. *We assume* $0\le\phi<\frac{\pi}{2}$.

Proof: We can distinguish several cases: endpoint 1 of the rod or endpoint 2 can be in contact with the slope, and the rod can be on the low or high energy side. We will prove the case where endpoint 1 is in contact and the rod is on the high energy side. The proof for the other cases is analogous. The case under consideration is shown in Figure 9(a). To roll counterclockwise over to the left side, $v^2>-2gh_1$. The distance h_1 is simply equal to $R(1+\sin\theta)$. If the rod rolls clockwise over to two-point contact and continues to roll over endpoint 2, the rod gains kinetic energy because the second contact point is lower than the first contact point. This gain is proportional to h_2.

One can easily verify that for two-point contact the relative orientations of contact points 1 and 2 are $\frac{3\pi}{2}-\frac{\alpha}{2}-\phi$ and $\frac{3\pi}{2}+\frac{\alpha}{2}-\phi$, respectively. The bound for rolling over endpoint 2 is therefore

$$\begin{aligned} v^2 &> -2g(h_3-h_2)\\ &= -2gR(1+\sin\theta-\cos(\tfrac{\alpha}{2}-\phi)+\cos(\tfrac{\alpha}{2}+\phi))\\ &= -2gR(1+\sin\theta-2\sin\tfrac{\alpha}{2}\sin\phi). \end{aligned}$$

If the center of mass is to the left of the contact point the last term will change sign. We can therefore combine the two bounds (one for rotating clockwise, and for rotating counterclockwise) into this bound:

$$\begin{aligned} v^2 > \min(&-2gR(1+\sin\theta),\\ &-2gR(1+\sin\theta+2\,\mathrm{sign}(\cos\theta)\sin\tfrac{\alpha}{2}\sin\phi))\\ =&-2gR(1+\sin\theta+(\mathrm{sign}(\cos\theta)-1)\sin\tfrac{\alpha}{2}\sin\phi). \end{aligned} \tag{3}$$

□

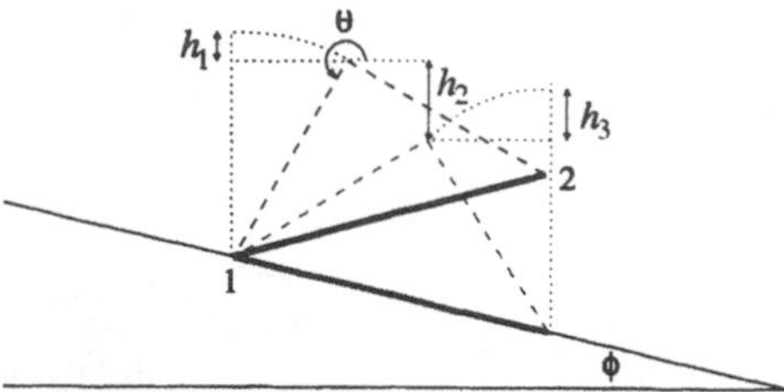

(a) Endpoint 1 in contact, high energy side

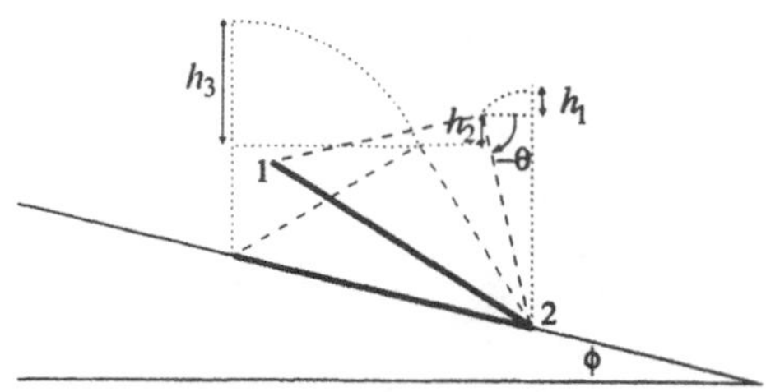

(b) Endpoint 2 in contact, high energy side

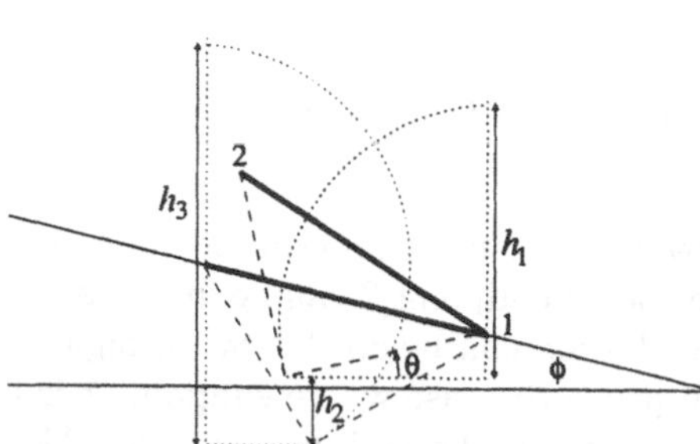

(c) Endpoint 1 in contact, low energy side

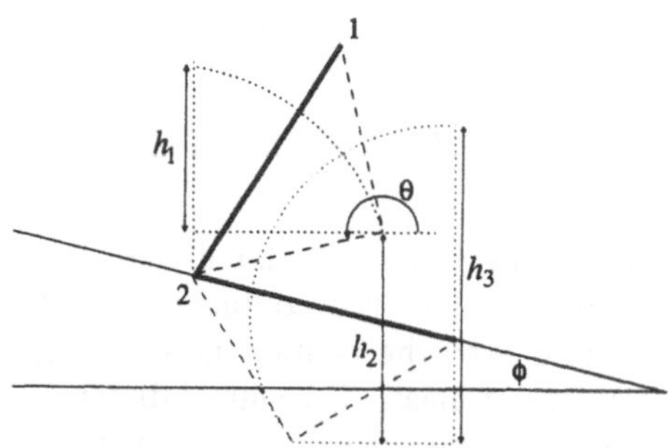

(d) Endpoint 2 in contact, low energy side

Figure 9: *Capture condition for rotation*

Theorem 4 *The rod with a velocity vector of length v and in contact with the surface is in a quasi-capture region if the following condition holds:*

$$v^2+\frac{2v\cos\xi\sin\phi}{\cos^2\phi}\left(v\sin(\xi+\phi)+\sqrt{v^2\sin^2(\xi+\phi)-2gd_n\cos\phi}\right)$$
$$-2g\left(\frac{d_n}{\cos\phi}+R\varepsilon\right)\leq -2gR\left(1+\cos\left(\frac{\alpha}{2}+\phi\right)\right),$$

where ξ is the direction of the velocity vector that will result in the largest increase of kinetic energy, $d_n = R(\cos\frac{\alpha}{2}-\sin(\theta+\phi))$ and $\varepsilon=\cos(\frac{\alpha}{2}+\phi)-\frac{\cos(\alpha/2)}{\cos\phi}+\max\left(\tan\phi,\ 2\sin\frac{\alpha}{2}\sin\phi\right)$.

Proof: The path of the center of mass during a bounce that increases the kinetic energy maximally is described by $\frac{1}{2}gt^2\cos\phi+v(\sin\xi\cos\phi+\cos\xi\sin\phi)t+d_n=0$. The smallest positive solution of this equation is

$$\begin{aligned}\hat{t}=&\frac{-v(\sin\xi\cos\phi+\cos\xi\sin\phi)}{g\cos\phi}\\&-\frac{\sqrt{v^2(\sin\xi\cos\phi+\cos\xi\sin\phi)^2-2gd_n\cos\phi}}{g\cos\phi}\\=&\frac{-v\sin(\xi+\phi)-\sqrt{v^2\sin^2(\xi+\phi)-2gd_n\cos\phi}}{g\cos\phi}.\end{aligned}\qquad(4)$$

The maximum change in v^2 is then bounded by

$$\begin{aligned}\Delta v^2=&2g\Delta h\leq 2g(\tfrac{1}{2}g\hat{t}^2+v(\sin\xi)\hat{t})\\=&\frac{2v\cos\xi\sin\phi}{\cos^2\phi}(v\sin(\xi+\phi)\\&+\sqrt{v^2\sin^2(\xi+\phi)-2gd_n\cos\phi})-\frac{2gd_n}{\cos\phi}.\end{aligned}\qquad(5)$$

After one bounce the orientation is assumed to be such that rod is parallel to the surface and the center of mass is below the rod, as this will result in the largest increase in kinetic energy according to lemma 1. This means that endpoint 1's relative orientation is equal to $\hat{\theta}=\frac{\pi}{2}-\frac{\alpha}{2}-\phi$. Substituting this value in Equation 3 of lemma 2 gives $-2gR(1+\sin\hat{\theta})$. In other words, if the kinetic energy after the bounce is less than $-2gR(1+\sin\hat{\theta})$ and the rod is in the ideal orientation, the rod cannot roll to the other side.

We can combine the two bounds to obtain a sufficient condition to determine whether the rod can rotate to the other side *if its new orientation after one bounce is equal to the ideal orientation.* Unfortunately this condition does not imply a similar condition for the general case where the new orientation is not necessarily equal to the ideal orientation.

Consider the case $v = 0^+$. Substituting this value in Equation 5 and expanding the definition of d_n shows that the maximum increase in kinetic energy is then

$$-\frac{2gd_n}{\cos\phi} = -\frac{2gR(\sin(\hat{\theta}+\phi)-\sin(\theta+\phi))}{\cos\phi}. \qquad (6)$$

Therefore, when $v = 0^+$ and the relative orientation of the contact point after the bounce is equal to ideal orientation the quasi-capture constraint is

$$-2gR\frac{\sin(\hat{\theta}+\phi)-\sin(\theta+\phi)}{\cos\phi} \leq -2gR(1+\sin\hat{\theta}). \qquad (7)$$

That is, if an upper bound on the kinetic energy after one bounce is less than the energy needed to rotate to the other side, the rod will not be able to rotate to the other side. Now suppose the new orientation is *not* equal to the ideal orientation. Then the increase of kinetic energy will be less, but the energy required to roll to the other side will be less, too. Let $\tilde{\theta}$ be the relative orientation of the contact point after the bounce. Equation 7 is of the form $f(\hat{\theta}) \leq g(\hat{\theta})$, where $f(\cdot)$ computes the kinetic energy after one bounce for a given new orientation and $g(\cdot)$ computes the energy needed to roll to the other side for a given orientation[3]. Unfortunately, this bound does not imply $\forall\tilde{\theta}.f(\tilde{\theta}) \leq g(\tilde{\theta})$. From the 'oracle argument' on page 139 it follows that $f(\tilde{\theta})$ is indeed an upper bound on the maximum increase of the kinetic energy. Substituting $\tilde{\theta}$ in lemma 2 shows that $g(\tilde{\theta})$ is a lower bound on the kinetic energy needed to roll to the other side. We would like to determine the smallest possible ε such that

$$f(\hat{\theta}) - 2gR\varepsilon \leq g(\hat{\theta}) \qquad \Rightarrow \qquad \forall\tilde{\theta}.f(\tilde{\theta}) \leq g(\tilde{\theta}).$$

It is not hard to see ε has to be equal to

$$\max_{\tilde{\theta}}(g(\hat{\theta}) - g(\tilde{\theta}) - f(\hat{\theta}) + f(\tilde{\theta}))/(-2gR).$$

The difference between $f(\hat{\theta})$ and $f(\tilde{\theta})$ is

$$-2gR\frac{\sin(\hat{\theta}+\phi)-\sin(\tilde{\theta}+\phi)}{\cos\phi}.$$

Similarly, the difference between $g(\hat{\theta})$ and $g(\tilde{\theta})$ is

$$-2gR\left(\sin\hat{\theta} - \sin\tilde{\theta} - (\mathrm{sign}(\cos\tilde{\theta}) - 1)\sin\tfrac{\alpha}{2}\sin\phi\right).$$

[3]Analagous to expression 3, $g(\tilde{\theta})$ equals $-2gR\left(1+\sin\tilde{\theta}+(\mathrm{sign}(\cos\tilde{\theta})-1)\sin\frac{\alpha}{2}\sin\phi\right)$.

The correction ε is therefore

$$\varepsilon = \max_{\tilde{\theta}}(\sin\hat{\theta} - \sin\tilde{\theta} - (\mathrm{sign}(\cos\tilde{\theta}) - 1)\sin\tfrac{\alpha}{2}\sin\phi - \tfrac{\sin(\hat{\theta}+\phi)-\sin(\tilde{\theta}+\phi)}{\cos\phi})$$

By differentiating the expression inside $\max(\cdot)$ with respect to $\tilde{\theta}$ we find that there is a local maximum at $\tilde{\theta} = 0$. Other local maxima occur when $\tilde{\theta}$ approaches $-\frac{\pi}{2}$ from below or $\frac{\pi}{2}$ from above. The correction ε therefore simplifies to

$$\begin{aligned}\varepsilon &= \max\left(\sin\hat{\theta} - \tfrac{\sin(\hat{\theta}+\phi)-\sin\phi}{\cos\phi},\ \sin\hat{\theta}+2\sin\tfrac{\alpha}{2}\sin\phi - \tfrac{\sin(\hat{\theta}+\phi)}{\cos\phi}\right)\\ &= \cos(\tfrac{\alpha}{2}+\phi) - \tfrac{\cos(\alpha/2)}{\cos\phi} + \max\left(\tan\phi,\ 2\sin\tfrac{\alpha}{2}\sin\phi\right).\end{aligned}$$

For $v \neq 0^+$ the difference between $f(\hat{\theta})$ and $f(\tilde{\theta})$ is even larger and $g(\tilde{\theta})$ does not depend on v, so the value for ε is an upper bound for all v. Combining all the bounds we arrive at the desired result. □

Note that for $\phi = 0$ this bound reduces to $v^2 \leq -2gR(1+\sin\theta)$. In other words, this bound is as tight as possible when the surface is horizontal.

For an arbitrary d_n it is not possible to compute the optimal ξ analytically. Fortunately, we *can* analytically solve for ξ if we assume that the bounce consists of pure translation. The resulting ξ can be used as an approximation. It can shown that the solution for ξ can be written as

$$\cos\xi = \frac{\cos\phi}{\sqrt{2}\sqrt{1-\sin\phi}} \ \wedge\ \sin\xi = \frac{\sqrt{1-\sin\phi}}{\sqrt{2}}.$$

Substituting these values in Equation 4, we find that the approximation for the bound for Δv^2 then simplifies to

$$\Delta v^2 \leq -\frac{2gd_n}{\cos\phi} + \frac{v^2\sin\phi}{1-\sin\phi}\left(1+\sqrt{1-4d_ng\frac{1-\sin\phi}{v^2\cos\phi}}\right)$$

The relative error in this approximation depends on ϕ, d_n, v and g and can be computed numerically. Somewhat surprisingly, the relative error appears to be constant in v, d_n and g. The relative error does vary significantly with ϕ, but is still fairly small (on the order of 10^{-2}).

Image Guided Surgery

Eric Grimson, *MIT, Cambridge, MA*
Michael Leventon, *MIT, Cambridge, MA*
Liana Lorigo, *MIT, Cambridge, MA*
Tina Kapur, *Visualization Technology Incorporated, Wilmington, MA*
Olivier Faugeras, *MIT, Cambridge, MA*
Ron Kikinis, *Brigham and Women's Hospital, Boston, MA*
Renaud Keriven, *Cermics, Ecole Nationale des Ponts et Chaussées, Paris, France*
Arya Nabavi, *Brigham and Women's Hospital, Boston, MA*
Carl-Fredrik Westin, *Brigham and Women's Hospital, Boston, MA*

Recent advances in image guided surgery are changing the manner in which surgeons are able to execute difficult procedures. By building detailed, patient-specific models of anatomy, and augmenting those models with other information, such as functional properties, the surgeon can better plan her procedure to optimally extract target tissue while avoiding nearby critical structures. By registering these models with the actual patient position in the operating theatre, and by tracking surgical instruments relative to the patient and the registered model, real-time feedback is provided about the position of the instrument and its relationship to nearby, hidden tissue.

A central aspect of IGS is creating accurate, detailed, patient-specific models from medical imagery. In this paper, we briefly outline an overall approach to image guided surgery, and present several examples of current methods for model building.

1 Introduction

Recent developments in computer vision and robotics are changing the manner in which modern surgery is being practiced. Image guided surgical methods are providing a surgeon with the ability to visualize internal structures and their geometric relationships, often in alignment with live imagery or direct views of the patient. Such visualization abilities enable a surgeon to plan minimally invasive procedures. Tracking methods further allow a surgeon to see the actual position of his instruments, and their physical relationship to critical or hidden structures, leading to faster, safer and more effective minimally invasive procedures.

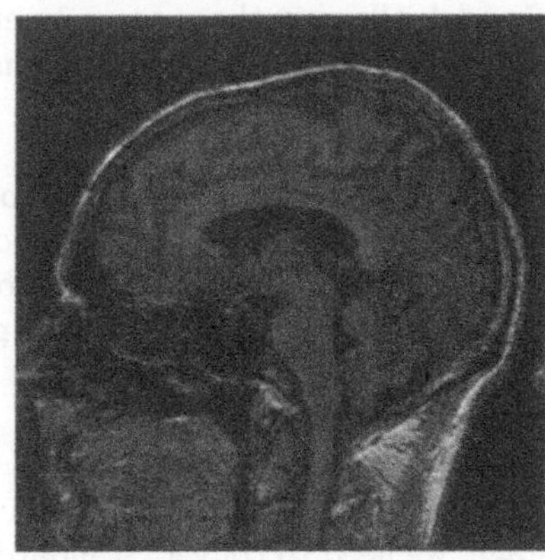

Figure 1: *Standard imagery, such as MRI, highlights different tissue responses.*

A standard framework for Image Guided Surgery consists of the following stages:

- A set of medical images (such as MRI or CT) are acquired of the relevant portion of the patient's anatomy (Figure 1).
- These images are segmented into distinct anatomical structures, yielding a 3D patient-specific model (Figure 2).
- The model is then registered to the actual position of the patient on the operating table. Surgical instruments are tracked relative to the patient

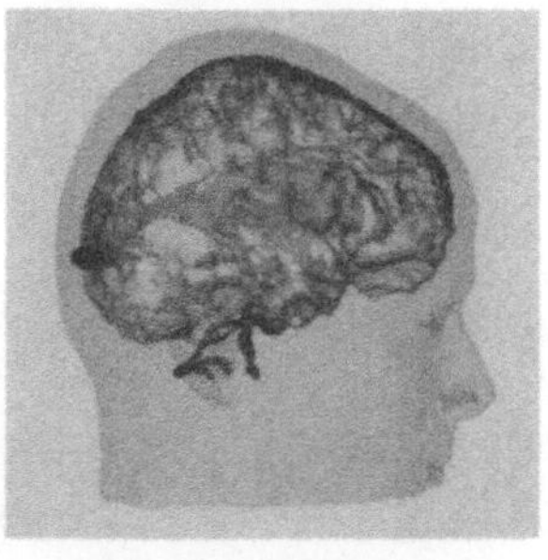

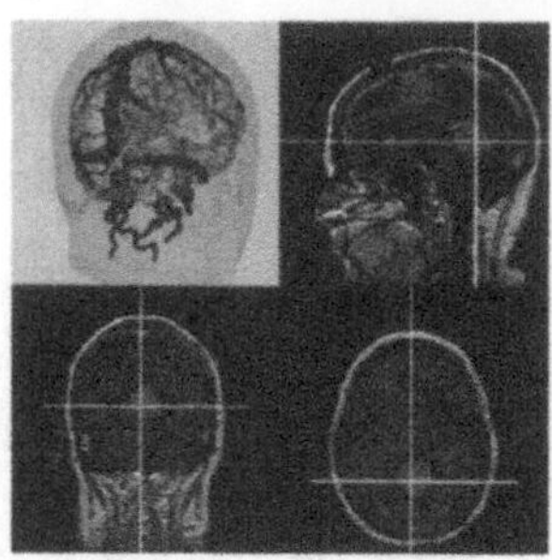

Figure 2: *These images are converted to patient-specific geometric models of anatomy. They can be used both for planning (left) and to support navigation by tracking surgical instruments (right).*

and the model, allowing the surgeon to effectively execute procedures while avoiding hidden, critical structures (Figure 2).

In this paper, we briefly outline current approaches to each of these stages, then examine in more detail the first stage – segmentation of medical scans into patient-specific models that can be used for surgical planning, surgical guidance, and surgical navigation.

2 Segmentation

In the first stage, we need to convert standard medical imagery into information that more directly reflects patient anatomy. As detailed in Section 5, we use a variety of algorithmic methods to label individual elements of a volumetric scan by tissue type, and to collect those labeled voxels into connected structures. We will discuss those methods in detail later, for now it suffices that patient specific geometric models can be created.

3 Registration

Given segmented models of the patient (see S ection 5), we want to use that information to assist the surgeon. One aspect of this is to allow the surgeon to visualize structures and their relationships, in order to support plan reliable procedures. To this end, it is often useful to augment the structure models created by segmenting MRI data, with functional and other information from other imagery (e.g., fMRI, PET, SPECT). This requires registration of two or more different modalities. Since there is often no set of corresponding landmarks between the scans, we have relied on a registration method that finds the best alignment between image volumes, based on the alignment of all the data in the volumes. In particular, our group has developed registration methods that maximize the Mutual Information between data sets, and have demonstrated that such methods can robustly align a wide range of medical imagery [29].

Once such augmented models are created, we also need to register them to the actual position of the patient. To do this, we use a surface matching algorithm. 3d data points from the skin surface of the patient are acquired using a trackable probe (typical methods involve locating synchronized LED's on the probe in a set of three cameras and using triangulation to locate the probe, or using electromagnetic coils tracking small metal balls attached to the probe, and similarly using triangulation to localize the probe). These data points are then matched to the skin surface of the reconstructed patient model to find the best fit, and thus to align the model with the actual patient position [13, 14]

4 Navigation

Finally, by tracking the position of any surgical instrument in the operating site, one can relate the position of the tip of the instrument to the corresponding point on the segmented model. By displaying this information on a monitor, the surgeon can rapidly navigate to desired target sites while avoid critical structures.

5 Segmentation

The first stage of this pipeline is key. One must be able to extract relevant information from the medical imagery, in order to most effectively present it to the surgeon. This becomes a problem of segmentation – labeling each voxel of the image with the associated tissue type, then agglomerating adjacent voxels with common labels into connected structures. In the following sections, we describe in more detail three current techniques for achieving this stage.

6 Statistical Labeling of Images

Our first method uses statistical properties of tissue response in MRI imagery to classify tissue type associated with each voxel [17, 18, 31]. The basis of the method is straightforward. Suppose one could identify a small set of voxels of each desired tissue type in the imagery, for example by using known anatomical landmarks. Record the intensity in the MRI data at such points. This gives us a set of samples of the intensity associated with each tissue type. In principle, a straightforward application of nearest neighbor labelling can then be used to identify the tissue type associated with all other points in the imagery.

Unfortunately, MRI imagery contains significant gain artifacts, which act as a nonlinear multiplier on the observed signal. This causes simple nearest-neighbor schemes to fail. As a consequence, our group has developed [17, 18, 31] methods that automatically account for the gain artifact, while correctly identifying the tissue type, using the Expectation/Maximization algorithm to iteratively solve for the gain field and the tissue labeling. The method proceeds by finding the best labeling of the voxels, given the current estimate of the gain field, then uses those labels to find the gain field estimate that maximizes the expected value of the field. This cycle iterates to convergence and provides accurate estimates of both the gain field and the tissue labelings.

The basic E/M method assumes a stationary prior on the intensities associated with different tissue types (which can be found by measuring relative volumes of labeled voxels in training images). Better use of anatomical knowledge can be made, however. In particular, one can use Markov Random Field methods to impose local continuity on the labeling, and one can use rough knowledge of the distribution of tissue types as a function of distance from known landmarks to create non-stationary prior probabilities [17, 18]. Such methods essentially incorporate coarse atlas information to guide the segmentation process.

7 Using Anatomical Knowledge

Because segmentation often requires distinguishing ambiguous information, having prior information about the expected shape of a structure can significantly aid in the segmentation process. For example, one can find corresponding points across training images, and use that to construct a statistical model of shape variation from the point positions. In this case, the best match of the model to the image is found by searching over the model parameters [7]. A second method [27] adds global shape information into the segmentation task by using an elliptic Fourier decomposition of the boundary and placing a Gaussian prior on the Fourier coefficients.

Our approach to object segmentation extends geodesic active contours [9, 20] by incorporating shape information into the evolution process. We first compute a statistical shape model over a training set of curves. To segment a structure from an image, we evolve an active contour both locally, based on image gradients and curvature, and globally to a maximum likelihood estimate of shape and pose.

8 Probability Distribution on Shapes

To incorporate shape information into the process of segmenting an object in an image, we consider a probabilistic approach, and compute a prior on shape variation given a set of training instances. To build the shape model, we choose a representation of curves, and then define a probability density function over the parameters of the representation.

8.1 Curve Representation

Each curve in the training dataset is embedded as the zero level set of a higher dimensional surface, u, whose height is sampled at regular intervals (say N^d samples, where d is the number of dimensions). The embedding function chosen is the signed distance function [26], where each sample encodes the distance to the nearest point on the curve, with negative values inside the curve (Figure 3). Each such surface (distance map) can be considered a point in a high dimensional space ($u \in \Re^{N^d}$). The training set, $\mathcal{T}$, consists of a set of surfaces $\mathcal{T} = \{u_1, u_2, \ldots, u_n\}$. Our goal is to build a shape model over this distribution of surfaces.

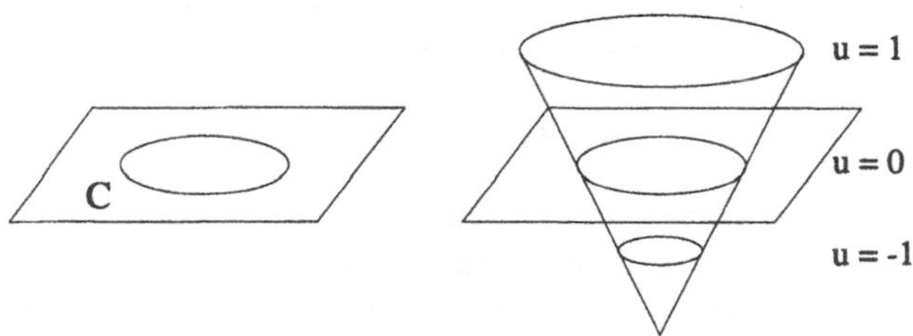

Figure 3: *Level sets of an embedding function u for a closed curve C in $\Re^2$.*

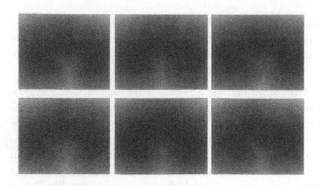

Figure 4: *Outlines of the corpus callosum of 6 out of 51 patients in the training set embedded as the zero level set of a higher dimensional surface, the signed distance function.*

The mean surface, μ, is computed by taking the mean of the signed distance functions, $\mu = \frac{1}{n}\sum u_i$. The variance in shape is computed using Principal Component Analysis (PCA). The mean shape, μ, is subtracted from each u_i to create an mean-offset map, $\hat{u}_i$. Each such map, $\hat{u}_i$, is placed as a column vector in an $N^d \times n$-dimensional matrix M. Using Singular Value Decomposition (SVD), the covariance matrix MM^T is decomposed as:

$$U\Sigma U^\mathsf{T} = MM^\mathsf{T} \tag{1}$$

where U is a matrix whose column vectors represent the set of orthogonal modes of shape variation and Σ is a diagonal matrix of corresponding singular values. An estimate of a novel shape, u, of the same class of object can be represented by k principal components in a k-dimensional vector of coefficients, α.

$$\alpha = U_k^\mathsf{T}(u - \mu) \tag{2}$$

where U_k is a matrix consisting of the first k columns of U that is used to project a surface into the eigen-space. Given the coefficients α, an estimate of the shape u, namely $\tilde{u}$, is reconstructed from U_k and μ.

$$\tilde{u} = U_k\,\alpha + \mu \tag{3}$$

Note that in general $\tilde{u}$ will not be a true distance function, since convex linear combinations of distance maps do not produce distance maps. However, the surfaces generally still have advantageous properties of smoothness, local dependence, and zero level sets consistent with the combination of original curves.

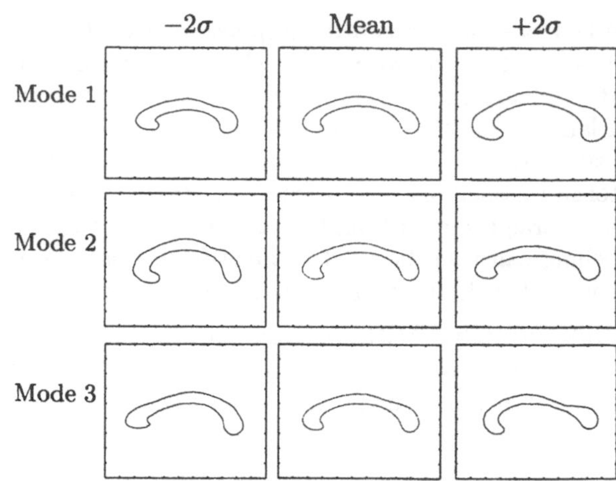

Figure 5: *The three primary modes of variance of the corpus callosum training dataset.*

Under the assumption of a Gaussian distribution of shape represented by α, we can compute the probability of a certain curve as:

$$P(\alpha) = \frac{1}{\sqrt{(2\pi)^k|\Sigma_k|}} \exp\left(-\frac{1}{2}\alpha^\mathsf{T}\Sigma_k^{-1}\alpha\right) \tag{4}$$

where Σ_k contains the first k rows and columns of Σ.

Figure 4 shows a few of the 51 training curves used to define the shape models of the corpus callosum. The original segmentations of the images are shown as red curves. The curves are overlaid on the signed-distance map. Before computing and combining the distance maps of these training shapes, the curves were aligned using centroids and second moments to approximate the correspondence. Figure 5 illustrates zero level sets corresponding to the means and three primary modes of variance of the shape distribution of the corpus callosum. Figure 6 shows the zero level set (as a triangle surface model) of seven rigidly aligned vertebrae of one patient used as training data. The zero level sets of the two primary modes are shown as well. Note that for both the corpus and the vertebrae, the mean shapes and primary modes appear to be reasonable representative shapes of the classes of objects being learned. In the case of the corpus callosum, the first mode seems to capture size, while the second mode roughly captures the degree of curvature of the corpus. The third

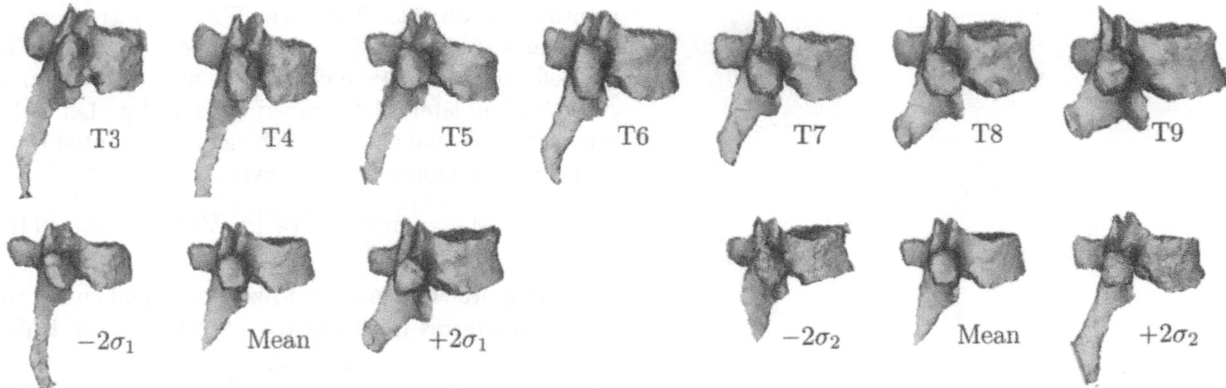

Figure 6: Top: *Three-dimensional models of seven thoracic vertebrae (T3-T9) used as training data.* **Bottom left and right**: *Extracted zero level set of first and second largest mode of variation respectively.*

mode appears to represent the shifting of the bulk of the corpus from front to back.

8.2 The Correspondence Problem

When measuring shape variance of a certain part of an object across a population, it is important to compare like parts of the object. For example, when looking at variances in the shape of the vertebrae, if two training examples are misaligned and a process of one is overlapping a notch of the other, then the model will not be capturing the appropriate anatomical shape variance seen across vertebrae.

One solution to the correspondence problem is to explicitly generate all point-wise correspondences to ensure that comparisons are done consistently, although this is difficult to automate and is manually tedious (see [7]). Another approach to correspondence is to roughly align the training data before performing the comparison and variance calculation. A rough alignment will not match every part of each training instance perfectly, so one must consider the robustness of the representation to misalignment.

Using the signed distance map as the representation of shape provides tolerance to slight misalignment of object features, in the attempt to avoid having to solve the general correspondence problem. In the examples presented here, the rough rigid alignment of the training instances resulted in the model capturing the shape variances inherent in the population due to the dependence of nearby pixels in the shape representation.

9 Shape Priors and Geodesic Active Contours

Given a curve representation (the k-dimensional vector α) and a probability distribution on α, the prior shape information can be folded into the segmentation process. This section describes adding a term to the level set evolution equation to pull the surface in the direction of the maximum likelihood shape and position of the final segmentation.

9.1 Geodesic Active Contours for Object Segmentation

The snake methodology defines an energy function $E(\mathcal{C})$ over a curve $\mathcal{C}$ as the sum of an internal and external energy of the curve, and evolves the curve to minimize the energy [19].

$$E(\mathcal{C}) = \beta \int |\mathcal{C}'(q)|^2 dq - \lambda \int |\nabla I(\mathcal{C}(q))| dq \qquad (5)$$

In [9], Caselles, et al. derive the equivalence of geodesic active contours to the traditional energy-based active contours (snakes) framework by first reducing the minimization problem to the following form:

$$\min_{\mathcal{C}(q)} \int g(|\nabla I(\mathcal{C}(q))|)\, |\mathcal{C}'(q)|\, dq \qquad (6)$$

where g is a function of the image gradient (usually of the form $\frac{1}{1+|\nabla I|^2}$). Using Euler-Lagrange, the following curve evolution equation is derived [9]:

$$\frac{\partial \mathcal{C}(t)}{\partial t} = g\kappa\mathcal{N} - (\nabla g \cdot \mathcal{N})\mathcal{N} \qquad (7)$$

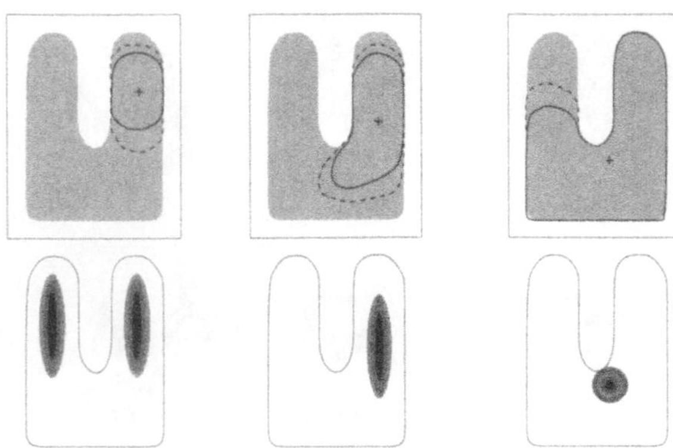

Figure 7: *Three steps in the evolution process. The evolving curve is shown in solid blue superimposed on the image (top row). The curve is matched to the expected curve to obtain a PDF over pose (bottom row). The next evolution step (based on pose and shape) is shown as the dotted blue line.*

where κ is the curvature and $\mathcal{N}$ is the unit normal. By defining an embedding function u of the curve $\mathcal{C}$, the update equation for a higher dimensional surface is [9]:

$$\frac{\partial u}{\partial t} = g\,(c+\kappa)\,|\nabla u| + \nabla u \cdot \nabla g \qquad (8)$$

where c is an image-dependent balloon force added to force the contour to flow outward [8, 22]. In this level set framework, the surface, u, evolves at every point perpendicular to the level sets as a function of the curvature at that point and the image gradient.

9.2 Estimation of Pose and Shape

In addition to evolving the level set based on the curvature and the image term, we include a term that incorporates information about the shape of the object being segmented. To add such a global shape force to the evolution, the pose of the evolving curve with respect to the shape model must be known (see Figure 7). Without an estimate of the pose, the shape model cannot adequately constrain or direct the evolution. Therefore, at each step of the curve evolution, we seek to estimate the shape parameters, α, and the rigid pose parameters, p, of the *final* curve using a maximum likelihood approach.

$$\langle \alpha_{ML}, p_{ML} \rangle = \operatorname*{argmax}_{\alpha,p} P(\alpha, p \mid u, \nabla I) \qquad (9)$$

In this equation, u is the evolving surface at some point in time, whose zero level set is the curve that is segmenting the object. The term ∇I is the gradient of the image containing the object to be segmented. By our definition of shape and pose, the final segmentation curve is completely determined by α and p. Let u^* be the estimated final curve, which can be computed from α and p. Therefore, we also have:

$$u^*_{ML} = \operatorname*{argmax}_{u^*} P(u^* \mid u, \nabla I) \qquad (10)$$

To compute the maximum likelihood final curve, we expand the terms from Equation 9 using Bayes' Rule.

$$\begin{aligned} P(\alpha, p \mid u, \nabla I) &= \frac{P(u, \nabla I \mid \alpha, p) P(\alpha, p)}{P(u, \nabla I)} \qquad (11) \\ &= \frac{P(u \mid \alpha, p) P(\nabla I \mid \alpha, p, u) P(\alpha) P(p)}{P(u, \nabla I)} \end{aligned}$$

We proceed by defining each term of Equation 11 in turn. We discard the normalization term in the denominator as it does not depend on shape or pose. More details are available in [21].

The first term in Equation 11 computes the probability of a certain evolving curve, u, given the shape and pose of the final curve, u^* (or $\langle \alpha, p \rangle$). Notice that this term does not include any image information whatsoever. We model this term as a Laplacian density function over $V_{outside}$, the volume of the curve u that lies outside the curve u^*.

$$P(u \mid \alpha, p) = \exp\left(-V_{outside}\right) \qquad (12)$$

This term assumes that any curve u lying inside u^* is equally likely. Since the initial curve can be located at any point inside the object and the curve can evolve along any path, we do not favor any such curve.

The second term in Equation 11 computes the probability of seeing certain image gradients given the current and final curves. Let $h(u^*)$ be the best fit Gaussian to the samples $(u^*, |\nabla I|)$. We model the gradient probability term as a Laplacian of the goodness of fit of the Gaussian.

$$P(\nabla I \mid u^*, u) = \exp\left(-\mid h(u^*) - |\nabla I| \mid^2\right) \qquad (13)$$

The last two terms in Equation 11 are based on our prior models, as described in Section 8. Our shape prior is a Gaussian model over the shape parameters, α, with shape variance Σ_k.

$$P(\alpha) = \frac{1}{\sqrt{(2\pi)^k |\Sigma_k|}} \exp\left(-\frac{1}{2}\alpha^T \Sigma_k^{-1} \alpha\right) \qquad (14)$$

In our current framework, we seek to segment one object from an image, and do not retain prior information on the likelihood of the object appearing in a certain location. Thus, we simply assume a uniform distribution over pose parameters, which can include any type of transformation, depending on application.

$$P(p) = \mathcal{U}(-\infty, \infty) \tag{15}$$

These terms define the maximum likelihood estimator of shape and pose, which estimates the final curve or segmentation of the object. For efficiency, these quantities are computed only in a narrow band around the zero level set of the evolving surface, and the ML pose and shape are re-estimated at each evolution step using simple gradient ascent on the log likelihood function in Equation 11. While each ascent may yield a local maximum, the continuous re-estimation of these parameters as the surface evolves generally results in convergence on the desired maximum. Next, we incorporate this information into the update equation commonly used in level set segmentation.

9.3 Evolving the Surface

Initially, the surface, u, is assumed to be defined by at least one point that lies inside the object to be segmented. Given the surface at time t, we seek to compute an evolution step that brings the curve closer to the correct final segmentation based on local gradient and global shape information.

The level set update expression shown in Equation 8 provides a means of evolving the surface u over time towards the solution to the original curve-minimization problem stated in Equation 6. Therefore, the shape of the surface at time t+1 can be computed from $u(t)$ by:

$$u(t+1) = u(t) + \lambda_1 \left(\mathrm{g}\,(c+\kappa)\,|\nabla u(t)| + \nabla u(t) \cdot \nabla g\right) \tag{16}$$

where λ_1 is a parameter defining the update step size.

By estimating the final surface u^* at a given time t, (Section 9.2), we can also evolve the surface in the direction of the maximum likelihood final surface:

$$u(t+1) = u(t) + \lambda_2 \left(u^*(t) - u(t)\right) \tag{17}$$

where $\lambda_2 \in [0, 1]$ is the linear coefficient that determines how much to trust the maximum likelihood estimate. Combining these equations yields the final expression for computing the surface at the next step.

$$u(t+1) = u(t) + \lambda_1 \left(g\,(c+\kappa)\,|\nabla u(t)| + \nabla u(t) \cdot \nabla g\right) + \lambda_2 \left(u^*(t) - u(t)\right) \tag{18}$$

The two parameters λ_1 and λ_2 are used to balance the influence of the shape model and the gradient-curvature model. The parameters also determine the overall step size of the evolution. The tradeoff between shape and image depends on how much faith one has in the shape model and the imagery for a given application. Currently, we set these parameters empirically for a particular segmentation task, given the general image quality and shape properties.

10 Results

We have tested the segmentation algorithm on synthetic and real shapes, both in 2D and in 3D. For controlled testing, a training set of rhombi of various sizes and aspect ratios was generated to define a shape model. Test images were constructed by embedding the shapes of two random rhombi with the addition of Gaussian speckle noise and a low frequency diagonal bias field.

Segmentation experiments were performed on 2D slices of MR images of the corpus callosum (Figure 8). The corpus callosum training set consisted of 49 examples like those in Figure 4. The segmentations of two corpora callosa are shown in Figure 8. Notice that while the ML shape estimator is initially incorrect, as the curve evolves, the pose and shape parameters converge on the boundary. The segmentations of the femur slices and the corpora all converged in under a minute on a 550 MHz Pentium III.

The vertebrae example illustrates the extension of the algorithm to 3D datasets. Figure 9 illustrates a few steps in the segmentation of vertebra T7. The training set in this case consisted of vertebrae T3-T9, with the exception of T7. The initial surface was a small sphere placed in the body of the vertebra. The red contour is a slice through the zero level set of the evolving hypersurface. The yellow overlay is the ML pose and shape estimate. Segmenting the vertebra took approximately six minutes.

To validate the segmentation results, we compute the undirected partial Hausdorff distance [15] between

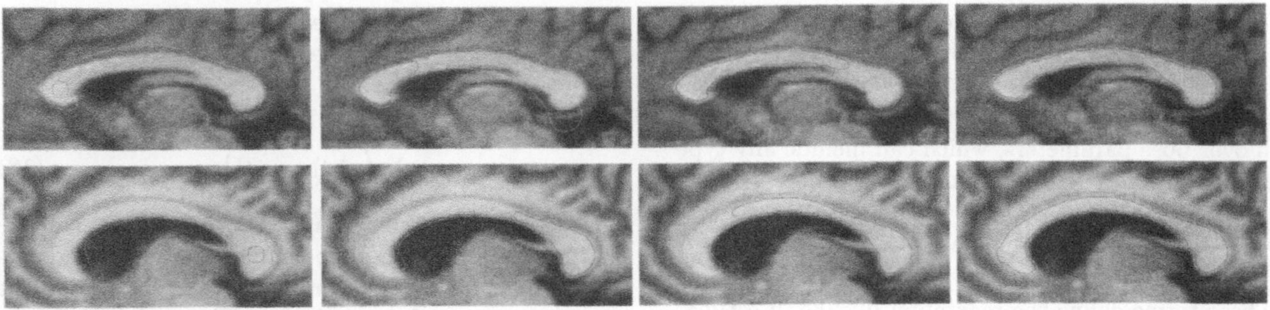

Figure 8: *Four steps in the segmentation of two different corpora callosa. The last image in each case shows the final segmentation in red. The cyan contour is the result of the standard evolution* without *the shape influence.*

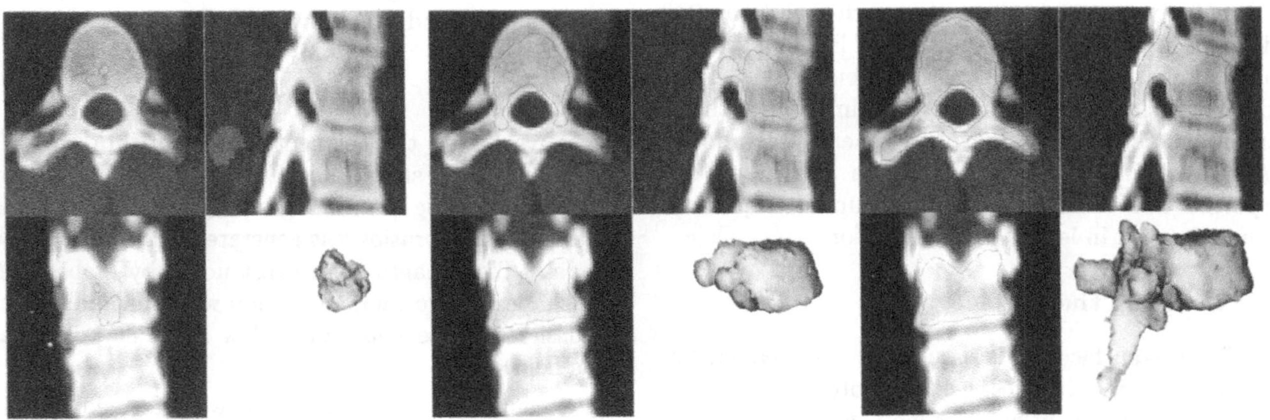

Figure 9: *Early, middle, and final steps in the segmentation of the vertebra T7. Three orthogonal slices and the 3D reconstruction are shown for each step. The red contour is a slice through the evolving surface. The yellow overlay is a slice through the inside of the ML final surface.*

the boundary of the computed segmentation and the boundary of the manually-segmented ground truth. The directed partial Hausdorff distance over two point sets A and B is defined as

$$h_K(A,B) = \mathrm{K}^{\mathrm{th}}_{a \in A} \min_{b \in B} ||a-b||$$

where K is a quantile of the maximum distance. The undirected partial Hausdorff distance is defined as $H_K(A,B) = \max(h_K(A,B), h_K(B,A))$. The results for the corpora and vertebra shown in Table 1 indicate that virtually all the boundary points lie within one or two voxels of the manual segmentation.

K	Corpus 1	Corpus 2	Vertebra
95%	1.3 mm	1.5 mm	2.7 mm
99%	1.6 mm	2.0 mm	4.4 mm

Table 1: *Partial Hausdorff distance between our segmentation and the manually-segmented ground truth.*

11 Evolution of Surfaces

The previous two methods for segmentation are designed to handle a wide range of structures, but work best for compact connected anatomical structures. Segmenting out tubelike structures, such as vessels often requires more specialized methods, and in this section, we outline our approach to such problems.

Curve-shortening flow is the evolution of a curve over time to minimize some distance metric. When this distance metric is based on image properties, it can be used for segmentation. The idea of *geodesic active contours* is to define the metric so that indicators of the object boundary, such as large intensity gradients, have a very small "distance" [4, 20, 2]. The minimization will attract the curve to such image areas, thereby segmenting the image, while preserving properties of the curve such as smoothness and connectivity. Geodesic active contours can also be viewed as a more mathematically sophisticated variant of classical snakes [19]. Further, they are implemented with level set methods [26] which are based on recent results in differential geometry [11, 5]. The method can be extended to evolve surfaces in 3D, for the segmentation of 3D imagery such as medical datasets [4, 32].

A limitation of the method was that it and its theoretical foundations applied only to hypersurfaces. More recent work in differential geometry, however, developed the equations necessary to evolve arbitrary dimensional manifolds in arbitrary dimensional space [1]. Subsequent work developed and analyzed a diffusion-generated motion scheme for co-dimension two curves [25]. We have developed the first application of geodesic active contours in 3D, based on [1]. Our system, CURVES, evolves 1D curves in 3D images and has been applied to automatic segmentation of blood vessels in medical images [23].

12 Background

Our approach is an extension of geodesic active contours research, also using a level set implementation. The extension was enabled by theoretical work on level set methods for arbitrary co-dimensional manifolds.

12.1 Geodesic Active Contours

The task of finding the curve that best fits the object boundary is posed [4, 2, 20] as a minimization problem over all closed planar curves $C(p) : [0,1] \to \mathbb{R}^2$. The objective function is:

$$\oint_0^1 g(|\nabla I(C(p))|)|C'(p)|dp$$

where $I : [0,a] \times [0,b] \to [0,\infty)$ is the image and $g : [0,\infty) \to \mathbb{R}^+$ is a strictly decreasing function such that $g(r) \to 0$ as $r \to \infty$, such as $g(|\nabla I|) = \frac{1}{1+|\nabla I|^2}$.

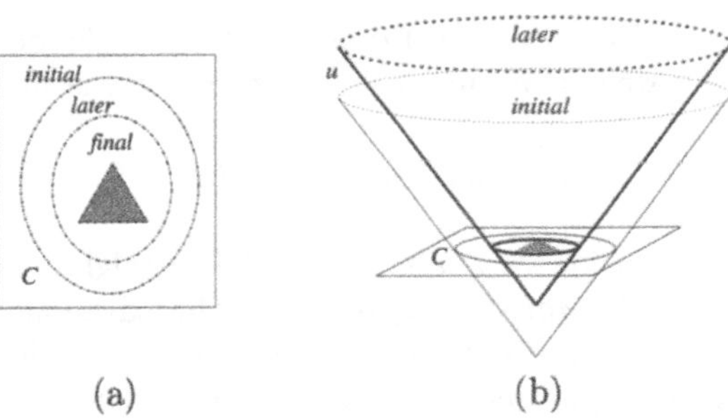

Figure 10: *Simple segmentation example: (a) Evolving curve. (b) Level set implementation of curve evolution.*

To minimize this objective function by steepest descent, consider C to be a function of time t as well as spatial parameter p. The Euler-Lagrange equations yield the curve evolution equation:

$$\vec{C}_t = g\kappa\vec{N} - (\nabla g \cdot \vec{N})\vec{N} \quad (19)$$

where κ is the Euclidean curvature and $\vec{N}$ is the unit inward normal. In the absence of image gradients, this equation causes the curve to shrink according to its curvature; the presence of image gradients causes the curve to stop on the object boundary (Figure 10a).

12.2 Level Set Method for Hypersurfaces

Level set methods increase the dimensionality of the problem from the dimensionality of the evolving manifold to the dimensionality of the embedding space [26]. For the example of planar curves, instead of evolving the one-dimensional curve, the method evolves a two-dimensional surface. Let $u : \mathbb{R}^2 \to \mathbb{R}$ be the signed distance function to curve C as in Figure 10b; C is thus the zero level-set of u, and u is an implicit representation of C. Let C_0 be the initial curve. It is shown in [11, 5] that evolving C according to:

$$\vec{C}_t = \beta\vec{N}$$

with initial condition $C(\cdot,0) = C_0(\cdot)$ for any function β, is equivalent to evolving u according to:

$$u_t = \beta|\nabla u|$$

with initial condition $u(\cdot,0) = u_0(\cdot)$ and $u_0(C_0) = 0$ in the sense that the zero level set of u is identical to the evolving curve for all time. Choosing

$\beta = g\kappa - (\nabla g \cdot \vec{N})$ as in Equation 19 gives the behavior illustrated in Figure 10b according to the update equation:

$$u_t = g\kappa|\nabla u| + \nabla g \cdot \nabla u.$$

The extension to surfaces in 3D is straightforward and is called *minimal surfaces* [4]. The advantages of the level set representation are that it is intrinsic (independent of parameterization) and that it is topologically flexible since different topologies of C are represented by the constant topology of u.

12.3 Level Set Method for Curves in Higher Co-dimension

For the task of evolving one-dimensional curves in three-dimensional space, however, the above level set relation does not hold. It is applicable only to hypersurfaces, that is, surfaces whose *co-dimension* is one: the *co-dimension* of a manifold is the difference between the dimension of the evolving space and the dimension of the manifold. The examples of a planar curve and a 3D surface have co-dimension one, but 1D curves in three dimensions have co-dimension two. Intuition for why the level set method above no longer holds is that there is not an "inside" and an "outside" to a manifold with co-dimension larger than one, so one cannot create the embedding surface u in the same fashion as for planar curves; a distance function must be everywhere positive, and is thus singular on the curve itself. Recently, however, more general level set equations were found for curvature-based evolution [1]. It is these equations that motivated the development of CURVES, which uses image information to create the auxiliary vector field.

Let $v : \mathbb{R}^3 \to [0, \infty)$ be an auxiliary function whose zero level set is identically $C(p) : [0,1] \to \mathbb{R}^3$, that is smooth near C, and such that ∇v is non-zero outside C. For a nonzero vector $\mathbf{q} \in \mathbb{R}^n$, define:

$$P_{\mathbf{q}} = I - \frac{\mathbf{q}\mathbf{q}^T}{|\,\mathbf{q}\,|^2}$$

where I is the identity matrix as the projector onto the plane normal to $\mathbf{q}$. Further define $\lambda(\nabla v(x,t), \nabla^2 v(x,t))$ as the smaller nonzero eigenvalue of $P_{\nabla v}\nabla^2 v P_{\nabla v}$. The level set evolution equation is [1]:

$$v_t = \lambda(\nabla v(x,t), \nabla^2 v(x,t)).$$

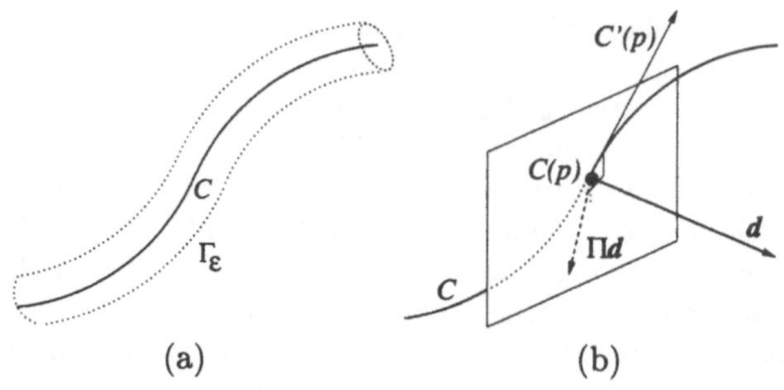

Figure 11: *Co-dimension two curve: (a) Tubular isolevel set Γ_ε of C. (b) The tangent to C at p, the normal plane, the external vector $\vec{d}$, and its projection onto the normal plane.*

That is, this evolution is equivalent to evolving C according to $\vec{C}_t = \kappa\vec{N}$ in the sense that C is the zero level set of v throughout the evolution. For intuition, let C be some curve in $\mathbb{R}^3$ and v the distance function to C. Consider then an isolevel set $\Gamma_\varepsilon = \{x | v(x) = \varepsilon\}$ of v where ε is small and positive, so Γ_ε is a thin tube around C (Figure 11a). The nonzero eigenvalues of $P_{\nabla v}\nabla^2 v P_{\nabla v}$ are equal to the principal curvatures of this tube. The larger principal curvature depends on ε while the smaller is related to the geometry of C. It is according to C that we want the evolution to proceed; thus, the smaller principal curvature is chosen.

Now assume there is an underlying vector field driving the evolution, so the desired evolution equation is:

$$\vec{C}_t = \kappa\vec{N} - \Pi\vec{d},$$

where Π is the projection operator onto the normal space of C (a vector space of dimension 2) and $\vec{d}$ is a given vector field in $\mathbb{R}^3$, (Figure 11). The evolution equation for the embedding space becomes [1]:

$$v_t = \lambda(\nabla v, \nabla^2 v) + \nabla v \cdot \vec{d}.$$

13 CURVES

The evolution equation we use follows directly from an energy-minimization problem statement.

13.1 Evolution Equation

For 1D structures in 3D images, we wish to minimize:

$$\oint_0^1 g(|\nabla I(C(p))|)|C'(p)|dp$$

where $C(p) : [0,1] \to \mathbb{R}^3$ is the 1D curve, $I : [0,a] \times [0,b] \times [0,c] \to [0,\infty)$ is the image, and $g : [0,\infty) \to \mathbb{R}^+$ is a strictly decreasing function such that $g(r) \to 0$ as $r \to \infty$ (analogous to [4]). For our current implementation, we use $g(r) = \exp(-r)$ because it works well in practice. The Euler-Lagrange equations give:

$$\vec{C}_t = \kappa\vec{N} - \frac{g'}{g}\Pi(\mathbf{H}\frac{\nabla I}{|\nabla I|}),$$

where $\mathbf{H}$ is the Hessian of the intensity function. The auxiliary vector field in the above equation is thus:

$$\vec{d} = \frac{g'}{g}\mathbf{H}\frac{\nabla I}{|\nabla I|},$$

so the equation for the embedding space is:

$$v_t = \lambda(\nabla v(x,t), \nabla^2 v(x,t)) + \frac{g'}{g}\nabla v(x,t) \cdot \mathbf{H}\frac{\nabla I}{|\nabla I|}$$

13.2 Features

Initial experiments required that the evolving volume be a distance function to the underlying curve; however, it was not clear how to robustly extract the zero level set or even evolve those points since the distance function was singular exactly there. For this reason, we developed the *ε-Level Set Method* which defines a thin tube of radius ε around the initial curve, then evolves that tube instead of the curve. ε does not denote a fixed value here, but means only that the evolving shape is a "tubular" surface of some unspecified and variable nonzero width. Thus, we are now evolving surfaces similar to [4], but that follow the motion of the underlying curve so they do not regularize against the high curvatures found in thin cylindrical structures such as blood vessels and bronchi. In addition to being more robust, this method better captures the geometry of such structures, which have nonzero diameter.

Next, we modified the update equation for the MRA segmentation application. To control the trade-off between fitting the surface to the image data and enforcing the smoothness constraint on the surface, we incorporate an image weighting term ρ which is set by the user or is pre-set to a default value. Second, because vessels in MRA and bronchi in CT appear brighter than the background, we weight the image term by the cosine of the angle between the normal to the surface and the gradient in the image. This cosine is given by the

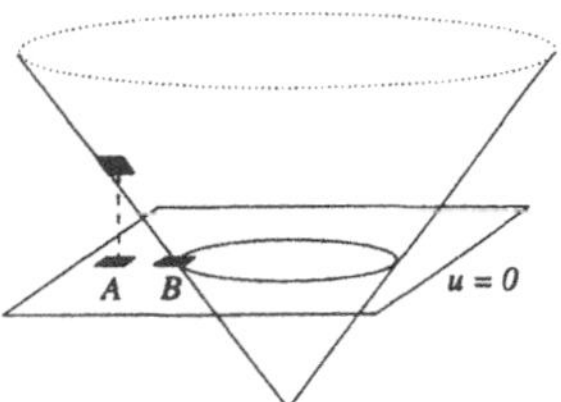

Figure 12: *To evolve a point on the distance function, CURVES chooses image information from A instead of B.*

dot product of the respective gradients of v and I, so the update equation becomes:

$$v_t = \lambda(\nabla v, \nabla^2 v) + \rho(\nabla v \cdot \nabla I)\frac{g'}{g}\nabla v \cdot \mathbf{H}\frac{\nabla I}{|\nabla I|} \quad (20)$$

A third aspect of our system is that we do not propagate the image information off the current object boundary to obtain the values near the boundary, as is customary in level set methods [4, 2, 20, 12]. Instead, we use directly the image information at each point in the image (Figure 12). Our choice has the advantage of enabling attraction to image gradients not on the current boundary, thereby reducing sensitivity to initialization, at the expense of requiring more frequent reinitializations of the distance function and losing the equivalence to the Lagrangian method (updating C directly instead of v) off the boundary.

In short, the CURVES system takes in a 3D image that contain thin curvilinear structures, such as an MRA image. An initial segmentation estimate is generated by thresholding the image. That estimate is used to generate an initial distance function, which is iteratively updated according to Equation 20. Convergence is detected automatically when volumetric change in the segmentation is very small over some number of iterations. Further detail was published in [23].

14 Results

We have run CURVES on over 20 medical datasets, primarily phase contrast magnetic resonance angiography (PC-MRA), of various resolutions and scanner types. We provide images of representative segmentations.

We first show successive boundary estimates in a segmentation of a cerebral MRA image to demonstrate the behavior of the algorithm over time, until convergence is reached. We then show CURVES segmentations of more cerebral MRA images compared to those

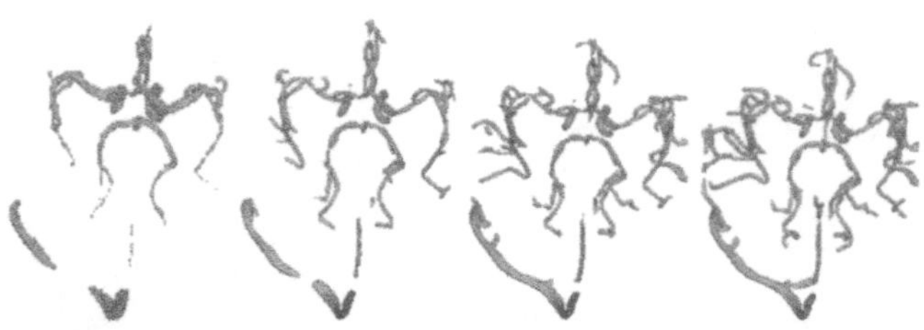

Figure 13: *Surface evolution over time: initialization, followed by successive boundary estimates.*

obtained with a manual segmentation technique used clinically at our institution. Finally, we illustrate the advantage of our system compared to co-dimension one surface evolution with an experiment involving the segmentation of bronchi in a computed tomography (CT) image of lung. The co-dimension one result was obtained using the mean curvature of the surface as the regularization force, as in previous level set segmentation schemes [4]; otherwise, all parameter settings were identical to those used in the CURVES experiment.

14.1 Example Evolution

Figure 13 illustrates the behavior of our system over time on a phase contrast MRA image of cerebral vessels. The initial surface is obtained by thresholding the raw dataset, then CURVES evolution produces the subsequent images.

14.2 Segmentations of Cerebral Vasculature

One specific practical motivation for our work is the use of surface models of cerebral vasculature as an aid in neurosurgical planning and procedure, especially in the context of the image-guided surgery program at our institution [13]. Currently the vessel models are obtained manually as follows. A neurosurgeon interactively chooses a threshold that is used to binarize the MRA dataset: all voxels brighter than that threshold are labeled as vessel, while all others are discarded. A "connectivity" program then partitions the set of labeled voxels into connected components. Each connected component appears in a distinct color on the user interface. The surgeon looks at individual slices and clicks on colored regions that correspond to vasculature. All connected components so chosen are stored as the final manual segmentation. The first drawback of this method is the expert user-interaction required, the second is that the thresholding step implies that all regions of image "noise" which adjoin vasculature are incorrectly labeled as vessel and small thin vessels which may appear broken or disconnected from larger structures will often be omitted. Thus, our goal is to reduce user interaction while increasing the ability to segment thin vessels.

Figure 14 shows CURVES segmentations (red) compared to segmentations acquired using the manual procedure just described (blue). The dataset shown here is PC-MRA acquired on a 1.5T scanner with voxel resolution of $1.171875 \times 1.171875 \times 0.8\text{mm}^3$ and a size of $256 \times 256 \times 84$ voxels. The raw dataset is shown in maximum intensity projection from three orthogonal viewpoints, as are the two segmentations. Notice that CURVES is able to capture more of the thin vessels than the manual procedure which is based on simple thresholding.

Acknowledgments

Support for our research was provided in part by the NSF under grant no. IIS-9610249, and in part by NSF ERC (Johns Hopkins University agreement) 8810-274.

References

[1] Ambrosio, L. and Soner, H.M.: Level set approach to mean curvature flow in arbitrary codimension. Journal of Differential Geometry **43** (1996) 693–737

[2] Caselles, V., Catte, F., Coll, T. and Dibos, F.: A geometric model for active contours. Numerische Mathematik **66** (1993) 1–31

[3] Caselles, V., Morel, J.M., Sapiro, G. and Tannenbaum, A.: Introduction to the special issue on partial differential equations and geometry-driven diffusion in image processing and analysis. IEEE Transactions on Image Processing **7(3)** (1998) 269–273

[4] Caselles, V., Kimmel, R. and Sapiro, G.: Geodesic active contours. Int'l Journal of Computer Vision **22(1)** (1997) 61–79

[5] Chen, Y.G., Giga, Y. and Goto, S.: Uniqueness and existence of viscosity solutions of generalized mean curvature flow equations. Journal of Differential Geometry **33** (1991) 749–786

[6] T. Cootes, C. Taylor, D. Cooper, and J. Graham. Active shape models - their training and application. *Computer Vision and Image Understanding* 1995.

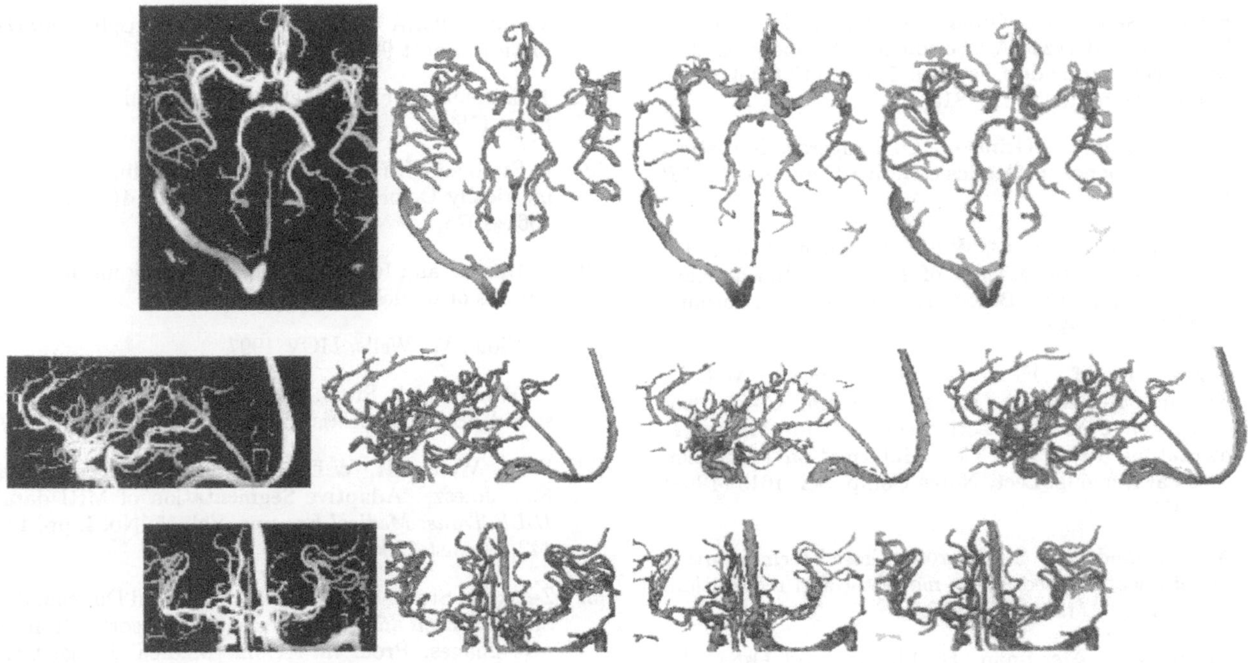

Figure 14: *The same MRA dataset is shown from three orthogonal viewpoints. For each viewpoint, the maximum intensity projection of the raw data is shown, followed by the CURVES segmentation (red), the manual segmentation (blue), and a combination image showing the differences between the segmentations.*

[7] T. Cootes, C. Beeston, G. Edwards, and C. Taylor Unified Framework for Atlas Matching Using Active Appearance Models. *Information Processing in Medical Imaging* 1999.

[8] L. Cohen On active contour models and balloons. *CVGIP: Image Understanding*, 53(2):211–218, 1991.

[9] V. Caselles, R. Kimmel, and G. Sapiro. Geodesic active contours. *Int'l Journal of Computer Vision* **22(1)** (1997) 61–79

[10] G. Christensen, R. Rabbitt, and M. Miller. Deformable templates using large deformation kinematics. *IEEE Trans. on Image Processing* **5(10)** (1996) 1435–1447.

[11] Evans, L.C. and Spruck, J.: Motion of level sets by mean curvature: I. Journal of Differential Geometry **33** (1991) 635–681

[12] Gomes, J., and Faugeras, O.: Reconciling distance functions and level sets.

[13] Grimson, W.E.L., Ettinger, G.J., Kapur, T., Leventon, M.E., Wells III, W.M., and Kikinis, R.: Utilizing Segmented MRI Data in Image-Guided Surgery. Int'l Journal Pattern Recognition and Artificial Intelligence (1996)

[14] W.E.L. Grimson, T. Lozano-Pérez, W.M. Wells III, G.J. Ettinger, S.J. White and R. Kikinis, An Automatic Registration Method for Frameless Stereotaxy, Image Guided Surgery, and Enhanced Reality Visualization, IEEE Trans. Medical Imaging, **15**(2), April 1996, pp. 129–140.

[15] D. Huttenlocher, G. Klanderman, W. Rucklidge, "Comparing images using the Hausdorff distance" in *IEEE Trans PAMI*, **15**:850-863, 1993.

[16] M. Jones and T. Poggio. Multidimensional Morphable Models. *Int'l Conf. Computer Vision*, 1998.

[17] T. Kapur, E. Grimson, R. Kikinis, W. Wells, "Enhanced spatial priors for segmentation of magnetic resonance imagery", Medical Image Computation and Computer Assisted Interventions, Boston, October 1998.

[18] T. Kapur, E. Grimson, W. Wells, R. Kikinis, "Segmentation of Brain Tissue from Magnetic Resonance Images", *Medical Image Analysis*, Vol. 1, Issue 2, 1997, pp 109–128.

[19] Kass, M., Witkin, A., and Terzopoulos, D.: Snakes: active contour models. Int'l Journal Computer Vision **1(4)** (1988) 321–331

[20] Kichenassamy, A., Kumar, A., Olver, P., Tannenbaum, A. and Yezzi, A.: Gradient flows and geometric active contour models. In Proc. IEEE Int'l Conf. Computer Vision (1995) 810–815

[21] M. Leventon, E. Grimson, O. Faugeras. Statistical Shape Influence in Geodesic Active Contours. *CVPR 2000*

[22] L. Lorigo, O. Faugeras, W.E.L. Grimson, R. Keriven, R. Kikinis. Segmentation of Bone in Clinical Knee MRI Using Texture-Based Geodesic Active Contours. In *MICCAI*, 1998.

[23] Lorigo, L. M., Faugeras, O., Grimson, W.E.L., Keriven, R., Kikinis, R., and Westin, C.-F.: Codimension 2 geodesic active contours for MRA segmentation. Proc. Int'l Conf. Information Procession Medical Imaging. Lect. Notes Comp. Sci. 1613 (1999) 126–139

[24] A. Pentland and S. Sclaroff. Closed-form solution for physically based shape modeling and recognition. *PAMI* **13(7)** (1991) 715–729.

[25] Ruuth, S.J., Merriman, B, Xin, J., and Osher, S.: Diffusion-generated motion by mean curvature for filaments. UCLA Computational and Applied Mathematics Report 98–47 (1998)

[26] J. Sethian. Level Set Methods. Cambridge University Press (1996)

[27] L. Staib and J. Duncan. Boundary Finding with Parametrically Deformable Models *PAMI* **14(11)**, (1992), 1061–1075.

[28] S. Ullman and R. Basri. Recognition by linear combinations of models. *CVPR 1991*

[29] P. Viola, W. Wells. IJCV 1997

[30] Y. Wang and L. Staib. Boundary Finding with Correspondence Using Statistical Shape Models *CVPR 1998*

[31] W.M. Wells, III, W.E.L. Grimson, R. Kikinis, and F.A. Jolesz, "Adaptive Segmentation of MRI data", *IEEE Trans. Medical Imaging*, Vol. 15, No. 4, pp. 429–442, August 1996.

[32] Zeng, X., Staib, L.H., Schultz, R.T., and Duncan, J.S.: Segmentation and measurement of the cortex from 3D MR images. Proc. Int'l Conf. Medical Image Computing Computer-Assisted Intervention. Lect. Notes Comp. Sci. 1496 (1998) 519–530

Pulling Motion Based Tactile Sensing

Makoto Kaneko, *Hiroshima University, Higashi-Hiroshima, Japan*
Toshio Tsuji, *Hiroshima University, Higashi-Hiroshima, Japan*

An algorithm for detecting the shape of a 2D concave surface by utilizing a tactile probe is discussed. Pulling a tactile probe whose tip lies on an object's surface can be easily achieved, while pushing it is more difficult due to stick-slip or blocking up with irregular surface. To cope with the difficulty of pushing motion on a frictional surface, the proposed sensing algorithm makes use of the pulling motion of tactile probe from a local concave point to an outer direction. The algorithm is composed of three phases, local concave point search, tracing motion planning, and infinite loop escape. The proposed algorithm runs until the tactile probe detects every surface which it can reach and touch. We show computer simulations obtained using the proposed algorithm.

1 Introduction

This paper focuses on the shape detection of a 2D surface by utilizing a tactile probe which can detect any contact point between itself and the environment. One emphasis of our research is to study how the probe motion should be planned for the object including concave surface. Suppose that the human moves his or her fingertip on the table as shown in Figure 1, where (a) and (b) denote the pushing and the pulling modes, respectively. It should be noted that pushing motion is not achieved smoothly due to the well-known stick-slip phenomenon and often blocked by a small irregularity on the surface, while pulling motion can be achieved even under significant friction. Considering this fact, we discuss the pulling motion-based algorithm where no pushing motion occurs. If the environment's surface is unknown, however, whether the resulting motion evolves as pulling or pushing strongly depends on the surface geometry and on the direction of the motion imparted to the tactile probe. In case that the sensing motion starts from an arbitrary point on the surface, pulling and pushing motions are expected to occur for outer and inner directions, respectively, as shown in Figure 2(a). If we can choose the local concave point as a starting point, however, we can expect pulling motions to occur for both directions, as shown in Figure 2(b).

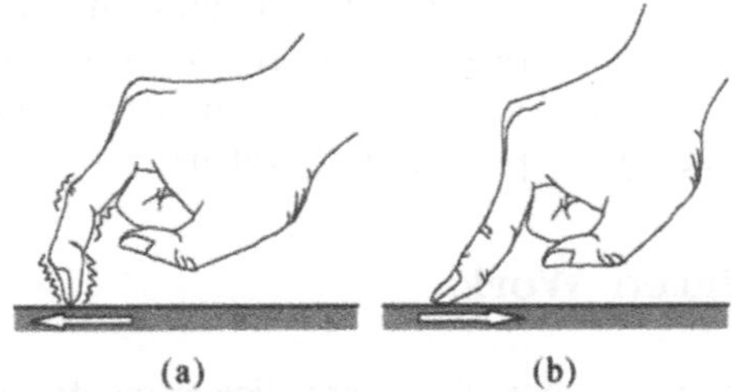

Figure 1: *Pulling and pushing motions by human hand.*

In order to utilize this advantage, the algorithm first searches for the local concave point by applying a bisection method (*local concave point search*). The tactile probe is then pulled outwards from the local concave point while keeping the tip of the probe in contact with the environment (*tracing motion planning*). During the tracing motion, however, the tip of the probe may break contact and touch again at a new contact point due to the existence of another local concave area, or the contact point may jump from the tip to an arbitrary point of the probe due to the existence of a local

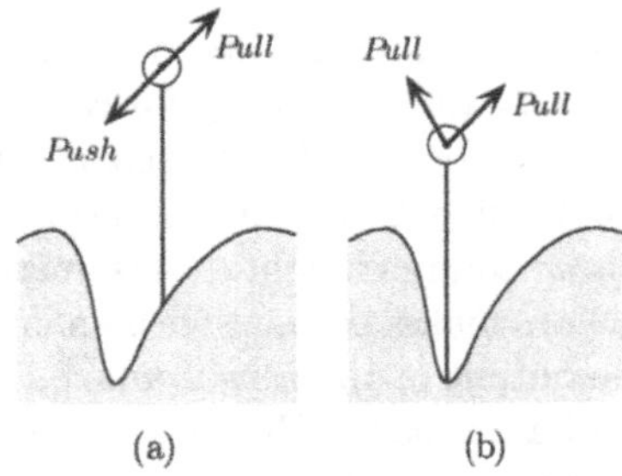

Figure 2: *Pulling and pushing motions by a sensor probe.*

convex area. When the tactile probe recognizes such a particular situation, the sensing system stores each contact point before and after jump. For such a non-traced area, the sensing motion is repeated recursively. There might be a failure mode in which the sensing motion results in repeating mode without finding any new contact point. To emerge from such an infinite loop, we prepare *infinite loop escape* by which the tactile probe can always find a new contact point if it exists. We show that the proposed algorithm can continue to run until the tactile probe detects every surface that it can reach and touch. Also, we show computer simulations obtained using the proposed algorithm.

2 Related Work

There are a number of papers discussing tactile and haptic perception linked with multi-fingered dexterous hands [1]–[10]. Caselli et al. [11] proposed an efficient technique for recognizing convex object from tactile sensing. They developed internal and external volumetric approximation of the unknown objects and exploited an effective feature selection strategy along with early pruning of incompatible objects to improve recognition performance. On the other hand, algorithms for tactile sensing have also been reported. Gaston and Lozano-Perez [12] and Grimson and Lozano-Perez [13] discussed object recognition and localization through tactile information under the assumption that the robot possesses the object models. Cole and Yap [14] have addressed the "Shape from probing" problem, where they discussed how many probes are necessary and sufficient for determining the shape and position of a polygon. They showed $3n-1$ probes are necessary and $3n$ are sufficient for any n-gon, where n is the number of probes. Most of these works [1]–[9], [11]–[14], however, deal with convex objects only and never discuss concave ones.

As far as we know, there are only a few papers [15]–[20] addressing tactile sensing for concave objects. Russell [15] designed a whisker sensor composed of an insensitive flexible beam, a potentiometer, a return spring and counterweight. Assuming that the whisker tip always makes contact with an environment, he showed the output of the sensor when scanning the concave bowl of a teaspoon. Roberts [17] discussed the strategy for determining the active sensing motion for the given set of convex and concave polyhedral model objects. Chen, Rink and Zhang [18] introduced an active tactile sensing strategy to obtain local object shape, in which they showed how to find the contact frame and how to find the local surface parameters in the contact frame. This work might be applied to a concave surface that can be described by the second order polynomial equation, although their experiment is limited to a convex object only. In these works, they picked up extremely simple concave objects as test examples but included no precise discussion on the inherent sensing algorithm for concave objects. This paper is an extended version of our former work [19].

3 Problem Formulation

3.1 Preliminary Definitions

Let $P(s)$ (or simply P) be a point on the environment's surface, where s is the coordinate along the surface as shown in Figure 3(a). We define $Dist(P(s_1), P(s_2))$, $C_{P_1}^{P_2}$ and $L_{P_1}^{P_2}$ as the distance, the environment's contour, and the line segment between $P(s_1)$ and $P(s_2)$, respectively. When $C_{P_1}^{P_2}$ becomes known by a probe tracing, we define $C_{P_1}^{P_2} \in W_1$, where W_1 is the assembly of the area traced by the probe. If $Dist(P(s_1), P(s_2)) < \varepsilon$ exists for a $\forall$ small $\varepsilon > 0$, we can regard $L_{P_1}^{P_2}$ as an approximate surface of $C_{P_1}^{P_2}$, and define $C_{P_1}^{P_2} \in W_2$, where W_2 is the assembly of approximately detected area by straight-line approximation. When the tactile probe recognizes the particular area between $P(s_1)$ and $P(s_2)$ where it can not reach and touch physically, we define the area as the non-reachable area and describe by $C_{P_1}^{P_2} \in W_3$, where W_3 is the assembly of the non-reachable area verified by a probing motion. For every concave object (or environment), we can make an equivalent convex shape by connecting common tangential lines. The outside

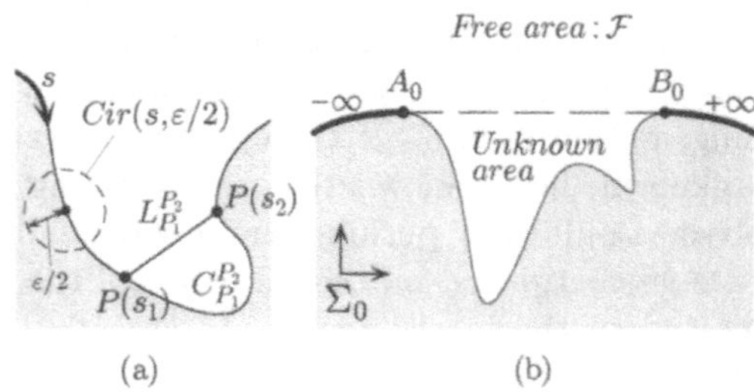

Figure 3: *Definition of symbols and problem notation.*

of the equivalent convex is defined as the free area $\mathcal{F}$ as shown in Figure 3(b) where it is guaranteed that there is no object (or environment). We also define $Area(\mathcal{G})$ as the area of $\mathcal{G}$, where $\mathcal{G}$ denotes the region constructed by two probe postures, while the precise definition of $\mathcal{G}$ will be given later.

3.2 Main Assumptions

Distributed sensing elements cover the entire tactile probe. The probe has negligible thickness and is connected with the end-joint of a robot arm having sufficient degrees of freedom so that the probe can achieve arbitrary position and pose in a 2D plane. The arm is assumed to have a joint torque and a joint position sensors in each joint. Both sensors are indispensable for achieving a compliant motion and determining the probe position. The torque sensor is further utilized for confirming whether a clockwise (or counter-clockwise) rotation of the probe is allowed or not. The probe is long enough to ensure that the end-joint of the arm always lies in the free area $\mathcal{F}$, which enables us to neglect any geometrical constraint imposed by the robot arm and to focus on the probe motion. The probe is sufficiently stiff to avoid bending. For the sake of simplicity, we give the following assumption on the shape of environment. Consider a small circle whose center and radius are given by $P(s)$ and $\varepsilon/2$, respectively, as shown in Figure 3(a). We also assume that there always are only two intersection points between the circle and the environment for an arbitrary $s \in [-\infty, +\infty]$ which means that the environment is smooth enough in ε level and does not include too much irregular surface. This assumption provides a valid reason for approximating the environment's contour as the straight-lined one, namely, $C_s^{s+\varepsilon} \cong L_s^{s+\varepsilon}$. We also assume that there exists only one object (or environment). The last assumption means that the objects (or environment) could be complicated but no island-like object in addition to the main object.

3.3 Problem Formulation and Nomenclature

Problem formulation : *Given $C_{P(-\infty)}^{A_0} \in W$ and $C_{B_0}^{P(+\infty)} \in W$, construct an algorithm such that $C_{A_0}^{B_0} \in W$ is achieved under the assumption in 3.2, where $W = \cup_{i=1}^{3} W_i$.*

$\mathcal{F}$: Free area
A_iB_i : Area between A_i and B_i
$\mathcal{G}_i$: Region constructed by two probe poses when the unknown area A_iB_i is found
$L_{A_i}^{B_i}$: Line segment between A_i and B_i
$\mathcal{V}_i$: Half plane that we can see in the right hand side when moving from B_i to A_i
$\mathcal{T}_i$: Semi-infinite region bounded by P_iA_i and P_iB_i, where P_i is the joint position

4 Pulling Motion Based Sensing

4.1 Outline of the Algorithm

Figure 4 shows an example explaining the sensing algorithm. The probe is first inserted from an arbitrary point in the free area $\mathcal{F}$ toward the unknown area until the tip makes contact with the environment. By monitoring the torque sensor output, it is checked whether a clockwise (or counter-clockwise) rotation of the probe is possible or not. After checking such a geometrical condition, the probe is rotated in the direction of rotation free until it again makes contact as shown by the dotted line in Figure 4(a), where φ is the rotational angle. Choosing the equally divided direction $\varphi/2$, we again insert the probe until it makes contact with the environment. By repeating this procedure, the tip can finally reach the local concave point, where the probe loses any rotational degree of freedom (*local concave point search*). Then, the probe is moved from the local concave point to the outer direction, while maintaining constant torque control for the last joint, where a clockwise torque is applied during the probe motion from D_0 to A_0 and a counter clockwise torque is imparted during the probe motion from D_0 to B_0. Figure 4(b) and (c) show two examples of tracing motion. By imparting the torque depending on the direction of tracing motion, it is ensured that the probe tip makes contact with the environment if the surface is smooth enough as shown in Figure 4(b) (*tracing motion planning*).

Thus, the tracing motion is executed by a pulling motion alone. This is the reason why we call the algorithm the pulling motion based sensing algorithm. If the environment includes another local concavity as shown in Figure 4(c), however, the tip will be once away from the surface and then make contact with another part of the environment due to the constant torque command imparted to the joint. When the probe recognizes such an additional concave area, we register A_1 and B_1 as

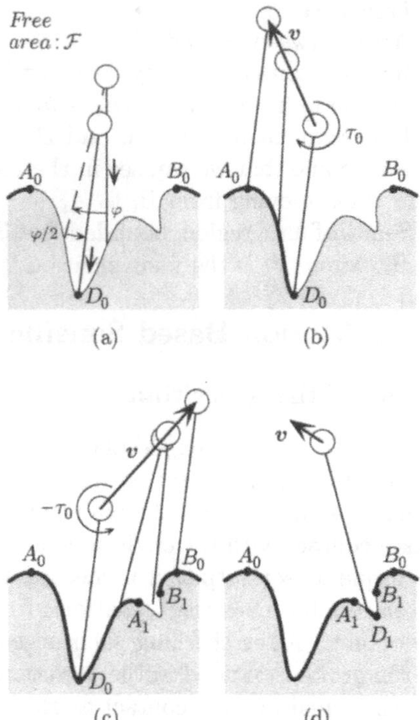

Figure 4: *Outline of the algorithm.*

a new pair indicating the unknown area. The sensing motion is repeated recursively for A_1B_1. During both *local concave point search* and *tracing motion planning,* the sensing motion may result in an infinite loop depending upon the environment's geometry. In order to emerge from such an infinite loop, we prepare the *infinite loop escape*, in which the probe is inserted so that it can reach a new contact point between the designated area. The *infinite loop escape* is not always necessary but called upon request when the same contact point is detected repeatedly.

4.2 Local Concave Point Search

4.2.1 Initial Pass Planning

Definition 1 *Consider two probe postures when the unknown area A_iB_i is found (see Figure 5(a)). $\mathcal{G}_i$ is defined as the region constructed by connecting A_i, B_i and each joint of the probe. If both A_i and B_i are detected by one probe posture, $\mathcal{G}_i$ is generated by rotating the probe around the tip till it makes contact with the environment as shown in Figure 6(a).*

All possible shapes for $\mathcal{G}_i$ can be classified into three groups, one polygon (triangle or quadrangle), two triangles (Figure 5(c) and (d)), and one line segment (Figure 6(a)), where the difference between (c) and (d) depends upon whether $L_{A_0}^{B_0}$ passes through $\mathcal{G}_i^{(1)}$ or $\mathcal{G}_i^{(2)}$. The quadrangle in Figure 5(a) or (b) results in a triangle when the line segment A_iB_i lies on the extended line of one of two probe postures. For a single probe detection as shown in Figure 6(a), $\mathcal{G}_i$ forms either a triangle or a line segment ($Area(\mathcal{G}_i) = 0$). $Area(\mathcal{G}_i) = 0$ means no more rotational degree of freedom around the tip. In other words, there exists no other pass except the current one resulting in the single probe posture as shown in Figure 6(b). Once $Area(\mathcal{G}_i) = 0$ is detected, the algorithm categorizes the A_iB_i into never touching area W_3 and leaves from the initial pass planning. In this subsection, we temporarily assume $Area(\mathcal{G}_i) \neq 0$.

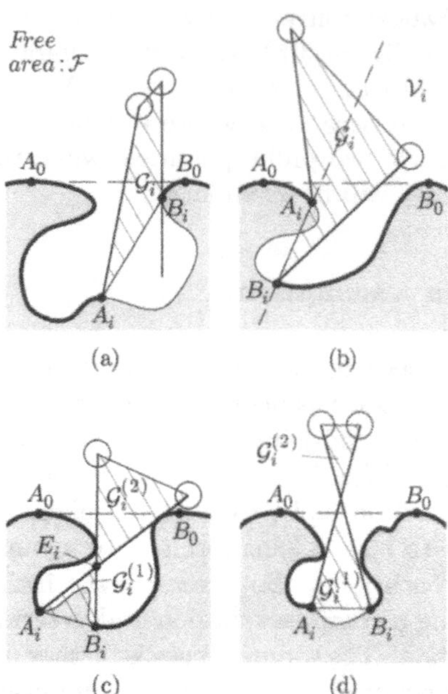

Figure 5: *Two probe postures when A_iB_i is detected.*

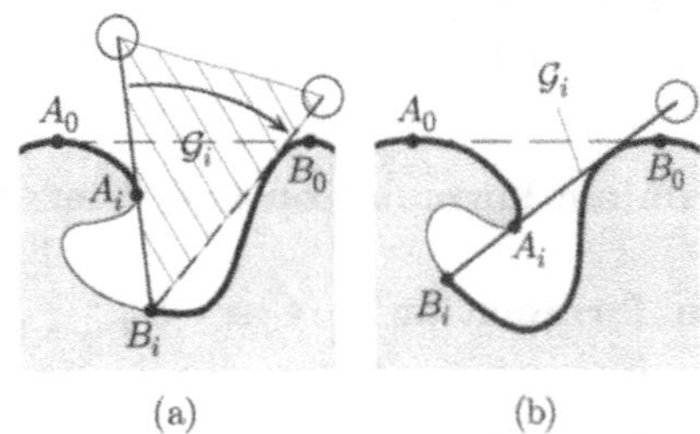

Figure 6: *An example of single probe detection.*

Under such an assumption, there exist only two patterns for $\mathcal{G}_i$ namely, one polygon and two triangles. Since $\mathcal{G}_i$ always links with the free area $\mathcal{F}$ under assumption in 3.2 (the probe is long enough to ensure that the joint never enters within the unknown area), it is guaranteed that if there is an environment within $\mathcal{G}_i$, its root must be connected with A_iB_i. If this does not hold true, the environment in $\mathcal{G}_i$ must be an island-like-object. Under the assumption of single object (or environment), however, such a situation never appears. In case of Figure 5(c), the probe trajectory should be strictly bounded, because it must be planned to pass through the common point for both triangles to absolutely avoid contacting with the environment whose root exists except A_iB_i. As for the environment classification, we give the following definitions.

Definition 2 *For the area A_iB_i, internal and external environments are defined as follows:*
$Int(A_iB_i)$: Internal environment is defined as the one whose root comes from the area between A_i and B_i.
$Ext(A_iB_i)$: External environment is defined by the area A_iB_i except $Int(A_iB_i)$.

Let us now define the area $\mathcal{V}_i$ which also restricts the initial pass to $Int(A_iB_i)$.

Definition 3 *Suppose an arbitrary point Q denoted by the vector $\boldsymbol{q}$ with respect to Σ_0. Let $\boldsymbol{a}_i$ and $\boldsymbol{b}_i$ be two vectors expressing the positions A_i and B_i with respect to Σ_0. Define $\mathcal{V}_i$ as an assemble of $\boldsymbol{q}$ satisfying $V_i(\boldsymbol{q}) = sgn\{(\boldsymbol{a}_i - \boldsymbol{q}), (\boldsymbol{b}_i - \boldsymbol{q})\} > 0$, where $sgn(\boldsymbol{x}, \boldsymbol{y})$ is given by*

$$sgn(\boldsymbol{x}, \boldsymbol{y}) = \frac{\boldsymbol{x} \otimes \boldsymbol{y}}{\|\boldsymbol{x} \otimes \boldsymbol{y}\|} \tag{1}$$

where, $\otimes$ denotes a vector product for two vectors $\boldsymbol{x} = (x_1, x_2)^{\mathrm{T}}$ and $\boldsymbol{y} = (y_1, y_2)^{\mathrm{T}}$.

Consider two half planes whose boundary line includes the line segment A_iB_i. By Definition 3, $\mathcal{V}_i$ denotes the half plane that we can see in the right-hand side when moving from B_i to A_i. Note that $\mathcal{G}_i \cap \mathcal{V}_i \neq \mathcal{G}_i$ for Figure 5(b), while $\mathcal{G}_i \cap \mathcal{V}_i = \mathcal{G}_i$ for Figure 5(a), (c), (d) and Figure 6(b). Another remark is that $\mathcal{G}_i$ forming a concave quadrangle (Figure 5(b)) is converted into convex one by $\mathcal{G}_i \cap \mathcal{V}_i$, while $\mathcal{G}_i$ having two triangles (Figure 5(c)) keeps the concave shape even under $\mathcal{G}_i \cap \mathcal{V}_i$. Supposing that the probe is inserted along its longitudinal direction, we now describe a sufficient condition for making the tip reach $Int(A_iB_i)$.

Theorem 1 *A sufficient condition for making the tip reach $Int(A_iB_i)$ is to move the probe along the line passing through one point on $L_{A_0}^{B_0} \cap \mathcal{G}_i \cap \mathcal{V}_i$ and one point on $L_{A_i}^{B_i}$, where if $\mathcal{G}_i \cap \mathcal{V}_i$ is composed of two triangles, one point on $L_{A_i}^{B_i}$ is replaced by the common point for both triangles.*

Proof : $L_{A_0}^{B_0} \cap \mathcal{G}_i$ expresses the line segment of $L_{A_0}^{B_0}$ within $\mathcal{G}_i$. Similarly, $L_{A_0}^{B_0} \cap \mathcal{G}_i \cap \mathcal{V}_i$ denotes the line segment of $L_{A_0}^{B_0}$ within $\mathcal{G}_i \cap \mathcal{V}_i$. Note that there are two possible shapes constructed by $\mathcal{G}_i \cap \mathcal{V}_i \cap \overline{\mathcal{F}}$, one is convex (Figure 5(a), (b), (d) and Figure 6(b)) and the other is concave (Figure 5(c)).

(i) $\mathcal{G}_i \cap \mathcal{V}_i \cap \overline{\mathcal{F}}$ is a convex polygon: Assume a half-straight line starting from an arbitrary point on $L_{A_0}^{B_0} \cap \mathcal{G}_i \cap \mathcal{V}_i$ toward $L_{A_i}^{B_i}$. Since $\mathcal{G}_i \cap \mathcal{V}_i \cap \overline{\mathcal{F}}$ is convex polygon, the half-straight line comes out only from the line segment A_iB_i without passing through the other line segments. Therefore, the tip comes out from $L_{A_i}^{B_i}$ or stops due to the existence of $Int(A_iB_i)$ before reaching $L_{A_i}^{B_i}$. Once the tip comes out from $L_{A_i}^{B_i}$, the only feasible case is that the tip makes contact with $Int(A_iB_i)$. In any case, the tip finally reaches on $Int(A_iB_i)$.

(ii) $\mathcal{G}_i \cap \mathcal{V}_i \cap \overline{\mathcal{F}}$ is a concave polygon: Figure 5(c) is the only example of this case. Since the polygon is composed of two triangles having the common top angle, the half-straight line starting from an arbitrary point on $L_{A_0}^{B_0} \cap \mathcal{G}_i \cap \mathcal{V}_i$ toward the common point always reaches $Int(A_iB_i)$ without coming out from the other line segment forming the polygon. Thus, the theorem holds true. □

The approaching strategy based on Theorem 1 guarantees finding a route for reaching a new contact point within the designated environment's surface. In the algorithm, we simply move the probe along the line connecting two central points on $L_{A_0}^{B_0} \cap \mathcal{G}_i \cap \mathcal{V}_i$ and $L_{A_i}^{B_i}$.

4.2.2 Bisection Method

Once the probe detects an initial contact point, the bisection method is continuously applied for finding a local concave point, while it is not always obtained. In this subsection, after a couple of definitions, we provide a theorem ensuring for always making the tip converge on $Int(A_iB_i)$. Suppose that the tip of the probe is already in contact with $Int(A_iB_i)$ by initial pass planning. Then, bisection method has the following procedure:

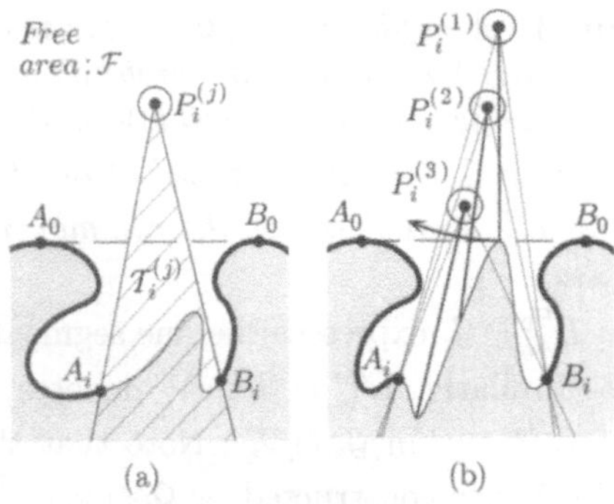

Figure 7: *Definition of* $\mathcal{T}_i^{(j)}$.

(i) Swing motion: The probe is rotated around the joint until either it makes contact with an environment or it exceeds a prescribed rotational angle.

(ii) Dividing: The angular displacement obtained during the swing motion is equally divided. Then, the probe is swung back by the equally divided angle.

(iii) Inserting motion: The probe is inserted along the longitudinal direction till the tip makes contact with an environment.

(iv) (i) through (iii) are repeated until the probe results in one of the following states, (a) The probe loses any rotational degree of freedom around the joint, or (b) The tip converges the intersection between the environment's surface and the boundary imparted as a constraint condition.

Definition 4 *Let the joint position be* P_i. *Define* $\mathcal{T}_i$ *as a semi-infinite region sandwiched by two lines* P_iA_i *and* P_iB_i.

$\mathcal{T}_i^{(j)}$ is shown by the hatched line in Figure 7(a), where the superscript (j) denotes the value after j-th insertion. Now let us consider an extreme case, where the initial contact is achieved at the top of the hill as shown in Figure 7(b). The bisection method starts by swinging the probe in the left direction (or right direction). Since the probe always goes into the safety area $\mathcal{F}$ in this particular case, the probe will not make contact with the environment any more. Therefore, we need a boundary, such that we can stop the swing motion. $\mathcal{T}_i^{(j)}$ provides a reasonable boundary, although it is still not a sufficient boundary for finally making the tip converge on $Int(A_iB_i)$.

Theorem 2 *Suppose that an initial contact with* $Int(A_iB_i)$ *is already completed. Also suppose that the maximum swing angle is determined such that the probe never comes out from the line segments* $P_i^{(j)}A_i$ *and* $P_i^{(j)}B_i$. *If there is no* $Ext(A_iB_i)$ *within* $\mathcal{T}_i^{(1)} \cap \mathcal{V}_i$, *the tip always reaches* $Int(A_iB_i)$ *through the bisection method.*

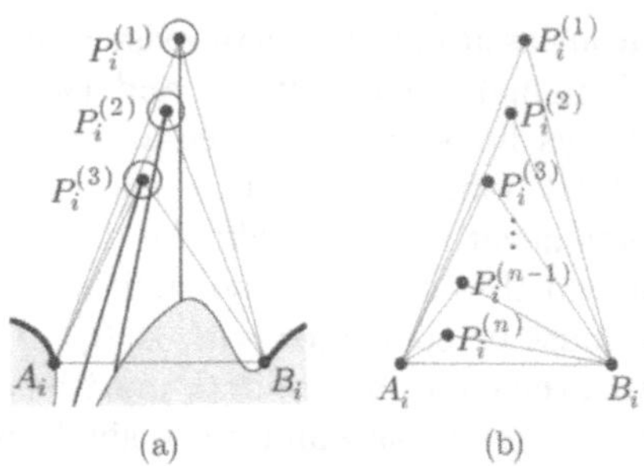

Figure 8: *Relationship between* $P_i^{(j)}$ *and* $\mathcal{T}_i^{(j)} \cap \mathcal{V}_i$.

Proof : By the assumption imparted to the maximum swing angle, the probe never comes out from the line segments $P_i^{(j)}A_i$ and $P_i^{(j)}B_i$ as shown in Figure 8(a). From Figure 8(b), we can easily show $(\mathcal{T}_i^{(1)} \cap \mathcal{V}_i) \supset (\mathcal{T}_i^{(2)} \cap \mathcal{V}_i) \supset \cdots \supset (\mathcal{T}_i^{(n)} \cap \mathcal{V}_i)$, which means if there is no $Ext(A_iB_i)$ within $\mathcal{T}_i^{(1)} \cap \mathcal{V}_i$, $\mathcal{T}_i^{(j)} \cap \mathcal{V}_i (j = 1, 2, \ldots, n)$ cannot include $Ext(A_iB_i)$ either. Therefore, the tip comes out from $L_{A_i}^{B_i}$ or stops due to the existence of $Int(A_iB_i)$ before reaching $L_{A_i}^{B_i}$. Once the tip comes out from $L_{A_i}^{B_i}$, the only feasible case is that it makes contact with $Int(A_iB_i)$. In any case, the tip finally reaches on $Int(A_iB_i)$. □

In order to utilize Theorem 2, we have to ensure that there is no $Ext(A_iB_i)$ within $\mathcal{T}_i^{(1)} \cap \mathcal{V}_i$. In case that there exists an external environment within $\mathcal{T}_i^{(1)} \cap \mathcal{V}_i$, however, the tip does not always reach $Int(A_iB_i)$ even though the bisection method is executed within $\mathcal{T}_i^{(j)} \cap \mathcal{V}_i$. Figure 9(a) may provide a good example. Suppose that the initial contact is achieved at the top of the hill and the bisection method is executed as shown in Figure 9(a). At the end of the bisection method, the tip will converge to the local concave point D_i belonging to $Ext(A_iB_i)$. To exclude the convergence to such an external environment, we again recall $\mathcal{G}_i$ where it is guaranteed that there is no $Ext(A_iB_i)$ in it. For example, if the bisection method is planned, such that the tip may move within $\mathcal{T}_i^{(j)} \cap \mathcal{V}_i \cap \mathcal{G}_i$ as shown in Figure 9(b), the convergence on $Int(A_iB_i)$ is always guaranteed. Theorem 3 provides a sufficient condition for making the tip always converge on $Int(A_iB_i)$.

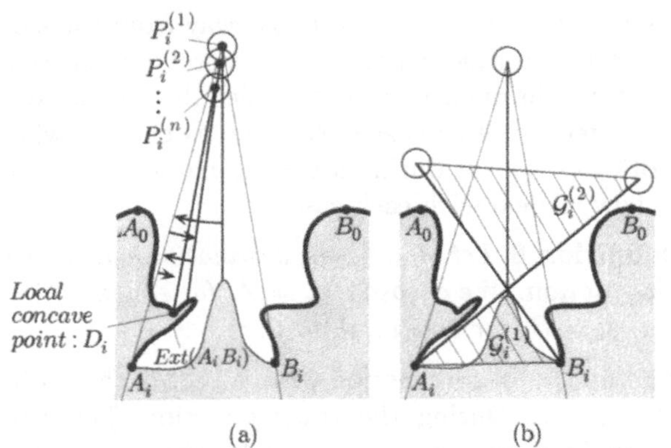

Figure 9: *An example of environment where there exists $Ext(A_iB_i)$ within $\mathcal{T}_i^{(1)} \cap \mathcal{V}_i$.*

Theorem 3 *Suppose that an initial contact with $Int(A_iB_i)$ is already completed. A sufficient condition for making the tip finally converge on $Int(A_iB_i)$ is to execute bisection method such that the tip may not come out from the boundary of $\mathcal{T}_i^{(j)} \cap \mathcal{V}_i \cap \mathcal{G}_i$ except the line segment A_iB_i.*

Proof : Since there is no $Ext(A_iB_i)$ within $\mathcal{T}_i^{(j)} \cap \mathcal{V}_i \cap \mathcal{G}_i$, the tip comes out from $L_{A_i}^{B_i}$ or stops due to the existence of $Int(A_iB_i)$ before reaching $L_{A_i}^{B_i}$. Therefore, by the similar logic that explained in the proof in theorem 2, we can show that the tip finally reaches on $Int(A_iB_i)$ in any case. □

4.3 Tracing Motion Planning

4.3.1 Realization of the Pulling Motion

When the *local concave search* comes to the end, there are basically two possible cases as shown in Figure 10, namely, the tip finally reaches a local concave point (a), and it stops at a non-local concave point ((b) and (c)). In Figure 10(b) and (c), the tip stops due to the boundary constraint (b) and due to multiple contacts (c), respectively.

Definition 5 *Y_1 denotes the final contact mode where $Y_1 = 1$ means that only the tip makes contact with the environment, and $Y_1 = 2$ means multiple contacts at the tip and other points. Y_2 denotes the rotational constraint around the joint, where $Y_2 = 0$, 1 and -1 are full constraint (Figure 10(a)(c)), single constraint for the clockwise direction, and single constraint for the counter clockwise direction (Figure 10(b)), respectively.*

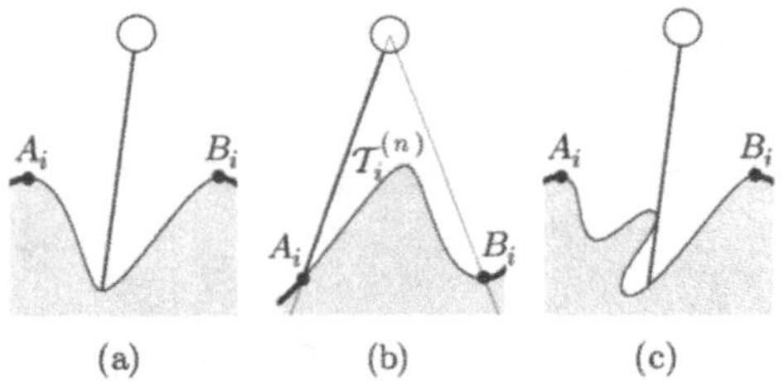

Figure 10: *Final states after bisection method: (a) Local concave point; (b) Intersection between the environment's surface and $\mathcal{T}_i^{(n)}$; (c) Multiple contacts.*

Definition 6 *Define $\mathcal{K}_1$ as an assemble of $\boldsymbol{v}$ satisfying $\boldsymbol{v}^{\mathrm{T}}\boldsymbol{t}^* < 0$, where $\boldsymbol{v}$ and $\boldsymbol{t}$ are vectors expressing the moving direction of the joint, and the longitudinal direction of the probe, respectively, and $*$ denotes the value just after the bisection method is completed. Also, define $\mathcal{K}_2$ as an assemble of $\boldsymbol{v}$ satisfying $sgn(\boldsymbol{v}, \boldsymbol{t}^*) < 0$.*

More simply, $\mathcal{K}_1$ denotes the half plane excluding $\boldsymbol{t}^*$ when the whole plane is divided at the joint by the line perpendicular to $\boldsymbol{t}^*$.

Definition 7 *$Face(right) = ON$ (or $Face(left) = ON$ or $Face(tip) = ON$) means that a part of the right side (or left side or tip) of the probe makes contact with an environment.*

Based on Definition 5, the three cases (a), (b), and (c) in Figure 10 can be classified by $(Y_1, Y_2) = (1, 0)$, $(Y_1, Y_2) = (1, -1$ or $1)$, $(Y_1, Y_2) = (2, 0)$, respectively. For each case, let us now consider a sufficient condition for realizing the pulling motion based tracing motion. Let $\mathcal{K}$ be the region where the joint can be moved without generating any pushing motion. In case (a), $\mathcal{K} = \mathcal{K}_1$ (the hatched area in Figure 11(a)). In case (b), if $Y_2 = -1$, $\mathcal{K} = \mathcal{K}_1 \cap \mathcal{K}_2$ (the double hatched area in Figure 11(b)) and if $Y_2 = 1$, $\mathcal{K} = \mathcal{K}_1 \cap \overline{\mathcal{K}_2}$. In case (c), if $Face(left) = ON$, $\mathcal{K} = \mathcal{K}_1 \cap \mathcal{K}_2$ (the double hatched area in Figure 11(c)) and if $Face(right) = ON$, $Y_2 = 1$, $\mathcal{K} = \mathcal{K}_1 \cap \overline{\mathcal{K}_2}$.

Theorem 4 *A sufficient condition for achieving a pulling motion based tracing is to determine the moving direction of the joint and the joint torque, such that $\boldsymbol{v}_i \in \mathcal{K}$ and:*

$$\tau = \tau_0 sgn(\boldsymbol{v}_i, \boldsymbol{t}^*) \tag{2}$$

where $\tau_0 (> 0)$ is the reference torque and the positive direction of τ is chosen in the clockwise direction, and $\mathcal{K}$ is given below.

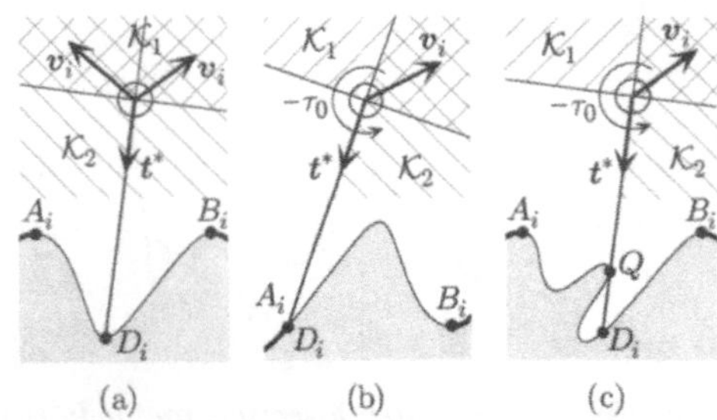

Figure 11: *Tracing motion planning.*

(**A**) : *If* $(Y_1, Y_2) = (1, 0)$, $\mathcal{K} = \mathcal{K}_1$.

(**B**) : *If* $(Y_1, Y_2) = (1, -1)$, $\mathcal{K} = \mathcal{K}_1 \cap \mathcal{K}_2$.

(**C**) : *If* $(Y_1, Y_2) = (1, 1)$, $\mathcal{K} = \mathcal{K}_1 \cap \overline{\mathcal{K}_2}$.

(**D**) : *If* $(Y_1, Y_2) = (2, 0)$ *and* $Face(left) = ON$, $\mathcal{K} = \mathcal{K}_1 \cap \mathcal{K}_2$.

(**E**) : *If* $(Y_1, Y_2) = (2, 0)$ *and* $Face(right) = ON$, $\mathcal{K} = \mathcal{K}_1 \cap \overline{\mathcal{K}_2}$.

Proof : (Omitted) □

In case (**A**), when $\boldsymbol{v}_i$ satisfying $\mathcal{K} = \mathcal{K}_1$ is chosen, only the right or the left tracing motion will be achieved. For completing the tracing motion for both directions, we further separate $\mathcal{K} = \mathcal{K}_1$ into $\mathcal{K} = \mathcal{K}_1 \cap \mathcal{K}_2$ and $\mathcal{K} = \mathcal{K}_1 \cap \overline{\mathcal{K}_2}$. Then, we execute the right and the left tracing motions by choosing $\boldsymbol{v}_i$ satisfying $\boldsymbol{v}_i \in \mathcal{K}_1 \cap \mathcal{K}_2$ and $\boldsymbol{v}_i \in \mathcal{K}_1 \cap \overline{\mathcal{K}_2}$, respectively.

4.3.2 All Possible Cases During Tracing Motion

Tracing motion is continued until the tip successfully traces from the initial point D_i to A_i (or B_i) as shown in Figure 12(a). However, the contact between the probe and the environment is not always guaranteed during the tracing motion. For example, the tip may be away from the environment at the top of the hill during the tracing motion from A_i to B_i, as shown in Figure 12(b). Once the probe is away from the surface, the joint torque sensor can not detect any reaction torque from the environment, by which the sensing system can recognize the case. Now let us classify what sort of cases appear during a tracing motion and consider how to deal with each case.

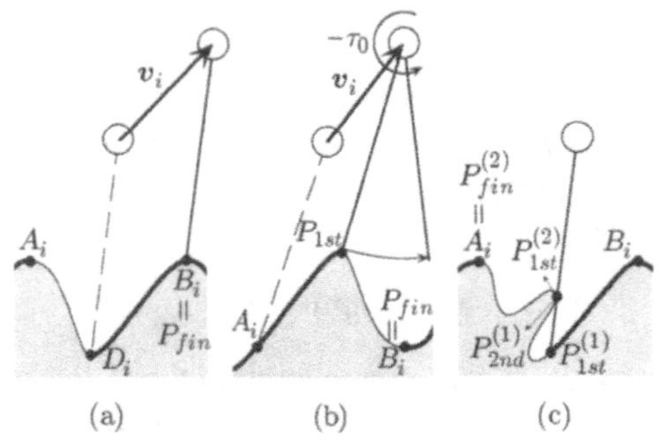

Figure 12: *Realization of tracing motion.*

Definition 8 *Let P_{fin} be the destination point for a tracing motion. We choose $P_{fin} = A_i$ for $sgn(\boldsymbol{v}_i, \boldsymbol{t}^*) > 0$ or $P_{fin} = B_i$ for $sgn(\boldsymbol{v}_i, \boldsymbol{t}^*) < 0$.*

We stop the tracing motion when a part of the probe reaches P_{fin}. During the tracing motion, however, there might appear a contact point jump due to the surface geometry. For such a contact point jump, we define two points P_{1st} and P_{2nd} as follows.

Definition 9 *Let P_{1st} and P_{2nd} be the contact points just before and after a contact point jump, respectively. In case of a single probe detection, the point closer to the tip is chosen as P_{1st}.*

All possible cases during a tracing motion can be classified as follows:

<Case 1> All contact points are continuously detected by the probe tip for the designated area.

<Case 2> At least, one contact point jump appears.

<Case 3> $\boldsymbol{v}_i{}^T\boldsymbol{t} > 0$ is detected.

<Case 4> The joint loses the moving degree of freedom in the direction given by $\boldsymbol{v}_i$.

All cases can be detected by utilizing the sensors implemented in the robot-probe system. For example, <Case 1> and <Case 2> are confirmed by the probe output. <Case 3> is certified by the sensors mounted in each joint of the robot arm. <Case 4> can be confirmed by the torque sensor.

Since <Case 1> means that the designated unknown area $Int(A_iB_i)$ becomes known, we can simply categorize such an area into W. For one of the three other cases, however, we have to memorize a part of $Int(A_iB_i)$ as a further non-detected area. Let us now consider which points we should register for <Case 2>, <Case 3> and <Case 4>. Just for our convenience, <Case 2> is further classified into the following three cases.

<Case 2-1> P_{2nd} exists on $Int(A_iB_i)$.

<Case 2-2> P_{2nd} exists on $Ext(A_iB_i)$.

<Case 2-3> P_{2nd} does not exist within $\overline{\mathcal{F}}$.

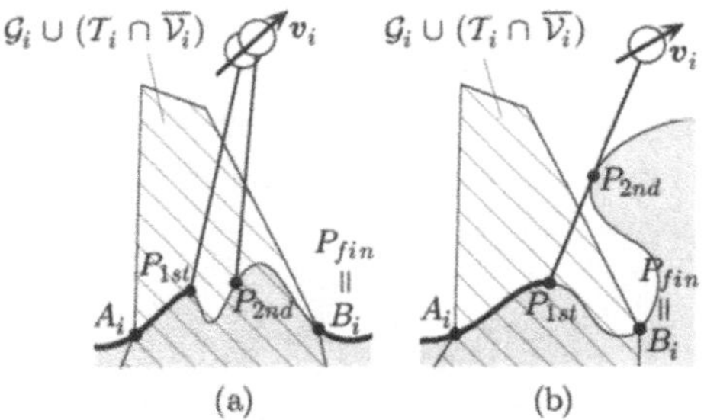

Figure 13: *P_{2nd} detected on $Int(A_iB_i)$ or $Ext(A_iB_i)$.*

<Case 2-1> or <Case 2-2> is distinguished by checking whether $P_{2nd} \in \mathcal{G}_i \cup (\mathcal{T}_i \cap \overline{\mathcal{V}_i})$ or $P_{2nd} \in \overline{\mathcal{G}_i} \cap (\overline{\mathcal{T}_i} \cup \mathcal{V}_i)$, respectively, as shown in Figure 13(a) and (b). <Case 2-3> happens when the probe enters $\mathcal{F}$ without any contact. In <Case 2-1>, the area between P_{1st} and P_{2nd} is registered as a non-detected area and then the tracing motion is continued for the remaining non-traced area. In both <Case 2-2> and <Case 2-3>, we stop any further tracing motion and register the area between P_{1st} and P_{fin} as a non-detected area. <Case 3> may occur because the condition realizing a pulling motion is satisfied only at the starting position. Once $\boldsymbol{v}_i{}^{\mathrm{T}}\boldsymbol{t} > 0$ is detected, we stop the tracing motion and register the area between P_{1st} and P_{fin} as a non-detected one, where P_{1st} is replaced by the point in which $\boldsymbol{v}_i{}^{\mathrm{T}}\boldsymbol{t} > 0$ is detected. <Case 4> may also occur depending on the surface geometry, the probe posture and the moving direction of the probe. Once the probe loses the degree of freedom in the direction $\boldsymbol{v}_i$, we stop the tracing motion and register the area between P_{1st} and P_{fin} as a non-detected one, where P_{1st} is replaced by the point in which the probe loses the degree of freedom. In case of Figure 11(c), according to Theorem 4, the tracing motion is not executed in the direction from D_i to A_i. In such a situation, we register $(P^{(1)}_{1st}, P^{(1)}_{2nd})$ and $(P^{(2)}_{1st}, P^{(2)}_{fin})$ as non-detected areas after regarding $(P^{(1)}_{1st}, P^{(1)}_{2nd}) = (D_i, Q)$ and $(P^{(2)}_{1st}, P^{(2)}_{fin}) = (Q, A_i)$ as shown in Figure 12(c).

4.3.3 Registering into A_jB_j

The non-detected area, such as (P_{1st}, P_{2nd}) has to be finally stored into (A_j, B_j) for executing the program recursively. In this subsection, we briefly describe the basic rule to obtain the set of (A_j, B_j).

<Rule to determine (A_j, B_j) > (see Figure 14)
(i) In case of $P_{fin} = A_i$: $(A_j, B_j) = (P_{2nd} \text{or} P_{fin}, P_{1st})$.

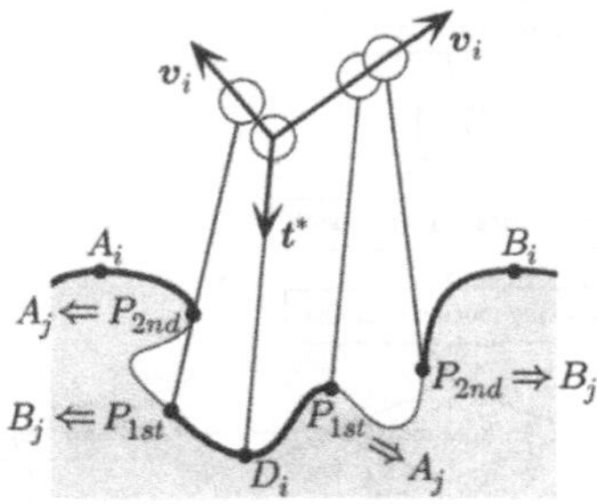

Figure 14: *How to store non-detected area.*

(ii) In case of $P_{fin} = B_i$:
$(A_j, B_j) = (P_{1st}, P_{2nd} \text{ or } P_{fin})$.

This rule is for keeping the characteristics of $\mathcal{V}_i$ defined in 4.2.1.

4.4 Infinite Loop Escape

There might be a particular state in which the tip can find the same point repeatedly during the *local concave point search* and the *tracing motion planning.* In order to avoid such an undesirable mode, we prepare the *infinite loop escape*, where the probe temporarily searches a new contact point by utilizing the same way taken in the initial pass planning. Since the initial pass planning ensures to find a new contact point between A_iB_i, we can separate the area A_iB_i into two new areas $A_{i+1}B_{i+1}$ and $A_{i+2}B_{i+2}$, as shown in Figure 15(a). After dividing the area, we leave from the *infinite loop escape* and come back to the normal mode given by *local concave point search* and *tracing motion planning.* Now, assume that an infinite mode appears every time after initial pass planning motion. In such an extreme case, $A_{i+1}B_{i+1}$ (or $A_{i+2}B_{i+2}$) is further separated into two unknown areas $A_{i+3}B_{i+3}$ and $A_{i+4}B_{i+4}$, and so forth, as shown in Figure 15(a). For

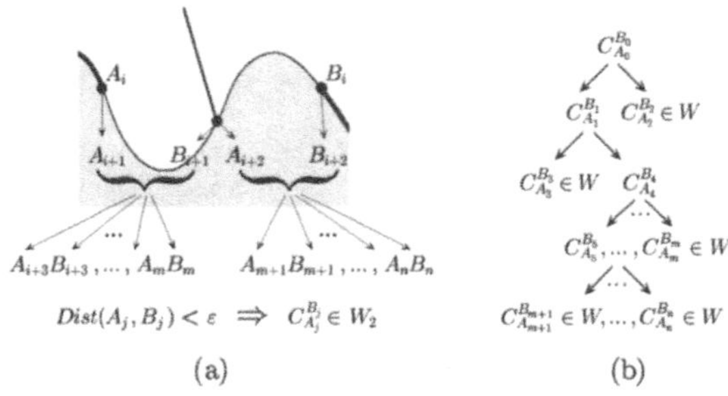

Figure 15: *Infinite loop escape.*

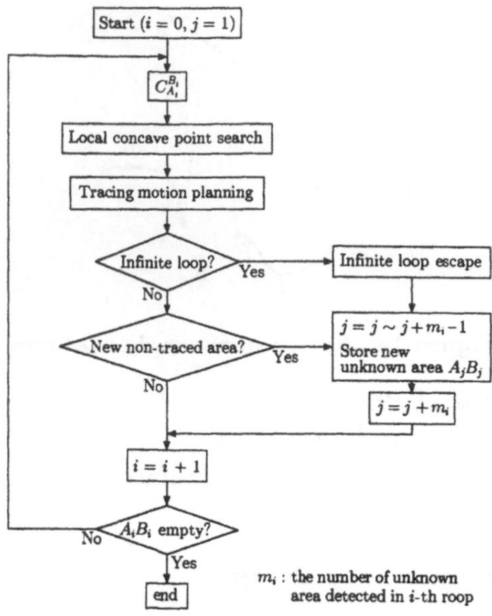

Figure 16: *Flowchart.*

every newly divided area, if $Dist(A_j, B_j) < \varepsilon$ is satisfied, any further separation between A_j and B_j is stopped and $C_{A_j}^{B_j} \in W_2$ is assigned. This implies that the algorithm brings the environment's shape in relief even when an infinite loop occurs continuously. Figure 16 shows the overall flowchart of the algorithm.

5 Simulation

Figure 17 and Figure 18 show simulation results, where the continuous and the dotted lines denote the known and the unknown areas, respectively, and the line segment passing through the joint expresses the moving direction of the joint, which is determined by the sufficient condition given by Theorem 4. Figure 17 is a simple example, where the tip can trace every surface eventually. Also, there is no $Ext(A_iB_i)$ within $\mathcal{T}_i^{(1)} \cap \mathcal{V}_i$ and, therefore, the local concave point can be found by utilizing either Theorem 2 or Theorem 3. Figure 18 shows an example including a couple of never-touching areas, where the tip cannot reach and touch. Although this example includes such a complicated surface, it can be seen that the proposed algorithm enables the tip to reach every reachable area step by step and, finally, it finds the contour of the surface except three never-touching areas. From these simulation results, it can be understood that the unknown area gradually decreases and finally disappears except the particular area which the probe cannot reach and touch.

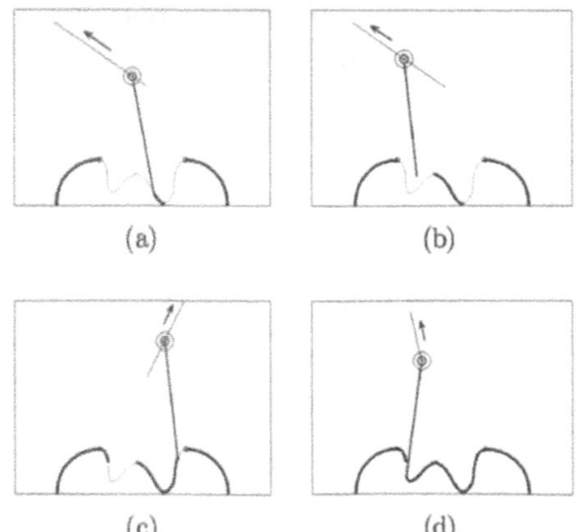

Figure 17: *Simulation result (example 1).*

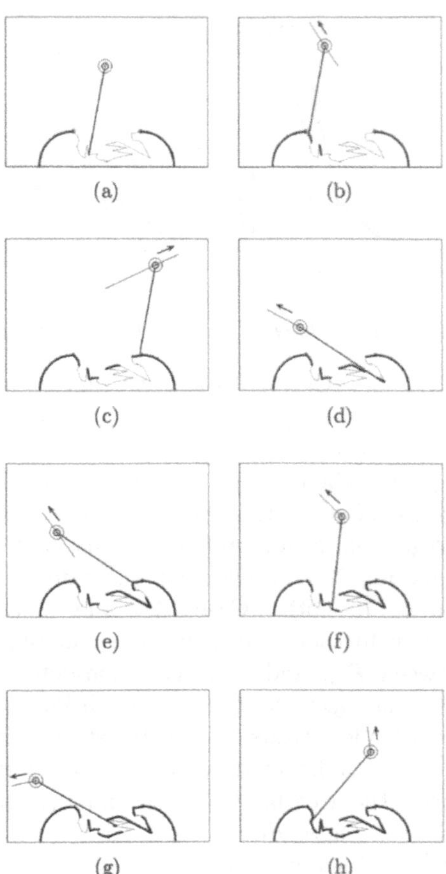

Figure 18: *Simulation result (example 2).*

6 Conclusions

A tactile sensing algorithm for detecting the concave surface was discussed. In our next step, we will relax this assumption so that the tip can reach much larger area. Also, we will try to extend the pulling-motion-based tactile sensing into a 3D concave environment. Since tracing every area for a 3D environment is not as feasible as that in a 2D one, the parametrized local surface function [18] may be a useful tool for expressing the surface shape with respect to a local contact frame. We believe, however, that the concept of the pulling-motion-based tracing should be still included even in the algorithm for surface sensing of a 3D environment.

Acknowledgments

This work has been supported by CREST of JST (Japan Science and Technology).

References

[1] Dario, P. and G. Buttazzo: An anthropomorphic robot finger for investigating artificial tactile perception, Int J. Robotics Research, vol.6, no.3, pp25–48, 1987.

[2] Fearing, R. S. and T. O. Binford: Using a cylindrical tactile sensor for determining curvature, Proc. of the IEEE Int. Conf. on Robotics and Automation, Philadelphia, pp765–771, 1988.

[3] Maekawa, H., K. Tanie, K. Komoriya, M. Kaneko, C. Horiguchi, and T. Sugawara: Development of a finger-shaped tactile sensor and its evaluation by active touch, Proc. of the IEEE Int. Conf. on Robotics and Automation, Nice, p1327, 1992.

[4] Salisbury, J. K.: Interpretation of contact geometries from force measurements, Proc. of the 1st Int. Symp. on Robotics Research, 1984.

[5] Brock, D.L. and S.Chiu: Environment perception of an articulated robot hand using contact sensors, Proc. of the IEEE Int. Conf. on Robotics and Automation, Raleigh, pp89–96, 1987.

[6] Kaneko, M., and K. Honkawa: Compliant motion based active sensing by robotic fingers, Preprints of the 4th IFAC Symp. on Robot Control, Capri, pp137–142, 1994.

[7] Allen, P. and K. S. Roberts: Haptic object recognition using a multi-fingered dexterous hand, Proc. of the IEEE Int. Conf. on Robotics and Automation, pp342–347, 1989.

[8] Bajcsy, R.: "What can we learn from one finger experiments?", Proc. of the 1st Int. Symp on Robotics Research, pp509–527, 1984.

[9] Bicchi, A., Salisbury J. K. and D. J. Brock: Contact sensing from force measurements, Int. J. of Robotics Research, vol.12, no.3, 1993

[10] Bays, J. S. : Tactile shape sensing via single- and multi-fingered hands, Proc. of the IEEE Int. Conf. on Robotics and Automation, PP 290–295, 1989.

[11] Caselli, S., C. Magnanini, F. Zanichelli, and E. Caraffi: Efficient exploration and recognition of convex objects based on haptic perception, Proc. of the IEEE Int. Conf. on Robotics and Automation, PP 3508–3513, 1996.

[12] Gaston, P. C., and T. Lozano-Perez: Tactile recognition and localization using object models, The case of polyhedra on a plane, MIT Artificial Intelligence Lab. Memo, no.705, 1983.

[13] Grimson, W. E. L. and T. Lozano-Perez: Model based recognition and localization from sparse three dimensional sensory data, AI Memo 738, MIT, AI Laboratory, Cambridge, MA, 1983.

[14] Cole, R., and C. K. Yap: Shape from probing, J. of Algorithms 8, pp19–38, 1987.

[15] Russell, R. A. : Using tactile whiskers to measure surface contours, Proc. of the 1992 IEEE Int. Conf. on Robotics and Automation, pp1295–1300, 1992.

[16] Tsujimura, T. and T. Yabuta: Object detection by tactile sensing method employing force/moment information, IEEE Trans. on Robotics and Automation, vol.5, no.4, pp444–450, 1988.

[17] Roberts, K. S. : Robot active touch exploration, Proc. of the IEEE Int. Conf. on Robotics and Automation, PP 980–985, 1990.

[18] Chen, N., R. Rink, and H. Zhang: Local object shape from tactile sensing, Proc. of the IEEE Int. Conf. on Robotics and Automation, PP 3496–3501, 1996.

[19] Kaneko, M., M. Higashimori, and T, Tsuji: Pulling motion based tactile sensing for concave surface, Proc. of the IEEE Int. Conf. on Robotics and Automation, pp2477–2484, 1997.

[20] Boissonnat, J. D. and M. Yvinec: Probing a scene of non-convex polyhedra, Algorithmica, vol.8, pp321–342, 1992.

Compensatory Grasping with the Parallel Jaw Gripper

Tao Zhang, *University of California, Berkeley, Berkeley, CA*
Gordon Smith, *University of California, Berkeley, Berkeley, CA*
Ken Goldberg, *University of California, Berkeley, Berkeley, CA*

For many industrial parts, their resting pose differs from the orientation desired for assembly. It is possible in many cases to compensate for this difference using a parallel-jaw gripper with fixed orientation. The idea is to arrange contact points on each gripper jaw so that the part is reoriented as it is grasped. We analyze the mechanics of compensatory grasping based on a combination of toppling, jamming, accessibility, and form closure and describe an algorithm for the design of such grasps based on the constrained toppling graph.

1 Introduction

Industrial parts on a flat worksurface will naturally come to rest in one of several stable orientations [1], but it is often necessary to rotate a part into a different (unstable) orientation for assembly [2].

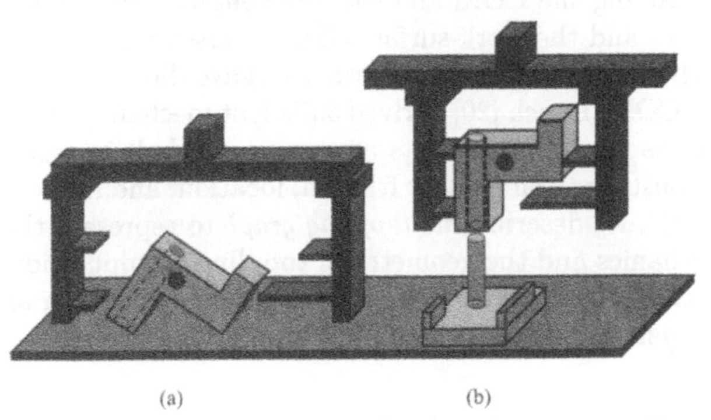

Figure 1: *A compensatory grasp.*

This paper proposes an inexpensive (minimalist) method for compensating for these differences in orientation by reorienting the part during grasping. As illustrated in Figure 1, the part is initially in stable orientation (a); it then is rotated by the gripper to orientation (b) for assembly onto the peg. We refer to this as a *compensatory grasp.*

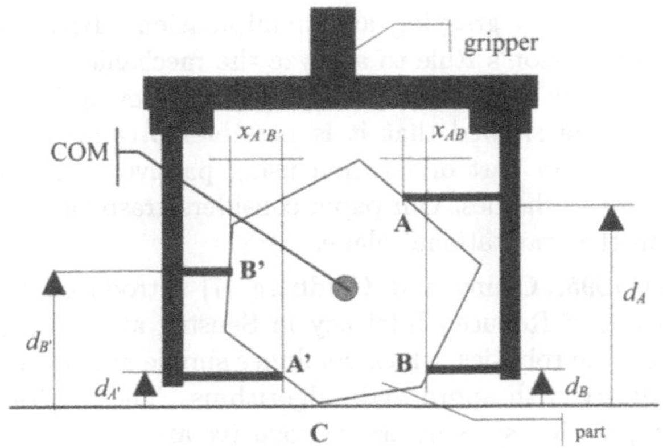

Figure 2: *Terminology.*

We achieve this using a simple parallel-jaw gripper with four tips as shown in Figure 2. First, toppling tip A and pushing tip A' make contact with the part and topple it from the initial orientation to the desired orientation. This process is referred to as *constrained toppling.* Then, as soon as the part reaches the desired orientation, left grasping tip B' and right grasping tip B make contact, stop the part's rotation, and securely grasp it. This process is simply referred to as *grasping.* Note that the pivot point, C, maintains contact with bottom surface at all times.

These four tips and the parallel jaw gripper are designed to be easily reconfigurable to handle different industrial parts, and low in cost, footprint and weight.

2 Related Work

Grasping is a fundamental issue in robotics; [3] provides a useful review of research on the topic.

The two most important classes of grasps are known as *force closure* and *form closure.* The difference between these two is that the latter is stable regardless of the external wrench applied to the object. In 1990, Markenscoff et al. [4] proved, by infinitesimal perturbation analysis, that four hard fingers are necessary and sufficient to achieve form closure of a 2D object in the absence of friction. The parallel jaw grippers we propose provide four contact points but the location of tips B and B' are dependent on the locations of A and A' respectively.

Mason [5] was the first to study the role of passive compliance in grasping and manipulation. Brost [6] applied Mason's Rule to analyze the mechanics of the parallel-jaw gripper for polygonal parts moving in the plane. He showed that it is possible to compensate for errors in part orientation using passive push and squeeze mechanics. Our paper considers grasp mechanics in the gravitational plane.

In 1995, Canny and Goldberg [7] introduced the concept of Reduced Intricacy in Sensing and Control (RISC) to robotics, which combines simple automation hardware with appropriate algorithms. The resulting manipulation systems are inexpensive and reliable.

Trinkle and Paul [8] studied how to orient parts in the gravitational plane by lifting them off work-surface using a gripper. Based upon the geometry of a grasp and quasi-static analysis, the authors generated *liftability regions*, which defines the qualitative motion of a squeezed object. They predicted the motion by solving the *forward object motion problem* that is the dual of a nonlinear program employing Peshkin's [9] minimum power principle. The pre-liftoff phase analysis of Trinkle and Paul's paper is related to our constrained toppling phase analysis in that we both applied graphic method to analyze the interaction among a planar object, a supporting surface, and a gripper in the plane containing gravity. One important difference is that we focus on the parallel-jaw gripper and consider how jaws can be designed to facilitate grasping using only translational motion; Trinkle and Paul considered two or three-point (contact) initial grasps, and employed both pusher and roll strategy to achieve form-closure grasps. Abell and Erdmann [10] studied how a planar polygon can be rotated in a gravitational field while stably supported by two frictionless contacts. Zumel and Erdmann [11][12] analyzed nonprehensile manipulation using two palms jointed at a central hinge. They also developed a sensorless approach to orient parts.

Another approach to reorienting parts is dextrous manipulation, where the part is reoriented as it is held in force closure using constrained slip. Rus [13] proposed a *finger tracking* technique to generate rotation of grasped objects with sliding. Hong *et al.* [14] developed a planing algorithm to acquire a desired grasp by using a *finger gait* technique, which allows reposition of fingers while maintaining a grasp. Fearing [15] considered both sliding and rolling manipulation, and developed grasp planning based upon local tactile feedback, geometry, and frictional constraints. Bicchi and Sorrentino [16] analyzed the effect of rolling. Compensatory grasps combine rolling and sliding. Another approach was studied by Rao et al. [17]. They proposed picking up a part using a parallel gripper with a pivoting bearing, allowing the part to pivot under gravity to rotate into a new configuration.

Our work was motivated by recent research in toppling manipulation. Zhang and Gupta [18] studied how parts can be reoriented as they fall down a series of steps. The authors derived the condition for toppling over a step and defined the transition height, which is the minimum step height to topple a part from a given stable orientation to another. Yu et al. [19] applied toppling technique to predict the location of an object's COM. The force and position information of the robot finger-tip was provided during *Tip (toppling) Operations* to estimate the *Gravity Equi-Effect Plane* containing the COM and the line contact between the object and the work-surface. The intersection of three such planes at different orientations gave the location of the COM. Lynch [20] derived sufficient mechanical conditions for toppling parts on a conveyor belt in terms of constraints on contact friction, location, and motion. In [21], we describe the *toppling graph* to represent the mechanics and the geometry of toppling manipulation. In this paper, we combine toppling mechanics with an analysis of jamming, accessibility and form closure.

3 Problem Definition

Given the planar projection of an n-sided convex polyhedral part P, how can we rotate the part to a desired orientation and grasp it securely? There are two phases involved in this process: constrained toppling and grasping. We are given as input: the part's center of mass (COM), uncertainty in vertex location, ε, the coefficient of friction between the part and the surface, μ_b, and the coefficient of friction between the part and the gripper, μ_t. Let θ denote the orientation of the part from the +X direction; initially $\theta = 0$. Let θ_d denote the angle at the desired orientation.

During the constrained toppling phase, only gripper tips A and A' make contact with the part. We require that these two tips cause the part to rotate counter-clockwise without causing C to lose contact with the surface. From the toppling analysis in section 4 we are able to show that on a given edge, the rolling conditions are more easily satisfied if A' is lower, i.e. $d_{A'}$ is smaller. This will in general produce a finite set of possible $d_{A'}$ candidates for the part. The height of A, d_A, can then be determined by the graphical analysis described in section 4.

The grasping phase occurs when the part has been toppled to $\theta = \theta_d$. During grasping the grasping gripper tips, B and B', make contact with the part. We require that the contacts corresponding to A, A', B, and B' create form-closure on the part, even when friction is disregarded. Since d_A and $d_{A'}$ are already known, we must determine d_B and $d_{B'}$ such that form-closure is achieved. We additionally require that the tips B and B' do not make contact with the part before $\theta = \theta_d$. This requirement is essentially a form of accessibility constraint and further limits the possible values of d_B and $d_{B'}$. Finally, because A and B are fixed on the same jaw of the gripper (and likewise A' and B') the geometry of the part along with the relative heights of A and B will determine the relative x offset between A and B, which we denote x_{AB}.

The final output of the analysis is the height of each of the four tips, d_A, $d_{A'}$, d_B, and $d_{A'}$, as well as the relative x offset between tips on each jaw, x_{AB} and $x_{A'B'}$ (see Figure 2). This set of variables determines the gripper design that will rotate the part to the desired orientation.

Figure 3 shows the notation used in our constrained toppling analysis. The part sits on a flat worksurface in a stable orientation. The worksurface friction cone half-angle is $\alpha_b = tan^{-1}\mu_b$, and the gripper tip friction half-angle is $\alpha_t = tan^{-1}\mu_t$. The contact point between the part and the worksurface is called a pivot vertex, denoted C in Figure 2, and taken to be at (0,0). The COM is a distance ρ from the origin and angle η from the +X direction at its initial orientation. The pushing tip, A', is a distance $d_{A'}$ from the bottom surface. We denote the vector at the left edge of the toppling tip friction cone as f_l and the right edge as f_r.

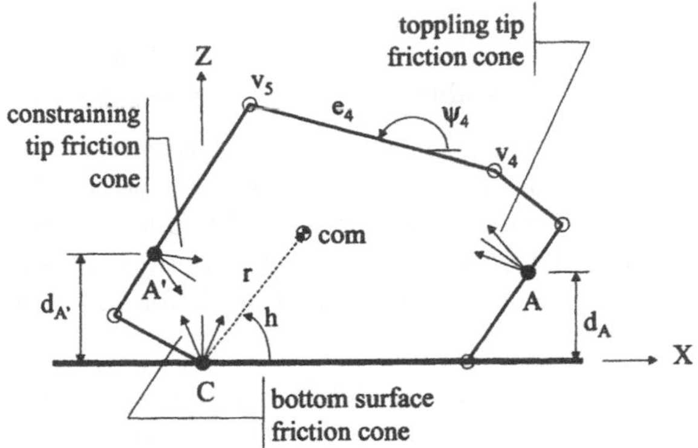

Figure 3: *Notation.*

We consider P as shown in Figure 3. Starting from the pivot, we consider each edge of the part in counter-clockwise order, namely e_1, e_2, ..., e_n. The edge e_i, with vertices v_i at (x_i, z_i) and v_{i+1} at (x_{i+1}, z_{i+1}), is in direction ψ_i from the +X axis.

We assume that the part and the gripper are rigid, and also that the part's geometry, the location of the COM, and the position of the jaws are known exactly. We also assume that the pushing tip and the toppling tip of the gripper contact the part simultaneously, and the motion of the part and is slow enough that we can ignore inertial effects.

4 Constrained Toppling Analysis

We divide constrained toppling into a *rolling* phase and a *settling* phase. Our analysis involves the graphical construction of a set of functions that represent the mechanics of these phases. In [21], we introduced the *toppling graph*, which includes the radius function, vertex height functions, constrained rolling height functions, and constrained jamming height functions.

The radius function, $R(\theta)$, is the height of the COM above the surface as the part is rotated through $\theta = 0 \rightarrow 2\pi$. The local minima of the radius function indicate the stable orientations of the part [22], while the local maxima are unstable equilibrium orientations. In this paper we will consider only the range of angles corresponding to rotation from one single stable orientation to the next. We assume that the part can be toppled into that orientation before grasping. This range consists of the angles $\theta = 0 \rightarrow \theta_n$ where θ_n is the angle of the ext stable orientation. Additionally, θ_t represents the unstable equilibrium angle in that range.

The vertex height function, $V_i(\theta)$, gives the height of vertex i above the surface as the part rotates. Each vertex of the part has a vertex height function. Using the vertex height functions we can determine which edge a gripper tip is in contact with for any θ and height.

The constrained rolling height function, $H_i(\theta)$, is the minimum height at θ that the toppling tip, A, in contact with edge e_i must be in order to roll the part given the height of A'. $H_i(\theta)$ is determined on the range $\theta = 0 \rightarrow \theta_t$. At $\theta = \theta_t$ the part is no longer being rolled, but it is now settling under its own weight. The constrained jamming height function, $H_i(\theta)$, indicates a range of d_A values that may cause jamming during the settling process. $H_i(\theta)$ is determined on the range $\theta = \theta_t \rightarrow \theta_n$.

All of these functions are dependant on θ and map from part orientation to distance: $S^1 \rightarrow \Re^+$, where S^1 is the set of planar orientations. The combination of these four functions forms the constrained toppling graph from which we can height of the toppling tip A required to guarantee toppling.

4.1 Constrained Rolling Height Function

During rolling, the part rotates about C. Friction between the part and the bottom surface must not prevent C from sliding to the right, and friction between the part and the tips must not prevent the part from slipping relative to the tips. Additionally, the system of forces on the part—the contact force at the bottom surface, the contact force at the tips, and the part's weight—must generate a positive moment on the part with respective to C.

The constrained rolling height function, $H_i(\theta)$, is the minimum height at θ and a given d that A in contact with edge e_i must be in order to roll a part. This height is determined as a function of θ based on the rolling conditions.

As the part rolls, A' could switch edges if a contact edge is not long enough. In this paper, we consider the situation that A' keeps contact with e_i during the entire rolling phase, and the same methodology can be

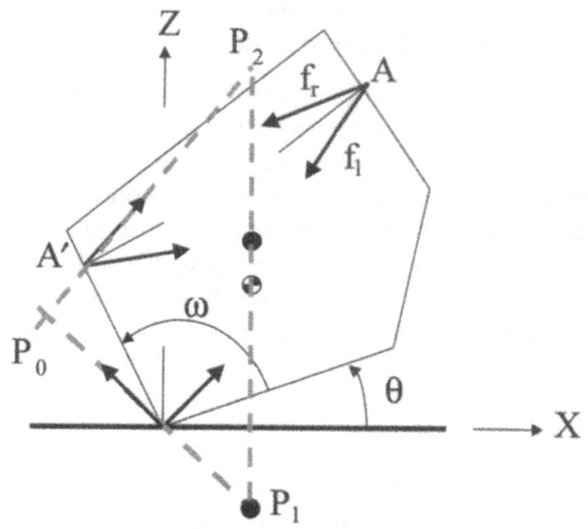

Figure 4: *Rolling conditions.*

applied for the case where the contact switches edges (see [23] for details).

Under the assumption that A' keeps contact with e_n, relative motion direction between the part and A' is uncertain, and it depends on an angle, $\omega + \theta$, where ω denotes the interior angle of a part at C as shown in Figure 4. We study the constrained rolling conditions based upon different $\omega + \theta$ ranges.

Let w_i be the distance along edge e_i as shown in Figure 4. Any point on e_i can be expressed as $(x_i + w_i\, cos\psi_i, z_i + w_i \sin\psi_i)$.

Consider the case where $\pi > \omega + \theta > \pi/2$. In such a case, rotation causes the contact between the part and A' to move away from C. To determine the constrained rolling height function, we begin by constructing a triangle as shown in Figure 4 with vertices P_0, P_1, and P_2. P_0 is determined by the intersection of the left edge of the bottom friction cone and the left edge of the pushing tip friction cone, and is at (x_{p0}, z_{p0}). P_1 is the intersection of the vertical line through the com and the left edge of the bottom friction cone, and is located at (x_{p1}, z_{p1}). P_2 is the intersection of the vertical line through the com and the left edge of the constrained tip friction cone, and is located at (x_{p2}, z_{p2}). From these definitions we have:

$$x_{p0} = s - t\sin\alpha_b, \quad (1)$$

$$z_{p0} = -s/\mu_b + t\cos\alpha_b, \quad (2)$$

$$x_{p1} = s, \quad (3)$$

$$z_{p1} = \frac{-s}{\mu_b}, \quad (4)$$

$$x_{p2} = s, \quad (5)$$

$$z_{p2} = \frac{d/\tan(\theta+\omega) - s}{\tan(\theta+\omega+\alpha_t)}, \quad (6)$$

where

$$s = \rho\cos(\eta+\theta), \quad (7)$$

$$t = \frac{(z_{p2}-z_{p1})\sin(\theta+\omega+\alpha_t)}{\sin(\theta+\omega+\alpha_t-\alpha_b)}. \quad (8)$$

Consider a region of the X-Z plane defined by linear edges. Let a *primary region* denote a region such that toppling is guaranteed if every force in the toppling tip friction cone makes a positive moment about every point in the primary region. The region is derived using a graphical method from Mason [24]. Therefore, the $P_0P_1P_2$ triangle is the primary region in this case.

For all forces in the toppling tip friction cone to generate a positive moment about the triangle, the left edge of the friction cone must pass above the triangle; all other vectors in the friction cone will pass higher. We find the height sufficient to roll the edge by projecting lines from P_0, P_1, and P_2 at the angle of the left edge of the pin friction cone, f_l, until they intersect the edge of the part. The intersection with the maximum height of those three is the minimum height sufficient to roll the part.

Let w_i^2 denote the edge contact on e_i where f_l passes exactly through point P_2. We can show through geometric construction that:

$$w_i^2(\theta) = (x_i\sin\theta + z_i\, cos\theta - z_{p2} - (x_i\cos\theta - z_i\, sin\theta - x_{p2})\tan(\beta_{il}+\theta))/(\cos(\theta+\psi_i)\tan(\beta_{il}+\theta) - \sin(\theta+\psi_i)). \quad (9)$$

where $\beta_{il} = \psi_i + \pi/2 + \alpha_t$.

Similarly, the edge contacts for f_l passing through P_0 and P_1 are given by w_i^0 and w_i^1.

The constrained rolling height function, $H_i(\theta)$, is based on w_i which is the maximum of w_i^0, w_i^1 and w_i^2 in the rolling region $0 < \theta < \theta_t$. w_i can be shown to be:

$$w_i = \begin{cases} w_i^2 & 0<\theta<\theta_{01} \text{ and } \psi_i<\omega \\ w_i^0 & 0<\theta<\theta_{01} \text{ and } \psi_i>\omega \\ w_i^1 & \theta_{01}<\theta<\theta_m \end{cases} \quad (10)$$

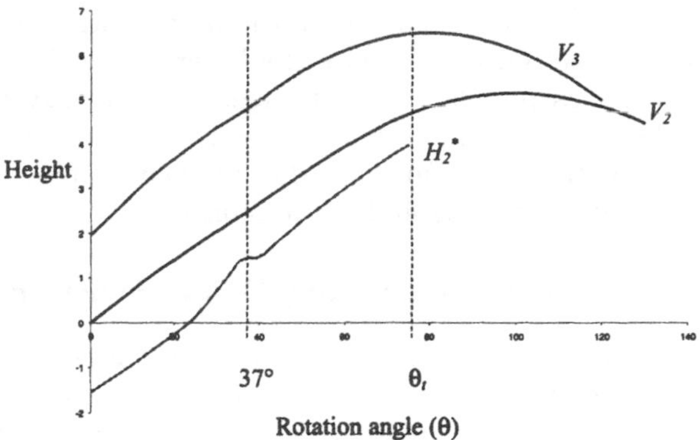

Figure 5: *$R(\theta)$ vs. $H_2^*(\theta)$, $V_2(\theta)$ and $V_3(\theta)$.*

where

$$\theta_{01} = \pi + \alpha_b - \alpha_t - \psi_i, \quad (11)$$

$$\theta_m = \min(\pi-\psi_i, \pi-\omega, \theta_t). \quad (12)$$

Thus, the constrained rolling height function within $0<\theta<\theta_m$ is given by:

$$H_i(\theta) = \begin{cases} H_i^*(\theta) & H_i^*(\theta) > V_i(\theta) \\ \emptyset & H_i^*(\theta) < V_i(\theta) \end{cases} \quad (13)$$

where

$$H_i^*(\theta) = x_i\sin\theta + z_i\, cos\theta + w_i\sin(\theta+\psi_i). \quad (14)$$

Following the same methodology, we find $H_i(\theta)$ under the condition $\theta+\omega=\pi/2$ and $\pi/2>\theta+\omega>0$ (see [23] for details).

Figure 5 illustrates the functions $R(\theta)$, $H_2(\theta)$, $V_2(\theta)$ and $V_3(\theta)$ for the part in Figure 3 with $\alpha_t = 5°$, $\alpha_b = 10°$ and $d_A = 0.9$cm. The kink at $\theta = 37°$ of $R(\theta)$ represents the orientation where e_6 is on the bottom surface. Notice that there are kinks at $\theta = 37°$ in $H_2(\theta)$, $V_2(\theta)$ and $V_3(\theta)$ too. When $\theta < 37°$, V_1 is C and x_2=4.1cm, z_2=0.0cm, $\psi_2 = 56°$, $\eta = 46°$, ρ=2.2cm, and $\omega = 53°$; when $\theta > 37°$, V_6 is C and x_2=4.6cm, z_2=2.4cm, $\psi_2 = 92°$, $\eta = 55°$, ρ=2.7cm, and $\omega = 89°$. At angle θ any toppling tip at a height, h, such that $\max(H_2(\theta),V_2(\theta)) < h < V_3(\theta)$ will instantaneously rotate the part. The graph indicates that A can roll the part at any contact on e_2 when $0<\theta<\theta_t$.

Given a certain contact edge, a lower A' results in a smaller primary region; thus, a lower A will be able to topple the part. Therefore, the best location of A' on a contact edge is (above the lower vertex, i.e. A' is as low as possible on that edge. This reduces $d_{A'}$ candidates to a finite set F ($|F| = n - 2$ in the worst case) for the part. For each $d_{A'}$, we may employ the graphical analysis to find the feasible d_A.

4.2 Constrained Jamming Height Function

We allow the part continue to rotate after it has reached θ_t if $\theta_d > \theta_t$. We call this process settling, and intend to determinate whether jamming may occur in this process. The part may jam while settling due to the frictional contact at the toppling tip. We will be conservative and eliminate any toppling tip height where jamming may occur even though we cannot be certain it will jam without further information. Note that we consider the statics (not full dynamics) of the settling process.

We only consider the situation where A' keeps contact with e_i during the entire settling phase. More general cases can be derived from the same methodology.

To determine the constrained jamming height function we begin by constructing a primary region as shown in Figure 6. Notice that there is no jamming if the constrained tip contact is left of the part gravity line, i.e.,

$$d \tan(\theta + \omega) \quad < \quad \rho \cos(\theta + \eta). \tag{15}$$

Otherwise, the primary region quadrilateral with vertices P_0, P_1, P_2 and P_3. P_0 is the intersection of the vertical line through the part's com and the right edge of the constrained tip friction cone, and is located at (x_{p0}, z_{p0}). P_1 is the constrained tip contact at (x_{p1}, z_{p1}). P_2 is the intersection of the left edge of the constrained friction cone and the left edge of the bottom friction cone, located at (x_{p2}, z_{p2}). And P_3 is the intersection of the vertical line through the part's com and the left edge of the bottom friction cone, located at (x_{p3}, z_{p3}).

$$\begin{aligned} x_{p0} &= s, && (16)\\ z_{p0} &= d - \frac{d\tan(\theta+\omega) - s}{\tan(\alpha_t - \theta - \omega)}, && (17)\\ x_{p1} &= d\tan(\theta+\omega), && (18)\\ z_{p1} &= d, && (19) \end{aligned}$$

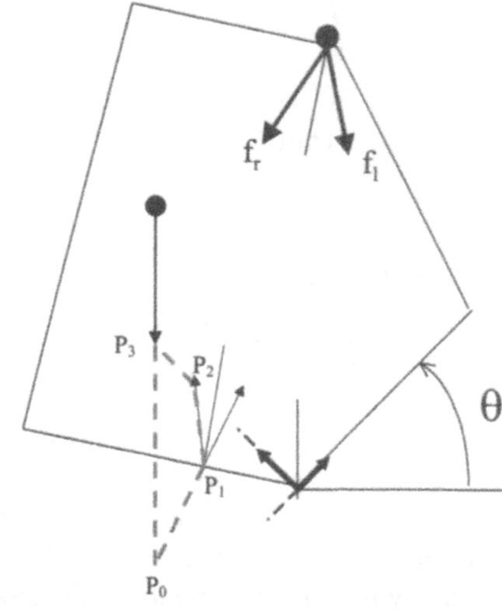

Figure 6: *Jamming conditions.*

$$\begin{aligned} x_{p2} &= \frac{-d\cos\alpha_t \sin\alpha_b}{\sin(\theta+\omega)\sin(\theta+\omega+\alpha_t-\alpha_b)}, && (20)\\ z_{p2} &= \frac{d\cos\alpha_t \cos\alpha_b}{\sin(\theta+\omega)\sin(\theta+\omega+\alpha_t-\alpha_b)}, && (21)\\ x_{p3} &= s, && (22)\\ z_{p3} &= \frac{-s}{\mu_b}. && (23) \end{aligned}$$

To guarantee that no jamming occurs, any force in the toppling tip friction cone must make a positive moment about the critical primary region then the left edge of the friction cone must make a positive moment. In other words, the left edge of the toppling tip friction cone determines the height at which jamming may occur.

Similar to the analysis of the toppling height function, we have $w_i^0(\theta)$, $w_i^1(\theta)$ and $w_i^2(\theta)$ and w_i, which is $\min(w_i^0, w_i^1, w_i^2)$ at a given θ. These functions are given by:

$$w_i = \begin{cases} w_i^0 & \theta < \theta_{32} \text{ and } \psi_i < \omega - 2\alpha_t \\ w_i^1 & \theta < \theta_{32} \text{ and } \omega - 2\alpha_t < \psi_i < \omega \\ w_i^2 & \theta < \theta_{32} \text{ and } \omega < \psi_i \\ w_i^3 & \theta_{32} < \theta < \pi - \omega_k \end{cases} \tag{24}$$

where

$$\theta_{32} = \pi - \alpha_t - \psi_i. \tag{25}$$

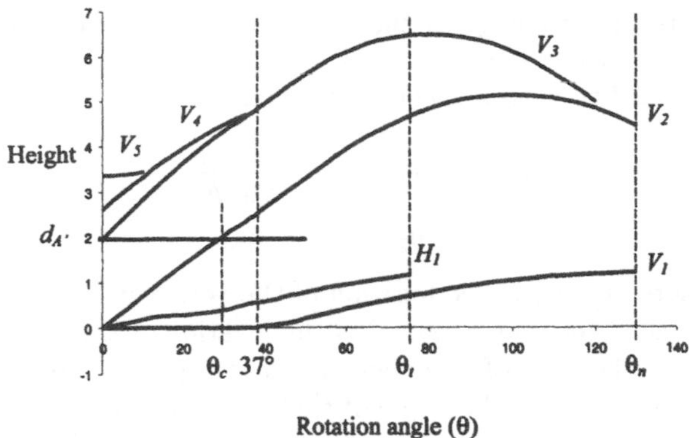

Figure 7: *Constrained toppling graph.*

The constrained jamming height function, $J_i(\theta)$, with $\theta_t < \theta < \theta_n$ is given by:

$$J_i(\theta) = \begin{cases} J_i^*(\theta) & H_i^*(\theta) > V_i(\theta) \\ \emptyset & H_i^*(\theta) < V_i(\theta) \end{cases} \tag{26}$$

where

$$J_i^*(\theta) = x_i \sin\theta + z_i \cos\theta + w_i \sin(\theta + \psi_i). \tag{27}$$

Therefore, for given θ and $d_{A'}$, jamming occurs if the heights of A is lower than $J_i(\theta)$.

4.3 The Constrained Toppling Graph

Figure 7 illustrates the entire constrained toppling graph that combines the vertex height, constrained rolling height, and constrained jamming height functions to represent the full mechanics of toppling. From the constrained toppling graph the necessary toppling height for A can be determined or shown to be nonexistent. Note that $H_i(\theta)$ must be bounded by the $V_i(\theta)$ and $V_{i+1}(\theta)$ and is truncated where it intersects them.

For toppling from an initial orientation to the desired orientation to be successful, there must exist a horizontal line from the angle of the initial orientation to the angle of the desired orientation at height h that has the following characteristics:

$1: h > H_i(\theta)$ for all θ, if $V_i(\theta) < h < V_{i+1}(\theta)$;

$2: h > J_i(\theta)$ for all θ, if $V_i(\theta) < h < V_{i+1}(\theta)$;

$3: h < \max(V_i(\theta))$ for all i at any $\theta < \theta_t$.

The first two criteria can be described as A must be above both the rolling height and the jamming height on the edge it is in contact with for all θ. Note that when the pin crosses a vertex height function it contacts a new edge and must then be above the rolling height and jamming height functions for that edge. The third criterion is that the pin must not lose contact with the part by passing over it during the rolling phase.

Figure 7 demonstrates the constrained toppling graph of the part shown in Figure 3. From the graph we can determine a toppling tip at $d_{A'} = 2.0$cm is capable to topple the part to any orientation with $0 < \theta < \theta_n$. Notice that A switches contact edge from e_2 to e_1 at θ_c.

5 Grasping

Once the part has been rotated to $\theta = \theta_d$, the grasping tips, B and B', must make contact with the part. Additionally, we require that the combination of the contacts corresponding to A, A', B, and B' generate a form-closure grasp on the part. There also exists an accessibility constraint on the locations of B and B' due to the requirement that they not make contact with the part until $\theta = \theta_d$. Therefore, we divide the grasping analysis into two sections corresponding to determining the accessibility constraint and meeting the form-closure requirement.

5.1 Accessibility

The accessibility constraint will limit the possible heights of the grasping tip, B or B', for a given height of A or A'. In order to determine the accessibility constraint we must consider the relative motion between the part and a jaw of the gripper. The rest of the accessibility discussion will be in a frame of reference fixed to the toppling tip A and will consider a contact B on edge e_j. The accessibility constraint requires that as the part rotates to the desired final orientation, that it is moving out towards the grasping tip and that at no previous angle has the part been as far out as the fixture tip and made contact.

To illustrate this situation, Figure 8 shows the rotation of a part with respect to the toppling tip. Note that at any height within the inaccessible range on edge e_2 indicated in the figure, the location of vertex v_2 would have contacted the grasping tip before the

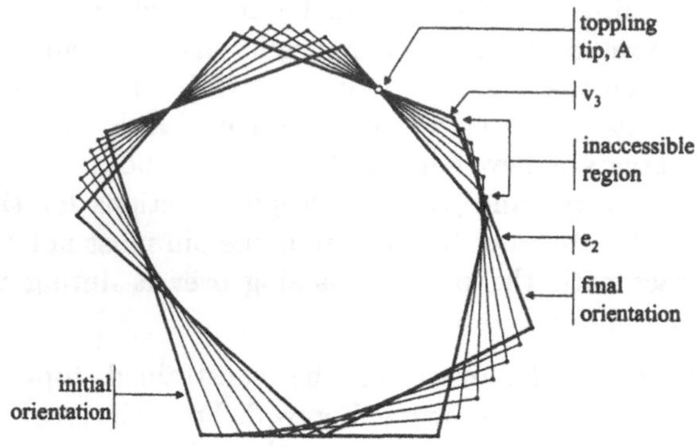

Figure 8: *Rotation of a part relative to the toppling tip A.*

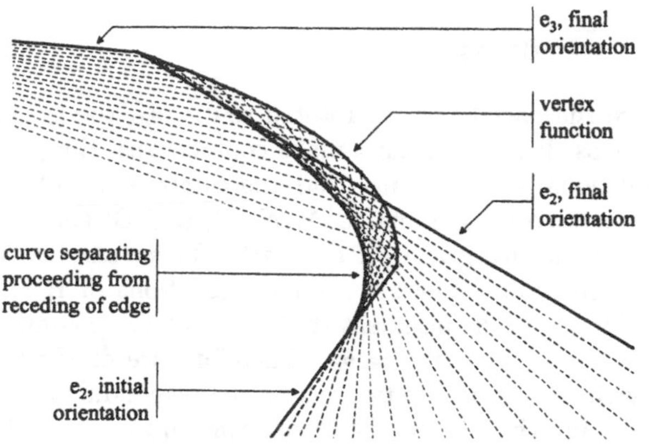

Figure 9: *A portion of the edge in the final orientation may be blocked in the positive x direction before the part reaches the final orientation. Additionally, a curve shows the separation between where the part is moving forward and where it is receding.*

part reached the desired orientation. Since this would prevent the part from reaching the desired final orientation, these heights are considered inaccessible.

By examining the inaccessible region more closely, as shown in Figure 9, we can see that there are two factors to be considered when determining the accessibility of an edge. The first factor is whether any portion of the edge in the final orientation is blocked from visibility in the positive X direction by the part as it rotates. The second factor is what portion of the edge is moving forward, in the +X direction, at the final orientation.

The first factor is taken into account by calculating a vertex function for the vertex at the top of edge e_j. The vertex function gives the location of the vertex with respect to the toppling tip as a function of rotation angle. The vertex function is given by:

$$x = x_j - x_i - \frac{d_A - y_i}{\tan\psi_i}, \tag{28}$$

$$y = y_j. \tag{29}$$

where (x_j, y_j) is the location of the vertex in the fixed frame of reference, (x_i, y_i) is the location of the vertex at the bottom of the edge that the toppling tip contacts, and ψ_i is the angle of that edge. We must check what portion of the edge is visible from the +X direction without being blocked by the combination of the vertex function and the initial orientation of the edge. Note that it is possible that lower edges and vertices may also block part of the final orientation of the edge, and must be checked as well. This check insures that the contact point is not blocked by the part during rotation.

The second consideration insures that the edge is moving out to the grasping tip at $\theta = \theta_d$. In other words, the relative X displacement between A and B, denoted x_{ab}, must be greater than the X location (with respect to the toppling tip) of the part at d_B for any $\theta < \theta_d$. We will denote the location of the part at a height d_B and angle θ as (x_c, d_B). Note that this is a different physical point at each θ.

At some critical height, h_c, the derivative of x_c with respect to θ is 0. All of the physical points on that edge below hc will meet the requirement of moving out to the part while those above will be receding. The accessibility constraint for this consideration is therefore, $d_B < x_c$ on a given edge.

The relative X displacement of a point of the part at height d_B can be shown geometrically to be:

$$x_c = x_j - x_i + \frac{d_B - y_j}{\tan\psi_j} - \frac{d_A - y_i}{\tan\psi_i}, \tag{30}$$

where (x_i, y_i) and (x_j, y_j) are the locations of the vertex at the bottom of the edge in contact with A and the edge in contact with B respectively, and ψ_i and ψ_i are the angles of those edges. Therefore the derivative of x_c with respect to θ is given by:

$$\frac{dx_c}{d\theta} = y_i - y_j - \frac{d_A - x_i}{\tan\psi_i} + \frac{d_A - y_i}{\sin^2\psi_j} + \frac{h_c - x_j}{\tan\psi_j} - \frac{h_c - y_j}{\sin^2\psi_j}, \tag{31}$$

and must be 0 at h_c. Setting (31) equal to 0 and solving for h_c yields:

$$h_c = (y_i - y_j - \frac{d_A - x_i}{\tan\psi_i} + \frac{d_A - y_i}{\sin^2\psi_j} - \frac{x_j}{\tan\psi_j} - \frac{y_j}{\sin^2\psi_j})/(\frac{1}{\tan\psi_j} - \frac{1}{\sin^2\psi_j}). \quad (32)$$

For a given edge, indicated by (x_j, y_j) and ψ_j, only heights less than h_c can be considered when determining the height for B in the form closure analysis. A corresponding procedure is used to determine a range of possible $d_{B'}$ for a given $d_{A'}$.

5.2 Form-Closure

At the end of the accessibility considerations we know d_A and $d_{A'}$, as well as ranges of possible values for d_B and $d_{B'}$. From these ranges we must determine values such that the four tips generate form-closure on the part. This is easily done using the method described in van der Stappen [25]. This method would entail determining the point at which the edge normals through A and A' meet, then selecting the locations of B and B' such that the edge normals through B and B' create opposite moments about that point. We omit the precise details for lack of space.

6 Physical Experiment

We conducted a physical experiment using an *AdeptOne* industrial robot and a parallel jaw gripper with tips designed using the methodology described in this paper. The part is a small lever from a standard videotape (*FUJI* serial number: 7410161160). Its planar convex hull is shown in Figure 3. The entire compensatory grasping is restricted in the X-Z plane due to the mechanical and the geometric property of the part.

As illustrated in Figure 1, the part begins at stable orientation (a). Its desired orientation is (b) where $\theta = 37°$. We choose A and A' at $d_A = 0.9$ cm and $d_{A'} = 2.0$ cm, respectively. The corresponding friction cone half angles are $\alpha_t = 5°$ and $\alpha_b = 10°$. When $\theta < 37°$, V_1 is C and x_2=4.1cm, z_2=0.0cm, $\psi_2 = 56°$, $\eta = 46°$, ρ=2.2cm, and $\omega = 53°$; when $\theta > 37°$, V_6 is C and x_2=4.6cm, z_2=2.4cm, $\psi_2 = 92°$, $\eta = 55°$, ρ=2.7cm, and $\omega = 89°$. The analysis yields the following tip values: $d_B = 2.7$ cm, $d_{B'} = 3.0$ cm, $x_{AB} = 0.0$ cm, and $x_{A'B'} = 2.1$cm. Figure 10 illustrates the compensatory grasp.

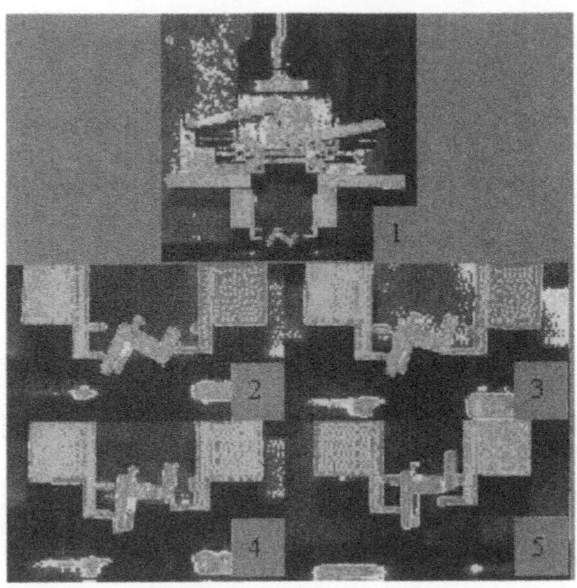

Figure 10: *Compensatory grasp experiment.*

7 Discussion and Future Work

In industrial practice, gripper jaw geometry is often custom-designed and machined for each part. Design has been ad-hoc and particularly challenging when the part's natural resting pose differs from the desired grip/insertion pose. In this paper we describe compensatory grasps, a new approach to this problem where 4 contact points on the jaws guide the part into alignment and hold it stably. The next step is to develop more sophisticated jaw shapes based on part trajectory (see Figure 11) and to address shape and position uncertainty, friction, and ultimately, 3D geometry. We are also interested in more efficient algorithms and knowing under what conditions a compensatory grasp exists.

Acknowledgments

This paper grew out of some practical experiments with a commercial assembly. We would like to thank Brian Carlisle and Randy Brost for suggesting such experiments, and Kevin Lynch for his elegant toppling analysis. We would also like to thank RP Berretty and Mark

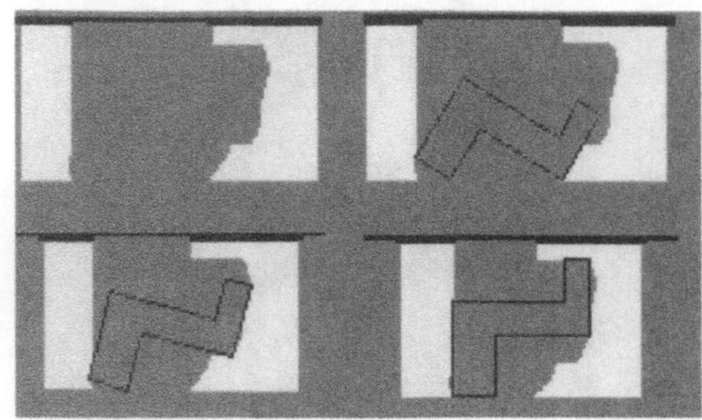

Figure 11: *Gripper Jaws based on compensatory grasp tips and complement of swept volume.*

Overmars for their contributions to our thinking about the toppling graph.

This work was supported in part by the National Science Foundation under CDA-9726389 and Presidential Faculty Fellow Award IRI-9553197.

References

[1] K. Goldberg, B. Mirtich, Y. Zhuang, J. Craig, B. Carlisle, and J. Canny. Part pose statistics: Estimators and experiments. *IEEE Transactions on Robotics and Automation. 15(5)*, pages 849–857, 1999.

[2] G. Causey and Roger Quinn. Design of a flexible parts feeding system. In *IEEE International Conference on Robotics and Automation*, 1997.

[3] Jocelyn Pertin-Troccaz. Grasping: A state of the art. In *The Robotics Review I*, pages 71–98. MIT Press, 1989.

[4] Xanthippi Markenscoff, Luqun Ni, and Christos H. Papadimitriou. The geometry of grasping. *International Journal of Robotics Research*, 9(1), February 1990.

[5] Matt Mason and J. Salisbury. *Robotic Hands and Mechanics of Manipulation.* MIT Press, 1985.

[6] Randy C. Brost. Automatic grasp planning in the presence of uncertainty. *The International Journal of Robotics Research*, December 1988.

[7] John Canny and Ken Goldberg. A risc approach to sensing and manipulation. *Journal of Robotic Systems*, 12(6):351–362, June 1995.

[8] J. C. Trinkle and R. P. Paul. Planning for dextrous manipulation with sliding contacts. *International Journal of Robotics Research*, June 1990.

[9] Michael A. Peshkin and Art C. Sanderson. The motion of a pushed, sliding workpiece. *IEEE Journal of Robotics and Automation*, 4(6), December 1988.

[10] T. Abell and M. Erdmann. Stably supported rotations of a planar polygon with two frictionless contacts. In *IEEE/RSJ International Conference on Intelligent Robots and Systems*, 1995.

[11] N. Zumel and M. Erdmann. Nonprehensile two plam manipulation with non-equilibrium transitions between stable states. In *IEEE International Conference on Robotics and Automation*, 1996.

[12] N. Zumel and M. Erdmann. Nonprehensile manipulation for orienting parts in the plane. In *IEEE International Conference on Robotics and Automation*, 1997.

[13] D. Rus. Dextrous rotations of polyhedra. In *IEEE International Conference on Robotics and Automation*, May 1992.

[14] J. Hong, G. Lafferriere, B. Mishra, and X. Tan. Fine manipulation with multifinger hands. In *International Conference on Robotics and Automation.* IEEE, May 1990.

[15] Ronald S. Fearing. Simplified grasping and manipulation with dextrous robot hands. *IEEE Journal of Robotics and Automation*, RA-2(4), December 1986.

[16] A. Bicchi and R. Sorrentino. Dextrous manipulation through rolling. In *International Conference on Robotics and Automation.* IEEE, 1995.

[17] A. Rao, D. Kriegman, and K. Goldberg. Complete algorithms for feeding polyhedral parts using pivot grasps. *IEEE Transactions on Robotics and Automation*, 12(6):331–42, April 1996.

[18] R. Zhang and K. Gupta. Automatic orienting of polyhedral through step devices. In *IEEE International Conference on Robotics and Automation*, 1998.

[19] Y. Yu, K. Fukuda, and S. Tsujio. Estimation of mass and center of mass of graspless and shape-unknown object. In *IEEE International Conference on Robotics and Automation*, 1999.

[20] Kevin Lynch. Toppling manipulation. In *IEEE International Conference on Robotics and Automation*, 1999.

[21] Tao Zhang, Gordon Smith, Robert-Paul Berretty, Mark Overmars, and Ken Goldberg. The toppling graph: Designing pin sequences for part feeding. In *IEEE International Conference on Robotics and Automation*, 2000.

[22] Ken Goldberg. Orienting polygonal parts without sensors. *Algorithmica*, 10(2):201–225, August 1993. Special Issue on Computational Robotics.

[23] Tao Zhang, Gordon Smith, and Ken Goldberg. Compensatory grasping with the parallel-jaw gripper. Technical Report, ALPHA Lab, UC Berkeley, 2000.

[24] Matt Mason. Two graphical methods for planar contact problems. In *IEEE/RSJ International Workshop on Intelligent Robots and Systems*, 1991.

[25] F. van der Stappen, C. Wentink, and M. Overmars. Computing form-closure configurations. In *IEEE International Conference on Robotics and Automation*, 1999.

Optimal Planning for Coordinated Vehicles with Bounded Curvature

Antonio Bicchi *Centro "E. Piaggio," University of Pisa, Pisa, Italy*
Lucia Pallottino *Centro "E. Piaggio," University of Pisa, Pisa, Italy*

In this paper we consider the problem of planning motions of a system of multiple vehicles moving in a plane. Each vehicle is modelled as a kinematic system with velocity constraints and curvature bounds. Vehicles can not get closer to each other than a predefined safety distance. For such system of multiple vehicles, we consider the problem of planning optimal paths in the absence of obstacles. The case when a constant distance between vehicles is enforced (such as when cooperative manipulation of objects is performed by the vehicle team) is also considered.

1 Introduction

In this paper we consider the problem of planning motions of a system of multiple vehicles moving in a plane. The motions of each vehicle are subject to some constraints: The velocity of the center of the vehicle is parallel to an axis fixed on the vehicle; the velocity is constant along such axis; the steering radius is bounded. Also, a minimum distance between vehicles must be enforced along trajectories.

The task of each vehicle is to reach a given goal configuration from a given start configuration. We consider optimal solutions in the sense of minimizing total completion time.

The literature on optimal path planning for vehicles of this type is very rich. The seminal work of Dubins [4] and the extension to vehicles that can back-up due to Reeds and Shepp [6], solved the single vehicle case by exploiting specialized tools. Later on, Sussmann and Tang [7], and Boissonnat et al. [2], reinterpreted these results as an application of Pontryagin's minimum principle [5]. Using these tools, Bui et al. [3] performed a complete optimal path synthesis for Dubins robots. The minimum principle framework used in these previous works is also fundamental in the developments presented here.

The paper is organized as follows. In Section 2 we describe the problem and introduce some notation. In Section 2.1 a formulation of the problem in a form amenable to application of optimal control theory is presented. Section 3 is devoted to the study of necessary conditions for extremal arcs. Finally, Section 4 describes a numeric algorithm to find solutions, which applies under some restrictions.

2 Problem Statement

Consider N vehicles in the plane, whose individual configuration is described by $q_i = (x_i, y_i, \theta_i) \in \mathbf{R}\times\mathbf{R}\times S^1$, with (x_i, y_i) coordinates in a fixed planar reference frame (o, x, y) and θ_i the heading angle of the vehicle with respect to the x axis. Each vehicle is assigned a task. In order to compute its task a vehicle starts in a configuration $q_{i,s}$ and moves in a final configuration $q_{i,g}$. We call these two particular configurations way-points. The initial way-point time is assigned and denoted by T_i^s. Assume vehicles are ordered such that $T_1^s \leq T_2^s \leq \cdots \leq T_N^s$. We denote by T_i^g the time at which the i-th vehicle reaches its goal, and let $T_i \stackrel{def}{=} T_i^g - T_i^s$. Motions of the i-th vehicle before T_i^s and after T_i^g are not of interest.

The i-th vehicle motion is subject to the constraint that its transverse velocity is zero, $\dot{x}_i \sin\theta_i - \dot{y}_i \cos\theta_i = 0$, $i = 1, \ldots, N$. Equivalently, this motion is described by the control system $\dot{q}_i = f_i(q_i, u_i, \omega_i)$. More explicitly:

$$\begin{pmatrix} \dot{x}_i \\ \dot{y}_i \\ \dot{\theta}_i \end{pmatrix} = \begin{pmatrix} u_i \cos\theta_i \\ u_i \sin\theta_i \\ \omega_i \end{pmatrix}, \tag{1}$$

where u_i and ω_i are the linear and angular velocity of the i-th vehicle, respectively. All vehicles are subject to the following additional constraints:

i) the linear velocity is constant: $u_i = \bar{u}_i$;

ii) the path curvature is bounded: $|\omega_i| \leq \Omega_i$, where $\Omega_i = \frac{\bar{u}_i}{R_i}$ and $R_i > 0$ denotes the minimum turning radius of the i-th vehicle;

iii) the distance between two vehicles must remain larger than, or equal to, a given separation limit: $D_{ij}(t) = (x_j(t)-x_i(t))^2+(y_j(t)-y_i(t))^2-d_{ij}^2 \geq 0$, at all times t ($d_{ii} = 0, i = 1, \ldots, N$).

We consider problems in which the goal is to minimize the total execution time:

$$\left\{\begin{array}{l} \min \sum_{i=1}^{N} T_i \\ \dot{q}_i = f_i(q_i, \bar{u}_i, \omega_i) \quad i = 1, \ldots, N \\ |\omega_i| \leq \frac{\bar{u}_i}{R_i} \quad i = 1, \ldots, N \\ D_{ij}(t) \geq 0, \quad \forall t, \; i,j = 1, \ldots, N \\ q_i(T_i^s) = q_{i,s}, \quad q_i(T_i^g) = q_{i,g}. \end{array}\right. \tag{2}$$

If separation constraints are disregarded, the minimum total time problem is clearly equivalent to N independent minimum length problems under the above constraints, i.e. to N classical Dubins' problems, for which solutions are well known in the literature ([4, 7, 2]). It should be noted that computation of the Dubins solution for any two given configurations is computationally very efficient.

2.1 Formulation as an Optimal Control Problem

The cost for the total time problem, $J = \sum_{i=1}^{N} T_i = \sum_{i=1}^{N} \int_{T_i^s}^{T_i^g} dt$, is not in the standard Bolza form. In order to use powerful results from optimal control theory, we rewrite the problem as follows. Let $h(t)$ denote the Heavyside function, i.e.:

$$h(t) = \left\{\begin{array}{ll} 0 & t < 0 \\ 1 & t \geq 0 \end{array}\right.,$$

and define the window function $w_i(t) = h(t-T_i^s) - h(t-T_i^g)$. Then the minimum total time cost is written as:

$$J = \int_0^\infty \sum_{i=1}^{N} w_i(t) dt. \tag{3}$$

Using the notation $\mathrm{col}_{i=1}^{N}(v_i) = [v_1^T, \ldots, v_N^T]^T$, define the aggregated state $q = \mathrm{col}_{i=1}^{N}(q_i)$, controls $u = \mathrm{col}_{i=1}^{N}(\bar{u}_i)$ and $\omega = \mathrm{col}_{i=1}^{N}(\omega_i)$, and define the admissible control sets Ω accordingly. In addition we define the separation vector $D = [D_{12}, \cdots, D_{1N}, D_{23}, \cdots, D_{N-1,N}]$, and the vector field $f(q, u, \omega) = \mathrm{col}_{i=1}^{N}(f_i w_i)$. Finally introduce matrices $\Gamma_i = \mathrm{col}_{j=1}^{N}(\sigma_{ij}\,[1\;1\;1]^T)$, with $\sigma_{ij} = 1$ if $i = j$, else $\sigma_{ij} = 0$, and functions $\gamma_i(q(t), \bar{q}) = \Gamma_i\,(q(t) - \bar{q})$. Our optimal control problem is then formulated as

Problem 1. *Minimize J subject to $\dot{q} = f(q, u, \omega)$, $\omega \in \Omega$, $D \geq 0$, and to the two sets of N interior-point constraints:*

$$\begin{array}{l} \gamma_i(q(t), q_i^s) = 0, \quad t = T_i^s \\ \gamma_i(q(t), q_i^g) = 0, \quad t = T_i^g (\text{unspecified}) \end{array}$$

3 Necessary Conditions

Necessary conditions for Problem 1 can be studied by adjoining the cost function with the constraints multiplied by unspecified Lagrange covectors. Omitting to write explicitly the extents of iterative operations when extending from 1 to N, let:

$$\begin{array}{rl} \hat{J} = & \sum_i \pi_i^s \gamma_i(q(T_i^s) - q_i^s) \\ & + \sum_i \pi_i^g \gamma_i(q(T_i^g) - q_i^g) \\ & + \int_0^\infty \sum_i w_i + \lambda^T(\dot{q} - f) + \nu^T D\, dt, \end{array} \tag{4}$$

with λ and ν costates of suitable dimension, and with $\nu_i = 0$ if $D_i > 0$, $\nu_i \geq 0$ if $D_i = 0$. Let the Hamiltonian be defined as:

$$H = \sum_i w_i + \lambda^T f + \nu^T D \tag{5}$$

Substituting Equation 5 in Equation 4, integrating by parts, and computing the variation of the cost, we get:

$$\begin{array}{rl} \delta\hat{J} = & \sum_i \left[\lambda^T(T_i^{s-}) - \lambda^T(T_i^{s+}) + \pi_i^s \frac{\partial\gamma_i}{\partial q(T_i^s)}\right] d\,q(T_i^s) \\ & + \sum_i \left[\lambda^T(T_i^{g-}) - \lambda^T(T_i^{g+}) + \pi_i^g \frac{\partial\gamma_i}{\partial q(T_i^g)}\right] d\,q(T_i^g) \\ & + \sum_i \left[H(T_i^{g-}) - H(T_i^{g+}) + \pi_i^g \frac{\partial\gamma_i}{\partial T_i^g}\right] dT_i^g \\ & + \int_0^\infty \left[\left(\dot{\lambda}^T + \frac{\partial H}{\partial q}\right)\delta q + \frac{\partial H}{\partial \omega}\delta\omega\right] dt \end{array} \tag{6}$$

(recall that $dT_i^s \equiv 0$). Therefore, we have the following necessary conditions for an extremal solution:

$$\lambda_i(T_i^{s-}) = \lambda_i(T_i^{s+}) + \Gamma_i^T \pi_i^s \tag{7}$$

$$\lambda_i(T_i^{g-}) = \lambda_i(T_i^{g+}) + \Gamma_i^T \pi_i^g \tag{8}$$

$$H(T_i^{g-}) = H(T_i^{g+}) \tag{9}$$

$$\dot{\lambda}^T = -\frac{\partial H}{\partial q} \quad (10)$$

$$\frac{\partial H}{\partial \omega}\delta\omega = 0 \;\; \forall \delta\omega \text{ admiss.} \quad (11)$$

Extremal trajectories for the i–th vehicle will be comprised in general of unconstrained arcs (with $D_{ij} > 0$, $\forall j \neq i$) and of constrained arcs, where the constraint is marginally satisfied ($\exists j : D_{ij} = 0$). We will proceed the discussion of necessary conditions by distinguishing constrained and unconstrained arcs.

3.1 Unconstrained Arcs

Suppose that, for the i–th vehicle, the separation constraints are not active in the interior of an interval $[t_i^a, t_i^b]$, $T_i^s \leq t_i^a < t_i^b \leq T_i^g$, i.e. $D_{ij}(t) > 0$, $j = 1, \ldots, N$, $t \in (t_i^a, t_i^b)$. Expanding Equation 10, we get:

$$\left[\dot{\lambda}_{i1}, \dot{\lambda}_{i2}, \dot{\lambda}_{i3}\right] = [0, 0, \lambda_{i,1}\bar{u}_i \sin\theta_i - \lambda_{i2}\bar{u}_i \cos\theta_i]. \quad (12)$$

The characterization of optimal solutions in the unconstrained case proceeds along the lines of the classical Dubins solution (see [4, 7, 2]). We report some results here for reader's convenience. By integrating Equation 12 one gets $\lambda_{i1}(t_i^a < t < t_i^b) = \bar{\lambda}_{i1}$, $\lambda_{i2}(t_i^a < t < t_i^b) = \bar{\lambda}_{i2}$, and $\lambda_{i3}(t_i^a < t < t_i^b) = \bar{\lambda}_{i1}y_i(t) - \bar{\lambda}_{i2}x_i(t) + \bar{\lambda}_{i3}$, with constant $\bar{\lambda}_{i,j}$, $j = 1, 2, 3$. In light of these relationships, the conditions in Equations 7 and 8 state that the costate components λ_{i1} and λ_{i2} are piecewise constant, with jumps possibly at the start and arrival time of the i–th vehicle. The addend in the Hamiltonian relative to the i–th vehicle can be written as $H_i = 1 + \bar{u}_i\rho_i\cos(\theta_i - \psi_i) + \lambda_{i3}\bar{u}_i\omega_i$, where $\rho_i = \sqrt{\bar{\lambda}_{i1}^2 + \bar{\lambda}_{i2}^2}$ and $\psi_i = \text{atan2}(\bar{\lambda}_{i2}, \bar{\lambda}_{i1})$. From Pontryagin's Minimum Principle (PMP), we know that $H_i(t) = \textit{const.} \leq 0$ along extremal unconstrained arcs and, being by assumption the way-points configurations unconstrained, it follows from Equation 9 that $H_i(t)$ is also continuous at $t = T_i^g$.

Extremals of H_i within the open segment $\{|\omega_i| < \bar{u}_i/R\}$ can only be obtained if:

$$\frac{\partial H_i}{\partial \omega_i} = \lambda_{i3} = \bar{\lambda}_{i1}y_i(t) - \bar{\lambda}_{i2}x_i(t) + \bar{\lambda}_{i3} = 0. \quad (13)$$

If the condition holds on a time interval of non-zero measure, then $\dot{\lambda}_{i,3} = 0$ on the interval. This implies $\rho_i\bar{u}_i\sin(\theta_i - \psi_i) = 0$, hence $\theta_i = \psi_i \bmod \pi$ and $\omega_i = 0$. In such an interval, the vehicle moves on the straight route (*the supporting line*) in the horizontal x, y plane described in Equation 13. Other extremals of H_i occur at $\omega = \pm\bar{u}_i/R$. The sign of the minimizing yaw rate ω_i is opposite to that of λ_{i3}. In other words, the supporting line also represent the switching locus for the yaw rate input. Trajectories corresponding to $\omega_i = \pm\bar{u}_i/R$ correspond to circles of minimum radius R followed counterclockwise or clockwise, respectively. It is important for our further developments to note that, along extremal arcs, the costates are completely determined by boundary configurations up to a multiplicative constant $\rho \neq 0$, which remains undetermined.

For each vehicle, extremal unconstrained arcs are concatenations of only two types of elementary arcs: line segments of the supporting line (denoted as "*S*"), and circular arcs of minimum radius (denoted by "*C*"). The latter type can be further distinguished between "*R*" clockwise arcs ($\omega_i = \bar{u}_i/R$), and "*L*" counterclockwise arcs ($\omega_i = -\bar{u}_i/R$). We will use subscripts to denote the length of rectilinear segments and the angular span of circular arcs.

Switchings of ω_i among 0, $\bar{u}_i/R$, and $-\bar{u}_i/R$ can only occur when the vehicle center is on the supporting line. As a consequence, all extremal unconstrained paths of each vehicle are written as $C_{u_1}S_{d_1}C_{u_2}S_{d_2}\cdots S_{d_n}C_{u_n}$, with $u_i = 2k\pi$, k integer, $i = 2, \ldots, n-1$.

In the case of a single vehicle, the discussion of optimal unconstrained arcs can be further refined by several geometric arguments, for which the reader is referred directly to the literature [4, 7, 2]. Optimal paths necessarily belong to either of two path types in the Dubins' sufficient family:

$$\{C_aC_bC_e\ ,\ C_uS_dC_v\} \quad (14)$$

with the restriction that:

$$b \in (\pi R, 2\pi R);\;\; a, e \in [0, b],\;\; u, v \in [0, 2\pi R),\;\; d \geq 0 \quad (15)$$

A complete synthesis of optimal paths for a single Dubins vehicle is reported in [3]. The length of Dubins paths between two configurations, denoted by $L_D(\xi_i^s, \xi_i^g)$, is then unique and defines a metric on $\mathbb{R}^2 \times S^1$. One simply has $L_D(\cdot,\cdot) = R(|a| + |b| + |c|)$ for a $C_aC_bC_e$ path, and $L_D(\cdot,\cdot) = R(|u| + |v|) + d$ for a $C_uS_dC_v$ path.

In our multivehicle problem, however, other extremal paths may turn out to be optimal, and therefore have to be considered. This may happen for instance for a path of type $C_a S_b C_{2k\pi} S_e C_f$ iff the corresponding Dubins' path $C_a S_{b+e} C_f$ (which is shorter), is not collision free. Arcs of the type $C_{2k\pi}$ can be interpreted as a circling maneuver that allows another vehicle to pass by and avoids collision[1]. Note explicitly that the length of two subpaths of type $\cdots C_{u_i} S_\alpha C_{2k\pi} S_\beta C_{u_{i+1}} \cdots$ and $\cdots C_{u_i} S_\gamma C_{2k\pi} S_\delta C_{u_{i+1}} \cdots$ are equivalent as long as $\alpha + \beta = \gamma + \delta$.

By "extremal trajectory" (Dubins' trajectory, respectively) we indicate henceforth a map $\mathbb{R}^+ \mapsto \mathbb{R}^2$ defined by $(x_i^D(t), y_i^D(t))$, denoting the position of the i–th vehicle at time t along an extremal (Dubins') path connecting q_i^s to q_i^g.

Remark 1. If a set of non–colliding Dubins' trajectories exists, then this is obviously a solution of the minimum total time problem. More interestingly, if there is a collision, the optimal solution will contain at least a constrained arc or at least one wait circle.

3.2 Constrained Arcs

Some further manipulation of the cost function is instrumental to deal with constrained arcs, i.e. arcs in which at least two vehicles are exactly at the critical separation ($D_{ij} = 0$, $i \neq j$). To simplify, consider a constrained arc involving only vehicles 1 and 2. Along a constrained arc, the derivatives of the constraint must vanish:

$$N = \begin{bmatrix} D_{12} \\ \dot{D}_{12} \end{bmatrix} = \begin{bmatrix} (x_2 - x_1)^2 + (y_2 - y_1)^2 - d^2 \\ 2(x_2 - x_1)(\dot{x}_2 - \dot{x}_1) + 2(y_2 - y_1)(\dot{y}_2 - \dot{y}_1) \end{bmatrix} = 0 \qquad (16)$$

with $d = d_{12}$. Let ϕ be the direction of the segment joining the two vehicles, so that:

$$\begin{aligned} x_2 - x_1 &= d\cos\phi, \\ y_2 - y_1 &= d\sin\phi, \end{aligned} \qquad (17)$$

From the second Equation in 16, we get:

$$\begin{aligned} &(x_2 - x_1)(\bar{u}_2\cos\theta_2 - \bar{u}_1\cos\theta_1) + \\ &(y_2 - y_1)(\bar{u}_2\sin\theta_2 - \bar{u}_1\sin\theta_1) = 0 \end{aligned} \qquad (18)$$

and, using Equation 17,

$$\bar{u}_1\cos(\phi - \theta_1) - \bar{u}_2\cos(\phi - \theta_2) = 0. \qquad (19)$$

[1]This is similar to the current practice in conflict resolution for air traffic control

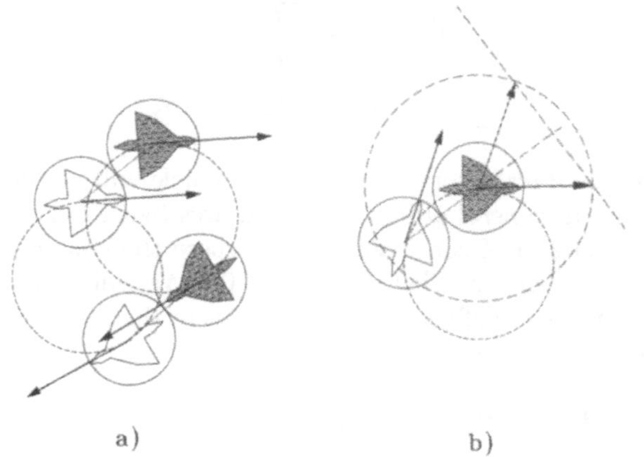

Figure 1: *Possible constrained arcs for two vehicles with the same velocity.*

When the constraint is active, the two vehicle envelopes are in contact, and the relative orientation of the two vehicles must satisfy Equation 19, which defines (for given $\bar{u}_1$, $\bar{u}_2$) two manifolds of solutions in the space $\{(\theta_1, \theta_2, \phi) \in S^1 \times S^1 \times S^1\}$ described as:

$$\textbf{a)}\quad \theta_2^a = \phi + \arccos\left(\frac{\bar{u}_1}{\bar{u}_2}\cos(\phi - \theta_1)\right); \qquad (20)$$

$$\textbf{b)}\quad \theta_2^b = \phi - \arccos\left(\frac{\bar{u}_1}{\bar{u}_2}\cos(\phi - \theta_1)\right). \qquad (21)$$

The two solutions correspond to two different types ("a" and "b") of relative configurations in contact. For instance, for: $\bar{u}_1 = \bar{u}_2$, one has:

$$\textbf{a)}\quad \theta_2^a = \theta_1; \qquad (22)$$

$$\textbf{b)}\quad \theta_2^b = 2\phi - \theta_1. \qquad (23)$$

In case a) the two vehicles have the same direction, while in case b) directions are symmetric with respect to the segment joining the vehicles (see Figure 3.2).

The two solutions of Equations 20, 21 coincide for:

$$\phi = \theta_1 \pm \arccos\left(\frac{\bar{u}_2}{\bar{u}_1}\right), \qquad (24)$$

such a ϕ exists only if $\frac{\bar{u}_2}{\bar{u}_1} \leq 1$. If we find the solution of Equation 19 in θ_1^a and θ_1^b, the solutions coincide for:

$$\phi = \theta_2 \pm \arccos\left(\frac{\bar{u}_1}{\bar{u}_2}\right), \qquad (25)$$

ϕ exists if $\frac{\bar{u}_1}{\bar{u}_2} \leq 1$. Hence, from Equations 24 and 25, ϕ exists if $\bar{u}_1 = \bar{u}_2$. In this case the solution of Equation 19 is:

$$\phi = \theta_1 = \theta_2.$$

In order to study constrained arcs of extremal solutions, it is useful to rewrite the cost function given by Equation 4 as:

$$\begin{aligned} \bar{J} = \ & \beta^T N \\ & + \sum_i \pi_i^s \gamma_i(q(T_i^s) - q_i^s) \\ & + \sum_i \pi_i^g \gamma_i(q(T_i^g) - q_i^g) \\ & + \int_0^\infty \sum_i w_i + \lambda^T(\dot{q} - f) + \mu \ddot{D}_{12} dt, \end{aligned} \tag{26}$$

with $\mu \geq 0$ along a constrained arc. The jump conditions at the entry point of a constrained arc, occurring at time τ, are now:

$$\lambda_i(\tau^-) = \lambda_i(\tau^+) + \beta \left. \frac{\partial N}{\partial q} \right|_\tau \tag{27}$$

$$H(\tau^-) = H(\tau^+) \tag{28}$$

$$\tag{29}$$

where $H = \sum_i w_i + \lambda^T f + \nu^T \ddot{D}_{12}$, and:

$$\left(\frac{\partial N}{\partial q} \right)^T = 2 \begin{bmatrix} (x_1 - x_2) & \bar{u}_1 \cos\theta_1 - \bar{u}_2 \cos\theta_2 \\ (y_1 - y_2) & \bar{u}_1 \sin\theta_1 - \bar{u}_2 \sin\theta_2 \\ 0 & d\bar{u}_1 \sin(\phi - \theta_1) \\ (x_2 - x_1) & \bar{u}_2 \cos\theta_2 - \bar{u}_1 \cos\theta_1 \\ (y_2 - y_1) & \bar{u}_2 \sin\theta_2 - \bar{u}_1 \sin\theta_1 \\ 0 & -d\bar{u}_2 \sin(\phi - \theta_2) \end{bmatrix}.$$

A further distinction among constrained arcs of zero and nonzero length should be done at this point.

3.2.1 Constrained Arcs of Zero Length

Consider first a constrained arc of zero length occurring at a generic contact configuration. This is completely described by the configuration of one vehicle (e.g., $q_c = q_1$), by the angle $\phi_c = \phi$, and by the contact type. Assume for the moment that there is only one constrained arc of zero length in the optimal path between way-points of the two vehicles. Equation 27, taking into account that costates of each vehicle are determined (once the way-points and contact configurations are fixed) up to constants $\rho_i(\tau^-)$, $\rho_i(\tau^+)$, provides a system of 6 equations in 6 unknowns of the

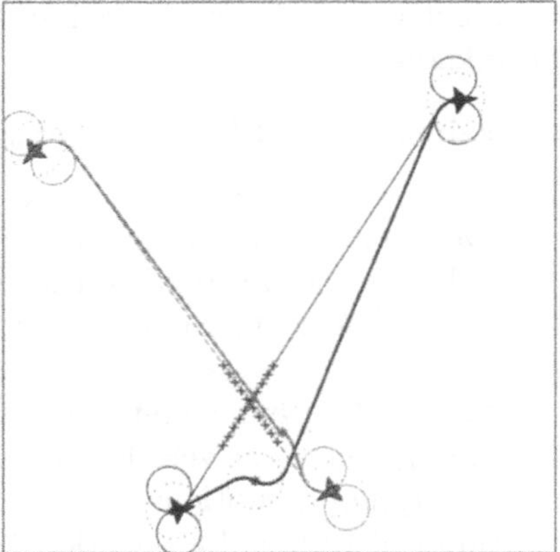

Figure 2: *A numerically computed solution to a two-vehicles minimum total time problem. Vehicles are represented as aircraft. Minimum curvature circles are reported at the start and goal configurations, along with safety discs of radius d/2 (dashed). The unconstrained Dubins' paths (thin lines) would achieve a cost of 88.75 units, but collide in this case (collisions are marked by "+" signs). The optimal solution consists of two unconstrained arcs for each vehicle, pieced together with a zero-length constrained arc of type b. Total cost is 92.25 units.*

form:

$$A(q_c, \phi_c) \begin{bmatrix} \rho_1(\tau^-) \\ \rho_1(\tau^+) \\ \rho_2(\tau^-) \\ \rho_2(\tau^+) \\ \beta_1 \\ \beta_2 \end{bmatrix} = 0,$$

where the explicit expression of matrix $A(q_c, \phi_c)$, for each contact type, can be easily evaluated in terms of $q_1^s, q_1^g, q_2^s, q_2^g$, and is omitted here for space limitations. Non-triviality of costates implies that (q_c, ϕ_c) must satisfy Equation $\det(\mathbf{A}) = 0$. A further constraint on contact configurations is implied by the equality of deplacement times from start to contact for the vehicles, which is expressed in terms of Dubins distances as:

$$L_D(\xi_1^s, \xi_c)/\bar{u}_1 = L_D(\xi_2^s, \xi_c')/u_2,$$

where ξ_c' denotes the configuration of vehicle 2 at contact. This is uniquely determined for each contact type. If m constrained arcs of zero length are present in an optimal solution, similar conditions apply (with way-points configurations suitably replaced by previous or successive contact configurations), yielding $2m$ equations in $4m$ unknowns.

3.2.2 Constrained Arcs of Nonzero Length

From this point on, we will make the assumption that forward velocity of all vehicles are equal to 1 ($\bar{u}_i = 1$). Consider an interval $[T_1, T_2]$ during which the constraint $D_{12} \equiv 0$. [2] A configuration of the two vehicles along such constrained arcs can be completely described by using only four parameters, for instance the configuration (x_1, y_1, θ_1) of one vehicle and the value of ϕ. Due to the tangency conditions on the constraint, the configuration is described by Equation 17 and one of Equations 22 or 23. Moreover, differentiating Equation 17, one finds:

$$\begin{aligned} \dot{x}_2 &= \dot{x}_1 - d\dot{\phi}\sin\phi, \\ \dot{y}_2 &= \dot{y}_1 + d\dot{\phi}\cos\phi, \end{aligned} \tag{30}$$

and

$$\dot{\phi} = \frac{1}{d}[\sin(\theta_2 - \phi) - \sin(\theta_1 - \phi)]. \tag{31}$$

Differentiating twice $D_{1,2}$ we obtain:

$$\begin{aligned} &\ddot{D}_{12}(q, \omega, t) = 0 = \\ &4 - 4\cos(\theta_1 - \theta_2) + 2\omega_1 d\sin(\theta_1 - \phi) - 2\omega_2 d\sin(\theta_2 - \phi). \end{aligned} \tag{32}$$

Constrained arcs of nonzero length that are part of an optimal solution must themselves satisfy necessary conditions, which can be deduced by rewriting the problem in terms of the reduced set of variables.

$$\begin{cases} \min 2(T_2 - T_1) \\ \dot{x}_i = \cos\theta_i \\ \dot{y}_i = \sin\theta_i \\ \dot{\theta}_i = \omega_i \\ \dot{\phi} = \frac{1}{d}[\sin(\theta_2 - \phi) - \sin(\theta_1 - \phi)] \\ \omega_1 \in \Omega_1, \omega_2 \in \Omega_2 \end{cases} \tag{33}$$

for $i = 1$ or $i = 2$, and for some initial and final specification of the variables $(x_i, y_i, \theta_i, \phi)$ and of the constrained arc type (a or b). Recall that $\theta_2 = \theta_1$ (arcs of type a), or $\theta_2 = 2\phi - \theta_1$ (type b).

[2] It should be pointed out that the study of constrained arcs of nonzero length is also useful to model cooperative manipulation of object by multiple vehicles, assuming that each vehicle supports the common load through a hinge joint.

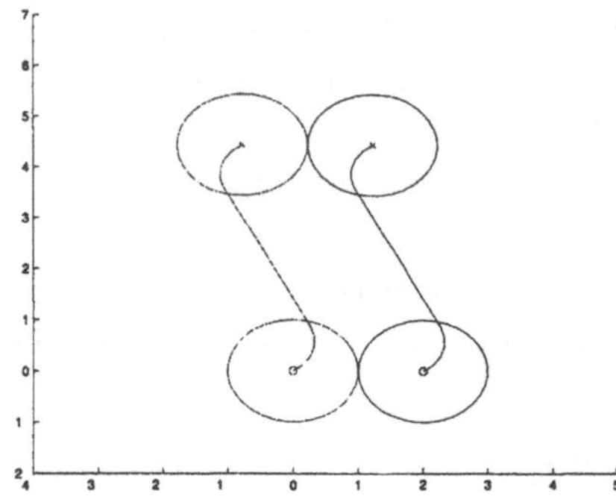

Figure 3: *Extremal constrained arcs of type a consist of two copies of a Dubins' path.*

Let us consider the two types of constrained arcs separately. Notice that two extremal constrained arcs of different type may be pieced together through a configuration with $\theta_1 = \theta_2 = \phi$, which is both of type a and b. **Type a)** From Equation 31, $\phi(t) \equiv \phi_0 = \arctan\frac{y_2(0) - y_1(0)}{x_2(0) - x_1(0)}$, hence:

$$\begin{aligned} \dot{x}_1 &= \dot{x}_2 \\ \dot{y}_1 &= \dot{y}_2 \\ \omega_1 &= \omega_2 \end{aligned} \tag{34}$$

Extremal constrained arcs of type a consist of a Dubins path for vehicle 1, and of a copy of the same path translated in the plane by $[d\cos\phi_0, d\sin\phi_0]^T$ for the other vehicle (see Figure 3).

Type b) In this case, using Equation 32, one obtains $\dot{\phi} = \frac{1}{2}(\omega_1 + \omega_2)$ in Equation 33. Introduce $\Lambda = (\lambda_1, \lambda_2, \lambda_3, \lambda_4)$, and $H = 2 + \lambda_1\cos\theta_1 + \lambda_2\sin\theta_1 + \lambda_3\omega_1 + \lambda_4(\omega_1 + \omega_2)/2$. Necessary conditions for optimality of solutions of Equation 33 are:

$$\dot{\Lambda} = -\Lambda \begin{pmatrix} 0 & 0 & -\sin\theta_1 & 0 \\ 0 & 0 & \cos\theta_1 & 0 \\ 0 & 0 & 0 & 0 \\ 0 & 0 & 0 & 0 \end{pmatrix}, \tag{35}$$

Hence, λ_1, λ_2 and λ_4 are constant. Letting $\lambda_1 = \rho\cos\psi$ and $\lambda_2 = \rho\sin\psi$, from Equation 35 we get:

$$\dot{\lambda_3} = \rho\sin(\theta_1 - \psi). \tag{36}$$

From P.M.P one also gets that, when $|\omega_1| < \Omega_1$ and $|\omega_2| < \Omega_2$, it is necessary for an optimal arc that $\frac{\partial H}{\partial \omega_1} = \frac{\partial H}{\partial \omega_2} = 0$, which implies $\lambda_3 = \lambda_4 = 0$. In this case, from

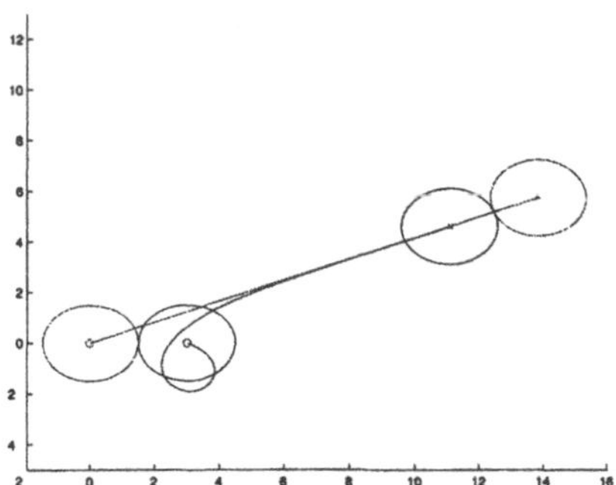

Figure 4: *Singular extremals in a constrained arc of type b.*

$\dot{\lambda}_3 = 0$ one easily gets $\theta_1 = \psi \pm \pi$, $\omega_1 = 0$. The direction of the segment joining the two vehicles varies as:

$$\begin{array}{l} \dot{\phi} = \frac{2}{d}\sin(\phi - \theta_1) \\ \phi(0) = \phi_0 \end{array} \tag{37}$$

Equilibria of Equation 37 at $\phi = \theta_1$ and $\phi = \theta_1 - \pi$ are respectively unstable and asymptotically stable. Hence, along a singular constrained arc of type b, one vehicle will be moving on a straight line, while the other will be trailing behind (see Figure 4).

Extremal constrained arcs may also obtain when a control variable is on the border of its domain, e.g. $\omega_1 = \pm\Omega_1$. In this case the motion of the two vehicles result in arcs such as those represented in Figure 5.

Notice that along such an arc, the steering velocity of vehicle 2 is uniquely determined by Equation 32. Hence, since $|\omega_2|$ is bounded by Ω_2, an arc of maximum curvature for vehicle one will be possible only until the limit curvature for vehicle two is reached. In other words, letting:

$$m_1(q) = \max\{-\Omega_2 - \frac{4}{d}\sin(\theta_1 - \phi), -\Omega_1\},$$

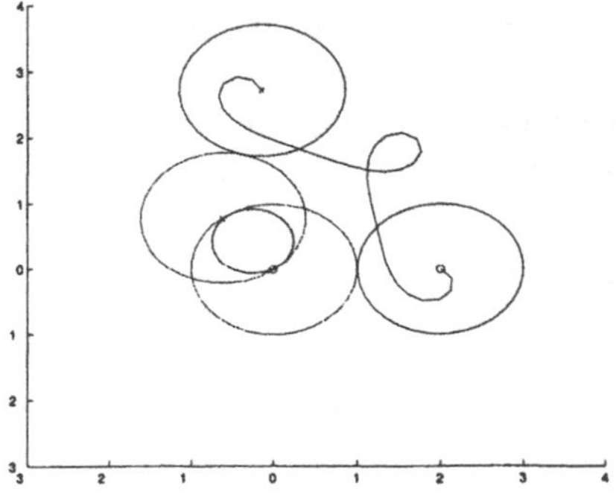

Figure 5: *An extremal constrained arc of type b.*

and:

$$M_1(q) = \min\{\Omega_2 - \frac{4}{d}\sin(\theta_1 - \phi), \Omega_1\},$$

we have that along constrained arcs of type b we must have $m_1 \leq \omega_1 \leq M_1$. Control ω_1 may equal M_1 if $\omega_1 = \Omega_1$ or if $\omega_1 = \Omega_2 - \frac{4}{d}\sin(\theta_1 - \phi)$. In the latter case, $\omega_2 = -\Omega_2$. A similar reasoning applies to vehicle two, for which we get $m_2 \leq \omega_2 \leq M_2$. In conclusion, along a nonsingular extremal constrained arc of type b, one of the vehicles moves along a circle of minimum radius, while the other follows a curve such as that described in Figure 5.

4 Numerical Computation of Solutions

The necessary conditions studied in this paper provide useful hints in the search for an optimal solution to the problem of planning trajectories of N vehicles in a common workspace. Although a complete synthesis has not been obtained so far, we will describe an algorithm that finds efficient solutions to the optimal planning problem in a reasonably short time.

Based on the discussion above, the optimal conflict resolution paths for multiple vehicles may include multiple waiting circles and constrained arcs of both zero and nonzero length. The algorithm presented in this section was developed to solve air traffic control problems [1], and is based on a few simplifying assumptions motivated by the application. Namely, we assume henceforth that

h1 all vehicles have equal geometric characteristics and equal (constant) speed;

h2 constrained arcs of nonzero length are not considered;

h3 multiple zero–length constrained arcs among the same vehicles are ruled out;

h4 the initial configurations of the vehicles are sufficiently separated.

Assumption **h4** implies that for each vehicle, the initial configuration are collision free and guarantee that wait circles at the initial configuration are collision free (this holds for instance if the distance between the initial position of vehicles i and j is larger than $2\pi R\frac{\bar{u}_j}{\bar{u}_i} + 2R + \frac{d_{ij}}{2}$).

Consider first the case of two vehicles. If the Dubins' trajectories joining the way-points configurations do not collide (i.e., $D(t) \geq 0, \forall t$), this is the optimal solution. Otherwise we compute the shortest contact–free solution with wait circles at the initial configurations, and let its length be L_f.

Hence we look for a solution with a concatenation of two Dubins' paths and a single constrained zero–length arc of either type a) or b) for both vehicles. Such a solution can be obtained by searching over a 2–dimensional submanifold of the contact configuration space ($\mathbb{R}^2 \times S^1 \times S^1$). The optimal solution can be obtained by using any of several available numerical constrained optimization routines. Computation is sped up considerably by using very efficient algorithms made available for evaluating Dubins' paths ([3]). The lenght L_c of such solution is compared with L_f, and the shorter solution is retained as the two–vehicle optimal conflict management path with at most a single constrained zero-length arc (OCMP21, for short). Some examples of OCMP21 solutions are reported in Figure 6.

If N vehicles move in a shared workspace, their possible conflicts can be managed with the following multilevel policy:

Level 0 Consider the unconstrained Dubins paths of all vehicles, which may be regarded as N single–vehicle, optimal conflict management paths, or OCMP10. If no collision occurs, the global optimum is achieved, and the algorithm stops. Otherwise compute the shortest contact–free paths (with wait circles) and go to next level;

Level 1 Consider the $M = 2\binom{N}{2}$ possible solutions with a single contact (of either type a or b), between two vehicles, and possibly wait circles for other vehicles, and compute the shortest path in this class. If this is longer than the shortest path obtained at level 0, exit. Otherwise, continue;

Level $m \geq 2$. Consider the $M \prod_{\ell=1}^{m-1}(M-2^{\ell})$ possible solutions involving m zero–length constrained arcs between different pairs of vehicles and (possibly) wait circles for other vehicles, and compute the shortest path in this class. If this is longer than the shortest path obtained at level $m-1$, exit. Otherwise, continue.

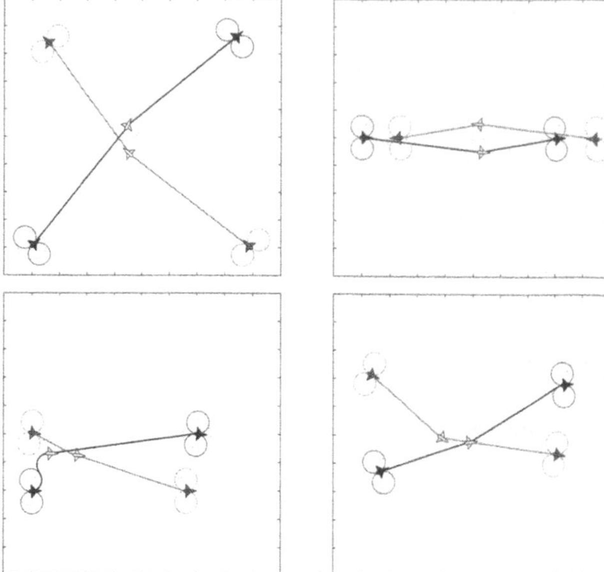

Figure 6: *Numerically computed solutions to optimal cooperative conflict resolution for two vehicles. Minimum curvature circles are reported at the way-point configurations, along with safety discs of radius d/2 (dashed). Optimal solutions consist of two unconstrained Dubins' trajectories for each vehicle, pieced together with a zero–length constrained arc.*

A few three–vehicle conflict resolution trajectories at different levels are reported in Figure 7.

When the number of vehicles increases, the number of optimization problems to be solved grows combinatorially. However, in practice, it is hardly to be expected that conflicts between more than a few vehicles at a time have to be managed. It was also observed in our simulations that, for vehicle parameters close to those of the kinematic model of commercial aircraft, solutions including wait circles are very rare.

5 Conclusion

In this paper, we have studied the problem of planning trajectories of multiple Dubins' vehicles in a plane. Necessary conditions have been derived for both free and constrained arcs. An algorithm for numerically finding solutions has been described.

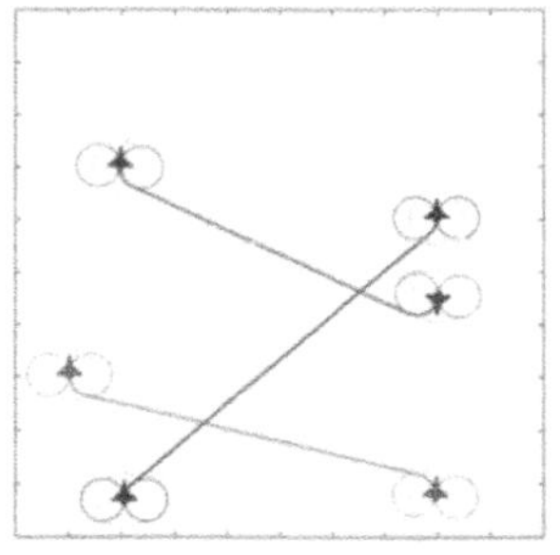
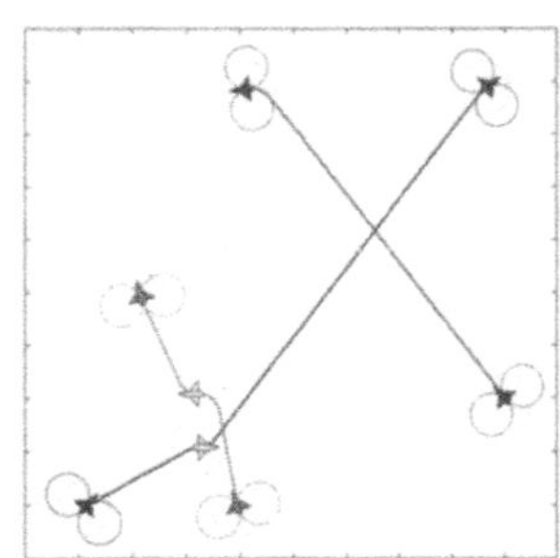
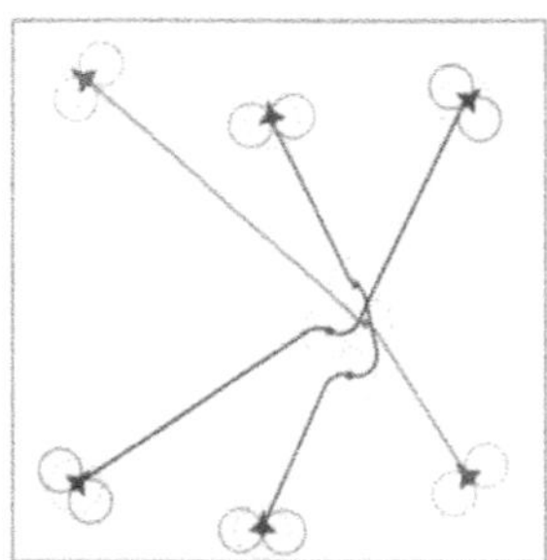
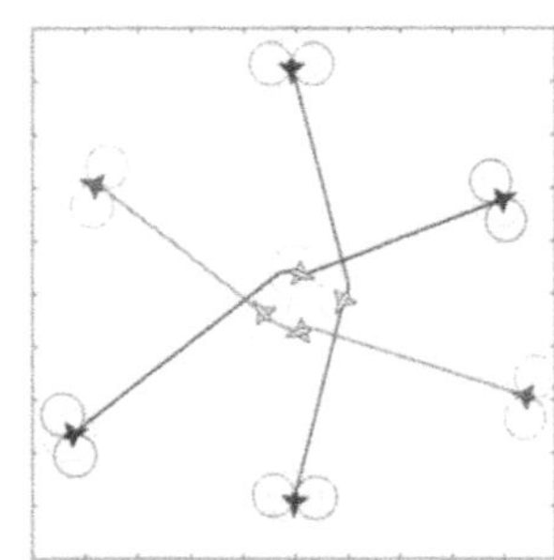

Figure 7: *Four cases of three–vehicles conflict resolution. Up left: the conflict is resolved at level 0. Up-right: a level 1 solution. Down left: a level 2 solution whereby the vehicle starting in the middle contacts first the one arriving on its right, and after the one arriving from left. Down right: a level 2 resolution that generates a roundabout–like maneuver.*

Future work on this topic will address the problem of finding a complete optimal synthesis at least for the simplest cases ($N = 2$). Further refinement of the algorithm will be sought, that could exploit more of the rich structure optimal solutions must satisfy. Finally, optimal paths of multiple agents at fixed distance will be studied in more detail, to address such problems as cooperative manipulation of objects by robotic vehicles and formation flight planning.

Acknowledgments

Authors wish to thank Gianfranco Parlangeli for help in an early phase of this work, and Jean–Paul Laumond for his useful remarks and suggestions on a previous version of this paper. This work was done with the support of NATO (CR no. 960750), ASI (ARS-96-170), and MURST (project RAMSETE).

References

[1] A. Bicchi, A. Marigo, G. Pappas, M. Pardini, G. Parlangeli, C. Tomlin, and S. S. Sastry. Decentralized air traffic management systems:performance and fault tolerance. In *Proc. Int. Works. on Motion Control*, pages 279–284. IFAC, 1998.

[2] J.D. Boissonnat, A. Cerezo, and J. Leblond. Shortest paths of bounded curvature in the plane. In *Proc. Int. Conf. on Robotics and Automation*, pages 2315–2320. IEEE, 1992.

[3] X.N. Bui, P. Souères, J-D. Boissonnat, and J-P. Laumond. Shortest path synthesis for Dubins non-holonomic robots. In *Proc. Int. Conf. on Robotics and Automation*, pages 2–7. IEEE, 1994.

[4] L. E. Dubins. On curves of minimal length with a constraint on average curvature and with prescribed initial and terminal positions and tangent. *American Journal of Mathematics*, 79:497–516, 1957.

[5] L. S. Pontryagin, V.G. Boltianskii, R.V. Gamkrelidze, and E.F. Mishenko. *The mathematical Theory of Optimal Processes*. Interscience Publishers, 1962.

[6] J. A. Reeds and R. A. Shepp. Optimal paths for a car that goes both forward and backward. *Pacific Journal of Mathematics*, 145(2), 1990.

[7] H. J. Sussmann and G. Tang. Shortest paths for the Reeds–Shepp car: a worked out example of the use of geometric techniques in nonlinear optimal control. Technical Report 91–10, SYCON, 1991.

An Efficient Approximation Algorithm for Weighted Region Shortest Path Problem

John Reif, *Duke University, Durham, NC*
Zheng Sun, *Duke University, Durham, NC*

In this paper we present an approximation algorithm for solving the following optimal motion planning problem: Given a planar space composed of triangular regions, each of which is associated with a positive unit weight, and two points s and t, find a path from s to t with the minimum weight, where the weight of a path is defined to be the weighted sum of the lengths of the sub-paths within each region.

Some prior algorithms (Lanthier et al. [9] and Aleksandrov et al. [1]) took a discretization approach by introducing m Steiner points on each edge. A discrete graph is constructed by adding edges connecting Steiner points in the same triangular region and an optimal path is computed in the resulting discrete graph using Dijkstra's algorithm. To avoid high time complexity, both [9] and [1] use a subgraph of the complete graph in each triangular region. As a result, in the discrete graph only an approximate optimal path can be achieved, whose error is proportional to the weight of the optimal path. This approximate optimal path then is used to approximate the optimal path in the original problem.

Our algorithm extends these algorithms by dynamically maintaining $O(m)$ edges in each region to compute the **exact** *shortest path in the discrete graph efficiently. The running time of this algorithm is $O(mn \log mn)$, where n is the number of triangular regions. Our algorithm provides an improvement over previous algorithms in the case when approximation error is to be bounded by an additive constant.*

Besides (additive) constant error bound, our algorithm also has the following three advantages: (a) it can compute exact solutions for a discrete case of this problem; (b) it can be applied to any discretization; and (c) it can be applied to a more general class of problems than the previous algorithms.

1 Introduction

1.1 Definition of the Problem

Robotics motion planning problems are some of the most fundamental problems in robotics research. An important category of these problems is to determine the optimal path (shortest path according to a user-defined metric on paths) between two points, s and t, in a 2D or 3D space under various conditions. This problem has drawn great attention from researchers in robotics as well as other areas such as computational geometry, geographical information systems (GIS), and graph theory, due to its direct applications. There have been numerous papers on the optimal path problem. Interested readers may refer to survey [11] and [12] for a comprehensive review.

In this paper we examine the weighted version of this problem in 2D space. More specifically, a weight function $F : \mathcal{R}^2 \rightarrow \mathcal{R}^+$ is defined on the space. For any path p from v_1 to v_2, the weighted length of p is defined to be $\int_{v_1}^{v_2} F(x)dx$, the integral of the weight function along the path. The objective of the minimum weighted path problem is to find the optimal path, i.e., the path with the minimum weighted length, from given source point s and a destination point t.

Figure 1 shows an optimal path in a 2D space with four triangular regions. In the figure, the darker a region is shaded, the more costly it is to move inside the region. (This designation will be used throughout this paper.)

Observe that the unweighted optimal path problem can be considered as a special case of the weighted optimal path problem: For any point in the "free space," the weight is defined to be 1; for any point in the "obstacle" (if there is any), the weight is defined to be $+\infty$.

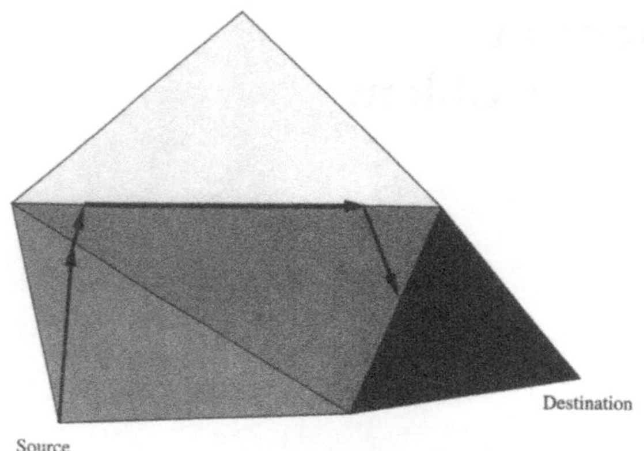

Figure 1: *Optimal path.*

Even though currently there is no complexity result regarding finding the exact solution of the weighted optimal path, it is generally considered to be very hard. (See Canny and Reif's work [5] for the NP-hardness proof on the unweighted optimal path problem in 3D space.) As a result, most previous research has focused on approximation algorithms on a special case of this problem where the planar space can be divided into polygons and the weight is uniform within each polygon. In some literature, this problem is named "homogeneous-cost-region-path-planning problem."

We define some notations that will be used in the rest of this paper. We use n to denote the number of vertices of polygonal regions in the plane, and L the length of the longest edge of all polygonal regions. For our convenience, we assume that all the coordinates of the vertices are integers, and N is the maximum coordinate. Among various weights of all regions, let W_{max} be the maximum weight and W_{min} be the minimum weight.

For any s and t, we use $D(s,t)$ to denote the weighted length of an optimal path from s to t and use $D'(s,t)$ to denote the weighted length of an approximate optimal path found by a given approximate algorithm. We decompose the error of an approximation into two components: The *absolute error* δ, which is bounded by a constant disregarding the value of $D(s,t)$; and the *relative error* ϵ, which is linear to $D(s,t)$. Therefore, we can represent the error of an approximation of the optimal path from s to t as $D'(s,t)-D(s,t) = \epsilon\cdot D(s,t)+\delta$.

We list in the following the performance parameters that will be used by various algorithms.

Performance Parameters

n	number of vertices
L	length of the longest edge
N	maximum coordinate of the vertices
W_{max} (W_{min})	the maximum (minimum) weight
ϵ	the user-defined relative error allowed

1.2 Previous Work

One of the early algorithms on this problem was given by Mitchell and Papadimitriou [13]. Their algorithm uses *Snell's Law* and "a continuous Dijkstra method" to give an optimal-path map for any given source point s. The time complexity of their algorithm is $O(n^8 M)$. Here M is a function of various input parameters including the (relative) error tolerance ϵ.

Mata and Mitchell [10] presented another approximation algorithm based on discretization, the "pathnet" algorithm. In their scheme, a "pathnet graph" of size $O(nk)$ is constructed and then an approximate optimal path with relative error of ϵ is computed, where $\epsilon = O(\frac{W_{max}/W_{min}}{k\theta_{min}})$. The time complexity, in terms of ϵ, n and other geometric parameters, is $O(\frac{n^3 W_{max}/W_{min}}{\epsilon\theta_{min}})$.

Lanthier et al. [9] and Aleksandrov et al. [1] adopted a natural approach to solving this problem. Both algorithms discretize the polygonal subdivision by placing m Steiner points along the edges of polygonal regions and then try to compute an optimal path in the resulting discrete graph using Dijkstra's algorithm. Instead of using a complete graph in each region by connecting all possible pairs of points (vertices and Steiner points) in the same region, which is costly in terms of both time and space complexity, they only selectively add $O(\frac{m}{\epsilon})$ ([9] or $O(m\log m)$ ([1]) edges in each region and compute an approximate optimal path with ϵ relative error.

Lanthier et al. [9] used a uniform discretization, which adds $m = O(n^2)$ points on each edge evenly, to compute approximate paths with a LW_{max}-absolute error and ϵ-relative error. The absolute error of LW_{max} is resulted from the discretization while the relative error of ϵ is resulted from using spanner to compute the approximate optimal path in the discrete graph. The relative error ϵ is treated as a constant in their complexity analysis. Including ϵ also as an input parameter, the complexity is $O(\frac{n^3}{\epsilon} + n^3 \log n)$, as in the discrete graph there will be $O(n^3)$ vertices and $O(\frac{n^3}{\epsilon})$ edges.

Aleksandrov et al. [1] proposed a logarithmic discretization that can guarantee a relative error of ϵ. The time complexity of this algorithm is $O(mn \log(mn))$, where $m = O(\log_\delta(L/r))$. Here r is the minimum distance from any point to the boundary of the regions adjacent to it, and $\delta = 1 + \frac{\epsilon W_{max}}{W_{min}} \sin \theta_{min}$, where θ_{min} is the minimum angle between any two edges. If all the points have integer coordinates and N is the maximum coordinate, we have $r = \Omega(1/N)$ and $\theta_{min} = \Omega(1/N^2)$, and thus the time bound is approximately $O(\frac{nN^2 W_{max}}{W_{min}\epsilon} \log(LN) \log(\frac{nN^2 W_{max}}{W_{min}\epsilon}))$.

In the following table, we list the time complexity and error bound of each of above-mentioned algorithm for a comparison.

Comparison of Algorithm Performance

Algorithm	Complexity	Error
Mitchell et al.	$O(n^8 \log \frac{nNW_{max}}{\epsilon W_{min}})$	ϵ-relative
Lanthier et al.	$O(\frac{n^3}{\epsilon} + n^3 \log n)$	ϵ-relative LW-absolute
Mata et al.	$O(\frac{n^3 W_{max}}{\epsilon \theta_{min} W_{min}}) =$ $O(\frac{n^3 N^2 W_{max}}{\epsilon W_{min}})$	ϵ-relative
Aleksandrov et al.	$O(nm \log(nm)) =$ $O(\frac{nN^2 W_{max}}{\epsilon W_{min}} \log(LN)$ $\log(\frac{nN^2 W_{max}}{\epsilon W_{min}}))$	ϵ-relative

For other related work, see [2, 3, 4, 8, 14, 15, 16, 17], or see survey [12].

1.3 Our Result

In this paper we introduce a wavefront-like algorithm that can compute optimal paths on any discretization (uniform or logarithmic) more effectively than the ones proposed in [9] and [1]. Instead of computing an approximate optimal path in the discretized space with ϵ-relative error, as the above-mentioned two algorithms do, our algorithm can give an exact optimal path in less (for the uniform discretization) or equivalent (for the logarithmic discretization) time. We achieve so by dynamically maintaining for each Steiner point a small set of edges that are to be considered in search for an optimal path. The total number of edges is $O(m)$ in each region.

For the purpose of solving the original (continuous) weighted shortest path problem, our algorithm improves Lanthier et al. [9]'s work in terms of both time complexity ($O(n^3 \log n)$ vs. $O(\frac{n^3}{\epsilon} + n^3 \log n)$) and accuracy (same absolute error but no relative error). It also matches Aleksandrov et al. [1]'s work ($O(mn \log(mn))$ time and ϵ-relative error) if logarithmic discretization is used. The reason our algorithm can not provide improvement in terms of time complexity over [1] is that during discretization a relative error of ϵ is already introduced and thus being able to compute the exact optimal path in the discretized problem does not improve the complexity result.

Although our algorithm is not superior for this particular problem that can be solved by specialized approach of [1] due to the nature of discretization, we found our algorithm interesting for the following reasons:

i. It provides an efficient way of computing the optimal path on the discretization graph. We refer to this problem as *Discrete Weighted Shortest Path Problem.*

> **Discrete Weighted Shortest Path Problem** A 2D space is composed of n triangular region, each of which is associated to a certain unit weight. On each edge e of each region, there are $m = f(e)$ Steiner points. For any given source point s and destination point t, find a weighted shortest path that only intersects region boundaries at vertices or Steiner points.

This is exactly the problem that [9], [1] and our algorithm solve after the discretization step. The other two algorithm only compute an approximate optimal path with time complexity greater than or equal to that of our algorithm.

ii. It can be applied to different discretizations to achieve different approximation goals. For example, if absolute error is to be limited, we can apply this algorithm on a uniform discretization. If restricting relative error is the main goal, the logarithmic discretization then can be used.

iii. It can be applied to a more general class of problems. As we will show later, our algorithm only assumes the following geometric property:

> Inside each region (not including the edges), the weighted length of a path p_1 is less than that of p_2 if and only if the Euclidean length of p_1 is less than that of p_2.

Therefore, our algorithm can be applied to optimal path problems on planar regions where weight of a path is defined differently but the above-mentioned property holds. A good example will be computing an optimal path in a space composed of triangular regions, each of which has a uniform flow.

2 Preliminaries

2.1 Overview of Our Approach

Let S be a polygonal subdivision of a planar space, and let n be the number of vertices of S. This implies the number of edges and faces (regions) is also $O(n)$. Without loss of generality, we assume that each face of S is a triangle. A triangulation can be performed in case there are non-triangular faces in the input subdivision. The resulting subdivision S' after triangulation will still have $O(n)$ vertices, edges and faces.

Each face f is associated with a non-negative real number w_f, the weight of f. For simplicity, we assume that $w_f \in [1, W_{max}]$. For each edge e, the weight w_e is defined to be $\min\{w_f, w_{f'}\}$, where f and f' are the two faces adjacent to e. W_{max} (W_{min}), as we defined in Section 1, is the maximum (minimum, respectively) weight of all regions. L is the length of the longest edge among the $O(n)$ edges.

A line segment $\overline{v_1v_2}$ in a triangular region f is called *edge-crawling* if v_1 and v_2 are on a same edge e of f; otherwise, it is called *face-crossing*. Refer to Figure 1, the third segment of the optimal path is edge-crawling and the other three segments are face-crossing. We use $|\overline{v_1v_2}|$ to denote its Euclidean length and $W(\overline{v_1v_2})$ to denote its weighted length which is defined to be $|\overline{v_1v_2}| \cdot w_f$. If v_1v_2 is edge-crawling, $W(\overline{v_1v_2})$ is $|\overline{v_1v_2}| \cdot w_e$. Here w_e might be different from w_f in case the other face f' adjacent to e has a smaller weight than that of f. For two paths p_1 and p_2, we use $p_1 + p_2$ to denote the concatenation of p_1 and p_2 and $p[v_1, v_2]$ to denote the part of path p between v_1 and v_2.

Let s and t be the source and the destination, respectively. We can assume that s and t are two vertices of S. (If otherwise, we can add a constant number of edges to construct from S a new subdivision S' which has s and t as vertices.) The problem is to find a shortest weighted length path (which will be called *optimal path*) from s to t.

It has been proved (see [13]) that an optimal path consists of $O(n^2)$ straight line segments. The endpoints of each segment are on the boundary of a triangular face.

Our algorithm is a general optimal path algorithm and can be applied to any discretization that places Steiner points on edges. Therefore, for now, we use $f(e)$ to denote the number of Steiner points added along a given edge e. Here $f(e)$ is dependent on e. It may also depend on n and some other parameters, as we shall see later in this paper. We use V to denote the set of vertices of S and use V_s to denote the set of all Steiner points. In the following discussion, a "point" refers to a point in $V \cup V_s$.

We first focus on a subspace P' of the original path space P. Each path in P' is composed of a number of line segments, each of whose two endpoints are points in $V \cup V_s$. Further, the two endpoints of such a line segment are in a same face of S. We call such a path a *discrete path*.

We first compute an optimal discrete path $p'_{opt}(s,t)$ from s to t in P' and use this path to approximate the optimal path $p_{opt}(s,t)$ from s to t in the original space.[1] Then, we show that $D'(s,t)$, the weighted length of $p'_{opt}(s,t)$, gives a good approximation to $D(s,t)$.[2]

Our algorithm is similar to Dijkstra algorithm. However, by utilizing some geometric properties of this problem, we achieve a time complexity lower than that of Dijkstra algorithm.

2.2 Data Structures

Before giving the description of our algorithm, we first present some basic data structures.

Our algorithm maintains a list QLIST of paths. Each path $p' \in$ QLIST is a *candidate optimal path* from s to some point $v \in V \cup V_s$, and is represented by a quadruplet $(v, v_{prev}, e_{prev}, l)$. Here v_{prev} is the previous point

[1] As a matter of fact, in a single run of our algorithm, an optimal discrete path $p'_{opt}(s,v)$ is computed for each $v \in V \cup V_s$.

[2] The accuracy of the approximation will depend on the discretization used.

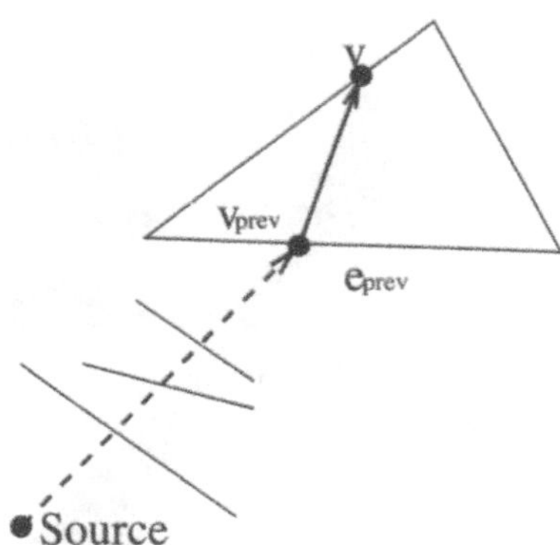

Figure 2: *Candidate optimal path.*

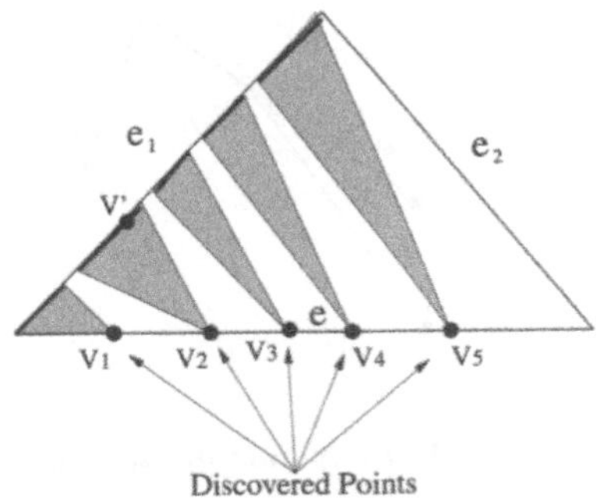

Figure 3: *Intervals.*

of v on p', e_{prev} is an edge that contains v_{prev}, and l is the weighted length of p'. Each p' is "almost optimal" in the sense that it is the concatenation of $p'_{opt}(s, v_{prev})$, the optimal discrete path from s to p_{prev}, and line segment $\overline{v_{prev}v}$. Thus, $l = D'(s, v_{prev}) + W(\overline{v_{prev}v})$.

QLIST is sorted in ascending order according to the weighted length of candidate optimal paths (i.e., value of l of the quadruplets). Possible operations on this list are insertions and removals. We can use an AVL tree (see [6]) to implement QLIST so that each operation costs only $O(\log |\text{QLIST}|)$ time.

Also, for each edge $e = \overline{v_i v_j}$ of S, we keep a list PLIST_e of *discovered* points on e. A point v is said to be discovered when the optimal discrete path from s to v is determined. The points in this list are ordered according to the distance to v_i. Possible operations on PLIST_e include insertions and finding the closest neighbors. Again, using an AVL tree can guarantee that each operation only costs $O(\log |\text{PLIST}_e|) = O(\log f(e))$ time.

Let f be a face adjacent to e, and let e_1 and e_2 be the other two edges of f. For each e_i and each discovered point v on e that is not an endpoint of e_i, we use I_{v,e,e_i} to denote the set of points on e_i with the following property: for any point v^* on e_i, $v^* \in I_{v,e,e_i}$ if and only if $W(\overline{vv^*}) + D'(s,v) < W(\overline{v'v^*}) + D'(s,v')$ for any other discovered point v' on e. We will show later that each such I_{v,e,e_i} is an interval of contiguous points on e_i, as indicated in Figure 3.

For any I_{v,e,e_i}, if the point $\hat{v}$ closest to v in I_{e,v,e_i} is not an endpoint of I_{v,e,e_i}, we split I_{v,e,e_i} into two intervals I'_{v,e,e_i} and I''_{v,e,e_i} in such a way that for each interval, the closest point to v is one of the two endpoints (and the farthest point to v is the other endpoint). This is to guarantee that the distance to v forms a monotonically increasing or decreasing sequence for points in I_{v,e,e_i} from left to right along edge e.

From the construction of I_{v,e,e_i}, it is easy to see that at most two intervals are associated with each point $v \in V \cup V_s$ due to interval splitting. For the sake of simplicity, in the following discussion we may assume that only one interval is associated with each point. Removing this assumption will only increase the time complexity of our algorithm by a constant factor.

We let v^l_{v,e,e_i} and v^r_{v,e,e_i} denote the left and right endpoint of I_{v,e,e_i}, respectively. As we shall see later, every point v' in I_{v,e,e_i} will be "processed," meaning that a candidate optimal path from s to v' which enters face f through v will be added to (and later removed from) QLIST. We let $\hat{v}_{v,e,e_i}$ be the closest point of I_{v,e,e_i} to v that has not yet been processed. Initially it is either one of the two endpoints of I_{v,e,e_i} whichever is closer to v. An interval I_{v,e,e_i} is said to be *active* if at least one point v^* in I_{v,e,e_i} has not been processed.

Figure 4 illustrates endpoints and closest unprocessed point of an interval. In the figure, we use solid circles to denote processed interval points and hollow circles to denote unprocessed interval points.

Let ILIST_{e,e_i} be the list of intervals I_{v,e,e_i} for all $v \in \text{PLIST}_e$. The intervals in ILIST_{e,e_i} are maintained in the order according to the order of v in PLIST_e. We also use an AVL tree to implement each ILIST_{e,e_i} to keep the cost for each operation on ILIST_{e,e_i} to be $O(\log |\text{ILIST}_{e,e_i}|) = O(\log f(e))$.

It is easy to see that, at any time, an edge vv' is considered to be added into an optimal path only if v'

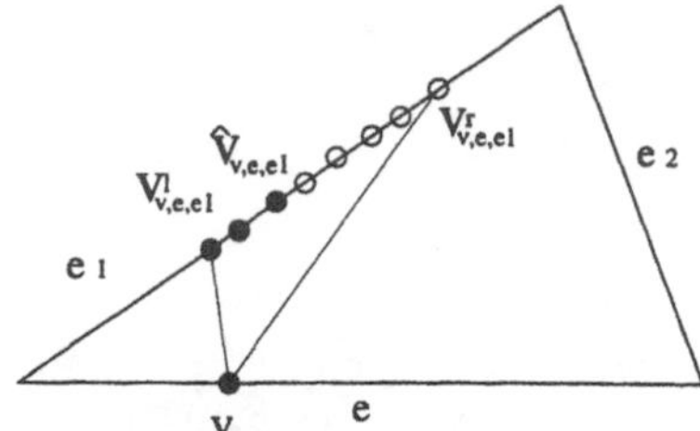

Figure 4: *Endpoints of interval.*

is in I_{v,e,e_i} for some e and e_i, and the total amount of such edges in each face f is bounded by $O(m)$. By eliminating edges that can not produce optimal paths, our algorithm is able to compute optimal paths more efficiently than Dijkstra's algorithm.

3 The Algorithm

We now give the formal specification of our algorithm.

Function *FindOptimalDiscrete*

[1] **(Initialization)** QLIST $= \{(s, s, \emptyset, 0)\}$. For each e, $\text{PLIST}_e = \emptyset$. For each edge e and each e_i that shares a face f with e, $\text{ILIST}_{e,e_i} = \emptyset$. For each point $v \in V \cup V_s$, $D'(s, v) = +\infty$.

[2] **(Main Loop)** While QLIST is not empty:

[a] Remove the first entry of QLIST. Let it be $(v, v_{prev}, e_{prev}, l)$.

[b] If $D'(s, v) = +\infty$, i.e., no other candidate optimal path p' (with weighted length at most l) has been removed from QLIST and thus v is not discovered yet:

[i] Set $D'(s, v) = l$ and set $p'_{opt}(s, v) = p'_{opt}(s, v_{prev}) + \overline{v_{prev}v}$.

[ii] Insert v into PLIST_e for each e that contains v.

[iii] For each edge e that contains v, and let f_1 and f_2 be the two faces adjacent to e. For each f_i, $i = 1, 2$, let $e_{i,1}$ and $e_{i,2}$ be the two edges of f_i other than e. Call *InsertInterval*$(v, e, e_{i,j})$ for $1 \le i \le 2$ and $1 \le j \le 2$ if v is not an endpoint of $e_{i,j}$.

[c] Call *Propagate*$(v, v_{prev}, e_{prev}, l)$.

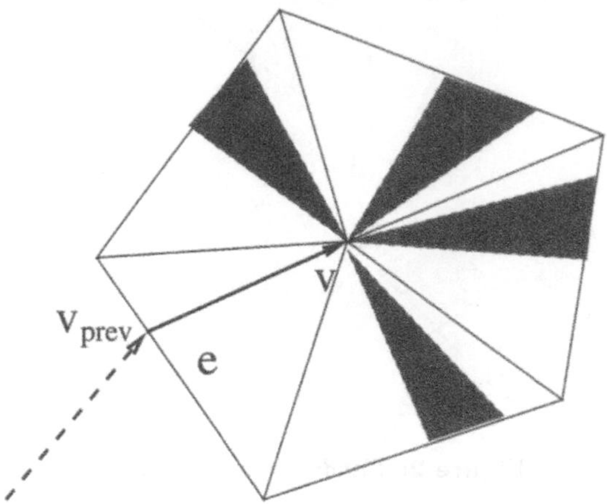

Figure 5: *Insert intervals for a discovered point.*

The main loop consists of two steps: the first step is to, in case v is a newly discovered point, update the discovered point list PLIST_e, and subsequently update the interval list ILIST_{e,e_i} for all appropriate e and e_i (by calling function *InsertInterval*). The second step is to propagate the path by one more line segment. This is done by calling function *Propagate*.

Figure 5 shows a special case of step [2.b.iii]. When the newly discovered point v is an endpoint of an edge (i.e. $v \in V$), we will need to insert new intervals to edges of all adjacent faces of v.

To update an interval list ILIST_{e,e_i} when a point v on e is discovered, function *InsertInterval* needs to determine v^l_{v,e,e_i} and v^r_{v,e,e_i}, the left and right endpoints of the new interval I_{v,e,e_i} associated with v, and to adjust the endpoints of other intervals in ILIST_{e,e_i} if necessary.

Function *InsertInterval*(v, e, e_i)

[1] Find the left and right neighbors, v_l and v_r, of v in PLIST_e.

[2] Initially, $v^l_{v,e,e_i} = \text{NULL}$, $v^r_{v,e,e_i} = \text{NULL}$.

[3] Do a binary search on the part of e_i to the right of (including) $v^r_{v_l,e,e_i}$ (the right endpoint of the interval associated with the left neighbor of v) until we find two adjacent points $\hat{v}_1$ and $\hat{v}_2$ such that, if $\hat{v}_1$ is in the interval associated with v^*_1 and $\hat{v}_2$ is in the interval associated with v^*_2, $D'(s, v) + W(v\hat{v}_1) < D'(s, v^*_1) + W(v^*_1\hat{v}_1)$ but $D'(s, v) + W(v\hat{v}_2) > D'(s, v^*_2) + W(v^*_2\hat{v}_2)$. In other words, to construct a path s to $\hat{v}_1$, it is advantageous to

extend $p'_{opt}(s, v)$ by adding one more line segment $v\hat{v}_1$ than to extend $p'_{opt}(s, v_1^*)$ by adding line segment $v_1^*\hat{v}_1$; however, to construct a path from s to $\hat{v}_2$, extending $p'_{opt}(s, v_2^*)$ is a better idea.

[4] Let $v^r_{v,e,e_i} = \hat{v}_1$.

[5] Similarly, compute v^l_{v,e,e_i}.

[6] Insert I_{v,e,e_i} into ILIST$_{e,e_i}$. For each interval I_{v',e,e_i} entirely covered by I_{v,e,e_i}, set $I_{v',e,e_i} = \emptyset$.

[7]Add $(\hat{v}_{v,e,e_i}, v, e, D'(s, v) + W(\overline{v, \hat{v}_{v,e,e_i}}))$ to QLIST.

[8] For any interval I_{v^*,e,e_i} that is partially covered by I_{v,e,e_i}, adjust its boundary by removing all points that are covered by I_{v,e,e_i}. If the closest unprocessed point to v^* in the original I_{v^*,e,e_i} is no longer inside the interval, add $(\hat{v}_{v^*,e,e_i}, v^*, e, D'(s, v^*)+ W(v^*, \hat{v}_{v^*,e,e_i}))$ into QLIST.

Step [7] is very important in function *InsertInterval*. It adds into QLIST a new candidate optimal path which enters face f through v and leads to the closest point $\hat{v}_{v,e,e_i}$ to v (in interval I_{v,e,e_i}). Here f is the face adjacent to both e and e_i. When later this path is processed and removed from QLIST, a similar optimal path that connects s to the subsequent point in interval I_{v,e,e_i} will be added into QLIST, as shown in function *Propagate*. Therefore, progressively, for each point $u \in I_{v,e,e_i}$, the candidate optimal path that enters f through v and leads to u will be tested if it is an optimal path.

Similarly, Step [8] is to ensure that, for any active interval I_{v^*,e,e_i} whose endpoints are changed because of the newly inserted interval, there is a candidate optimal path in QLIST that enters face f through v^* and leads to the closest unprocessed point in I_{v^*,e,e_i}.

Function *Propagate* is used to add candidate optimal paths to QLIST after $(v, v_{prev}, e_{prev}, l)$ is removed from the list.

Function *Propagate*$(v, v_{prev}, e_{prev}, l)$

[1] If v is a Steiner point:

[a] If v is not on e_{prev}: suppose f is the face that contains both e_{prev} and v, and e is the edge of f that contains v.

[a.1] If there has been another quadruplet $(v, v'_{prev}, e_{prev}, l')$ removed from QLIST previously (i.e., another candidate optimal path in the form of $p'_{opt}(s, v'_{prev}) + \overline{v'_{prev}v}$ has been found with a shorter weighted length, where v'_{prev} is also on e_{prev}), **return**.

[a.2] Let v_{new} be the left or right neighbor of v on e such that $|\overline{v_{prev}v_{new}}| > |\overline{v_{prev}v}|$. If $W(\overline{v_{prev}v_{new}}) \geq W(\overline{v_{prev}v}) + W(\overline{vv_{new}})$,[3] insert $(v_{new}, v, e, D'(s, v) + W(\overline{vv_{new}}))$ into QLIST; otherwise insert $(v_{new}, v_{prev}, e_{prev}, D'(s, v_{prev}) + W(\overline{v_{prev}v_{new}}))$.

[b] If v is on e_{prev}:

[b.1] If there has been another quadruplet $(v, v_{prev}, e_{prev}, l')$ removed from QLIST, **return**

[b.2] Otherwise, let v_{new} be the neighbor of v on e_{prev} that is farther from v_{prev} than v is. Insert $(v_{new}, v, e_i, D'(s, v) + W(\overline{vv_{new}}))$ into QLIST.

[2] If v is a vertex of the polygonal subdivision:

Let $e_1, e_2, \cdots, e_k$ be the edges adjacent to v, and let v_i be the closest point $\in V \cup V_s$ that is on e_i for $i = 1, 2, \cdots, k$. Insert $(v_i, v, e_i, D'(s, v) + W(\overline{(vv_i)}))$ into QLIST for each $i = 1, 2, \cdots, k$. This is to extend the candidate optimal path along each e_i.

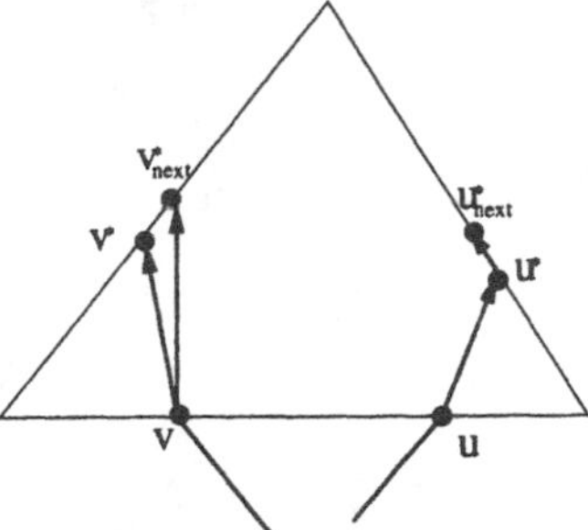

Figure 6: *Two ways of propagating candidate optimal paths.*

The two cases in step [a.2] correspond to the two ways of propagate candidate optimal paths as shown in Figure 6. On the left, a candidate optimal path to

[3]It is possible as the weight of e may be smaller than the weight of f.

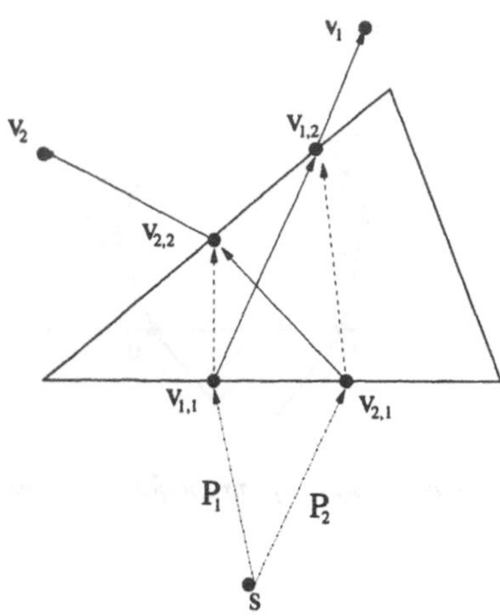

Figure 7: *Two optimal paths can not intersect.*

v^*_{next} is generated by replacing the last line segment vv^* of the candidate optimal path to v^* by line segment vv^*_{next}. We call this type of propagation *face-crossing-propagation.* On the right, a candidate optimal path to u^*_{next} is generated by adding line segment $u^*u^*_{next}$ to the candidate optimal path from s to u^*. We call this type of propagation *edge-crawling-propagation.* Edge-crawling-propagation is to take advantage of the smaller weight of the other face.

4 Correctness and Error Bound

In the first subsection we present the correctness proof for our algorithm of finding an optimal discrete path $p'_{opt}(s, v)$ for every $v \in V \cup V_s$. Then we show how an optimal discrete path can provide an approximate solution to the original problem.

4.1 Correctness Proof

We first give a lemma that is key to our algorithm.

Lemma 1 *Any two discrete optimal paths cannot intersect in the interior of any triangular face.*

Proof Assume that, as shown in Figure 7, two optimal path $p'_{opt}(s, v_1)$ and $p'_{opt}(s, v_2)$ intersect at b inside a triangular face f. For each $i = 1, 2$, let $\overline{v_{i,1}v_{i,2}}$ be the segment of $p'_{opt}(s, v_i)$ that contains b. Thus, $v_{i,j}$ are points on the boundary of f. Let $p_{i,1}$ be the portion of $p'_{opt}(s, v_i)$ between s and $v_{i,1}$ and let $p_{i,2}$ be the portion of $p'_{opt}(s, v_i)$ between $v_{i,2}$ and v_i. Therefore, $p_{2,1} + \overline{v_{2,1}v_{1,2}} + P_{1,2}$ is a path from s to v_1 and $p_{1,1} + \overline{v_{1,1}v_{2,2}} + P_{2,2}$ is a path from s to v_2. However, the total weighted length of the above two paths is shorter than that of $p'_{opt}(s, v_1)$ and $p'_{opt}(s, v_2)$, as $W(\overline{v_{1,1}v_{1,2}}) + W(\overline{v_{2,1}v_{2,2}}) > W(\overline{v_{1,1}v_{2,2}}) + W(\overline{v_{2,1}v_{1,2}})$. Therefore, one of $p'_{opt}(s, v_1)$ and $p'_{opt}(s, v_2)$ must not be optimal. A contradiction. □

With exactly the same argument, we can prove the following lemma:

Lemma 2 *For any two points v_1 and v_2 in $PLIST_e$ that are not endpoint of e, each point in $I_{v_1,e,f}$ is to the right (left) of each point in $I_{v_2,e,f}$ if and only if v_1 is to the right (left, respectively) of v_2 on e.*

It directly follows from this lemma that each $I_{v,e,f}$ is an interval of contiguous points on $e_1 \cup e_2$. Thus, a binary search can be used in step [3] of function *InsertInterval* to find the left and right endpoints of the new interval I_{v,e,e_i}.

Lemma 3 *For any active I_{v,e,e_i}, let $\hat{v} \in I_{v,e,e_i}$ be the closest point (in terms of Euclidean distance) to v that has not been processed. Then, there must be a point $\hat{v}' \in I_{v,e,e_i}$ with $W(\overline{v\hat{v}'}) \leq W(\overline{v\hat{v}})$ that satisfies at least one of the following two conditions: (i) Quadruplet $(\hat{v}', v, e, D'(s, v) + W(\overline{v\hat{v}'})) \in QLIST$; (ii) Quadruplet $(\hat{v}', \hat{v}'', e_1, D'(s, \hat{v}'') + W(\overline{\hat{v}''\hat{v}'})) \in QLIST$ and $D'(s, \hat{v}'') + W(\overline{\hat{v}'\hat{v}''}) < D'(s, v) + W(\overline{v\hat{v}})$. Here $\hat{v}''$ is either the left neighbor or right neighbor of $\hat{v}'$ on e_1. We call a point that satisfies condition (i) (condition (ii)) a "handle point" of Type A (B, respectively). Observe that there might be more than one handle point for an interval.*

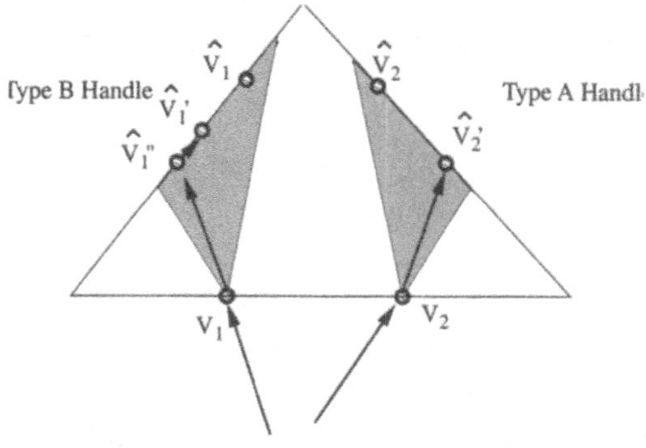

Figure 8: *Interval handles.*

Proof After an interval I_{v,e,e_i} is first created (and thus inserted into ILIST_{e,e_i}) when the optimal path from s to v is decided, the function *Propagate* is called which will add $(\hat{v}_{v,e,e_i}, v, e, D'(s,v)+W(\overline{v\hat{v}_{v,e,e_i}}))$ into the list QLIST. Here $\hat{v}_{v,e,e_i}$ is the closest point to v in I_{v,e,e_i} and thus is a handle point.

There are two events that could possibly make I_{v,e,e_i} lose handle points: (a) a quadruplet is removed from QLIST; (b) The boundary of I_{v,e,e_i} is adjusted (as in step [7] of function *InsertInterval*) so that the handle points are now in another interval I_{v',e,e_i}. In the following we prove that in any of the above two cases, at least one handle point will be generated unless I_{v,e,e_i} becomes inactive.

(a) Suppose I_{v,e,e_i} loses a Type A handle point $\hat{v}'$ after a quadruplet $(\hat{v}', v, e, D'(s,v)+W(\overline{v\hat{v}'}))$ is removed from QLIST. The fact that $\hat{v}' \in I_{v,e,e_i}$ excludes the possibility that another quadruplet $(\hat{v}', v', e, l')$ removed from QLIST previously. Therefore, step [1.b] of function *Propagate* is executed and quadruplet $(\hat{v}'_{new}, v, e, D'(s,v) + W(\overline{v\hat{v}'_{new}}))$ is inserted into QLIST if $W(\overline{v\hat{v}'_{new}}) \leq W(\overline{v\hat{v}'}) + W(\overline{\hat{v}'\hat{v}'_{new}})$; otherwise, quadruplet $(\hat{v}'_{new}, \hat{v}', e_1, D'(s,\hat{v}') + W(\overline{\hat{v}'\hat{v}'_{new}}))$ is inserted. Here $\hat{v}'_{new}$ is the (left or right) neighbor of $\hat{v}'$ on e_1 such that $W(\overline{v\hat{v}'_{new}}) > W(\overline{v\hat{v}'})$.

If $\hat{v}'_{new}$ is not in I_{v,e,e_i}, then $\hat{v}'$ must be an endpoint of I_{v,e,e_i}. Furthermore, it must be the point of I_{v,e,e_i} that is farthest from v. Therefore, all the points in I_{v,e,e_i} must have already been processed. Thus, I_{v,e,e_i} is not active anymore. If $\hat{v}'_{new}$ is in I_{v,e,e_i}, we claim that it must be a handle point of I_{v,e,e_i}. Observe that, for any unprocessed point $\hat{v} \in I_{v,e,e_i}$, $\hat{v} \neq \hat{v}'$, $W(\overline{v\hat{v}}) > W(\overline{v\hat{v}'})$ and thus $W(\overline{v\hat{v}}) \geq W(\overline{v\hat{v}'_{new}})$. Thus, in case quadruplet $(\hat{v}'_{new}, v, e, D'(s,v) + W(\overline{v\hat{v}'_{new}}))$ is inserted into QLIST, $\hat{v}'_{new}$ is a Type A handle point. In case quadruplet $(\hat{v}'_{new}, \hat{v}', e_1, D'(s,\hat{v}') + W(\overline{\hat{v}'\hat{v}'_{new}}))$ is inserted into QLIST, we have

$$\begin{aligned} & D'(s,\hat{v}') + W(\overline{\hat{v}'\hat{v}'_{new}}) \\ \leq\ & D'(s,v) + W(\overline{v\hat{v}'}) + W(\overline{\hat{v}'\hat{v}'_{new}}) \\ \leq\ & D'(s,v) + W(\overline{v\hat{v}'_{new}}) \\ \leq\ & D'(s,v) + W(\overline{v\hat{v}}). \end{aligned}$$

Hence, $\hat{v}'_{new}$ is a Type B handle point.

Similarly, we can prove that if I_{v,e,e_i} loses a Type B handle point $\hat{v}'$, another point $\hat{v}'_{new} \in I_{v,e,e_i}$ will become a new Type B handle point, unless I_{v,e,e_i} is not active anymore.

(b) Suppose interval I_{v,e,e_i} loses all of its handle points as a result of a boundary adjustment. However, as specified in step [7] of *InsertInterval*, a new quadruplet $(\hat{v}_{v,e,e_i}, v, e, D'(s,v) + W(\overline{v\hat{v}_{v,e,e_i}}))$ is inserted into the list QLIST. Here $\hat{v}_{v,e,e_i}$ is the closest point to v in the (adjusted) I_{v,e,e_i} and thus is a handle point.

Therefore, at any time, as long as I_{v,e,e_i} is active, there is at least one handle point in I_{v,e,e_i}. □

A Type A handle point ensures that a candidate optimal path to $\hat{v}$ will be generated and added into *QLIST* by face-crossing-propagation, while a Type B handle point ensures an edge-crawling-propagation to $\hat{v}$.

With the above lemmas, we are able to prove the following theorem:

Theorem 4 *Function FindOptimalDiscrete determines an optimal discrete path* $p'_{opt}(s,v)$ *for every point* $v \in V \cup V_s$.

Proof It suffices to prove that, at each time step [b] of function *FindOptimalDiscrete* is executed, the optimal path $p'_{opt}(s,v) = p'_{opt}(s,v_{prev}) + \overline{v_{prev}v}$ is determined.

Suppose for the sake of contradiction that p^* is an optimal discrete path from s to v with $W(p^*) < W(p'_{opt}(s,v))$. Let v^* be the discovered point on p^* that is closest to v. v^* can not be v as v is not discovered before quadruplet (v, v_{prev}, e, f) is removed from QLIST. Let v^*_{next} be the successor of v^* on p^*.

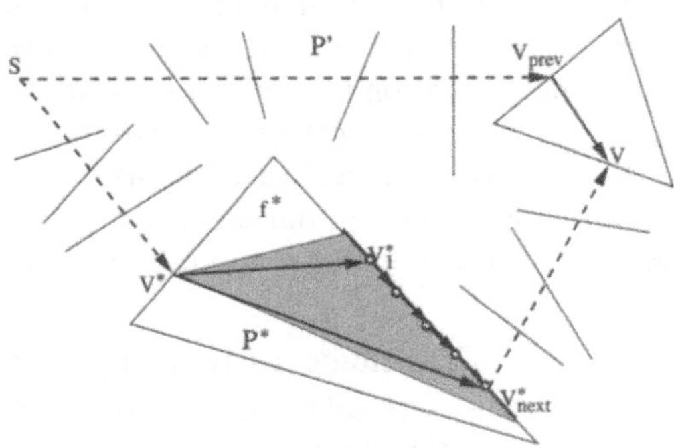

Figure 9: *Proof by contradiction.*

First assume $v^*v^*_{next}$ is a face-crossing line segment. Let f^* be the face that contains both v^* and v^*_{next}, let e^* be the edge of f^* that contains v^* and let e^*_{next} be the edge that contains v^*_{next}. Since p^* is an optimal path from s to v, $p^*[s, v^*_{next}]$ is an optimal path from s to v^*_{next}. Thus, v^*_{next} must be in $I_{v^*,e^*,e^*_{next}}$.

From Lemma 3 we know that there is at least one handle point in $I_{v^*,e^*,e^*_{next}}$. Let v' be a handle point of Type A. (The proof for the case when v' is a Type B handle point is similar.) Let $v_1^* = v', v_2^*, \cdots, v_k^* = v^*_{next}$ be the points of I_{v^*,e^*,f^*} between v' and v^*_{next}. Since $D'(s,v^*) + W(\overline{v^*v_i^*}) \leq D'(s,v^*) + W(\overline{v^*v^*_{next}}) < W(p^*) < D'(s,v_{prev}) + W(\overline{v_{prev}v})$, for $i = 1, 2, \cdots k$, (equality holds only when $v' = v_k^*$,) quadruplet $(v_i^*, v^*, e^*, D'(s,v^*) + W(\overline{v^*v_i^*}))$ should in turn be inserted into (and then removed from) QLIST before quadruplet $(v, v_{prev}, e, D'(s,v_{prev}) + W(\overline{v_{prev}v}))$ is removed, for $i = 1, 2, \cdots, k$. Therefore, $v_1^*, v_2^*, \cdots, v_k^*$ are already discovered before v is discovered. A contradiction to the assumption that v^*_{next} is not discovered before v is discovered.

The case when $v^*v^*_{next}$ is an edge-crawling line segment can be handled similarly. Therefore, there can not be such an optimal path p^* with $W(p^*) < p'_{opt}(s,v)$, and hence $p'_{opt}(s,v) = p'_{opt}(s,v_{prev}) + \overline{v_{prev}v}$ is an optimal path. □

4.2 Approximating the Optimal Path

As for each $v \in V \cup V_s$ our algorithm uses an optimal path $p'_{opt}(s,v)$ in the discretized path space P' to approximate the optimal path $p_{opt}(s,v)$ in the original path space P, the accuracy of our approximation is totally determined by the discretization on which our algorithm is based upon.

If the uniform discretization as in [9] is used, which puts $O(n^2)$ Steiner points on each edge, the absolute error of our approximation is bounded by a constant of LW_{max}. However, since we calculate the exact optimal path in the discretized space instead of using spanner to approximate it (as in [9]), our algorithm eliminates the extra relative error of ϵ introduced by the usage of spanner.

If we adopt the logarithmic discretization proposed by [1] so that $m = O(\log_\delta(L/r))$ Steiner points are placed on each edge, the approximation will have a relative error of ϵ. Although our algorithm is able to compute the exact optimal path in the discretized space, as compared to [1] where only an approximate optimal path is computed, it does not make a difference in this particular case as a relative error of the same magnitude has already been introduced in the first phase of the algorithm (discretization).

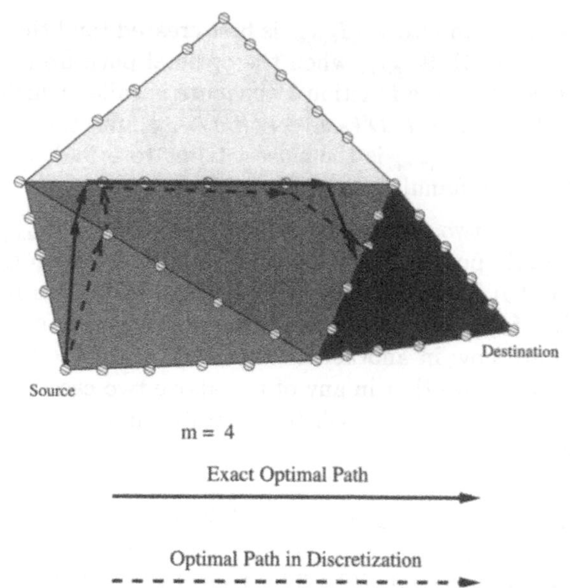

Figure 10: *Approximating optimal path by discretization.*

5 Complexity Analysis

In the following complexity analysis, we assume $f(e) = m$ for all e in the polygonal subdivision S for simplicity. (Thus, the number of Steiner points added to each edge is the same.)

To analyze the complexity of our algorithm, we need to examine the number of quadruplets inserted into QLIST. Observe that, when a quadruplet is removed from QLIST, there may be two occasions when new quadruplets are inserted into QLIST. One is occurred within function *Propagate* and the other is occurred within function *InsertInterval*.

Quadruplets added through ***InsertInterval*:**
For each $v \in V \cup V_s$, when the optimal discrete path $p'_{opt}(s,v)$ is determined, function $InsertInterval(v, e, e_i)$ is executed once for each face-sharing edge pair (e, e_i) such that $v \in e$. For each edge e, there are at most four face-sharing

edge e_i as e is adjacent to two triangular faces. In *InsertInterval*(v, e, e_i), a new interval is added into ILIST$_{e,e_i}$ and a quadruplet (candidate optimal path) is inserted into QLIST, as indicated by line [7] of function *InsertInterval*. Also, a new quadruplet is inserted into QLIST for each interval partially covered by the new interval I_{v,e,e_i}, as indicated by line [8] of *InsertInterval*. Since there are at most two intervals partially covered by the new interval, at most three quadruplets are inserted into QLIST for each I_{v,e,e_i}. Thus the total number of quadruplets added for point v is $12E(v)$, with $E(v)$ defined as the number of edges that contains v. For $v \in V_s$, $E(v) = 2$. For $v \in V$, $E(v)$ varies but we have $\sum_{v \in V} E(v) = O(n)$. Therefore, the total number of quadruplets added in executions of function *InsertInterval* is $\sum_{v \in V_s} 12E(v) + \sum_{v \in V} 6E(v) = O(|V_s|) + O(n) = O(nm)$.

Quadruplets added through *Propagate*: Now we calculate the number of new quadruplets inserted through executions of function *Propagate*$(v, v_{prev}, e_{prev}, l)$ for each $v \in V \cup V_s$. First we assume that v is a Steiner point, i.e., $v \in V_s$. Let e be the edge that contains v and let f_1, f_2 be the two faces adjacent to e. Assume $e_{i,1}, e_{i,2}$ are the other two edges of face f_i, for $i = 1, 2$. Observe that e_{prev} is either e, or one of $e_{i,j}$s. There are at most six quadruplets in the form of $(v, *, *, *)$ whose removal from QLIST can lead to an insertion of a new quadruplet into QLIST. The six possible quadruplets are $(v, v_{left}, e, D'(s, v_{left}) + W(\overline{v_{left}v}))$, $(v, v_{right}, e, D'(s, v_{left}) + W(\overline{v_{right}v}))$, and $(v, v_{prev}, e_{i,j}, D'(s, v_{prev}) + W(\overline{v_{prev}v}))$ for $i = 1, 2, j = 1, 2$. Here v_{left} and v_{right} are the left and right neighbors of v on e, respectively. This is guaranteed by line [1.a] and [2.a] of function *Propagate*. Similarly, for $v \in V$, the number of quadruplets in the form of $(v, *, *, *)$ whose removal can lead to an insertion of a new quadruplet is bounded by $O(E(v))$. Therefore, the total number of quadruplets added through *Propagate* is $\sum_{v \in V_s} O(1) + \sum_{v \in V} O(E(v)) = O(|V_s|) + O(n)$, which is again $O(nm)$.

From the above discussion, we can conclude that the total number of quadruplets inserted into QLIST is $O(nm)$. The time cost for maintaining QLIST is thus $O(nm \log(nm))$. Also, as there are $O(nm)$ different intervals $I_{v,e,f}$, and it takes $O(\log m)$ time to create such an interval, (observe that a binary search is needed in step [3] of *InsertInterval*,) the total time cost for creating these intervals is $O(nm \log m)$. Hence, our algorithm takes $O(nm \log(nm))$ time in total.

For the uniform discretization used in [9], we have $m = n^2$ and thus the time complexity is $O(n^3 \log n)$. For the logarithmic discretization used in [1], the time complexity is $O(mn \log(mn))$ which matches [1]'s result, here $m = O(\log_\delta(L/r))$.

6 Implementation Issues

We have implemented our algorithm in C++. We used several lists in our algorithm, each of which needs to be able to perform operations such as insertion, deletion, and find next (or previous) node efficiently. As mentioned previously, using AVL tree to implement the lists may be more efficient than using an ordinary binary tree. In particular, the discovered point list, PLIST$_e$, for each edge e, might be extremely skewed if implemented by an ordinary binary tree, as the discovering of points on an edge e might start at a point in the middle of e and spreads towards both ends of e. We implemented AVL tree using a C++ template so that it can handle different types of data structure, such as discovered point, candidate path and interval. Due to the complexity of AVL tree (i.e., the large constant in the analytical complexity), we expect to use large-scale data sets to demonstrate the efficiency of our algorithm.

As mentioned earlier, the real strength of our algorithm is the ability to compute efficiently an exact optimal path on a discretization of the original space. Both Lanthier's[9] and Aleksandrov's[1] are based on Dijkstra's algorithm, but they construct a relatively sparse graph in each triangular region and thus only an approximation can be achieved. To compute an exact optimal path by Dijkstra's algorithm, a complete graph has to be constructed in each region, which means connecting each pair of points by an edge. The total number of edges is thus $O(m^2n)$ and the total number of points is $O(mn)$. Here again m is the number of Steiner points placed on each edge of the original space and n is the number of triangular regions. Therefore the time complexity of Dijkstra' algorithm is $O(m^2n + mn\log(mn))$ to solve the exact optimal path problem, compared to $O(mn\log(mn))$ of our algorithm.

We implemented two versions of Dijkstra's algorithm, one based on binary tree and the other based on Fibonacci heap. The binary tree based Dijkstra's algorithm is the traditional one and has a time complexity of $O((|E| + |V|)\log|V|)$. The Fibonacci heap based one is provided by Fredman and Tarjan [7] and has an improved complexity of $O(|E| + |V|\log|V|)$. Some preliminary experiments with relatively small data inputs show that our algorithm performs better than both versions of Dijkstra's algorithm in finding exact optimal paths in discretized space.

For large scale experiments, we plan to use standard data sets for experiment inputs to make comparison. If standard data sets are not available, we will try to use computer-generated data sets.

7 Conclusion

In this paper we present a new approximation algorithm to solve the weighted shortest path problem. Compared to some of the previous work, our algorithm provide a more effective way of finding optimal paths in the discretized space (resulted from either uniform or non-uniform discretization). Our algorithm has the following three advantages over previous algorithms: (a) can compute exact solutions for a discrete case of this problem; (b) can be applied to any discretization; and (c) can be applied to a more generic class of problems.

We expect to find more applications for our algorithm.

Acknowledgements

This work was supported in part by Grants NSF/DARPA CCR-9725021, CCR-96-33567, NSF IRI-9619647, ARO contract DAAH-04-96-1-0448, and ONR contract N00014-99-1-0406.

References

[1] L. Aleksandrov, M. Lanthier, A. Maheshwari, and J.-R. Sack. An ϵ-approximation algorithm for weighted shortest paths on polyhedral surfaces. *Lecture Notes in Computer Science*, 1432:11–22, 1998.

[2] R. Alexander. Construction of optimal-path maps for homogeneous-cost-region path-planning problems. Ph.D. Thesis, Computer Science, U.S. Naval Postgraduate School, Monterey, CA, 1989.

[3] R. Alexander and N. Rowe. Geometrical principles for path planning by optimal-path-map construction for linear and polygonal homogeneous-region terrain. Technical report, Computer Science, U.S. Naval Postgraduate School, Monterey, CA, 1989.

[4] R. Alexander and N. Rowe. Path planning by optimal-path-map construction for homogeneous-cost two-dimensional regions. In *IEEE Int. Conf. Robotics and Automation*, 1990.

[5] J. Canny and J. Reif. New lower bound techniques for robot motion planning problems. In A. K. Chandra, editor, *Proceedings of the 28th Annual Symposium on Foundations of Computer Science*, pages 49–60, Los Angeles, CA, Oct. 1987. IEEE Computer Society Press.

[6] T. H. Cormen, C. E. Leiserson, and R. L. Rivest. *Introduction to Algorithms*. MIT Press, 1990.

[7] M. L. Fredman and R. E. Tarjan. Fibonacci heaps and their uses in improved network optimization algorithms. In *25th Annual Symposium on Foundations of Computer Science*, pages 338–346, Los Angeles, Ca., USA, Oct. 1984. IEEE Computer Society Press.

[8] M. Kindl, M. Shing, and N. Rowe. A stochastic approach to the weighted-region problem, II: Performance enhancement techniques and experimental results. Technical report, Computer Science, U.S. Naval Postgraduate School, Monterey, CA, 1991.

[9] M. Lanthier, A. Maheshwari, and J. Sack. Approximating weighted shortest paths on polyhedral surfaces. In *6th Annual Video Review of Computational Geometry, Proc. 13th ACM Symp. Computational Geometry*, pages 485–486. ACM Press, 4–6 June 1997.

[10] C. Mata and J. Mitchell. A new algorithm for computing shortest paths in weighted planar subdivisions. In *Proceedings of the 13th International Annual Symposium on Computational Geometry (SCG-97)*, pages 264–273, New York, June4–6 1997. ACM Press.

[11] J. S. B. Mitchell. Shortest paths and networks. In J. E. Goodman and J. O'Rourke, editors, *Handbook of Discrete and Computational Geometry*, chapter 24, pages 445–466. CRC Press LLC, Boca Raton, FL, 1997.

[12] J. S. B. Mitchell. Geometric shortest paths and network optimization. In J.-R. Sack and J. Urrutia, editors, *Handbook of Computational Geometry*. Elsevier Science Publishers B.V. North-Holland, Amsterdam, 1998.

[13] J. S. B. Mitchell and C. H. Papadimitriou. The weighted region problem: Finding shortest paths through a weighted planar subdivision. *Journal of the ACM*, 38(1):18–73, Jan. 1991.

[14] N. Papadakis and A. Perakis. Minimal time vessel routing in a time-dependent environment. *Transportation Science*, 23(4):266–276, 1989.

[15] N. Papadakis and A. Perakis. Deterministic minimal time vessel routing. *Operations Research*, 38(3):426–438, 1990.

[16] N. C. Rowe. Roads, rivers, and obstacles: optimal two-dimensional route planning around linear features for a mobile agent. *International Journal of Robotics Research, Vol 9(6), Dec 1990.*, pages 67–74, 1990.

[17] N. C. Rowe and R. F. Richbourg. An efficient snell's law method for optimal-path planning across multiple two dimensional, irregular, homogeneous-cost regions. *International Journal Of Robotics Research, Vol 9, No 6, Dec 1990. Page 48-66*, 1990.

Toward the Regulation and Composition of Cyclic Behaviors

E. Klavins *University of Michigan, Ann Arbor, MI*
D.E. Koditschek *University of Michigan, Ann Arbor, MI*
R. Ghrist *Georgia Institute of Technology, Atlanta, GA*

Many tasks in robotics and automation require a cyclic exchange of energy between a machine and its environment. Since most environments are "underactuated" — that is, there are more objects to be manipulated than actuated degrees of freedom with which to manipulate them — the exchange must be punctuated by intermittent repeated contacts. In this paper, we develop the appropriate theoretical setting for framing these problems and propose a general method for regulating coupled cyclic systems. We prove for the first time the local stability of a (slight variant on a) phase regulation strategy that we have been using with empirical success in the lab for more than a decade. We apply these methods to three examples: juggling two balls, two legged synchronized hopping and two legged running — considering for the first time the analogies between juggling and running formally.

1 Introduction

A robot is a source of programmable work. Robot programming problems arise when a mechanism designed with certain directly actuated degrees of freedom is required to exchange energy with its environment in such a fashion that some useful work — its "task," involving the imposition of specified forces over specified motions — is accomplished. Typically, the environment is not directly actuated and has its own preferred natural dynamics whose otherwise uninfluenced motions would be at least indifferent and, possibly, inimical to the task. The prior century's end has witnessed the practical triumph and emerging formal understanding of programs for information exchange and manipulation. There does not yet seem to exist a programming paradigm that can specify similarly goal oriented work exchange at any reasonable level of generality with any reasonable likelihood of successful implementation (much less, of formal verification).

For reasons we have discussed elsewhere at length [7, 21], we construe "task" to mean any behavior that can be encoded as the limit set of the closed loop dynamical system resulting from coupling the robot up to its environment. By "programming" is meant (at the very least) a means of composing from extant primitive task behaviors new, more specialized, or elaborate capabilities. A decade's research by the second author and colleagues has yielded the beginnings of a compositional methodology for tasks that can be encoded as point attractors [29, 28, 7, 14, 19]. In the present paper, we take the first steps toward a formal foundation for tasks that can be encoded as the next simplest class of steady state dynamical systems behavior — limit cycles.

1.1 Contributions of the Paper

In this paper we are able to prove for the first time the partial correctness of a (slight variant on) a phase regulation strategy that we have been using with empirical success in the lab for more than a decade [5, 31]. The object of study is a discrete dynamical control system on a co-dimension one subset of the tangent bundle over the two-torus, $\Sigma \subseteq \mathbb{T}^2$. The theoretical result is the proof of local asymptotic stability for a specified fixed point on this subset, $x^* \in \Sigma$. To illustrate the potential implications of the emerging theory, we introduce three example systems that move from juggling toward phase coordination strategies for legged machines.

We show in our first example that this discrete system corresponds to the parametrized family of return maps that arise when a one degree of freedom actuated piston strikes two otherwise unactuated one degree of freedom balls falling in gravity. This abstraction presumes sufficient control of the piston to track a suitably distorted image — a "mirror law" [6] — of the trajectory described by the two balls, along the lines of the empirical setup in [5, 30]. Under these assumptions, the torus bundle represents the *phase coordinate representation* of the two degree of freedom hybrid flow formed when the paddle repeatedly but intermittently strikes one or the other ball. The functional freedom afforded by the choice of "mirror law" yields the parametrization of the available closed loop return maps whose domain, Σ, is now interpreted as the phase condition at which an impact is made. The preliminary analysis presented here develops a set of sufficient conditions on the mirror law that guarantees the local asymptotic stability of a limit cycle corresponding to the the desired two-juggle. We suspect, but have not yet proven, that the desired limit cycle is essentially globally asymptotically stable.

We have not yet formalized the notion of behavioral composition (as we have begun to do for behaviors encoded as point attractors [7, 19]) but it represents a strong unifying theme throughout the paper. The two-juggle mirror law may be seen as a kind of informal "interleaving" of two one-juggle functions. However, because we desire a more general compositional notion not tied to the (effective but very costly in sensory effort) mirror constructions, we next apply our phase regulation method to "interleave" a very different style of individual controller. Specifically, with the appropriate notion of phase coordinates described above, we are now able to consider for the first time the analogies between juggling and running.

Our second example concerns two vertical Raibert hoppers [27], each of whose leg springs has an adjustable stiffness. Now, although both legs are partially actuated, the contact with ground is no longer instantaneous and we abandon mirror laws in favor of a Raibert style energy management strategy coordinated over repeated intermittent stance modes so as to nudge the total vertical energy toward that value which encodes the desired behavior. When the legs are decoupled from each other, arguments nearly identical to those we have developed in past work [20] yield essential global asymptotic stability of the two independent vertical "gaits". Note that the reference energy is achieved asymptotically, rather than by a deadbeat one step correction as was assumed in the first example. Nevertheless, applying the identical phase regulation scheme yields a closed loop system that exhibits in simulation the same striking coordinated behavior as we have proven to hold true (at least locally) in the case of the two-juggle. We suspect, but have not yet proven, that the coordinated bipedal vertical gait is once again (essentially globally) asymptotically stable.

The final example represents our first and still rather tentative efforts to interleave constituent cyclic behaviors that arise in systems possessed of more than one degree of freedom. Raibert's running machines combined in parallel, for each leg, three independent and decoupled controllers that operated in three very strongly coupled degrees of freedom, with excellent empirical success. Moreover, he devised a notion of "virtual leg" that successfully coordinated the relative phases of the "vertical" components of the physical legs without damaging their other degrees of freedom. In this paper, we are content simply to extend our emerging notion of "phase" to a pair of two-degree of freedom pogo sticks (the "Spring Loaded Inverted Pendulum" [35]) and assume that their individual phase regulation mechanisms are deadbeat. Once again, simulations suggest that this is the appropriate generalization, but we remain cautious regarding the larger implications pending more realistic constituent models.

1.2 Motivation and Relation to Existing Literature

Coupled oscillators have long been used to model complex physical and biological settings wherein phase regulation of cyclic behavior is paramount [15]. The biological reality of neural central pattern generators (CPGs) — oscillatory signals that arise spontaneously from appropriate intercommunication between neurons — seems by now to have been conclusively demonstrated in organisms ranging from insects [26, 12] to lampreys [9]. Mathematical models proposed to explain the manner in which families of coupled dynamical systems can stimulate a sustained oscillation and stably entrain a desired phase relationship have become progressively more biologically detailed [8, 13, 16]. But the work presented here has relatively little overlap with that literature. While we are intrigued by the capabilities of purely "clock driven" systems [36, 32], it seems clear that no significant level of autonomy can be developed in the absence of perceptual feedback. The present investigation cleaves to the opposite (i.e., perceptually driven) end of the sensory spectrum in adopting the device of a "mirror law" [6] with its commitment to profligate sensory dependence [30]. In this sense, the present work bears a closer relationship to the biological literature concerned with reflex modulated phase regulation [11].

Many tasks in robotics and automation entail a cyclic exchange of energy between a machine and its environment. This is evidently the case for legged locomotion systems as well as for many less obvious examples wherein a mechanism repeatedly changes its local "shape" so as to effect some global "progress" [24]. When viewed from an appropriate geometric perspective, the recourse to repetitive self-motion may be interpreted as a means of "rectification" — exercising indirectly the unactuated degrees of freedom through the influence of the actuated degrees of freedom arising from an interaction between symmetries and constraints [2]. Because our notion of a task is so completely bound up with a closed loop dynamical interaction between the robot and its environment, this invaluable geometric control perspective provides no solution but merely a complete account of the (open loop) setting within which our search for stabilizing feedback controllers can begin. Since the dynamics in question are inevitably nonlinear, the relation between open loop controllability properties and feedback stabilizability properties is far from clear.

In our understanding, the most relevant connection to date remains the nearly two decade old observation of Brockett [4], precluding smooth feedback stabilization to a point in the face of conditions known [3] to characterize the nonholonomic constraints that appear in the present underactuated setting [22]. At the very least, this fact necessitates the appearance of hybrid controllers — feedback laws whose resulting closed loops make non-smooth transitions triggered by state — in the case of tasks encoded as point attractors [22]. In the present situation, when tasks are encoded as limit cycles, we are aware of no similar necessary conditions. Nevertheless, the feedback solutions we construct have a strong hybrid character. Since the nonholonomic constraints in our setting arise from the "underactuated" nature of the problem [21], it seems intuitively clear that the robot's work on the components of its environment must be punctuated by intermittent repeated contacts.

One last influence on the present work that bears some comment concerns the possibility of composition. Since good regulation mechanisms are hard to find, there is considerable interest in developing techniques for composing existing behaviors to get new ones. However, as the degrees of freedom increase, the burdens of high dimensionality make centralized control schemes prohibitively expensive. There is considerable interest in developing cyclic behaviors that are as decoupled as possible, promoting decentralized regulation. Our present model for pursuing this desideratum is provided by our initial work on concurrent composition of point attractors [19]. Since our reference flows have gradient-like cross-section maps, we harbor some hope that the connection may be forthcoming.

2 Preliminary Discussion

We start in Section 2.1 by defining phase coordinates that enable us to re-cast physical equations involving potential and kinetic energy as geometric equations relating progress around a circle and its velocity. In the examples at hand, the physical control variables are used to adjust the energy of the unactuated degrees of freedom upon their intermittent contacts with the actuated components. In phase coordinates, the phase velocity of each constituent subsystem is subject to control at each impact, and effects a corresponding resetting of the various relative phases.

Having arrived at a convenient model space, the torus, we next examine in Section 2.2 the notion of a "reference field" — a family of limit cycle generating vector fields on the k-torus whose return maps on the $(k-1)$-torus admit as a Lyapunov function a "Navigation Function" [23] down to the fixed point. The topologically unavoidable repellors can be identified with the application as phase pairs that are to be avoided (e.g., when both balls come down at exactly the same of time). Although our ultimate constructions appear as maps of an appropriate cross section (so the topological constraints appear to lose their force) these toral maps are classical objects and yield very convenient and workable targets.

2.1 Controlling Phase

Let $f^t : \mathbb{R} \times X \to X$ be a flow on X. We are concerned with flows that are cyclic in the sense that a global cross section can be found. Formally, a global cross section Σ is a connected submanifold of X which transversally intersects every flowline. For any point $x \in X$, define the **time to return** of x to be:

$$t^+(x) = min\{t > 0 \mid f^t(x) \in \Sigma\} \qquad (1)$$

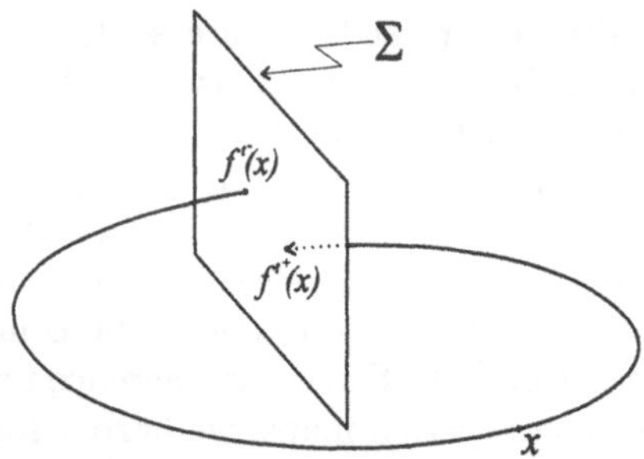

Figure 1: *The relationship of* $t^-(x)$ *and* $t^+(x)$ *to* x.

and define the **time since return** of x to be

$$t^-(x) = min\{t \geq 0 \mid f^{-t}(x) \in \Sigma\} . \tag{2}$$

The first return map, $P : \Sigma \to \Sigma$, is the discrete, real valued map given by $P(x) = f^{t^+}(x)$.

Let $s(x) = t^+(x) + t^-(x)$. s is the time it takes the system starting at the point $f^{t^-}(x) \in \Sigma$ to reach Σ again. Now, define the **phase** of a point x by:

$$\phi(x) = \frac{t^-(x)}{s(x)} \tag{3}$$

Notice that the rate of change of phase, $\dot{\phi}$, is equal to $1/s$. The relationship of these functions to Σ is shown in Figure 1. Therefore, $\dot{\phi}$ is constant or piecewise constant, changing only when the state passes through Σ.

In Section 3, we give a one-dimensional example (Juggling) where $h : X \to Y$, defined by $h(x, \dot{x}) = (\phi, \dot{\phi})$, is actually a change of coordinates where $Y = S^1 \times \mathbb{R}^+$. We use the section $\Sigma \in X$ defined by $x = 0$ which corresponds to the set of states where the robot may contact (and thereby actuate) the system. The image of this section $h(\Sigma)$ will be given by the set $\mathcal{C} = \{(0, \dot{\phi}) \mid \dot{\phi} \in \mathbb{R}^+\}$. Because we consider intermittent control situations, it is only in this section that $\dot{\phi}$ may be altered by the control input u. That is, we change $\dot{\phi}$ according to a control policy u to get the return map $P' : \mathcal{C} \to \mathcal{C}$ given by $P'(0, \dot{\phi}) = (0, u(\dot{\phi}))$[1]. We design the controller so that there is a unique and stable fixed point at some desired phase velocity $\dot{\phi}^* = \omega$.

[1] In Section 3.1, deadbeat control of the phase velocity is possible. In Section 3.2, the control of phase velocity is asymptotically stable. Our analysis in Section 4 depends on the former. We believe a similar treatment will eventually apply to the latter.

Of course we really want to control the system so that the return map P has a stable fixed point at some x^*. Whether or not $h^{-1}(0, \omega) = (0, \dot{x}^*)$ depends on the dimension of Σ. If $dim\ \Sigma = 1$, as it will be in the examples we supply, then the preimage of ω is indeed a point.

The main contribution of this paper concerns the composition or interleaving of two such systems. That is, we suppose that we have the system $(x_1, \dot{x}_1, x_2, \dot{x}_2) \in X^2$ with corresponding phase coordinates $(\phi_1, \dot{\phi}_1, \phi_2, \dot{\phi}_2) \in Y^2$. As before, system i may only be actuated when $\phi_i = 0$. In the examples we will consider, we suppose that the systems cannot be actuated simultaneously. Thus the set of states where $\phi_1 = \phi_2 = 0$ should be repelling. We will design a controller such that the attracting limit cycle is given by:

$$\begin{aligned} \mathcal{G} = \{(\phi_1, \dot{\phi}_1, \phi_2, \dot{\phi}_2) \quad | \quad & \phi_1 = \phi_2 + \frac{1}{2} \ (\mathrm{mod}\ 1) \\ \wedge \quad & \dot{\phi}_1 = \dot{\phi}_2 = \omega\} . \end{aligned} \tag{4}$$

The constraint $\phi_1 = \phi_2 + \frac{1}{2}$ (mod 1) encodes our desire to have the pair of phases as far from the situation $\phi_1 = \phi_2 = 0$ as possible. In fact, we will consider the more general case wherein the phase velocities are controlled to $(\kappa A, \kappa B)$ for some $A, B \in \mathbb{Z}$ and scaling factor κ.

To analyze and control such a system, we restrict our attention to the sections $\Sigma_1 \subseteq Y^2$ and $\Sigma_2 \subseteq Y^2$ defined by $\phi_1 = 0$ and $\phi_2 = 0$ respectively. Suppose that the flow alternates between the two sections. Let $G^t = H \circ F^t \circ H^{-1}$ be the flow in Y^2 conjugate to the flow in X^2 where $F = (f, f)$ and $H = (h, h)$ and $\tau_i(w) = \min\{\tau > 0 \mid H \circ F^\tau \circ H^{-1}(w) \in \Sigma_{3-i}\}$. Start with a point $w \in \Sigma_1$. Let $w' = G^{\tau_1}(w)$ and $w'' = G^{\tau_2}(w')$. We have $w' \in \Sigma_2$ and $w'' \in \Sigma_1$, so we have defined the return map on Σ_1. Now since G is parameterized by the control inputs u_1 and u_2 we get:

$$\begin{aligned} w = (0, \dot{\phi}_1, \phi_2, \dot{\phi}_2) \quad & \mapsto \quad w' = (\phi_1', u_1, 0, \dot{\phi}_2) \\ & \mapsto \quad w'' = (0, u_1, \phi_2', u_2) . \end{aligned}$$

Thus, the phase velocity updates $u_1(w)$ and $u_2(w')$ must be found so that (4) is achieved. We do this with two examples in Section 3 and prove the stability of our method in Section 4.

Notice that a single phase describes a circle S^1 and two phases describe a torus $\mathbb{T}^2 = S^1 \times S^1$. In the next section, we define a "reference" vector field on the k-dimensional $\mathbb{T}^k$ which encodes the ideal behavior of the system as though it were fully actuated. Then, we show how to use the field to generate velocity updates as above.

2.2 Construction of a Reference Flow on $\mathbb{T}^k$

The problem of composing dynamical systems with point-goal attractors is relatively straightforward, due in no small part to the convenient topological fact that the product of a zero-dimensional set (a point goal) with a zero-dimensional set is again a zero-dimensional set: point-goals are well-behaved with respect to Cartesian products. This is not so for the case of systems with an attracting periodic orbit. The Cartesian product of k such continuous systems gives rise to a flow with an attracting k-torus $\mathbb{T}^k \triangleq S^1 \times \cdots \times S^1$. The desired behavior for a flow on this set is again an attracting periodic orbit; however, such mode locking can occur only if the oscillators are coupled. More unfortunately, such dynamics arise only through a relatively careful tuning of the individual systems and their mutual couplings. Baesens et al. [1] carefully explore the intricacies of this problem, illustrating the prevalence of complexity in both the dynamics and the bifurcation structures of flows on the attracting $\mathbb{T}^k$ in the (ostensibly simple) case $k = 3$.

An important measure of complexity for the dynamics of a flow on a torus $\mathbb{T}^k$ is the set of *winding vectors*. Choose a lift $\tilde{\phi}_t$ of the flow ϕ_t on $\mathbb{T}^k$ to the universal cover $\mathbb{R}^k$ of $\mathbb{T}^k$. Then, consider for each $x \in \mathbb{T}^k$ with lift $\tilde{x} \in \mathbb{R}^k$ the vector $\tilde{\phi}_t(\tilde{x})$ normalized to unit length: denote this by $w_t(\tilde{x})$. This vector lies in the unit $(k-1)$-sphere $S^{k-1} \subset \mathbb{R}^k$ of directions in $\mathbb{R}^k$. Define $w(x) \subset S^{k-1}$ to be the set limit points of $w_t(\tilde{x})$ as $t \to \infty$. This set (independent of the lift $\tilde{x}$ in the case of a nonsingular flow) defines the *winding vectors* of x. The union of $w(x)$ over all $x \in \mathbb{T}^k$ comprises the *winding set* of the flow. Winding vectors/sets are the continuous analogues of the rotation vectors/sets defined for torus homeomorphisms[2]: cf. the discussions in [1, 25] in the context of coupled oscillators and [34] for a topological generalization to arbitrary spaces.

The systems we consider have specific constraints on the winding vectors. In order to have a single mode-locked attracting periodic orbit, the winding set must consist of a unique winding vector. In Section 3.1 we present a system consisting of a piston which must vertically juggle two balls so that the first bounces A times for every B times the second bounces (See Figure 4), where A and B are integers: the winding vector is thus (A, B) (rescaled to unit length). The generalization of this situation to n juggled items requires a winding vector of integers $(A_1, A_2, \ldots, A_n)$.

Our goal is to couple systems with unique attractors satisfying the above restrictions in such a manner that the product system remains in the same dynamical class: a single attractor with appropriate winding vector. In addition, the existence of unstable invariant sets (in general forced by topological considerations) is desirable for setting up "walls" of repulsion in the configuration space. For example, in juggling it may be desirable for the configuration wherein both balls hit the paddle simultaneously to be a repellor. For both attractors and repellors, the freedom to "tune" these invariant sets geometrically is a necessity. We thus turn to a brief exposition of two appropriate classes of reference flows on the k-torus $\mathbb{T}^k$ which will serve as skeletons for the goal dynamics of the control schemes to be constructed.

The flows we consider on $\mathbb{T}^k$ will all have global cross sections Σ homeomorphic to $\mathbb{T}^{k-1}$. To obtain a unique attracting periodic orbit for the flow, we specify the appropriate dynamics on the cross-section and accordingly suspend to a flow: the flow is then determined by the dynamics of the return map and the desired winding vector.

Consider the diffeomorphism which is the time-1 map of the gradient field $-\nabla V$, where:

$$V : \Sigma \to \mathbb{R} \qquad (\theta_1, \theta_2, \ldots, \theta_{k-1}) \mapsto \kappa \sum_{i=1}^{k-1} \cos(2\pi\theta_i). \tag{5}$$

Here $\Sigma \cong S^1 \times \cdots \times S^1$ is parameterized by $k-1$ angles $\theta_i \in [0, 1]$ and $\kappa > 0$ is an amplitude which controls the rate of attraction. The dynamics of this return map decouples into the cross-product of the circle maps which have the "north pole" ($\theta_i = 0$ for all i) as a repellor and the "south pole" ($\theta_i = 1/2$ for all i) as an attractor. It thus follows that (5) has exactly $(k-1)$-choose-j hyperbolic fixed points whose unstable manifold is of dimension j. There is thus a unique attracting fixed point, and V defines a navigation function [29] for the

[2] More specifically, for those homeomorphisms which are continuously deformable to the identity map.

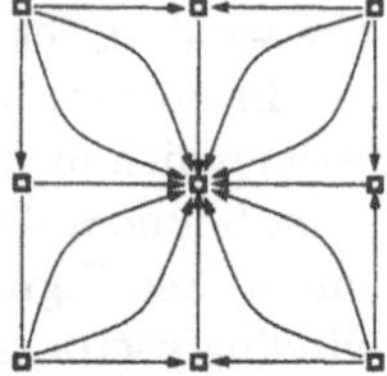
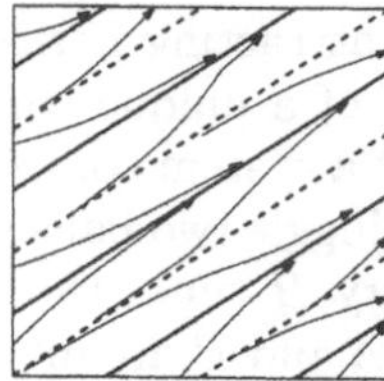

Figure 2: *(left) The ideal reference dynamics on a $\mathbb{T}^2$ cross-section to a flow on $\mathbb{T}^3$ having a single attracting fixed point. Here, the 2-torus is represented as a square with opposite sides identified. (right) A reference flow on $\mathbb{T}^2$ with winding vector* $(3,2)$. *The repelling orbit passes through the origin. The appropriate cross-section here is the circle along the "diagonal" of the square.*

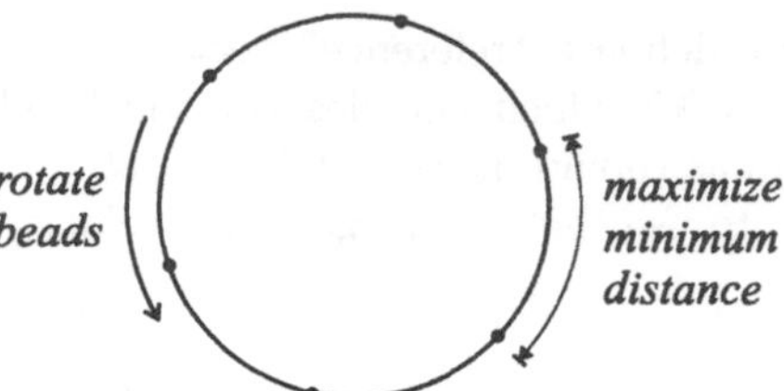

Figure 3: *Embedding distinct phases as the "beads on a circle" problem. The beads must rotate around the circle while maximally avoiding their neighbors.*

Poincaré return map of the flow. See Figure 2(left) for an illustration of the dynamics in the case $k = 3$.

The existence, quantity, and placement of the unstable invariant manifolds in the dynamics of (5) are governed by Morse-theoretic considerations (see [29] for applications of Morse theory to the design of navigation functions). Of particular interest is the forced existence of repelling unstable invariant manifolds of all dimensions. This is extremely relevant to the control problem in that the "obstacles" in the configuration space (where the "paddle" is forced to contact several distinct elements simultaneously) can be of variable dimension. The prevalence of unstable manifolds in the dynamics of (5) is a valuable resource when one wants to "tune" the dynamics on the configuration space.

Consider the problem of designing a vector field on $\mathbb{T}^k$ such that all orbits possess a unique winding vector w which points in the direction $(A_1, \ldots, A_k) \in \mathbb{Z}^k$, assuming that the A_i are all relatively prime. The cross-sectional dynamics of this system will be conjugate to (5) for an appropriate cross-section: Namely, the cross-section which is the orthogonal complement to τ, where τ is an integer vector satisfying $\tau \cdot (A_1, \ldots, A_k) = 1$, see [1, App. A]. To obtain a reference flow with the desired winding vector, we may suspend (5) to a flow on $\mathbb{T}^k$ and then change coordinates so that the attracting orbit in the new coordinates has slope $(A_1, \ldots, A_k)$. Supposing we start with the slope $(0, 0, ..., 1)$, we desire a linear map M on $\mathbb{R}^k$ (the covering space for $\mathbb{T}^k$) such that $M \cdot (0, 0, ..., 1) = (A_1, \ldots, A_k)$ and so that M, when projected onto the torus is a change of coordinates. This amounts to requiring that $M \in SL(n, \mathbb{Z})$ with its last column given by $(A_1, \ldots, A_k)$.

This construction can be tuned so that the attracting orbit does not pass through the pairwise "obstacle" where two phases are identical. However, in the resulting system the obstacles become dynamically neutral — neither attracting nor repelling. In applications where these obstacles are not physically meaningful, we may use this construction. Otherwise we must design a reference field wherein these obstacles are dynamically repelling.

One manner of generalizing (5) to a flow on $\mathbb{T}^k$ which avoids determining a complicated coordinate change and which may be suitable in applications where the obstacles are important is as follows. We imagine the phases of the system as coming from k distinct point on a circle. Each point must rotate around the circle with some velocity and the distance between any two consecutive points must be maximized, as in Figure 3. The potential function:

$$\begin{aligned} V &: \mathbb{T}^k \to \mathbb{R} \\ (\phi_1, \ldots, \phi_k) &\mapsto \kappa \sum_{i<j} \cos\left(2\pi[A_i\phi_j - A_j\phi_i]\right) \end{aligned} \quad (6)$$

is used to accomplish the latter task. Here the coordinates $\phi_i \in S^1 = [0,1]/\{0 \sim 1\}$ are angular coordinates on T^k. The function V attains a global maximum along the straight line (mod 1) through the origin tangent to the winding vector $(A_1, \ldots, A_k)$. The function attains its global minimum along a shifted parallel line (mod 1). The addition of a global drift term in the winding direction gives a realization of the desired flow.

$$\dot{\phi}_i = \kappa_1 A_i - (\nabla V)_i \quad (7)$$

See Figure 2 (right) for an illustration of the two-dimensional case which is used in the examples in this

With this field the obstacles are indeed repelling, but there are several attracting orbits, one for each of the $(k-1)!$ arrangements of k beads on a circular wire. For applications such as juggling any of these orbits will represent a viable juggling behavior. For other applications, such as controlling the gait of a legged machine, further tuning to achieve the correct order will be required. In the case of $k = 2$ that we specifically consider in this paper, the attracting orbit is unique and maximally bounded away from the origin.

3 From Juggling to Running

In this section we examine in detail the task of regulating the phases of just two cyclic processes. We will consider intermittent contact systems. For example, a ball being bounced on a controllable paddle is an intermittent contact system where the phase velocity corresponds to the energy of the ball between contacts. Another type of intermittent contact system is one that has a *stance mode*, that is, $\dot{\phi}_i$ is controllable only when $\phi_i \in [a, b] \subseteq [0, 1]$ for some a and b. A hopping robot, for example, is in its stance mode when it is touching the ground. Only in stance mode may the controller change the energy of the robot. We will not consider stance systems in general but instead show how to consider certain models of hopping robots as though their phase velocities were determined by their phase velocities at the single point $\phi_i = a = 0$, thereby reducing the problem to one apparently involving instantaneous contact.

We are concerned with regulating the two systems so that (1) the rate of change of each phase is some desired value (i.e. the first system oscillates A times for every B times the second does) and (2) the phases are maximally separated. That is, we require that:

$$\begin{pmatrix} \dot{\phi}_1 \\ \dot{\phi}_2 \end{pmatrix} = \kappa_1 \begin{pmatrix} A \\ B \end{pmatrix} \text{ and } A\phi_2 = B\phi_1 + \frac{1}{2} \text{ (mod 1)} \tag{8}$$

where κ_1 scales the phase velocities A and B to values reasonable for the system.

We construct a reference vector field on $\mathbb{T}^2$ with this circle as a limit cycle such that $(\dot{\phi}_1, \dot{\phi}_2) = \kappa_1(A, B)$ along the cycle as described in Section 2. This field encodes the ideal behavior of the system as though it were fully actuated. The potential function is:

$$V(\phi_1, \phi_2) = \cos(2\pi[A\phi_2 - B\phi_1]) \, . \tag{9}$$

and the field is:

$$\mathfrak{R}(\phi_1, \phi_2)^T = \kappa_1 \begin{pmatrix} A \\ B \end{pmatrix} - \kappa_2 \nabla V(\phi_1, \phi_2), \tag{10}$$

the two dimensional instantiation of (6) and (7). Here κ_2 is an adjustable gain which controls the rate of convergence to the limit cycle. The lines $A\phi_2 = B\phi_1$ and $A\phi_2 = B\phi_1 + \frac{1}{2}$ are equilibrium orbits. The first is unstable, the second is stable.

3.1 Juggling

Consider the system wherein a paddle with position p controls a single ball with position b to bounce, repeatedly, to a prespecified apex. We suppose the paddle always strikes the ball at $p = b = 0$ and instantaneously changes its velocity according to the rule:

$$\dot{b}_{new} = -\alpha \dot{b} + (1 + \alpha)\dot{p} \tag{11}$$

where α is the coefficient of restitution in a simple ball and paddle collision model. We suppose that the velocity of the paddle is unchanged by collisions. Evidently, a paddle velocity of $\dot{p} = (\alpha - 1)/(\alpha + 1)\dot{b}$ will set $\dot{b}_{new} = -\dot{b}$. Now define $\eta = \frac{1}{2}\dot{b}^2 + bg$ to be the total energy of the system. By conservation of energy, $\dot{\eta} = 0$ between collisions. Set η^* to be the desired energy (corresponding to a desired apex). Define a reference trajectory for the paddle to follow by $\mu = cb$ where:

$$c = \frac{\alpha - 1}{\alpha + 1} + \kappa(\eta - \eta^*)$$

is constant between collisions. μ is called a *mirror law* because it defines a distorted "mirror" of the ball's trajectory. As the ball goes up the paddle goes down and *vice versa.* κ is a gain that adjusts how aggressively the controller minimizes the energy error. The analysis of this system in [5] proceeds, roughly, as follows. A "return map," mapping the energy just before a collision to the energy just before the next collision, is derived. It is shown that the discrete, real valued dynamic system that results is globally asymptotically stable by showing that the map is unimodal [10] with parameter κ adjusting the period of the map.

3.1.1 Phase Regulation of Two Balls

To control *two* balls to bounce on the paddle so that one hits exactly when the other is at its highest point as in Figure 4(a), we will use the reference field (10).

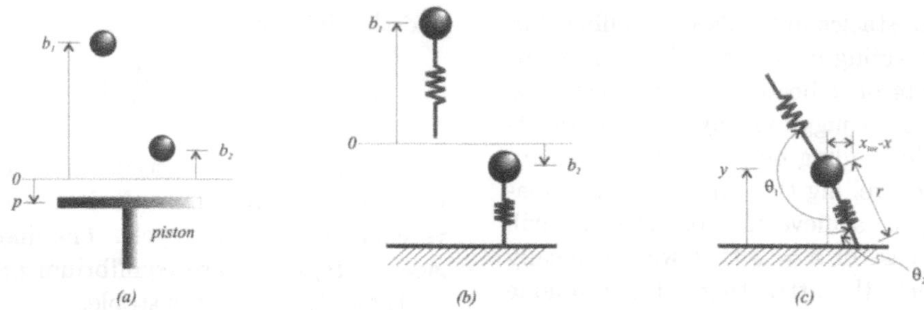

Figure 4: *Models considered in this paper: (a) two balls juggled on a piston; (b) two Raibert style hoppers, hopping out of phase; (c) two legged spring loaded inverted pendulum (SLIP) model of a biped.*

This represents a point of departure from earlier work on juggling two balls where a *phase error* term was combined with two mirror laws somewhat informally.

We construct a mirror law for each system separately and then combine the laws into a single mirror law using an "attention function." First define the phase of a ball according to the discussion in Section 2.1. Suppose a ball rebounds from a collision with the paddle with velocity $\dot{b}_0$. By integrating the dynamics $\ddot{b} = -g$ and noting that collisions occur when $b = 0$, we obtain the time since the last impact and the time between impacts, a computationally effective instance of (1) and (2), as:

$$t^- = \frac{\dot{b}_0 - \dot{b}}{g} \text{ and } s = t^- + t^+ = \frac{2\dot{b}_0}{g} \qquad (12)$$

respectively. The change of coordinates $h : (\mathbb{R}^+ \times \mathbb{R}) - (0,0) \to S^1 \times \mathbb{R}^+$ from ball coordinates to phase coordinates is given by $h(b, \dot{b}) = (\phi, \dot{\phi})$ where, following the recipe (3), we take:

$$\phi = \frac{t^-}{s} = \frac{\dot{b}_0 - \dot{b}}{2\dot{b}_0} \text{ and } \dot{\phi} = \frac{g}{2\dot{b}_0}. \qquad (13)$$

In this manner, for a two ball system with ball positions b_1 and b_2, we obtain two phases ϕ_1 and ϕ_2. Notice that $\phi_i \in [0,1]$ and that $\dot{\phi}_i$ is constant between collisions between ball i and the paddle as established in Section 2.1. The velocity $\dot{\phi}_i$ is reset instantaneously upon collisions, corresponding to the update rule (11).

In the rest of this section we will take advantage of the fact that the flow $G^t = H \circ F^t \circ H^{-1}$, described in Section 2 and instantiated here, has the very simple form $(y_1, \dot{y}_1, y_2, \dot{y}_2) \mapsto (y_1 + \dot{y}_1 t, \dot{y}_1, y_2 + \dot{y}_2 t, \dot{y}_2)$ between collisions.

We define reference trajectories μ_1 and μ_2 based on the mirror law idea. Given any pair (ϕ_1, ϕ_2) we define a lookahead function $C_1 : [0,1]^2 \to [0,1]^2$ for ball one which gives the phase of ball two at the next ball one collision. Thus:

$$C_1(\phi_1, \phi_2) = \frac{\dot{\phi}_2}{\dot{\phi}_1}(1 - \phi_1) + \phi_2.$$

We desire that after this collision, $\dot{\phi}_{1,new} = u_1 = \mathcal{R}_1 \circ C_1(\phi_1, \phi_2)$ (i.e. the control input u_i follows $\mathcal{R}$). Since $\dot{\phi}_{1,new} = -g/\sqrt{2\eta_{1,new}}$ and $\eta_{1,new} = \frac{1}{2}\dot{b}_{1,new}^2$ (since $b_1 = 0$ at the collision), we have, using (11),

$$\dot{\phi}_{1,new} = \frac{-g}{-\alpha\dot{b}_1 + (1+\alpha)c_1\dot{b}_1} = \mathcal{R}_1 \circ C_1(\phi_1, \phi_2) \quad (14)$$

where c_1 is the coefficient in the mirror law trajectory $\mu_1 = c_1 b_1$. Solving for c_1 and using the fact that at $b_1 = 0$, $\dot{b}_1 = \sqrt{2\eta_1}$, gives:

$$c_1 = \frac{1}{(1+\alpha)\sqrt{2\eta_1}}\left[\alpha\sqrt{2\eta_1} - \frac{g}{\mathcal{R}_1 \circ C_1(\phi_1, \phi_2)}\right]. \quad (15)$$

A similar expression for c_2 can be obtained in terms of $\mathcal{R}_2 \circ C_2$. This gives us a mirror law for each ball. However, we have only one paddle so we need an attention function $s : [0,1]^2 \to [0,1]$ and a new reference trajectory composed of μ_1 and μ_2:

$$\mu = s\mu_1 + (1-s)\mu_2.$$

Such a function is fairly easy to define for specific instances of A and B. A more complete treatment of attention functions can be found in [18].

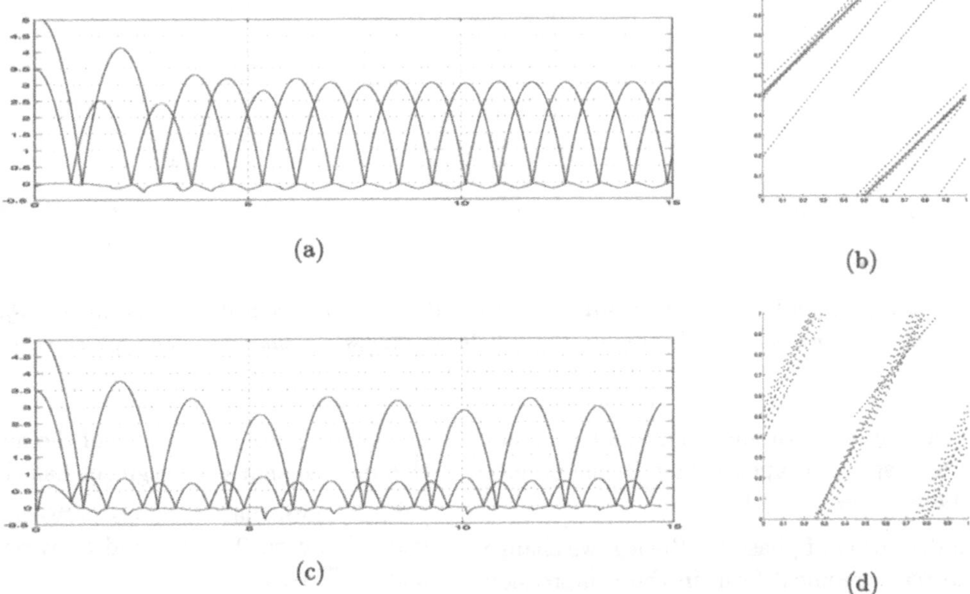

Figure 5: *(a) The positions of the two balls and the paddle vs. time for $A:B = 1:1$ starting from a randomly chosen initial condition. (b) The phase plot $\phi_1(t)$ vs. $\phi_2(t)$ for the same run. Note the limit cycle where $\phi_1 = \phi_2 + \frac{1}{2}$. (c) and (d) show the same information for $A:B = 1:2$.*

We have simulated this system and have found it to work as expected. Figures 5(a) and (b) show a run of the system with $A : B = 1 : 1$ and Figures 5(c) and (d) show a run of the system with $A : B = 2 : 1$. In both cases the paddle regulates the phases to very near the limit cycle within two or three hits of the balls. After presenting two more examples of phase regulated systems, we give an analysis of this controller for the 1:1 case in Section 4.

3.2 Synchronized Hoppers

In this section we examine a model of a bouncing point mass reminiscent of Raibert's hopper [27]. A single, vertical hopping leg is modeled by a mass $m = 1$ attached to a massless spring leg. The hybrid dynamics has three discrete modes: flight (the leg is not touching the ground), compression (the leg has landed and is compressing), and decompression (the mass has reached its lowest point and is being pushed upward). These latter modes each have the dynamics of a linear, damped spring. Flight mode is entered again once the leg has reached its full extension. The equations of motion, borrowed and altered somewhat from [20], are as follows:

$$\ddot{x} = \begin{cases} -g & \text{if } x > 0 \quad \text{flight} \\ -\omega^2(1+\beta^2)x - 2\omega\beta\dot{x} & \text{if } x < 0 \wedge \dot{x} < 0 \quad \text{compression} \\ -\omega_2^2(1+\beta_2^2)x - 2\omega_2\beta_2\dot{x} & \text{if } x < 0 \wedge \dot{x} > 0 \quad \text{decompression} \end{cases} \tag{16}$$

where ω and β are parameters which determine the spring stiffness $\omega^2(1+\beta^2)$ and damping $2\omega\beta$. During decompression, $w_2 \triangleq \omega\tau$ and $\beta_2 \triangleq \beta/\tau$. We define τ, the thrust, by $\tau \triangleq \kappa/(1+x_b^2)$ where x_b is the lowest point reached by the mass the last time the decompression mode was entered and κ is a gain which determines the energy of the leg (and which we will use as a control input in the next section). Notice that the damping during decompression, $2\omega_2\beta_2 = 2\omega\beta$, is the same as during compression. The spring stiffness during decompression is $\omega_2^2(1+\beta_2^2) = \omega^2(\tau^2+\beta^2)$ and is thus proportional to the magnitude of the thrust. The trajectory in the position/velocity plane of a simulation of this system is shown in Figure 6 (a). That this is a stable system follows from an argument similar to that made in [20] wherein similar systems are shown

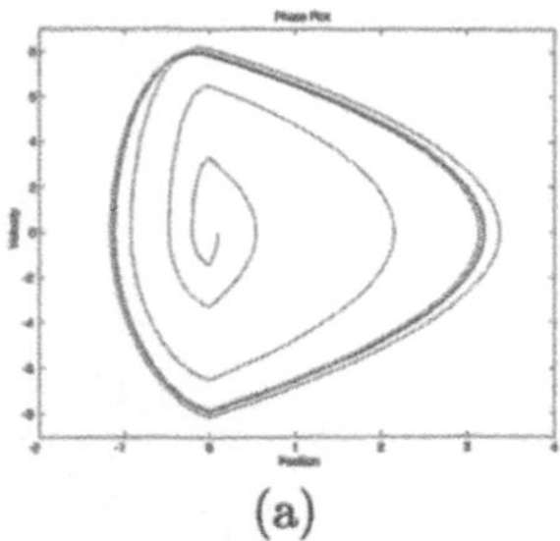

(a)

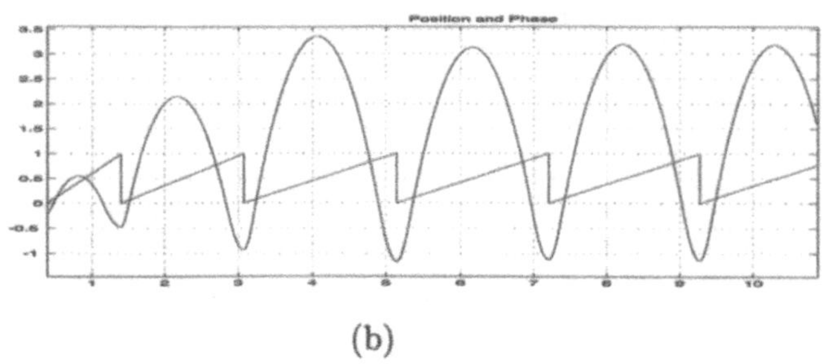

(b)

Figure 6: *(a) A plot of the velocity vs. the position of a simulated one legged hopper emphasizing the stable limit cycle. (b) A plot of the position and phase of the hopper vs. time. Between minimal points of the hop, phase velocity is constant.*

to have global asymptotic stability. Thus, a leg cannot be controlled to a specified apex height in one bounce as could the balls in Section 3.1.

To obtain a definition of phase for the leg, we change coordinates to real canonical form in the compression and decompression modes. This is demonstrated in [20]. We arrive at systems $(\dot{E}_c, \dot{\theta}_c) = (-2\omega\beta E_c, -\omega)$ and $(\dot{E}_d, \dot{\theta}_d) = (-2\omega\beta E_d, -\omega\tau)$ for compression and decompression respectively. Note that the changes in the second coordinates of these systems are constant. They can therefore be used as phase variables. In compression, for example, $\dot{\theta}_c = \omega$ and θ_c varies between $\theta_{c,max}$ and 0. Thus:

$$\theta_c = \omega t = \theta_{c,max}\frac{t}{t_c} \tag{17}$$

where t_c is the duration of the compression mode. This expression is a scalar multiple of the general phase definition described in Section 2.1. However, to control two physically unconnected legs to hop in a synchronized manner as in Figure 4(b), our present constructive approach requires constant phase velocity throughout each cycle of a hop as in Figure 6(b). We must transform three piecewise linear phases so that each has the same rate of change, as assumed in the discussion of Section 2.1. We begin with the phases during compression and decompression, θ_c and θ_d and the phase during flight, $\theta_f \triangleq \frac{\dot{x}_0 - x}{2\dot{x}_0}$.

The construction of a suitable phase definition from θ_d, θ_c and θ_f is in two steps. First, apply affine transformations so that the decompression phase varies from 0 to 1/3, the flight phase from 1/3 to 2/3 and the compression phase from 2/3 to 1. This gives a piecewise linear map from interval of time connecting the time of one lowest point to the time of the next. Next, smooth this map so that each segment has the same slope.

First, note that θ_d varies between 0 and $\theta_{d,min}$, θ_f varies between 0 and 1 and θ_c varies between $\theta_{c,max}$ and 0. Thus:

$$\begin{array}{ll} \tilde{\theta}_d \triangleq -\dfrac{\theta_d}{3\theta_{d,min}}, & \tilde{\theta}_f \triangleq \dfrac{\theta_f}{3} + \dfrac{1}{3} \\ \text{and} & \tilde{\theta}_c \triangleq -\dfrac{\theta_c}{3\theta_{c,max}} + 1 \end{array} \tag{18}$$

is the first transformation.

Second, think of each of these phases as a linear homogeneous transformation $P_d : [0, t_1] \to [0, 1/3]$, $P_f : [t_1, t_2] \to [1/3, 2/3]$ and $P_c : [t_2, t_3] \to [2/3, 1]$ where t_1, t_2 and t_3 are the durations of the liftoff, touchdown and bottom modes, respectively. We desire the phase to increase from 0 to 1 linearly as time increases from 0 to t_3. The following, final definition of phase for the hopping leg system, has this property:

$$\theta \triangleq \begin{cases} HP_d^{-1}(\tilde{\theta}_d,\ 1) & \text{if } x < 0 \wedge \dot{x} > 0 \\ HP_f^{-1}(\tilde{\theta}_f,\ 1) & \text{if } x > 0 \\ HP_c^{-1}(\tilde{\theta}_c,\ 1) & \text{if } x < 0 \wedge \dot{x} < 0\,. \end{cases} \tag{19}$$

Here H is the transformation $(x, 1) \mapsto (x/t_3, 1)$. This results in a phase definition for a single leg that is equivalent to $t/(t_1 + t_2 + t_3)$ as in (3). Figure 6 (b) illustrates how θ changes with the position of the mass in simulation.

For a system with two physically unconnected legs modeled by (16) with states $(x_1, \dot{x}_1)$ and $(x_2, \dot{x}_2)$, let θ_1 and θ_2 be the phases of the legs. Since each leg actuates itself, there is no need for an attention function. Once again, we wish to control the legs so that they are hopping to a prespecified height and so that they

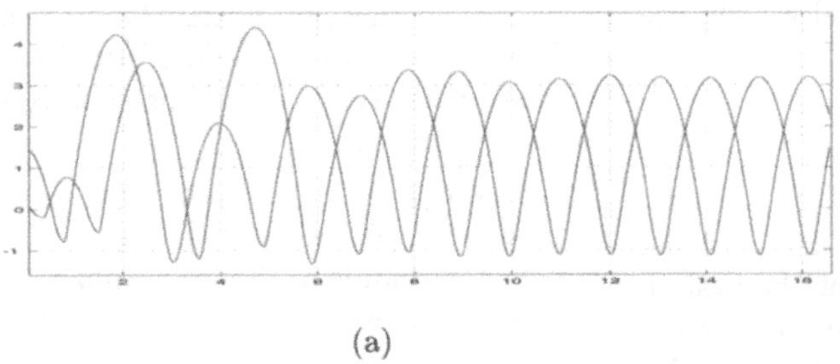

(a)

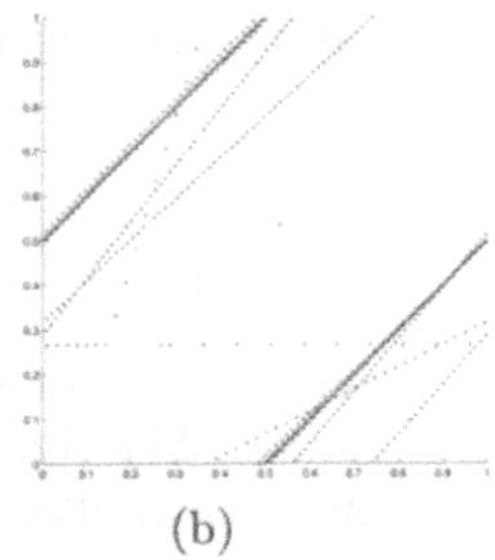

(b)

Figure 7: *(a) The positions of the leg's centers of mass versus time for $A\!:\!B = 1\!:\!1$ starting from a randomly chosen initial condition. (b) The phase plot $\theta_1(t)$ vs. $\theta_2(t)$ for the same run. Note the limit cycle where $\theta_1 = \theta_2 + \frac{1}{2}$.*

are out of phase: one leg is at its lowest point when the other is at its highest. We use a reference field $\mathfrak{R}$ with $A\!:\!B = 1\!:\!1$.

Recall that the thrust, τ_i, supplied by a leg is constant through the decompression phase. The gain for each thrust, κ_i, controls the phase velocity. A larger thrust gives a smaller phase velocity (because the leg goes higher and takes longer to return to its lowest point). Thus, we simply reset the thrust gains, κ_i to be $1/\mathfrak{R}_i(\theta_1, \theta_2)$ whenever $\theta_i = 0$, $i = 1, 2$.

Figures 7(a) and (b) show simulations of the system with $A\!:\!B = 1\!:\!1$. In all runs with various reasonable values of the parameters the legs regulate their phases to very near the limit cycle within two or three stance modes.

3.3 Two Legged SLIP

One obvious shortcoming of the synchronized hopping model of walking is that the two legs are physically unconnected. In this section we examine the *Spring Loaded Inverted Pendulum* (SLIP) model [35] of a hopping leg which we have modified to have two legs as shown in Figure 4(c). The SLIP model has a single point mass (which we assume to be 1 in this paper) in the plane and a massless, spring loaded leg. When it is not touching the ground, its dynamics are ballistic. When the toe of the leg is touching the ground, the spring of the leg exerts force on the mass along the direction of the leg. Our slightly different model consists of a mass with two rotating, massless legs which we call the *roadrunner*. In an alternating gait, the roadrunner uses one leg to hop and then the other. When a leg is touching the ground, it is in *stance* mode – which again consists of compression and thrust phases. Between hops, a leg must swing around the mass to ready itself for the next hop. This is called the *swing phase.* The task is to construct a controller for the legs that realizes this gait.

This example is different from our previous examples in that we start with a specification of the aggregate behavior of the system and decompose it into controllers for the individual subsystems. In the juggling and hopping examples, the phase regulated systems are themselves cyclic. For example, in the $1:1$ case, the aggregate phase of a phase regulated system at equilibrium is

$$\phi_{agg}(\phi_1, \phi_2) = \begin{cases} \frac{1}{2}(\phi_1 + \phi_2 - \frac{1}{2}) \text{ if } \phi_1 < \phi_2 \\ \frac{1}{2}(\phi_1 + \phi_2 + \frac{1}{2}) \text{ otherwise}. \end{cases} \quad (20)$$

In the juggling example, ϕ_{agg} is a measurement of the phase of the paddle. In this example, ϕ_{agg} will be a measurement of the phase of the underlying SLIP model which the legs will then "service" according to the pseudo-inverse of ϕ_{agg}:

$$\phi^{\dagger}_{agg}(\phi) = (\phi, \phi + \frac{1}{2}) \pmod 1 . \quad (21)$$

That is, the phases of the legs are functions of the phase of the underlying SLIP model. It remains to define phase for the SLIP model.

As shown in Figure 4(c) the system is described by variables x, y, θ_1, and θ_2 where (x, y) is the position of the body and θ_1 and θ_2 are the angles of the legs. When in stance mode, the position of the body will also be described by the distance r from the body to the toe (given by (x_{toe}, y)) and the angle θ of the leg touching the ground. When a leg is compressing, $\dot{r} < 0$, the stiffness of the spring is k_1 and when it is decompressing,

$\dot{r} > 0$, the stiffness is k_2. The spring model we will use has potential:

$$U(r) = \frac{k}{2}\left(\frac{1}{r^2} - \frac{1}{l^2}\right)$$

where l is the natural length of the leg and k is the current spring stiffness. The dynamics of the system can be derived from the Hamiltonian as in [35] for the cases with or without gravity in stance. We consider only the case without gravity. We have during flight:

$$\begin{pmatrix} \ddot{x} \\ \ddot{y} \end{pmatrix} = \begin{pmatrix} 0 \\ -g \end{pmatrix} \text{ and } \begin{pmatrix} \dot{\theta}_1 \\ \dot{\theta}_2 \end{pmatrix} = \begin{pmatrix} u_1 \\ u_2 \end{pmatrix}$$

where g is the force of gravity and u_1 and u_2 are velocity inputs the legs. During stance, suppose that leg 1 is touching the ground and leg 2 is not so that $\theta = \theta_1$. Then we have:

$$\begin{pmatrix} \ddot{r} \\ \ddot{\theta} \end{pmatrix} = \begin{pmatrix} r\dot{\theta}^2 - \frac{1}{m}\nabla U(r) \\ -2\dot{r}\dot{\theta}/r \end{pmatrix} \text{ and } \begin{pmatrix} \ddot{\theta}_1 \\ \dot{\theta}_2 \end{pmatrix} = \begin{pmatrix} \ddot{\theta} \\ u_2 \end{pmatrix}$$

where u_2 is a control input. The equations for when the legs are reversed are similar. We do not consider the case where both legs are touching the ground or when the mass hits the ground—situations we would like to avoid. In [33], the control of the SLIP model is discussed. We do not repeat this discussion here, but simply assume that upon liftoff that θ_{td}, k_1 and k_2 are given by the controller.

The phase of the virtual leg will once again be composed of the phases of flight, compression and decompression. In the rest of the section, variables subscripted with l represent the state at liftoff and those subscripted with td represent that state a touchdown. As in the previous example, the phase is obtained from a piecewise linear transformation on the phases during the various modes. We use the results from [35] wherein the systems are integrated to obtain the durations of the flight, compression and decompression modes, t_f, t_c and t_d respectively. For a given state w of the leg, equation (2) defines $t^-(w)$ to be the time since the last lift off. Then the phase is:

$$\phi = \frac{t^-}{t_f + t_c + t_d} .$$

Notice that the phase varies between 0 and 1. Since each leg will service every other stance mode, we could define the phase of the SLIP model so that it completes two cycles between 0 and 1 instead of one cycle so that (21) makes sense in the present context. We neglect this detail here.

Now define the position of the legs during their swing phases as a function of the phase. Let θ_{top} be an angle near the middle of the swing phase, such as π. We give each leg a discrete state s_i defined by:

$$s_i = \begin{cases} 0 \text{ if } \theta_l < \theta_i < \theta_{top} \\ 1 \text{ if } \theta_{top} < \theta_i < \theta_{td} \\ 2 \text{ otherwise (leg is touching the ground)} . \end{cases} \tag{22}$$

Thus, a leg is characterized by a sequence such as $\langle 0, 1, 2, 0, 1, 2... \rangle$. We define reference maps (functions of phase, which is in turn a function of the state of the body) $\theta_{ref,0}$ and $\theta_{ref,1}$ as functions of the leg phase which give the ideal trajectory of a leg during each of the discrete states 1 and 2. $\theta_{ref,0}$ varies between θ_l and θ_{top} as ϕ varies from 0 to 1 so that $\dot{\theta}_{ref,0}$ is equal to $\dot{\theta}$ at liftoff. $\theta_{ref,1}$ varies from θ_{top} to θ_{td} as ϕ varies from 0 to its value at touchdown. These may be smoothed in various ways to minimize, for example, the velocity of the toe relative to the ground at touchdown. There is no reference phase during stance because when a leg is in stance it is not actuated. If the discrete states of the legs are initially different, they will alternately service the stance mode of the robot.

4 Analysis of the Phase Regulation Algorithm

We have presented three examples of phase regulation that differ in several important respects. In the juggling controller, we are assured that the reference field can be followed closely because of the deadbeat nature of the ball control. That is, within the limits of the actuator, we can achieve any desired ball energy by striking it with the paddle using (11). Therefore, to analyze the stability of the control method, we need only consider the system in terms of the phase states and velocities. We do so in this section. With the synchronized hopping example, we do not have deadbeat leg control, but only asymptotic stability. Thus, to analyze the stability of the hoppers, we would need to take in to account the rate of convergence of a single leg to the reference phase velocity dictated by the reference field controller. We have not yet performed this analysis. However, because of the fast rate of convergence of the single leg controller in practice, the analysis in this section is likely appropriate. The two legged SLIP controller, in a sense, needs no further analysis. If we assume that the legs can follow the reference trajectory accurately, the model is the same as the original SLIP model [35].

4.1 Analysis

Consider the phase regulated system $(\phi_1, \phi_2, \dot{\phi}_1, \dot{\phi}_2) \in \mathbb{T}^2 \times \mathbb{R}^2$ where $\dot{\phi}_i$ is constant except for discrete jumps made when $\phi_i = 0$. These jumps are governed by the reference field (10). That is, when $\phi_i = 0$, $\dot{\phi}_i$ becomes $\mathcal{R}(\phi_1, \phi_2)$. Notice that when $A : B = 1 : 1$, then $\mathcal{R}(0, \phi_2) = \mathcal{R}(\phi_1, 0)$. To simplify notation in this section, we redefine $\mathcal{R} : S^1 \to \mathbb{R}$ to be the reference field restricted to $\phi_1 = 0$. Therefore, with $A{:}B = 1{:}1$, $\mathcal{R}(\phi) = \kappa_1 - \kappa_2 \sin(2\pi\phi)$.

To analyze the dynamics of this system, we consider the Poincaré sections Σ_1 and Σ_2 of $\mathbb{T}^2 \times \mathbb{R}^2$ given by $\phi_1 = 0$ and $\phi_2 = 0$ respectively. We suppose that adjustments to the phase velocities alternate between the two phases (i.e. the system is near the limiting behavior). We construct the return map from Σ_1 into Σ_1 as follows. Start with a point $w \in \Sigma_1$, integrate the system forward to obtain a point in Σ_2, then integrate again to get a point in $f(w) \in \Sigma_1$.

A point in Σ_1 has the form $w = (0, \phi_2, \dot{\phi}_1, \dot{\phi}_2)$. This maps to the point $w' = (C_1, 0, \mathcal{R}(\phi_2), \dot{\phi}_2) \in \Sigma_2$ where C_1 is the phase of the first system when the trajectory of the total system first intersects Σ_2. w' in turn maps to the point $f(w) = (0, C_2, \mathcal{R}(\phi_1), \mathcal{R}(C_1))$ where C_2 is the phase of the second system when the trajectory next intersects Σ_1. The phases C_1 and C_2, which can be obtained via the point-slope formula for a line (in the ϕ_1, ϕ_2 plane), are given by:

$$C_1 = \frac{\mathcal{R}(\phi_2)}{\dot{\phi}_2}(1 - \phi_2) \text{ and } C_2 = \frac{\mathcal{R}(C_1)}{\mathcal{R}(\phi_2)}(1 - C_1)\,. \quad (23)$$

Let $(x, y, z) = (\phi_2, \dot{\phi}_1, \dot{\phi}_2)$. Then, expanding $f(w)$, we obtain a discrete, real valued map on Σ_2 given by

$$\begin{aligned} x_{k+1} &= \frac{\mathcal{R}\left[\frac{\mathcal{R}(x_k)}{z_k}(1 - x_k)\right]}{\mathcal{R}(x_k)}\left[1 - \frac{\mathcal{R}(x_k)}{z_k}(1 - x_k)\right] \\ y_{k+1} &= \mathcal{R}(x_k) \\ z_{k+1} &= \mathcal{R}\left[\frac{\mathcal{R}(x_k)}{z_k}(1 - x_k)\right]\,. \end{aligned} \quad (24)$$

Since the x and z advance functions are not functions of y, we can treat y as an output of this system. Thus, analytically, it will suffice to treat (24) as an iterated map of the the variables $(x, z) \in S^1 \times \mathbb{R}^+$ given by $F(x_k, z_k) = (x_{k+1}, z_{k+1})$. We have the following fixed point conditions:

Proposition 4.1 *$F(x, z) = (x, z)$ if and only if $\mathcal{R}(x) = \mathcal{R}(1 - x) = z$.*

We omit the proof, which is straightforward algebra (note that the values of x are always taken modulo 1 since $x \in S^1$). For the reference field we are using, we have:

Corollary 4.1 *If $\mathcal{R}(\phi) = \kappa_1 - \kappa_2 \sin(2\pi\phi)$, then the only fixed points of F are $(1/2, \kappa_1)$ and $(0, \kappa_1)$.*

We wish to show that the first fixed point, $(1/2, \kappa_1)$, is stable, since it corresponds to the situation where the two subsystems are out of phase and at the desired velocity. To do this, we examine the Jacobian. Suppose that the fixed point condition we desire is $F(1/2, v) = (1/2, v)$ where v is the desired phase velocity. Then:

$$J_{(\frac{1}{2}, \kappa_1)}F = \begin{pmatrix} \frac{1}{2v^2}\left(\frac{m}{2} - 1\right)m - \frac{m}{v} + 1 & \frac{1}{2v} - \frac{m}{4v^2} \\ \frac{1}{v}\left(\frac{m}{2} - 1\right)m & -\frac{m}{2v} \end{pmatrix}. \quad (25)$$

Here, $m = \mathcal{R}'(1/2)$ is the slope of $\mathcal{R}$ at $1/2$. F is stable at $(1/2, v)$ if the eigenvalues of the Jacobian lie within the unit circle. Values for m and v which guarantee this are not difficult to find. For example,

Proposition 4.2 *If $m = 2v - 2$ then the eigenvalues of $J_{(\frac{1}{2}, \kappa_1)}F$ are 0 and $\frac{2}{v^2} - 1$ which implies that $(1/2, v)$ is a stable fixed point of F whenever $v > 1$.*

Once again, the proof is just a calculation: simplify (25) using $m = 2v - 2$ and compute the eigenvalues. With the reference field we are using, $m = 2\pi\kappa_1$. Thus, for a given value of v, we set $\mathcal{R}(\phi) = v - \frac{m}{2\pi} sin(2\pi\phi)$. In practice, it is not difficult to find other parameters which make F stable. For a given v, we first choose m to be quite small and increase it slowly until the controller is aggressive, yet still stable.

5 Conclusion

In this paper we have taken the first steps toward a formal treatment of phase regulation for underactuated environments that must be repeatedly and intermittently contacted by an actuated robot. We have introduced a variant of the two-juggle controller [5, 30] and, by re-writing the system in phase coordinates, exhibited sufficient conditions for local asymptotic stability of a 1:1 mode-locked rhythm. The obvious next step concerns the extent of the domain of attraction. Here, there is a natural hybrid structure — the order of "contact events" (i.e., the sequence of balls hit) — whose desired sequences might be seen as a pattern to be regulated against disturbances. Moreover, there is a "forbidden" set in phase space — where both balls must be hit at the same time — that must be shown to be a repeller. We have also suggested the manner in which this 1:1 "juggling" framework carries over to simple problems in legged locomotion. Because the effective input enters through an additional dynamical lag in such problems, our present sufficient conditions for asymptotic stability will need to be modified in order to address them. We have not dealt at all with the problem of regulating more general A:B mode-locking, but we believe that similar methods can be used to achieve such behaviors.

Although the applications focus of this paper is limited to locomotion systems, we are intrigued by the prevalence of phase regulation problems in more abstract settings such as factory automation [19, 17] and will seek to apply these ideas in that context as well.

Acknowledgments

We thank Bill Rounds for providing many insights concerning the compositional semantics of dynamic systems. This work is supported in part by DARPA/ONR under grant N00014-98-1-0747 and in part by the NSF under grant IRI-9510673 at the University of Michigan. It is supported in part by the NSF under grant DMS-9971629 at the Georgia Institute of Technology.

References

[1] C. Baesens, J. Guckenheimer, S. Kim, and R. MacKay. Three coupled oscillators: Mode-locking, global bifurcations and toroidal chaos. *Physica D*, 49(3):387–475, 1991.

[2] A.M. Bloch, P.S. Krishnaprasad, J.E. Marsden, and R. Murray. Nonholonomic mechanical systems with symmetry. *Archive for Rational Mechanics and Analysis*, 136:21–99, 1996.

[3] A.M. Bloch and N.H. McClamroth. Control of mechanical systems with classical non-holonomic contraints. In *Proc. 28th IEEE Conf. on Decision and Control*, pages 201–205, Tampa, FL, Dec 1989.

[4] R.W. Brockett. Asymptotic stability and feedback stabilization. In R.W. Brockett, R.S. Millman, and H.J. Sussman, editors, *Differential Geometric Control*, chapter 3, pages 181–191. Birkhäuser, 1983.

[5] M. Buehler, D. E. Koditschek, and P.J. Kindlmann. Planning and control of robotic juggling and catching tasks. *International Journal of Robotics Research*, 13(2), April 1994.

[6] M. Bühler and D.E. Koditschek adn P.J. Kindlmann. A family of control strategies for intermittent dynamical environments. *IEEE Control Systems Magazine*, 10(2):16–22, 1990.

[7] R. Burridge, A. Rizzi, and D.E. Koditschek. Sequential composition of dynamically dexterous robot behaviors. *International Journal of Robotics Research*, 18(6):534–55, 1999.

[8] A. H. Cohen, P. J. Holmes, and R. H. Rand. The nature of the coupling between segmental oscillators of the lamprey spinal generator for locomotion: A mathematical model. *J. Math. Biology*, 13:345–369, 1982.

[9] Avis H. Cohen, Serge Rossignol, and Sten Grillner (eds.). *Neural Control of Rhythmic Movements in Vertebrates*. Wiley Inter-Science, NY, 1988.

[10] P. Collet and J.P. Eckmann. *Iterated Maps on the Interval as Dynamical Systems*. Birkhäuser, Boston, 1980.

[11] H. Cruse. What mechanisms coordinate leg movement in walking arthropods? *Trends in Neurosciences*, 13:15–21, 1990.

[12] F. Delcomyn. Neural basis of rhythmic behavior in animals. *Science*, 1980.

[13] G.B. Ermentrout and N. Kopell. Inhibition-produced patterning in chains of coupled nonlinear oscillators. *SIAM Journal of Applied Mathematics*, 54:478–507, 1994.

[14] R. Ghrist and D.E. Koditschek. Safe cooperative robot patterns via dynamics on graphs. In *Robotics Research*, pages 81–92, 1998.

[15] J. Guckenheimer and P. Holmes. *Nonlinear Oscillations, Dynamical Systems, and Bifurcations of Vector Fields*. Springer-Verlag, New York, 1983.

[16] R. H. Harris-Warrick, F. Nagy, and M. P. Nusbaum. Neuromodulation of stomatogastric networks by identified neurons and transmitters. In Harris-Warrick, Marder, Selverston, and Moulins, editors, *Dynamic Biological Networks*, pages 87–137. MIT Press, 1992.

[17] E. Klavins. Automatic compilation of concurrent hybrid factories form product assembly specifications. In *Hybrid Systems: Computation and Control Workshop, Third International Workshop*, Pittsburgh, PA, 2000.

[18] E. Klavins. The construction of attention functions for phase regulation. Technical report, University of Michigan, 2000.

[19] E. Klavins and D.E. Koditschek. A formalism for the composition of concurrent robot behaviors. In *Proceedings of the IEEE Conference on Robotics and Automation*, 2000.

[20] D. E. Koditschek and M. Bühler. Analysis of a simplified hopping robot. *International Journal of Robotics Research*, 10(6):587–605, December 1991.

[21] D.E. Koditschek. Task encoding: Toward a scientific paradigm for robot planning and control. *Robotics and Autonomous Systems*, 9:5–39, 1992.

[22] D.E. Koditschek. An approach to autonomous robot assembly. *Robotica*, 12:137–155, 1994.

[23] D.E. Koditschek and E. Rimon. Robot navigation functions on manifolds with boundary. *Advances in Applied Mathematics*, 11, 1990.

[24] P.S. Krishnaprasad. Motion, control and geometry. In *Board of Mathematical Sciences, National Research Council Motion, Control and Geometry: Proceedings of a Symposium*, pages 52–65. National Academy Press, 1997.

[25] R. S. MacKay. Chaos, order, and patterns (lake como, 1990). In *NATO Adv. Sci. Inst. Ser. B Phys.*, volume 280, pages 35–76, Plenum, New York, 1991.

[26] K. Pearson. The control of walking. *Scientific American*, 235(6):72–86, December 1973.

[27] M.H. Raibert, H.B. Brown, and M. Chepponis. Experiments in balance with a 3D one-legged hopping machine. *International Journal of Robotics Research*, 3:75–92, 1984.

[28] E. Rimon and D.E. Koditschek. The construction of analytic diffeomorphisms for exact robot navigation on star worlds. *Transactions of the American Mathematical Society*, 327:71–115, 1991.

[29] E. Rimon and D.E. Koditschek. Exact robot navigation using artificial potential fields. *IEEE Transactions on Robotics and Automation*, 8(5):501–518, October 1992.

[30] A. Rizzi and D.E. Koditschek. An active visual estimator for dexterous manipulation. *IEEE Transactions on Robotics and Automation*, 12(5):697–713, October 1996.

[31] A.A. Rizzi, L.L. Whitcomb, and D.E. Koditschek. Distributed real-time control of a spatial robot juggler. *IEEE Computer*, 25(5):12–26, May 1992.

[32] U. Saranli, M. Buehler, and D.E. Koditschek. Design, modeling and preliminary control of a compliant hexapod robot. In *Proceedings of the IEEE Conference on Robotics and Automation*, 2000.

[33] U. Saranli, W.J. Schwind, and D.E. Koditschek. Toward the control of a multi-jointed monoped runner. In *Proc. IEEE Intl. Conf. on Robotics and Automation*, pages 2676–2682, 1998.

[34] S. Schwartzman. Asymptotic cycles. *Annals of Mathematics*, 2(66):270–284, 1957.

[35] W.J. Schwind and D.E. Koditschek. Approximating the stance map of a 2 dof monoped runner. *Journal of Nonlinear Science*, 2000. To appear.

[36] P. J. Swanson, R. R. Burridge, and D. Koditschek. Global asymptotic stability of a passive juggling strategy: A parts possible feeding strategy. *Mathematical Problems in Engineering*, 1(3), 1995.

A Framework for Steering Dynamic Robotic Locomotion Systems

James P. Ostrowski, *University of Pennsylvania, Philadelphia, PA*
Kenneth A. McIsaac, *University of Pennsylvania, Philadelphia, PA*

We seek to formulate control and motion planning algorithms for a class of dynamic robotic locomotion systems. We consider mechanical systems that involve some type of interaction with the environment and have dynamics that possess rotational and translational symmetries. Research in nonholonomic systems and geometric mechanics has led to a single, simplified framework that describes this class of systems, which includes examples such as wheeled mobile robots, bicycles, and the snakeboard robot; undulatory robotic and biological locomotion systems, such as paramecia, inchworms, snakes, and eels; and the reorientation of satellites and underwater vehicles with attached robotic arms. We explore a hybrid systems approach in which small amplitude, periodic inputs, or **gaits**, *are used to yield simplified approximate motions. These motions are then treated as abstract control inputs for a simplified, kinematic representation of the locomotion system. We describe the application of such an approach as applied to two examples: the snakeboard robot and an eel-like, underwater robot.*

1 Introduction

The field of robotic motion planning has generally focused on the study of wheeled vehicles, and very often this has implied kinematic representations of such systems [9, 14]. When dynamics are treated, they are generally directly actuated, e.g., through wheel torques applied to the wheels, and do not involve unactuated motions. In such cases, the kinematic representation of such a system can be proven to be fully adequate [13].

More recently, the motion planning field has branched out to explore a wider range of topics, including motion planning for flexible or articulated objects, parts assembly, and biological molecules [10]. Researchers have also begun to study underactuated mechanical systems, where some or even most of the dynamics of the system can only be actuated *indirectly*, through control inputs on other parts of the system. It is this class of systems in which we are interested, and in particular to a sub-class that is characterized by robotic systems that move through their environment, or ***robotic locomotion*** systems. Some examples of such systems include flying robots (satellites with thrusters, blimps, helicopters, etc.), the snakeboard, bicycles, swimming robots, and even legged systems. Our current work does not include legged robots, which possess intermittent contacts, though extensions to such systems are certainly possible [5].

To study control and motion planning for such systems, we utilize a geometric framework developed in previous work [19] to capture the general form of most locomotion systems. The basic concept is to decompose the system into a part defining the motion (and momentum) in the ***position*** of the robot, and a complementary part describing the internal ***shape*** of the system. Internal shape controls can then be used not only to change the position of the system, but to generate velocities and hence truly *locomote* in a dynamic sense. The current study focuses on introducing methods for controlling and planning trajectories (steering) for such dynamic systems.

In this paper, we use perturbation methods to derive approximate expressions for the effect of internal shape changes on the motion of the system. By assuming small amplitude inputs we can introduce a scaling parameter into the system. Such perturbation techniques have proven useful in a variety of control contexts [2, 12, 18, 22].

In order to develop the appropriate motion plans, we utilize a framework for abstraction of the system dynamics that is used in hybrid systems. We note that the motion planning algorithms we are proposing for dynamic underactuated systems have a natural realization as hybrid systems, since they involve systems with continuous time dynamics for which the controls are generated over discrete intervals (input periods), and with discrete mode-switching.

2 Characterization of the Steering Problem

2.1 Background and Problem Formulation

In working with locomotion problems in general, and steering problems in particular, there is generally a natural decomposition of the state space (say, Q) into a ***position space***, represented by a Lie group, G, and a ***shape space***, given by a general manifold, M. A more detailed description of this is given in [19], but some easy examples to visualize the shape variables include the internal bending of a snake, the rotational position of the wheels on a mobile robot, or the leg positions on a walking robot.

Denoting the configuration variables by $q = (g, r) \in G \times M = Q$, the general formulation of the system dynamics is found via Lagrange's equations, perhaps with undetermined multipliers representing the constraints. Another important characteristic of robotic locomotion systems, however, is that the motion of such systems is generally independent of the actual position of the system, g. Stated another way, the system's dynamics are ***invariant*** with respect to position. This leads to a reduced representation when viewed in the appropriate body-fixed frame. Using this invariance we write the (possibly constrained) dynamics in terms of a body velocity, $\xi = g^{-1}\dot{g}$, and a body momentum, p (see [1, 17] for more formal definitions):

$$\xi = g^{-1}\dot{g} = -\mathbb{A}(r)\dot{r} + I^{-1}(r)p, \tag{2.1}$$

$$\dot{p} = \dot{r}^T\alpha(r)\dot{r} + p^T\beta(r)\dot{r} + p^T\gamma(r)p + \tau_g, \tag{2.2}$$

$$\nabla_{\dot{r}}\dot{r} = \tau_r. \tag{2.3}$$

Although the derivation of these equations is quite involved, for our purposes here the basic form of them is all that is critical (to gain a much better insight into these equations, the reader is referred to the paper by Bloch et al. [1], and examples from robotics explored in [8, 19]). Equations 2.1 and 2.3 are the ***fiber*** and ***base*** equations, respectively. They define the body velocity of the system and the dynamics of the shape space, respectively (∇ represents an ***affine connection***, which we will use to denote the second order ODE given by Lagrange's equations). Thus, Equation 2.1 governs the motion of the system in position space, based on changes in shape and the momentum in the position (group) directions. Equation 2.3 describes the evolution of the internal shape of the robot in terms of the internal applied forces, τ_r. Lastly, Equation 2.2 is called the ***generalized momentum equation***, where p is a momentum vector associated with the momentum along each of the kinematically unconstrained fiber directions. This equation describes the dynamics of the system, via the momentum, in the position, or group, directions. We have included a forcing term in this equation, τ_g, to allow us to model control inputs or external forces, such as those arising from fluid drag models or viscous damping. Notice that by using the symmetries and writing the equations in a body-fixed frame we can pull the position (group) variable g out of the equation— this greatly helps to simplify the necessary calculations.

2.2 Analogies and Abstractions to Kinematic Systems

Our main goal in this paper is to develop a motion planner for dynamic locomotion systems by using equivalent representations, or simplified ***abstractions***, for the system. This will enable us to compute paths for the simplified systems that are then tracked by the full dynamic system. Figure 1 shows our general approach to the system representation at various levels of abstraction. Also see Figure 2 for an overlayed comparison between the motion plans.

The key idea here is to start with the full dynamics, shown at the lowest level in Figure 1, and build simplified approximations of this system at various levels. In the example we have shown, the full dynamics are first approximated by an equivalent dynamic system in which the inputs have been restricted to a class of simple cyclic inputs. This leads to an "averaged" set of dynamics in which we assume direct control over the momentum terms (the derivation for this is done in Section 3 below). The next level of abstraction involves replacing the dynamic model with a kinematic approximation (also shown in Section 3). In this paper, the kinematic approximation (the second level from the top) is all that is needed, since we use steering algorithms based on a kinematic model of this form. However, we can consider the use of an even coarser approximation to this kinematic model in which all of

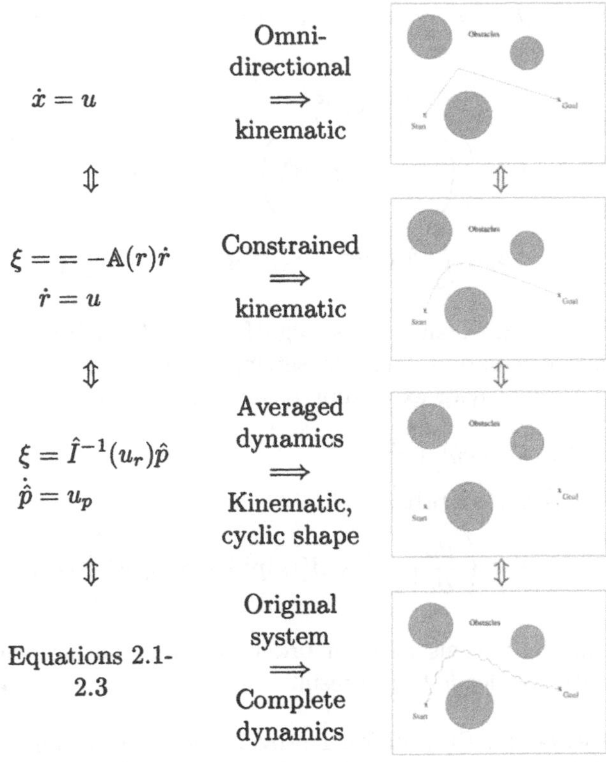

Figure 1: *A hierarchical description of the different models used.*

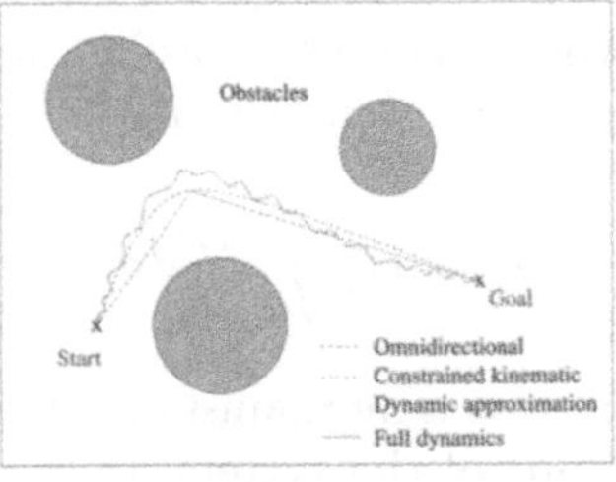

Figure 2: *Comparison of the paths planned at different levels.*

the directions of motion are directly controlled via the inputs.

Of course, formulating the different levels of abstraction is only half of the work. The reason for abstracting at the various levels is to simplify the motion planning effort at each stage. However, motion plans formulated at the highest levels of abstraction may not be implementable at the lower levels (though this is certainly a goal of the abstraction). Thus, a motion planning process that works its way down the hierarchy also must occur, in which motion plans are generated with the simplest level of dynamics at the top, and filtered down through the hierarchy until plans for the original system can be generated. In the case that the choice of abstractions is constructive, the motion plans at one level will implicitly define motion plans at the next lowest level. They may also serve to define an initial guess as to what the motion plans should be. Regardless, this overall scheme will likely require some type of negotiation between motion planners at the various levels of abstraction, an issue that we do not address here.

There is an obvious compromise here between building sufficiently simple higher-level abstractions and losing the fidelity to model the characteristics of the more complex dynamic system. For example, in an omnidirectional model, right angle (piecewise continuous ($\mathcal{C}^0$) paths) turns are allowed, while this is obviously not feasible at many of the lower levels (which generally require at least $\mathcal{C}^1$ continuity). On the other hand, dealing with the full details of the original dynamic model is generally not tractable, and even special methods such as those based on optimal controls or randomized motion planners tend to require a great deal of machinery and computation to compute plans for even relatively simple systems [21]. In the next section, we show a mechanism for approximating a dynamic locomotion system with cyclic inputs, and in Section 3.3 we show that this leads to a simplified representation using a kinematic, car-like robot model. Then, in Sections 4 and 5, we apply these algorithms to the steering of snakeboard and eel robotic systems.

3 Algorithms for Generating Motion

3.1 Cyclic Inputs and Locomotive Gaits

One of the goals discussed above is to form a simplified approximate model of the system. One way in which we can do this is by working from a set of pre-defined inputs. This technique is commonly used in underactuated and nonholonomic systems [16, 7, 2, 11], and for a restricted class of inputs is sometimes referred to as "steering using sinusoids." Since we are investigating locomotion systems, we derive an analogy from biological systems where a very common observation is that motions are most often generated by ***cyclic*** shape changes. The motion takes on a characteristic form, called a ***gait***, which we can loosely define as a specified cyclic pattern of internal shape changes (inputs) which couple to produce a net motion. For each species, there usually exist at most a handful of gaits, often tailored for specific needs or environments. For instance, a human will walk or run, depending on the desired speed, but can also hop or skip.

3.2 A Perturbation Approach to Momentum Generation

Motivated by recent work by Leonard, Krishnaprasad, and Bullo [2, 12], we use a perturbation approach to analyze the response of dynamic locomotion systems to small amplitude cyclic inputs (see Khalil [6] for more details on the use of perturbation methods). We begin with the general form of the equations for a dynamic nonholonomic system, but with the restriction that the terms in the momentum equation that are quadratic in momenta be zero (that is, $\gamma = 0$ from Equation 2.2). We re-write these equations here using indicial notation, where repeated indices imply summation over that index, since this will be important for making sense of the results given below:

$$\xi^a = (g^{-1}\dot{g})^a = -\mathbb{A}^a_i(r)\dot{r}^i + (I^{-1})^{ab}(r)p_b, \qquad (3.4)$$

$$\dot{p}_a = \alpha_{aij}(r)\dot{r}^i\dot{r}^j + \beta^b_{ai}(r)\dot{r}^i p_b. \qquad (3.5)$$

It can be shown that in all cases where the generalized momentum is one-dimensional the term γ will be identically zero [1]. This is true for many rolling robotic systems including the bicycle. The case in which γ is the only nonzero term in the momentum equation is treated implicitly in [2].

Next, consider inputs of the form $r(t) = r_0 + \epsilon u(t)$, where ϵ represents a small-amplitude parameter used in the perturbation expansion. We will be most interested in the case where $u(t)$ is cyclic, and $r(t)$ follows a cyclic loop in the shape space about the base point r_0. We include the possibility of a nonzero initial shape, r_0, to allow flexibility in the modeling of the inputs. The time derivative of r is just $\dot{r} = \epsilon\dot{u}$. A simple perturbation analysis of the momentum equation leads to the following result, which we use to develop motion plans for the snakeboard system below. We remark that the form of the eel dynamics is slightly different, and so a separate derivation is provided in Section 5.

Proposition 1 *Given a system with initial momentum $p(0) = p_0$ and inputs $r(t)$ of the form $r(t) = r_0 + \epsilon u(t)$, the momentum to second order in ϵ is given by:*

$$\begin{aligned} p_a(t) &= p_{0a} + \epsilon\beta^b_{ai}p_{0b}u^i|^t_0 \\ &+ \epsilon^2\left(\left(\frac{\partial\beta^b_{aj}}{\partial r^i} + \beta^c_{aj}\beta^b_{ci}\right)p_{0b}\int_0^t u^i\dot{u}^j d\tau \right. \\ &\left. +\alpha_{aij}\int_0^t \dot{u}^i\dot{u}^j d\tau - \alpha^c_{aj}\alpha^b_{ci}p_{0b}u^i(0)u^j|^t_0\right) + \ldots. \end{aligned}$$

Proof: This result is a straightforward, though extensive, calculation, found by setting $p = p^0 + \epsilon p^1 + \epsilon^2 p^2 + \ldots$, and Taylor expanding α and β about r_0:

$$\begin{aligned} \dot{p} &= \dot{p}^0 + \epsilon\dot{p}^1 + \epsilon^2\dot{p}^2 + \ldots \\ &= 0 + \epsilon\beta(r_0)p^0\dot{u} \\ &+ \epsilon^2\left(\left.\frac{\partial\beta}{\partial r}\right|_{r_0} p^0 + \beta(r_0)p^1\dot{u} + \alpha(r_0)\dot{u}\dot{u}\right) + \ldots \end{aligned}$$

Equating terms at each order of ϵ and noting that $p(0) = p_0$ leads to the result. ∎

Remarks: Proposition 1 provides very interesting insights into the generation of momentum for this type of system. For cyclic inputs, there are only two terms that lead to net changes in momentum, the terms stemming from α and β, and they are both at order ϵ^2. Furthermore, when one uses sinusoids that are driven in-phase,[1] the change in momentum is:

$$\Delta p_a = \alpha_{aij}\int_0^t \dot{u}^i\dot{u}^j d\tau.$$

This directly gives us a mechanism for building up the speed (momentum) of the system when starting from rest. Note that this also tells us how to stop the system, using the same in-phase inputs that are used to start the system, but with the sign switched on one of the inputs, i.e., using a 180° change in phase.

To be a little more explicit about this, let $u^i = a^i \sin 2\pi\omega t$. This leads to a net change of momentum after one cycle of $\Delta p = 2\pi^2\omega^2\alpha_{ij}a^i a^j$. In general, one would like to move from $p = p_0$ to p^d, using small magnitude inputs. To achieve this, it may

[1] An important distinction of the eel robot is that the interaction with the environment implies that zero momentum is generated using in-phase inputs, and instead the gaits require slightly out-of-phase inputs.

be necessary to break the motion up into several repeated cycles, say N cycles. Thus, a suitable method for choosing inputs to control to a desired component of the momentum, $(p^d)_b$, is to choose a magnitude of inputs, a^i and a^j, and a number of cycles, N, such that $2N\pi^2\omega^2\alpha_{aij}a^i a^j = (p^d - p_0)_b$.

3.3 Steering for "Car-like" Systems

In the previous section, we derived results for controlling the individual momenta of a dynamic nonholonomic system. In this section, we explore the implication of this for steering, using an analogy to wheeled mobile robots. While the results are fairly straightforward applications of the above algorithms to motion planning, they are useful because they allow us to control the robot over a non-local region of the state-space. In other words, instead of being forced to patch together purely local results as is generally done in steering algorithms for nonholonomic systems using cyclic inputs, one can determine motion plans for more non-local control goals. This has a similar feel to the steering of a mobile robot or an airplane, where one pieces together very simple trajectories to generate global (not STLC) controllability.

First, recall the results of Dubins [4] for kinematic mobile robots. It was shown that the optimal path for a car steering in $SE(2)$ with limits on the turning radius is given by two circular arcs (one arc tangent to each of the initial and final headings) connected by a straight line segment. See Figure 3. In this section we utilize these paths as the backbone for steering dynamic nonholonomic systems. In order to make the abstraction shown in Figure 1 from the dynamic system to the approximate dynamic system (and hence further to the kinematic, car-like system) work, the

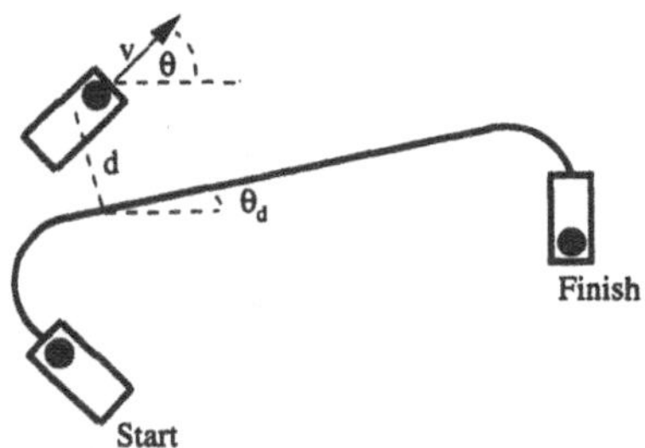

Figure 3: *Car-like steering using linear and circular arcs.*

approximation must lead to a system that is "steerable" in the sense of a wheeled vehicle. We characterize as ***steerable*** any system for which we can use the shape variables to generate body velocities (c.f., Equation 2.1) of the form $\xi = g^{-1}\dot{g} = (v, 0, v/R)^T$, where v is the forward velocity, and $R_{\min} < R < \infty$ is a control variable representing the turning radius (and $R_{\min}$ a turning radius constraint). This effectively means that we can create generalized circles, using either the shape inputs directly (turning the wheels of the snakeboard), or through the momentum of the system (as is done with the eel robot). Generalizations to this type of steering (e.g., to 3D motions) are certainly possible, as long as the same type of analogy can be made between the full dynamic system and its kinematic representation.

The steering algorithm, then, is as follows:

1. **Momentum generation:** As shown above in Proposition 1, momentum can be generated by the simple mechanism of in-phase inputs. For steering in $SE(2)$, the momentum is assumed to be one-dimensional. Thus, this step takes the system from $(0, 0, 0, 0)$ to $(x_1, y_1, \theta_1, p^d)$, for some desired momentum, p^d.

2. **Steering to the desired setup point:** Once the desired momentum has been achieved, the system can be steered along any desired path (by assumption). This involves computing the necessary arcs and line segments to connect (x_1, y_1, θ_1) to (x_d, y_d, θ_d). In order to setup for the final step of reducing the momentum, steer the system to the point $(x_d - x_s, y_d - y_s, \theta_d - \theta_s)$, where $T_s = (x_s, y_s, \theta_s)$ is the approximate change in position that will result from the stopping motion in Step 3. Assuming the appropriate steering can be executed, the setup point is reached with a momentum of p^d.

3. **Zero the momentum:** Execute the opposite maneuver to that chosen in Step 1 in order to reduce the momentum to zero.

4. **Cancel errors in position and momentum:** Since the above analysis is just an approximation, it is expected that the final state after Step 3 will not be exactly $(x_d, y_d, \theta_d, 0)$, but instead some $(\hat{x}, \hat{y}, \hat{\theta}, \hat{p})$. The final step, then, is to make corrections to zero the momentum and repeat the

process to move to the final state. For systems such as the snakeboard, it may be possible to use other gaits to make small corrections in position, as is shown in [18].

This steering algorithm is based solely on the use of feedforward terms to control the motion. We explore the use of feedback terms in discussing steering for the eel in Section 5. We also remark that if one is interested strictly in configuration controllability (moving between configurations only, without regards to velocity) [3], then the motion can be achieved using Steps 1 and 2 only.

4 Steering with a Feedforward Approach: The Snakeboard Example

The *Snakeboard* is a commercial variant of the skateboard that allows for independent rotation of the wheel trucks. The simplified model of the *Snakeboard* (referred to as the snakeboard model) is shown in Figure 4, along with a robotic version that was constructed in order to verify theoretical simulations. We briefly recall the description of the snakeboard as developed in [20, 21]. The basic premise is that by properly coupling the rotating flywheel (modeling the swinging of a human torso) with the turning of the wheels, one can generate a variety of motions, including a forward serpentine-like gait reminiscent of a snake.

As a mechanical system the snakeboard has a configuration manifold given by $Q = SE(2) \times \mathbb{S} \times \mathbb{S}$. As coordinates for Q we shall use $(x, y, \theta, \psi, \phi)$ where (x, y, θ) describes the position of the board with respect to a reference frame, ψ is the angle of the rotor with respect to the board, and ϕ denotes the angle of the back wheel truck with respect to the board (and the opposite of the angle of the front wheels, which is assumed to move through equal and opposite rotations). Note

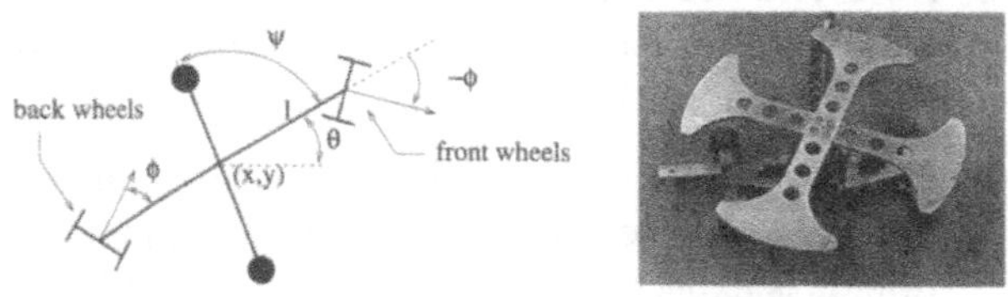

Figure 4: *The simplified model of the* Snakeboard, *along with a demo prototype.*

that the wheels themselves are allowed to spin freely, just as with a traditional skateboard.

Following along the lines of Ostrowski et al. [21], we use a non-dimensional set of variables, normalized by the board length and rotor momentum, so that the fiber equations reduce to the form of Equation 2.1 with:

$$\mathbb{A} = \begin{pmatrix} -\hat{J}\sin\phi\cos\phi & 0 \\ 0 & 0 \\ \hat{J}\sin^2\phi & 0 \end{pmatrix} \quad \text{and} \quad I^{-1} = \begin{pmatrix} -\frac{\hat{J}}{2} \\ 0 \\ \frac{\hat{J}}{2}\tan\phi \end{pmatrix},$$

and $\hat{J}$ a non-dimensional ratio between the rotor inertia and the board inertia. The generalized momentum equation, Equation 2.2, is given by:

$$\dot{p} = 2\cos^2\phi\,\dot{\phi}\dot{\psi} - \tan\phi\,\dot{\phi}p.$$

Using this, we can explore the relationship between the integrals of Proposition 1 and the appearance of cyclic inputs. Using inputs of the form $\psi(t) = \frac{\pi}{2} - a_\psi \sin 2\pi t$ and $\phi(t) = a_\phi \sin 2\pi t$, the α term in the momentum equation leads to changes of momentum equal to:

$$\begin{aligned}\Delta p &= \alpha_{aij}(r_0)\int_0^1 \dot{u}^i\dot{u}^j d\tau = 2\int_0^1 \dot{\phi}\dot{\psi}d\tau \\ &= -2(2\pi)^2 a_\phi a_\psi \int_0^1 \cos^2 2\pi\tau\, d\tau = -4\pi^2 a_\phi a_\phi. \end{aligned} \tag{4.6}$$

A set of simulations for this pair of inputs is presented in Figure 5. The right-hand plot in this figure shows the time evolution of the momentum. For each

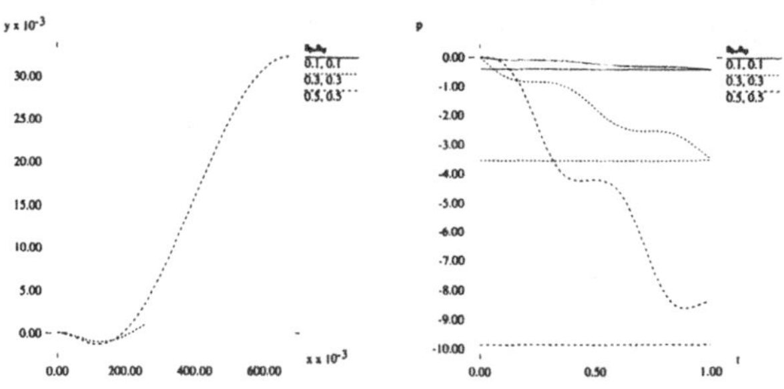

Figure 5: *Forward motion during momentum generation, and momentum versus time. The approximated value of the momentum is shown in horizontal lines.*

set of inputs, the value of the momentum after one cycle predicted by Equation 4.6 is shown as a horizontal line. It should be noted that the prediction for the small inputs works quite well up to an amplitude (in radians) of approximately $a_\phi = a_\psi = 0.3$. Larger amplitude inputs ($a_\phi = a_\psi = 0.5$) are also plotted to show that while some discrepancy does exist between the predicted and simulated final values of momentum, the error is relatively small (on the order of 15%).

For the snakeboard example, steering comes from controlling ϕ as it affects the term $I^{-1} = (-\hat{J}/2, 0, \hat{J}\tan\phi/2)^T$. To see how the steering algorithm could be implemented, we provide in Figure 6 a sample plot of motion control for the snakeboard moving from $(x, y, \theta) = (0, 0, 0)$ to $(0, -5.0, \pi)$, along with a plot of the momentum as a function of time. Notice that in this plot the system first builds up momentum using a serpentine-like snakeboard "drive" gait in Step 1, and then executes a turn, a straight coast, and a final turn in Step 2. The turns are executed at the minimum turning radius, which we have constrained to be $\phi_{max} = \pm 1.0$rad. For this simulation, we execute the same set of input motions at the end of the trajectory as was used at the start, except that the wheel angles are turned through opposite rotations of those used originally. These in-phase cyclic gaits used to build up and to reduce momentum at the start and finish of the motion are run for the same length of time. As is clear from the the right-hand graph in Figure 6, this results in a near exact cancelation of the momentum that was built up during the initial motions.

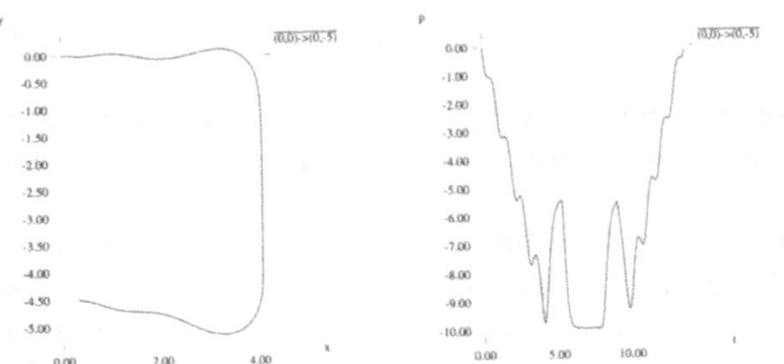

Figure 6: *Steering using the full algorithm, along with associated changes in momentum.*

5 Adding a Feedback Term: The Snake/Eel Robot

In [15], we studied anguilliform locomotion using a simplified physical model of a snake (we use the term "snake" interchangeably with "eel") to be used as a platform to test various locomotive gaits. We model the snake as a planar, serial chain of identical links with mass m and inertia J. We assume full control of the internal shape of the snake (the joint angles ϕ_i) which allows us to solve the dynamic equations in terms of the unknown configuration variables (x, y, θ)—the position and orientation of the first link. Since our mechanical robot (the REEL eel [15]) is composed of four links, all simulations in this paper have been performed for a four link model of a snake. For simplicity and symmetry, however, the analytical derivations have been performed on a three link model.

The crucial elements in this model are the drag force terms, which generate the locomotion. To simulate the forces in the water, we adopt a simple fluid mechanical model. We assume that the Reynolds number is high enough that inertial forces dominate over viscous effects—a reasonable approximation for smooth bodies in an inviscid fluid. We also assume that the fluid is stationary, so the force of the fluid on a given link is due only to the motion of that link. The pressure differential created by an object moving in a fluid causes a drag force opposing the motion. Under the assumptions above, the drag force developed takes the form $F \propto \mu_w v^2$. Here, v is the forward speed of the link and μ_w is a drag coefficient for the water, determined by the formula $\mu_w = \rho AC/2$, where A is the effective area of the object, ρ is the density of water, and C is a shape coefficient.

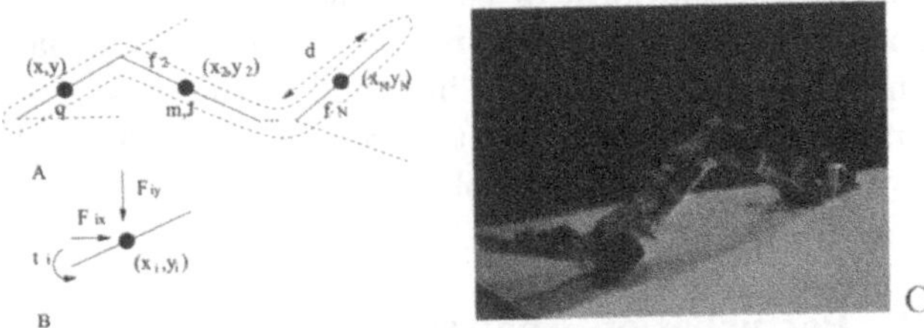

Figure 7: *(A) Model of snake, (B) Forces and torques on link i, (C) The REEL eel robot.*

5.1 Force Approximation and Momentum Equation

In our simulations, we assume that pressure differentials in the directions parallel to the moving body are decoupled from pressure differentials perpendicular to the body, to yield:

$$F_i^{\perp} = -\mu_w \text{sgn}(v_i^{\perp}) \cdot (v_i^{\perp})^2 \tag{5.7}$$

where $v_i^{\perp}$ is the projection of the vector $(\dot{x}_i, \dot{y}_i)$ along a direction perpendicular to the link. We exclude drag forces parallel to the link because they were determined in simulation to have negligible effects. For all numerical simulations, the full forcing term in Equation 5.7 is used. However, the discontinuity in $\text{sgn}(v)$ means that this expression is not tractable for use in calculations. Therefore, in our analytical derivations we utilize an approximation to this function, which turns out to be linear in v: $F_{\text{approx}} = \mu v^{\perp}$, where μ is defined by a least squares fit over some small range around $v = 0$. We also note that this force model can be interpreted as a viscous damping model, as might be encountered with a snake moving over soft sand.

Using this expression for the drag forces and exploiting the invariance of the system with respect to changes in position and orientation, the momentum equation becomes:

$$\begin{aligned} \dot{p}_i = \gamma^{kl} p_k p_l - \beta_l^k p_k \dot{r}^l + \lambda_{ij} (I^{-1})^{jl} p_l \\ + (\eta_{il} - \lambda_{ij} \mathbb{A}_l^j) \dot{r}^l, \end{aligned}$$

where λ and η are transformations related to the drag forces. This momentum equation differs from the equation introduced in Section 2.1 because the frictional forces must enter the dynamics explicitly, rather than implicitly in the form of constraint forces. A direct impact of this is that the form of the momentum equation is no longer quadratic in p and $\dot{r}$, but now also contains linear terms. For this reason, the results of Proposition 1 no longer apply, but similar conclusions can still be drawn.

5.2 Perturbation Analysis

We proceed as before, setting $r(t) = r_0 + \epsilon u(t)$, and solving for $p_i = p_{i0} + \epsilon p_{i1} + \epsilon^2 p_{i2} + \ldots$. A traveling wave of the form $r_i = \phi_i = \epsilon \sin(\omega t + i\phi_s)$ can be used to drive the snake in the forward and backward directions by appropriately choosing the sign of ϕ_s. With the assumption that $p(0) = 0$, we can determine that the momentum at both the zeroth order (p_{i0}) and first order (p_{i1}) in ϵ are both zero after any complete cycle of the gait.

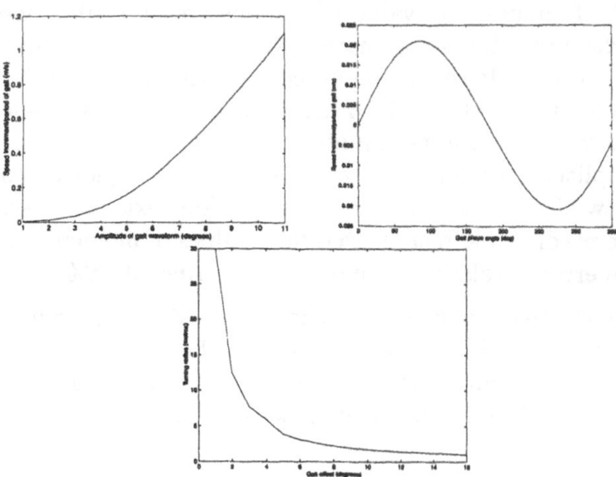

Figure 8: *Simulation of speed changes per cycle with respect to amplitude, ϵ (top left), and phase, ϕ_s (top right). Also shown (bottom) is turning radius versus the gait offset parameter, ϕ_{offs}.*

Examination of the second order momentum term reveals a single term that drives the momentum in the forward direction. For the case of the three link snake (the simplest configuration capable of locomotion), we can express the average forward momentum (over one cycle) with:

$$\Delta p_1 = -\epsilon^2 K(\phi_0) \omega \sin(\phi_s), \tag{5.8}$$

where the momentum gained over one cycle is proportional to the square of the gait amplitude and the sine of the gait phase angle (K is a constant of proportionality that is a function of the robot parameters). Although this is only an approximation, we find that the simulation results show a very good agreement with the approximation, even for fairly large amplitude inputs, as shown in Figure 8. For longer snakes (with more than three links), the simple sinusoid is replaced by a more complex sum of harmonics of the phase angle. Equation 5.8 provides a theoretical justification for the open loop inputs, presented above, for generating forward momentum. Intuition suggests that a similar expression exists describing the evolution of p_3

(angular momentum) when unbalanced (turning) gaits are used, by adding an offset, $\phi_0 = \phi_{\text{off}}$ to the joint angles. Initial numerical simulations show that this is the case, but we are continuing to investigate these issues further analytically.

5.3 Controller Development

There are three basic components to the steering algorithm discussed above in Section 3: building up speed, steering, and stopping. We have just derived an expression that can be used in driving the system to a steady-state speed. Next, we examine a mechanism for steering with feedback.

5.3.1 Steering Controller

In our previous work [15], we presented results from open loop control using this model. One of the important points that can be taken from this work involves the ability to steer. We have examined the addition of a biasing term, ϕ_{offs}, to the wave gait, which causes the snake to follow a circular path. There is an inverse relationship between turning radius and gait offset, as shown in Figure 8.

It should also be noted that when performing feedback on a system using periodic gaits, the control is based on discrete sampling by averaging the center of mass position and velocity over one cycle. This places a theoretical (Nyquist) limit on the controller dynamics of $F_S/2$. In practice, for a 1Hz drive gait, controller responses of under five seconds are likely not achievable.

We use a feedforward law based on the (curvature of the) path generated by the kinematic motion planner. In addition to this, we note that since the steering is only approximate, the addition of a feedback term is desirable. If we define d as the signed, perpendicular distance to the desired trajectory (with d negative when the desired trajectory is to the right of the actual trajectory), the motion of the cart is described by $\dot{d} = v\sin(\theta_d(t) - \theta(t))$ (see Figure 3). Using a proportional feedback control law for the cart of the form $\dot{\theta}(t) = k_\theta(\theta_d(t) - \theta(t)) - k_d(d)$, one can show that the linearized equations of motion can be easily stabilized about a straight path. The characteristic equation for the linearized system is by $s^2 + k_\theta s - vk_d = 0$. The condition for stability of this system, therefore is $k_\theta > 0$ and $k_d < 0$ which matches our intuition about steering.

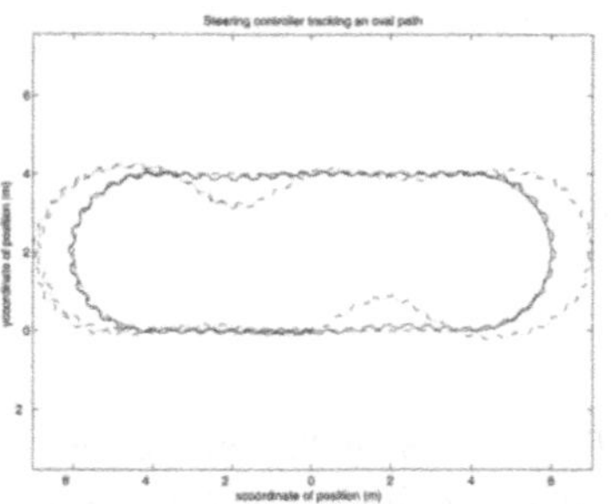

Figure 9: *Steering around an oval shaped path*

Figure 9 shows the path of the eel tracking an oval shaped path that integrates both circular and straight line segments. With our simple kinematic tracking (feedback only) approach, (outer path–dot-dashed) there is considerable overshoot where the controller attempts to transition from circular to straight line motion and *vice versa* due to accumulated linear and angular momentum. This overshoot can be eliminated by the use of a feedforward look-ahead control (inner path–solid), allowing tracking of path curvature. The desired path is the dashed curve.

5.3.2 Stopping Controller

The problem of stopping given an initial forward momentum is considered independently of the problem of steering. In general, a stopping controller that drives the robot's velocity to zero will result in a repeatable motion, possibly including both a translation and a rotation during some stopping time, t_s. Unlike the snakeboard, the first order approximation to stopping does not lead to the opposite motion as that found in starting, largely due to the dissipation found in the eel.

Let us define a ***stopping transformation***, T_s, for a given initial velocity. Then we can use our steering controller to steer to a configuration that can reach the goal through the transformation T_s. Using the stopping controller to drive the velocity to zero from that point will result in the robot having a final configuration at the goal point with zero velocity. The transformation, T_s, depends primarily on the momentum of the robot before initiating the stopping gait, which would make the determination of T_s an untractable problem if it were necessary to consider the entire momentum space. However, since we have limited ourselves to straight line trajectories and circular arcs, we

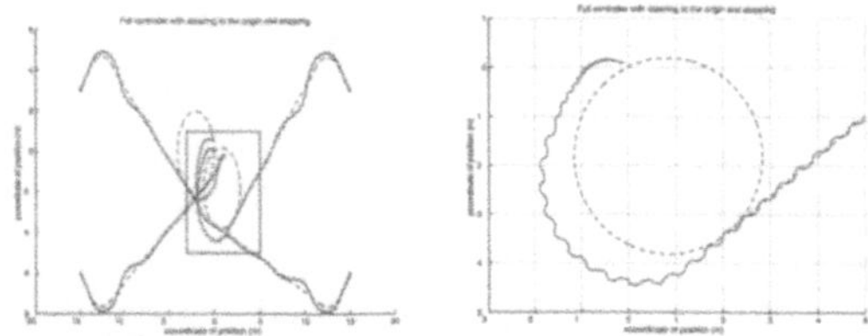

Figure 10: *The motion of the eel tracking from all four quadrants, along with a focused plot of the motion from the first quadrant*

need only consider three cases: stopping after straight line motion and stopping after motion along clockwise and counterclockwise circles.

5.3.3 Path Planning

Figure 10 shows the results of the proposed motion planning algorithm for initial conditions in each quadrant, with differing initial orientation. Desired paths are shown dashed and actual paths solid. Also shown in Figure 10 is the boxed area in greater detail for the initial position in the first quadrant. The robot stops within 50cm of the origin in all four cases, with orientation errors of less than 30 degrees. We are currently exploring "station keeping" gaits to correct these small errors.

6 Conclusions and Future Work

This paper establishes easily implemented feedforward and feedback control algorithms for mechanical systems with Lie group symmetries, external forces, and nonholonomic constraints. We focus our attention on systems that use small amplitude, periodic shape inputs, or gaits, for momentum generation and steering. We have proposed a motion planning algorithm based on Dubins' work on the control of kinematic wheeled vehicles. The motion planning problem reduces to a momentum generation phase (with a complementary momentum dissipation phase) and a steering phase in which the complex dynamics of the system can be approximated using a kinematic steering model. We develop these abstract (kinematic) steering models using a hierarchical, hybrid-systems approach to the problem, in which we look for commonalities among a wide range of dynamic systems– in this case, the commonality is that the system must be *steerable*.

We present results from simulations in which we applied our motion planning algorithm to two different types of mobile robots: the snakeboard and the eel. Although these systems differ in the type of interaction with the environment (non-holonomic versus viscous drag forces), we show that the same abstract motion planning algorithm can be used in both cases, with only slight modifications required to overcome the presence of dissipation. Since our algorithms build on results from kinematic path planners, we believe that it will be natural to extend this work to include obstacle avoidance and other motion planning issues.

Acknowledgments

The authors gratefully acknowledge the support of NSF grants IRI-9711834 and IIS-9876301, and ARO grants P-34150-MA-AAS, DAAH04-96-1-0007, and DURIP DAAG55-97-1-0064.

References

[1] A. M. Bloch, P. S. Krishnaprasad, J. E. Marsden, and R. M. Murray. Nonholonomic mechanical systems with symmetry. *Archive for Rational Mechanics and Analysis*, 136:21–99, December 1996.

[2] F. Bullo, N. E. Leonard, and A. D. Lewis. Controllability and motion algorithms for underactuated Lagrangian systems on Lie groups. Submitted to IEEE *Transactions on Automatic Control*, February 1998.

[3] F. Bullo and A. D. Lewis. Configuration controllability for mechanical systems on Lie groups. In *Symposium on Mathematical Theory of Networks and Systems*, St. Louis, June 1996.

[4] L. E. Dubins. On curves of minimal length with a constraint on average curvature and with presribed initial and terminal positions and tangents. *American Journal of Mathematics*, 79:497–516, 1957.

[5] B. Goodwine and J. W. Burdick. Trajectory generation for kinematic legged robots. In *Proc. IEEE Int. Conf. Robotics and Automation*, pages 2689–2696, Albuquerque, NM, April 1997.

[6] H. Khalil. *Nonlinear Systems*. Macmillan Publishing Co., 1992.

[7] P. S. Krishnaprasad and D. P. Tsakiris. Oscillations, $SE(2)$–snakes and motion control. In *IEEE Conf. Decision and Control*, pages 2806–2811, New Orleans, LA, December 1995.

[8] P. S. Krishnaprasad and D. P. Tsakiris. *G*-snakes: Nonholonomic kinematic chains on Lie groups. In *Proc.* 33rd *IEEE Conf. Decision and Control*, pages 2955–2960, Lake Buena Vista, FL, December 1994.

[9] J.-C. Latombe. *Robot Motion Planning*. Kluwer, Boston, 1991.

[10] J.-C. Latombe. Motion planning: A journey of robots, molecules, digital actors, and other artifacts. *International Journal of Robotics Research*, 18(11):1119–1128, December 1999.

[11] N. E. Leonard. Periodic forcing, dynamics and control of underactuated spacecraft and underwater vehicles. In *Proc. IEEE Conf. Decision and Control*, pages 1131–1136, New Orleans, LA, December 1995.

[12] N. E. Leonard and P. S. Krishnaprasad. Motion control of drift-free, left-invariant systems on Lie groups. *IEEE Trans. on Automatic Control*, 40(9):1539–1554, September 1995.

[13] Andrew D. Lewis. When is a mechanical control system kinematic? In *Proc. IEEE Conf. on Decision and Control*, pages 1162–1167, December 1999.

[14] Z. Li and J. F. Canny, editors. *Nonholonomic Motion Planning*. Kluwer, 1993.

[15] K. McIsaac and J. Ostrowski. A geometric formulation of underwater snake-like locomotion: Simulation and experiments. In *IEEE Conf. on Robotics and Automation*, pages 2843–2848, Detroit, May 1999.

[16] R. M. Murray and S. S. Sastry. Nonholonomic motion planning: Steering using sinusoids. *IEEE Transactions on Automatic Control*, 38(5):700–716, 1993.

[17] J. P. Ostrowski. *The Mechanics and Control of Undulatory Robotic Locomotion*. Ph.D. thesis, California Institute of Technology, Pasadena, CA, 1995. Available electronically at http://www.cis.upenn.edu/~jpo/papers.html.

[18] J. P. Ostrowski. Steering for a class of dynamic nonholonomic systems. To appear in IEEE *Transactions on Automatic Control*, September 1999.

[19] J. P. Ostrowski and J. W. Burdick. The geometric mechanics of undulatory robotic locomotion. *International Journal of Robotics Research*, 17(7):683–702, July 1998.

[20] J. P. Ostrowski, A. D. Lewis, R. M. Murray, and J. W. Burdick. Nonholonomic mechanics and locomotion: The snakeboard example. In *Proc. IEEE Int. Conf. Robotics & Automation*, pages 2391–7, San Diego, May 1994.

[21] J. Ostrowski, J. P. Desai, and V. Kumar. Optimal gait selection for nonholonomic locomotion systems. *International Journal of Robotics Research*, 19(3):225–237, March 2000.

[22] J. Radford and J. Burdick. Local motion planning for nonholonomic control systems evolving on principal bundles. *Conf. Mathematical Theory of Networks and Systems*.

A Kinematics-Based Probabilistic Roadmap Method for Closed Chain Systems

Li Han, *Texas A&M University, College Station, TX*
Nancy M. Amato, *Texas A&M University, College Station, TX*

In this paper we consider the motion planning problem for closed chain systems. We propose an extension of the PRM *methodology which uses the kinematics of the closed chain system to guide the generation and connection of closure configurations. In particular, we break the closed chains into a set of open subchains, apply standard* PRM *random sampling techniques and forward kinematics to one subset of the subchains, and then use inverse kinematics on the remaining subchains to enforce the closure constraints. This strategy preserves the PRM sampling philosophy, while addressing the fact that the probability that a random configuration will satisfy the closure constraints is zero, which has proven problematical in previous attempts to apply the PRM methodology to closed chain systems.*

Another distinguishing feature of our approach is that we adopt a two-stage strategy, both of which employ the PRM *framework. First, we disregard the environment, fix the position and orientation of one link (the "virtual" base) of the system, and construct a* kinematic roadmap *which contains different self-collision-free closure configurations. Next, we populate the environment with copies of the kinematic roadmap (nodes and edges), and then use rigid body planners to connect configurations of the same closure type. This two-stage approach enables us to amortize the cost of computing and connecting closure configurations.*

Our results in 3-dimensional workspaces show that good roadmaps for closed chains with many links can be constructed in a few seconds as opposed to the several hours required by the previous purely randomized approach.

1 Introduction

Closed chain mechanisms arise in many practical problems, such as the Stewart Platform [22], closed molecular chains [21], reconfigurable robots [14, 20], and the closed chain system formed by multiple robots grasping an object [13] (see Figure 1). Closed chains are sometimes called parallel chains since they can be viewed as consisting of two or more serial/open chains that provide parallel linkages between two points. While closed chains can offer advantages over open chains in terms of the rigidity of the mechanism, motion planning and control of closed chains is complicated by the need to maintain the closed chain structure, the so-called *closure constraint.*

Figure 1: *The Stanford Assistant Mobile Manipulator [12]. (Photo Courtesy of Prof. O. Khatib.)*

In this paper we consider the motion planning problem for closed chain systems. The motion planning problem is to find a collision-free path that takes the closed chain from one configuration to another. A real world example is to find a collision-free path for a multi-fingered robotic hand, to move a grasped part from one station to another for machining. For some tasks, the robot might need to regrasp the object, i.e., to change the grasp points and grasp fingers, so as to accommodate the workspace limits of the robot or to avoid collisions. In general, it is not easy to move and regrasp the object simultaneously. One approach [16] proposed for this problem is to interleave *transit paths,* which only implement the regrasp without moving the object, and *transform paths*, which only transfer the object using fixed grasps. A manipulation system on a transform path with a fixed grasp can be viewed as a system with closed chains.[1] While the transform path planning problem has been studied for simple robot manipulation systems, the general problem remains open and is one motivation for our work. Other motivational applications of closed chain motion planning include animation, virtual reality, and training.

Motion planning [15] is a challenging problem which involves complicated physical constraints and high-dimensional configuration spaces. The fastest existing deterministic planner [6] takes time exponential in the number of degrees of freedom of the robot. On the other hand, a class of randomized planners proposed during the last decade have successfully solved many previously unsolved problems. In particular, *Probabilistic Roadmap Methods (PRMs)* [2, 3, 4, 5, 10, 11, 24] have been used successfully in high-dimensional configuration spaces. The general methodology of PRMs is to construct a graph (the roadmap) during preprocessing that represents the connectivity of the robot's free configuration space (C-space), and then to query the roadmap to find a path for a given motion planning task (see Figure 2). Roadmap vertices are (generated from) randomly sampled configurations which satisfy feasibility requirements (e.g., collision free), and roadmap edges correspond to connections between "nearby" vertices found with simple local planning methods. For the most part, the major successes for PRMs have been limited to rigid bodies or articulated objects without closed chains. Recently, some efforts have been made to apply the PRM paradigm to closed chain systems [18] and to flexible objects [9].

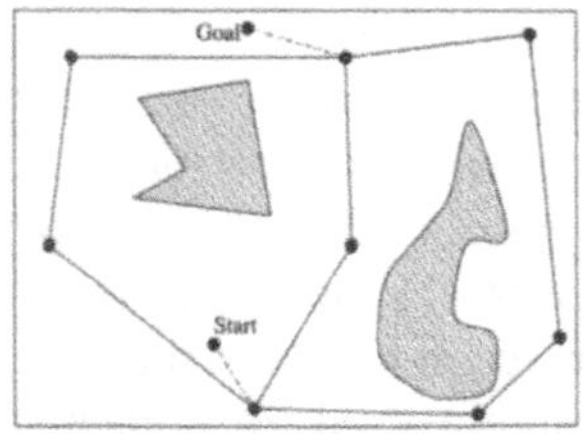

Figure 2: *A probabilistic roadmap (C-space).*

[1]Besides the collision-free and closure constraints, more manipulation constraints need to be taken into account for regrasp planning, which will be addressed in a follow up paper.

The PRM planner for closed chain mechanisms proposed in [18] builds a roadmap in the portion of the configuration space that satisfies the closure constraints. The roadmap vertices are generated by first sampling points from the entire configuration space, and then performing a randomized gradient descent to try to transform them into configurations satisfying the closure constraints. Roadmap vertices are connected by a randomized gradient descent traversal of the constraint surface. When applied to closed chain linkages composed of line segments in the plane, this approach required several hours of computation to generate a well connected roadmap. Thus, while this work represents a crucial first step towards extending the PRM methodology to this important class of problems, the methods do not yet lead to efficient solutions for many practical problems.

1.1 Our Approach

In this paper, we propose a new approach for planning the motion of kinematic chains with closure constraints. Like [18], we believe the PRM methodology can be extended to closed chains. However, since the probability that a randomly generated node lies on the constraint surface is zero [18], we believe that the purely randomized philosophy of the PRM must be augmented with more deliberate techniques developed in the robotics community to deal with the closure constraints. In particular, we advocate the use of kinematics to guide the generation and connection of closure configurations. Briefly, the kinematics of an articulated object (such as a robotic finger) describes the relationship between the configuration/motion of the joints of the linkage and the resulting configuration/motion of the rigid bodies which form the linkage. Our planner uses both forward and inverse kinematics to generate closure configurations. In particular, we break the closed chain into a set of open subchains, apply standard PRM random sampling techniques and forward kinematics to one subset of the subchains, and then use inverse kinematics on the remaining subchains to enforce the closure constraints.

Another distinguishing feature of our approach is that we adopt a two-stage strategy, both of which employ the PRM framework. First, we disregard the environment, fix the position and orientation of one link of the chain (which can be viewed as a "virtual" base

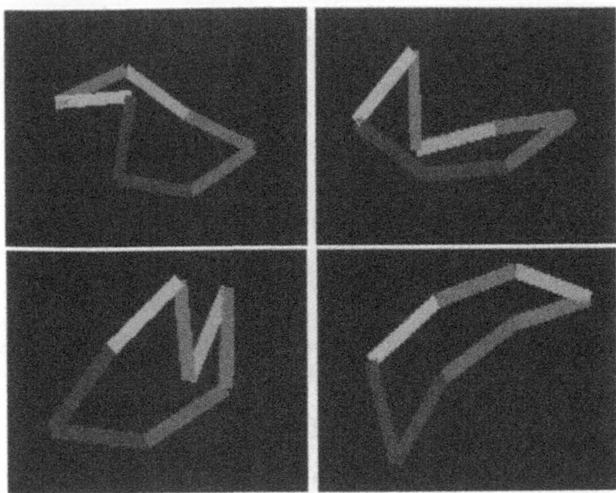

Figure 3: *Snapshots of a kinematic roadmap path for a 7-link closed chain.*

of the system)[2] , and construct a roadmap which contains different self-collision-free closure configurations. We call this roadmap a ***kinematic roadmap*** since it deals solely with the robot's kinematics and utilizes both forward and inverse kinematics in its construction. Figure 3 shows snapshots from a path contained in a kinematic roadmap found by our planner. Next, we populate the environment with copies of (portions of) the kinematic roadmap. This stage again employs the PRM strategy. In particular, we select random configurations (position and orientation) in the environment for the base in the kinematic roadmap, and retain those portions of the kinematic roadmap that are collision-free. Finally, local planning methods for rigid bodies are used to connect configurations of the same closure type. Note that this strategy restricts the connection of different closure configurations to the kinematic roadmap, i.e., the only edges between different closure types in the main roadmap are copied from the kinematic roadmap.

Our motivation for this two-stage approach is that it amortizes the cost of computing and connecting the closed chain configurations. Another benefit of this strategy is that we do not waste time trying to connect unconnectable closure configurations when constructing the final roadmap. Indeed, our experimental results in 3D workspaces show that good roadmaps for closed chains with many links can be constructed in a few seconds as opposed to the several hours required by the previous purely randomized approach [18].

This paper is outlined as follows. In Section 2, we describe the related work in more detail. We formally define the closed chain motion planning problem in Section 3. The details of our approach are described in Sections 4 through 6. We present some experimental results in Section 7, and some concluding remarks in Section 8.

For simplicity and clarity, two-dimensional figures are used throughout this paper to illustrate the closed chain motion planning problem and our kinematics-based PRM motion planning approach. It should be noted that the general discussion and planning framework are applicable to both 2D and 3D workspaces.

2 Related Work

PRMs. As mentioned in Section 1, the success of PRMs for rigid body and articulated open chain robots [2, 3, 4, 5, 10, 11, 24], has motivated the extension of the PRM strategy to planning for elastic objects (FPRM) [9] and closed chains [18]. The major challenge here is in finding an effective way to deal with the additional constraints imposed on the feasible robot configurations. In particular, while the only constraint on feasible configurations for rigid bodies and open chains is that they be collision free (constraints on the joint variables of an open chain linkage can generally be encoded in the robot's configuration space), elastic objects can only achieve deformations with (local) minimum elastic energy and closed chains need to satisfy the closure constraints.

As previously mentioned, while [18] pioneered the use of PRMs on closed chain systems, the relative inefficiency of the randomized gradient descent technique used to generate closure configurations from randomly sampled configurations, and to connect closure configurations, emphasized the need for better techniques. The fundamental problem is that since the probability that a sampled node lies on a constraint surface is zero [18], it is very difficult to find (and connect) configurations satisfying the closure constraints using purely randomized techniques. This is what motivates our kinematic roadmap, whose construction employs both forward and inverse kinematics. We note that although

[2]The virtual base can be chosen in other ways. For example, in a hand-object manipulation system, we can choose the object as the virtual base of the system.

it was not used, the possibility of placing pre-computed closure configurations at different locations in the environment was mentioned in [18].

FPRM [9] deals with the energy requirement for elastic objects in a manner similar to the kinematic roadmap we use for closed chains: minimal energy deformations are computed *a priori*, disregarding the obstacles in the environment, and then copies of these deformations are placed at randomly selected locations (positions and orientations) in the environment. However, a difference between FPRM and our approach is that we also use the PRM strategy to construct the kinematic roadmap and populate kinematic roadmap edges, which correspond to connections between closure configurations. In contrast, FPRM populates minimal energy deformations only and performs node connection directly in the environment. We note that our two-stage PRM strategy can be applied to flexible object motion planning. More specifically, we can first generate a "deformation" roadmap with connections between minimal energy deformations in a clear environment and then populate it to the real environment containing obstacles.

Other Methods. A random exploration strategy, based on the *Ariadne's Clew Algorithm* [1], has been used to solve point-to-point inverse kinematics problems for redundant manipulators. Given an initial configuration of a robot, the problem was to find a reachable configuration that corresponds to a desired position and orientation of the robot end-effector and to find a feasible path connecting the initial and the goal configuration. Central to their approach was the construction of a roadmap[3] which took into account constraints due to joint limits, self-collision, and collision with environment obstacles. A point-to-point inverse kinematics problem was then solved by querying the roadmap.

In [25], the recently proposed *Rapidly-exploring Random Trees (*RRT*)* [17] strategy was used to greatly decrease the computation time required for several of the examples studied in [18].

[3]While the roadmap in [1] is called a *kinematic roadmap*, it is different from our kinematic roadmap which we construct without any knowledge of the obstacles in the environment.

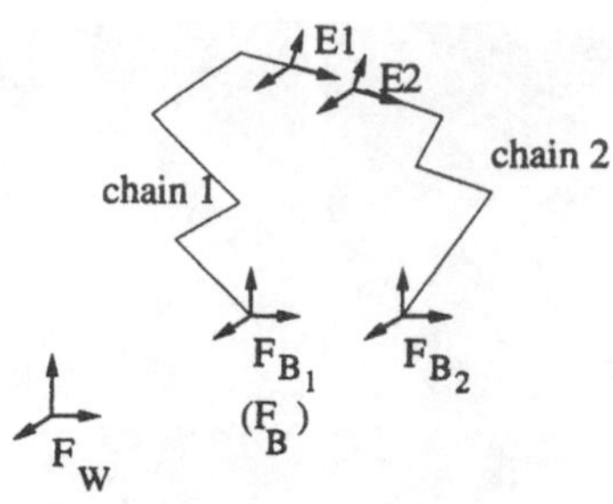

Figure 4: *Breaking a Closed Chain into 2 Open Chains.*

3 Problem Formulation

In this section, we study the configuration spaces of multi-link chains and discuss the effect of closure constraints on these spaces. When a linkage system involves multiple closed chains, the overall closed chain constraint is satisfied if and only if each closed chain constraint is satisfied. Therefore, we discuss in detail only the case for one closed chain, with the understanding that the problem involving multiple closed chains can be similarly formulated and handled by our planner.

We first note that a closed chain system can alternatively be viewed as a linkage system consisting of a collection of open chains, where we "break" each closed chain, and then satisfy the closure constraints, if any, by forcing the break points to coincide. For example, in Figure 4, chain 1 and chain 2 form a closed chain where the frames E_1 and E_2 attached to the breakpoint (the "end effector") must coincide to satisfy the closure constraints.

Consider a closed chain system that can be broken into k open chains. The configuration of an open chain i can be specified by its base configuration and its joint variables. In particular, the configuration of the base can be specified by the Euclidean (rigid body) transformation from the world frame F_W to the body frame F_{Bi}: $g_{wb_i} = (p_{wb_i}, R_{wb_i}) \in SE(d)$ where $d \in \{2,3\}$ is the dimension of the workspace, $p_{wb_i} \in \mathrm{R}^d$ and $R_{wb_i} \in SO(d)$ are, respectively, the position and orientation of F_{Bi} relative to F_W, and $SE(d)$ denotes the special Euclidean group. Denote by $\beta_i = (\beta_{i1}, \ldots, \beta_{in_i})$, the vector of joint variables for chain i. For a revolute joint, the joint variable is an angle $\beta_{ij} \in [0, 2\pi)$, with the angle 2π equated to angle 0, which is naturally associated with a unit circle in the plane, denoted by S^1, and hence we write

$\beta_{ij} \in S^1$. A prismatic joint is described by a linear displacement $\beta_{ij} \in R$ along a directed axis. In summary, the *configuration space* of a multi-link robot can be represented as:

$$\mathcal{C} = \{(g_{wb_1}, \beta_1, \cdots, g_{wb_k}, \beta_k)| \\ g_{wb_i} \in SE(d), \beta_i \in S^{r_i} \times R^{p_i}, i = 1, \cdots, k\}. \quad (1)$$

where r_i and p_i are the number of revolute joints and prismatic joints for link i, respectively.

One of the reasons PRMs work well for systems without closed chains is that any configuration q sampled from $\mathcal{C}$ is a *valid* configuration (when collision constraints are ignored). However, this is not true for systems involving closed chains where q must also satisfy the closure constraints. For example, the two end-effector frames in Figure 4 must coincide:

$$g_{we_1} = g_{wb_1} g_{b_1e_1}(\beta_1) = g_{wb_2} g_{b_2e_2}(\beta_2) = g_{we_2} \quad (2)$$

where $g_{b_ie_i}(\beta_i), i = 1, 2$, is the *forward kinematic transformation* [7, 19] of chain i which determines the end-frame configuration based on joint variables. (*Inverse kinematics* solves the inverse problem of determining proper joint variables to achieve some specified end-frame configuration.)

The closure constraint is often expressed in the form $f(q) = 0$ (e.g., $g_{we_1} - g_{we_2} = 0$). When multiple closed chains exist in a linkage system, each closed chain imposes one closure constraint; the lth such constraint is denoted by f_l, and we use $f(q) = 0$ to denote $f_l(q) = 0$, for all $1 \leq l \leq K$, where K is the number of closed chain constraints. In general, the valid configurations of a closed chain system lie in the set:

$$\mathcal{C}_{closure} = \{q | q \in \mathcal{C} \text{ and } f(q) = 0\}. \quad (3)$$

Notice that closure constraints such as Equation 2 can be transformed to the zeros of polynomials using the projective transformation. Then, the valid configurations of the system, $\mathcal{C}_{closure}$, define a lower-dimensional *algebraic variety* embedded in the higher-dimensional configuration space $\mathcal{C}$. This is roughly analogous to embedding a 2D surface or a 1D curve in a 3D space. The fact that the volume measure of a low-dimensional entity in a high-dimensional ambient space is zero is why the probability that a random configuration $q \in \mathcal{C}$ will satisfy the closure constraint is zero. While closure constraints pose difficulties for standard PRM planners, as we will see, the structure of $\mathcal{C}_{closure}$ can be utilized to guide the generation and connection of closure configurations.

Finally, for both open and closed chain systems, feasible configurations should not involve collision between the robot and an obstacle, or self-collision among the links. We denote by $\mathcal{C}_{free}$ the set of robot configurations $q \in \mathcal{C}$ which do not cause any collision in the system.

Using the notation defined above, the closed-chain motion planning problem can be defined as follows:

Problem 1 *Given a start configuration q_0 and a goal configuration q_1, the objective of the planner is to find a path $q(t), t \in [0,1]$, such that $q(0) = q_0$, $q(1) = q_1$, and $\forall t \in [0,1], q(t) \in \mathcal{C}_{closure} \cap \mathcal{C}_{free}$.*

4 A Kinematics-Based PRM

In this section, we describe the high-level strategy of our kinematics-based PRM for closed chain systems.

We begin by noting that the closure constraint in Equation 2 can be reduced to:

$$g_{wb_1}(g_{b_1e_1}(\beta_1) - g_{b_1b_2} g_{b_2e_2}(\beta_2)) = 0 \quad (4)$$

$$g_{b_1e_1}(\beta_1) - g_{b_1b_2} g_{b_2e_2}(\beta_2) = 0 \quad (5)$$

where $g_{b_ie_i}, i = 1, 2$, are the end-frame configurations described in the body frame F_{Bi}, and $g_{b_1b_2}$ is the transformation from the base link of chain 1 to the base link of chain 2. If we think of the base link of the first chain as the virtual (mobile) base of the system, its configuration g_{wb_1} can be interpreted as a rigid body motion on the system. In other words, Equation 4 reveals one important property of closure configurations: *Rigid body transformations preserve closure configurations.* Equation 5 further shows that closure constraints can be defined independent of the base configuration.

In the following discussion, we will use g_{wb} to denote the configuration of the virtual base of the system. In addition, we will treat the transformation from the system base to the base of any chain, say chain i, as a virtual link with joint variables being the parameterization of the transformation g_{bb_i}, e.g., a position vector (prismatic joints) $p_{bb_i} \in R^d$ and an orientation vector (revolute joints) $\alpha_{bb_i} \in S^{\frac{d(d-1)}{2}}$. Thus, every open chain can be viewed to be virtually extended to the system base. As a result, we can define a joint variable for the extended chain as $\theta_i = (p_{bb_i}, \alpha_{bb_i}, \beta_i), i = 1, \cdots, k$. For simplicity, we will call θ_i the joint variable of sub-chain i, with the understanding that it includes the virtual joint variables, when applicable. Define the joint variable of the system to be $\theta = (\theta_1, \cdots, \theta_k) \in S^r \times R^p$, where p and r are the total number of prismatic joints and revolute joints, respectively. Then the system configuration space can be rewritten as:

$$\mathcal{C} = \{(g_{wb}, \theta) | g_{wb} \in SE(d), \theta \in S^r \times R^p\}. \quad (6)$$

Since the closure constraint (Equation 2) or its general form $f(q) = 0$, in fact, does not depend on the base configuration and only specifies constraints with respect to joint variables, we can define the equivalent constraints of f on the joint variables as:

$$\tilde{f}(\theta) = 0 \quad (7)$$

Furthermore, define:

$$\tilde{\mathcal{C}}_{closure} = \left\{\theta | \theta \in S^r \times R^p, \tilde{f}(\theta) = 0\right\}. \quad (8)$$

Now, the subset of $\mathcal{C}$ corresponding to closure configurations can be expressed as:

$$\mathcal{C}_{closure} = \left\{(g_{wb}, \theta) | g_{wb} \in SE(d), \theta \in \tilde{\mathcal{C}}_{closure}\right\} \quad (9)$$

The following two observations summarize the above discussion, and form the basis of our two-stage PRM closed chain planner:

Observation 1 *Only the joint variables θ determine if a configuration $q = (g_{wb}, \theta)$ is closure or not. The closure constraint defines an algebraic variety (Equation 8) on $\theta \in S^r \times R^p$, which can be parameterized almost everywhere. In other words, θ can be partitioned into θ_a and θ_p, $\theta = \theta_a \times \theta_p$, where θ_p can be determined from θ_a based on the closure constraints (Equation 7).*

Observation 2 *A given "closure configuration" θ, can be combined with different base configurations $g_{wb} \in SE(d)$, which corresponds to placing the same closure configuration at different locations in the environment.*

The first observation suggests an efficient way to generate closure configurations: Ignore the environment (obstacles) and set a nominal base configuration, randomly generate θ_a and determine the corresponding θ_p by solving the closure constraints (Equation 7), and retain the self-collision free closure configurations. For example, consider the closed chain shown in Figure 4. Suppose we select the joint variables of chain 2, θ_2, as the *active variables* θ_a, and randomly generate values for them. We then use *forward kinematics* to determine the end-frame configuration g_{be_2}, and then use *inverse kinematics* to compute joint variables of chain 1 (the *passive variables* θ_p) which will make g_{be_1} coincide with g_{be_2} and satisfy the closure constraints.

The closure configurations generated will be the vertices of a (small) roadmap which records paths connecting self-collision-free closure configurations (again, with no dependence on the base configuration g_{wb}). The edges in this roadmap can be generated using straight-line, or any other simple local planner to connect the active variables θ_a of the two closure configurations, and then computing the corresponding passive variables θ_p along the local path. As with any PRM, a self-collision-free local path is recorded as a roadmap edge. By a slight abuse of terminology, we call such a roadmap a *kinematic roadmap* since it reveals the kinematic connectivity of the closure structures, and its construction involves the computation of both forward and inverse kinematics. Figure 5(a) shows a three-node kinematic roadmap for a 4-link closed chain.

The second observation suggests that we place multiple copies of the same closure configuration in the environment. There are two advantages to this approach. First, we quickly populate the environment with closure configurations and amortize the cost of computing a closure configuration. Second, we can treat configurations with the same closure structure as rigid body configurations, which can then be connected by efficient rigid body PRM local planning methods.

While one could generate a roadmap simply from configurations with *one* closure structure, the query process would have to connect any start configuration and goal configuration to this closure structure, which might become very hard, or might not be possible at all. Hence, instead of copying only one closure configuration, we copy an entire kinematic roadmap to the environment and retain all vertices and edges that are collision free (see Figure 5(b)). Next, we group configurations by closure structure, and attempt to make (rigid body) connections within each group (see Figure 5(c)).

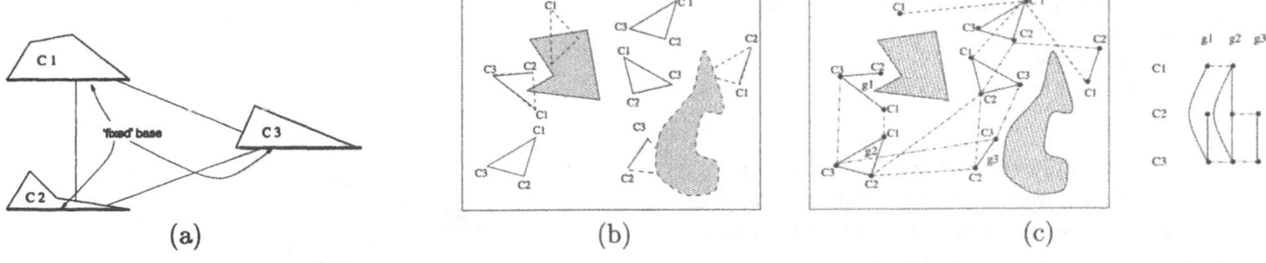

Figure 5: *(a) A kinematic roadmap (Workspace), (b) copying the kinematic roadmap to different base configurations (C-space), and (c) connecting configurations with the same closure structure (C-space).*

5 The Kinematic Roadmap

This section discusses a PRM planner to construct a kinematic roadmap. Recall that the kinematic roadmap encodes the connectivity of the closure configurations $\theta \in \tilde{C}_{closure}$ and does not depend on the environment.

<u>KINEMATIC ROADMAP CONSTRUCTION</u>
1. NODE GENERATION
   (find self-collision-free closure configurations)
2. CONNECTION
   (connect nodes and save paths with edges)
   (repeat as desired)

5.1 Node Generation

The task here is to generate joint variables θ which satisfy the closure constraints and retain self-collision-free closure configurations as kinematic roadmap nodes. Recall that θ can be partitioned into active and passive variables, θ_a and θ_p, where the value of θ_p can be determined for a given θ_a value based on closure constraints.

From an algorithmic point of view, θ_a needs to be chosen such that for a given value of θ_a, the corresponding value of θ_p satisfying the closure constraints, if any, can be computed efficiently. While solutions are not known for inverse kinematics problems for general linkages, the closed-form inverse kinematic solutions for simple chains (such as 4-link chains) and most industrial robots do exist. Therefore, we are most interested in choosing θ_a and θ_p as consecutive joint variables. More specifically, for a system involving K closed chains, each closed chain is broken into two open chains: one an "active chain" with joint variables θ_{la} and the other a "passive chain" with joint variables θ_{lp}, where $l = 1, \ldots, K$, is the index of the closed chain. We also need to ensure that the chain corresponding to the passive joint variables θ_p has a closed-form inverse kinematic solution. This can always be done, for example, by choosing any three consecutive joint variables in each closed chain as θ_p.

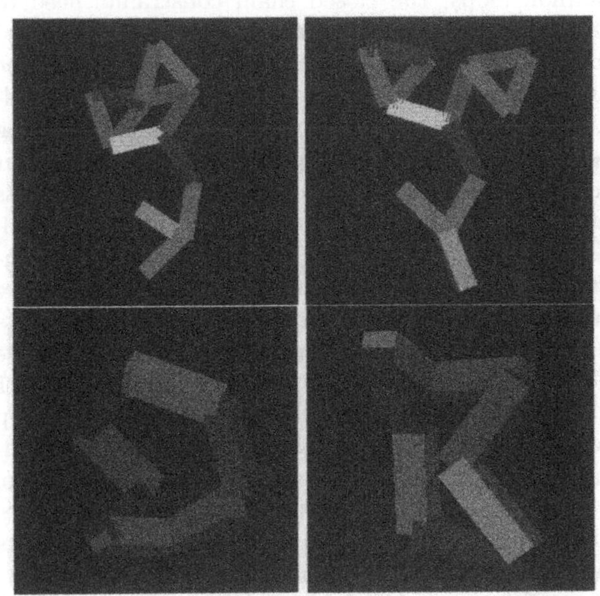

Figure 6: *Closure configurations for some kinematic chains generated by our planner.*

<u>NODE GENERATION FOR KINEMATIC ROADMAP</u>
1. Randomly generate θ_a
2. Use forward kinematics for active chains to compute end-frame configurations $g_{la}, l = 1, \ldots, K$, at the break point of each closed chain;
3. Use inverse kinematics for passive chains to compute joint variables θ_p to achieve the end-frame configurations computed in Step 2.

4. If a solution is found in Step 3 (closure exists)
5. If closure configuration $\theta = (\theta_a, \theta_p)$ is self-collision free
6. retain θ as a kinematic roadmap node

In Step 3, if multiple solutions exist for the inverse kinematics problem, we can either keep all solutions or randomly choose one. Figure 6 shows some closure configurations generated by our planner.

Finally, we note that when a link or joint is involved in multiple closed chains, i.e., when the system involves common loops, the closed chain constraints need to be carefully handled to guarantee that different closed chains will result in the same link configuration or the same joint variable. In particular, we can use the algorithm above to close loops one by one, by choosing joint values to avoid breaking loops while creating other loops. The first loop can choose arbitrary values for its active joint variables θ_a^1. If the loop cannot be closed, i.e., there does not exist θ_p^1 satisfying the closure constraints, then discard it. Otherwise, continue working on its neighboring loop, say loop 2. Assume the joints θ_c^{12} are common for loop 1 and loop 2. Then we will use the values of θ_c^{12} that have been computed from loop 1 as part of the active joint values θ_a^2 of loop 2 (to keep loop 1) and then compute the corresponding θ_p^2. In general, if a loop shares joint variables with other loops, it has to keep the values of the common joint variables that have been determined from the closure constraints of other loops. Clearly, when a loop has more determined joint values, it is more constrained, and thus it is more difficult to close the loop. One heuristic for node generation of common loops is to start from the loop with the largest number of common joints. For the example shown in Figure 7, it would be better to start from the center loop.

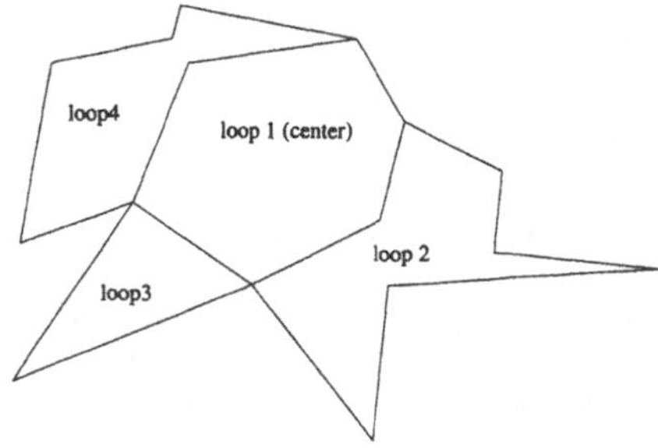

Figure 7: *A System with Common Loops.*

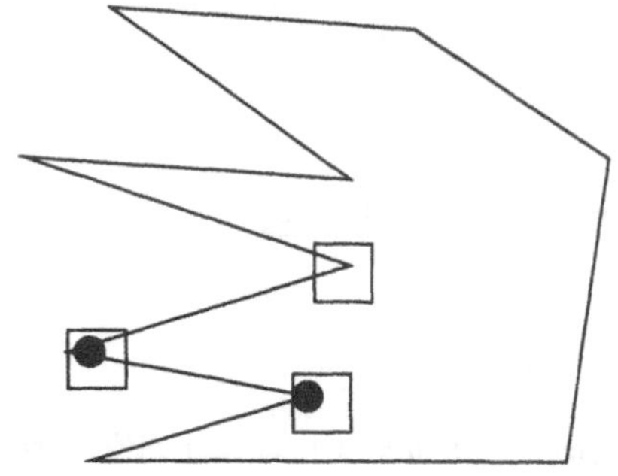

Figure 8: *Different joint partition schemes.*

For the efficiency of the PRM node generation, it is important to choose θ_a and θ_p in a way that the inverse kinematics for the passive chains, corresponding to the θ_p joint variables, has a closed form solution. It is also important to maximize the probability that a closure configuration can be obtained given a randomly generated θ_a. For example, consider the closed chain shown in Figure 8. Two possible selections of θ_p are shown: (i) the two joints marked with black circles, or (ii) the three joints marked with squares. In this case, the three joint option would be preferable, since the intersection of the workspaces of the active and passive chains is larger. In the following, we make this argument more precise.

Recall that the workspace of an open chain with joint variables θ is defined as:

$$W = \{g_{be} \in SE(d) | \exists \theta \in Q, s.t. g_{be} = g_{be}(\theta)\} \quad (10)$$

where Q is the joint space, i.e., the set of all possible joint variable values with joint limits taken into account, g_{be} is the end-frame configuration with respect to the base, and $g_{be}(\theta)$ is the forward kinematics of the open chain.

For one closed chain, we denote the end-frame configuration of the active chain by g_{ba}, and the workspaces of the active chain and passive chains by W_a and W_p, respectively.

Observation 3 *A randomly generated value of θ_a can result in a closure configuration, i.e., there exist passive joint variable values satisfying the closure constraint,* if and only if *the end-frame configuration of the active chain, g_{ba}, falls in the workspace of the passive chain, i.e. $g_{ba} \in W_p$.*

Therefore, a rough estimate of the probability that a randomly generated θ_a results in a closure configurations is:

$$\text{prob(closure)} \approx \frac{Volume(W_p \cap W_a)}{Volume(W_a)} \qquad (11)$$

where $Volume$ denotes the volume of the workspace measured in $SE(d)$.

Remark 1 *If both W_p and W_a can be computed, then a random closure configuration $\theta = (\theta_a, \theta_p)$ can be obtained by randomly choosing a configuration in $W_p \cap W_a$ and using inverse kinematics of both the active chain and the passive chain to compute θ_a and θ_p. This is one of the most effective ways to generate a random closure configuration. However, it is not easy to compute the workspaces. We will still use θ_a and θ_p in the following discussion with the understanding that the knowledge of the workspaces, when available, should be exploited to improve the effectiveness of node generation.*

Since multiple joint variable values may result in the same end-frame configuration, the above probability measure is not accurate because it does not take the multiplicity into account. Denote by $g^{-1}(g_{ba})$ the inverse kinematic solutions for one end-frame configuration of the active joints. Then, the probability of obtaining a closure configuration from a randomly generated θ_a is:

$$\text{prob(closure)} = \frac{Volume(g^{-1}(W_p \cap W_a))}{Volume(g^{-1}(W_a))} \qquad (12)$$

where $g^{-1}(W)$ denotes set of inverse kinematic solutions for each configuration in W, and where the volume is computed in the active joint space instead of $SE(d)$ as in Equation 11.

As different joint partition schemes may result in different probabilities of successfully generating a closure configuration, the joint partition scheme should ideally be chosen based on this probability. However, the computation of the probability measure (Equation 12) involves the computation of the workspaces, inverse kinematics, and volume integrals, which is probably too complicated to be practical for most linked systems. Nevertheless, the probability measure provides us with insight that can be used to develop heuristics to guide the partition of the joint variables. For example, balancing the lengths of the active and passive chains is one possible heuristic.

Finally, we note that increasing the probability that a closure configuration exists will in many cases complicate the inverse kinematics for the passive chain. For example, the length heuristic mentioned above might result in longer passive chains with more complicated inverse kinematics than the 3-joint/4-link chains we have selected.

5.2 Node Connection

An edge between two closure configurations in the kinematic roadmap consists of a sequence of intermediate closure configurations. Since it is relatively expensive to generate closure configurations, the edges of the kinematic roadmap are saved for future use. Node connection in the kinematic roadmap follows the standard PRM framework.

<u>NODE CONNECTION FOR KINEMATIC ROADMAP</u>

1. For any two "nearby" closure configurations θ_i and θ_j
2. Use (simple) local planner to find path from θ_{ia} to θ_{ja}: $\theta_a(t), t \in [0,1]$, where $\theta_a(0) = \theta_{ia}$, $\theta_a(1) = \theta_{ja}$
3. For each intermediate point on the path $\theta_a(t)$
4. If inverse kinematics determines that no $\theta_p(t)$ exists to satisfy the closure constraints
5. return no-edge
6. Choose the closure configuration $\theta(t)$ that is continuous from previous step
7. If $\theta(t)$ involves self-collision, then return no-edge
8. endfor
9. save the edge (and with it the path $\theta(t), t \in [0,1]$)
10. return edge-exist

The connection of common loop configurations has to be processed similarly as for the node generation of common loops. In general, any connection strategy and local planner for rigid body robots or serial chains, such as the nearest neighbors connection strategy and the C-space straight line local planner, can be used to choose pairs of closure configurations for potential edge generation and to connect the active joint variables. When the system involves complicated closed chain structures, more sophisticated techniques such as the Jacobian method or a point-to-point inverse kinematic solver [1] can also be used to generate kinematic roadmap edges. The distance metric on the set $\tilde{C}_{closure}$ can be, e.g., Euclidean distance (Equation 13) with the modification that the distance between two joint angles is at most π.

$$dist(\theta_1 - \theta_2) = \|\theta_1 - \theta_2\| \qquad (13)$$

6 Building a Roadmap from a Kinematic Roadmap

The kinematic roadmap provides us with a set of self-collision-free closure configurations $\theta \in \tilde{C}_{closure}$ (Equation 8) and connections between them. Therefore, by randomly generating base configurations g_{wb}, we can "populate" the environment with kinematic roadmap nodes and edges—this will only require collision detection with environment obstacles since we save the paths associated with the kinematic roadmap edges. Furthermore, roadmap nodes generated from the same closure configuration can be treated as rigid body configurations during roadmap connection.

<u>PROTOTYPE KINEMATICS-BASED PRM</u>

I. <u>POPULATE ENVIRONMENT WITH KINEMATIC ROADMAP</u>
    generate random base configurations and retain collision-free parts of kinematic roadmap in roadmap

II. <u>ADDITIONAL CONNECTION OF ROADMAP NODES</u>
    connect roadmap nodes with the same closure structure using rigid body planners

6.1 Populating the Environment with Copies of the Kinematic Roadmap

<u>POPULATING ENV. WITH COPIES OF KINEMATIC ROADMAP</u>

```
1.  Choose random vertex θ from the kinematic roadmap
2.      Generate random base configuration g_wb
3.      If the configuration (g_wb, θ) is collision-free
4.          Retain (g_wb, θ) as a roadmap vertex
5.          For each neighbor of θ, say θ̄, in the kinematic
            map (repeat with their neighbors as needed)
6.          If (g_wb, θ̄) is collision-free
7.              Retain (g_wb, θ̄) as a roadmap vertex
8.              Retrieve the path θ(t) connecting θ and θ̄
                from the kinematic map
9.              If (g_wb, θ(t)) is collision-free for all intermediate
                closure configurations along the path
10.                 Add an edge between (g_wb, θ) and (g_wb, θ̄)
(repeat as desired)
```

The generation of the random base configuration in Step 2 can be implemented with any node generation strategy developed for rigid body robots such as PRM [11], PRM with Gaussian filter [5], OBPRM [2], and the medial-axis PRM [23]. We next note that since the kinematic roadmap's nodes and edges are known to be self-collision free, the only collision checks needed in this stage are between environment obstacles and the robot, i.e., it is not necessary to check the links for self-collision. Therefore, the reuse of the closure configurations and their connection edges can significantly reduce the total number of collision detection calls, which represent the major computation cost at this stage.

6.2 Connecting Same Closure Nodes

Closed-chain configurations with the same closure configurations can be viewed as configurations of a rigid body. Therefore, we can use rigid body PRM methods to connect them. More specifically,

<u>CONNECTING NODES OF SAME CLOSURE TYPE</u>

```
1. For each closure configuration θ in kinematic roadmap
2.     Collect all roadmap nodes with this closure
       configuration in a set
3.     Use rigid body PRM connection methods to connect
       configurations in the set
4.     Add the edges generated in Step 3 to the roadmap
5. endfor
```

Figure 9 shows a portion of the roadmap, being progressively built by connecting configurations with closure structures $C1, C2$, and $C3$, respectively.

7 Experimental Results

7.1 Implementation Details

Our prototype closed chain PRM planner was developed on top of the C++ OBPRM software package developed by the robotics group at Texas A&M University [2, 4]. This strategy was taken because the construction of the kinematic roadmap and the connection of the roadmap nodes with the same closure configuration are both basically simple PRM planners. It turned out to be fairly easy to incorporate the closed-chains into the PRM framework due to the object oriented design of the code. All experimental results reported in this section were performed on an SGI Octane and used the RAPID [8] package for 3D collision detection.

7.2 Experiments

While our planner can handle complicated three-dimensional closed chains, the results presented here are for single loop closed chains in three-dimensional environments. In particular, the chains we consider consist of m identical links, all joints are revolute, and we partition the m joint angles into 3 consecutive passive angles and $m - 3$ active angles.

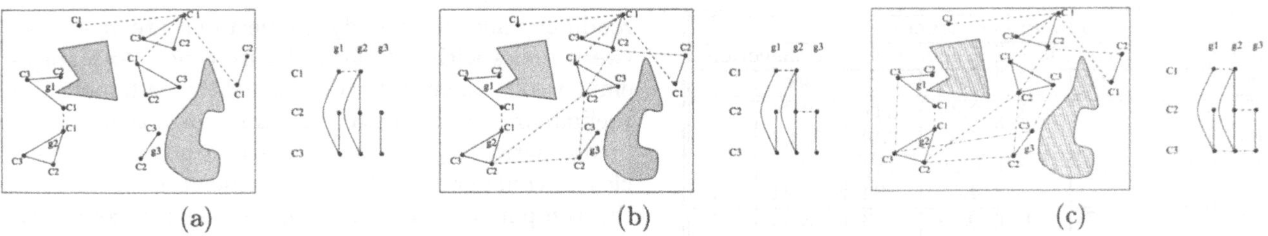

(a) (b) (c)

Figure 9: *Roadmap connection in three rigid body phases: connecting configurations of type (a) C1, (b) C2, and (c) C3.*

Kinematic Roadmap Construction				
Chain	Generation		Connection	
Links	sec	cfg	sec	CC
6(P)	0.84	203	24.66	2
7(P)	0.90	122	9.88	2
11(P)	1.24	16	0.48	4
15(P)	1.63	3	0.01	3
6(S)	0.97	28	0.88	4
7(S)	1.09	20	0.41	6
11(S)	1.76	1	0.00	1

Table 1: *Kinematic roadmap construction times (seconds) and statistics for k-link closed chains, $k = 6, 7, 11$ and 15. In the table, planner and spatial chains are labeled respectively with P and S following their link numbers, and cfg and CC denote the number of nodes and the number of connected components respectively in the resulting kinematic roadmap. Note that the roadmap for a single planar closed chain should have two connected components.*

We first study the effectiveness of our method for generating the kinematic roadmap, which involves generating and connecting closure configurations in the absence of obstacles in the environment. We used the algorithm presented in Section 5.1 to check if it is feasible for a randomly generated active joint angle to achieve a closure configuration. As seen in Figures 3 and 6, this method was effective in generating and connecting closure configurations.

However, as predicted in Section 5.1, our success in generating closure configurations is closely tied to the "balance" of the partition of the joints into the active and passive sets. Table 1 shows the kinematic roadmap statistics for closed chains with various numbers of links (all cases contained three passive joint angles). The nodes were generated from 2000 trials using the algorithm sketched in Section 5.1. The edges were generated by trying to connect each node to its 20 nearest neighbors using a C-space straight line planner. The

Figure 10: *The "Walls" Environment.*

results shown in the table are consistent with the analysis of Section 5.1. Namely, for both planar and spatial types of chains, the planner generated more closure nodes for the chains with fewer links, which are also those for which the partition into the active and passive chains is most equal, and where their workspaces overlap the most. In addition, the results for planner chains are better than their spatial counterparts, again due to their larger overlapping workspaces. Clearly, the simple PRM kinematic roadmap planner does not work as well for the 15-link planar chain and 11-link spatial chain. We are working on incorporating more sophisticated planning methods, such as the Ariadne's Clew algorithm [1], into our system which should help in such situations.

We now analyze the benefit of the kinematic preprocessing. That is, the benefit of using the kinematic roadmap as opposed to simply generating closure configurations directly in the environment (as was done in [18]). To study this issue, we compare roadmaps constructed with and without the kinematic preprocessing. We used the environment shown in Figure 10. Table 2 shows the statistics for 7 and 9-link chains. The roadmaps without kinematic preprocessing were generated using the PRM planner implemented in the

Roadmap Construction								
Chain	Kinematic Map			Generation		Connection		
Links	sec	cfg	CC	sec	cfg	sec	CC	
7(P)	1.25	31	2	2.44	341	12.56	2	
7(P)	–	–	–	5.25	338	62.07	6	
9(P)	0.44	13	3	0.99	132	9.27	3	
9(P)	–	–	–	4.63	104	11.94	8	
7(S)	0.31	7	6	0.09	43	3.93	7	
7(S)	–	–	–	3.52	39	5.61	16	

Table 2: *Roadmap construction times (seconds) and statistics for roadmaps constructed with and without the kinematic roadmap. In the table, planner and spatial chains are labeled respectively with P and S following their link numbers, and cfg and CC denote the number of nodes and connected components in the roadmap (there are two), respectively.*

OBPRM software package. The nodes were generated from 4000 attempts, using the generation method outlined in Section 5.1, the only difference being that now the base configuration was randomly generated and collision was checked with the obstacles in the environment as well. The edges were generated using the straight line planner to connect each node to its 20 nearest neighbors. The kinematic map nodes were generated from 400 attempts for planer chains and 500 attempts for the spatial 7-link chain, and, again, the 20 nearest neighbors were checked using the straight line planner. Then, five different base configurations were generated for each closure node when populating the environment with copies of the kinematic roadmap. The final rigid body connections between configurations with the same closure type also used the straight line planner with the 20 nearest neighbors. As can clearly be seen from the table, the roadmaps constructed using kinematic preprocessing are superior in all aspects: faster computation and improved roadmap quality (fewer connected components).

8 Conclusion

This paper presents a kinematics-based probabilistic roadmap planner for closed chains. The two-stage construction of our roadmap first builds a (small) kinematic roadmap that deals solely with the robot's kinematics and utilizes both forward and inverse kinematics in its construction. In the second stage, the environment is populated with copies of the kinematic roadmap, and rigid body connections are made between nodes with the same closure type. Both stages employ PRM planners to construct their roadmaps. Our preliminary experimental results indicate that the use of kinematics to guide the generation and connection of closure configurations is very beneficial, both reducing the computation costs and improving the connectivity of the resulting roadmap as compared to previous purely randomized approaches.

Therefore, we believe that augmenting the randomized philosophy of PRMs with more deliberate techniques developed in the robotics community is a promising direction to pursue for motion planning problems involving additional constraints.

While our results are promising, there are still many issues to be addressed. First, we intend to more fully exercise our current closed chain PRM planner on more complex linkages. We anticipate that the computational costs will grow, so that additional optimization will be required. One area we intend to explore in this regard is parallelization, which has been shown to be effective for traditional PRMs [3]. We also plan to study more complicated manipulation planning problems such as regrasp planning.

Acknowledgments

We would like to thank Drs. Steve Wilmarth, Jeff Trinkle, and Peter Stiller for valuable discussions. We are also grateful to the robotics group at Texas A&M, especially Lucia Dale and Guang Song, for their help regarding the OBPRM software package.

This research was supported in part by NSF CAREER Award CCR-9624315, NSF Grants IIS-9619850, EIA-9805823, and EIA-9810937, and by the Texas Higher Education Coordinating Board under grant ARP-036327-017.

References

[1] J. M. Ahuactzin and K. Gupta. The kinematic roadmap: A motion planning based global approach for inverse kinematics of redundant robots. *IEEE Trans. Robot. Automat.*, RA-15(4):653–670, 1999.

[2] N. M. Amato, O. B. Bayazit, L. K. Dale, C. V. Jones, and D. Vallejo. OBPRM: An obstacle-based PRM for 3D workspaces. In *Proc. Int. Workshop on Algorithmic Foundations of Robotics (WAFR)*, pages 155–168, 1998.

[3] N. M. Amato and L. K. Dale. Probabilistic roadmap methods are embarrassingly parallel. In *Proc. IEEE Int. Conf. Robot. Autom. (ICRA)*, pages 688–694, 1999.

[4] N. M. Amato and Y. Wu. A randomized roadmap method for path and manipulation planning. In *Proc. IEEE Int. Conf. Robot. Autom. (ICRA)*, pages 113–120, 1996.

[5] V. Boor, M. H. Overmars, and A. F. van der Stappen. The gaussian sampling strategy for probabilistic roadmap planners. In *Proc. IEEE Int. Conf. Robot. Autom. (ICRA)*, pages 1018–1023, 1999.

[6] J. F. Canny. On computability of fine motion plans. In *Proc. IEEE Int. Conf. Robot. Autom. (ICRA)*, pages 177–182, 1989.

[7] John J. Craig. *Introduction to Robotics: Mechanics and Control, 2nd Edition.* Addison-Wesley Publishing Company, Reading, MA, 1989.

[8] S. Gottschalk, M.C. Lin, and D. Manocha. Obb-tree: A hierarchical structure for rapid interference detection. Technical Report TR96-013, University of N. Carolina, Chapel Hill, CA, 1996.

[9] L. Kavraki, F. Lamiraux, and C. Holleman. Towards planning for elastic objects. In *Proc. Int. Workshop on Algorithmic Foundations of Robotics (WAFR)*, 1998.

[10] L. Kavraki and J. C. Latombe. Randomized preprocessing of configuration space for fast path planning. In *Proc. IEEE Int. Conf. Robot. Autom. (ICRA)*, pages 2138–2145, 1994.

[11] L. Kavraki, P. Svestka, J. C. Latombe, and M. Overmars. Probabilistic roadmaps for path planning in high-dimensional configuration spaces. *IEEE Trans. Robot. Automat.*, 12(4):566–580, August 1996.

[12] O. Khatib. Stanford Robotic Manipulation Group, robotics.Stanford.EDU/groups/manips/projects.

[13] O. Khatib, K. Yokoi, K. Chang, D. Ruspini, R. Holmberg, and A. Casal. Vehicle/arm coordination and multiple mobile manipulator decentralized cooperation. In *Proc. IEEE Int. Conf. Robot. Autom. (ICRA)*, pages 546–553, 1996.

[14] K. Kotay, D. Rus, M. Vona, and C. McGray. The self-reconfiguring robotic molecule: Design and control algorithms. In *Proc. Int. Workshop on Algorithmic Foundations of Robotics (WAFR)*, pages 375–386, 1998.

[15] J. C. Latombe. *Robot Motion Planning.* Kluwer Academic Publishers, Boston, MA, 1991.

[16] J.-P. Laumond and M.H. Overmars. Algorithms for robotic motion and manipulation. In *Proc. Int. Workshop on Algorithmic Foundations of Robotics (WAFR)*, 1996.

[17] S. M. LaValle and J. J. Kuffner. Randomized kinodynamic planning. In *Proc. IEEE Int. Conf. Robot. Autom. (ICRA)*, pages 473–479, 1999.

[18] S.M. LaValle, J.H. Yakey, and L.E. Kavraki. A probabilistic roadmap approach for systems with closed kinematic chains. In *Proc. IEEE Int. Conf. Robot. Autom. (ICRA)*, 1999.

[19] Richard M. Murray, Zexiang Li, and S. Shankar Sastry. *A Mathematical Introducation to Robotic Manipulation.* CRC Press, Boca Raton, FL, 1994.

[20] A. Nguyen, L. J. Guibas, and M. Yim. Controlled module density helps reconfiguration planning. In *Proc. Int. Workshop on Algorithmic Foundations of Robotics (WAFR)*, 2000.

[21] A.P. Singh, J.C. Latombe, and D.L. Brutlag. A motion planning approach to flexiable ligand binding. In *7th Int. Conf. on Intelligent Systems for Molecular Biology (ISMB)*, pages 252–261, 1999.

[22] D. Stewart. A platform with six degrees of freedom. *Proc. of the Institute of Mechanical Engineering*, 180(part I(5)):171–186, 1954.

[23] S. A. Wilmarth, N. M. Amato, and P. F. Stiller. MAPRM: A probabilistic roadmap planner with sampling on the medial axis of the free space. In *Proc. IEEE Int. Conf. Robot. Autom. (ICRA)*, pages 1024–1031, 1999.

[24] S. A. Wilmarth, N. M. Amato, and P. F. Stiller. Motion planning for a rigid body using random networks on the medial axis of the free space. In *Proc. ACM Symp. on Computational Geometry (SoCG)*, pages 173–180, 1999.

[25] Jeffery H. Yakey. Randomized path planning for linkages with closed kinematic chains. Master's thesis, Iowa State Univ, Computer Science Dept., 1999.

Randomized Kinodynamic Motion Planning with Moving Obstacles

David Hsu, *Stanford University, Stanford, CA*
Robert Kindel, *Stanford University, Stanford, CA*
Jean-Claude Latombe, *Stanford University, Stanford, CA*
Stephen Rock, *Stanford University, Stanford, CA*

A randomized motion planner is presented for robots that must avoid collision with moving obstacles under kinematic and dynamic constraints. This planner samples the robot's state×time space by picking control inputs at random and integrating the equations of motion. The result is a probabilistic roadmap, i.e., a collection of sampled state×time points, called milestones, connected by short admissible trajectories. The planner does not precompute the roadmap; instead, for each planning query, it generates a new roadmap to connect the input initial and goal state×time points. This paper shows that the probability that the planner fails to find a trajectory when one exists quickly goes to 0 as the number of milestones grows. The planner has been tested successfully in both simulated and real environments. In the latter case, a vision module estimates the obstacle motions just before planning, and the planner is then allocated a small amount of time to compute a trajectory. If a change in obstacle motion is detected while the robot executes the planned trajectory, the planner re-computes a trajectory on the fly.

1 Introduction

In its most basic form, motion planning is a purely geometric problem: Given the geometry of a robot and static obstacles, compute a collision-free path of the robot between two configurations. This formulation ignores several key aspects of the real world. Robot motions are subject to kinematic and dynamic constraints that, unlike obstacles, cannot be represented by forbidden regions in the configuration space. The environment may also contain moving obstacles requiring that computed paths be parametrized by time. In this paper, we consider motion planning problems with both kinodynamic constraints and moving obstacles. We propose an efficient algorithm for this class of problems.

Our work extends the probabilistic roadmap (PRM) framework originally developed for planning collision-free geometric paths [18]. A PRM planner samples the robot's configuration space at random and connects the generated free samples, called *milestones*, by simple local paths, typically straight-line segments in configuration space. The result is a undirected graph called a probabilistic roadmap. Multi-query PRM planners precompute the roadmap [18], while single-query planners compute a new roadmap for each query [11]. It has been proven that, under reasonable assumptions, a small number of milestones picked uniformly at random are sufficient to capture the free space's connectivity with high probability [11, 17]. However, with nonholonomic and/or dynamic constraints, straight local paths are not feasible. Moreover, allowing obstacles to move requires indexing milestones by the times when they are attained.

For each planning query, our algorithm builds a new roadmap in the collision-free subset of the robot's state×time space, where a state typically encodes both the configuration and velocity of the robot (Section 3). Each milestone is obtained by selecting a control input at random in the set of admissible controls and integrating the motion induced by this input over a short duration of time, from a previously-generated milestone. The local trajectory thus obtained automatically satisfies the motion constraints, and if it does not collide with the obstacles, its endpoint is added to the roadmap as a new milestone. This iterative process yields a tree-shaped roadmap rooted at the initial state×time and oriented along the time axis. It ends when a milestone falls in an "endgame" region from which it is known how to reach the goal.

In Section 4, we show that the probability that our algorithm fails to find a trajectory when one exists converges toward 0 (probabilistic completeness), with a convergence rate that is exponential in the number of generated milestones. We have also tested the algorithm in both simulated and real environments. In simulation (Sections 5 and 6), we have verified that it can solve tricky problems. In the hardware robot testbed (Section 7), we have checked that the planner operates properly despite various uncertainties and delays associated with an integrated system. In this testbed, a vision module measures obstacle motions, and the planner has a short, predefined amount of time to compute a trajectory (real-time planning). The vision module monitors the obstacles while the robot executes the computed trajectory. If an obstacle deviates from its predicted trajectory, the planner re-computes the robot's trajectory on the fly.

2 Previous Work

Motion planning by random sampling This approach was originally proposed to solve geometric path-planning problems for robots with many degrees of freedom (dofs) [2, 3, 11, 16, 18]. Sampling replaces the prohibitive computation of an explicit representation of the free space by collision checking operations. Proposed techniques differ mainly in their sampling strategies. An important distinction is between *multi-query* planners that precompute a roadmap (e.g., [18]) and *single-query* planners that don't (e.g., [11]). Single-query planners build a new roadmap for each query by constructing trees of sampled milestones from the initial and goal configurations [11, 21]. Our planner falls in this second category.

It has been shown in [11, 15, 17, 29] that, under some assumptions, a multi-query PRM path planner that samples milestones uniformly from the configuration space is probabilistically complete and converges quickly. More formally, the probability that it fails to find a path when one exists converges toward 0 exponentially with the number of milestones. In [11], this result is established under the assumption that the free space verifies a geometric property called *expansiveness*. In this paper, we generalize this property to state×time space and prove that our new randomized planner for kinodynamic planning is also probabilistically complete with a convergence rate exponential in the number of sampled milestones. No formal guarantee of performance had previously been established for single-query planners.

Kinematic and dynamic constraints Kinodynamic planning refers to problems in which the robot's motion must satisfy nonholonomic and/or dynamic constraints. With few exceptions (e.g., [9]), previous work has considered these two types of constraints separately.

Planning for nonholonomic robots has attracted considerable interest [4, 19, 20, 22, 23, 26, 28]. One approach [19, 20] is to first generate a collision-free path, ignoring the nonholonomic constraints, and then to break this path into small pieces and replace them by feasible canonical paths (e.g., Reeds and Shepp curves [24]). This approach works well for simple robots (e.g., car-like robots), but requires the robots to be locally controllable [4, 23]. A related approach [26, 28] uses a multi-query PRM algorithm that connects milestones by canonical path segments such as the Reeds and Shepp curves. Another approach, presented in [4], generates a tree of sampled configurations rooted at the initial configuration. At each iteration, a chosen sample in the tree is expanded into a few new samples, by integrating the robot's equation of motion over a short duration of time with deterministically picked controls. A space partitioning scheme limits the density of samples in any region of the configuration space. This approach has been shown to be also applicable to robots that are not locally controllable. Our new planner has many similarities with this approach, but picks controls at random. It is probabilistically complete whether the robot is locally controllable or not.

Algorithms for dealing with dynamic constraints are comparable to those developed for nonholonomic constraints. In [5, 27] a collision-free path is first computed, ignoring the dynamic constraints; a variational technique then deforms this path into a trajectory that both conforms the dynamic constraints and optimizes a criterion such as minimal execution time. No formal guarantee of performance has been established for these planners. The approach in [7] places a regular grid over the robot's state space and searches the grid for a trajectory using dynamic programming. It offers provable performance guarantees, but is only applicable to robots with few dofs (typically, 2 or 3), as the size of the grid grows exponentially with the number of dofs. Our planner is related to this second approach, but randomly discretizes the state×time space, instead of placing a regular grid over it. The planner in [21] resembles ours, but no guarantee of performance has been established for it.

Moving obstacles When obstacles are moving, the planner must compute a trajectory parametrized by time. This problem has been proven to be computationally hard, even for robots with few dofs [25]. Heuristic algorithms [8, 10, 14] have been proposed, but they usually do not consider constraints on the robot's motion other than an upper bound on its velocity. The technique in [14] first ignores the moving obstacles and computes a collision-free path of the robot among the static obstacles; it then tunes the robot's velocity along this path to avoid colliding with moving obstacles. The resulting planner is clearly incomplete. The planner in [10] tries to reduce incompleteness by generating a network of paths. The planner in [9] is a rare example of planner dealing concurrently with kinodynamic constraints and moving obstacles. It extends the approach of [7] to state×time space and thus is also limited to robots with few dofs.

3 Planning Framework

Our algorithm builds a probabilistic roadmap in the collision-free subset $\mathcal{F}$ of the state×time space of the robot [9]. The roadmap is computed in the connected component of $\mathcal{F}$ that contains the robot's initial state×time point.

3.1 State-Space Formulation

Motion constraints We consider a robot whose motion is governed by an equation of the form:

$$\dot{s} = f(s, u), \tag{1}$$

where $s \in \mathcal{S}$ is the robot's state, $\dot{s}$ is its derivative relative to time, and $u \in \Omega$ is a control input. The set $\mathcal{S}$ and Ω are the robot's *state space* and *control space*, respectively. Given a state at time t and a control function u: $[t, t'] \rightarrow \Omega$, where $[t, t']$ is a time interval, the solution of Equation (1) is a function s: $[t, t'] \rightarrow \mathcal{S}$ describing the robot's state over $[t, t']$. We assume that $\mathcal{S}$ and Ω are bounded manifolds of dimensions n and m ($m \leq n$), respectively. By defining appropriate charts, we can treat $\mathcal{S}$ and Ω as subsets of $\mathbb{R}^n$ and $\mathbb{R}^m$.

Equation (1) can represent both nonholonomic and dynamic constraints. A nonholonomic robot is constrained by k independent, non-integrable scalar equations of the form $F_i(q, \dot{q}) = 0$, $i = 1, 2, \ldots, k$, where q and $\dot{q}$ denote the robot's configuration and velocity, respectively. Let the robot's state be $s = q$. It is shown in [4] that, under appropriate mathematical conditions,

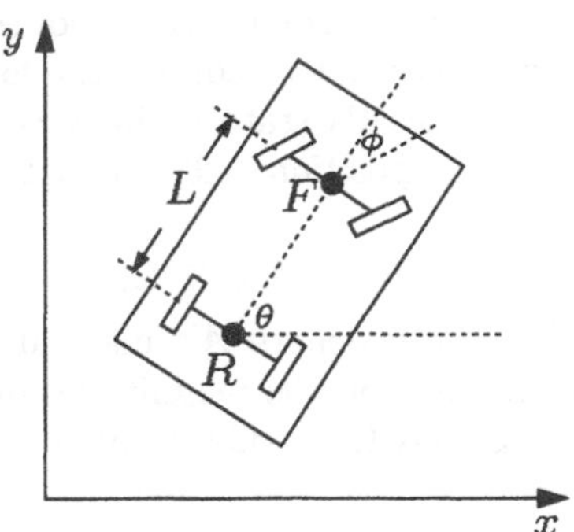

Figure 1: *Model of a car-like robot.*

the constraints $F_i(s, \dot{s}) = 0$ are equivalent to Equation (1) in which u is a vector in $\mathbb{R}^m = \mathbb{R}^{n-k}$. Reciprocally, Equation (1) can be rewritten into $k = n - m$ independent equations of the form $F_i(s, \dot{s}) = 0$. Furtheremore, in Lagrangian mechanics, dynamics equations are of the form $G_i(q, \dot{q}, \ddot{q}) = 0$, where q, $\dot{q}$, and $\ddot{q}$ are the robot's configuration, velocity, and acceleration, respectively. Defining the robot's state as $s = (q, \dot{q})$, we can rewrite the dynamics equations in the form $F_i(s, \dot{s}) = 0$, which, as in the nonholonomic case, is equivalent to Equation (1).

Robot motions can also be constrained by inequalities of the forms $F_i(q, \dot{q}) \leq 0$ and $G_i(q, \dot{q}, \ddot{q}) \leq 0$. Such constraints restrict the sets of admissible states and controls to subsets of $\mathbb{R}^n$ and $\mathbb{R}^m$.

Examples We illustrate these notions with two examples that will be useful later in the paper:

Nonholonomic car-like robot. Consider a car A modeled as shown in Figure 1. Let (x, y) be the position of the midpoint R between A's rear wheels and θ be the orientation of the rear wheels with respect to the x-coordinate axis. We encode A's configuration by $(x, y, \theta) \in \mathbb{R}^3$. The nonholonomic constraint $\tan\theta = \dot{y}/\dot{x}$ is equivalent to the system:

$$\dot{x} = v\cos\theta, \quad \dot{y} = v\sin\theta, \quad \dot{\theta} = (v/L)\tan\phi,$$

which has the same form as Equation (1). This reformulation corresponds to defining the state of A to be its configuration and choosing the vector (v, ϕ), where v and ϕ denote the car's linear velocity and steering angle, as the input control. Bounds on v and ϕ define Ω as a subset of $\mathbb{R}^2$.

Point-mass robot. For a point-mass robot A moving in a plane, we typically want to control the force applied to A. So, we define A's state to be $s = (x, y, \dot{x}, \dot{y})$, where (x, y) is A's position. The equations of motion are:

$$\ddot{x} = u_x/m, \qquad \ddot{y} = u_y/m,$$

where m and (u_x, u_y) denote A's mass and the force applied to it. Bounds on the magnitudes of $(\dot{x}, \dot{y})$ and (u_x, u_y) define $\mathcal{S}$ and Ω as subsets of $\mathbb{R}^4$ and $\mathbb{R}^2$, respectively.

Planning query Let $\mathcal{ST}$ denote the state×time space $\mathcal{S} \times [0, +\infty)$. Obstacles in the robot's workspace are mapped into this space as forbidden regions. The *free space* $\mathcal{F} \subseteq \mathcal{ST}$ is the set of all collision-free points (s, t). A trajectory $s\colon [a, b] \rightarrow \mathcal{S}$ is *admissible* if, for all $t \in [a, b]$, $s(t)$ is an admissible state and $(s(t), t)$ is collision-free.

A planning query is specified by an initial state×time (s_b, t_b) and a goal state×time (s_g, t_g). A solution to this query is a function $u\colon [t_b, t_g] \rightarrow \Omega$ that produces an admissible trajectory from state s_b at time t_b to state s_g at time t_g. In the following, we consider piecewise-constant functions $u(t)$ only.

3.2 The Planning Algorithm

Like the planner in [11], our algorithm iteratively builds a tree-shaped roadmap T rooted at $m_b = (s_b, t_b)$. At each iteration, it picks at random a milestone (s, t) from T, a time $t' \leq t_g$, and a control function $u\colon [t, t'] \rightarrow \Omega$. The trajectory from (s, t) induced by u is computed by integrating Equation (1). If this trajectory is admissible, its endpoint (s', t') is added to T as a new milestone; an arc is created from (s, t) to (s', t'), and u is stored with this arc. The kinodynamic constraints are thus naturally enforced in all trajectories represented in T. The planner exits with success when the newly generated milestone lies in an "endgame" region that contains (s_g, t_g).

Milestone selection The planner assigns a weight $w(m)$ to each milestone m in T. The weight of m is the number of other milestones contained in a neighborhood of m. So $w(m)$ represents how densely the neighborhood of m has already been sampled. At each iteration, the planner picks an existing milestone m in T at random with probability $\pi_T(m)$ inversely proportional to $w(m)$. Hence, a milestone lying in a sparsely sampled region has a greater chance of being selected than a milestone lying in an already densely sampled region. This technique avoids oversampling any particular region of $\mathcal{F}$.

Control selection Let $\mathcal{U}_\ell$ be the set of all piecewise-constant control functions with at most ℓ constant pieces. Hence, for any $u \in \mathcal{U}_\ell$, there exist $t_0 < t_1 < \ldots < t_{\ell-1} < t_\ell$ such that $u(t)$ is a constant $c_i \in \Omega$ over the time interval (t_{i-1}, t_i), for $i = 1, \ldots, \ell$. We also require $t_i - t_{i-1} \leq \delta_{\max}$, where $\delta_{\max}$ is a constant. Our algorithm picks a control $u \in \mathcal{U}_\ell$, for some prespecified ℓ and $\delta_{\max}$, by sampling each constant piece of u independently. For each piece, c_i and $\delta_i = t_i - t_{i-1}$ are selected uniformly at random from Ω and $[0, \delta_{\max}]$, respectively. The choices of the parameters ℓ and $\delta_{\max}$ will be discussed in Subsection 4.4.

Endgame connection The above "control-based" sampling technique does not allow us to reach the goal (s_g, t_g) exactly. We need to "expand" the goal into an *endgame* region that the sampling algorithm will eventually attain with high probability. There are several ways of creating such an endgame region.

For some robots, it is possible to analytically compute one or several canonical control functions that exactly connect two given points while obeying the kinodynamic constraints. The Reeds and Shepp curves developed for nonholonomic car-like robots are an example of such functions [24]. If such control functions are available, one can test if a milestone m belongs to the engame region by checking that a canonical function generates an admissible trajectory from m to (s_g, t_g).

A more general technique is to build a secondary tree T' of milestones rooted at the goal, in the same way as that for the primary tree T, except that Equation (1) is integrated backwards in time. The endgame region is then the union of small neighborhoods of the milestones in T': The planner exits with success when a milestone $m \in T$ falls in the neighborhood of a milestone $m' \in T'$. The trajectory following the appropriate arcs of T and T' contains a small gap between m and m', but this gap can often be dealt with in practice. For example, beyond m, one can use a PD controller to track the trajectory extracted from T'.

Algorithm in pseudo-code The planning algorithm is formalized in the following pseudo-code. We will refer to it as KDP.

Algorithm 1

```
1.  Initialize T with m_b; i ← 1
2.  repeat
3.     Pick a milestone m from T with probability π_T(m)
4.     Pick a control function u from U_ℓ uniformly at random
5.     m' ← PROPAGATE(m, u)
6.     if m' ≠ nil then
7.        Add m' to T; i ← i + 1
8.        Create an arc e from m to m'; store u with e
9.        if m' ∈ ENDGAME then exit with SUCCESS
10.       if i = N then exit with FAILURE
```

In line 5, PROPAGATE(m, u) integrates the equations of motion from m with control u. It returns a new milestone m' if the computed trajectory is admissible; otherwise it returns *nil*. If there exists no admissible trajectory from $m_b = (s_b, t_b)$ to (s_g, t_g), the algorithm cannot detect it. Therefore, in line 10, we bound the total number of milestones by a constant N.

4 Analysis of the Planner

The experiments that will be described in Sections 5–7 demonstrate that KDP provides an effective approach for solving difficult kinodynamic motion planning problems. Nevertheless some important questions cannot be answered by experiments alone: what is the probability γ that the planner fails to find a trajectory when one exists? Does γ converge toward 0 as the number of milestones increases? In this section we show that the failure probability γ decreases exponentially with the number of sampled milestones. Hence, with high probability, a relatively small number of milestones are sufficient to capture the connectivity of the free space and answer the query correctly. Our analysis is based on a generalization of the notion of expansiveness proposed in [11].

4.1 Expansive State×Time Space

Expansiveness characterizes the difficulty of finding a path between two points in a given space by random sampling. To be concrete, let us first consider the simple example in Figure 2. The free space $\mathcal{F}$ consists of two subsets S_1 and S_2 connected by a narrow passage.

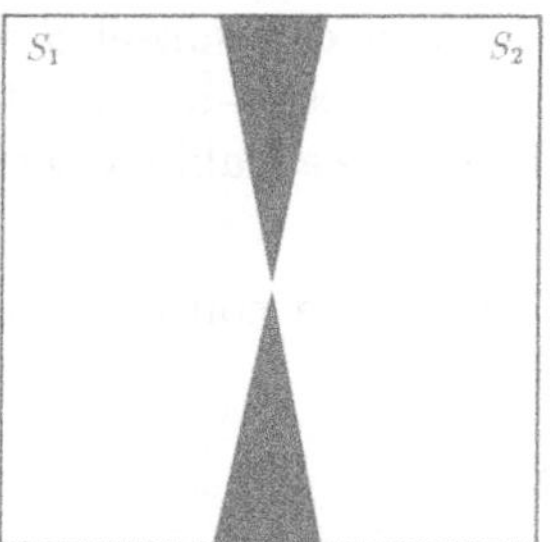

Figure 2: *A free space with a narrow passage.*

Let us say that two points in $\mathcal{F}$ *see* each other (or are mutually *visible*) if the straight line segment between them lies entirely in $\mathcal{F}$ (no kinodynamic constraint is considered in this example). A classical PRM planner samples $\mathcal{F}$ uniformly at random and connects any two milestones that see each other. Let the *lookout* of S_1 be the subset of all points in S_1 that sees a large fraction of S_2. If the lookout of S_1 were large, the planner would easily pick a milestone in S_1 and another in S_2 that see each other. However, due to the narrow passage between S_1 and S_2, S_1 has a small lookout. Consequently, it is difficult for the planner to generate a connection between S_1 and S_2. In [11], $\mathcal{F}$ is said to be expansive if every subset $S \subset \mathcal{F}$ has a large lookout. It is shown that in an expansive space, the convergence rate of a classical PRM planner is exponential in the number of milestones.

KDP generates a different kind of roadmap, in which trajectories between milestones may neither be straight, nor reversible. This leads us to generalize the notion of visibility to that of reachability. Given two points (s, t) and (s', t') in $\mathcal{F} \subset \mathcal{ST}$, (s', t') is said to be *reachable* from (s, t) if there exists a control function $u \colon [t, t'] \to \Omega$ that induces an admissible trajectory from (s, t) to (s', t'). If (s', t') remains reachable from (s, t) by using $u \in \mathcal{U}_\ell$, a piecewise-constant control with at most ℓ segments, then we say that (s', t') is *locally reachable*, or *ℓ-reachable*, from (s, t). Let $\mathcal{R}(p)$ and $\mathcal{R}_\ell(p)$ denote the set of points reachable and ℓ-reachable from p, respectively; we call them the *reachability* and the *ℓ-reachability set* of p. For any subset $S \subset \mathcal{F}$, we define:

$$\mathcal{R}(S) = \bigcup_{p \in S} \mathcal{R}(p) \quad \text{and} \quad \mathcal{R}_\ell(S) = \bigcup_{p \in S} \mathcal{R}_\ell(p).$$

We define the lookout of a subset S of $\mathcal{F}$ as the subset of all points in S whose ℓ-reachability sets overlap significantly with their reachability sets outside S:

Definition 1 Let β be a constant in $(0, 1]$. The β-*lookout* of $S \subset \mathcal{F}$ is:

$$\beta\text{-LOOKOUT}(S) = \{p \in S \mid \mu(\mathcal{R}_\ell(p)) \setminus S) \geq \beta\, \mu(\mathcal{R}(S) \setminus S)\},$$

where $\mu(Y)$ denote the volume of any subset $Y \subset \mathcal{R}(S)$ relative to $\mathcal{R}(S)$.

The free space $\mathcal{F}$ is expansive if every subset $S \subset \mathcal{F}$ has a large lookout.

Definition 2 Let α and β be two constants in $(0, 1]$. For any $p \in \mathcal{F}$, $\mathcal{R}(p)$ is (α, β)-*expansive* if for every connected subset $S \subseteq \mathcal{R}(p)$, $\mu(\beta\text{-LOOKOUT}(S)) \geq \alpha\, \mu(S)$. The free space $\mathcal{F}$ is (α, β)-*expansive* if for every $p \in \mathcal{F}$, $\mathcal{R}(p)$ is (α, β)-*expansive.*

Think of p in Definition 2 as the initial milestone m_b and S as the ℓ-reachability set of a set of milestones produced by KDP after some iterations. If α and β are both reasonably large, then KDP has a good chance to sample a new milestone whose ℓ-reachability set adds significantly to the size of S. In fact, we show below that with high probability, the ℓ-reachability set of the milestones sampled by KDP expands quickly to cover most of $\mathcal{R}(m_b)$; hence, if the goal (s_g, t_g) lies in $\mathcal{R}(m_b)$, then with high probability, the planner will quickly find an admissible trajectory.

4.2 Probabilistic Convergence of the Planner

Let $\mathcal{X} = \mathcal{R}(m_b)$ be the reachability set of m_b. Suppose that $\mathcal{X}$ is (α, β)-expansive. We establish below an upper bound on the number of milestones needed to guarantee that a milestone lies in the endgame region E with high probability, if $E \cap \mathcal{X}$ has non-zero volume relative to $\mathcal{X}$. For convenience, we scale all volumes so that $\mu(\mathcal{X}) = 1$.

Let us assume for now that there is an ideal sampler IDEAL-SAMPLE that picks a point uniformly at random from the ℓ-reachability set $\mathcal{R}_\ell(M)$ of any set of milestones M. The procedure IDEAL-SAMPLE replaces lines 3-5 in KDP. We will discuss how to approximate IDEAL-SAMPLE in Subsection 4.3.

Let $M = (m_0, m_1, m_2, \ldots)$ be a sequence of milestones generated by KDP with IDEAL-SAMPLE ($m_0 = m_b$), and let M_i denote the first i milestones in M. The milestone m_i is called a *lookout point* if it lies in the β-lookout of $\mathcal{R}_\ell(M_{i-1})$. Lemma 3 states that the ℓ-reachability set of M spans a large volume if it contains enough lookout points, and Lemma 4 estimates the probability that this happens. Together, they imply that with high probability, the ℓ-reachability set of a relatively small number of milestones spans a large volume in $\mathcal{X}$.

Lemma 3 *If a sequence of milestones M contains k lookout points, then $\mu(\mathcal{R}_\ell(M)) \geq 1 - e^{-\beta k}$.*

Proof: Let $(m_{i_0}, m_{i_1}, \ldots, m_{i_k})$ be the subsequence of lookout points in M. For any $i = 1, 2, \ldots,$ we have:

$$\mu(\mathcal{R}_\ell(M_i)) = \mu(\mathcal{R}_\ell(M_{i-1})) + \mu(\mathcal{R}_\ell(m_i) \setminus \mathcal{R}_\ell(M_{i-1})). \tag{2}$$

Thus $\mu(\mathcal{R}_\ell(M_j)) \geq \mu(\mathcal{R}_\ell(M_i))$, for any $i \leq j$. In particular:

$$\mu(\mathcal{R}_\ell(M)) \geq \mu(\mathcal{R}_\ell(M_{i_k})). \tag{3}$$

Using (2) with $i = i_k$ in combination with the fact that m_{i_k} is a lookout point, we get:

$$\mu(\mathcal{R}_\ell(M_{i_k})) \geq \mu(\mathcal{R}_\ell(M_{i_k - 1})) + \beta\, \mu(\mathcal{X} \setminus \mathcal{R}_\ell(M_{i_k - 1})).$$

Let $v_i = \mu(\mathcal{R}_\ell(M_i))$. We observe:

$$\mu(\mathcal{X} \setminus \mathcal{R}_\ell(M_{i_k - 1})) = \mu(\mathcal{X}) - \mu(\mathcal{R}_\ell(M_{i_k - 1})) = 1 - v_{i_k - 1}.$$

Hence, $v_{i_k} \geq v_{i_k - 1} + \beta\,(1 - v_{i_k - 1})$, which can be rewritten as:

$$v_{i_k} \geq v_{i_{k-1}} + \beta\,(1 - v_{i_{k-1}}) + (1 - \beta)(v_{i_k - 1} - v_{i_{k-1}}).$$

Since $i_k - 1 \geq i_{k-1}$, it follows that $v_{i_k - 1} - v_{i_{k-1}} \geq 0$. Therefore, the previous inequality yields:

$$v_{i_k} \geq v_{i_{k-1}} + \beta\,(1 - v_{i_{k-1}}).$$

Setting $w_k = v_{i_k}$ leads to the recurrence $w_k \geq w_{k-1} + \beta\,(1 - w_{k-1})$, with the solution:

$$w_k \geq (1-\beta)^k w_0 + \beta \sum_{j=0}^{k-1} (1-\beta)^j = 1 - (1-\beta)^k (1 - w_0).$$

As $w_0 \geq 0$ and $1 - \beta \leq e^{-\beta}$, we get $w_k \geq 1 - e^{-\beta k}$. Combined with (3), it yields:

$$\mu(\mathcal{R}_\ell(M)) \geq 1 - e^{-\beta k}.$$

□

Lemma 4 *A sequence of r milestones contains k lookout points with probability at least* $1 - ke^{-\alpha\lfloor r/k \rfloor}$.

Proof: Let M be the sequence of r milestones and L be the event that M contains k lookout points. Assume that r is an integer multiple of k. We divide M into k subsequences of r/k consecutive milestones. Denote by L_i the event that the ith subsequence contains at least one lookout point. Since the probability of M having k lookout points is greater than the probability of every subsequence having at least one lookout point, we have:

$$\Pr(L) \geq \Pr(L_1 \cap L_2 \ldots \cap L_k),$$

which implies:

$$\Pr(\overline{L}) \;\leq\; \Pr(\overline{L}_1 \cup \overline{L}_2 \ldots \cup \overline{L}_k) \;\leq\; \sum_{i=0}^{k} \Pr(\overline{L}_i).$$

Since each milestone picked by IDEAL-SAMPLE has probability α of being a lookout point, the probability $\Pr(\overline{L}_i)$ of having no lookout point in the ith subsequence is at most $(1-\alpha)^{r/k}$. Hence:

$$\Pr(L) = 1 - \Pr(\overline{L}) \geq 1 - k(1-\alpha)^{r/k}.$$

If r is not an integer multiple of k, we divide M into $k-1$ subsequences of length $\lfloor r/k \rfloor$ and a kth subsequence of length $r - k\lfloor r/k \rfloor$. We get:

$$\Pr(L) \geq 1 - ke^{-\alpha\lfloor r/k \rfloor}.$$

□

We are now ready to state our main result which establishes an upper bound on the number of milestones needed to guarantee that the planner finds a trajectory with high probability, if one exists.

Theorem 5 *Let* $g > 0$ *be the volume of the endgame region* E *in* $\mathcal{X}$ *and* γ *be a constant in* $(0,1]$. *A sequence* M *of* r *milestones contains a milestone in* E *with probability at least* $1-\gamma$, *if* $r \geq (k/\alpha)\ln(2k/\gamma) + (2/g)\ln(2/\gamma)$, *where* $k = (1/\beta)\ln(2/g)$.

Proof: Divide $M = (m_0, m_1, m_2, \ldots, m_r)$ into two subsequences M' and M'' such that M' contains the first r' milestones and M'' contains the next r'' milestones with $r' + r'' = r$. By Lemma 4, M' contains k lookout points with probability at least $1-k(1-\alpha)^{r'/k}$. If there are k lookout points in M', then by Lemma 3, $\mathcal{R}_\ell(M')$ has volume at least $1 - g/2$, provided that $k \geq 1/\beta \ln(2/g)$. As a result, $\mathcal{R}_\ell(M')$ has a non-empty intersection I with E of volume at least $g/2$, and so does the ℓ-reachability set of every subsequence $M_i \supset M'$.

Since IDEAL-SAMPLE picks a milestone uniformly at random from the ℓ-reachability set of the existing milestones, every milestone m_i in M'' falls in I with probability $(g/2)/\mu(\mathcal{R}_\ell(M_{i-1}))$. Since $\mu(\mathcal{R}_\ell(M_{i-1})) \leq 1$ for all i, and the milestones are sampled independently, M'' contains a milestone in I with probability at least $1 - (1-g/2)^{r''} \geq 1 - e^{-r''g/2}$.

If M fails to have a milestone in E, then either the ℓ-reachability set of M' does not have a large enough intersection I with E (event A), or no milestone of M'' lands in I (event B). We know that $\Pr(A) \leq \gamma/2$ if $r' \geq (k/\alpha)\ln(2k/\gamma)$ and $\Pr(B) \leq \gamma/2$ if $r'' \geq (2/g)\ln(2/\gamma)$. Choosing $r \geq (k/\alpha)\ln(2k/\gamma) + (2/g)\ln(2/\gamma)$ guarantees that $\Pr(A \cup B) \leq \Pr(A) + \Pr(B) \leq \gamma$. Substituting $k = (1/\beta)\ln(2/g)$ into the inequality bounding r, we get:

$$r \geq \frac{\ln(2/g)}{\alpha\beta} \ln\frac{2\ln(2/g)}{\beta\gamma} + \frac{2}{g}\ln\frac{2}{\gamma}.$$

□

If KDP returns FAILURE, either the query admits no solution, i.e., $(s_g, t_g) \notin \mathcal{X}$, or the algorithm has failed to find one. The latter event, which corresponds to returning an incorrect answer to the query, has probability less than γ. Since the bound in Theorem 5 contains only logarithmic terms of γ, the probability of an incorrect answer converges toward 0 exponentially in the number of milestones.

The bound given by Theorem 5 also depends on the expansiveness parameters α, β and the volume g of the endgame region. The greater α, β, and g, the smaller the bound. In practice, it is often possible to establish a lower bound for g. However, α and β are difficult or impossible to estimate, except for trivial cases. This prevents us from determining the parameter N (maximal number of milestones) for KDP *a priori*. This is not different from previous analyses of PRM path planners [11, 15, 17, 29].

4.3 Approximating IDEAL-SAMPLE

One way of implementing IDEAL-SAMPLE would be to use rejection sampling [13]: Generate many samples and throw away a fraction of them in the more densely sampled regions. However, this would lead KDP to generate and then discard many potential milestones.

KDP seeks to approximate IDEAL-SAMPLE in a more efficient way. Note that every new milestone m' created in line 5 of Algorithm 1 tends to be relatively close to m, because long trajectories induced by controls picked at random are often in collision. Therefore, if we selected milestones uniformly in line 3, the resulting distribution would be very uneven; indeed, with high probability, at line 3, the planner would pick a milestone in an already densely sampled region, which would yield a new milestone in that same region in line 5. The distribution $\pi_T(m) \sim 1/w(m)$ used at line 3 contributes to the diffusion of milestones over $\mathcal{R}(m_b)$ and avoids oversampling. In general, maintaining the weights $w(m)$ as the roadmap is being built has a much smaller computational cost than performing rejection sampling.

There is a slightly greater chance of generating a new milestone in an area where the ℓ-reachability sets of several milestones already in T overlap. However, milestones in T with overlapping ℓ-reachability sets are more likely to be close to one another than milestones with no such overlapping. Therefore, the use of π_T at line 3 keeps the problem under control by preventing it from worsening as the number of milestones grows.

There is yet another issue to consider. Though line 4 selects u uniformly at random from $\mathcal{U}_\ell$, the distribution of m' in $\mathcal{R}_\ell(m)$ is not uniform in general, because the mapping from $\mathcal{U}_\ell$ to $\mathcal{R}_\ell(m)$ may not be linear. In many cases, one may precompute a distribution π_U such that picking u from $\mathcal{U}_\ell$ with probability $\pi_U(u)$ yields a uniform distribution of m' in $\mathcal{R}_\ell(m)$. In other cases, rejection sampling can be used locally as follows: In line 4, pick several control functions u_i; in line 5, compute the corresponding m'_i, throw away some of them to achieve a uniform distribution among the remaining m'_i, and pick a remaining m'_i at random.

4.4 Choice of ℓ and $\delta_{\max}$

In theory, the parameter ℓ must be chosen such that for any $p \in \mathcal{R}(m_b)$, $\mathcal{R}_\ell(p)$ has the same dimension as $\mathcal{R}(m_b)$. Otherwise, $\mathcal{R}_\ell(p)$ has zero volume relative to $\mathcal{R}(m_b)$, and $\mathcal{R}(m_b)$ cannot be expansive even for arbitrarily small values of α and β. This can only happen when some dimensions of $\mathcal{R}(m_b)$ are not directly spanned by constant controls in Ω. But these dimensions can then be generated by combining several controls in Ω using Lie-brackets [4]. The mathematical definition of a Lie bracket can be interpreted as an infinitesimal "maneuver" involving two controls. Spanning all the dimensions of $\mathcal{R}(m_b)$ may require combining more than two controls of Ω, by imbricating multiple Lie brackets. At most $n-2$ Lie brackets are needed, where n is the dimension of $\mathcal{S}$. Hence, it is sufficient in all cases to choose $\ell = n - 2$.

To simplify the implementation, however, one may choose $\ell = 1$, since a path passing through several consecutive milestones in T corresponds to applying a sequence of constant controls. In general, the larger ℓ, the greater α and β, hence the smaller the number of milestones needed according to our analysis, but also the more costly the generation of each milestone. The choice of $\delta_{\max}$ is somewhat related. A larger $\delta_{\max}$ results in greater α and β, but also leads the planner to integrate longer trajectories that are more likely to be non-admissible. Experiments show that ℓ and $\delta_{\max}$ can be selected in rather wide intervals without significant impact on the performance of the planner. However, if the values for ℓ $\delta_{\max}$ are too large, it may be difficult to approximate IDEAL-SAMPLE well.

5 Nonholonomic Robots

5.1 Robot Description

We implemented KDP for two different robot systems. One consists of two nonholonomic carts connected by a telescopic link and moving among stationary obstacles. The other system is an air-cushioned robot that is actuated by air thrusters and operates among moving obstacles on a flat table. It is subject to strict dynamic constraints. In this section, we discuss the implementation of KDP for the nonholonomic robot. In the next two sections, we will do the same for the air-cushioned robot.

(a) (b)

Figure 3: *Two-cart nonholonomic robots.*

Wheeled vehicles are a classical example for nonholonomic motion planning. The robot considered here is a new variation on this theme. It consists of two independently-actuated carts moving on a flat surface (Figure 3). Each cart obeys a nonholonomic constraint and has non-zero minimum turning radius. In addition, the two carts are connected by a telescopic link whose length is lower and upper bounded. This system has been inspired by two scenarios. One is the mobile manipulation project in the GRASP Laboratory at the University of Pennsylvania [6]; the two carts are each mounted with a manipulator arm and must remain within a certain distance range so that the two arms can cooperatively manipulate an object (Figure 3(*a*)). The manipulation area between the two carts must be clear of obstacles. In the other scenario, two carts patrolling an indoor environment must remain in a direct line of sight of each other (Figure 3(*b*)), within some distance range, in order to allow visual contact or simple directional wireless communication.

We project the geometry of the carts and the obstacles onto the horizontal plane. For $i = 1, 2$, let R_i be the midpoint between the rear wheels of the ith cart and F_i be the midpoint between the front wheels. Let L_i be the distance between R_i and F_i. We define the state of the system by $s = (x_1, y_1, \theta_1, x_2, y_2, \theta_2)$, where (x_i, y_i) are the coordinates of R_i, and θ_i is the orientation of the rear wheels of cart i relative to the x-axis (Figure 1). The distance constraint between the two carts is expressed as $d_{\min} \leq \sqrt{(x_1 - x_2)^2 + (y_1 - y_2)^2} \leq d_{\max}$.

Each cart has two scalar controls: the magnitude u_i of the velocity of R_i and the steering angle ϕ_i (the orientation of F_i's velocity relative to the rear wheels). The equations of motion for the system are:

$$\begin{array}{lllllll} \dot{x}_1 & = & u_1 \cos\theta_1 & \quad & \dot{x}_2 & = & u_2 \cos\theta_2 \\ \dot{y}_1 & = & u_1 \sin\theta_1 & & \dot{y}_2 & = & u_2 \sin\theta_2 \\ \dot{\theta}_1 & = & (u_1/L_1)\tan\phi_1 & & \dot{\theta}_2 & = & (u_2/L_2)\tan\phi_2 \end{array}$$

The control space is restricted by $|u_i| \leq U_i$ and $|\phi| \leq \Phi_i$, which bound the carts' velocities and steering angles.

5.2 Implementation Details

Since all obstacles are stationary, the planner builds a roadmap T in the robot's 6-D state space.

Computing the weights To compute the weight $w(m)$ of a milestone m, we define the neighborhood of m to be a ball of radius ρ centered at m. Our implementation uses a naive method that checks every new milestone m' against all the milestones currently in T. Thus, for every new milestone, updating w takes linear time in the size of T. More efficient range search techniques [1] would improve the planner's running time for problems requiring very large roadmaps.

Implementing PROPAGATE Given a milestone m and a control function u, PROPAGATE (m, u) uses the Euler method with a fixed step size to integrate the trajectory τ of the robot. It then discretizes τ into a sequence of states and returns *nil* if any of these states is in collision. For each cart, a 3-D bitmap that represents the collision-free configurations of the cart is precomputed prior to planning. PROPAGATE then takes constant time to check whether a configuration

(a)

(b)

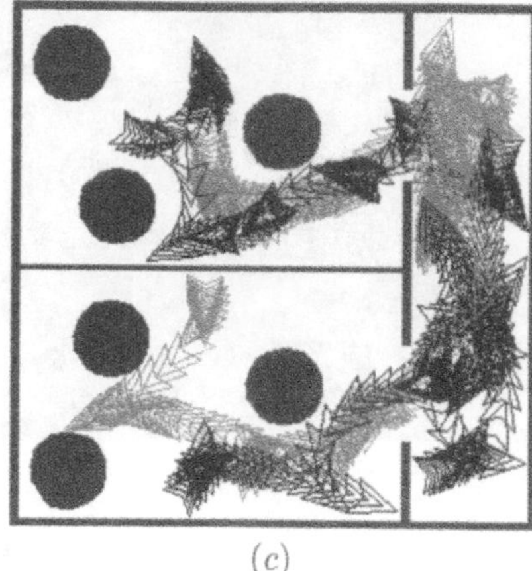

(c)

Figure 4: *Examples of nonholonomic paths computed by the planner.*

is in collision, or not. In our experiments, we used a $128 \times 128 \times 64$ bitmap.

Endgame region We obtain the endgame region by generating a secondary tree T' of milestones rooted at s_g.

5.3 Experimental Results

We experimented with the planner in many workspaces. Each is a 10 m×10 m square region with static obstacles. The two carts are identical, each represented by a polygon contained in a circle of radius 0.4 m. $L_1 = L_2 = 0.5$ m. Each cart's speed ranges from −3 m/s to 3 m/s, and its steering angle ϕ varies between −30° and 30° The distance between R_1 and R_2 ranges between 1.4 m and 3.3 m.

Figure 4 shows three computed examples. Workspace (*a*) is a maze; the robot must navigate from one side of it to the other. Workspace (*b*) contains two large obstacles separated by a narrow passage. The two carts, which are initially parallel to one another, change formation and proceed in a single file through the passage, before becoming parallel again. Workspace (*c*) consists of two rooms cluttered with obstacles and connected by a hallway. The carts need to move from the bottom one to the the top one. The maximum steering angles and the size of the circular obstacles conspire to increase the number of required maneuvers.

We have run the planner for several different queries in each workspace shown in Figure 4. For every query,

E	Q	time (sec)		coll.	mil.	prop.
		mean	std			
(*a*)	1	1.39	0.91	62,400	2473	21316
	2	0.74	0.65	43,600	1630	15315
	3	0.54	0.41	36,000	1318	12815
	4	0.55	0.44	38,400	1310	14066
(*b*)	1	4.45	3.92	126,100	4473	45690
(*c*)	1	14.09	7.42	287,800	9123	107393
	2	0.92	0.51	56,400	1894	20250

Table 1: *Planning statistics for the nonholonomic robot.*

we ran the planner 30 times independently with different random seeds. The results are collected in Table 1. Every row of the table corresponds to a particular query. Columns 3-7 list the average running time, its standard deviation, the average number of collision checks, the average number of milestones generated, and the average number of calls to PROPAGATE. The running times range from less than a second to a few seconds. The first query in environment (*c*) takes longer because the carts must perform several maneuvers in the hallway before reaching the goal (see Figure 4(*c*)). The planner was written in C++, and the running times were collected on a 195 Mhz SGI Indigo2 with an R10000 processor.

The standard deviations in Table 1 are larger than we would like. Figure 5 plots a histogram of more than 100 independent runs for a given query. In most runs, planning time is well under the mean or slightly above. This indicates that our planner performs well most of the time. The large deviation is caused by a few runs that take as long as four times the mean. The long and

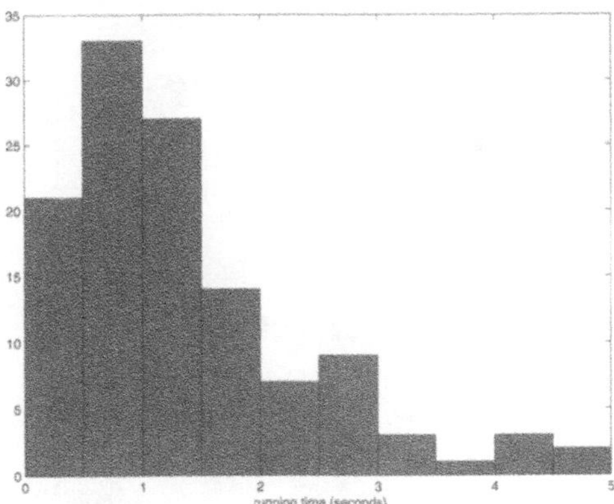

Figure 5: *Histogram of planning times for more than 100 runs on a particular query. The average time is 1.4 sec, and the four quartiles are 0.6, 1.1, 1.9, and 4.9 sec.*

thin tail of the distribution is typical of the tests that we have performed.

6 Air-Cushioned Robots

6.1 Robot Description

The second robot system used to evaluate our algorithm was developed in the Stanford Aerospace Robotics Laboratory for testing space robotic technologies. This robot (Figure 6) moves frictionlessly on an air bearing on a flat granite table. Air thrusters provides omni-directional motion capability, but the thrust available is small compared to the robot's mass, resulting in tight acceleration limits. We represent the workspace by a 3 m×4 m rectangle, the robot by a disc of radius 0.25 m, and the obstacles by discs of radii between 0.1 and 0.15 cm. Each obstacle moves along a straight path at constant velocity ranging between 0 and 0.2 m/s (more complex trajectories will be considered in Subsection 7.4). In the simulation environment, collisions among obstacles are ignored. So two obstacles may temporarily overlap without changing courses. When an obstacle reaches the workspace's boundary, it leaves the workspace and is no longer considered a threat to the robot.

We define the robot's state to be $(x, y, \dot{x}, \dot{y})$, where (x, y) are the coordinate of the robot's center. The equations of motion are:

$$\ddot{x} = \frac{1}{m}u\cos\theta \quad \text{and} \quad \ddot{y} = \frac{1}{m}u\sin\theta,$$

Figure 6: *The air-cushioned robot.*

where m is for the robot's mass, and u and θ denote the magnitude and direction of the force generated by the thrusters. We have $u/m < 0.025$ m/s^2, and θ varies freely between 0° and 360°. We also bound the robot's velocity by 0.18 m/s.

6.2 Implementation Details

The planner builds a roadmap T in the robot's 5-D state×time space, The initial state×time is of the form $(s_b, 0)$ and the goal is of the form (s_g, t_g) where t_g can be any time less than a given $t_{\max}$. The planner is given the obstacle trajectories, and unlike in the experiments with the real robot in the next section, planning time is not limited. This is equivalent to assuming that the world is frozen until the planner returns a trajectory.

Computing the weights The 3-D configuration-×time space of the robot is partitioned into an 8×11×10 array of identically sized rectangles called bins. When a milestone is inserted in T, the planner adds it to a list of milestones associated with the bin in which it falls. In line 3 of KDP, the planner picks at random a bin containing at least one milestone and then a milestone from within this bin. Both choices are made uniformly at random. This corresponds to picking a milestone with a probability approximately proportional to the inverse of the density of samples in the robot's configuration×time space (rather than its state×time space). We did some experiments with higher-dimensional bin arrays, but the results were not improved significantly.

Implementing PROPAGATE The simple equations of motion make is possible to compute trajectories an-

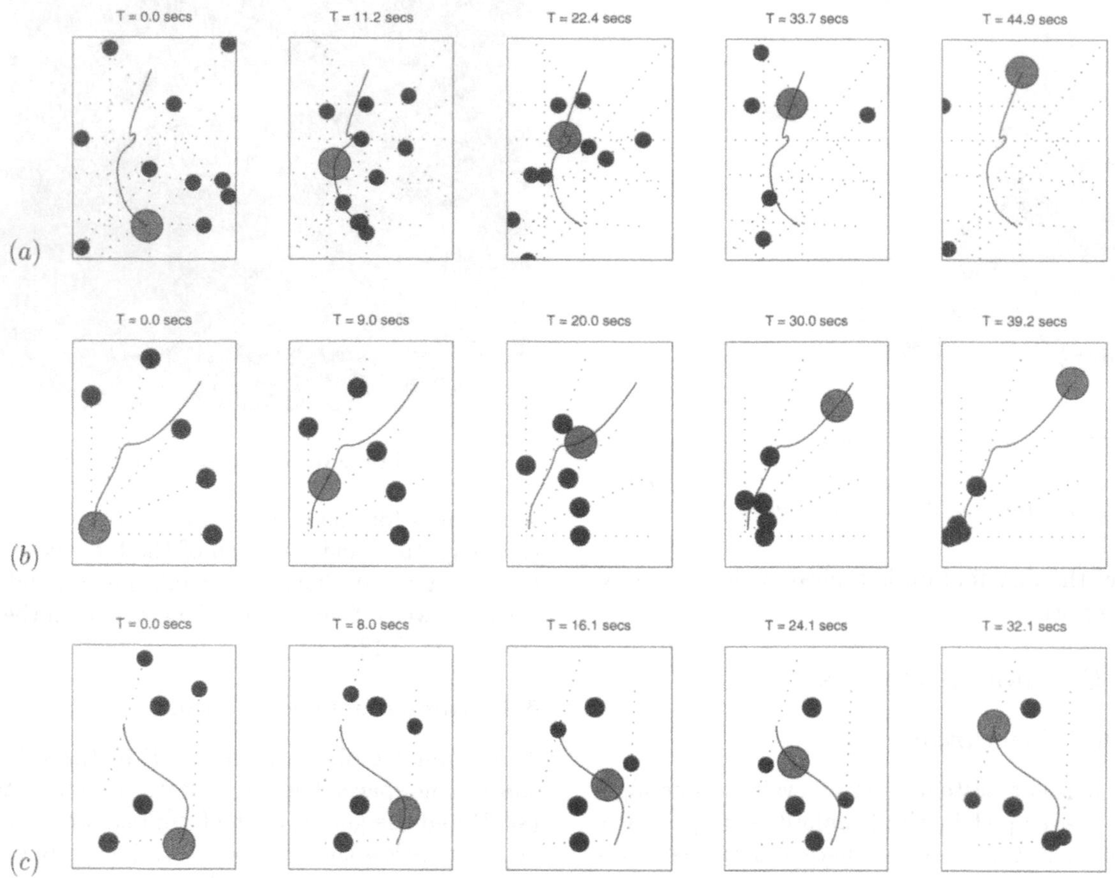

Figure 7: *Trajectories produced by the planner for the air-cushioned robot.*

alytically. Trajectories are discretized and at each discretized state×time the robot disc is checked for collision with each obstacle disc. This naive technique works reasonably well when the number of obstacles is small. It could easily be improved.

Endgame region When a milestone m is added to T, it is checked for a connection to k goal points (s_g, t_g). Each of the k values of t_g is picked uniformly at random in the interval $[t_{\min}, t_{\max}]$, where $t_{\min}$ is an estimate of the earliest time when the robot may reach s_g given its maximal velocity and assuming no obstacles. For each value of t_g, the planner computes the third-order spline connecting m to (s_g, t_g). It then verifies that the spline is collision free and satisfies the velocity and acceleration bounds. If all the tests succeed, then m lies in the endgame region. In all the experiments below, k is set to 10.

6.3 Experimental Results

We performed numerous experiments in more than one hundred simulated environments. In each case, planning time was limited to five minutes. For a small number of queries, the planner failed to return a trajectory, but for none of them were we able to show that an admissible trajectory exists. On the other hand, the planner successfully solved several queries for which we initially thought there were no solution.

Three trajectories computed by the planner are shown in Figure 7. For every example, we display five snapshots labeled by time. The large disc is the robot; the smaller discs are the obstacles. The solid and dotted lines mark the trajectories of the robot and the obstacles, respectively. For each query, we ran the planner 100 times independently with different random seeds. The planner successfully returned a trajectory

E	time (sec)		milestones	
	mean	std	mean	std
(a)	0.249	0.264	2008	2229
(b)	0.270	0.285	1946	2134
(c)	0.002	0.005	22	25

Table 2: *Planning statistics for the air-cushioned robot.*

in all runs. In Example (a), the duration of the trajectory varied from 40 to 70 sec. Table 2 lists the means and standard deviations of the planning time and number of milestones for each example. The reported times are based on a planner written in C and running on a Pentium-III PC with a 550 Mhz processor and 128M of memory.

In the first two examples, the moving obstacles create narrow passages through which the robot must pass to reach the goal. Yet planning time remains much under 1 second. The fact that the planner never failed in 100 runs indicates its reliability. The configuration×time space for Example (b) is shown in Figure 8. The robot maps to a point (x, y, t), and the obstacles to cylinders. The velocity and acceleration constraints force any solution trajectory to pass through a small gap between cylinders. Example (c) is much simpler. There are two stationary obstacles obstructing the middle of the workspace and three moving obstacles. Planning time is well below 0.01 second, with an average of 0.002 second. The number of milestones is also small, confirming the result of Theorem 5 that in the absence of narrow passages, KDP is very efficient. As in the experiments on nonholonomic robot carts, the running time distribution of the planner tends to have a long and thin tail.

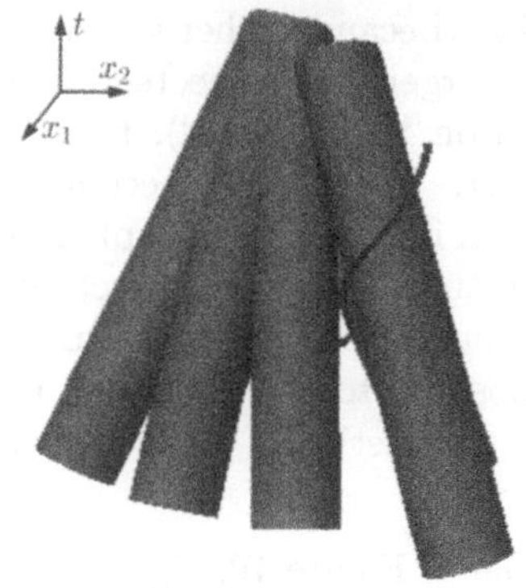

Figure 8: *Configuration×space for Example (b).*

7 Experiments with the Real Robot

We connected the planner of the previous section to the air-cushioned robot of Figure 6 in order to verify that KDP remains useful in a system integrating control and sensing modules over a distributed architecture and operating in a physical environment with uncertainties, time delays, and real-time constraints.

7.1 Testbed Description

The robot in Figure 6 moves frictionlessly on an air bearing within the limits of a 3 m×4 m table. Obstacles, also on air-bearings, translate without friction on the table. The positions of the robot and the obstacles are measured at 60 Hz by an overhead vision system thanks to LEDs placed on each moving object. The measurement is accurate to 5 mm. Velocity estimates are derived from position data.

The robot is untethered. Gas tanks provide compressed air for both the air-bearing and thrusters. An onboard Motorola ppc2604 computer performs motion control at 60 Hz. The planner runs on on a 333 Mhz Sun Sparc 10. The robot communicates with the planner and the vision module over the radio Ethernet.

The obstacles have no thrusters. They are initially propelled by hand from various locations, and then move at constant speed until they reach the boundary of the table, where they stop due to the lack of air bearing.

7.2 System Integration

Implementing the planner on the hardware testbed raises a number of new challenges:

Delays Various computations and data exchanges produce delays between the instant when the vision module measures the trajectories of the robot and the obstacles and the instant when the robot starts executing the planned trajectory. Ignoring these delays would lead the robot to begin executing the planned trajectory behind the start time assumed by the planner. It then may not be able to catch up with the planned trajectory before collision occurs. To deal with this issue, the planner computes a trajectory assuming that the robot will start executing it 0.4 second into the future. It extrapolates the positions of the obstacles accordingly, as well as that of the robot if its initial velocity is non-zero. The 0.4 second contains all the delays in the system (not just the time needed for planning). It could be further reduced by running the planner on a machine faster than the Sun Sparc 10 that we are using currently.

Path optimization Because robot starts executing the trajectory 0.4 second after planning begins, the planner exploits any extra time after obtaining a first trajectory to generate additional milestones and keeps track of the best trajectory generated. The cost function used to compare trajectories is $\sum_{i=1}^{i=k}(u_i + b)\delta_i$, where k is the number of segments in the trajectory, u_i is the magnitude of the force exerted by the thrusters along the ith segment, δ_i is the duration of the ith segment, and b is a constant. This cost combines fuel consumption with execution time. A larger b yields a faster motion, while a smaller b yields less fuel consumption. In our experiments, the cost of trajectories was reduced, on average, by 14% with this simple improvement.

Sensing errors The obstacle trajectories are assumed to be straight lines at constant velocities. However, inaccuracy in the measurements by the vision module and assymmetry in air-bearings cause actual trajectories to be slightly different. The planner deals with these errors by increasing the radius of each moving obstacle by an amount $\xi V_{\max} t$, where t denotes time, $V_{\max}$ is the measured velocity of the obstacle, and ξ is a constant.

Safe-mode planning If the planner does not find a trajectory to the goal within the allocated time, we find it useful to compute an *escape* trajectory. The endgame region E_{esc} for the escape trajectory consists of all the reachable, collision-free states (s_e, t_e) with $t_e \geq T_{\text{esc}}$ for some T_{esc}. An escape trajectory corresponds to any acceleration-bounded, collision-free motion in the workspace for a small duration of time. In general, E_{esc} is very large, and so generating an escape trajectory often takes little time. To ensure collision-free motion beyond T_{esc}, a new escape trajectory must also be computed long before the end of the current escape trajectory so that the robot can escape collision despite the acceleration constraints. We modified the planner to compute concurrently both a normal and an escape trajectory. In our experiments, the modification slowed down the planner by about 2%.

Trajectory tracking The trajectory received by the robot specifies the position, velocity, and acceleration of the robot at all times. A PD-controller with feedforward is used to track this trajectory. Maximum tracking errors are 0.05 m and 0.02 m/s. The size of the disc modeling the robot is increased by 0.05 m to ensure safe collision checking by the planner.

7.3 Experimental Results

The planner successfully produced complex maneuvers of the robot among static and moving obstacles in various situations, including obstacles moving directly toward the robot, as well as perpendicular to the line connecting its initial and goal positions. The tests also demonstrated the ability of the system to wait for an opening to occur when confronted with moving obstacles in the robot's desired direction of movement and to pass through openings that are less than 10 cm larger than the robot. In almost every trial, a trajectory was computed within the allocated time. Figure 9 shows snapshots of the robot during one test.

Several constraints limited the complexity of the planning problems which we could test. Two are related to the testbed itself: The size of the table relative to the robot and the obstacles, and the robot's small acceleration. The other two constraints result from the design of our system: The requirement that obstacles move along straight lines and hence do not collide with each other, and the relatively high uncertainty on their movements, which forces the planner to grow the obstacles and thus reduce free space. To eliminate the last two constraints, we introduced on-the-fly replanning.

7.4 On-the-Fly Replanning

Whenever an obstacle leaves the disc in which the planner believes it lies (because either the error on the predicted motion is larger than expected, or the obstacle's direction of motion has changed), the vision module alerts the planner. The planner recomputes a trajectory on the fly as if it were a normal planning operation within the same time limit, by projecting the state of the world 0.4 second into the future. On-the-fly replanning makes it possible to perform much more complex experiments in the testbed. We show two examples below.

In the example in Figure 10, eight replanning operations are performed over the 75 seconds taken by the robot's motion. Initially, the robot moves to the left

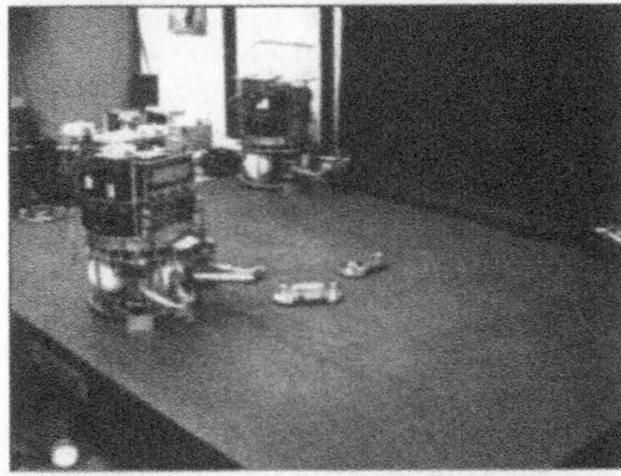

Figure 9: *Snapshots of the robot executing a trajectory.*

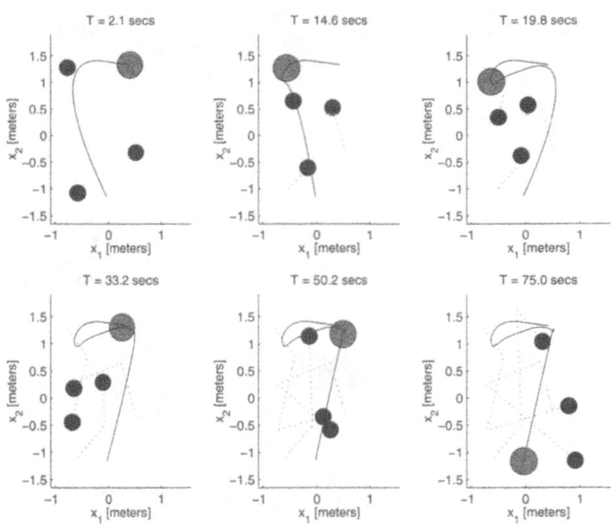

Figure 10: *Simulation example with replanning.*

to reach the goal at the bottom middle. At 14 seconds, the upper-left obstacle changes course, forcing a replan (snapshot 2). Soon after, the motion of the upper-right obstacle changes, forcing the robot to reverse direction and approach the goal from the other side of the workspace (snapshot 3). In the remaining time, new changes in obstacle motion cause the robot to pause (tight angle in snapshot 5) before a straight trajectory to the goal is possible (snapshot 6).

The efficacy of the replanning procedure in the testbed is demonstrated by the example in Figure 11. In snapshot 1, the middle obstacle is stationary, while the two outer obstacles are moving towards the robot. The robot attempts to move from the back middle to the front middle of the workspace. In snapshot 2, the robot dodges the faster-moving obstacle from the left, in order to let it pass by and then proceeds toward the goal. In snapshot 3, that obstacle is redirected to block the trajectory of the robot, causing it to slow down and stay behind the obstacle to avoid collision. Snapshot 5 shows the leftmost obstacle being redirected again, this time towards the robot's goal. The robot follows this obstacle and move slowly towards the goal. Before snapshot 7, however, the rightmost obstacle is directed back towards the robot, forcing the robot to wait and let it pass (snapshot 8). Finally, in the last snapshot, the robot attains the goal. The entire motion lasted 40 seconds. Throughout this example, other replanning operations were performed due to errors in the

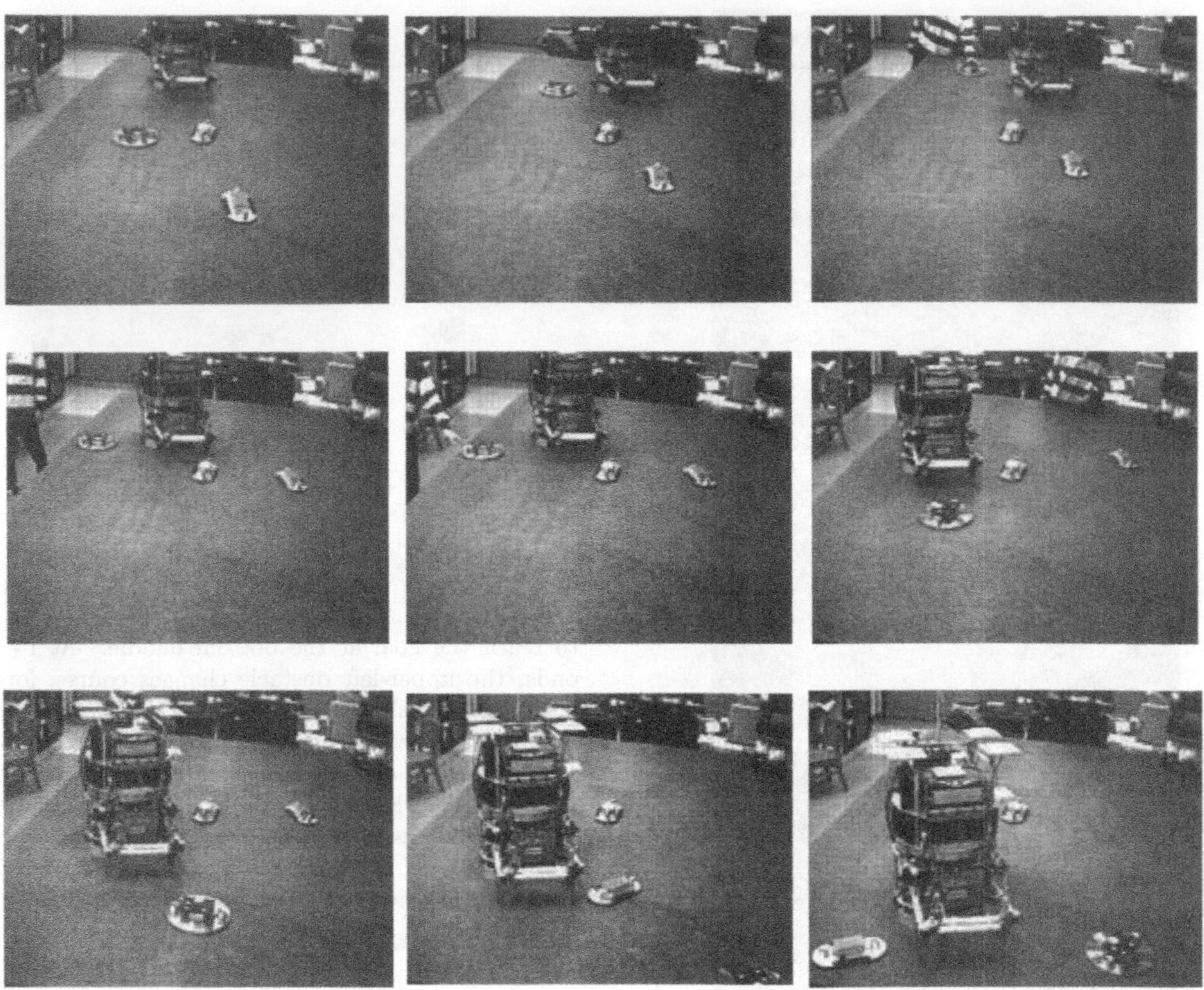

Figure 11: *An example with the real robot using replanning.*

measurement of the obstacle trajectories (We set ξ such that sensing errors caused replanning to occur, on average, every 5 seconds). However, none resulted in a major redirection of the robot.

8 Conclusion

We have presented a simple, efficient randomized planner for kinodynamic motion planning problems in the presence of moving obstacles. Under the expansiveness assumption, we have formally proven that the planner is probabilistically complete and converges quickly when a solution exists. This proof also applies to robots that are not locally controllable. The planner was tested successfully with simulated and real robots. The experiments in the hardware robot testbed demonstrate that the planner remains effective despite various delays and uncertainties inherent to an integrated system interacting with the physical world. They also show that the planner can be used in real-time when obstacle trajectories are not known in advance.

In the future, we plan to apply the planner to environments with more complex geometry. Geometrical complexity essentially increases the cost of collision-checking, but hierarchical techniques deal with this issue well. In [11], a similar, but simpler planner was successfully applied to compute geometric disassembly paths with CAD models having up to 200,000 triangles. Another issue that we would like to investigate is the reduction of the standard deviation of the planning time. We suspect that the long thin tail shown in Figure 5 is typical of all PRM planners developed so far. But it seems more critical to reduce it for single-query planners, since such planners are more likely to be used interactively or in real-time than multi-query ones.

Acknowledgments

This work was supported by ARO MURI grant DAAH04-96-1-007, NASA TRIWG Coop-Agreement NCC2-333, Real-Time Innovations, and the NIST ATP program. David Hsu has also been the recipient of a Microsoft Graduate Fellowship, and Robert Kindel, the recipient of an NSF Graduate Fellowship.

References

[1] P. K. Agarwal. Range searching. In *Handbook of discrete and computational geometry*, J.E. Goodman and J. O'Rourke (eds.), CRC Press, Chap. 31, p. 575–598, 1997.

[2] N. M. Amato, O. B. Bayazit, L. K. Dale, C. Jones, and D. Vallejo. OBPRM: An obstacle-based PRM for 3D workspaces. In *Robotics: The Algorithmic Perspective*, P.K. Agarwal, L.E. Kavraki, and M.T Mason (eds.), A K Peters, p. 155–168, 1998.

[3] J. Barraquand and J. C. Latombe. Robot motion planning: a distributed representation approach. *Int. J. of Robotics Research*, MIT Press, 10(6):628-649, 1991.

[4] J. Barraquand and J. C. Latombe. Nonholonomic multibody mobile robots: controllability and motion planning in the presence of obstacles. *Algorithmica*, 10(2-4):121–155, 1993.

[5] J. E. Bobrow, S. Dubowsky, and J. Gibson. Time-optimal control of robotic manipulators along specified paths. *Int. J. of Robotics Research*, 4(3):3–17, 1985.

[6] J.P. Desai and V. Kumar. Motion planning for cooperating mobile manipulators. *J. of Robotic Systems*, 16(10):557–579, 1999.

[7] B. Donald, P. Xavier, J. Canny, and J. Reif. Kinodynamic motion planning. *J. of the ACM*, 40(5):1048–1066, 1993.

[8] P. Fiorini and Z. Shiller. Time optimal trajectory planning in dynamic environments. *Proc. IEEE Int. Conf. on Robotics and Autom.*, p. 1553–1558, 1996.

[9] T. Fraichard. Trajectory planning in a dynamic workspace: a 'state-time space' approach. *Advanced Robotics*, 13(1):75-94, 1999.

[10] K. Fujimura. Time-minimum routes in time-dependent networks. *IEEE Tr. on Robotics and Autom.*, 11(3):343–351, 1995.

[11] D. Hsu, J. C. Latombe, and R. Motwani. Path planning in expansive configuration spaces. *Int. J. Comp. Geometry and Applications*, 9(4-5):495–512, 1999.

[12] P. Jacobs and J. Canny. Planning smooth paths for mobile robots. In *Proc. IEEE Int. Conf. on Robotics and Autom.*, p. 2–7, 1989.

[13] M.H. Kalos and P.A. Whitlock. *Monte Carlo Methods*, Vol. 1, John Wiley & Son, 1986.

[14] K. Kant and S. W. Zucker. Toward efficient trajectory planning: The path-velocity decomposition. *Int. J. of Robotics Research*, 5(3):72–89, 1986.

[15] L.E. Kavraki, M. Kolountzakis, and J. C. Latombe. Analysis of probabilistic roadmaps for path planning. *IEEE Tr. Robotics and Autom.*, 14(1):166–171, 1998.

[16] L.E. Kavraki and J. C. Latombe. Randomized preprocessing of configuration space for fast path planning. *Proc. IEEE Int. Conf. on Robotics and Autom.*, p. 2138–2145, 1994.

[17] L.E. Kavraki, J. C. Latombe, R. Motwani, and P. Raghavan. Randomized query processing in robot motion planning. *J. Computer and System Sciences*, 57(1):50–60, 1998.

[18] L.E. Kavraki, P. Švestka, J. C. Latombe, and M. Overmars. Probabilistic roadmaps for path planning in high-dimensional configuration spaces. *IEEE Tr. Robotics and Autom.*, 12(4):566–580, 1996.

[19] J.P. Laumond. Feasible trajectories for mobile robots with kinematic and environmental constraints. *Proc. Int. Conf. on Intelligent Autonomous Systems*, p. 346–354, 1986.

[20] J.P. Laumond, P.E. Jacobs, M. Taïx., and R.M. Murray. A motion planner for nonholonomic mobile robots. *IEEE Tr. on Robotics and Autom.*, 10(5):577–593, 1994.

[21] S. LaValle and J. Kuffner. Randomized kinodynamic planning. *Proc. IEEE Int. Conf. on Robotics and Autom.*, p. 473–479, 1999.

[22] Z. Li, J. F. Canny, and G. Heinzinger. Robot motion planning with nonholonomic constraints. In *Robotics Research: The 5th Int. Symp.*, H. Miura et al. (eds.), MIT Press, p. 309–316, 1989.

[23] K. M. Lynch and M. T. Mason. Stable pushing: mechanics, controllability, and planning. *Int. J. of Robotics Research*, 15(6):533–556, 1996.

[24] J.A. Reeds and L.A. Shepp. Optimal paths for a car that goes forwards and backwards. *Pacific J. of Mathematics*, 145(2):367-393, 1990.

[25] J. Reif and M. Sharir. Motion planning in the presence of moving obstacles. *Proc. IEEE Symp. on Foundations of Computer Science*, p. 144–154, 1985.

[26] S. Sekhavat, P. Švestka, J.-P. Laumond, and M. Overmars. Multi-level path planning for nonholonomic robots using semi-holonomic subsystems. *Algorithms for Robotic Motion and Manipulation*, J.P. Laumond and M. Overmars (eds.), A.K. Peters, p. 79–96, 1997.

[27] Z. Shiller and S. Dubowsky. On computing the global time-optimal motions of robotic manipulators in the presence of obstacles. *IEEE Tr. on Robotics and Autom.*, 7(6):785–797, 1991.

[28] P. Švestka and M. H. Overmars. Motion planning for car-like robots using a probabilistic learning approach. Tech. Rep. RUU-CS-94-33, Dept. of Computer Science, Utrecht Univ., The Netherlands, 1994.

[29] P. Švestka and M. Overmars. Probabilistic path planning: robot motion planning and control. *Lecture Notes in Control and Information Sciences*, 229, Springer, p. 255–304, 1998.

On Random Sampling in Contact Configuration Space

Xuerong Ji, *University of North Carolina, Charlotte, NC*
Jing Xiao, *University of North Carolina, Charlotte, NC*

Random sampling strategies play critical roles in randomized motion planners, which are promising and practical for motion planning problems with many degrees of freedom (dof). In this paper, we address random sampling in a constrained configuration space – the contact configuration space between two polyhedra, motivated by the need for generating contact motion plans. Given a contact formation (CF) between two polyhedra A and B, our approach is to randomly generate configurations of A satisfying the contact constraints of the CF. Key to the approach is to guarantee that sampling happens only in the constrained space to be efficient, which has not been addressed in the literature. We first describe our random sampling strategy for configurations constrained by CFs consisting of one or two principal contacts (PCs) and then present implementation results.

1 Introduction

Contact motions are important in automatic assembly processes, not only because they happen frequently when clearance between objects is tight, but also because they reduce degrees of freedom (dof), thus reduce uncertainties [14, 16]. Contact motion occurs on the boundary of configuration space obstacles (C-obstacles) [13], but computing C-obstacles remains a formidable task to date. While there were exact descriptions of C-obstacles for polygons [2, 4], there were only approximations for polyhedra [5, 11]. Contact motions, however, require exactness of contact configurations. Hence, some researchers started exploring methodologies on contact motion planning without explicitly computing C-obstacles (see, for example, [7]).

Recently the authors introduced a general divide-and-merge approach for automatically generating a contact state graph between arbitrary polyhedra [9, 20]. Each node in the graph denotes a contact state, indicated by a *contact formation* (CF) [17] and a representative configuration of the CF. Each edge denotes the neighboring relationship between the two nodes connected by the edge. With this approach, the problem of contact motion planning is effectively simplified as graph search at high-level for state transitions and motion planning at low-level within the set of contact configurations constrained by the *same* contact state[1]. However, even for such reduced-dimension and reduced-scope motion planning, the dimensionality or dof can still be quite high for less-constrained contact states, such as those consisting of only a single *principal contact* (PC) [17] (see Figure 1 for PCs between two polyhedra). Thus, randomized motion planning is desirable for planning contact motions, which requires random sampling of *contact* configurations.

There are promising randomized planners for collision-free motions, such as those based on probabilistic roadmaps (PRM) [12, 15]. In the PRM model random sampling consists of generating arbitrary configurations of the considered object/robot. The sampled configurations may or may not be collision-free. To make sampling more efficient, several researchers introduced different methods targeting at producing more samples in certain critical areas that tend to be close to C-obstacles and sparse samples in other areas[1, 3, 8]. However, sampling *exactly* on the C-obstacle surface, or in the *contact* configuration space, has not been addressed in the literature.

In this paper, we extends the random sampling strategy for a single PC reported in [10] to contact formations (CFs) of two PCs. Our approach takes advantage of our work on automatic generation of contact state graphs by building sampling on the knowledge of a contact formation and a representative contact configura-

[1]Note that a general contact motion crossing several contact states consists of segments of motion in each contact state.

tion under the contact formation (obtainable from such a graph). Particularly, given a CF and a *seed* configuration satisfying the CF, the goal is to randomly sample configurations satisfying the CF.

The paper is outlined as follows. In Section 2, we review the notion of CFs and analyze the dof for each kind of single-PC or two-PC CF between two arbitrary polyhedra. In Section 3, we present the random sampling algorithms. In Section 4, we provide some experimental results of the sampling strategy. Section 5 concludes the paper.

2 Contact Formations and Degrees of Freedom (dof)

In this section, we first introduce notations related to contact formations used throughout the paper and then analyze the dof for different types of contact formations. Knowledge of the dof is needed for our random sampling strategy.

2.1 Notations

Consider two arbitrary polyhedra A and B. Assume that A is moveable and B is static. The faces, edges, and vertices of each object are the object's topological *elements*. The *boundary elements* of a face are its edges and vertices, and the boundary elements of an edge are its vertices.

As defined in [17], a *principal contact* (PC) between A and B describes a single contact between a pair of contacting elements which are *not* the boundary elements of other contacting elements. There are 10 types of PCs (Figure 1): face-face (f-f), face-edge(f-e)/edge-face(e-f), face-vertex (f-v)/vertex-face(v-f), edge-edge-cross (e-e-c), edge-edge-touch (e-e-t), edge-vertex (e-v)/vertex-edge (v-e), and vertex-vertex (v-v), of which e-e-t, e-v/v-e and v-v PCs between convex elements are *degenerate* PCs. We denote a PC as u_A-u_B, where u_A and u_B denote the contacting elements of A and B respectively. Given this notion of PCs, an arbitrary contact between A and B can be characterized by the set of PCs formed, called a *contact formation* (CF), denoted by $\{PC_1,...,PC_n\}$. Since degenerate PCs rarely happen in practice, in this paper, we only consider CFs formed by *non-degenerate* PCs, referred to simply as PCs.

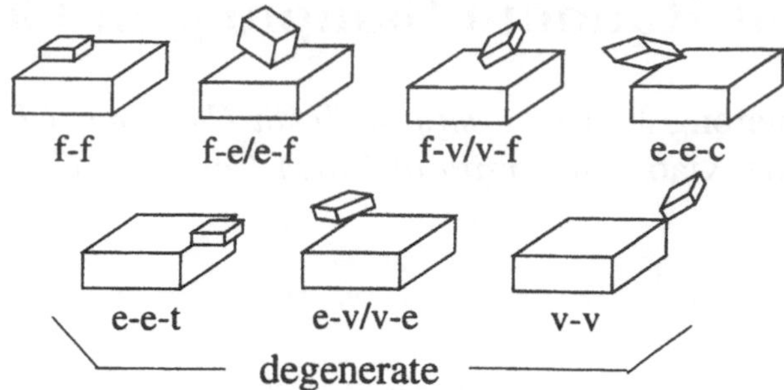

Figure 1: *Principal Contacts (PCs)*

The two contacting elements of a non-degenerate PC uniquely determine a plane, which we call a *contact plane* (CP). Based on the region of contact on the contact plane, we can further classify the PCs into the following three types:

- **plane PC**: f-f, where the contacting elements intersect at a planar region on the contact plane,
- **line PC**: e-f and f-e, where the contacting elements intersect at a line, called a *contact line.*
- **point PC**: v-f, f-v, and e-e-c, where the contacting elements intersect at a point, called a *contact point.*

A *contact configuration* is defined as the configuration of A relative to that of B when A and B are in contact. Thus, the geometric interpretation of a PC is the region of contact configurations (on the contact plane) where the PC holds, and that of a CF is the intersection of the regions of contact configurations of the participating PCs.

2.2 Degrees of Freedom

The dof of a PC is expressed by the constraints it imposes on contact configurations. To represent such contact constraints, for an arbitrary polyhedron P, we attach coordinate system (or frame) to it. Moreover, for each element (vertex, edge or face) of the object P, we attach a coordinate system as follows:

- vertex: the coordinate system v has its origin at the vertex, and the orientation is the same as that of P.
- edge: the coordinate system e has its origin at one of its bounding vertices, the direction of $+X$ is along the edge pointing to the other bounding vertex, the direction of $+Z$ is along the *outward normal*

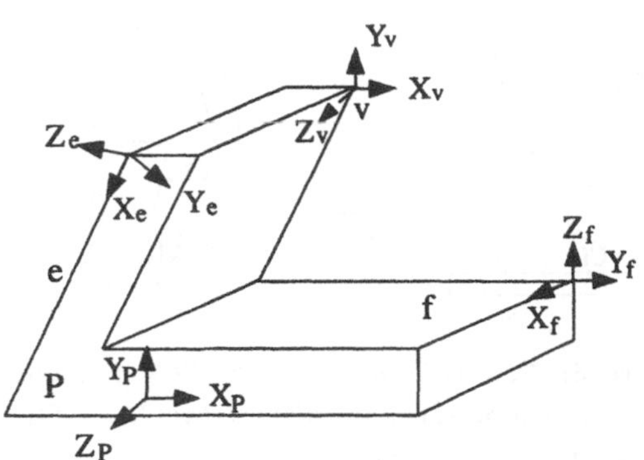

Figure 2: *Coordinate systems for an object P and for its vertex v, edge e and face f*

of the edge, defined as the sum of the outward normals of the faces forming the edge, and the direction of $+Y$ is determined by the right-hand rule.

- face: the coordinate system f has its origin at one of its bounding vertices, the direction of $+Z$ is defined as the outward normal of the face, the direction of $+X$ is along one of the bounding edges of the face, and the direction of $+Y$ is determined by the right-hand rule.

Figure 2 illustrates the coordinate systems of object P and some of its elements.

We can present the contact constraint equations for each type of PC between A and B in terms of expressions for contact configurations ${}^{B}T_{A}$ (homogeneous transformation matrix) (extending [19]), where for A, B and their elements, we attach coordinate systems by the above definitions. In each expression, the underlined symbols are independent variables, of which the Greek symbols are rotational variables in Euler angles.

- *v-f*:
  ${}^{B}T_{A} = {}^{B}T_{fB} \cdot T_{trans}(\underline{x}, \underline{y}, 0) \cdot T_{rotzyx}(\underline{\alpha}, \underline{\beta}, \underline{\gamma}) \cdot {}^{A}T_{vA}^{-1}$
- *f-v*:
  ${}^{B}T_{A} = {}^{B}T_{vB} \cdot T_{rotzyx}(\underline{\alpha}, \underline{\beta}, \underline{\gamma}) \cdot T_{trans}(\underline{x}, \underline{y}, 0) \cdot {}^{A}T_{fA}^{-1}$
- *e-e-c*:
  ${}^{B}T_{A} = {}^{B}T_{eB} \cdot T_{trans}(\underline{x_1}, 0, 0) \cdot T_{rotzyx}(\underline{\alpha}, \underline{\beta}, \underline{\gamma}) \cdot T_{trans}(\underline{x_2}, 0, 0) \cdot {}^{A}T_{eA}^{-1}$
- *e-f*:
  ${}^{B}T_{A} = {}^{B}T_{fB} \cdot T_{trans}(\underline{x}, \underline{y}, 0) \cdot T_{rotz}(\underline{\alpha}) \cdot T_{rotx}(\underline{\gamma}) \cdot {}^{A}T_{eA}^{-1}$
- *f-e*:
  ${}^{B}T_{A} = {}^{B}T_{eB} \cdot T_{rotx}(\underline{\gamma}) \cdot T_{rotz}(\underline{\alpha}) \cdot T_{trans}(\underline{x}, \underline{y}, 0) \cdot {}^{A}T_{fA}^{-1}$
- *f-f*:
  ${}^{B}T_{A} = {}^{B}T_{fB} \cdot T_{trans}(\underline{x}, \underline{y}, 0) \cdot T_{rotx}(\pi) \cdot T_{rotz}(\underline{\alpha}) \cdot {}^{A}T_{fA}^{-1}$

where ${}^{A}T_{vA}$, ${}^{A}T_{eA}$ and ${}^{A}T_{fA}$ (or ${}^{B}T_{vB}$, ${}^{B}T_{eB}$ and ${}^{B}T_{fB}$) represent the transformation matrices from the frame of a vertex, edge and face of A (or B) to the frame of A (or B) respectively. $T_{trans}(*,*,*)$ is the 4×4 translational matrix. $T_{rotz}(*)$, $T_{roty}(*)$, and $T_{rotx}(*)$ are the 4×4 rotational matrix about z, y and x axis respectively, and $T_{rotzyx}(*,*,*)$ is the combination of the three rotational matrices.

The dof for each type of PC equals to its number of independent variables. Using "t" to indicate translational dof and "r" to indicate rotational dof, we summarize the dof for single-PC CFs below:

- **plane PC**: **dof** = 3, (2t and 1r)
- **line PC**: **dof** = 4, (2t and 2r)
- **point PC**: **dof** = 5, (2t and 3r)

Since a free-flying polyhedron has 6 dof, it is easy to see that a point PC reduces the dof of the object by 1, a line PC by 2, and a plane PC by 3.

The constraint equations for a two-PC CF are the set of equations for the PCs involved. The dof of a two-PC CF depends not only on the topological types of the PCs but also on the geometrical relation between the PCs and their corresponding contact regions. We summarize them below:

CFs where the two contact planes are not parallel:

- **two plane PCs {f-f, f-f}**:
  dof = 1, (1t)
- **line PC and plane PC {e-f/f-e, f-f}**[2]:

$$\textbf{dof} = \begin{cases} 2,\ \text{(1t and 1r)} \\ \quad \text{if } CL_2 \perp CL \\ 1,\ \text{(1t)} \\ \quad \text{otherwise} \end{cases}$$

Without losing generality, we denote PC_1 to be the plane PC with CP_1 being the contact plane, PC_2 to be the line PC with CP_2 being the contact plane and CL_2 being the contact line. We further denote CL to be the intersection line of CP_1 and CP_2.

[2]The notion "e-f/f-e" means either e-f or f-e. The same explanation applies to all "/" used in the paper.

Clearly there is one translational dof along CL. In addition, the only possible kind of rotation maintaining (or compliant to) the plane PC PC_1 is along a normal of CP_1, and the possible rotation (or combined rotation and translation) maintaining PC_2 is either along an axis parallel to CL_2, or along a normal of CP_2, or along an axis on the plane passing CL_2 and perpendicular to CP_2, denote it as P_{CL_2}. Since CP_1 and CP_2 are not parallel, only when $CL_2 \perp CL$ (the first case in the above formula), as shown in Figure 3(a), there exists a common axis to allow rotations or combined rotation and translation (where the translation depends on the rotation) to maintain both PCs, i.e., there is a rotational dof (see Figure 3(b)). Otherwise, such as shown in Figure 3(b), there is no rotational dof.

- **point PC and plane PC {v-f/f-v/e-e-c, f-f}**: **dof** = 2, (1t and 1r)
- **two line PCs {e-f/f-e, e-f/f-e}**:

$$\textbf{dof} = \begin{cases} 3, \text{(1t and 2r)} \\ \quad \text{if } CL_1 \perp CL \text{ and } CL_2 \perp CL \\ 2, \text{(1t and 1r)} \\ \quad \text{otherwise} \end{cases}$$

In the above, CL_1 and CL_2 are the contact lines of PC_1 and PC_2 respectively, CP_1 and CP_2 are the contact planes of PC_1 and PC_2, and CL is the intersecting line of CP_1 and CP_2.
Clearly there is one translational dof along the CL. Additionally, there can be one or two rotational dof depending on the contact geometry. Denote the plane passing CL_1 and perpendicular to CP_1 as P_{CL_1} and the plane passing CL_2 and perpendicular to CP_2 as P_{CL_2}. Figure 4 depicts cases of different relations between P_{CL_1} and P_{CL_2}:

- In (a) and (b), P_{CL_1} and P_{CL_2} intersect at line r_1, which is the only possible instantaneous rotation axis for CF-compliant motion (of combined rotation and translation), and thus there is only one rotational dof.
- In (c) and (d), P_{CL_1} and P_{CL_2} are coplanar, thus any line on the plane can be an instantaneous rotation axis for CF-compliant combined motion, and one can pick a pair of orthogonal axes r_1 and r_2; hence there are 2 rotational dof.
- In (e) and (f), P_{CL_1} and P_{CL_2} are parallel, a pair of orthogonal rotation axes r_1 and r_2 that enable CF-compliant motion can be found on either P_{CL_1} or P_{CL_2}. On P_{CL_1}, for example, r_1 and r_2 can be selected to satisfy $r_1 \perp CL_2$ and $r_2 \parallel CL_2$. Thus, there are also 2 rotational dof.

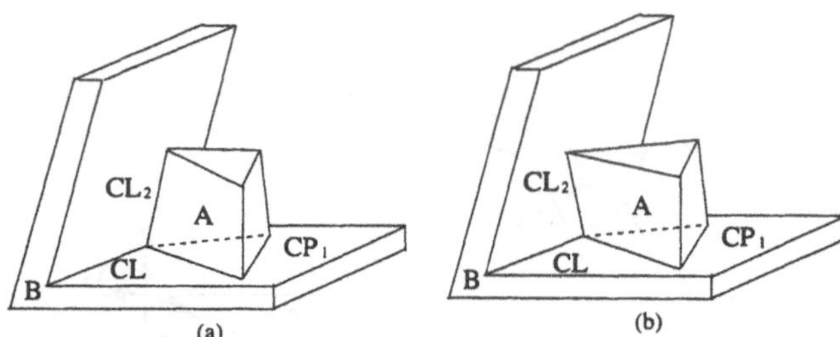

Figure 3: *CFs with one plane PC PC_1 and one line PC PC_2 of non-parallel contact planes: (a) $CL_2 \perp CL$ (2 dof), (b) $CL_2 \neg \perp CL$ (1 dof).*

Note that for cases (c)-(f), after any CF-compliant rotation or combined rotation and translation, the cases become the general case (a), that is, the condition ($CL_1 \perp CL$ and $CL_2 \perp CL$) no longer holds. Note also that in a CF-compliant combined rotation and translation, the translation depends on the rotation so that there is no additional translational dof.

- **point PC and line PC {v-f/f-v/e-e-c, e-f/f-e}**: **dof** = 3, (1t and 2r)
- **two point PCs {v-f/f-v/e-e-c, v-f/f-v/e-e-c}**: **dof** = 4, (1t and 3r)

For CFs where the contact planes are parallel (see Figure 5):

$$\textbf{dof} = \begin{cases} 4, \text{(2t and 2r)} \\ \quad \text{if two point PCs, else} \\ 4, \text{(2t and 2r)} \\ \quad \text{if contact point(s)/line(s) are collinear} \\ 3, \text{(2t and 1r)} \\ \quad \text{otherwise} \end{cases}$$

3 Random Sampling Strategy

As mentioned in Section 1, the authors introduced a general divide-and-merge approach [9, 20] to generate contact state graphs automatically. This approach reduces the contact motion planning problem to a graph search problem at the high level and the problem of

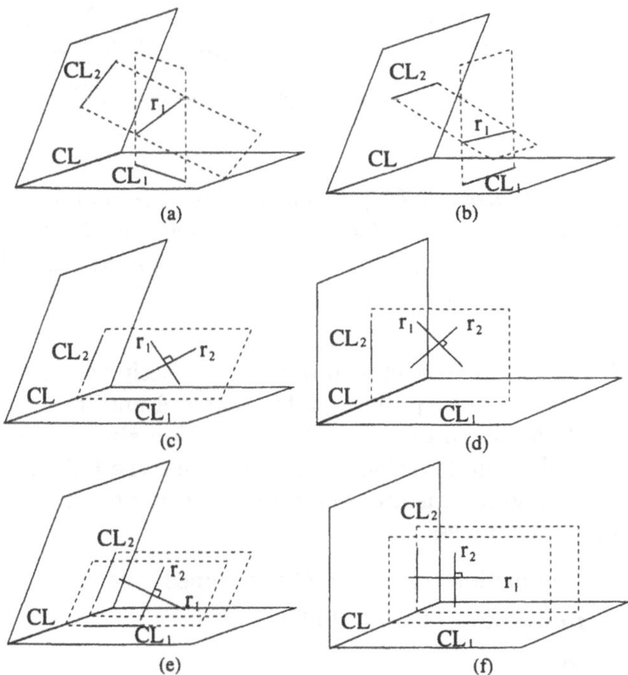

Figure 4: *CFs with two line PCs of non-parallel contact planes: (a) and (b), $CL_1, CL_2 \neg\perp CL$ (2 dof), (c) and (d), $CL_1, CL_2 \perp CL$ and CL_1 and CL_2 are coplanar (3 dof), (e) and (f), $CL_1, CL_2 \perp CL$ and CL_1 and CL_2 are not coplanar (3 dof). The instantaneous rotational axes r_1 (and r_2) are shown. For clarity, only the elements of the objects in contact are shown. The planes P_{CL_1} and P_{CL_2} are shown in dotted lines.*

planning of contact motions within the *same* contact formation at the low level, which we call *CF-compliant* motion planning. Our random sampling strategy aims at CF-compliant motion planning and takes advantage of the known information provided by the divide-and-merge approach: A CF and a (seed) contact configuration satisfying the CF.

For each CF, our strategy generates contact configurations which satisfy the CF and are guaranteed no *local penetration*, that is, *no penetration between the two objects through elements u_A and u_B of each PC or through an element directly connected to u_A (or u_B) and u_B (or u_A).* We call such configurations *valid* contact configurations for the CF. Figure 6 shows two configurations with the same {e-f, e-f} CF, where (a) is a valid configuration, and (b) is invalid because it has a

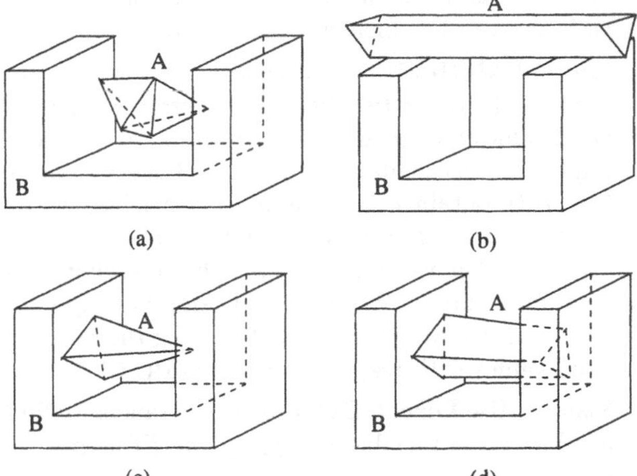

Figure 5: *CFs with parallel contact planes: (a) two point PCs (4 dof), (b) two line PCs with collinear contact lines (4 dof), (c) one point and one plane PCs (3 dof), (d) two plane PCs (3 dof)*

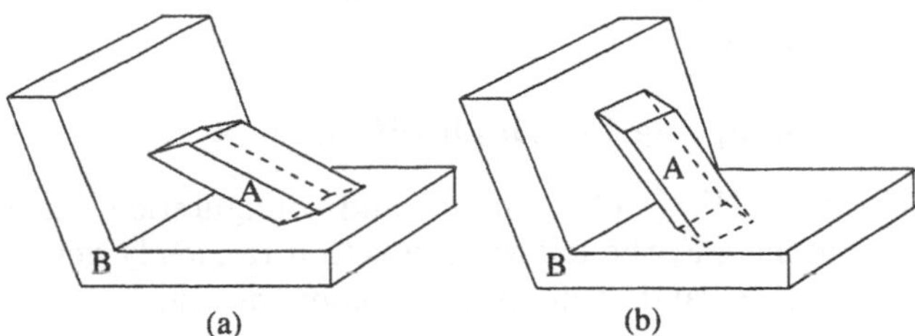

Figure 6: *Two configurations of the same {e-f, e-f} CF: (a) valid with no local penetration, (b) invalid with local penetration*

local penetration between an adjacent face of the edge and the face of B in the bottom PC.

We use two general methods to randomly generate a valid configuration:

- **Direct Calculation:** this method first calculates the valid range for the values of each independent variable[3] and then randomly selects a value within the range for the variable. In this way, all sampled configurations are valid ones. For single-PC CFs and some two-PC CFs (see Section 3.1 and 3.2 for details), it is a good method for sampling since

[3] A valid range refers to the range of values for a variable which satisfy the contact constraints of the CF and do not cause local penetration between the two objects.

in those cases, the value ranges of all independent variables can be efficiently calculated.

- **Hybrid Method:** this method is for sampling regarding the other two-PC CFs where the range of valid values for certain variables are hard or nearly impossible to calculate. It first uses **Direct Calculation** to obtain valid samples for variables whose ranges can be efficiently computed. Then, if there are still other variables, it uses the following two-step procedure to obtain a valid random sample for each such variable without calculating the range of valid values with respect to the two-PC CF:

  Step 1: Use **Direct Calculation** to randomly find a value of the variable *satisfying one PC only.*

  Step 2: If the value sampled does not result in a configuration satisfying the other PC as well, simply discard the value and repeat Step 1, which we call **resampling**. Alternatively, a **convergent iteration** strategy can be used to modify the invalid value iteratively until a valid value is resulted (i.e., it leads to a valid configuration satisfying both PCs).

3.1 Sampling for Single-PC CFs

Given a $CF=\{PC_1\}$, and a valid configuration C_{seed} under the CF, the following function randomly generates a new valid configuration under the CF:

func $random\_sample\_1PC(C_{seed}, PC_1)$
begin
  // randomly translate A from C_{seed} to get C
  $C \leftarrow trans\_1PC(C_{seed}, PC_1)$;
  // randomly rotate A from C to get new C
  $C \leftarrow rotate\_1PC(C, PC_1)$;
  return C;
end.

The above function calls two subfunctions, which we explain in turn now.

Function $trans\_1PC()$ is used to translate A randomly along the contact plane of PC_1 to new valid configurations. Note that finding explicitly the valid ranges for the translational variables needs to calculate the Minkowski sum of the two contacting elements of the PC. In this function, we use a simple and efficient method to achieve the same sampling effects without calculating the Minkowski sum. The function randomly picks two points on the two contacting elements u_A and u_B of PC_1 and then translates A until the two points meet. In this way, the translation always leads to a valid configuration (i.e., maintaining the PC), and the configurations are evenly sampled due to the uniform randomness.

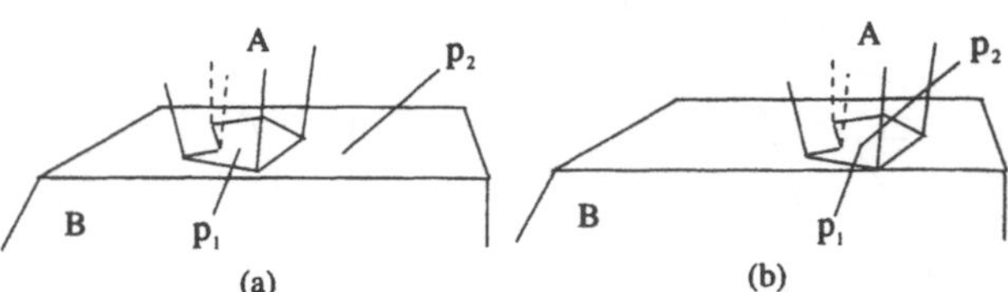

Figure 7: *Example {f-f} CF and the sampling procedure of the translational variables: (a) pick p_1 and p_2 randomly on the two contacting faces of A and B respectively, and (b) translate A so that p_1 and p_2 meet.*

The function $trans\_1PC()$ is outlined below:

func $trans\_1PC(C_t, PC_1)$
begin
  // u_A is the contacting element of A in PC_1
  randomly sample a point p_1 on u_A;
  // u_B is the contacting element of B in PC_1
  randomly sample a point p_2 on u_B;
  $C \leftarrow$ translate A from C_t so that $p_1 = p_2$;
  return C;
end.

Note that if any of u_A and u_B is a vertex, the randomly sampled point on it is the vertex itself; if any of them is an edge or a face, the randomly sampled point is a random point *inside* the bounded edge or face. It is easy to pick a point randomly inside a bounded edge or a bounded convex face: The point is some convex combination of the boundary vertices [4]. If the face is concave, it is first decomposed into several convex parts (e.g., triangles), and then the point is sampled inside the convex parts.

Function $rotate\_1PC()$ generates a random rotation (of A) *compliant to* the contact constraints of the CF to achieve a new valid contact configuration. It pro-

[4] The convex combination of vertices p_i $(i = 1, 2, ..., n)$ is $\sum_{i=1}^{n} \alpha_i * p_i$, where $\sum_{i=1}^{n} \alpha_i = 1$. To sample a point inside the face, we randomly pick up a vertex p_i and randomly sample α_i in $[0, 1-s]$, where s is the sum of α_j already sampled.

duces the random or arbitrary rotation through a sequence of rotations with respect to each independent rotational variable. For each such variable, it first calls a function *get_axis*() to get its axis r and then calls *find_angle_range*() to determine the valid range of values for the variable (to satisfy the CF and cause no local penetration). Next it randomly picks an angle inside the range and makes a rotation about r by the angle. The function is outlined below.

```
func rotate_1PC(C_r, PC_1)
begin
  C ← C_r;
  d ← rotational dof of the CF;
  for l = 1 to d do begin
    r ← get_axis(C, l, PC_1);
    (θ_1,θ_2) ← find_angle_range(C, r, PC_1);
    θ ← randomly sampled angle in (θ_1,θ_2);
    C ← rotate A by θ about r from C;
  end;
  return C;
end.
```

The function *get_axis*() works based on the type of the CF. The axes are determined to facilitate the calculation of valid value ranges for the rotations: the first axis is along the normal of the contact plane of the PC, the second axis (if exists) is either along the contact line (for line PCs) or any line on the contact plane of the PC and passing through the contact point (for point PCs), and the third axis (if exists) is determined by the right-hand rule from the other two. The angle range for the first rotation is $(-\pi, \pi]$, and the ranges for the other two (if exist) are returned by function *find_angle_range*(), which uses the algorithm described in [18] to calculate angle ranges.

3.2 Sampling for Two-PC CFs

For two-PC CFs with only one translational degree of freedom, **Direct Calculation** is sufficient for sampling valid configurations, while for other two-PC CFs, the **Hybrid Method** introduced earlier (in the beginning of Section 3) is used to sample valid configurations. Nevertheless, we can combine the sampling processes for all two-PC CFs in a general function as described below.

From a given seed configuration C_{seed}, function *random_sample_2PC*() randomly generates a valid configuration for an arbitrary two-PC CF. Without losing generality, we designate *PC_1 to be the PC with fewer dof if PC_1 and PC_2 have different dof.* Let CP_1 and CP_2 denote the contact planes of PC_1 and PC_2 respectively, CL denote the intersecting line of CP_1 and CP_2 if they are not parallel, and $\vec{cl}$ denote a unit vector along either direction of CL. The function is outlined as follows:

```
func random_sample_2PC(C_seed, PC_1, PC_2)
begin
  if (CF has 1 translational dof) then
    C ← trans_2PC(C_seed, PC_1, PC_2, cl⃗);
  else begin
    v⃗_0 ← random unit vector on CP_1;
    C ← trans_2PC(C_seed, PC_1, PC_2, v⃗_0);
  end;
  for l = 1 to 3 do
    if (CF has l-th rotational dof) then
      C ← rotate_1dof(l, C, PC_1, PC_2);
  return C;
end.
```

In the above function, independent translational variables are sampled by **Direct Calculation** with guaranteed valid values. Function *trans_2PC*() implements a random translation satisfying the two-PC CF. With the axis $\vec{v}$ as an input, the function starts from a given valid configuration C_t and calls a procedure *find_trans_range*() to calculate the valid range of translations along $\vec{v}$ *relative to the given configuration* C_t, and then it randomly generates an increment of translation for object A within the valid range and obtains a new valid configuration, as outlined below.

```
func trans_2PC(C_t, PC_1, PC_2, v⃗)
begin
  // find translational ranges along -v⃗ and v⃗
  // while maintaining PC_1 and PC_2. d_1, d_2 ≥ 0
  [-d_1, d_2] ← find_trans_range(C_t, v⃗, PC_1, PC_2);
  d ← randomly sampled value in [-d_1, d_2];
  //if d < 0, translate along -v⃗ by |d|
  C ← translate A along v⃗ by d from C_t;
  return C;
end.
```

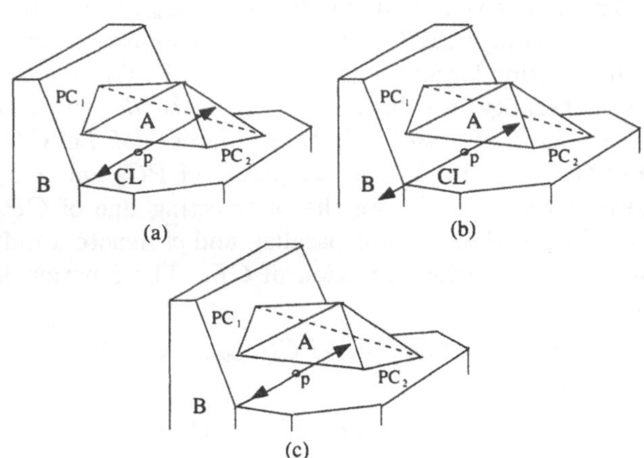

Figure 8: *The calculation of valid range of the translation variable along CL for a CF with two non-parallel line PCs: (a) calculate the range for PC_1 along CL, (b) calculate the range for PC_2 along CL, (c) find intersection of the ranges.*

The implementation of function $find\_trans\_range()$ involves computing the shortest separation distance between the two contacting elements of the PC involved, which can be two faces, a face and an edge, a face and a vertex, or two edges, along a given tangential direction of the contact plane. Figure 8 shows how $find\_trans\_range()$ works in the case where CP_1 and CP_2 are non-parallel by an example. The function first projects an arbitrary point of A to CL, denoted as p, and then calculates the valid ranges to translate A along $\vec{cl}$ and $-\vec{cl}$ without breaking PC_1 and PC_2 respectively. Next it finds the intersection of the two ranges and returns it as the final range. In the case of a given random unit vector, the function works similarly.

To sample rotational variables, we need to first determine rotation axes. From analyzing all kinds of two-PC CFs, we discover the following three general characterizations of rotation axes, each corresponding to one rotational variable (if it exists), which are sufficient for all possible rotations constrained by two-PC CFs. We simply index them by $l = 1, 2, 3$ in $random\_sample\_2PC()$:

- $l = 1$: the rotation axis is denoted by X and defined as passing through one point on PC_2 (i.e., one point on both contacting elements of PC_2) along the normal of CP_1.
- $l = 2$: the rotation axis is denoted by Y and defined as along the contact line of PC_1 (if PC_1 is a line PC), or passing through the contact point of PC_1 and parallel to CL (if PC_1 is a point PC).
- $l = 3$: the rotation axis is denoted by Z and defined as passing through the two contact points (for two point PCs) or along one contact line (for two collinear line PCs, or one point and one line PCs with the point on the line).

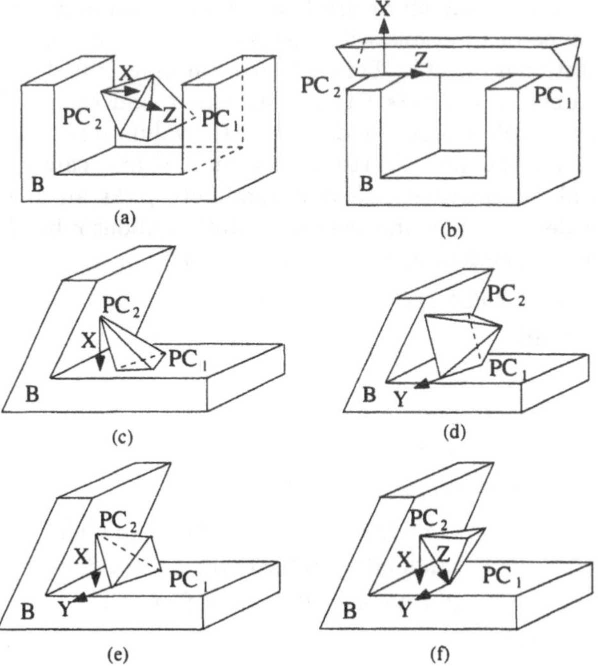

Figure 9: *The axes of the three rotational variables for various CFs, (a) two point PCs with parallel CPs, (b) two line PCs with parallel CPs and collinear contact lines, (c) one point and one plane PCs, (d) two lines PCs, (e) one point and one line PCs, (f) two point PCs.*

Figure 9 illustrates these three kinds of rotation axes in various examples. Note that in all cases, the rotation axis Y is actually for a combined rotation and translation in order to maintain the CF. As the rotation and translation are mutually dependent, there is only one independent variable. We will explain its sampling later.

Clearly, depending on the rotational dof, not all rotations about these axes are always possible. Table 1 summarizes all kinds of two-PC CFs, their rotational dof and corresponding rotational axes.

CF type		rot. dof	X	Y	Z
$CP_1 \parallel CP_2$	two point PCs	2	$\surd$	x	$\surd$
	collinear contact points/lines	2	$\surd$	x	$\surd$
	others	1	$\surd$	x	x
$CP_1 \neg \parallel CP_2$	2 plane PCs	0	x	x	x
	1 line, 1 plane PCs	0/1	x/$\surd$	x	x
	1 point, 1 plane PCs	1	$\surd$	x	x
	two line PCs	1/2	x/$\surd$	$\surd$	x
	1 point, 1 line PCs	2	$\surd$	$\surd$	x
	2 point PCs	3	$\surd$	$\surd$	$\surd$

Table 1: *Two-PC CFs, their rotational dof, and corresponding rotational axes*

Now we explain the sampling strategies regarding rotational variables about X, Y, and Z respectively in more detail.

Rotation About X, i.e., $l = 1$:

We use the **Hybrid Method** introduced earlier. The function $rotate\_1dof(1, C, PC_1, PC_2)$ first calculates the angle range about X which satisfies PC_2 by calling $find\_angle\_range()$ (see Section 3.1). It then samples an angle θ inside the range, and next rotates A about X by θ to get a new configuration C. If C also satisfies PC_1, i.e., forms a valid configuration, it is returned; otherwise, either **resampling** or **convergent iteration** can be used (as introduced before Section 3.1). Here convergent iteration is to modify θ by $k\theta$, i.e., $\theta \leftarrow k\theta$, where $0 < k < 1$, repeatedly until θ results in a valid configuration, i.e., a convergence to the valid value range is achieved.

Rotation About Y, i.e., $l = 2$:

Sampling again uses the **Hybrid Method**. As mentioned earlier, the motion here is a combined translation and rotation with one independent variable. Although the *instantaneous* rotational axis for the variable can be found (see Figure 4), it is generally difficult to be used for sampling because the axis is not fixed and it is hard to analytically describe the axis's motion. Therefore, we use another strategy to sample this variable, which is given below.

In all cases where this motion is possible (Figure 9 gives some examples), CP_1 and CP_2 are not parallel, and they intersect at line CL. The function $rotate\_1dof(2, C, PC_1, PC_2)$ uses a translational variable d along an axis $\vec{v}$ on CP_1 and perpendicular to CL as the independent variable for the combined motion. It first randomly samples d with a value satisfying PC_1 by **Direct Calculation** (in a procedure similar to but simpler than $trans\_2PC()$, since only one PC needs to be satisfied). Next it checks whether PC_2 can also be satisfied by a *guarded* rotation about Y, with the angle calculated, which depends on the value of d. If so, the function returns a valid configuration C; otherwise, again, either **resampling** or **convergent iteration** on the value d (i.e., $d \leftarrow kd$, where $0 < k < 1$, repeatedly until d results in a valid configuration) can be used.

Figure 10 shows how a rotation about Y is sampled for a CF with two line PCs of parallel contact lines (and non-parallel contact planes).

Figure 11 shows the sampling for a CF with two line PCs of non-parallel contact lines and contact planes. Besides the similar steps as those for the case with parallel contact lines as shown in Figure 10, here an extra step is needed as shown in Figure 11(d) to rotate about X axis, which is a guarded rotation trying to satisfy PC_2.

Rotation About Z, i.e., $l = 3$:

Unlike the two previous cases about X and Y, here the rotational variable about Z is sampled using **Direct Calculation.** The function $rotate\_1dof(3, C, PC_1, PC_2)$ first calculates the rotational angle ranges $(\theta_{11}, \theta_{12})$ and $(\theta_{21}, \theta_{22})$ about Z for PC_1 and PC_2 respectively by calling $find\_angle\_range()$ (see Section 3.1), and then finds the intersection of the ranges as the valid range of the rotational variable. Finally, an angle θ is sampled randomly inside the range, and A is rotated about Z by θ to generate a valid configuration.

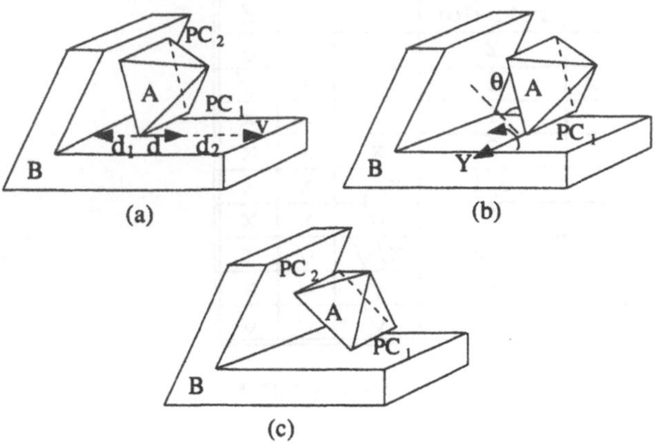

Figure 10: *Sampling a combined motion with rotation about Y for a CF with two parallel line PCs and non-parallel contact planes: (a) seed configuration, translational range $[-d_1, d_2]$ along $\vec{v}$, and sampled value d in $[-d_1, d_2]$; (b) configuration after translation along $\vec{v}$ by d, and the guarded rotation angle θ about Y; (c) configuration after the guarded rotation about Y by θ to form PC_2.*

In summary, we have presented a general function *random_sample_2PC*() to produce random configurations satisfying any given two-PC CFs. Note that only in two steps regarding rotations about X and Y (i.e., $l = 1, 2$), the **Hybrid Method** is used, and **Direct Calculation** is used in sampling all the other variables to maximize efficiency.

In the **Hybrid Method**, the alternative to **Direct Calculation** is either **resampling** or **convergent iteration**. Resampling for a single variable is simply to repeat the sampling process for that variable until a value which results in a valid configuration is found or after some pre-determined number of tries.

Convergent iteration, on the other hand, guarantees to find a valid value for the variable or makes it converge to the valid value range. If there is more than one connected valid value range, which may happen in the cases where the two-PC CF has multiple connected regions of contact configurations, caution is needed to make the convergence to each valid range equally likely in order to ensure the even distribution of samples. This, however, can be achieved by always using the newly randomly sampled configuration as the seed configuration for the next sample.

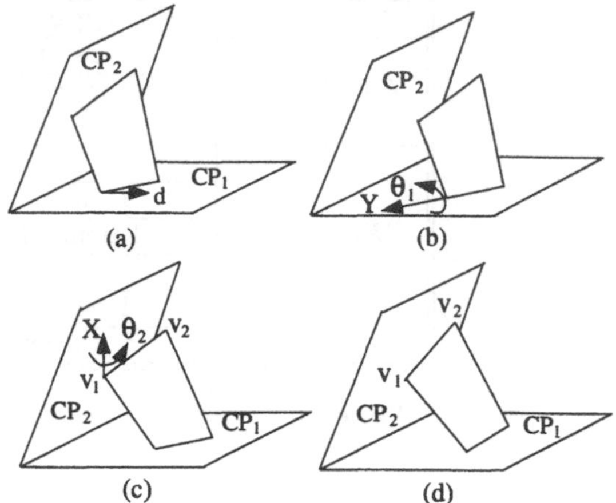

Figure 11: *Sampling a combined motion with rotation about Y for a CF with two non-parallel line PCs and non-parallel contact planes: (a) sampled distance vector d on CP_1; (b) configuration after translation by d with PC_1 maintained but PC_2 broken, and the guarded rotational angle θ_1 about Y so that at least one vertex of the contact edge of PC_2 can be on CP_2 after the rotation; (c) configuration after the guarded rotation about Y by θ_1, with only one vertex v_1 of the contact edge of PC_2 on CP_2, and guarded rotational angle θ_2 about X so that the contact edge v_1v_2 can be on CP_2; (d) configuration after rotation about X by θ_2, where both PC_1 and PC_2 are satisfied. Note that though the figure shows the two contact lines are coplanar, the non-coplanar cases are handled similarly.*

4 Experimental Results

The random sampling strategy for single-PC and two-PC CFs has been implemented in *C*. The program runs on a SUN Ultra10 workstation. The machine is rated at 12.1 SPECint95 and 12.9 SPECfp95. The input to the program is a CF and a valid contact configuration satisfying the CF. The output are random configurations of the same CF which are guaranteed no local penetration. We use VRML as the output format.

Figure 12 shows the sampling results for several single-PC CFs between a cube A and an L-shape B. Figure 13 shows the sampling results for two-PC CFs between different shapes of objects. The running time (in seconds) for generating 1000 samples of the examples in Figure 12 and Figure 13 are summarized in Table 2.

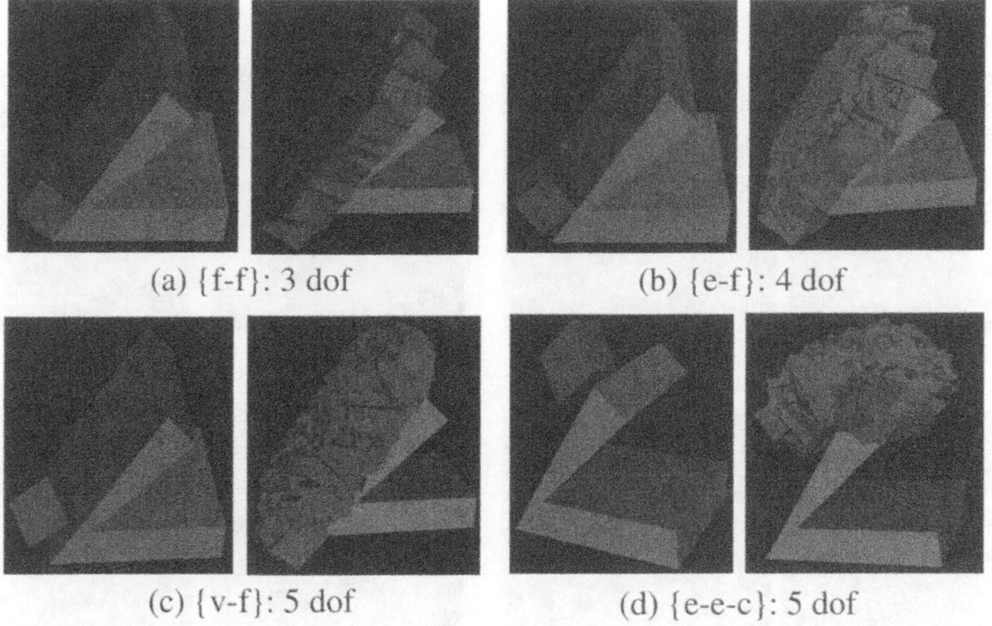

Figure 12: *Examples for single-PC CFs: seed configurations and the results for 1000 samples*

The results summarized in Table 2, show clearly that it takes much shorter time to generate samples for single-PC CFs. This is because, though usually single-PC CFs have higher dof, sampling is done by **Direct Calculation**. For two-PC CFs of the same objects, usually the higher dof the CF has, the more time is needed to sample the same number of configurations, although the time also depends on the geometry of the objects. For the same CF, the running time of the algorithm is nearly proportional to the number of samples generated.

In the last three rows of Table 2, we show the running times of the examples using **convergent iteration** and **resampling** respectively. It seems that **convergent iteration** runs faster in most cases. Our experiments show that for **convergent iteration** with $k = 0.5$, after at most 13 iterations, a valid configuration can be obtained for all the four examples. For **resampling**, on the other hand, after at most 73 resampling iterations, a valid configuration can be obtained.

Our experiments also show that among the three rotational variables, the sampling of the second one about Y with dependent translation is the most expensive, followed by that of the first variable, and sampling for the third variable is the fastest because of **Direct Calculation**.

5 Conclusion

This paper addresses random sampling of configurations constrained by contact, which is not only necessary for planning contact motions with certain randomized planners but also useful for planning collision-free motions since the sampled contact configurations probabilistically characterize the C-obstacles. An efficient random sampling strategy is implemented for sampling configurations constrained by single-PC or two-PC CFs, which satisfy contact constraints of the CF without causing local penetration. The strategy is characterized by directly computing valid samples wherever possible to maximize efficiency. In the next step, we intend to apply such a strategy to randomized contact motion planning.

Acknowledgments

This work has been supported by the National Science Foundation under grants IIS-9700412 and CDC-9726424. The authors appreciate early inspiration from

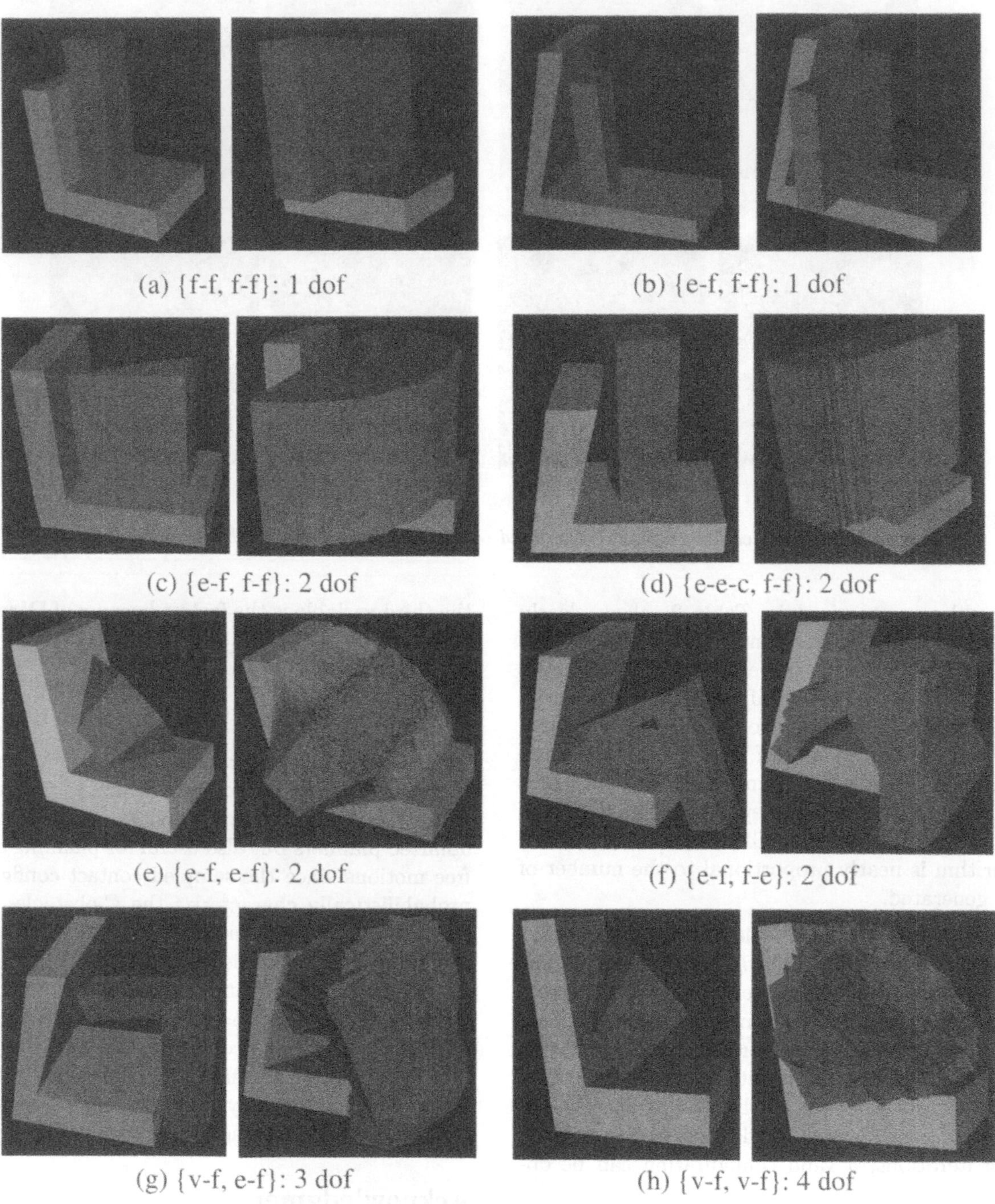

Figure 13: *Examples for two-PC CFs: seed configurations and the results for 1000 samples*

Method	CF	dof	time(s)	CF	dof	time(s)
Direct	{f-f}, Figure 12(a)	3	0.23	{e-f}, Figure 12(b)	4	0.25
Direct	{v-f}, Figure 12(c)	5	0.45	{e-e-c}, Figure 12(d)	5	0.45
Direct	{f-f, f-f}, Figure 13(a)	1	6.7	{e-f, f-f}, Figure 13(b)	1	3.5
Hybrid	{e-f, f-f}, Figure 13(c)	2	3.3 or 3.4	{e-e-c, f-f}, Figure 13(d)	2	3.7 or 3.8
Hybrid	{e-f, e-f}, Figure 13(e)	2	49.5 or 54.6	{e-f, f-e}, Figure 13(f)	2	61.1 or 56.1
Hybrid	{v-f, e-f}, Figure 13(g)	3	50.4 or 50.9	{v-f, v-f}, Figure 13(h)	4	34.7 or 40.4

Table 2: *Examples in Figure 12 and Figure 13 and their running times for 1000 samples. In the last three rows, the two numbers given for each case under* **times(s)** *correspond to the running times using* **convergent iteration** *and* **resampling** *respectively.*

discussions with Jean-Paul Laumond, Jean-Claude latombe, David Hsu. The authors are also grateful to comments by Matt Mason and Mark Yim.

References

[1] N. Amato, O. Bayazit, L. Dale, C. Jones, D. Vallejo, "OBPRM: An Obstacle-Based PRM for 3D Workspaces", *Workshop Algor. Found. of Robotics*, pp. 155-168, Mar. 1998.

[2] F. Avnaim, J. D. Boissonnat, and B. Faverjon, "A practical exact motion planning algorithm for polygonal objects amidst polygonal obstacles", *IEEE Int. Conf. Robotics & Automation*, pp. 1656-1661, Apr. 1988.

[3] V. Boor, M. H. Overmars, and A. F. Stappen, "The Gaussian Sampling Strategy for Probabilistic Roadmap Planners", *IEEE Int. Conf. Robotics & Automation*, May 1999.

[4] R. Brost, "Computing Metric and Topological Properties of C-space Obstacles", *IEEE Int. Conf. Robotics & Automation*, pp. 170-176, May 1989.

[5] B. Donald, "A Search Algorithm for Motion Planning with Six Degrees of Freedom", *Artificial Intelligence*, pp. 295-353, 31(3), 1987.

[6] A. Farahat, P. Stiller, and J. Trinkle, "On the Algebraic Geometry of Contact Formation Cells for Systems of Polygons", *IEEE Trans. Robotics & Automation*, 11(4), pp. 522-536, Aug. 1995.

[7] H. Hirukawa, "On Motion Planning of Polyhedra in Contact", *Workshop Algor. Found. of Robotics*, 1996.

[8] D. Hsu, L. Kavraki, J.C. Latombe, R. Motwani, and S. Sorkin, "On Finding Narrow Passages with Probabilistic Roadmap Planners", *Workshop Algor. Found. of Robotics*, pp. 141-153, Mar. 1998.

[9] X. Ji and J. Xiao, "Automatic Generation of High-Level Contact State Space," *IEEE Int. Conf. Robotics & Automation*, pp. 238-244, May 1999.

[10] X. Ji and J. Xiao, "Towards Random Sampling with Contact Constraints", to appear in *ICRA*'2000.

[11] L. Joskowicz, R. H. Taylor, "Interference-Free Insertion of a Solid Body Into a Cavity: An Algorithm and a Medical Application", *Int. J. Robotics Res.*, 15(3):211-229, June 1996.

[12] L.E. Kavraki, P. Svestka, J.C. Latombe, and M. Overmars, "Probabilistic Roadmaps for Path Planning in High-Dimensional Configuration Spaces", *IEEE Trans. Robotics & Automation*, 12(4):566-580, 1996.

[13] T. Lozano-Pérez, "Spatial Planning: A Configuration Space Approach", *IEEE Trans. Comput.*, C-32(2):108-120, 1983.

[14] M. T. Mason, "Compliant Motion", *Robot Motion: Planning and Control*, MIT Press, 1982.

[15] C. Nissoux, T. Simeon, and J. P. Laumond, "Visibility based probabilistic roadmaps", *IEEE Int. Conf. Intell. Robots and Systems*, 1999.

[16] D. E. Whitney, "Historical Perspective and State of the Art in Robot Force Control", *IEEE Int. Conf. Robotics & Automation*, pp. 262-268, 1985.

[17] J. Xiao, "Automatic Determination of Topological Contacts in the Presence of Sensing Uncertainties," *IEEE Int. Conf. Robotics & Automation*, pp. 65-70, May 1993.

[18] J. Xiao and L. Zhang, "Computing Rotation Distance between Contacting Polytopes," *IEEE Int. Conf. Robotics & Automation*, pp. 791-797, Apr. 1996.

[19] J. Xiao and L. Zhang, "Contact Constraint Analysis and Determination of Geometrically Valid Contact Formations from Possible Contact Primitives", *IEEE Trans. Robotics and Automation*, 456-466, Jun. 1997.

[20] J. Xiao and X. Ji, "A Divide-and-Merge Approach to Automatic Generation of Contact States and Planning of Contact Motion", to appear in *ICRA*'2000.

Randomized Path Planning for a Rigid Body Based on Hardware Accelerated Voronoi Sampling

Charles Pisula, *University of North Carolina, Chapel Hill, NC*
Kenneth Hoff III, *University of North Carolina, Chapel Hill, NC*
Ming C. Lin, *University of North Carolina, Chapel Hill, NC*
Dinesh Manocha, *University of North Carolina, Chapel Hill, NC*

Probabilistic roadmap methods have recently received considerable attention as a practical approach for motion planning in complex environments. These algorithms sample a number of configurations in the free space and build a roadmap. Their performance varies as a function of the sampling strategies and relative configurations of the obstacles. To improve the performance of the planner through narrow passages in configuration space, some researchers have proposed algorithms for sampling along or near the medial axis of the free space. However, their usage has been limited because of the practical complexity of computing the medial axis or the cost of computing such samples.

In this paper, we present efficient algorithms for sampling near the medial axis and building roadmap graphs for a free-flying rigid body. We use a recent algorithm for fast computation of discrete generalized Voronoi diagrams accelerated by graphics hardware [10]. We initially compute a bounded error discretized Voronoi diagram of the obstacles in the workspace and use it to generate samples in the free space. We use multi-level connection strategies and local planning algorithms to generate roadmap graphs. We also utilize the distance information provided by our Voronoi algorithm for fast proximity queries and sampling the configurations. The resulting planner has been applied to a number of free flying rigid bodies in 2D (with 3-dof) and 3D (with 6-dof) and compared with the performance of earlier planners using a uniform sampling of the configuration space. Its performance varies with different environments and we obtain 25% to over 1000% speed-up.

1 Introduction

Motion planning is one of the important, classical problems in algorithmic robotics [18]. Besides robotics, it has applications in many areas, including animation of digital actors or autonomous agents [16], maintainability studies [4], drug design [7] and robot-assisted medical surgery [22, 24].

This problem has been extensively studied over the last three decades. A number of analytical methods and approximate techniques have been proposed. However, due to the computational complexity of complete motion planning algorithms, most practical algorithms are based on randomized motion planning algorithms, such as randomized potential field methods (RPP) or probablistic roadmap methods (PRMs).

The simplest PRM algorithms generate a set of configuration in the free space. The planning algorithm involves three main steps [26, 14, 15, 21]:

1. Generate a list of samples in the configuration space. The simplest algorithms use uniform sampling techniques.
2. Use collision detection or distance computation algorithms to select the samples lying in the free space.
3. Try connecting the samples in the free space by local planning methods and build a roadmap graph.

The roadmap graph is used to generate a path between the start and goal configurations. However, the efficiency of these planners can degrade in configurations containing narrow passages or cluttered environments. In such cases, a significant fraction of randomly generated configurations are not in the free space. Moreover, it is hard to generate a sufficient number of samples in the narrow passages or connect all the nodes in the free space using local planning methods. Many approaches have been proposed in the literature to overcome these problems. These include:

Dilating the Free Space[13]: The main idea is to dilate the free space by allowing the rigid body to penetrate the obstacles by a small amount. The areas near these nodes are resampled to find collision free configurations. However, dilation can alter the topology of the free space. Furthermore, no good practical (in terms of efficiency and robustness) algorithms are known for penetration depth computation between polyhedral models. Some of the public domain implementations like I-COLLIDE and V-Clip for convex polytopes, based on Lin-Canny distance computation algorithm, provide only an approximation to the penetration depth.

Sampling near the Obstacle Boundaries [1]: It samples the nodes from the contact space, the configurations where the robot just touches one of the obstacles. It works well in many cases, but its performance is difficult to analyze.

Information about the Environment [12, 21]: These algorithms make use of information known about the environment. These include executing random reflections at the C-obstacle surfaces [12] and adding "geometric" nodes for non-articulated robots near the boundaries of the obstacles in the workspace [21].

Sampling Based on the Medial Axis [27, 26, 9]: The main idea is to generate nodes that lie on the medial axis of the workspace or the free space. Intuitively speaking, the medial axis corresponds to a set of points that are furthest from the obstacle boundaries and have maximum clearance. It has been used for a number of motion planning algorithms [20, 3, 18, 6, 5]. However, its practical use has been limited because of the difficulties in accurately computing the medial axis or generalized Voronoi diagrams. Wilmart et al. [27, 26] generate random configurations and retract them onto the medial axis without explicitly computing the medial axis. They use a search algorithm based on distance to the obstacles for the retraction. They have applied the resulting algorithm to a few configurations with narrow passages and obtained considerable speedups over uniform sampling. Guibas et al. [9] use point approximations on the boundary, compute their Voronoi diagrams to approximate the medial axis and use it for motion planning of flexible objects.

1.1 Our Results

In this paper, we present improved algorithms for sampling based on medial axis and use them for efficiently constructing the Voronoi roadmap. We make use of bounded error discretized Voronoi diagrams, computed using graphics rasterization hardware [10]. More specifically, we render distance functions for each primitive of the obstacle and use the frame buffer output to identify the Voronoi boundaries upto a given resolution. Furthermore, the depth buffer provides the distance information which is used to speed up the proximity queries and sampling the configuration space. This computational framework enables insightful analysis of the workspace to identify different types of narrow passages and efficient generation of sampling points with guaranteed distribution on the medial axis. We use multi-stage connection strategies along with local planning algorithms to build the roadmap graphs. The resulting PRM has been applied to a number of complex environments composed of 3-dof (in 2D) and 6-dof rigid bodies (in 3D). As compared to uniform sampling of the configuration space, it can considerably improve the performance of the planner. Our initial experiments demonstrate 25% to over 1000% speed-up in running times. Since the cost of generating the samples is relatively small, the planner never underperforms as compared to uniform sampling.

Some of the main advantages of our approach include:

- **Efficiency**: The resulting algorithms for generating the bounded error approximation of the Voronoi diagram and samples on the medial axis run relatively fast. For environments composed of thousands of polygons, we can generate Voronoi diagrams at $128 \times 128 \times 128$ resolution at in a few seconds or minutes (depending on the size of the environment) on a SGI workstation. Furthermore, the distance buffer helps us speed up the collision queries by a factor of two. Based on our sampling strategies, a high fraction of the nodes generated are in the free space.

- **Simplicity**: The resulting algorithm is relatively simple to implement and doesn't suffer from robustness or degeneracies. The basic PRM planner is a rather simple approach and our sampling algorithm doesn't introduce any complications.

- **Identification and Characterization of Narrow Passages**: The distance buffer information gives us information about narrow passages in the workspace. Using a combination of distance metric for dimension analysis and retraction methods, it can also be used to identify narrow passages in many cases.
- **Global connectivity information based on Voronoi Roadmaps**: The connectivity information provided by the discretized generalized Voronoi diagram is used in constructing the roadmap and accelerating performance of the local planner.

The rest of the paper is organized as follows: In Section 2, we highlight our notation, provide a brief overview of the medial axis and the fast approximate to compute a discretized approximation using graphics hardware. We describe the planner in Section 3 and present the multi-stage connection strategies to build the roadmap. Section 4 presents our sampling strategies based on medial axis of the workspace.

Section 5 discusses various implementation issues, including collision culling, computations of Voronoi vertices and graphs, and other data structures. We also highlight the performance of our planner on different benchmarks. Finally we compare it with related approaches in Section 6.

2 Background

In this section, we highlight our notation and provide a quick overview of medial axis and fast computation of discretized Voronoi diagrams using graphics hardware.

2.1 Notation and Representation

In this paper, we restrict ourselves to dealing with non-articulated rigid bodies in 2D or 3D. We use the symbol W to represent the workspace. We will denote the robot as R and the set of obstacles as O. We assume that the robot and each obstacle is a closed and bounded set. Our current implementation assumes that each obstacle is represented as a collection of triangles. It is relatively simple to extend them to other primitives (e.g., spline models) based on our current algorithmic framework. The configuration space (denoted by C) of R is given by the set of all positions and orientations. We use (x, y, z) coordinates to represent the position and a unit quaternion to represent the orientation. The (x, y, z) coordinates will be referred to as the *translation component* of a configuration and the quaternion is the *rotation component* of a configuration. Furthermore, C can be partitioned into *free space, F* and *blocked space, B*. The blocked space corresponds to configurations, where the robot R collides with at least one of the obstacle. The rest of the configuration space corresponds to the free space. In other words, $F = C \backslash B$. Furthermore, we denote the boundary of the free space as ∂F and ∂O represents the union of boundaries of all the obstacles.

2.2 Medial Axis and Voronoi Diagrams

The medial axis of a solid object provides shape analysis in terms of its boundary elements. It is a skeletal representation that can be formulated as the locus of the center of a maximal sphere as it rolls around the object interior. It is closely related to the Voronoi diagram of a solid and for a suitably defined boundary, it can be computed from the Voronoi diagram and vice versa. The concept of medial axis was first proposed by Blum [2] for biological shape measurement and since been used for mesh generation, feature recognition and molding simulation. Medial axis and Voronoi diagrams have been long used for robot motion planning [20, 3, 18, 6]. The medial axis of the free space F, denoted by MA(F) has lower dimension than F but is still a complete representation for planning the motion. Strictly speaking, MA(F) is a strong deformation retract of F, implying that F can be continuously deformed into MA(F) and maintain its topology structure [20, 3, 6, 27].

Algorithms to compute Voronoi diagrams and medial axis have been extensively studied in computational geometry and solid modeling. Good theoretical and practical algorithms are known for point primitives. However, the boundaries of the Voronoi regions for higher order primitives (e.g., lines, triangles) correspond to high-degree algebraic curves and surfaces. No good practical algorithms are known for computing them efficiently and robustly. As a result, their applications have been limited.

Given the practical complexity of computing Voronoi diagrams, many researchers have proposed approximate approaches. Some common approximations include generating point approximations of the boundaries and computing their Voronoi diagrams [23]. However, it is hard to give any guarantees on the accuracy of the resulting Voronoi diagram. Other approaches use spatial subdivision techniques [25, 19]. While these algorithms can be used to generate bounded error approximations of Voronoi diagrams, they can be rather slow for large environments.

2.3 Bounded Error Approximation of Voronoi Diagrams Using Graphics Hardware

We compute a bounded error approximation of the Voronoi diagram of the obstacles in the workspace, W. If any of the obstacles is a closed and bounded solid, we do not compute the Voronoi diagram inside that solid. Each boundary triangle, edge, and vertex of an obstacle is treated as a separate site. We compute a discrete generalized Voronoi diagram by rendering a three-dimensional distance mesh for each site. The 3D polygonal distance mesh is a bounded-error approximation of a possibly non-linear distance function over a plane. Each site is assigned a unique color, and the corresponding distance mesh is rendered in that color using parallel projection. The graphics system performs a depth test for each pixel in order to resolve the visibility of surfaces. The depth buffer keeps a running minimum depth as polygons are rendered. When the minimum depth is updated, the frame buffer is also updated with the pixel's color. Thus, the rasterization provides, for each pixel, the identity of the nearest site (encoded as a color) and the distance to that site. The error in the mesh is bounded to be smaller than the distance between two pixels, in order to maintain an accurate Voronoi diagram. More details are given in [10]. This algorithm runs very fast in practice.

2.3.1 Motion Planning Using Discretized Voronoi Diagrams

Many motion planning algorithms have been proposed based on discretized Voronoi diagrams [25, 10, 11]. Based on the color and distance buffer information, [10, 11] define a potential field and use it to navigate the robot. Since the distance buffer information is computed at interactive rates, this planner has also been applied to dynamic 2D environments (3-dof robots).

3 Path Planning Based on Discretized Voronoi Diagrams

In this section, we describe our path planning algorithm. Initially we present algorithms for computing the discretized Voronoi diagrams, followed by use of boundary finding algorithms that extract the Voronoi graph. We use a multi-stage strategy to build the roadmap. It generates configurations in the free space and connect them using local planning algorithms. The first stage selects nodes using the medial axis of the workspace and connects them using simple local planning algorithms. Next it attempts to connect different components. Finally, it tries to estimate narrow passages based on local characteristics of the Voronoi diagram and generates more samples in those regions. More details on improved sampling algorithms are given in Section 4.

3.1 Discretized Voronoi Diagram of the Workspace

Given a bound on the discretization error, the algorithm computes the Voronoi diagram of the obstacles O slice-by-slice (along the z-axis). It renders the distance function for each vertex, edge and triangular face of each obstacle. Each slice is generated using the graphics rasterization hardware. We read back the color buffer and the depth-buffer for each slice. The color buffer gives the index of the nearest obstacle to each sample point in the slice. The distance buffer gives the distance to that obstacle. It generates a 3D voxel grid that corresponds to a uniform sampling of the space containing the geometric primitives. This 3D image gives a volumetric representation of the generalized Voronoi diagram of the primitives. The resolution of the Voronoi graph can have significant impact on the performance of the planner (as highlighted in Section 5). In our current implementation, we take uniform samples along the z-axis. In theory, it is possible to vary the step size adaptively using bisection. For PRM, we are only interested in computing an accurate approximation of the Voronoi boundaries.

3.2 Extracting the Voronoi Graph

The Voronoi diagram is in the form of two 3D images: a color image corresponding to the IDs of the closest site to each sample point and a depth image giving the related distances to the closest sites. Since we are using a volumetric representation, the actual continuous boundaries of the Voronoi graph are described implicitly as lying between sample points of different colors (sample points that have different closest primitives).

Our goal is to extract the boundary graph structure in order to bias the randomized motion planning sampling. In 3D workspace, the Voronoi graph structure is composed of Voronoi vertices, edges, and faces. However, for our applications, we only need the vertices and the edges. This will give us a graph structure forming maximally clear paths from the obstacles in the workspace. Voronoi vertices are the set of points equidistant among four or more primitives, and Voronoi edges are the set of points equidistant among three primitives. To extract these features we use continuation methods, which are very similar to common iso-surface extraction techniques commonly used in volume rendering. Our goal is to continuously "bracket" the boundary curves in a $2 \times 2 \times 2$ region of sample points and then walk along the boundary one voxel at a time. We only step to the next $2 \times 2 \times 2$ region if it is part of the same boundary. In this manner, we only touch 3D sample points that are close to the boundary. Since we are effectively growing the entire boundary from some starting "seed" point, we are able to form a correctly connected graph structure easily and efficiently. The seed point is found along the boundaries of the voxel grid since the boundary graph must typically intersect the bounding volume of the workspace.

3.3 Multi-Stage Roadmap Construction

Given the Voronoi graph, we present a multi-stage algorithm to build the roadmaps. The goal is to initially generate portions of the roadmap using the medial axis and simple local planning algorithms. Next we connect different roadmaps generated using more sophisticated techniques. Finally, we make use of the Voronoi graph and the distance buffer to estimate narrow passages and generate additional samples along these narrow passages and try to connect them with the other sampled nodes. Our overall strategy in using a multi-stage algorithm is similar to that highlighted in [1]. However, we choose samples based on the medial axis and estimate narrow passages, as opposed to generating samples in the contact space. Our roadmap construction algorithm proceeds in three stages.

1. **Preprocessing:** As part of a pre-process, we generate a number of nodes using points near the medial axis of the workspace. We assign configurations where the position of the robot corresponds to a point lying on or near the medial axis. The orientation is assigned either randomly or based on the local characteristics of the medial axis (see Section 4). Since the points on the medial axis have the maximal clearance from the obstacles, these nodes will bias the robot to plan a path near the medial axis. We use an adaptive strategy to select the nodes. Initially, we select nodes corresponding to the vertex locations in the Voronoi graph. We use a local planning algorithm to check whether we can generate an edge of the roadmap between those nodes. If not, we select a midpoint location along the Voronoi edge and repeat the process for each Voronoi edge in the roadmap graph. This process is repeated until both nodes are connected. At the end of this phase, the roadmap may consist of one or more connected components and its topology is similar to that of the Voronoi graph. The local planning algorithm is similar to the *rotate-at-s* algorithm highlighted in [1].

2. **Connecting Multiple Connected Components:** If the roadmap has more than one connected component, we try to connect them. This requires trying different set of nodes between each component and trying to find a path using the local planner. Or we generate more nodes that retain the same positions close to the medial axis of the workspace, but with different orientations for the robot. Then, we try to connect these nodes to different components of the roadmap.

3. **Sampling along Narrow Passages:** We estimate narrow passages based on the medial axis and the distance buffer. More details are given in Section 4. We generate nodes near these narrow passages by using a combination of uniform sampling, Gaussian sampling and biased angle sampling to connect them with the roadmap.

4 Sampling Strategies Based on Voronoi Graphs

In this section, we present strategies to estimate narrow passages and improved sampling algorithms based on the medial axis of the workspace.

4.1 Estimation of Narrow Passages

Our algorithm computes a discretized Voronoi diagram of the workspace. With each voxel, we associate a color that corresponds to an ID of the closest obstacle and a distance value, which stores the distance to that obstacle. By comparing this value against the dimensions of the robot, we can estimate the narrow passage in many cases. Given a robot R, let r_{out} be the radius of minimum enclosing sphere and r_{in} be the radius of the maximum inscribed sphere. If we are given a subset of voxels on the Voronoi boundary, whose distance value is less than r_{out}, then all the configurations whose corresponding positions in workspace (translation components) are close the location of these voxels may contain "narrow passages." In such cases, it may be difficult to plan the motions of the robot through these narrow passages, even when only *translational displacements* are required to navigate through the narrow corridors. Furthermore, if any collection of voxels on the Voronoi boundary have a distance value less than r_{in}, then all configurations whose translation components (or the corresponding positions in workspace) are close to the location cannot be contained in the free space.

Furthermore, the degrees and types of "tightness" in a cluttered environment are determined by the dimensions of the robot with respect to the dimensions of narrow corridor in the workspace (defined based on some type of distance metric). Normally, the medial representation computed is only the generalized Voronoi diagram or medial axis of the "workspace." Ideally we wish to compute the medial axis of the free space and identify the narrow passages more precisely by considering the rotational components of robot configurations. Unfortunately, no fast and practical algorithm is known for computing the configuration space.

Our algorithm computes an approximate medial axis of the workspace. Therefore, the scheme we mention to estimate the location of "narrow passages" is only an approximation technique. In the rest of this section, we describe some techniques that perform dimension analysis by comparing the dimensions of the robot with the distance values of the voxels that represent the discretized medial axis of the workspace.

Let us assume that the tightest-fitting arbitrarily oriented bounding box (OBB) is given for the robot. Moreover, the box's local coordinate frame and its dimensions with respect to the global coordinate frame are known. Algorithms based on covariance matrix to approximate the "principle component directions" of a given collection of polygons can be used to compute a tight fitting OBB [8]. Let us further assume that the exact dimensions of the "potential" narrow corridors are given. Based on this information, we define, the following types of narrow passages:

Type-T Narrow Passages: Robot can move through narrow passages by translational motion. Rotations are *not* required for the robot to navigate through the narrow corridors. This occurs when all the dimensions of the robot's OBB are equal or slightly smaller than the dimensions of the narrow corridors. Images 1-3 are examples of such narrow passages where merely careful translational movements are sufficient to navigate the robot through the narrow corridors.

Type-R Narrow Passages: Rotations are required for the robot to navigate through the corridors. This occurs often when the largest dimension of the robot's OBB exceeds those of the narrow corridors. Images 3 and 6 highlight examples of two narrow passages that require rotations to plan the path for the robot and the piano.

4.2 Random Sampling Near the Medial Axis

Once we can identify different types of narrow passages in the configuration space, we can design better sampling strategies for handling these scenarios.

In general, if the robot encounters a type-T narrow passage, the robot can easily move through the narrow corridors by placing the geometric center of the robot's OBB along the edges of the Voronoi graph. In such cases taking samples on the medial axis (the first step in our multi-level strategy) has considerable benefits.

For handling type-R narrow passages, sampling *along* the Voronoi graph with random sampling of angles can lead to poorer performance. Instead, we propose to sample *near* the Voronoi graph using a *combination* of the following strategies:

- **Simple Uniform Sampling:** Place more sampling points uniformly inside the narrow passages, once they have been identified. This technique basically increases the density of sampled nodes inside the narrow passages everywhere.
- **Gaussian distribution:** Sample near the medial axis with a Gaussian distribution to randomly position the points with higher distribution density *near* the medial axis while uniformly sampling in the angular space.
- **Angular Bias:** Biasing the angular sampling by positioning the local coordinate axis along which the robot's OBB has its smallest dimension so that the axis is perpendicular to the tangent direction of the Voronoi graph with Gaussian distribution. Intuitively, this sampling strategy attempts to orient the robot to intelligently adapt to the change in curvatures of the Voronoi graph, as it moves through the narrow passages.

We have experimented with a combination of these strategies on several different benchmark environments and have observed good performance improvement over uniform sampling. More details of our implementation results are given in the next section.

5 Implementation and Performance

We describe the implementation of our planner and its application to different benchmarks.

5.1 Implementation

We have implemented a preliminary version of the planning algorithm for 2D (3-dof) and 3D (6-dof) environments on an SGI IRIX workstation. The planner proceeds in distinct stages: Pre-processing, roadmap construction, and path planning query.

In the pre-processing phase, the planner computes information about the environment. First, it computes the discretized Voronoi diagram. With each voxel of the 3D image, we store information about the closest obstacle and the distance computed using the depth buffer. The performance of the Voronoi algorithm on different scenarios has been highlighted in Table 1. It is a linear function of the number of voxels in the grid and increases as we subdivide a voxel into 8 sub-voxels. We expect this performance can be improved by using adaptive grid size selection. We are currently investigating the implementation issues related to adaptive grid size.

We perform collision and distance queries between the robot and the obstacles using PQP [17]. PQP uses a hierarchy of swept sphere volumes and works efficiently on general polygonal models. A typical planner spends a high percentage of its time in performing collision and distance queries. To speed up these queries, we enclose the robot R with a bounding sphere. Let its radius be r_{out}. We compare the radius with the distance value associated with the voxel that contains the position of the center of the enclosing sphere. If r_{out} is less than the distance value, it implies that the given configuration lies in the free space and we don't have to perform explicit collision check between the robot and all the obstacles. Otherwise, we use PQP routines for exact collision detection. We refer to this acceleration technique as *QR collision check* for quick rejections. Given the color buffer, we compute the boundary graph of the Voronoi diagram using a marching technique. We next compute an approximation to the medial axis, and identify regions to be considered as narrow passages in the workspace (as described in Section 4).

In the roadmap construction phase, the planner attempts to build a network of configuration nodes that can be used for the path planning queries. We use the multi-stage algorithm described in Section 3. We represent the roadmap as a graph with multiple connected components. Given an initial and goal configuration, we combine the roadmap and query phase into one step so that each query adds new connectivity information to the roadmap. The planner builds the roadmap by iteratively growing from initial configuration C_i and the goal configuration C_g. We use a growth strategy similar to the one highlighted in [15, 13]. To grow a component c of the roadmap, the algorithm randomly selects a node N_e in c to expand. N_e is chosen from the set of previously generated samples in c, giving priority to those near the medial axis, and the ones in the low density areas. The main goal is to bias the planner towards the nodes whose translational component lies

Model Size (polygons)	16	32	64	128	256	512
46 (Benchmark 5)	.02	.05	.12	.48	3.22	25.33
2180 (Benchmark 6)	.16	.44	1.74	9.82	68.01	557.36
10900	.77	2.09	8.49	48.47	336.83	2897.43

Table 1: The cost of computing the discretized Voronoi diagram in seconds as a function of the resolution. The table lists the number of triangles against Voronoi diagram resolution. All timings are computed on a SGI Infinite Reality (Onyx2). Each represents a different grid size ($16 \times 16 \times 16$ to $512 \times 512 \times 512$).

close to the medial axis and to select new configurations in free space.

To expand a node, we generate a number of random configurations N_r in the neighborhood of N_e using a Gaussian distribution. The size of the neighborhood is defined by the distance to the nearest obstacle and is computed using the distance information associated with each voxel. After that we check if N_e is in the free space by performing a collision query. We try to insert free configurations into the component c by checking for a valid path from N_e to N_e using the local planner. Finally, after each component has completed its expansion stage, an attempt is made to connect the unconnected components using local planning algorithms. This process is repeated until a path has been found between the initial and the goal configuration.

5.2 Benchmark Environments

We have tested the performance of the planner on several 2D and 3D benchmarks. We also compared its performance with Stanford PRM developed by David Hsu, J. Latombe et al. [13]. It uses uniform sampling in the configuration space to generate the configurations.

We have chosen a suite of environments and scenarios to test the effectiveness of our sampling scheme over basic uniform sampling. Most of them have either Type-T narrow passage or Type-R narrow passage. The set of benchmarks used include:

- **Benchmark 1:** A 2D environment requiring the robot to navigate a narrow passage to move from the open area on the left to the open area on the right (Image 1(a-b)). The environment consists of two wide open areas connected by one narrow channel.
  - Scenario (a): Type-T narrow passage, robot dimension of 2.5×14.0.
  - Scenario (b): Type-R narrow passage, robot dimension of 4.0×40.0.
  - Size of environment : 400×400
  - Width of narrow passage: 20
  - Number of line segments : 158
- **Benchmark 2:** A 2D environment requiring the robot to traverse a long Type-T narrow passage resembling a maze (Image 2).
  - Size of environment : 400×400
  - Width of narrow passage: 15
  - Number of triangles : 1044
  - Robot dimensions : 10×10
- **Benchmark 3:** A complex 2D environment consisting of chairs, pianos and a music stand, each projected into the XY plane (Image 3). The robot (a music stand) must navigate a Type-T narrow passage.
  - Size of environment : 15.5×17.5
  - Number of triangles : 12054
  - Robot dimensions : 1.0×1.0
- **Benchmark 4:** A 3D environment with two open areas connected by a single channel (Image 4(a-c)). The robot (a block) must move from the initial configuration to the goal by traversing a narrow tunnel.
  - Scenario (a): Type-T narrow passage, robot dimension of $.025 \times .025 \times .025$.
  - Scenario (b): Type-T narrow passage, robot dimension of $.08 \times .08 \times .08$.
  - Scenario (c): Type-R narrow passage, robot dimension of $.025 \times .025 \times .125$.
  - Size of environment: $1.0 \times 1.0 \times 1.0$
  - Width of narrow passage: $0.1 \times 0.1 \times 0.1$
  - Number of triangles: 28

Benchmark	Sampling Type	Time	C-Space Samples	Free Samples	Full Coll. Checks	QR Coll.	Connect Comp. Calls
1(a)	Uniform	36.10	5062	3414	21740	0	2655585
1(a)	Voronoi Based	1.50	1239	1213	3202	9864	36447
1(b)	Uniform	80.22	14843	8636	20756	0	2360384
1(b)	Voronoi Based	20.47	8297	3365	8753	6201	201654

Table 2: Benchmark 1, environment shown in Image 1. The size of the robot varies so that in (a) No rotation is required. (b) Rotation is required.

- **Benchmark 5:** A 3D environment similar to benchmark 4, except that the passage is not a simple straight channel. The channel "spirals" through space. The tunnel itself provides a narrow passage in the workspace. The sharp corners further complicate planning by causing a Type-R narrow passage (Image 5).
  - Scenario (a): Type-R narrow passage, robot dimension of $.02 \times .02 \times .125$.
  - Scenario (b): Type-R narrow passage, robot dimension of $.07 \times .07 \times .07$.
  - Size of environment : $1.0 \times 1.0 \times 1.0$
  - Width of narrow passage: $0.1 \times 0.1 \times 0.1$
  - Number of triangles : 46
- **Benchmark 6:**

  A 3D environment consisting of chairs, a table, and a piano (Image

  6). The goal is to move the piano through the window. The

  environment provides Type-R narrow passage, as the piano must

  rotate to fit through the window. This benchmark has been provided to us by Jean-Paul Laumond at LAAS, Toulouse.
  - Size of environment : $6000 \times 6000 \times 3000$
  - Dimension of narrow passage: 2000×1000 (W $\times$ H)
  - Robot dimension (with legs): $1000 \times 1210 \times 850$
  - Robot dimension (sans legs): $1000 \times 1200 \times 350$

5.3 Performance Data

We highlight the performance of the planner on different benchmarks. The different columns in the table used are:

- Sampling Type—We have compared our Voronoi Based sampling to Stanford's planner that uses Uniform sampling [13]. The basic structure of the implementations are the same. The main difference is in the sampling strategy.
- Time—The running time to service the benchmark query. The time to compute the Vornoi diagram is not included, but is listed in a separate table. But it includes the time to build the roadmap.
- C-Space Sample—The total number of random C-Space configuration generated by the planner. This number represents all configurations, even those in contact space, and those that are in free space that fail to connect to a node in the road map.
- Free Samples—The total number of samples which lie in free space and successfully become part of the roadmap.
- Full Coll. Checks—The number of times a full PQP collision query was performed.
- QR Coll.—The number of times a full collision query was avoided, by doing a quick rejection.
- Connect Comp. Calls—The number of times the planner attempted to connect two nodes from distinct components.

6 Analysis and Comparison

Overall the performance of our planning algorithm varies with the environment. In all our current benchmarks, it significantly performs uniform sampling in Type-T narrow passages. It also improves the performance of the planner when there are Type-R narrow passages in the configuration space. In all our benchmark comparisons, we try to make minimal changes to the connection strategies. Based on the medial axis, we have also proposed a novel multi-stage strategy to build the roadmap. It leads to considerable improvement in the performance.

The idea of using the medial axis for sampling is not novel. Other authors have proposed using medial axis based sampling [27, 26, 9]. The major benefit in our approach comes from the fact that we have bounded error approximation of the medial axis computed using graphics hardware. Wilmarth et al. [27, 26] do not compute a medial axis. They take random configurations and retract them to the medial using iterative approaches. However, they either cannot guarantee sufficient number of samples in all the narrow passages, or it will take many more random configurations (followed by retraction) to generate sufficient samples in some of the challenging scenarios.

Guibas et al. [9] take point samples on the boundary and compute their Voronoi diagram to estimate samples on the medial axis of the workspace. Again, in this case, it is difficult to guarantee any bounds on the medial axis or quality of the resulting samples.

Overall, having a bounded error approximation of the medial axis of the workspace, along with the distance information computed using the depth buffer, helps us in identifying the narrow passages and characterizing them in many cases. Our proposed algorithms seem to perform well in all the benchmark cases we have tried. Clearly, there is no one universal sampling strategy that will work well in all cases. Therefore, we are continuing to experiment with various benchmark examples to find the worst case scenarios, as well as investigating a theoretical analysis that relates a medial axis in a workspace to a Voronoi roadmap in the corresponding configuration space. We believe that a formal analysis and characterization will give us a more rigorous method to distinguish between different types of narrow passages and enable us to intelligently choose a combination of sampling strategies for adapting to different regions of the environment.

7 Conclusion

We have presented techniques to improve the performance of PRM in configurations with narrow passages. We use a bounded error approximation of the generalized Voronoi diagram to improve the sampling and estimate narrow passages in the free space. It is also used to improve the performance of proximity queries. We have applied it to a number of 2D and 3D benchmarks and the preliminary results are promising.

Next, we would like to investigate the theoretical analysis of "Voronoi skeleton" or medial axis for Type-R narrow passages and develop more rigorous computational techniques using graphics hardware. It will be useful to conduct more extensive performance comparisons between our approach accelerated by graphics hardware against other PRMs based on medial-axis sampling [27, 26, 9]. We would like to further extend our approach to motion planning with constraints and planning of articulated objects or manipulators.

Acknowledgments

We are grateful to David Hsu and Jean-Claude Latombe for a version of their planner and many useful discussions related to this problem. We are also grateful to Jean-Paul Laumond for providing us with the model of Benchmark 6. Supported in part by ARO Contract DAAH04-96-1-0257, NSF Career Award CCR-9625217, NSF grants EIA-9806027 and DMI-9900157, ONR Young Investigator Award, DOE ASCI Contract and Intel.

References

[1] N. Amato, O. Bayazit, L. Dale, C. Jones, and D. Vallejo. Obprm: An obstacle-base prm for 3d workspaces. *Proceedings of WAFR98*, pages 197-204, 1998.

[2] H. Blum. A transformation for extracting new descriptors of shape. In W. Whathen-Dunn, editor, *Models for the Perception of Speech and Visual Form*, pages 362-380. MIT Press, 1967.

[3] J. Canny and B. R. Donald. Simplified Voronoi diagrams. In *Proc. 3rd Annu. ACM Sympos. Comput. Geom.*, pages 153-161, 1987.

[4] H. Chang and T. Li. Assembly maintainability study with motion planning. In *Proceedings of Internationa Conference on Robotics and Automation*, 1995.

[5] H. Choset. Nonsmooth analysis, convex analysis and their applications to motion planning. *International Journal of Computational Geometry and Applications*, 1997.

[6] H. Choset, I. Konukseven, and A. Rizzi. Sensor based planning: Using a honing strategy and local map method to implement the generalize voronoi graph. *SPIE Mobile Robotics*, 1997.

[7] P. W. Finn, L. E. Kavraki, J. C. Latombe, R. Motwani, C. Shelton, S. Venkatasubramanian, and A. Yao. Rpid: Randomized pharmacophore identification for drug design. *Proc. of 13th ACM Symp. on Computational Geometry (SoCG'97)*, 1997. A revised version of this paper also appeared in Computational Geometry: Theory and Applications, 10, pp. 263-272, 1998.

[8] S. Gottschalk, M. Lin, and D. Manoch. Obb-tree: A hierarchical structure for rapid interference detection. In *Proc. of ACM Siggraph '96* pages 171-180, 1996.

[9] L. Guibas, C. Holleman, and L. Kavraki. A probabilistic roadmap planner for flexible objects with a workspace medial-axis-based sampling approach. In *Proc. of IROS*, 1999.

[10] K. Hoff, T. Culver, J. Keyser, M. Lin, and D. Manocha. Fast computation of generalized voronoi diagrams using graphics hardware. *Proceedings of ACM SIGGRAPH 1999*, 1999.

[11] K. Hoff. T. Culver, J. Keyser, M. Lin, and D. Manocha. Interactive motion planning using hardware accelerated comutation of generalized voronoi diagrams. Techincal report, Department of Computer Science, University of North Carolina, 1999. To appear in IEEE Conference on Robotics and Automation.

[12] T. Horsch, F. Schwarz, and H. Tolle. Motion planning for many degrees of freedon-random reflections at c-space obstacles. *Proc. of IEEE International Conf. on Robotics and Automation*, pages 3318-3323, 1994.

[13] D. Hsu, L. Kavraki, J. Latombe, R. Motwani, and S. Sorkin. On finding narrow passages with probabilistic roadmap planners. *Proc. of 3rd Workshop on Algoricthmic Foundations of Robotics*, 1998.

[14] L. Kavraki and J. C. Latombe. Randomized preprocessing of configuration space for fast path planning. *IEEE Conference on Robotics and Automation*, pages 2138-2145, 1994.

[15] L. Kavraki, P. Svestka, J. C. Latombe, and M. Overmars. Probabilistic roadmaps for path planning in high-dimensional configuration spaces. em IEEE Trans. Robot. Automat., pages 12(4):566-580, 1996.

[16] Yoshito Koga, Koichi Kondo, James Kuffner, and Jean-Claude Latombe. Planning motions with in tentions. In Andrew Glassner, editor, *Proceedings of SIGGRAPH '94 (Orlando, Flordia, July 24-29, 1994)*, Computer Graphics Proceedings, Annual Confrence Series, pages 395-408. ACM SIGGRAPH, ACM Press, July 1994. ISBN 0-89791-667-0.

[17] E. Larsen, S. Gottschalk, M. Lin, and D. Manocha. Fast proximity queries with swept sphere volumes. Technical Report TR99-018, Department of Computer Science, University of North Carolina, 1999.

[18] J. C. Latombe. *Robot Motion Planning*. Kluwer Academic Publishers, 1991.

[19] D. Lavender, A. Bowyer, J. Davenport, A. Wallis, and J. Woodwark. Voronoi diagrams of set-theoretic solid models. *IEEE Computer Graphics and Applications*, pages 69-77, September 1992.

[20] C. O'Dunlaing, M. Sharir, and C. K. Yap. Retraction: A new approach to motion-planning. In *Proc. 15th Annu. ACM Sympos. Theory Comput.*, pages 207-220, 1983.

[21] M. H. Overmars and P Svestka. A probabilistic learning approach to motion planning. In *Algorithmic Foundations of Robotics*, Boston, MA 1995. A. K. Peters.

[22] A. Schweikard, R. Tombropoulos, L. E. Kavraki, J. R. Adler, and J. C. Latombe. Treatment planning for a radiosurgical system with general kinematics. *IEEE Confrence on Robotics and Automation*, 1994.

[23] D. J. Sheehy, C. G. Armstrong, and D. J. Robinson. Computing the medial surface of a solid from a domain delaunay triangulatio. In Chris Hoffman and Jarek Rossignac, editors, *Solid Modeling '95*, pages 201-212, May 1995.

[24] R. Tombropouloas, J. R. Adler, and J. C. Latombe. Carabeamer. A treatment planner for a robotic radiosurgical system with general kinematics. *Medical Image Analysis*, pages 3(3):1-28, 1999.

[25] J. Vleugels and M. Overmars. Approximating voronoi diagrams of convex sites in any dimension. *International Journal of Computational Geometry and Applications*, 8:201-222, 1997.

[26] Steven A. Wilmarth, Nancy M. Amato, and Peter F. Stiller. Maprm: Aprobabilisitc roadmap planner with sampling on the medial axis of the free space. *IEEE Conference on Robotics and Automation*, 1999.

[27] Steven A. Wilmarth, Nancy M. Amato, and Peter F. Stiller. Motion planning for a rigid body using random networks on the medial axis of the free space. *Proc. of the 15th Annual ACM Symposium on Computational Geometry (SoCG'99)*, 1999.

Benchmark	Sampling Type	Time	C-Space Samples	Free Samples	Full Coll. Checks	QR Coll.	Connect Comp. Calls
2	Uniform	625.67	30692	3479	267045	0	2423896
2	Voronoi Based	213.89	10018	1393	86902	78251	308528

Table 3: Benchmark 2, maze environment shown in Image 2. The narrow passage is long with respect to the robot dimensions.

Benchmark	Sampling Type	Time	C-Space Samples	Free Samples	Full Coll. Checks	QR Coll.	Connect Comp. Calls
3	Uniform	540.44**	36631	17812	101087	0	104032
3	Voronoi Based	114.5	23680	8739	33646	24907	54986

Table 4: Benchmark 3, house environment shown in Image 3. Notice that the Uniform based sampling does not even complete in the allocated time.

Benchmark	Sampling Type	Total Time	C-Space Samples	Free Samples	Full Coll. Checks	QR Coll.	Connect Comp. Calls
4(a)	Uniform	8.28	4669	2714	10148	0	375736
4(a)	Voronoi Based	3.48	471	378	3098	2908	12685
4(b)	Uniform	7892.55	126189	15079	212859	0	55216398
4(b)	Voronoi Based	721.28	35328	2707	33344	21088	1803603
4(c)	Uniform	14179.06**	313014	54593	488821	0	57429630
4(c)	Voronoi Based	2956.83	93723	12718	93528	67087	4634587

Table 5: Benchmark 4, a simple tunnel shown in Image 4. The size of the robot varies so that in (a) The robot is small, and no rotation is required. (b) Rotation is required. (c) The robot is large, but no rotation is required. Notice that in (c) the Uniform based sampling does not even complete in the allocated time.

Benchmark	Sampling Type	Total Time	C-Space Samples	Free Samples	Full Coll. Checks	QR Coll.	Connect Comp. Calls
5(a)	Uniform	296.65	61675	10985	98704	0	2028713
5(a)	Voronoi Based	160.99	22021	6087	52956	32091	715593
5(b)	Uniform	19223.3	733663	13335	845172	0	14035450
5(b)	Voronoi Based	15434.1	543401	9874	618233	152987	10329874
5(c)	Uniform	296.65	61675	10985	98704	0	2028713
5(c)	Voronoi Based	12.45	5672	472	2732	982	Not Applicable

Table 6: Benchmark 5, spiral tunnel shown in Image 5. In (a) and (b) the robot size varies. In (c) we compute the path with our boundary graph approach. (a) No rotation is required. (b) Rotation is required. (c) The same scenario as in (a) with the different connection scheme that takes advantage of medial representations. We observe substantial performance improvement by using a better connection strategy that exploits the Voronoi graph structures.

Benchmark	Sampling Type	Total Time	C-Space Samples	Free Samples	Full Coll. Checks	QR Coll.	Connect Comp. Calls
6	Uniform	3948.86	91264	43542	250043	0	452459428
6	Voronoi Based	2983.24	58265	38098	220023	10821	351247399

Table 7: Benchmark 6, piano environment shown in Image 6. A very challenging Type-R narrow passage still shows 25% speedup.

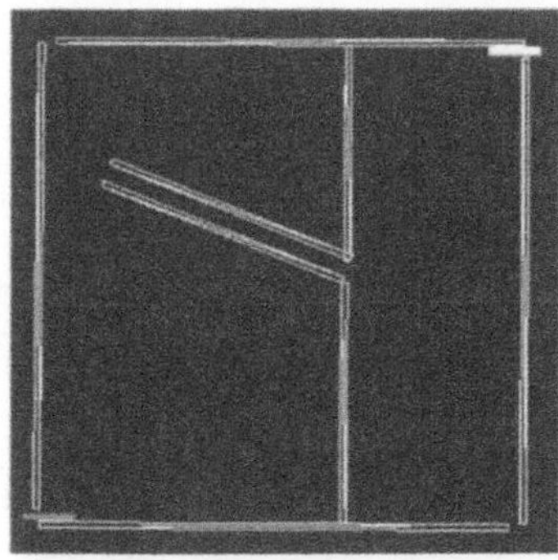
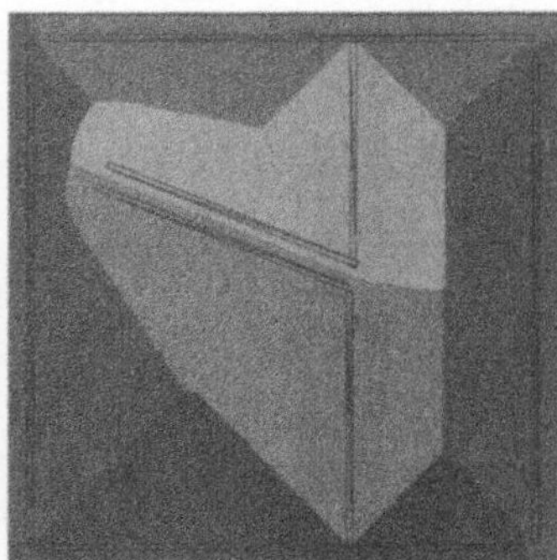
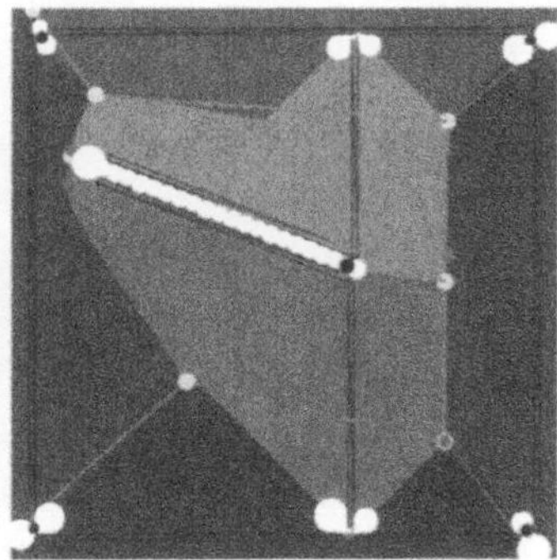

Image 1: Benchmark 1 showing (a) initial (red) and goal (green) configurations, (b) computed voronoi diagram, (c) boundary graph and narrow passage identification.

Image 2: The maze, Benchmark 2.

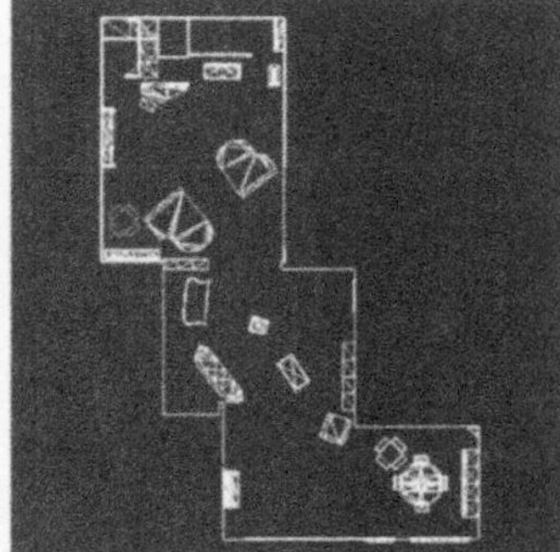

Image 3: The house, Benchmark 3.

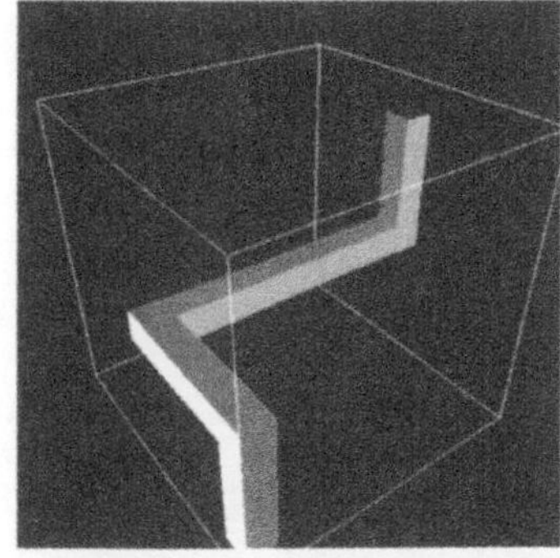
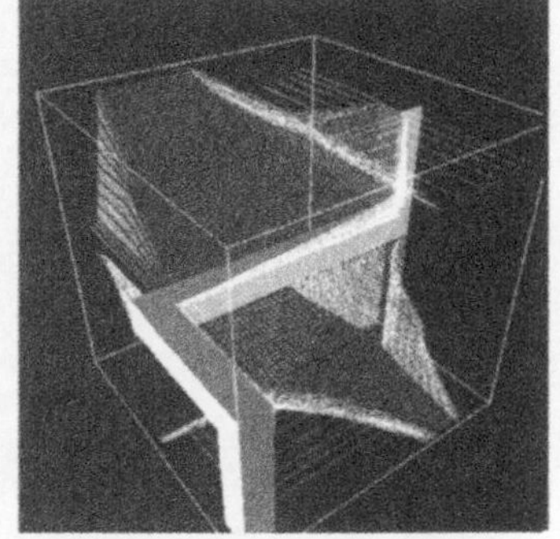
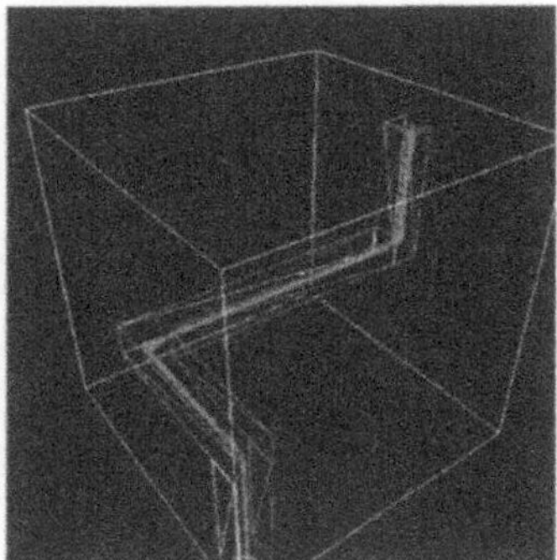

Image 4: Benchmark 4 showing (a) planning environment, (b) computed voronoi surface, (c) extracted core through workspace.

Image 5: A simple tunnel, Benchmark 5.

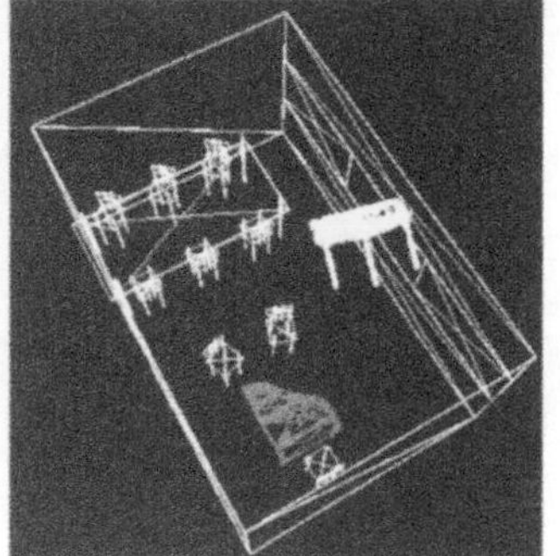

Image 6: Piano Environment, Benchmark 6. Red shows the initial position. The Green position shows an R Narrow Passage.

Rapidly-Exploring Random Trees: Progress and Prospects

Steven M. LaValle, *Iowa State University, Ames, IA*
James J. Kuffner, Jr., *University of Tokyo, Tokyo, Japan*

We present our current progress on the design and analysis of path planning algorithms based on Rapidly-exploring Random Trees (RRTs). The basis for our methods is the incremental construction of search trees that attempt to rapidly and uniformly explore the state space, offering benefits that are similar to those obtained by other successful randomized planning methods. In addition, RRTs are particularly suited for problems that involve differential constraints. Basic theoretical properties of RRT-based planners are established. Several planners based on RRTs are discussed and compared. Experimental results are presented for planning problems that involve holonomic constraints for rigid and articulated bodies, manipulation, nonholonomic constraints, kinodynamic constraints, kinematic closure constraints, and up to twelve degrees of freedom. Key open issues and areas of future research are also discussed.

1 Introduction

Given the vast, growing collection of applications that involve the design of motion strategies, the successes of motion planning algorithms have just begun to scratch its surface. The potential for automating motions is now greater than ever as similar problems continue to emerge in seemingly disparate areas. The traditional needs of roboticists continue to expand in efforts to automate mobile robots, manipulators, humanoids, spacecraft, etc. Researchers in computer graphics and virtual reality have increasing interests in automating the animations of life-like characters or other moving bodies. In the growing field of computational biology, many geometric problems have arisen, such as studying the configuration spaces of flexible molecules for protein-ligand docking and drug design. Virtual prototyping is a rapidly-expanding area that allows the evaluation of proposed mechanical designs in simulation, in efforts to avoid the costs of constructing physical prototypes. Motion planning techniques have already been applied to assembly problems in this area [9]. As the power and generality of planning techniques increase, we expect that more complicated problems that include differential constraints can be solved, such as the evaluation of vehicle performance and safety through dynamical simulation conducted by a virtual "stunt driver."

As we approach applications of increasing difficulty, it becomes clear that planning algorithms need to handle problems that involve a wide variety of models, high degrees of freedom, complicated geometric constraints, and finally, differential constraints. Although existing algorithms address some of these concerns, there is relatively little work that addresses all of them simultaneously. This provides the basis for the work presented in this paper, which presents randomized, algorithmic techniques for path planning that are particularly suited for problems that involve differential constraints.

We present an overview of the progress on our development of Rapidly-exploring Random Trees (RRTs) [29]. The results and discussion presented here summarize and extend the work presented in [30, 22]. In [30], we presented the first randomized approach to kinodynamic trajectory planning, and applied it to problems that involve up to twelve-dimensional state spaces with nonlinear dynamics and obstacles (for problems with moving obstacles, a similar randomized approach has been proposed more recently, in [20]). In [22], we presented and analyzed a holonomic path planner that gives real-time performance for many challenging problems. RRTs build on ideas from optimal control theory [5], nonholonomic planning (see [27] for an overview), and randomized path planning [1, 19, 36]. The basic idea is to use control-theoretic representations, and incrementally grow a search tree from an initial state by applying control inputs over short time intervals to reach new states. Each vertex in the tree represents a state, and each directed edge represents an input that was applied to reach the new state from a previous state. When a vertex reaches a desired goal region, an open-loop trajectory from the initial state is represented.

For problems that involve low degrees of freedom, classical dynamic programming ideas can be employed to yield numerical optimal control solutions for a broad class of problems [5, 24, 28]. Since control theorists have traditionally preferred feedback solutions, the representation takes the form of a mesh over which cost-to-go values are defined using interpolation, enabling inputs to be selected over any portion of the state space. If open-loop solutions are the only requirement, then each cell in the mesh could be replaced by a vertex that represents a single state within the cell. In this case, control-theoretic numerical dynamic programming techniques can often be reduced to the construction of a tree grown from an initial state (referred to as forward dynamic programming [24]). This idea has been proposed in path planning literature for nonholonomic [4] planning and kinodynamic planning in [13]. Because these methods are based on dynamic programming and systematic exploration of a grid or mesh, their application is limited to problems with low degrees of freedom.

We would like to borrow some of the ideas from numerical optimal control techniques, while weakening the requirements enough to obtain methods that can apply to problems with high degrees of freedom. As is common in most of path planning research, we forego trying to obtain optimal solutions, and attempt to find solutions that are "good enough," as long as they satisfy all of the constraints. This avoids the use of dynamic programming and systematic exploration of the space; however, a method is needed to guide the search in place of dynamic programming.

Inspired by the success of randomized path planning techniques and Monte-Carlo techniques in general for addressing high-dimensional problems, it is natural to consider adapting existing planning techniques to our problems of interest. The primary difficulty with many existing techniques is that, although powerful for standard path planning, they do not naturally extend to general problems that involve differential constraints. The randomized potential field method [3], while efficient for holonomic planning, depends heavily on the choice of a good heuristic potential function, which could become a daunting task when confronted with obstacles, and differential constraints. In the probabilistic roadmap approach [1, 19], a graph is constructed in the configuration space by generating random configurations and attempting to connect pairs of nearby configurations with a local planner that will connect pairs of configurations. The assumption is that the same roadmap will be used for *multiple queries.* For planning of holonomic systems or steerable nonholonomic systems (see [27] and references therein), the local planning step might be efficient; however, in general the connection problem can be as difficult as designing a nonlinear controller, particularly for complicated nonholonomic and dynamical systems. The probabilistic roadmap technique might require the connections of thousands of configurations or states to find a solution, and if each connection is akin to a nonlinear control problem, it seems impractical many problems with differential constraints.

2 Problem Formulation

The class of problems considered in this paper can be formulated in terms of six components:

1. **State Space:** A topological space, X
2. **Boundary Values:** $x_{init} \in X$ and $X_{goal} \subset X$
3. **Collision Detector:** A function, $D : X \to \{true, false\}$, that determines whether global constraints are satisfied from state x. This could be a binary-valued or real-valued function.
4. **Inputs:** A set, U, which specifies the complete set of controls or actions that can affect the state.
5. **Incremental Simulator:** Given the current state, $x(t)$, and inputs applied over a time interval, $\{u(t')|t \leq t' \leq t + \Delta t\}$, compute $x(t + \Delta t)$.
6. **Metric:** A real-valued function, $\rho : X \times X \to [0, \infty)$, which specifies the distance between pairs of points in X.

Path planning will generally be viewed as a search in a state space, X, for a continuous path from an initial state, x_{init} to a goal region $X_{goal} \subset X$ or goal state $x_{goal} \in X$. It is assumed that a complicated set of global constraints is imposed on X, and any solution path must keep the state within this set. A collision detector reports whether a given state, x, satisfies the global constraints. We generally use the notation X_{free} to refer to the set of all states that satisfy the global constraints. Local, differential constraints are imposed through the definition of a set of inputs (or controls) and an incremental simulator. Taken together, these two components specify possible changes in state. The incremental simulator can be considered as the response of a discrete-time system (or a continuous-time system that is approximated in discrete time). Finally, a metric is defined to indicate the closeness of pairs of points in the state space. This metric will be used in Section 3, when the RRT is introduced. It will generally be assumed that a single path planning query is presented, as opposed to performing precomputation for multiple queries, as in [1, 19].

Basic (Holonomic) Path Planning Path planning can generally be viewed as a search in a configuration space, $\mathcal{C}$, in which each $q \in \mathcal{C}$ specifies the position and orientation of one or more geometrically-complicated bodies in a 2D or 3D world [33, 25]. The path planning task is to compute a continuous path from an initial configuration, q_{init}, to a goal configuration, q_{goal}. Thus, $X = \mathcal{C}$, $x_{init} = q_{init}$, $x_{goal} = q_{goal}$, and $X_{free} = \mathcal{C}_{free}$, which denotes the set of configurations for which these bodies do not collide with any static obstacles in the world. The obstacles are modeled completely in the world, but an explicit representation of X_{free} is not available. However, using a collision detection algorithm, a given configuration can be tested. (To be more precise, we usually employ a distance-computation algorithm that indicates how close the geometric bodies are to violating the constraints in the world. This can be used to ensure that intermediate configurations are collision free when discrete jumps are made by the incremental simulator.) The set, U, of inputs is the set of all velocities $\dot{x}$ such that $\|\dot{x}\| \leq c$ for some positive constant c. The incremental simulator produces a new state by integration to obtain, $x_{new} = x + u\Delta t$, for any given input $u \in U$.

Nonholonomic Path Planning Nonholonomic planning [26] addresses problems that involve nonintegrable constraints on the state velocities, in addition to the components that appear in the basic path planning problems. These constraints often arise in many contexts such as wheeled-robot systems [27], and manipulation by pushing [35]. A recent survey appears in [27]. The constraints often appear in the implicit form $h_i(\dot{q}, q) = 0$ for some i from 1 to $k < N$ (N is the dimension of $\mathcal{C}$). By the implicit function theorem, the constraints can also be expressed in control-theoretic form, $\dot{q} = f(q, u)$, in which u is an input chosen from a set of inputs U. Using our general notion, x replaces q to obtain $\dot{x} = f(x, u)$. This form is often referred to as the *state transition equation* or *equation of motion*. Using the state transition equation, an incremental simulator can be constructed by numerical integration (using, for example Runge-Kutta techniques).

Kinodynamic[1] Path Planning For kinodynamic planning, constraints on both velocity and acceleration exist, yielding implicit equations of the form $h_i(\ddot{q}, \dot{q}, q) = 0$ [6, 8, 10, 11, 13, 12, 14, 45, 46]. A state, $x \in X$, is defined as $x = (q, \dot{q})$, for $q \in \mathcal{C}$. Using the state space representation, this can be simply written as a set of m implicit equations of the form $G_i(x, \dot{x}) = 0$, for $i = 1, \ldots, m$ and $m < 2N$. The implicit function theorem can again be applied to obtain a state transition equation. The collision detection component may also include global constraints on the velocity, since $\dot{q}$ is part of the state vector.

Other Problems A variety of other problems fit within our problem formulation, and can be approached using the techniques in this paper. In general, any open-loop trajectory design problem can formulated because the models are mostly borrowed from control theory. For example, the planner might be used to compute a strategy that controls an electrical circuit, or an economic system. In some applications, a state transition equation might not be known, but this does not present a problem. For example, a physical simulator might be developed by engineers for simulating a proposed racing car design. The software might simply accept control inputs at some sampling rate, and produce new states. This could serve directly as the incremental simulator for our approach. Other minor variations of the formulation can be considered. Time-varying problems can be formulated by augmenting the

[1]In nonlinear control literature, kinodynamic planning for underactuated systems is encompassed by the definition of nonholonomic planning. Using control-theoretic terminology, the task is to design open-loop trajectories for nonlinear systems with drift.

state space with a time. State-dependent inputs sets can also be considered. For example, a robot engaged in a grasping task might have different inputs available than while navigating. Depending on the state, different decisions would have to be made. Problems that involve kinematic closure constraints can also be addressed; an example is shown in Figure 14.

3 Rapidly-Exploring Random Trees

The Rapidly-exploring Random Tree (RRT) was introduced in [29] as a planning algorithm to quickly search high-dimensional spaces that have both algebraic constraints (arising from obstacles) and differential constraints (arising from nonholonomy and dynamics). The key idea is to bias the exploration toward unexplored portions of the space by sampling points in the state space, and incrementally "pulling" the search tree toward them. At least two other randomized path planning techniques have been proposed that generate an incremental search tree in the configuration space (for holonomic path planning): The Ariadne's clew algorithm [37, 36] and the planners in [18, 50]. Intuitively, these planners attempt to "push" the search tree away from previously-constructed vertices, contrasting the RRT, which uses the surrounding space to "pull" the search tree, ultimately leading to uniform coverage of the state space. To the best of our knowledge, a randomized search tree approach has not been proposed previously for nonholonomic or kinodynamic planning. Perhaps the most related approaches are [47, 44], in which the probabilistic roadmap method is combined with nonholonomic steering techniques to plan paths for wheeled mobile robot systems.

The basic RRT construction algorithm is given in Figure 1. A simple iteration in performed in which each step attempts to extend the RRT by adding a new vertex that is biased by a randomly-selected state. The EXTEND function, illustrated in Figure 2, selects the nearest vertex already in the RRT to the given sample state. The "nearest" vertex is chosen according to the metric, ρ. The function NEW_STATE makes a motion toward x by applying an input $u \in U$ for some time increment Δt. This input can be chosen at random, or selected by trying all possible inputs and choosing the one that yields a new state as close as possible to the sample, x (if U is infinite, then an approximation or analytical technique can be used). In the case of holonomic planning, the optimal value for u can be chosen

```
BUILD_RRT(x_init)
    T.init(x_init);
    for k = 1 to K do
        x_rand ← RANDOM_STATE();
        EXTEND(T, x_rand);
    Return T

EXTEND(T, x)
    x_near ← NEAREST_NEIGHBOR(x, T);
    if NEW_STATE(x, x_near, x_new, u_new) then
        T.add_vertex(x_new);
        T.add_edge(x_near, x_new, u_new);
        if x_new = x then
            Return Reached;
        else
            Return Advanced;
    Return Trapped;
```

Figure 1: *The basic RRT construction algorithm.*

easily by a simple vector calculation. NEW_STATE also implicitly uses the collision detection function to determine whether the new state (and all intermediate states) satisfy the global constraints. For many problems, this can be performed quickly ("almost constant time") using incremental distance computation algorithms [16, 31, 38] by storing the relevant invariants with each of the RRT vertices. If NEW_STATE is successful, the new state and input are represented in x_{new} and u_{new}, respectively. Three situations can occur: *Reached*, in which the new vertex reaches the sample x (for the nonholonomic planning case, we might instead have a threshold, $\|x_{new} - x\| < \epsilon$ for a small $\epsilon > 0$); *Advanced*, in which a new vertex $x_{new} \neq x$ is added to the RRT; *Trapped*, in which NEW_STATE fails to produce a state that lies in X_{free}. The left column of Figure 3 shows an RRT for a holonomic planning problem, constructed in a 2D square space. The right column shows the Voronoi diagram of the RRT vertices; note that the probability that a vertex is selected for extension is proportional to the area of its Voronoi region.

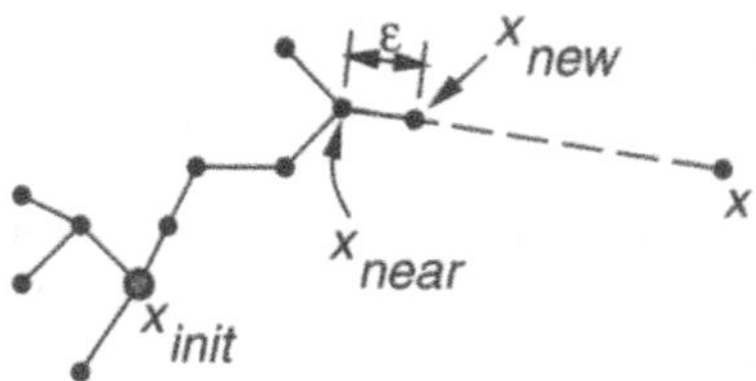

Figure 2: *The EXTEND operation.*

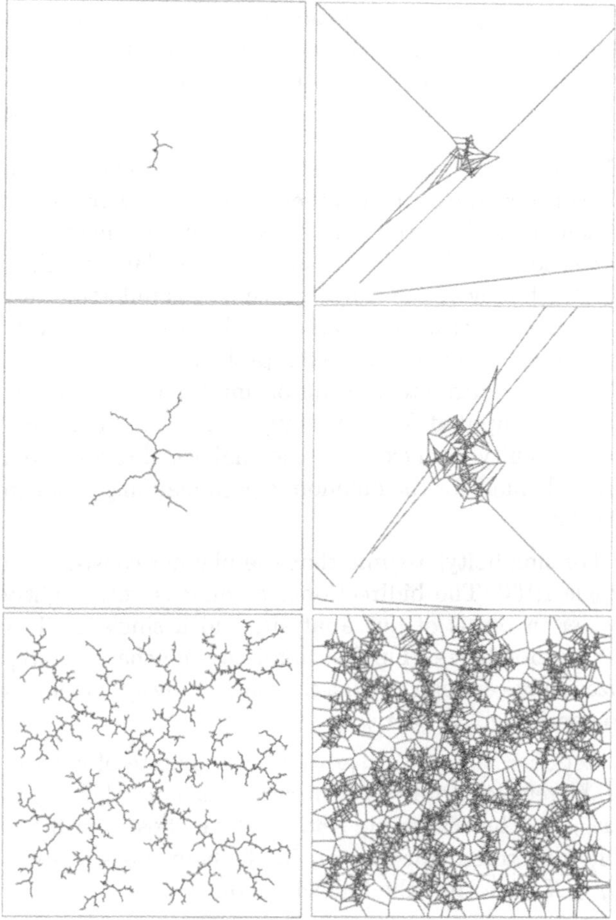

Figure 3: *An RRT is biased by large Voronoi regions to rapidly explore, before uniformly covering the space.*

This biases the RRT to rapidly explore. In Section 4 it is shown that RRTs also arrive at a uniform coverage of the space.

4 Analysis of RRTs

This section provides some analysis of RRTs, and indicates several open problems for future investigation. A key result shown so far is that the distribution of the RRT vertices converges to the sampling distribution, which is usually uniform. This currently has been shown for holonomic planning in a nonconvex state space. We have also verified the results through simulations and chi-square tests. We have generally had many experimental successes, indicated in Section 6, that far exceed our current analysis capabilities. Considerable effort remains to close the gap between our experimental success, and the analysis that supports the success.

The limiting distribution of vertices Let $D_k(x)$ denote a random variable whose value is the distance of x to the closest vertex in G, in which k is the number of vertices in an RRT. Let d_k denote the value of D_k. Let ϵ denote the incremental distance traveled in the EXTEND procedure (the RRT step size).

Consider the case of a holonomic planning problem, in which $\dot{x} = u$ (the incremental simulator permits motion in any direction). The first lemma establishes that the RRT will (converging in probability) come arbitrarily close to any point in a convex space.

Lemma 1 *Suppose X_{free} is a convex, bounded, open, n-dimensional subset of an n-dimensional state space. For any $x \in X_{free}$ and positive constant $\epsilon > 0$, $\lim_{k\to\infty} P[d_k(x) < \epsilon] = 1$.*

The proofs for all propositions, except Theorems 5-7, appear in [22].

The next lemma extends the result from convex spaces to nonconvex spaces.

Lemma 2 *Suppose X_{free} is a nonconvex, bounded, open, n-dimensional connected component of an n-dimensional state space. For any $x \in X_{free}$ and positive real number $\epsilon > 0$, then $\lim_{n\to\infty} P[d_n(x) < \epsilon] = 1$.*

For holonomic path planning, this immediately implies the following:

Theorem 3 *Suppose x_{init} and x_{goal} lie in the same connected component of a nonconvex, bounded, open, n-dimensional connected component of an n-dimensional state space. The probability that an RRT constructed from x_{init} will find a path to x_{goal} approaches one as the number of RRT vertices approaches infinity.*

This establishes probabilistic completeness, as considered in [19], of the basic RRT.

The next step is to characterize the limiting distribution of the RRT vertices. Let $\mathbf{X}$ denote a vector-valued random variable that represents the sampling process used to construct an RRT. This reflects the distribution of samples that are returned by the RANDOM_STATE function in the EXTEND algorithm. Usually, $\mathbf{X}$ is characterized by a uniform probability density function over X_{free}; however, we will allow $\mathbf{X}$ to be characterized by any continuous probability density function. Let $\mathbf{X}_k$ denote a vector-valued random variable that represents the distribution of the RRT vertices after k iterations.

Theorem 4 *$\mathbf{X}_k$ converges to $\mathbf{X}$ in probability.*

We now consider the more general case. Suppose that motions obtained from the incremental simulator are locally constrained. For example, they might arise by integrating $\dot{x} = f(x, u)$ over some time Δt. Suppose that the number of inputs to the incremental simulator is finite, Δt is constant, no two RRT vertices lie within a specified $\epsilon > 0$ of each other according to ρ, and that EXTEND chooses the input at random. It may be possible eventually to remove some of these restrictions; however, we have not yet pursued this route. Suppose x_{init} and x_{goal} lie in the same connected component of a nonconvex, bounded, open, n-dimensional connected component of an n-dimensional state space. In addition, there exists a sequence of inputs, u_1, u_2, $\ldots$, u_k, that when applied to x_{init} yield a sequence of states, $x_{init} = x_0$, x_1, x_2, $\ldots$, $x_{k+1} = x_{goal}$. All of these states lie in the same open connected component of X_{free}.

The following establishes the probabilistic completeness of the nonholonomic planner.

Theorem 5 *The probability that the RRT initialized at x_{init} will contain x_{goal} as a vertex approaches one as the number of vertices approaches infinity.*

Overview of Proof: The argument proceeds by induction on i. Assume that the RRT contains x_i as a vertex after some finite number of iterations. Consider the Voronoi diagram associated with the RRT vertices. There exists a positive real number, c_1, such that $\mu(Vor(x_i)) > c_1$ in which $Vor(x_i)$ denotes the Voronoi region associated with x_i. If a random sample falls within $Vor(x_i)$, the vertex will be selected for extension, and a random input is applied; thus, x_i has probability $\mu(Vor(x_i))/\mu(X_{free})$ of being selected. There exists a second positive real number, c_2 (which depends on c_1), such that the probability that the correct input, u_i, is selected is at least c_2. If both x_i and u_i have probability of at least c_2 of being selected in each iteration, then the probability tends to one that the next step in the solution trajectory will be constructed. This argument is applied inductively from x_1 to x_k, until the final state $x_{goal} = x_{k+1}$ is reached. ■

Convergence Rate Theorems 6 and 7 express the rate of converge of the planner. These results represent a significant first step towards gaining a complete understanding of behavior of RRT-based planning algorithms; however, the convergence rate is unfortunately expressed in terms of parameters that cannot be easily measured. It remains an open problem to characterize the convergence rate in terms of simple parameters that can be computed for a particular problem. This general difficulty even exists in the analysis of randomized path planners for the holonomic path planning problem [18, 23].

For simplicity, assume that the planner consists of a single RRT. The bidirectional planner is only slightly better in terms of our analysis, and a single RRT is easier to analyze. Furthermore, assume that the step size is large enough so that the planner always attempts to connect x_{near} to x_{rand}.

Let $\mathcal{A} = \{A_0, A_1, \ldots, A_k\}$ be a sequence of subsets of $\mathbf{X}$, referred to as an *attraction sequence*. Let $A_0 = \{x_{init}\}$. The remaining sets must be chosen with the following rules. For each A_i in $\mathcal{A}$, there exists a *basin*, $B_i \subseteq \mathbf{X}$ such that the following hold:

1. For all $x \in A_{i-1}$, $y \in A_i$, and $z \in \mathbf{X} \setminus B_i$, the metric, ρ, yields $\rho(x, y) < \rho(x, z)$.
2. For all $x \in B_i$, there exists an m such that the sequence of inputs $\{u_1, u_2, \ldots, u_m\}$ selected by the EXTEND algorithm will bring the state into $A_i \subseteq B_i$.

Finally, it is assumed that $A_k = X_{goal}$.

Each basin B_i can intuitively be considered as both a safety zone that ensures an element of B_i will be selected by the nearest neighbor query, and a potential well that attracts the state into A_i. An attraction sequence should be chosen with each A_i as large as possible and with k as small as possible. If the space contains narrow corridors, then the attraction sequence will be longer and each A_i will be smaller. Our analysis indicates that the planning performance will significantly degrade in this case, which is consistent with analysis results obtained for randomized holonomic planners [17]. Note that for kinodynamic planning, the choice of metric, ρ, can also greatly affect the attraction sequence, and ultimately the performance of the algorithm.

Using μ to represent measure, let p be defined as:

$$p = \min_i \{\mu(A_i)/\mu(\mathbf{X}_{free})\},$$

which corresponds to a lower bound on the probability that a random state will lie in a particular A_i.

The following theorem characterizes the expected number of iterations required to find a solution.

Theorem 6 *If a connection sequence of length k exists, then the expected number of iterations required to connect q_{init} to q_{goal} is no more than k/p.*

Proof: If an RRT vertex lies in A_{i-1}, and a random sample, x, falls in A_i, then the RRT will be connected to x. This is true because using the first property in the definition of a basin, it follows that one of the vertices in B_i must be selected for extension. Using the second property of the basin, inputs will be chosen that ultimately generate a vertex in A_i.

In each iteration, the probability that the random sample lies in A_i is at least p; hence, if A_{i-1} contains an RRT vertex, then A_i will contain a vertex with probability at least p. In the worst-case, the iterations can be considered as Bernoulli trials in which p is the probability of a successful outcome. A path planning problem is solved after k successful outcomes are obtained because each success extends the progress of the RRT from A_{i-1} to A_i.

Let $C_1, C_2, \ldots, C_n$ be i.i.d. random variables whose common distribution is the Bernoulli distribution with parameter p. The random variable $C = C_1 + C_2 + \cdots + C_n$ denotes the number of successes after n iterations. Since each C_i has the Bernoulli distribution, C will have a binomial distribution,

$$\binom{n}{k} h^k (1-h)^{n-k},$$

in which k is the number of successes. The expectation of the binomial distribution is n/p, which also represents an upper bound on the expected probability of successfully finding a path. ∎

The following theorem establishes that the probability of success increases exponentially with the number of iterations.

Theorem 7 *If an attraction sequence of length k exists, for a constant $\delta \in (0,1]$, the probability that the RRT finds a path after n iterations is at least $1 - exp(-np\delta^2/2)$, in which $\delta = 1 - k/(np)$.*

Proof: The random variable C from the proof of Theorem 6 has a binomial distribution, which enables the application of a Chernoff-type bound on its tail probabilities. The following theorem [39] is directly applied to establish the theorem. If C is binomially distributed, $\delta \in (0,1]$, and $\mu = E[C]$, then $P[C \leq (1-\delta)\mu] < exp(\mu\delta^2/2)$. ∎

An RRT in a Large Disc In the limit as the number of iterations approaches infinity, the RRT becomes uniformly distributed, but what happens when the RRT is placed in a "large" space? Intuitively, it seems that the best strategy would be to grow the tree away from the initial state as quickly as possible. To determine whether this occurs, we performed many simulation experiments (each with hundreds of thousands of iterations) to characterize how an RRT grows in the limit case of a disc with a radius that approaches infinity. Consider the case of a 2D state space and holonomic planning. Figure 4(a) shows a typical result, in which the RRT has three major branches, each roughly 120 degrees apart. This behavior was repeatedly observed for the 2D case, although the orientation of the branches is random. In higher dimensions, we have observed that the RRT makes $n+1$ branches in an n-dimensional space. The branches also have equal separation from each other (they appear to touch the vertices of a regular $(n+1)$-simplex, centered at the origin. This gives experimental evidence that in the expected sense, the RRT grows outward from the origin at a rate that is linear in the number of iterations, and decreases moderately with the number of dimensions. It remains an open question to confirm these observations by proving the number and directions of these branches.

Relationship to Optimality Another observation that we have made through simulation experiments is that the paths in a holonomic RRT, while jagged, are not too far from the shortest path (recall Figure 3). This is not true for paths generated by a simpler technique, such as Brownian motion. For paths in the plane, we have performed repeated experiments that compare the distance of randomly-chosen RRT vertices

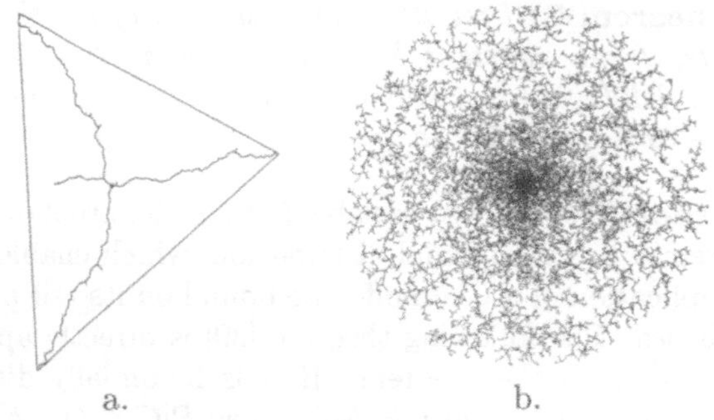

Figure 4: *a) The convex hull of an RRT in an "infinitely" large disc; b) a 2D RRT that was constructed using biased sampling.*

to the root by following the RRT path to the Euclidean distance to the root. Experiments were performed in a square region in the plane. The expected ratio of RRT-path distance to Euclidean distance is consistently between 1.3 and 1.7. It remains an open question to prove an upper bound on the expected path length, with respect to optimal solutions (e.g., such as a ratio bound of two).

5 Designing Path Planners

Sections 3 and 4 introduced the basic RRT and analyzed its exploration properties. Now the focus is on developing path planners using RRTs. We generally consider the RRT as a building block that can be used to construct an efficient planner, as opposed to a path planning algorithm by itself. For example, one might use an RRT to escape local minima in a randomized potential field path planner [3]. In [48], an RRT was used as the local planner for the probabilistic roadmap planner. We present several alternative RRT-based planners in this section. The recommended choice depends on several factors, such as whether differential constraints exist, the type of collision detection algorithm, or the efficiency of nearest neighbor computations.

Single-RRT Planners In principle, the basic RRT can be used in isolation as a path planner because its vertices will eventually cover a connected component of X_{free}, coming arbitrarily close to any specified x_{goal}. The problem is that without any bias toward the goal, convergence is often slow. An improved planner, called RRT-GoalBias, can be obtained by replacing RANDOM_STATE in Figure 2 with a function that tosses a biased coin to determine what should be returned. If the coin toss yields "heads", then x_{goal} is returned; otherwise, a random state is returned. Even with a small probability of returning heads (such as 0.05), RRT-GoalBias usually converges to the goal much faster than the basic RRT. If too much bias is introduced; however, the planner begins to behave like a randomized potential field planner that is trapped in a local minimum. An improvement called RRT-GoalZoom replaces RANDOM_STATE with a decision, based on a biased coin toss, that chooses a random sample from either a region around the goal or the whole state space. The size of the region around the goal is controlled by the closest RRT vertex to the goal at any iteration. The effect is that the focus of samples gradually increases around the goal as the RRT draws nearer. This planner has performed quite well in practice; however, it is still possible that performance is degraded due to local minima. In general, it seems best to replace RANDOM_STATE with a sampling scheme that draws states from a nonuniform probability density function that has a "gradual" bias toward the goal. Figure 4(b) shows an example of an RRT that was constructed by sampling states from a probability density that assigns equal probability to concentric circular rings. There are still many interesting research issues regarding the problem of sampling. It might be possible to use some of the sampling methods that were proposed to improve the performance of probabilistic roadmaps [1, 7].

One more issue to consider is the size of the step that is used for RRT construction. This could be chosen dynamically during execution on the basis of a distance computation function that is used for collision detection. If the bodies are far from colliding, then larger steps can be taken. Aside from following this idea to obtain an incremental step, how far should the new state, x_{new} appear from x_{near}? Should we try to connect x_{near} to x_{rand}? Instead of attempting to extend an RRT by an incremental step, EXTEND can be iterated until the random state or an obstacle is reached, as shown in the CONNECT algorithm description in Figure 5. CONNECT can replace EXTEND, yielding an RRT that grows very quickly, if permitted by collision detection constraints and the differential constraints. One of the key advantages of the CONNECT function is that a long path can be constructed with

```
CONNECT(T, x)
  repeat
      S ← EXTEND(T, x);
  until not (S = Advanced)
  Return S;
```

Figure 5: *The CONNECT function.*

only a single call to the nearest-neighbor algorithm. This advantage motivates the choice of a greedier algorithm; however, if an efficient nearest-neighbor algorithm (e.g., [2]) is used, as opposed to the obvious linear-time method, then it might make sense to be less greedy. After performing dozens of experiments on a variety of problems, we have found CONNECT to yield the best performance for holonomic planning problems, and EXTEND seems to be the best for nonholonomic problems. One reason for this difference is that CONNECT places more faith in the metric, and for nonholonomic problems it becomes more challenging to design good metrics.

Bidirectional Planners Inspired by classical bidirectional search techniques [41], it seems reasonable to expect that improved performance can be obtained by growing two RRTs, one from x_{init} and the other from x_{goal}; a solution is found if the two RRTs meet. For a simple grid search, it is straightforward to implement a bidirectional search; however, RRT construction must be biased to ensure that the trees meet well before covering the entire space, and to allow efficient detection of meeting.

Figure 5 shows the RRT_BIDIRECTIONAL algorithm, which may be compared to the BUILD_RRT algorithm of Figure 1. RRT_BIDIRECTIONAL divides the computation time between two processes: 1) exploring the state space; 2) trying to grow the trees into each other. Two trees, $\mathcal{T}_a$ and $\mathcal{T}_b$ are maintained at

```
RRT_BIDIRECTIONAL(x_init, x_goal)
1  T_a.init(x_init); T_b.init(x_goal);
2  for k = 1 to K do
3      x_rand ← RANDOM_STATE();
4      if not (EXTEND(T_a, x_rand) = Trapped) then
5          if (EXTEND(T_b, x_new) = Reached) then
6              Return PATH(T_a, T_b);
7      SWAP(T_a, T_b);
8  Return Failure
```

Figure 6: *A bidirectional RRT-based planner.*

all times until they become connected and a solution is found. In each iteration, one tree is extended, and an attempt is made to connect the nearest vertex of the other tree to the new vertex. Then, the roles are reversed by swapping the two trees. Growth of two RRTs was also proposed in [30] for kinodynamic planning; however, in each iteration both trees were incrementally extended toward a random state. The current algorithm attempts to grow the trees into each other half of the time, which has been found to yield much better performance.

Several variations of the above planner can also be considered. Either occurrence of EXTEND may be replaced by CONNECT in RRT_BIDIRECTIONAL. Each replacement makes the operation more aggressive. If the EXTEND in Line 4 is replaced with CONNECT, then the planner aggressively explores the state space, with the same tradeoffs that existed for the single-RRT planner. If the EXTEND in Line 5 is replaced with CONNECT, the planner aggressively attempts to connect the two trees in each iteration. This particular variant was very successful at solving holonomic planning problems. For convenience, we refer to this variant as RRT-ExtCon, and the original bidirectional algorithm as RRT-ExtExt. Among the variants discussed thus far, we have found RRT-ExtCon to be most successful for holonomic planning [22], and RRT-ExtExt to be best for nonholonomic problems. The most aggressive planner can be constructed by replacing EXTEND with CONNECT in both Lines 4 and 5, to yield RRT-ConCon.

Through extensive experimentation over a wide variety of examples, we have concluded that, when applicable, the bidirectional approach is much more efficient than a single RRT approach. One shortcoming of using the bidirectional approach for nonholonomic and kinodynamic planning problems is the need to make a connection between a pair of vertices, one from each RRT. For a planning problem that involves reaching a goal region from an initial state, no connections are necessary using a single-RRT approach. The gaps between the two trajectories can be closed in practice by applying steering methods [27], if possible, or classical shooting methods, which are often used for numerical boundary value problems.

Other Approaches If a dual-tree approach offers advantages over a single tree, then it is natural to ask whether growing three or more RRTs might be even better. These additional RRTs could be started at random states. Of course, the connection problem will become more difficult for nonholonomic problems. Also, as more trees are considered, a complicated decision problem arises. The computation time must be divided between attempting to explore the space and attempting to connect RRTs to each other. It is also not clear which connections should be attempted. Many research issues remain in the development of this and other RRT-based planners.

It is interesting to consider the limiting case in which a new RRT is started for every random sample, x_{rand}. Once the single-vertex RRT is generated, the CONNECT function from Figure 5 can be applied to every other RRT. To improve performance, one might only consider connections to vertices that are within a fixed distance of x_{rand}, according to the metric. If a connection succeeds, then the two RRTs are merged into a single graph. The resulting algorithm simulates the behavior of the probabilistic roadmap approach to path planning [19]. Thus, the probabilistic roadmap can be considered as an extreme version of an RRT-based algorithm in which a maximum number of separate RRTs are constructed and merged.

6 Implementations and Experiments

In this section, results for four different types of problems are summarized: 1) holonomic planning, 2) nonholonomic planning, 3) kinodynamic planning, and 4) planning for systems with closed kinematic chains. Presently, we have constructed two planning systems based on RRTs. One is written in Gnu C++ and LEDA, and experiments were conducted on a 500Mhz Pentium III PC running Linux. This implementation is very general, allowing many planning variants and models to be considered; however, it is limited to planar obstacles and robots, and performs naive collision detection. The software can be obtained from *http://janowiec.cs.iastate.edu/~lavalle/rrt/*. The second implementation is written in SGI C++ and SGI's OpenInventor library, and experiments were conducted on a 200MHz SGI Indigo2 with 128MB. This implementation considers 3D models, and was particularly designed inclusion in a software platform for automating the motions of digital actors [21]. Currently, we are constructing a third implementation, which is expected to be general-purpose, support 3D models, and be based on freely-available collision detection and efficient nearest neighbor libraries.

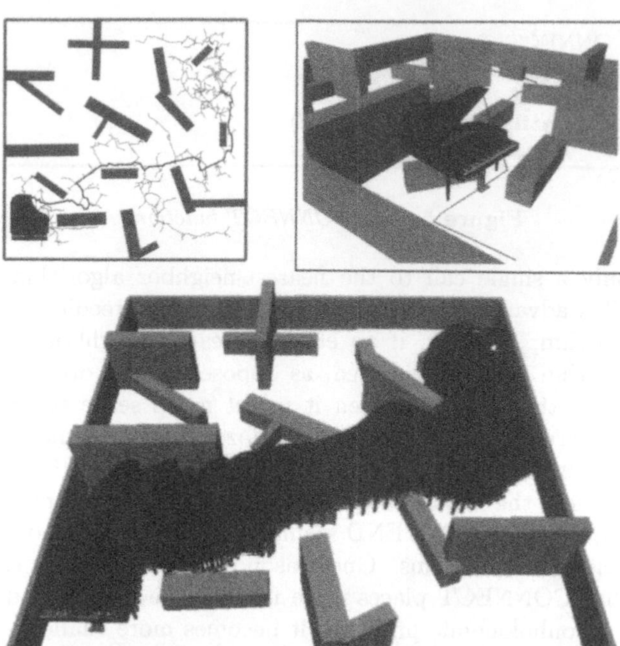

Figure 7: *Moving a Piano.*

Holonomic planning experiments Through numerous experiments, we have found RRT-based planners to be very efficient for holonomic planning. Note that our planners attempt to find a solution without performing precomputations over the entire state space, and are therefore suited for single-query path planning problems, in contrast to the probabilistic roadmap method. It is easy to construct single-query examples on which an RRT-based planner will be superior by terminating before covering the entire state space, and it is easy to construct multiple-query problems in which the probabilistic roadmap method will be superior by repeatedly using its precomputed roadmap.

Most of the experiments in this section were conducted on the 200MHz SGI Indigo2. More holonomic planning experiments are presented in [22]. The CONNECT function is most effective when one can expect relatively open spaces for the majority of the planning queries. We first performed hundreds of experiments on over a dozen examples for planning the motions of rigid objects in 2D, resulting in 2D and 3D configu-

Figure 8: *Playing a game of virtual chess.*

ration spaces. Path smoothing was performed on the final paths to reduce jaggedness. Figure 7 depicts a computed solution for a 3D model of a grand piano moving from one room to another amidst walls and low obstacles. Several tricky rotations are required of the piano in order to solve this query.

Figure 8 shows a human character playing chess. Each of the motions necessary to reach, grasp, and reposition a game piece on the virtual chess board were generated using the RRT-ExtCon planner in an average of 2 seconds on the 200 MHz SGI Inidigo2. The human arm is modeled as a 7-DOF kinematic chain, and the entire scene contains over 8,000 triangle primitives. The 3D collision checking software used for these experiments was the RAPID library based on OBB-Trees developed by the University of North Carolina [32]. The speed of the planner allows for the user to interact with the character in real-time, and even engage in an interactive game of "virtual chess".

The final holonomic planning example, shown in Figure 9, was solved using the RRT-ExtExt planner. This problem was presented in [7] as a test challenge for randomized path planners due to the narrow passages that exist in the configuration space when the "U"-shaped object passes through the center of the world. In the example shown, the RRT does not explore to much of the surrounding space (some of this might be due to the lucky placement of the corridor in the center of the world). On average, about 1500 nodes are generated,

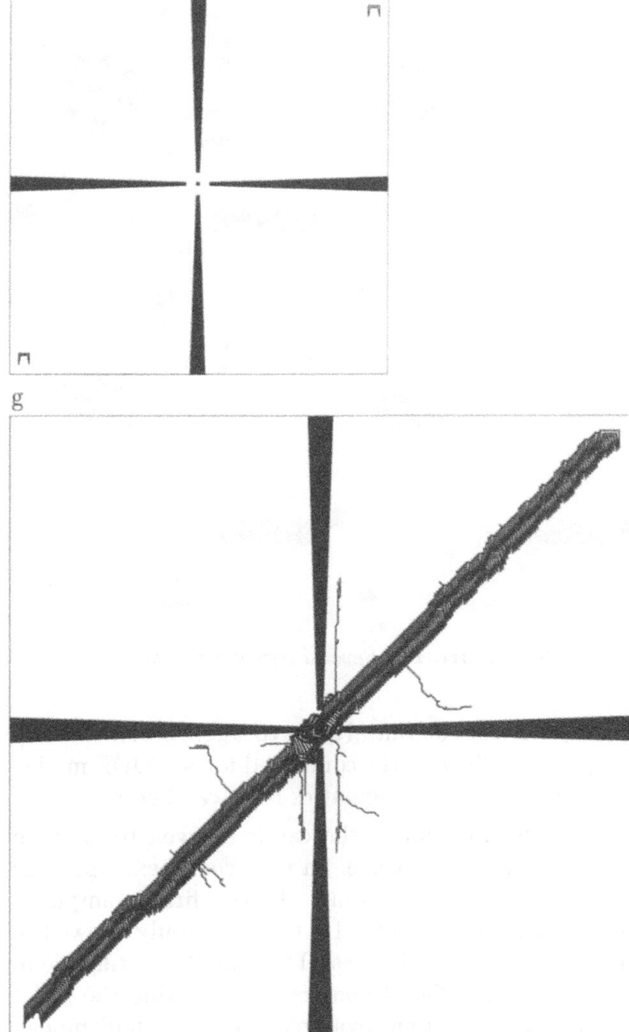

Figure 9: *A narrow-corridor example.*

and the problem is solved in two seconds on the PC using naive collision checking.

Nonholonomic planning experiments Several nonholonomic planning examples are shown in Figures 10 and 11. These examples were computed using the RRTExtExt planner, and the average computation times were less than five seconds on the PC using naive collision detection. The four examples in Figure 10 involve car-like robots that moves at constant speed under different nonholonomic models. A 2D projection of the RRTs is shown for each case, along with the com-

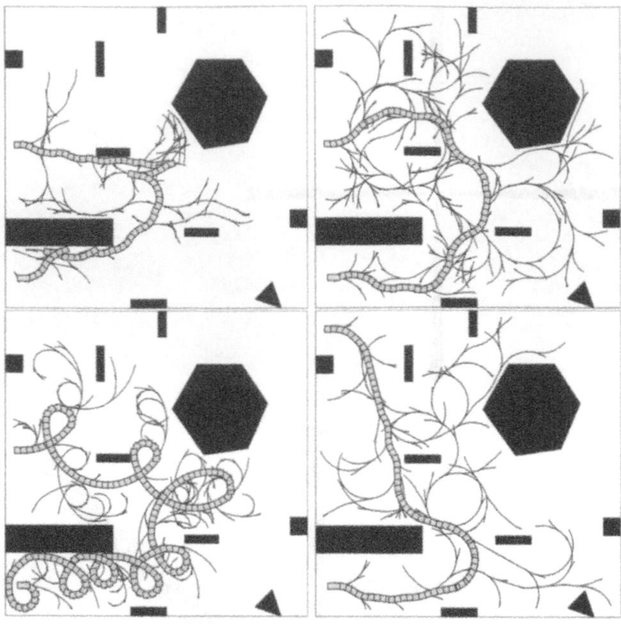

Figure 10: *Several car-like robots.*

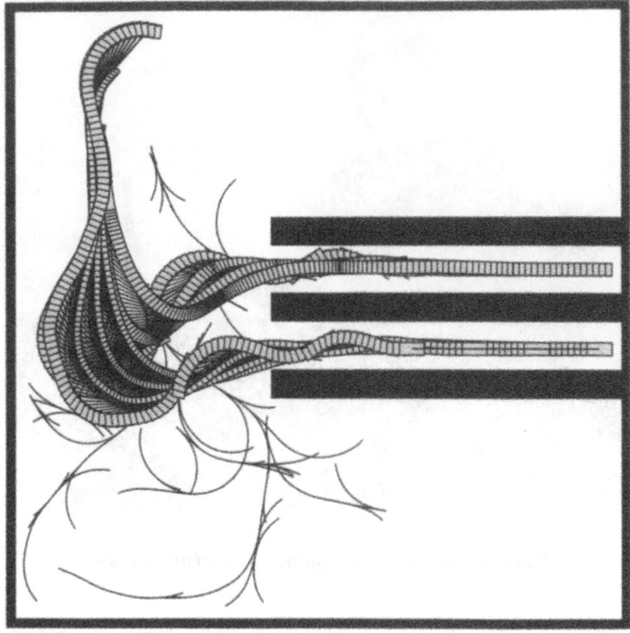

Figure 11: *A 7-DOF nonholonomic planning example.*

puted path from an initial state to a goal state. The top two pictures show paths computed for a 3-DOF model, using the standard kinematics for a car-like robot.

In the first example, the car is allowed to move in both forward and reverse. In the second example, the car can move forward only. In the first example in the second row in Figure 10, the car is only allowed to turn left in varying degrees! The planner is still able to overcome this difficult constraint and bring the robot to its goal. The final example uses a 4-DOF model, which results in continuous curvature paths [43].

The final nonholonomic planning problem involves the 4-DOF car pulling three trailers, resulting in a 7-DOF system. The kinematics are given in [40]. The goal is to pull the car with trailers out of one stall, and back it into another. The RRTs shown correspond to one of the best executions; in other iterations the exploration was much slower due to metric problems.

Kinodynamic planning experiments Several kinodynamic planning experiments have been performed for both non-rotating and rotating rigid objects in 2D and 3D worlds with velocity and acceleration bounds obeying L^2 norms. For the 2D case, controllability issues were studied recently in [34]. All experiments utilized a simple weighted L^2 metric on X, and were performed using variants of the RRT_BIDIRECTIONAL planner. More experiments were presented in [30]. Recently, RRTs have been applied to automating the flight of helicopters in complicated 3D simulations that contain obstacles [15].

Consider the case of a fully-orientable satellite model with limited translation. The satellite is assumed to have momentum wheels that enable it to orient itself along any axis, and a single pair of opposing thruster controls that allow it to translate along the primary axis of the cylinder. This model has a 12-dimensional state space. The task of the satellite, modeled as a rigid cylindrical object, is to perform a collision-free docking maneuver into the cargo bay of the space shuttle model amidst a cloud of obstacles. Figure 12 shows the candidate solution found after 23,800 states were explored. The total computation time was 8.4 minutes on the SGI.

The final result, given in Figure 13, shows a fictitious, underactuated spacecraft that must maneuver through two narrow gates and enter a hangar that has a small entrance. There are five inputs, each of which applies a thrust impulse. The possible motions are: 1) forward, 2) up, 3) down, 4) clockwise roll, 5) counterclockwise roll. There are 4000 triangles in the model. Path plan-

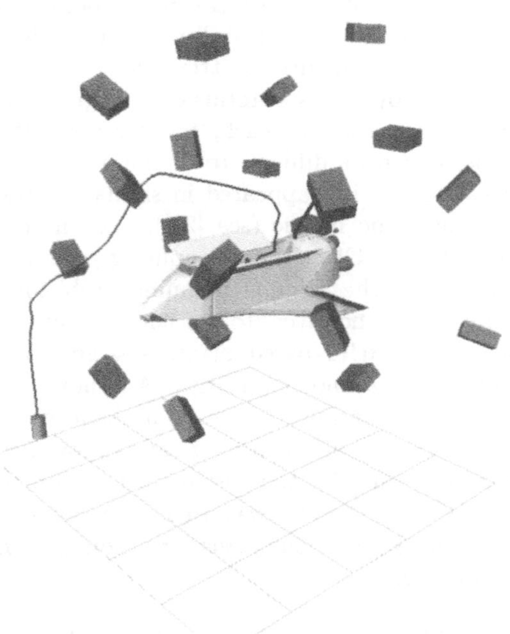

Figure 12: *A 12-DOF kinodynamic planning example.*

ning is performed directly in the 12-dimensional state space. A typical run requires about 12 minutes on an R12000.

Planning for closed kinematic chains Figure 14 shows a problem that involves a kinematic closure constraint that must be maintained in addition to performing holonomic path planning. Many more examples and experiments are discussed in [49]. In the initial state, the closure constraint is satisfied. The incremental simulator performs local motions that maintain with closure constraint within a specified tolerance.

7 Discussion

We have presented a general framework for developing randomized path planning algorithms based on the concept of Rapidly-exploring Random Trees (RRTs). After extensive experimentation, we are satisfied with the results obtained to date. There is, however, significant room for improvement given the complexity of problems that arise in many applications. To date, we believe we have presented the first randomized path planning techniques that are particularly designed for

Figure 13: *An underactuated spacecraft that performs complicated maneuvers. The state space has twelve dimensions, and there are five inputs.*

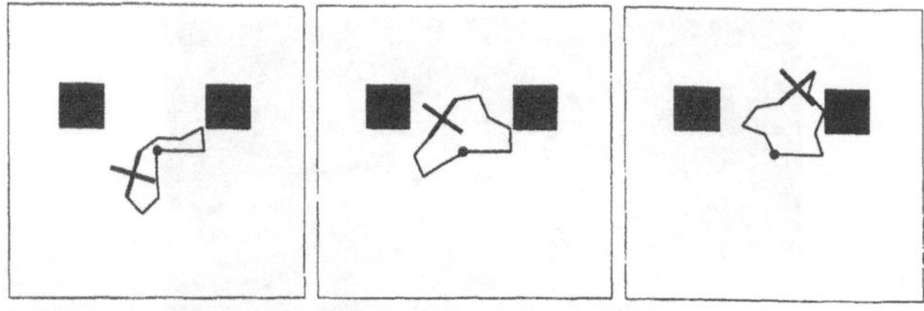

Figure 14: *Two manipulators transport a cross-shaped object while maintaining kinematic closure.*

handling differential constraints (without necessarily requiring steering ability). RRTs have also led to very efficient planners for single-query holonomic path planning. Several issues and topics are mentioned below, which are under current investigation.

Designing Metrics The primary drawback with the RRT-based methods is the sensitivity of the performance on the choice of the metric, ρ. All of the results presented in Section 6 were obtained by assigning a simple, weighted Euclidean metric for each model (the same metric was used for different collections of obstacles). Nevertheless, we observed that the computation time varies dramatically for some problems as the metric is varied. This behavior warrants careful investigation into the effects of metrics. This problem might seem similar to the choice of a potential function for the randomized potential field planer; however, since RRTs eventually perform uniform exploration, the performance degradation is generally not as severe as a local minimum problem. Metrics that would fail miserably as a potential function could still yield good performance in an RRT-based planner.

In general, we can characterize the ideal choice of a metric (technically this should be called a pseudometric due to the violation of some metric properties). Consider a cost or loss functional, L, defined as:

$$L = \int_0^T l(x(t), u(t))dt + l_f(x(T)).$$

As examples, this could correspond to the distance traveled, the energy consumed, or the time elapsed during the execution of a trajectory. The optimal cost to go from x to x' can be expressed as

$$\rho^*(x, x') = \min_{u(t)} \left\{ \int_0^T l(x(t), u(t))dt + l_f(x(T)) \right\}.$$

Ideally, ρ^* would make an ideal metric because it indicates "closeness" as the ability to bring the state from x to x' while incurring little cost. For holonomic planning, nearby states in terms of a weighted Euclidean metric are easy to reach, but for nonholonomic problems, it can be difficult to design a good metric. The ideal metric has appeared in similar contexts as the nonholonomic metric (see [27]), and the cost-to-go function [5]. Of course, computing ρ^* is as difficult as solving the original planning problem! It is generally useful, however, to consider ρ^* because the performance of RRT-based planners seems to generally degrade as ρ and ρ^* diverge. An effort to make a crude approximation to ρ^*, even if obstacles are neglected, will probably lead to great improvements in performance. In [15], the cost-to-go function from a hybrid optimal controller was used as the metric in an RRT to generate efficient plans for a nonlinear model of a helicopter.

Efficient Nearest-Neighbors One of the key bottlenecks in construction of RRTs so far has been nearest neighbor computations. To date, we have only implemented the naive approach in which every vertex is compared to the sample state. Fortunately, the development of efficient nearest-neighbor for high-dimensional problems has been a topic of active interest in recent years (e.g., [2]). Techniques exist that can compute nearest neighbors (or approximate nearest-neighbors) in near-logarithmic time in the number of vertices, as opposed to the naive method which takes linear time. Implementation and experimentation with nearest neighbor techniques is expected to dramatically improve performance. Three additional concerns must be addressed: 1) any data structure that is used for efficient nearest neighbors must allow incremental insertions to be made efficiently due to the incremental construction of an RRT, and 2) the method must support whatever metric, ρ, is chosen, and 3) simple adaptations must be made to account for the topology of the state space (especially in the case of S^1 and P^3, which arise from rotations).

Collision Detection For collision detection in our previous implementations, we have not yet exploited the fact that RRTs are based on incremental motions. Given that small changes usually occur between configurations, a data structure can be used that dramatically improves the performance of collision detection and distance computation [16, 31, 38, 42]. The incorporation of such approaches into our RRT-based planners should cause dramatic performance benefits.

Acknowledgments

This work has benefitted greatly from discussions with Nancy Amato, Jim Bernard, Francesco Bullo, Peng Cheng, Bruce Donald, Yan-Bin Jia, Lydia Kavraki, Jean-Claude Latombe, Jean-Paul Laumond, Kevin Lynch, Ahmad Masoud, and Jeff Yakey. The authors thank Valerie Boor for supplying the problem shown in Figure 9. LaValle is supported in part by NSF CAREER Award IRI-9875304 and Honda. Kuffner is supported in part by a jointly-funded NSF-JSPS (Japan) Post-Doctoral Fellowship.

References

[1] N. M. Amato and Y. Wu. A randomized roadmap method for path and manipulation planning. In *IEEE Int. Conf. Robot. & Autom.*, pages 113–120, 1996.

[2] S. Arya, D. M. Mount, N. S. Netanyahu, R. Silverman, and A. Y. Wu. An optimal algorithm for approximate nearest neighbor searching. *Journal of the ACM*, 45:891–923, 1998.

[3] J. Barraquand, B. Langlois, and J. C. Latombe. Numerical potential field techniques for robot path planning. *IEEE Trans. Syst., Man, Cybern.*, 22(2):224–241, 1992.

[4] J. Barraquand and J.-C. Latombe. Nonholonomic multibody mobile robots: Controllability and motion planning in the presence of obstacles. In *IEEE Int. Conf. Robot. & Autom.*, pages 2328–2335, 1991.

[5] R. E. Bellman. *Dynamic Programming.* Princeton University Press, Princeton, NJ, 1957.

[6] J. E. Bobrow, S. Dubowsky, and J. S. Gibson. Time-optimal control of robotic manipulators along specified paths. *Int. J. Robot. Res.*, 4(3):3–17, 1985.

[7] V. Boor, N. H. Overmars, and A. F. van der Stappen. The gaussian sampling strategy for probabilistic roadmap planners. In *IEEE Int. Conf. Robot. & Autom.*, pages 1018–1023, 1999.

[8] J. Canny, A. Rege, and J. Reif. An exact algorithm for kinodynamic planning in the plane. *Discrete and Computational Geometry*, 6:461–484, 1991.

[9] H. Chang and T. Y. Li. Assembly maintainability study with motion planning. In *IEEE Int. Conf. Robot. & Autom.*, pages 1012–1019, 1995.

[10] M. Cherif. Kinodynamic motion planning for all-terrain wheeled vehicles. In *IEEE Int. Conf. Robot. & Autom.*, 1999.

[11] C. Connolly, R. Grupen, and K. Souccar. A Hamiltonian framework for kinodynamic planning. In *Proc. of the IEEE International Conf. on Robotics and Automation (ICRA'95)*, Nagoya, Japan, 1995.

[12] B. Donald and P. Xavier. Provably good approximation algorithms for optimal kinodynamic planning: Robots with decoupled dynamics bounds. *Algorithmica*, 14(6):443–479, 1995.

[13] B. R. Donald, P. G. Xavier, J. Canny, and J. Reif. Kinodynamic planning. *Journal of the ACM*, 40:1048–66, November 1993.

[14] Th. Fraichard and C. Laugier. Kinodynamic planning in a structured and time-varying 2d workspace. In *IEEE Int. Conf. Robot. & Autom.*, pages 2: 1500–1505, 1992.

[15] E. Frazzoli, M. A. Dahleh, and E. Feron. Robust hybrid control for autonomous vehicles motion planning. Technical Report LIDS-P-2468, Laboratory for Information and Decision Systems, Massachusetts Institute of Technology, 1999.

[16] L. J. Guibas, D. Hsu, and L. Zhang. H-Walk: Hierarchical distance computation for moving convex bodies. In *Proc. ACM Symposium on Computational Geometry*, pages 265–273, 1999.

[17] D. Hsu, L. E. Kavraki, J.-C. Latombe, R. Motwani, and S. Sorkin. On finding narrow passages with probabilistic roadmap planners. In et al. P. Agarwal, editor, *Robotics: The Algorithmic Perspective*, pages 141–154. A.K. Peters, Wellesley, MA, 1998.

[18] D. Hsu, J.-C. Latombe, and R. Motwani. Path planning in expansive configuration spaces. *Int. J. Comput. Geom. & Appl.*, 4:495–512, 1999.

[19] L. E. Kavraki, P. Svestka, J.-C. Latombe, and M. H. Overmars. Probabilistic roadmaps for path planning in high-dimensional configuration spaces. *IEEE Trans. Robot. & Autom.*, 12(4):566–580, June 1996.

[20] R. Kindel, D. Hsu, J.-C. Latombe, and S. Rock. Kinodynamic motion planning amidst moving obstacles. In *IEEE Int. Conf. Robot. & Autom.*, 2000.

[21] J. J. Kuffner. *Autonomous Agents for Real-time Animation.* PhD thesis, Stanford University, 1999.

[22] J. J. Kuffner and S. M. LaValle. RRT-connect: An efficient approach to single-query path planning. In *Proc. IEEE Int'l Conf. on Robotics and Automation*, 2000.

[23] F. Lamiraux and J.-P. Laumond. On the expected complexity of random path planning. In *IEEE Int. Conf. Robot. & Autom.*, pages 3306–3311, 1996.

[24] R. E. Larson. A survey of dynamic programming computational procedures. *IEEE Trans. Autom. Control*, 12(6):767–774, December 1967.

[25] J.-C. Latombe. *Robot Motion Planning*. Kluwer Academic Publishers, Boston, MA, 1991.

[26] J.-P. Laumond. Finding collision-free smooth trajectories for a non-holonomic mobile robot. In *Proc. Int. Joint Conf. on Artif. Intell.*, pages 1120–1123, 1987.

[27] J. P. Laumond, S. Sekhavat, and F. Lamiraux. Guidelines in nonholonomic motion planning for mobile robots. In J.-P. Laumond, editor, *Robot Motion Plannning and Control*, pages 1–53. Springer-Verlag, Berlin, 1998.

[28] S. M. LaValle. Numerical computation of optimal navigation functions on a simplicial complex. In P. Agarwal, L. Kavraki, and M. Mason, editors, *Robotics: The Algorithmic Perspective*. A K Peters, Wellesley, MA, 1998.

[29] S. M. LaValle. Rapidly-exploring random trees: A new tool for path planning. TR 98-11, Computer Science Dept., Iowa State University. <http://janowiec.cs.iastate.edu/papers/rrt.ps>, Oct. 1998.

[30] S. M. LaValle and J. J. Kuffner. Randomized kinodynamic planning. In *Proc. IEEE Int'l Conf. on Robotics and Automation*, pages 473–479, 1999.

[31] M. C. Lin and J. F. Canny. Efficient algorithms for incremental distance computation. In *IEEE Int. Conf. Robot. & Autom.*, 1991.

[32] M. C. Lin, D. Manocha, J. Cohen, and S. Gottschalk. Collision detection: Algorithms and applications. In J.-P. Laumond and M. Overmars, editors, *Algorithms for Robotic Motion and Manipulation*, pages 129–142. A K Peters, Wellesley, MA, 1997.

[33] T. Lozano-Pérez. Spatial planning: A configuration space approach. *IEEE Trans. on Comput.*, C-32(2):108–120, 1983.

[34] K. M. Lynch. Controllability of a planar body with unilateral thrusters. *IEEE Trans. on Automatic Control*, 44(6):1206–1211, 1999.

[35] K. M. Lynch and M. T. Mason. Stable pushing: Mechanics, controllability, and planning. *Int. J. Robot. Res.*, 15(6):533–556, 1996.

[36] E. Mazer, J. M. Ahuactzin, and P. Bessière. The Ariadne's clew algorithm. *J. Artificial Intell. Res.*, 9:295–316, November 1998.

[37] E. Mazer, G. Talbi, J. M. Ahuactzin, and P. Bessière. The Ariadne's clew algorithm. In *Proc. Int. Conf. of Society of Adaptive Behavior*, Honolulu, 1992.

[38] B. Mirtich. V-Clip: Fast and robust polyhedral collision detection. Technical Report TR97-05, Mitsubishi Electronics Research Laboratory, 1997.

[39] R. Motwani and P. Raghavan. *Randomized Algorithms*. Cambridge University Press, 1995.

[40] R. M. Murray and S. Sastry. Nonholonomic motion planning: Steering using sinusoids. *Trans. Automatic Control*, 38(5):700–716, 1993.

[41] I. Pohl. Bi-directional and heuristic search in path problems. Technical report, Stanford Linear Accelerator Center, 1969.

[42] S. Quinlan. Efficient distance computation between nonconvex objects. In *IEEE Int. Conf. Robot. & Autom.*, pages 3324–3329, 1994.

[43] A. Scheuer and Ch. Laugier. Planning sub-optimal and continuous-curvature paths for car-like robots. In *IEEE/RSJ Int. Conf. on Intelligent Robots & Systems*, pages 25–31, 1998.

[44] S. Sekhavat, P. Svestka, J.-P. Laumond, and M. H. Overmars. Multilevel path planning for nonholonomic robots using semiholonomic subsystems. *Int. J. Robot. Res.*, 17:840–857, 1998.

[45] Z. Shiller and S. Dubowsky. On computing time-optimal motions of robotic manipulators in the presence of obstacles. *IEEE Trans. on Robotics and Automation*, 7(7), December 1991.

[46] K. Shin and N. McKay. Open-loop minimum-time control of mechanical manipulators and its application. In *Proc. of American Control Conf.*, pages 296–303, San Diego, CA, 1984.

[47] P. Svestka and M. H. Overmars. Coordinated motion planning for multiple car-like robots using probabilistic roadmaps. In *IEEE Int. Conf. Robot. & Autom.*, pages 1631–1636, 1995.

[48] D. Vallejo, C. Jones, and N. Amato. An adaptive framework for "single shot" motion planning. Texas A&M, October 1999.

[49] J. H. Yakey. Randomized path planning for linkeages with closed kinematic chains. Master's thesis, Iowa State University, Ames, IA, 1999.

[50] Y. Yu and K. Gupta. On sensor-based roadmap: A framework for motion planning for a manipulator arm in unknown environments. In *IEEE/RSJ Int. Conf. on Intelligent Robots & Systems*, pages 1919–1924, 1998.

Encoders for Spherical Motion Using Discrete Optical Sensor

Edward R. Scheinerman, *Johns Hopkins University, Baltimore, MD*
Gregory S. Chirikjian, *Johns Hopkins University, Baltimore, MD*
David Stein, *Johns Hopkins University, Baltimore, MD*

We develop a methodology for absolute encoding of spherical motion. This is accomplished by painting the surface of a moving ball in two colors and sensing the color at a finite set of points. We show how the painting of a ball in two colors should be performed, and how the point measurements can resolve an arbitrary rotation to within a range depending on the number of sensors and the painting of the ball.

1 Introduction

We consider the following problem: A ball is held in a housing, but is free to rotate arbitrarily. How can we determine the orientation of the ball?

We solve this problem by painting the surface of the ball with two colors (black and white). Fixed point sensors are located in the housing and we determine the orientation based on the feedback from these sensors. In this paper we present a technique for painting the surface of the ball, and how to decode the orientation of the ball based on the sensor readings.

The need to determine the orientation of a ball undergoing spherical motion arises in several scenarios. One application area is the feedback control of camera pointing devices used in robot vision [2]. Another application area is in the design of an optical mouse/trackball (see [1, 18] and references therein). Figure 1 illustrates another application in which an array of spherical motors is used as a distributed manipulation device. Other mechanisms for distributed manipulation have been considered (see, e.g., [9, 17]). For this application to become a reality, both robust and inexpensive spherical motors and encoders must become available.

Spherical motors have been studied for quite some time. The basic operating principles of spherical DC induction motors are described in [7, 28]. Kaneko et al. [14] developed a spherical DC servo motor. The design and implementation of spherical variable-reluctance motors has been studied by Kok-Meng Lee and coworkers for a number of years [16, 21, 22, 29]. Toyama and coworkers have developed spherical ultrasonic motors [25, 26]. Most recently, mathematical issues in the design and commutation of spherical stepper motors with full spherical motion has been addressed in [6]. Techniques from noncommutative harmonic analysis are useful in this regard (see e.g., [5] for an introduction). Despite the rather large literature on spherical motors, the study of optical encoders for spherical motion appears not to have been approached from a mathematical or algorithmic perspective. In this paper we fill this gap by providing a general technique

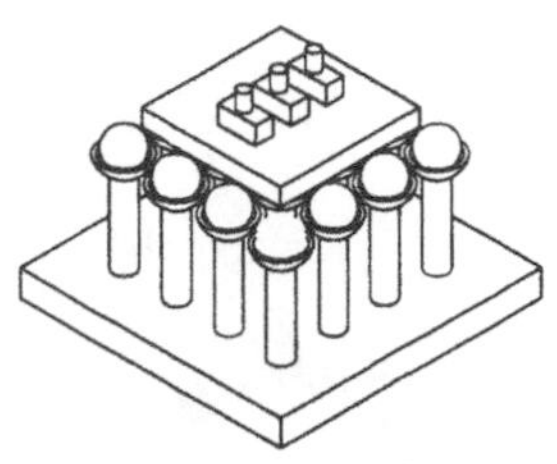

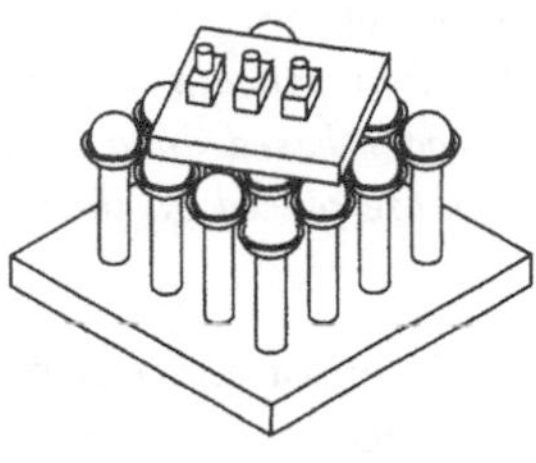

Figure 1: *An application of spherical motors in a distributed manipulation device.*

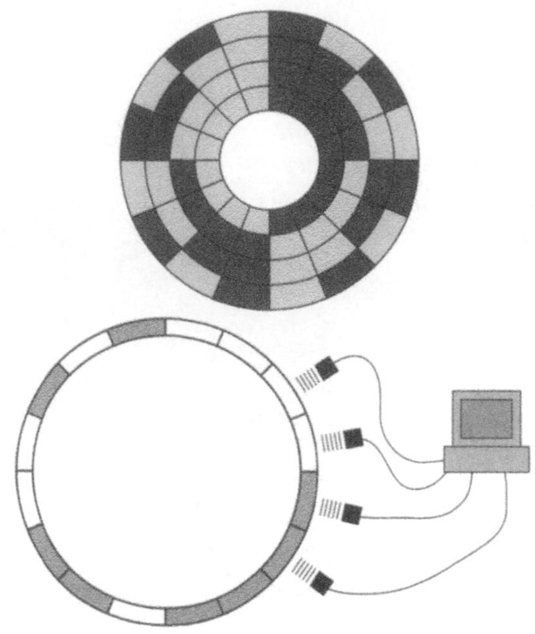

Figure 2: *The angular orientation of a disk can be determined using optical sensors. Above we see a standard encoding in which the four sensors are positioned radially. Below, we see an encoder based on a de Bruijn sequence; the colors and sensors are placed on the edge of the disk.*

for ball painting and sensor placement. This is analogous to the optical encoder for axial rotations shown in the upper portion of Figure 2. The main difference is that in spherical motion, we cannot take advantage of a dimension perpendicular to motion. That is, the painting and sensor measurements must be performed on the surface of the ball. This is analogous to only circumferential painting and sensor measurements in the axial case. See Figure 2, bottom.

2 Voronoi Surface Painting

Our method of determining the orientation of a ball is inspired by the following technique for determining the orientation of a disk that is free to rotate about a single axis. We paint the disk and locate optical sensors to determine the angle of rotation. For example, the upper portion of Figure 2 shows a standard optical encoding. However, for this encoder, the sensors must be able to see the interior of the disk. An alternative method, based on de Bruijn sequences, allows us to determine the angle of the disk using colors and sensors along the edge of the disk.

Let n be a positive integer. A *de Bruijn sequence* is a sequence of 2^n zeros and ones arranged so that each of the 2^n possible n-bit binary numbers appears (cyclically) exactly once in the sequence. For example, the sequence

$$0\ 0\ 0\ 0\ 1\ 1\ 1\ 1\ 0\ 1\ 1\ 0\ 0\ 1\ 0\ 1$$

contains all 2^4 length-4 binary numbers. The binary number 0011 appears starting at the third element of the sequence, and the binary number 0100 appears starting at the next-to-last element of the sequence (and wrapping around to the start). By painting the edge of the disk according to a de Bruijn sequence (as in the lower portion of Figure 2), all 2^n bit patterns appear exactly once and we can resolve 2^n different rotations of the disk. See [4, 27]. Our purpose is to present a technique to use optical sensors to establish the orientation of the ball. We begin by describing a scheme for painting the surface of the ball. Unfortunately, a simple, regular painting of the ball is unacceptable; such paintings often have symmetries and, consequently, different orientations of the ball produce identical sensor readings. On the other hand, a highly irregular coloring might be difficult to describe and be inefficient to handle computationally. We desire a coloring that is irregular (highly asymmetrically) but for which the following problem is efficiently solved: Given a point on the surface of the ball, determine the color of that point.

We call the coloring method we employ a *random Voronoi coloring*. Let m be a positive integer (say, between 50 and 100). We choose m *anchor points* independently and uniformly at random on the surface of the sphere; call these points $\mathbf{a}_1, \mathbf{a}_2, \ldots, \mathbf{a}_m$. (We assume these points lie on a unit sphere, so all $\mathbf{a}_i \in \mathbb{R}^3$ and $\|\mathbf{a}_i\| = 1$.) We assign a color to each of these points; independently, we color point $\mathbf{a}_i$ either black or white, each with probability $\frac{1}{2}$. Let $c(\mathbf{a}_i)$ be the color of $\mathbf{a}_i$. It is convenient to take $c(\mathbf{a}_i) = 1$ if $\mathbf{a}_i$ is black and $c(\mathbf{a}_i) = -1$ if the point is white.

Now let $\mathbf{x}$ be an arbitrary point on the surface of the sphere. We color $\mathbf{x}$ to match the color of its nearest anchor point. That is, choose i so that $d(\mathbf{x}, \mathbf{a}_i) \leq d(\mathbf{x}, \mathbf{a}_j)$ for all j. Let $c(\mathbf{x}) = c(\mathbf{a}_i)$. Such a coloring is mildly ambiguous. If $\mathbf{x}$ is nearest to two (or more) points of different colors, we assign $c(\mathbf{x})$ arbitrarily (e.g., to

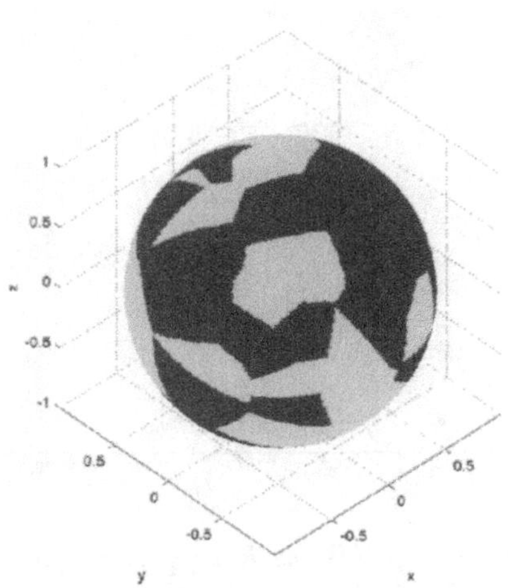

Figure 3: *A random Voronoi painting of a ball.*

match the nearest anchor point of lowest index). Such a coloring of the ball is shown in Figure 3.

The anchor points $\mathbf{a}_1, \ldots, \mathbf{a}_m$ partition the surface of the sphere into regions based on nearest neighbor. That is, each point is assigned to a region based on which of the anchor points is nearest to it. The boundaries of these regions (points that are nearest to two or more anchor points) are arcs of great circles. Such a partitioning of the sphere is known as a *Voronoi* decomposition. Thus, given random anchor points, we independently color each region of the Voronoi decomposition black or white (each with probability $\frac{1}{2}$).

The random Voronoi coloring of the ball achieves the irregularity we desire, but what of the computational efficiency? That is, given a test point $\mathbf{x}$, how quickly can we determine $c(\mathbf{x})$? A simple method is to calculate the m inner products $\mathbf{x} \cdot \mathbf{a}_i$ and then select the index of the maximum such value. This requires $O(m)$ calculations. Because m is of modest size, the color of any single point can be found rapidly. However, in our method, we repeatedly compute the colors of many points, and so a more efficient algorithm is needed.

Fortunately, such algorithms already exist [8, 10, 20]. One can construct (via a one-time calculation) a data structure which supports nearest neighbor query in time $O(\log m)$. That is, given a test point $\mathbf{x}$ on the surface of the sphere, we can determine which of the fixed anchor points $\mathbf{a}_1, \ldots, \mathbf{a}_m$ is closest to $\mathbf{x}$ in time $O(\log m)$.

3 Sensors

We determine the orientation of the ball by reading the color at various points on its surface. That is, we place an array of sensors in the device's housing. Each sensor is capable of detecting the color (black or white) at a point on the surface of the ball. In our laboratory, we use infrared photosensors that have both an emitter and a detector in the same package. The sensors use digital logic. When a black or an absorbent surface it placed in front of it, the output is in one state. When a white or reflective surface is placed in front of it the output switches state. The output of these sensors is fed into a data acquisition board.

The locations of the sensors correspond to points on the sphere. If we have n sensors, let $\mathbf{s}_1, \mathbf{s}_2, \ldots, \mathbf{s}_n$ be the points on the sphere (so $\mathbf{s}_i \in \mathbb{R}^3$ and $\|\mathbf{s}_i\| = 1$) corresponding to the sensors.

Fix a particular orientation of the ball as a home orientation. Every orientation of the ball can then be specified by a rotation matrix $A \in SO(3)$.

If the ball has been rotated from the home position via a rotation A, then the color detected by sensor i is $c(A^T\mathbf{s}_i)$. We define the *color vector* of a rotation A to be the following vector (whose entries are ± 1):

$$\mathbf{c}(A) = \begin{bmatrix} c(A^T\mathbf{s}_1) & c(A^T\mathbf{s}_2) & \cdots & c(A^T\mathbf{s}_n) \end{bmatrix}^T.$$

Of course, the color vector depends not only on the rotation A, but also on the painting of the ball c and the placement of the sensors $\{\mathbf{s}_i\}$. However, only the orientation A changes; the painting and sensor placement are fixed.

The problem we want to solve is: Given a color vector $\mathbf{c}(A)$, determine A. The problem does not have a unique solution. Because there are only n sensors, there are (at most) 2^n possible color vectors, but there are infinitely many different rotation matrices A. Therefore, the problem we actually solve is, given a color vector $\mathbf{c}(A)$, find a rotation matrix B so that $\mathbf{c}(A) = \mathbf{c}(B)$, and B is close to A. We can measure the discrepancy between A and B as an angle. Since $AB^T \in SO(3)$, we can consider AB^T as a rotation about an axis; the angle of that rotation is our measure of closeness. Thus, the "distance" between A and B is $\cos^{-1}\left(\frac{\mathrm{tr}(AB^T)-1}{2}\right)$.

How many sensors should we use and where should they be placed? Our experiments indicate that the *number* of sensors is more important than their placement. As long as the sensors are reasonably distributed around the surface of the sphere we achieve nearly identical performance of our decoding method. The more sensors we use, the more accurately can we determine A from $\mathbf{c}(A)$. However, the computational burden rises accordingly. With 50 sensors, we usually find A to within 2 or 3 degrees. With 200 sensors we typically find A to within half a degree.

4 Local Solution

Our problem is: Given $\mathbf{c}(A)$, find a $B \in SO(3)$ so that $\mathbf{c}(B) = \mathbf{c}(A)$. To this end, we define a function $f : SO(3) \to \mathbb{R}$ by:

$$f(B) = \frac{1}{\sqrt{n}} \|\mathbf{c}(B) - \mathbf{c}(A)\|.$$

The factor $1/\sqrt{n}$ is not strictly necessary, but simply rescales the image of the function so that it always lies in $[0, 2]$ regardless of the size of n (since the entries of $\mathbf{c}(A)$ and $\mathbf{c}(B)$ are ± 1, the value of $\|\mathbf{c}(B) - \mathbf{c}(A)\|$ is at most $2\sqrt{n}$).

The problem can then be restated: Find $B \in SO(3)$ so that $f(B) = 0$. Equivalently, since f is non-negative, we can think of this problem as searching for a (global) minimizer of f.

In this section we present an iterative method that leads to a solution provided we are given an initial B_0 that is close (say, within 10 degrees) to the desired solution. This assumption is justifiable for two reasons. First, as the ball turns in its housing, the sensors continuously track its progress. We can use the previous orientation of the ball as an initial guess for the current orientation. Second, the method described in the next section can provide us with an initial guess B_0 reasonably close to the solution value.

We begin by examining a plot of $f(B)$ where B is restricted to a single axis of rotation; such a plot is presented in Figure 4.

Notice that the graph of f is piecewise flat. This happens because, as the ball rotates, sensors cross the boundaries between the Voronoi regions and suddenly change state. This causes a step-jump in the value of f.

Also notice that within 10° of the actual position, the graph of f decreases (stepwise) to the minimum value. Thus, a sensible approach to minimizing $f(B)$ would be to follow a steepest-descent trajectory. However, to make this precise, we need to be clear on what we mean by the gradient of f, and then deal with the fact that wherever the gradient is defined, its value is zero.

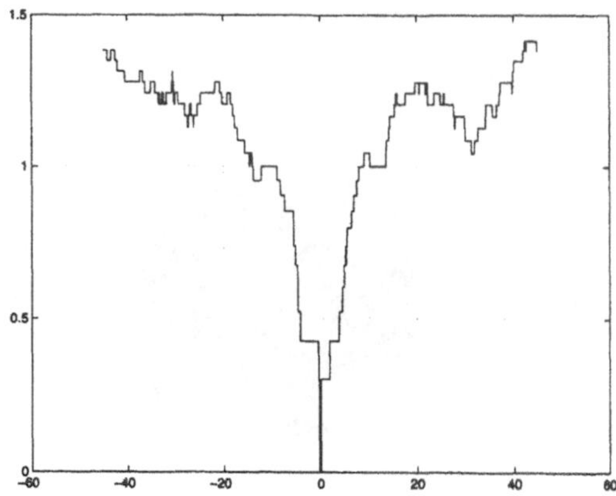

Figure 4: *Plot of $f(B)$ as B varies through a single axis of rotation. On the horizontal axis, 0° corresponds to the home orientation. (For this figure, we used a ball with 75 Voronoi regions and 44 sensors.)*

Because the domain of f is $SO(3)$, we use concepts from Lie groups/algebras [5, 11, 12, 19] to speak carefully of the gradient of f; see [24]. Let X be a 3×3 skew-symmetric matrix. We set:

$$(X^R f)(B) = \left. \frac{df\left(Be^{tX}\right)}{dt} \right|_{t=0}.$$

We can think of X^R as a (right) directional derivative of f in the direction X.

Define the (right) gradient of f at B to be the vector:

$$[E_1^R f \quad E_2^R f \quad E_3^R f]$$

where

$$E_1 = \begin{bmatrix} 0 & 0 & 0 \\ 0 & 0 & -1 \\ 0 & 1 & 0 \end{bmatrix},$$

$$E_2 = \begin{bmatrix} 0 & 0 & 1 \\ 0 & 0 & 0 \\ -1 & 0 & 0 \end{bmatrix}, \quad \text{and}$$

$$E_3 = \begin{bmatrix} 0 & -1 & 0 \\ 1 & 0 & 0 \\ 0 & 0 & 0 \end{bmatrix}.$$

The gradient points in the direction of steepest ascent, so by following a path in the opposite direction, we are led to a (local) minimum of the function. However, for the function we are considering, the gradient of f at B is $\mathbf{0}$ for almost all $B \in SO(3)$, and is undefined otherwise.

Our solution is to use approximate gradients. We replace the right derivatives $E_j^R f$ (with $j = 1, 2, 3$) with finite difference approximations:

$$\frac{f(Be^{tE_j}) - f(Be^{-tE_j})}{2t}.$$

We begin with a specific step size for t (e.g., $t = 0.01$). If the approximate gradient we calculate is $\mathbf{0}$, we increase t by a constant factor. We do this until the approximate gradient is some nonzero vector $\mathbf{x}$. We then update B by a step of size t in the negative gradient direction, i.e.,

$$B \leftarrow B \exp\left\{-t\begin{bmatrix} 0 & x_1 & x_2 \\ -x_1 & 0 & x_3 \\ -x_2 & -x_3 & 0 \end{bmatrix}\right\}$$

where x_j is the finite difference approximation to $(E_j^R f)(B)$.

We then iterate this procedure. At each successive step, we decrease t by a constant factor so we do not overrun the minimum.

Ideally, we continue this procedure until we reach a B at which $f(B) = 0$. Of course, we could get trapped at a local minimum. In this case, if $f(B)$ is fairly small (say $f(B) < 0.2$) we are actually quite close to the minimum and may stop there.

5 Global Solution

In the previous section we presented a technique to find a $B \in SO(3)$ that solves the equation $\mathbf{c}(B) = \mathbf{c}(A)$ provided we are given a starting B_0 sufficiently near A (within 10°). However, a steepest-descent method fails if we start too far from the solution because of the presence of many local minima and the highly nonsmooth nature of the function. See Figure 5.

Our strategy is simple. Since $SO(3)$ is compact and since the descent method just needs a starting value that is reasonably close to the correct value, we can evaluate $f(B)$ over a discrete subset of $SO(3)$. That

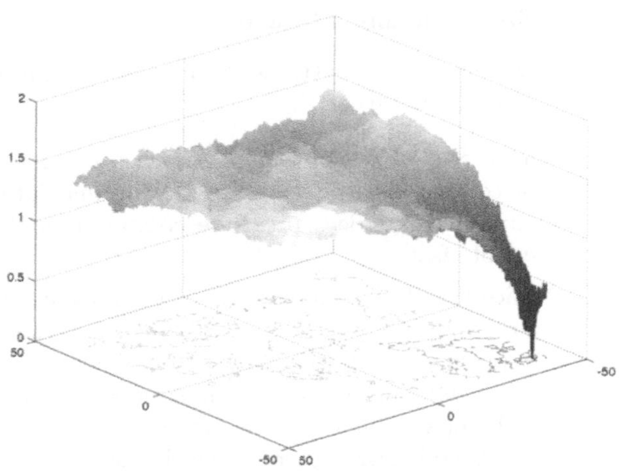

Figure 5: *Plot of $f(A)$ as A varies along two orthogonal axes.*

is, we can choose a threshold (say 10°) and find a finite subset $Z \subset SO(3)$ so that for all $A \in SO(3)$ there is a $B \in Z$ so that the rotational difference between A and B is below the threshold.

Our method is equally simple. We precompute the set $Z \subset SO(3)$. We record in a table the vector $\mathbf{c}(B)$ for all $B \in Z$. Suppose $Z = \{B_1, B_2, \dots, B_k\}$. We create a $k \times n$ matrix M whose i,j-entry is $c(B_i^T \mathbf{s}_j)$; i.e., this entry is the color observed by the jth sensor if the ball is rotated by B_i from its home position.

Let $A \in SO(3)$ be arbitrary. Then we choose B_i in $SO(3)$ so that $\mathbf{c}(A) \cdot \mathbf{c}(B_i)$ is as large as possible (best possible match). This can be done by a single matrix-vector multiply, $M\mathbf{c}(A)$ and finding the coordinate with largest index.

This calculation is reasonably fast, but the matrix M is quite large (for 100 sensors and 10000 saved orientations we need one million entries). One idea that saves some memory and speeds up the calculation is to replace M by a reduced-rank approximation.

Let $M = U\Sigma V^T$ be M's singular value decomposition [13]. Here U is a $k \times k$ unitary matrix, V is an $n \times n$ unitary matrix, and Σ is a $k \times n$ diagonal matrix whose diagonal entries are real, nonnegative, and in decreasing order. Let r be a modest positive integer (e.g., $r = 10$) and let:

- $\hat{\Sigma}$ be the $r \times r$ upper left corner of Σ,

- $\hat{U}$ be the $k \times r$ matrix formed by choosing just the first r columns of U, and
- $\hat{V}$ be the $n \times r$ matrix formed by choosing just the first r columns of V.

Then $M \approx \hat{U}\hat{\Sigma}\hat{V}^T$. This approximate decomposition can be computed without finding the full singular value decomposition. It is a one-time computation that is reasonably fast.

Notice that these three matrices consume a total of $rk + r^2 + rn$ storage. For $k = 10000$, $n = 100$ and $r = 10$, this is about 100,000 which is significantly less than one million. We can then approximate $M\mathbf{c}(A)$ by $\hat{U}\hat{\Sigma}\hat{V}^T\mathbf{c}(A)$, and select the largest component(s) to give reasonable starting values to the descent method. Our experience has been that we need to try a few of the largest starting values of the approximation to $M\mathbf{c}(A)$ to get a good starting value.

Acknowledgments

Edward Scheinerman is supported, in part, by an Interdisciplinary Grant in the Mathematical Sciences from the National Science Foundation, DMS-9971696. Gregory Chirikjian and David Stein are supported by a grant from the Robotics and Human Augmentation program of the National Science Foundation, IIS-9731720.

References

[1] X. Arreguit, F.A. van Schaik, F.V. Bauduin, M. Bidiville, and E. Raeber. A CMOS motion detector system for pointing devices. *IEEE Journal of Solid-State Circuits*, 31(12):1916–1921, December 1996.

[2] B.B. Bederson, R.S. Wallace, and E.L. Schwartz. A miniature pan-tilt actuator: the spherical pointing motor. *IEEE Trans. on Robotics and Automation*, 10(3):298–308, June 1994.

[3] K.-F. Böhringer, B.R. Donald, and N.C. MacDonald. Programmable vector fields for distributed manipulation, with applications to MEMS actuator arrays and vibratory parts feeders. *International Journal of Robotics Research*, 18(2):168–200, February 1999.

[4] J.A. Bondy and U.S.R. Murty. *Graph Theory with Applications.* North-Holland, 1976.

[5] G.S. Chirikjian and A.B. Kyatkin. *Engineering Applications of Noncommutative Harmonic Analysis.* CRC Press, to appear.

[6] G.S. Chirikjian and D. Stein. Kinematic design and commutation of a spherical stepper motor. *IEEE/ASME Transactions on Mechatronics*, 4(4):342–353, December 1999.

[7] K. Davey and G. Vachtsevanos. The analysis of fields and torques in a spherical induction motor. *IEEE Trans. Magn.*, MAG-23, March 1987.

[8] M. de Berg, M. van Kreveld, M. Overmars, and O. Schwarzkopf. *Computational Geometry: Algorithms and Applications.* Springer-Verlag, 1997.

[9] B.R. Donald, J. Jennings, and D. Rus. Information invariants for distributed manipulation. *International Journal of Robotics Research*, 16(5):673–702, October 1997.

[10] H. Edelsbrunner. *Algorithms in Combinatorial Geomery.* Springer-Verlag, 1987.

[11] I.M. Gel'fand, R.A. Minlos, and Z.Ya. Shapiro. *Representations of the rotation and Lorenz groups and their applications.* Macmillan, 1963.

[12] S. Helgason. *Differential geometry, Lie groups, and symmetric spaces.* Academic Press, 1978.

[13] R.A. Horn and C.R. Johnson. *Matrix Analysis.* Cambridge University Press, 1985.

[14] Y. Kaneko, I. Yamada, and K. Itao. A spherical DC servo motor with three degrees-of-freedom. *ASME Dynamic Systems and Controls Division*, 11:398–402, 1988.

[15] K.-M. Lee. Orientation sensing system and method for a spherical body, #5,319,577. U.S. Patent Office, 1994.

[16] K.-M. Lee and C.-K. Kwan. Design concept development of a spherical stepper for robotic applications. *IEEE Trans. on Robotics and Automation*, 7(1):175–181, February 1991.

[17] J. Luntz, W. Messner, and H. Choset. Parcel manipulation and dynamics with a distributed actuator array: The virtual vehicle. In *Proc. 1997 IEEE International Conference on Robotics and Automation*, 1997.

[18] R.F. Lyon. The optical mouse and an architectural methodology for smart digital sensors. In *CMU Conf. VLSI Systems Computations*, pages 1–19, 1981.

[19] W. Miller, Jr. *Lie theory and special functions.* Academic Press, 1968.

[20] J. O'Rourke. *Computational Geometry in C.* Cambridge University Press, second edition, 1998.

[21] J. Pei. *Methodology of Design and Analysis of Variable-Reluctance Spherical Motors*. PhD thesis, Georgia Institute of Technology, 1990.

[22] R.B. Roth and K.-M. Lee. Design optimization of a three degrees-of-freedom variable-reluctance spherical wrist motor. *ASME J. Engineering for Industry*, 117:378–388, August 1995.

[23] D. Stein and G.S. Chirikjian. Experiments in the commutation and motion planning of a spherical stepper motor. preprint.

[24] C.J. Taylor and D.J. Kriegman. Minimization on the Lie group $SO(3)$ and related manifolds. Technical Report 9405, Yale Center for Systems Science, April 1994.

[25] S. Toyama, S. Sugitani, G. Zhang, Y. Miyatani, and K. Nakamura. Multi degree of freedom spherical ultrasonic motor. In *Proc. 1995 IEEE Int. Conf. on Robotics and Automation*, pages 2935–2940, Nagoya, Japan, 1995.

[26] S. Toyama, G. Zhang, and O. Miyoshi. Development of new generation spherical ultrasonic motor. In *Proc. 1996 IEEE Int. Conf. on Robotics and Automation*, pages 2871–2876, Minneapolis, Minnesota, April 1996.

[27] D.B. West. *Introduction to Graph Theory*. Prentice-Hall, 1996.

[28] F. Williams, Laithwaire, and G.F. Eastham. Development and design of spherical induction motors. *Proc. of the IEEE*, pages 471–484, December 1959.

[29] Z. Zhou and K.-M. Lee. Real-time motion control of a multi-degree-of-freedom variable reluctance spherical motor. In *Proc. 1996 IEEE Int. Conf. on Robotics and Automation*, pages 2859–2864, Minneapolis, Minnesota, April 1996.

Notes on Visibility Roadmaps and Path Planning

J.-P. Laumond, *LAAS-CNRS, Toulouse, France*
T. Siméon, *LAAS-CNRS, Toulouse, France*

This paper overviews the probabilistic roadmap approaches to path planning whose surprising practical performances attract today an increasing interest. We first comment on the configuration space topology induced by the methods used to steer a mechanical system. Topology induces the combinatorial complexity of the roadmaps tending to capture both coverage and connectivity of the collision-free space. Then we introduce the notion of optimal coverage and we provide a paper probabilistic scheme in order to compute what we called visibility roadmaps [26]. Reading notes conclude on recent results tending to better understand the behavior of these probabilistic path planning algorithms.

1 Introduction

The framework of this work lies in the tentative to provide path planners working for large classes of mechanical systems. Such a generality is imposed by an increasing number of path planning applications that extend the robotics area where the research has been initially conducted [19].

In our case, we are interested in providing CAD systems with path planning facilities in the context of logistics and operation in industrial installations. A typical scenario is the following one. An operator has to define a maintenance operation involving the moving of an heavy freight. He has CAD software including all the geometric details and facilities to display the plant together with catalogue containing lists of available tools to perform the maintenance task. He should choose suitable handling devices among cranes, rolling bridges, carts... and validate his choice by simulating the task within the CAD system. As the other facilities offered by the CAD software, the use of path planners should be as easy as possible. The operator does not care about path planning algorithms.

Probabilistic path planners, and among them, probabilistic roadmap algorithms, appeared as a promising route to face the constraints imposed by such a scenario.

This paper does not contain new algorithms, new theorems nor new original results. It constitutes a set of remarks and working notes on probabilistic roadmaps algorithms introduced at the beginning of the '90s [13, 25, 15] and now attracting numerous research groups, including ours.

The first section starts at the beginning: It deals with the controllability of mechanical systems independently from any computational perspective. In the next section we put emphasis on the choice of the methods used to steer a mechanical system from a configuration to another one. The set of configurations reachable from a given configuration has various shapes depending on the considered steering method. Paving the space with reachable sets asks for combinatorial topology issues which are discussed. The roadmaps induced by such pavements capture both coverage and connectivity of the space and allow to apply a retraction approach to path planning. At this stage, we introduce the notions of optimal coverage and visibility roadmaps. Section 4 addresses the computational point of view: Probabilistic algorithms are today the only effective way to pave the space with reachable sets. After presenting the principle of the probabilistic roadmap algorithms we show how visibility roadmaps may be computed. An interest of the algorithm lies in its control. Another one is the small size of the computed roadmaps. The work related to probabilistic roadmap algorithms is reported and commented in Section 5.

2 Controllability and CS Topology

Examples of mechanical systems Consider the mechanical systems displayed in Figure 1 together with examples of collision-free paths. Modeling a mechanical system in the context of path planning addresses two issues.

The first one deals with the placement constraints. We should identify the configuration space CS of the system, i.e., a minimal set of parameters locating the system in its workspace. This is an easy task for the robot arm, the mobile robot, and the rolling bridge, all of them being open kinematic chains. The system constituted by the two holonomic mobile robots manipulating an dumbbell is a closed kinematic chain. The placement parameters of both robots are linked by equations modeling the grasping of the dumbbell and giving rise to holonomic constraints. For this special example it is possible to select five independent parameters defining the configuration space properly, the other placement parameters being deduced by explicit equations involving only the five independent parameters. For more general closed loop chains, characterizing CS properly is not an easy task.

Second, we should consider the kinematic constraints. There is no special constraint for the robot arm, any configuration parameter being a degree of freedom directly controlled by a motor independently from any other one. Any path in CS is admissible. The same property holds for the coordinated path of the two holonomic mobile robots. The control of the rolling bridge may impose special constraints, like moving a degree of freedom at once. In that case the only admissible paths in CS are Manhattan paths. The motion of the mobile manipulator is submitted to the constraint of rolling without sliding. This is a nonholonomic constraint that affects the range of admissible paths, but not the dimension of the reachable configuration space.

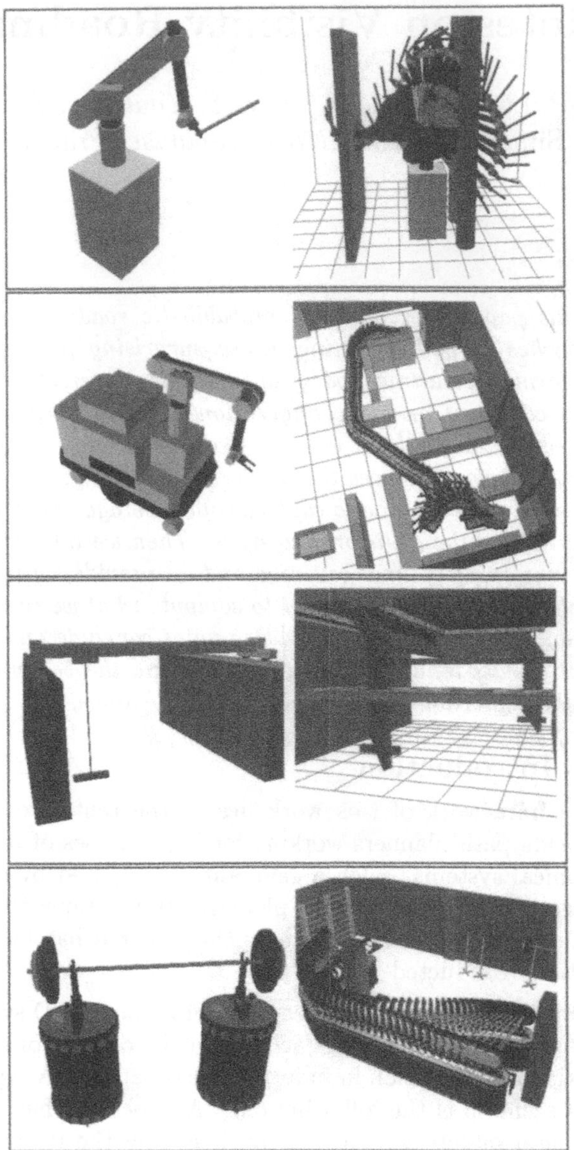

Figure 1: *Examples of mechanical systems and collision-free paths.*

Controllability All the previous systems are small-time controllable: Starting from any configuration, the set of configurations reachable before any given time contains a neighborhood of the starting configuration. Let us translate this property in geometric terms: The set of configurations reachable by all the admissible paths not escaping a given neighborhood of the starting configuration contains a neighborhood of the starting configuration (Figure 2).

Small-time controllability plays a central role in obstacle avoidance. Indeed, for small-time controllable systems, any collision-free path in CS may be approximated by a finite sequence of collision-free paths respecting the kinematic constraints. In other words, there exists an admissible collision-free path between two configurations if and only if both of them lie in the same connected component of the collision-free configuration space[1] CS_{free}. This means that the knowledge of the connected components of CS_{free} is sufficient to

[1]Configurations touching an obstacle are not considered as collision-free. CS_{free} is an open domain in CS.

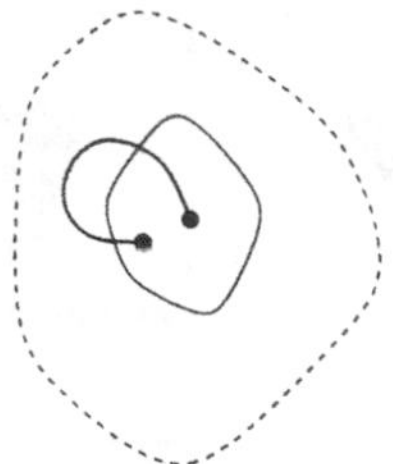

Figure 2: *Small-time controllable systems: The domain reachable without escaping a given neighborhood contains a neighborhood of the origin.*

prove the existence of an admissible collision-free path between two given configurations. The route is open to devise deterministic and complete algorithms to solve the path planning problem.

The small-time controllability property does not hold for any system. A car moving only forward is locally controllable (the reachable set from a configuration contains a neighborhood of the configuration). It is not small-time controllable and deciding on the existence of an admissible collision-free path is still an open problem for this system.

Complexity Small-time controllability allows to forget the kinematic constraints to decide on the existence of an admissible path in CS_{free}. Nevertheless the kinematic constraints affect the combinatorial complexity of the admissible paths. Consider that CS_{free} is reduced to a long straight-line tube. Moving within the tube is easy for the robot arm: A single straight-line path is sufficient. Depending on the orientation of the tube in CS_{free}, the same task will be more difficult for the rolling bridge: The number of elementary pieces of the Manhattan paths may grow linearly when the width of the tube decreases. It will be even more difficult for a nonholonomic robot: The number of elementary pieces of admissible paths may grow exponentially with its degree of nonholonomy when the width of the tube decreases (e.g., the number of maneuvers required to park a car along a sidewalk varies quadratically with the inverse of the free space size).

3 Steering Methods, Visibility, and Roadmaps

Steering methods The decision part of the path planning problem (existence of a path) depends only on the controllability of the considered system. Now, to solve the problem completely (compute a path) we have to devise effective ways to steer the system. What we call a steering method is a procedure computing an admissible path between any two configurations in the absence of obstacles[2].

Any steering method respecting the kinematic constraints of the system is not necessarily a good one with respect to obstacle avoidance. For instance, a steering method for a car moving only forward applies also to a car moving both forward and backward. Nevertheless, it is impossible to park a car by using only forward motions. A steering method induces a topology in CS that should account for the topology induced by the controllability property of the mechanical system. A steering method accounting for small-time controllability should verify the property illustrated in Figure 2. It is said to be **stc**. Devising **stc** steering methods is not necessarily a trivial problem (especially for nonholonomic systems). Moreover the choice is not unique. It may affect the combinatorial complexity of the path planning algorithm.

This section aims to illustrate the importance of this choice through two examples of **stc** steering methods for a point moving on a plane. Both induce the same topology in CS. We will see that they differ by the combinatorics of the pavements induced on CS_{free}.

The first one (Steer_{lin}) consists in computing a straight-line segment between two points. The second one is Steer_{man}. Two points being given, Steer_{man} computes first an horizontal path from the left point until the right point is reachable by a vertical path. Both steering methods are symmetric and **stc**. Figure 3(a) shows the structure of the sets of points reachable with paths of fixed length.

A symmetric **stc** steering method **Steer** being given, a configuration is said to be *visible* from another one if the path computed between them by **Steer** lies in CS_{free}. The set of configurations visible from a given configuration constitutes its *visibility set*. The configuration is said to be the *guard* of its visibility set. Figure 3(b) shows visible sets of the same configuration, in the same environment, for both Steer_{lin} and Steer_{man}.

Covering CS_{free} with visibility sets Let us consider a small-time controllable system together with a

[2]In the path planning literature, authors usually refer to the notion of *local methods*.

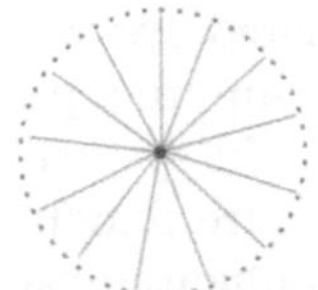
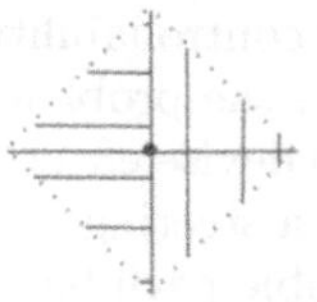

a: Reachable sets with paths of fixed length.

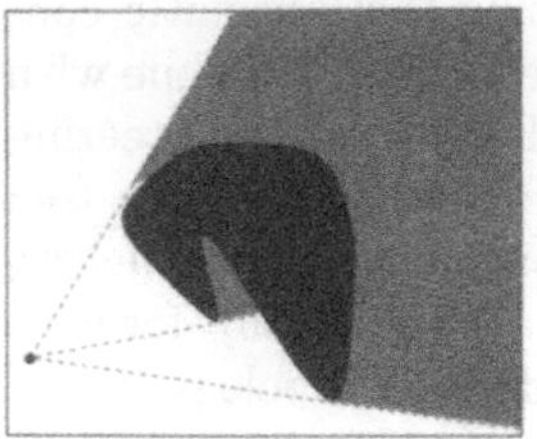
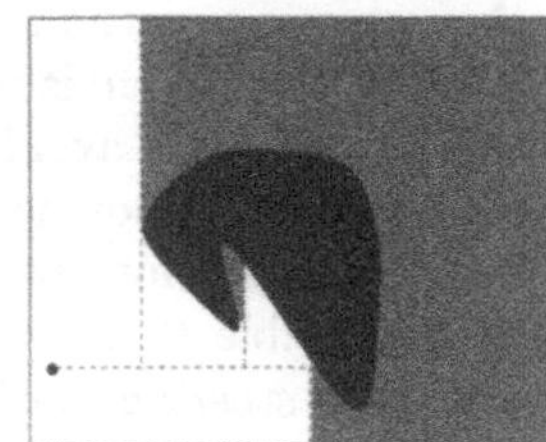

b: Visible sets in the presence of an obstacle (grey domains are not reachable).

Figure 3: *Visible sets of a configuration with* Steer_{lin} *(left) and* Steer_{man} *(right).*

stc steering method. A set of guards is said to be *covering* CS_{free} if any configuration in CS_{free} is visible from at least one guard.

We first address a question: Is there a finite set of guards covering CS_{free} ? The answer to this question is difficult and there exists today no general result. It depends on both the shape of CS_{free} and the considered steering method.

Let us illustrate the problem via four examples of 2-dimensional CS_{free} (see Figure 4). First consider Steer_{lin}. For the cases of Figures 4(a) and 4(b), we may provide coverage of CS_{free}, e.g., by putting guards sufficiently close to convex vertices of the obstacles. The cases 4(c) and 4(d) contain an obstacle whose boundary piece is circular and tangent to a straight-line segment: There is no way to provide a finite number of guards covering CS_{free} for the case 4(c), while this is possible for the case 4(d) (in this later example it is necessary to put at least one guard on the dashed line).

Consider the same environments with Steer_{man}. Finite coverage exists for the cases 4(e), 4(g), and 4(h) while it does not exist for the case 4f.

Optimal coverage Now, we impose an additional constraint. We wish to provide coverage such that any guard does not see any other one. Withdrawing a single guard does not provide coverage anymore. For that reason we say that the set of guards provides *optimal coverage*.

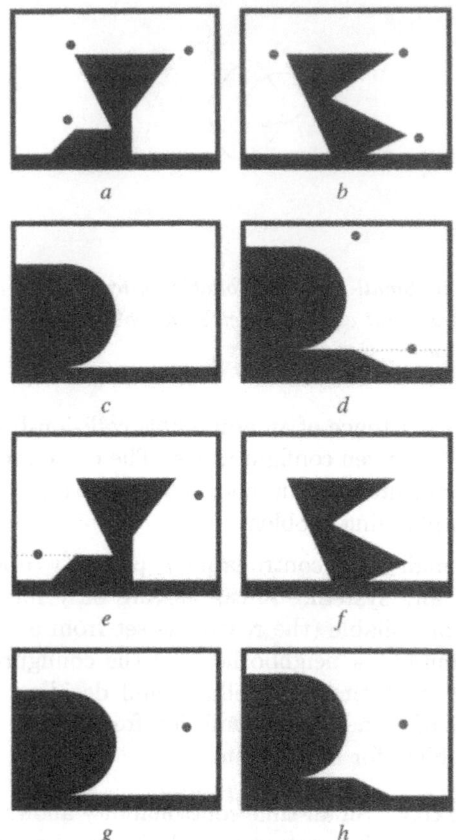

Figure 4: *Finite coverage using* Steer_{lin} *exists for the cases a, b, and d. It does not for the case c. Finite coverage using* Steer_{man} *exists for the cases e, g, and h. It does not for the case f.*

Clearly, the set of guards providing optimal coverage is not unique. Taking a combinatorial point of view, a second question is: If there exists finite optimal coverage, are all the sets of guards achieving optimal coverage necessarily finite? The answer is "no" in general.

The answer is "yes" if we consider Steer_{lin} and a polygonal CS_{free}: The maximal number of guards is bounded by the number of straight-line segments[3]. The answer is "no" if we consider a single circular obstacle (Figure 5).

[3]Remark: computing the minimal number of guards constitutes the classical art gallery problem [23].

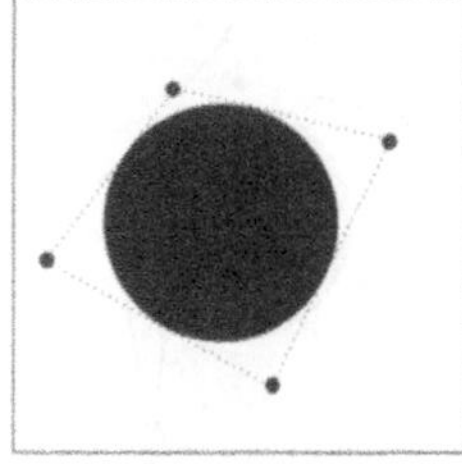
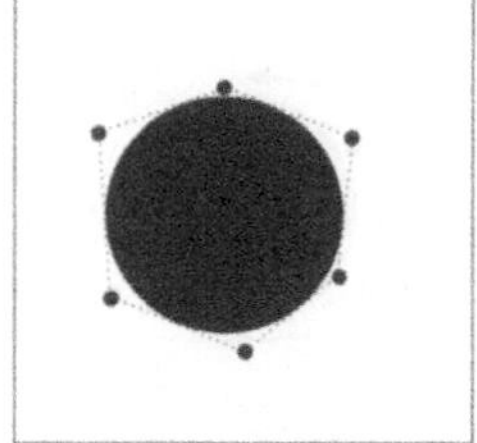

Figure 5: Steer_{lin}: *Examples of optimal coverage with 4 and 6 guards. The number of guards achieving optimal coverage is not a constant number. It may be unbounded.*

Also note that one cannot necessarily extract optimal coverage from any given set of guards providing coverage. More than that, finite coverage may exist while finite optimal coverage does not. In the example of Figure 6 finite coverage (with Steer_{lin}) necessarily requires two guards on the dashed lines. It is impossible to make one of them not visible from the other one.

Coverage and robustness In several examples above (Figures 4(d), 4(e), 5, and 6) some guards should belong to a domain whose dimension is strictly lower than the dimension of CS_{free}. In that sense, the corresponding set of guards is not robust: The probability to select such guards randomly is null.

Capturing CS_{free} connectivity from coverage In this section we show that coverage induces connectivity. Let **Steer** be a symmetric **stc** steering method in CS_{free}. Consider a given (finite or not) set of guards achieving coverage of CS_{free} together with a (finite length) path. The path is covered. Visibility sets are open sets. The path being a compact set, there is a *finite* number of guards whose visibility sets cover that path[4].

This property means that it is always possible to build a (finite or not) undirected graph accounting for the connectivity of CS_{free}. Two guards are connected if they are visible by **Steer**. At this level there does not exist necessarily a one to one mapping between the connected components of the graph and the connected components of CS_{free}. Nevertheless, it is always possible to complete the graph. When two guards belong to the same connected component of CS_{free} but not to the same component of the graph, and when their visibility sets intersect, we chose a configuration in the intersection set and we add it as a new node to the graph. Such a new configuration does not increase coverage, it just makes the number of connected components of the graph decreasing. It is called a *connector*.

[4]This gives rise to a notion of complexity of a path with respect to a given set of guards: the complexity of a path is the minimal number of guards necessary to cover the path.

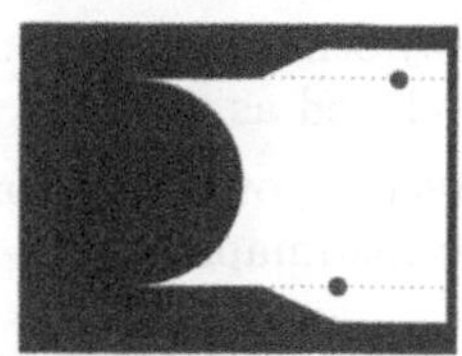

Figure 6: Steer_{lin}: *Finite coverage may exist while finite optimal coverage does not.*

By this way, it is always possible to build a graph capturing the connectivity of CS_{free}, i.e., such that there is a one-to-one mapping between the connected components of the graph and the connected components of CS_{free}.

Everything is now in place to define a retraction method solving the path planning problem. The retraction function applied to a configuration just consists in selecting a guard seeing it. This is a well defined function since the set of guards covers CS_{free}. Finally some starting and goal configurations being given there exists an admissible collision-free path between them if and only if their retractions belong to a same connected component of the graph. The continuous dimension of the path planning problem is then transformed into a computational one.

The graphs above have been introduced in path planning literature with a computational point of view and especially with a probabilistic one. Indeed computing such roadmaps with complete deterministic algorithms is a difficult problem: For given CS_{free} and **Steer**, we do not know *a priori* if there exists finite coverage, and even if there exists, we do not know how to compute one of them in a deterministic way. The probabilistic approaches relax the completeness property and tend to cover CS_{free} at the best. Experiences show that they behave well for practical problems. This is the key of the success of the so-called *probabilistic roadmaps*.

When considering optimal coverage, the roadmap resulting from the construction above is a bipartite graph: The connector nodes are adjacent only to guard nodes and, from the definition of optimal coverage, a guard cannot see any other guard. Such roadmaps have been recently introduced as *visibility roadmaps* in [26].

The following section overviews probabilistic algorithms to compute roadmaps.

4 Probabilistic Visibility Roadmaps

Computing probabilistic roadmaps Consider a small-time controllable mechanical system. Computing a probabilistic roadmap for the system requires a collision checker and a symmetric stc steering method Steer that accounts for the kinematic constraints of the system.

The basic version of a probabilistic roadmap algorithm generates configurations randomly. As soon as a collision-free configuration is found, it is added as a new node to the graph. Then the algorithm checks if the current configurations previously generated are visible by Steer from the new configuration. Each time one of them is visible, a new edge is added; otherwise, the new node creates a new connected component. A less expensive version consists in restricting the connectivity test to connect the new configuration to only one node for each connected component of the current graph. In that case, the resulting graph is a tree.

The more time consuming step of the algorithm is the call of Steer. Indeed the power of the algorithm is to avoid the computation of the visibility domains. Checking whether a configuration is visible from another one is done by checking if the path computed by Steer between them is collision-free. The number of calls to Steer then appears as a critical parameter that measures the combinatorial complexity of the construction.

Computing probabilistic visibility roadmaps The general principle above should be adapted to compute visibility roadmaps. We have seen that visibility roadmaps are bipartite graphs with two classes of nodes: The guards and the connectors. When a new collision-free configuration is (randomly) found, three cases may arise:

1. either it is not visible from any existing guard: Then it is added as a new guard to the graph (it creates a new connected component),
2. or it is visible by guards belonging to separate connected components of the current graph: then it is added as a new connector and the corresponding connected components are merged by adding the edges between the new connector and the guards seeing it,
3. otherwise it is visible only by guards belonging to a same connected component and then it is rejected.

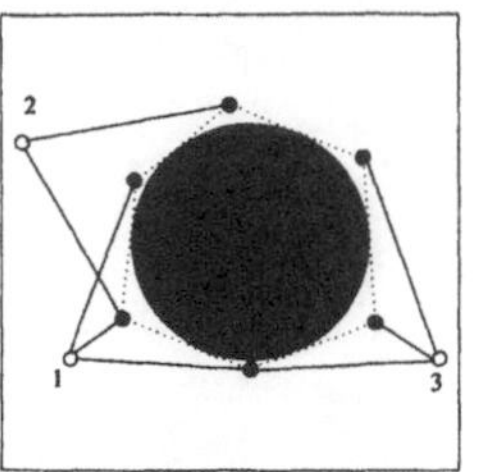

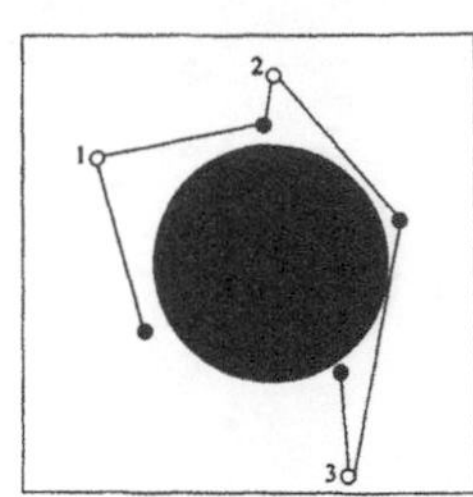

Figure 7: *The apparition order of the connectors are numbered. The visibility roadmap is not robust when connectors are not allowed to increase coverage (left): the probability to find such a roadmap is null. The variant of the visibility roadmap algorithm provides a robust roadmap (right).*

From a practical point of view, we have proposed a variant of the algorithm [26] allowing connectors to increase coverage of CS_{free}: Connectors are added to the list of the guards. By this way covering CS_{free} is faster. Moreover, an advantage of this variant is to increase the robustness of the visibility roadmaps. In Figure 7 we show that the visibility roadmap built from guards achieving optimal coverage is not robust, while the roadmap computed by the variant of the algorithm is robust.

Comparison Figure 8 shows two examples of roadmaps computed for the same environment (polygonal CS_{free} with Steer_{lin}) with the same list of 250 random configurations. The first one is a basic roadmap. The second one is a visibility roadmap. In both cases the trees cover CS_{free} and capture its connectivity. On this example, the visibility roadmap algorithm ran 6 times faster than the basic roadmap algorithm.

This simple example suggests a better performance of the visibility roadmap algorithm. They both cover and capture the connectivity of CS_{free}. The difference between the performances is due to the fact that visibility roadmaps keep a number of nodes smaller than basic roadmaps. The number of calls to Steer in the main loop of the algorithm is then smaller. This is illustrated on Figure 9.

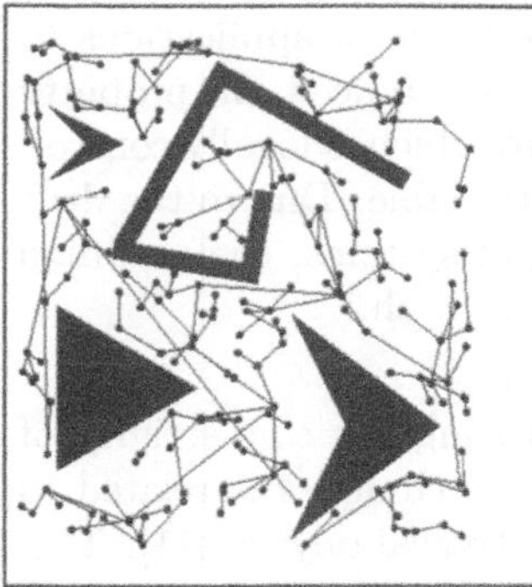

Figure 8: *A basic roadmap and a visibility roadmap.*

However, we do not succeed in proving formally a better behavior. Visibility roadmap algorithms may fail (in probability) in capturing the connectivity of CS_{free}. Guards may induce artificial narrow passages as shown in Figure 10. In that case, the probability to put a connector belonging to visibility sets of both guards is very low. With the same random sampling of CS_{free}, the visibility roadmap will certainly fail in capturing the connectivity of the space, while the basic roadmap will certainly succeed. Nevertheless, this behavior may appear as a side-effect of the algorithm, since the probability to fall in this special case is clearly low (see [24] for details).

Control of the algorithm A critical problem of the probabilistic algorithms is first to stop them, and second to get some information about the quality of the roadmap in terms of both coverage and connectivity. The structure of the visibility roadmap algorithm allows to control the quality of the roadmap in term of coverage.

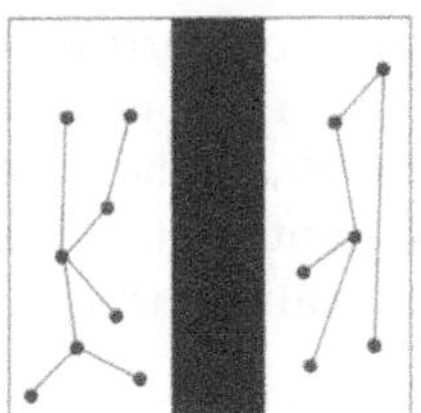
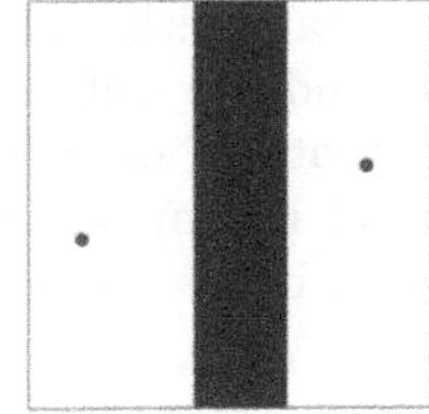

Figure 9: *Example of CS_{free} with two connected components. Any new collision-free configuration should be checked to know if it belongs to both connected components of CS_{free}. Checking this is much more expansive in the case of the basic roadmap (leftt) than in the case of the visibility roadmap (right).*

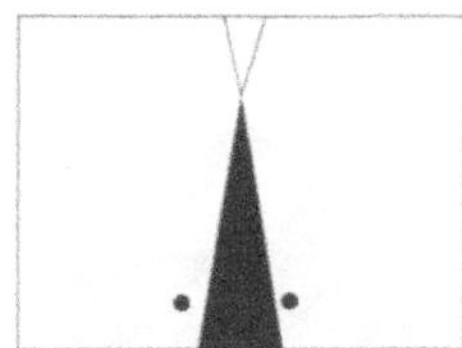

Figure 10: *A possible side-effect of the visibility roadmaps.*

We have seen that the algorithm consists in rejecting the configurations which are reachable only from nodes belonging to a same connected component of the roadmap. Therefore, the expansion of the visibility set of a connected component of the visibility roadmaps may be *a priori* slower than those of the basic roadmaps. However, in addition to the computational gain discussed above, rejecting those configurations has an advantage: The number $\#try$ of failures in adding a new guard to the visibility roadmap allows to compute an estimation of the volume of the free-space remaining non covered. We model the percentage of the non covered free-space as the inverse of the number of failures before adding a new guard. The plain curve in Figure 11 shows the evolution of this estimated percentage of the covered free-space for a free-flying body. The horizontal axis represents the number of guards, while the vertical one represents $(1 - \frac{1}{\#try})$. For the dotted curve, the vertical axis represents the real percentages of the covered free-space (the curve is monotonic and increasing). The estimated percentage converges to the real one when the number of guards increases.

Guards achieve good coverage when the curve becomes horizontal. Such behavior may be detected within the algorithm by looking at the evolution of the number $\#try$. The algorithm stops when $\#try$ becomes constant (in probability). Of course we may also bound $\#try$ by a huge integer (a detailed version of the algorithm appears in [26]). In both cases, when the algorithm stops we get an estimation of the percentage of the non covered free-space volume.

5 Reading Notes and Comments

Probabilistic roadmaps were introduced at the beginning of the '90s at both Stanford and Utrecht University [13, 25, 15]. Ph.D. Theses of L. Kavraki [12] and P. Švestka [27] state with details the seminal foundations of these new algorithms together with the first experimental results which are at the origin of their success.

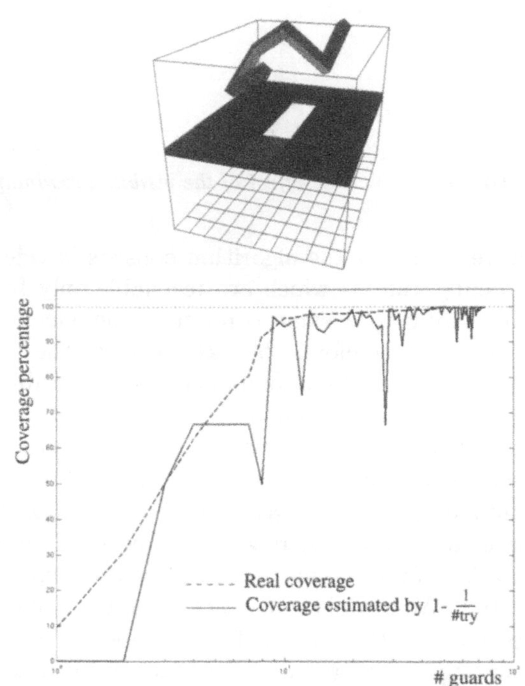

Figure 11: *Convergence of the visibility roadmap algorithm for a free flying body.*

In this section we comment on several issues tending to understand their behavior on the basis of recently published papers.

Devising adequate steering methods In our presentation, we assumed that the considered steering methods are both symmetric and **stc**. The symmetry property allows to compute undirected roadmaps, while the **stc** property guarantees the convergence of the probabilistic algorithms.

Devising **stc** steering methods is not necessarily an easy task especially when we consider nonholonomic systems. An overview on this aspect may be found in [21].

In his PhD thesis, P. Švestka relaxes the **stc** property. He shows that the steering methods may verify the following topological property: The set of configurations reachable by paths lying in an arbitrarily constrained domain contains a neighborhood in that domain (and not necessarily a neighborhood of the starting configuration as for the **stc** property). This property is important because it opens the range of applications to some locally controllable systems. Indeed, the property holds for a car moving forward which is locally controllable but not small-time controllable. Due to the drift usually present in that kind of systems, the roadmap can no more be an undirected graph.

Algorithm analysis Capturing the connectivity of geometric spaces by sampling techniques is related to the percolation problem is statistical physics [11]. This problem asks deep and challenging questions in mathematics. Analyzing the probabilistic roadmap algorithms is a difficult problem. Several authors, mainly around Latombe's team, gave pertinent insights on this issue.

The notion of ϵ-goodness is introduced in [5]. It is related to the coverage property. CS_{free} is ϵ-good if the volume of the visibility set of any configuration in CS_{free} is greater than some fixed percentage $(1-\epsilon)$ of the total volume of CS_{free}. The authors prove that if CS_{free} is ϵ-good the probability that a (basic) roadmap does not cover CS_{free} decreases exponentially with the number of nodes. Moreover the number of nodes increases moderately when ϵ increases. Notice that the notion of ϵ-goodness is not strictly related to the existence of a finite set of guards achieving coverage. Indeed, for the cases illustrated in Figures 4(a), 4(b), 4(g) and 4(h), CS_{free} is ϵ-good, while it is not for the cases 4(c), 4(d), 4(e) and 4(f). Finite coverage may exist for non ϵ-good spaces. Moreover, the number of guards achieving optimal coverage may be unbounded even for ϵ-good spaces (Figure 5).

The notion of expansiveness introduced in [8] deals with connectivity. The proposed model is rather technical. It deals with the notion of narrow passages in CS_{free} and the difficulty to go through them. It is shown that for expansive CS_{free} the probability for a (basic) roadmap not to capture the connectivity of CS_{free} decreases exponentially with the number of nodes.

The clearance of a path is also a pertinent factor. In [14] a bound on the number of nodes required to capture the existence of a path of given clearance is provided. This bound depends also on the length of the path. In [28] the dependence on the length is replaced by the dependence on the number of visibility sets required to cover the path. Again, the probability to fail in capturing the existence of a path decreases exponentially with the number of roadmap nodes.

Finally, the dependence on the dimension of CS is discussed in [10].

All these results are based on parameters characterizing the geometry of CS_{free}. The knowledge of these parameters would allow to control the algorithms. Unfortunately they are a priori unknown. As commented in [10], "*one could be tempted to use Monte Carlo technique to estimate the values of* [such parameters] *in a given free space, and hence obtain an estimate of the number of* [nodes] *needed to get a roadmap that adequately represents* [CS_{free}]. *But it seems that a reliable estimation would take at least as much time as building the roadmap itself*". An idea could be to try to estimate these parameters *during* the construction of the roadmaps. This is the underlying principle to control the visibility roadmap algorithm above. Generalizing it to estimate the ϵ-goodness, the expansiveness of the spaces or to detect narrow passages seems to be a promising issue to be investigated. The main challenge is to not degrade the performance of the algorithm with sophisticated computations.

On heuristics guiding probabilistic searches The same challenge appears to improve the basic roadmap algorithm behavior when facing difficult problems.

Capturing the connectivity of CS_{free} in the presence of narrow passages is one of them. In such contexts a first improvement has been introduced in [16]: The algorithm concentrates the search around small isolated components of the roadmaps. Another idea consists in sampling CS_{free} with configuration close to the obstacles. This can be done for free-flying systems by analyzing the contact configuration space, and by sampling contact subspaces defined by some geometric constraints like polyhedron vertex on polyhedron facet [3].

Contact space may be randomly sampled without using any explicit constraints. In [7] the method generates a pair of configurations separated by a random distance. When both configurations are either both collision-free or both in the collision space, they are rejected. Otherwise, the collision-free configuration is kept.

Another way to capture narrow passages is to exploit distance computations (an operation more expensive than a simple collision-checking): In a first step, configurations inducing small penetrations between bodies are accepted, then they are progressively moved to finally be collision-free [9].

Sampling close to the obstacles is a good strategy to capture the connectivity of narrow passages. Nevertheless, it may induce undesirable side effects when the space is not very cluttered: In the example shown in Figure 5, sampling close to the obstacle requires more configurations than sampling far from the obstacle.

Experiences conduced on the visibility roadmaps show that the time gained by constructing small roadmaps may be used to concentrate the search on regions not connected to the roadmap. In other words, for a given fixed running time, the visibility roadmap algorithm will explore more space than the basic one. Visibility roadmaps behave rather well to capture narrow passages even if they do not have been devised to this end. The visibility roadmap algorithm runs 20 times faster than the basic roadmap algorithm to capture the connectivity of CS_{free} in the example of Figure 12 (see [26] for details).

On the choice of steering methods By definition, stc steering methods verify the topological property illustrated in Figure 2. This means that any path computed with a given method may be approximated by sub-paths computed with another one. Nevertheless, we have seen that the combinatorics of a roadmap depends on the considered steering method. Depending on the shape of CS_{free} a steering method may perform better than another one. In Figure 13 we can see that, for the same example of environment, Steer_{lin} induces a smaller roadmap than Steer_{man}. Moreover, the construction of the roadmap is much faster with

Figure 12: *Translating square (grey) with* Steer_{lin}*: The visibility roadmap algorithm (right) performs 20 times better than the basic roadmap algorithm (left) in capturing the connectivity of* CS_{free}.

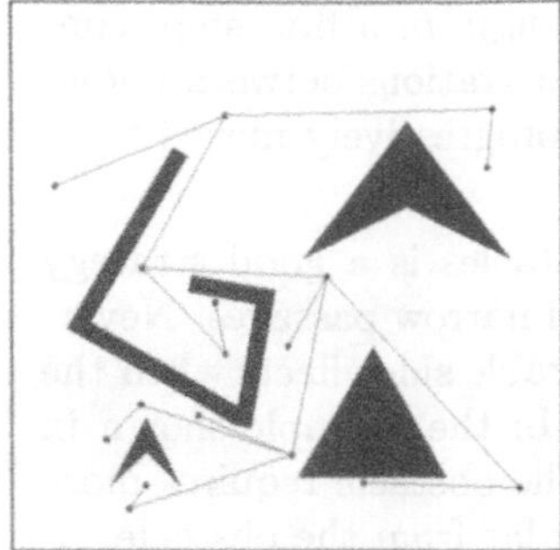
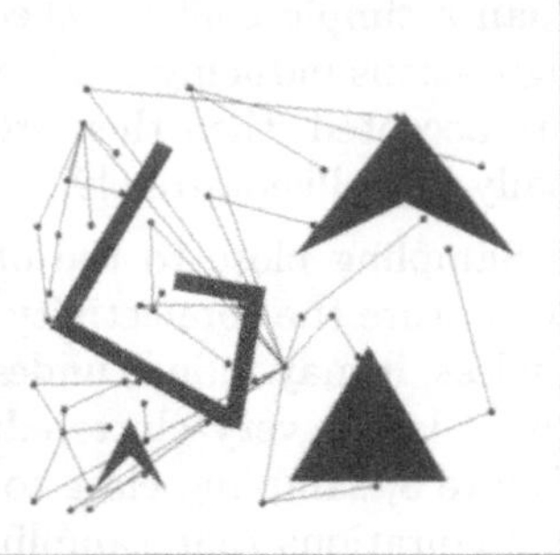

Figure 13: *Two visibility roadmaps computed with* Steer_{lin} *(left) and* Steer_{man} *(right).*

Steer_{lin} than with Steer_{man}. However, such a behavior is specific to the example. In the case of the rolling bridge in Figure 1, our experiments show a performance gain of 10 when using Steer_{man} instead of Steer_{lin}. This is a consequence of the special geometric structure of the workspace to be explored by a rolling bridge. Steer_{man} is more suitable than Steer_{lin} in that case.

Therefore there is *a priori* no arguments to conclude on general results guiding the choice of a suitable steering method. This choice should be done after practical experiments on special classes of systems and environments. For instance, in [4] several steering methods working for free-flying rigid bodies are compared from a practical point of view. In particular a steering method consisting in decoupling translation and rotation is shown performing well.

6 Current Directions

We began the integration of various probabilistic roadmap algorithms within a same software platform[5] Move3D. All the examples displayed in Figure 1 have been computed within Move3D by using the visibility roadmap algorithm. The generality constraint introduced as the main motivation of this work has been respected.

From a practical point of view, it remains to show that such techniques may face real size problems. The environment of Figure 14 represents a canonical example we are working on. The industrial installation is a stabilizer (subset of a plant in chemical industry). The geometric model was translated from PDMS geometric data structures[6]. The crane was added within Move3D. Its workspace *a priori* covers all the environment. The model contains around 300.000 polygonal facets. Here the difficulty is to face the geometrical complexity of the environment in a reasonable time.

Figure 14: *A real-size problem.*

We are working on new hybrid collision-checkers combining the techniques developed in [22] and allowing to process polyhedra together with volumic primitives (e.g., spheres, tubes, torus...) as in [6].

On the other hand we are testing incremental approaches to roadmap constructions. The principle consists in partitioning the CS space of the crane into elementary regions. Then for each corresponding elementary workspace, we apply a filter on the geometric database to select the bodies to be handled by the collision-checker. Elementary roadmaps are computed in each CS regions and then merged together.

Another challenging issue is to provide the operator with facilities for combining several handling devices to carry freights. Path planning should be extended to handling task planning. This problem is often referenced as the manipulation planning problem. Its geo-

[5] http://www.laas.fr/~nic/Move3D

[6] PDMS is a product of Cadcentre Ltd, partner of LAAS-CNRS, EDF and Utrecht University within the MOLOG project.

metrical formulation [2] allows to apply probabilistic approaches. Results already appeared in [18] in the domain of digital actors animation and in [1] in the context of manufacturing.

A priori such issues do not ask for new fundamental research. They require incremental research extending at best the existing state of the art.

From a theoretical point of view, the formal analyses referenced in Section 5 have opened the route toward a better understanding of the behavior and performance of probabilistic roadmap algorithms. Moreover additional work remains to be done to better control the probabilistic roadmap algorithms. A lot of information is usually lost when running probabilistic roadmap algorithms. For instance the configurations belonging to an obstacle are rejected. Counting them may give an estimation of CS_{free} volume. Memorizing their location may help in estimating the shape of the obstacles. At the same time, it may be possible to estimate on line the expansiveness of a node. The problem is to avoid increasing the cost of the algorithm by expensive operations. The visibility roadmap algorithm commented in this paper is an attempt in this research issue.

Acknowledgments

We thank Carole Nissoux who has been the kingpin of Move3D.

This work is supported by the European Esprit Project 28226 MOLOG (http://www.laas.fr/molog).

References

[1] J.M. Ahuactzin, K.K. Gupta and E. Mazer. Manipulation planning for redundant robots: a practical approach. *Practical Motion Planning in Robotics*, K.K. Gupta and A.P. Del Pobil (Eds), J. Wiley, 1998.

[2] R. Alami, J.P. Laumond and T. Siméon. Two manipulation planning algorithms. *Algorithmic Foundations of Robotics (WAFR94)*, K. Goldberg et al (Eds), AK Peters, 1995.

[3] N. Amato, O Bayazit, L. Dale, C. Jones and D. Vallejo. OBPRM: an obstacle-based PRM for 3D workspaces. *Robotics: The Algorithmic Perspective (WAFR98)*, P. Agarwal and all (Eds), AK Peters 1998.

[4] N. Amato, O Bayazit, L. Dale, C. Jones and D. Vallejo. Choosing good distance metrics and local planners for probabilistic roadmap methods. *IEEE International Conference on Robotics and Automation*, Leuven (Belgium), 1998.

[5] J. Barraquand, L. Kavraki, J.C. Latombe, T.Y. Li, R. Motvani and P. Raghavan. A random sampling scheme for path planning. *International Journal of Robotics Research*, Vol 16 (6), 1997.

[6] G. van den Bergen. Efficient Collision Detection of Complex Deformable Models using AABB Trees. *Journal of Graphics Tools*, Vol 2 (4), 1997.

[7] V. Boor, M. Overmars and A. Frank van der Stappen. The Gaussian sampling strategy for probabilistic roadmap planners. *IEEE International Conference on Robotics and Automation*, Detroit (USA), 1999.

[8] D. Hsu, J.C. Latombe and R. Motwani. Path planning in expansive configuration spaces. *IEEE International Conference on Robotics and Automation*, Albuquerque (USA), 1997.

[9] D. Hsu, L. Kavraki, J.C. Latombe, R. Motwani and S. Sorkin. On finding narrow passages with probabilistic roadmap planners. *Robotics: The Algorithmic Perspective (WAFR98)*, P. Agarwal et al (Eds), AK Peters, 1998.

[10] D. Hsu, L. Kavraki, J.C. Latombe and R. Motwani. Capturing the connectivity of high-dimensional geometric spaces by parallelizable random sampling techniques. *Advances in Randomized Parallel Computing*, P.M. Pardalos and S. Rajasekaran (Eds.), Combinatorial Optimization Series, Kluwer Academic Publishers, Boston, 1999.

[11] G. Grimmet. *Percolation*. Springer Verlag, 1989.

[12] L. Kavraki. Random networks in configuration spaces for fast path planning. PhD Thesis, Dept of Computer Science, Stanford University, 1995.

[13] L. Kavraki and J.C. Latombe. Randomized preprocessing of configuration space for fast path planning. *IEEE International Conference on Robotics and Automation*, San Diego (USA), 1994.

[14] L. Kavraki, L. Kolountzakis and J.C. Latombe. Analysis of probabilistic roadmaps for path planning. *IEEE International Conference on Robotics and Automation*, Minneapolis (USA), 1996.

[15] L. Kavraki, P. Švestka, J.C. Latombe and M.H. Overmars. Probabilistic Roadmaps for Path Planning in High-Dimensional Configuration Spaces, *IEEE Transactions on Robotics and Automation*, Vol 12 (4), 1996.

[16] L. Kavraki and J.C. Latombe. Probabilistic roadmaps for robot path planning. *Practical Motion Planning in Robotics*, K. Gupta and A. del Pobil (Eds), Wiley Press, 1998.

[17] L. Kavraki, M. Kolountzakis and J.C. Latombe. Analysis of Probabilistic Roadmaps for Path Planning, *IEEE Transactions on Robotics and Automation*, Vol 14 (1), 1998.

[18] Y. Koga, K. Kondo, J. Kuffner and J.C. Latombe. Planning motion with intentions. *ACM SIGGRAPH*, Orlando (USA), 1994.

[19] J.C. Latombe. Motion planning: a journey of robots, molecules, digital actors, and other artifacts. *The International Journal of Robotics Research*, Vol 18 (11), 1999.

[20] J.P. Laumond (Ed.). *Robot Motion Planning and Control.* LNCS 229, Springer Verlag, 1998 (out of print, available freely at `http://www.laas.fr/~jpl/`).

[21] J.P. Laumond, F. Lamiraux and S. Sekhavat. Guidelines in nonholonomic motion planning. In [20].

[22] M. Lin, D. Manocha, J. Cohen and S. Gottschalk. Collision detection: Algorithms and applications. *Algorithms for Robotic Motion and Manipulation (WAFR96)*, JP. Laumond and M. Overmars (Eds), AK Peters, 1997.

[23] J. Goodman and J. O'Rourke. *Handbook of Discrete and Computational Geometry.* CRC Press, 1997.

[24] C. Nissoux. Visibilit et mthodes probabilistes pour la planification de mouvement en robotique. *PdD Thesis (in french), University Paul Sabatier, Toulouse, 1999.*

[25] M. Overmars and P. Švestka. A Probabilistic learning approach to motion planning. *Algorithmic Foundations of Robotics (WAFR94)*, K. Goldberg et al (Eds), AK Peters, 1995.

[26] T. Siméon, J.P. Laumond and C. Nissoux. Visibility-based probabilistic roadmaps for motion planning. (Submitted to Advanced Robotics Journal. A short version appeared in IEEE IROS, 1999).

[27] P. Švestka. Robot motion planning using probabilistic road maps. PhD Thesis, Dept of Computer Science, Utrecht University, 1997.

[28] P. Švestka and M. Overmars. Probabilistic path planning. In [20]

AutoBalancer: An Online Dynamic Balance Compensation Scheme for Humanoid Robots

Satoshi Kagami, *University of Tokyo, Tokyo, Japan*
Fumio Kanehiro, *University of Tokyo, Tokyo, Japan*
Yukiharu Tamiya, *Namco Ltd, Tokyo, Japan*
Masayuki Inaba, *University of Tokyo, Tokyo, Japan*
Hirochika Inoue, *University of Tokyo, Tokyo, Japan*

Algorithms for maintaining dynamic stability are central to legged robot control. Recent advances in computing hardware have enabled increasingly sophisticated physically based simulation techniques to be utilized for the offline generation of dynamically-stable motions for complex robots, such as humanoid robots. However, in order to design humanoid robots that are reactive and robust, a low-level online balancing scheme is required.

This paper presents an online algorithm for automatically generating dynamically stable compensation motions for humanoid robots. Given an input motion trajectory, the "AutoBalancer" software reactively generates a modified dynamically-stable motion for a standing humanoid robot. The system consists of two parts: A planner *for state transitions derived from contacts between the robot and the ground, and a dynamic* balance compensator *which formulates and solves the balance problem as a constrained, second order nonlinear programming optimization problem. The balance compensator can be made to compensate for deviations in the centroid position and tri-axial moments of any standing motion for a humanoid robot, using all joints of the body in real-time. The complexity of the AutoBalancer algorithm is $O((p+c)^3)$, where p is number of DOFs and c is the number of constraint equations.*

We describe results obtained by an experimental implementation of the AutoBalancer algorithm that has been applied to 16-DOF and 30-DOF humanoid robots, controlled via a master-slave puppet interface.

1 Introduction

Algorithms for maintaining dynamic stability are fundamental to efforts to develop practical humanoid robots that can successfully operate in the real world. A humanoid robot may have to perform several concurrent tasks, for example: walking, carrying an object in its arms, and maintaining visibility of a target. In order to achieve such tasks, both the desired high-level behaviors and the low-level dynamic stability should be satisfied simultaneously.

Developing techniques for bipedal locomotion has been an active area of research for decades. Many approaches have been proposed to bipedal walking pattern generation [1, 2, 3]. This work can be divided into two major directions: (1) torque-based online trajectory generation methods, and (2) position based offline trajectory generation methods. The torque-based approach has many advantages, including conceptual simplicity and adaptability to rough terrain. However, in order to achieve torque-based control, the dynamic equations of the system must be solved. Since this is usually difficult, most existing solutions use model simplification assumptions such as ignoring the weight of the leg.

Position-based offline trajectory generation has been adopted mostly using ZMP[4] constraints since a humanoid type robot has many degrees of freedom. The current methods for computing dynamically stable trajectories can be categorized in three ways: (1) heuristic search such as GA (genetic algorithm), (2) problem optimization such as gradient-descent methods, and (3) solving the problem using a simplified model and iteratively calculating solutions to more detailed models

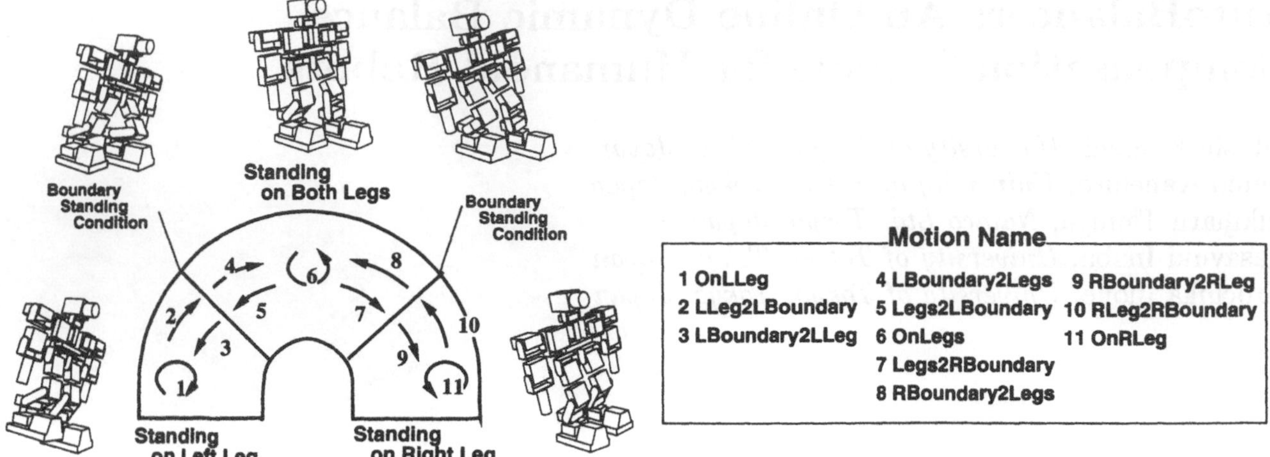

Figure 1: *Classification of Standing Motion States and their Transitions.*

(converging towards the actual model). These schemes have been successfully used to generate walking patterns for humanoid type robots [5, 6, 7].

This paper describes an online dynamic balance stabilization algorithm for humanoid robots that works by enumerating possible contact state changes, and compensating for dynamic instabilities using all joints of the body in real-time. We formulate and solve the balance compensation problem as an optimization problem. Our method can be applied to full-body motions of humanoid robots that stand on one or two legs. Although our experiments use bipedal humanoid-type robots, the algorithm can generally be applied to any type of legged robot.

The paper is organized as follows: Section 2 outlines the design of the AutoBalancer algorithm. Section 3 explains the balance compensation method in detail. Section 4 describes experiments using a 16-DOF remote-brained robot and 30-DOF child-size humanoid robot. Section 5 contains some concluding remarks and discussion.

2 Design of AutoBalancer

"AutoBalancer" makes the following assumptions:

1. A filter function is used for maintaining dynamic balance in real-time.
2. The robot tries to follow the input motion while satisfying the dynamic balance constraints.
3. All joints or selected joints can be used in order to maintain balance.
4. No assumptions are made regarding the dynamic stability of the input motion.
5. Information regarding the future input motion is unavailable, aside from the current posture.
6. The floor and the soles of the robot's feet are both planar.
7. Disturbances from the environment are assumed to be slow.

AutoBalancer is designed as a two-layered architecture: the *Planner* manages transitions between states, while the *Compensator* keeps the robot balanced in the current state.

2.1 States and Transitions for a Standing Humanoid Robot

We identify the nature of contacts between the robot and the ground by enumerating states. The robot moves by making transitions between these states. Figure 1 describes the five states and eleven transitions used by our system. Boundary condition states represent the special cases when both soles of the feet are in contact with the ground, but the entire weight of the robot is supported by only a single foot. Transitions between the single-leg contact states and the dual-leg contact state must pass through the boundary condition states. Additional transitions include: positioning the robot centroid above a sole, and lifting up or lowering a free leg to the ground. AutoBalancer implements balance compensation functions for every state transition, and switches between these functions according to the robot state.

2.2 Planner

The planner controls the robot state and transitions according to Figure 1. In this paper, we assume that direct transitions can only be made between adjacent states. Therefore we have five states and eleven transitions (including self-transitions which return to the current state). Four behaviors are introduced in order to control a humanoid robot as follows:

1. Single-leg behavior
2. Dual-leg behavior
3. Lifting up a foot behavior
4. Lowering a foot behavior

The planner controls the state transition from a given input by using these four behaviors. The planner manages the state transitions to maintain stability even when the input state changes very quickly, so it can cope with highly dynamic situations. The basis of our dynamic balance compensation method and the details for each behavior are given in the next section.

3 Dynamic Balance Compensation

3.1 Center of Gravity and Moment of Inertia Constraints

Basic Control Strategy for Dynamic Balance

The moment of inertia M and center of gravity of the robot C are utilized in order to maintain dynamic balance. We introduce the desired point of the projection of center of gravity to the ground, and name it DPCGG. The following schemes are the basic constraints for dynamic balance using DPCGG:

- Constrain the center of gravity C to lie along the perpendicular to the pre-defined DPCGG, inside the sole which is supporting the body.

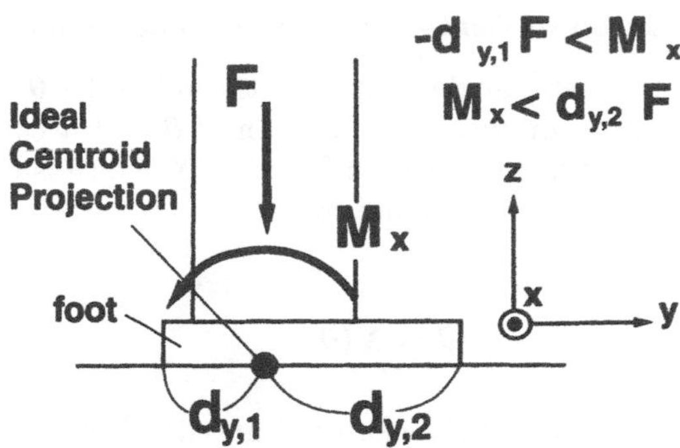

Figure 2: *Limitation of Moment.*

- Constrain the moment of inertia M around the supporting sole by a predefined tolerance value which is calculated from the shape of the sole and the leg state.

In this paper, the shape of the sole is assumed to be rectangular, and the constraints are defined as in Figure 2.

The constraints can be formalized by using the current joint angle $\boldsymbol{\theta}$ and the joint angle difference $\delta\boldsymbol{\theta}$ at each control cycle as follows:

$$\delta C_x(\boldsymbol{\theta}, \delta\boldsymbol{\theta}) = 0, \quad (1)$$

$$\delta C_y(\boldsymbol{\theta}, \delta\boldsymbol{\theta}) = 0, \quad (2)$$

$$-d_{y,1}F < M_x(\boldsymbol{\theta}, \delta\boldsymbol{\theta}) < d_{y,2}F, \quad (3)$$

$$-d_{x,1}F < M_y(\boldsymbol{\theta}, \delta\boldsymbol{\theta}) < d_{x,2}F, \text{and} \quad (4)$$

$$-\mu F < M_z(\boldsymbol{\theta}, \delta\boldsymbol{\theta}) < \mu F. \quad (5)$$

The axial components of the center of gravity and moment of inertia are denoted by $\delta C_x, \delta C_y, M_x, M_y, M_z$. Let F be a force from the ground, and $d_{x,1}, d_{x,2}, d_{y,1}, d_{y,2}$ be the distances from the edge of the supporting sole along each axis. Let μ be the coefficient of static friction. Then Equation 1 and 2 constrain the center of gravity to lie along the perpendicular line of DPCGG, given that DPCGG already lies inside the support polygon. Equation 3 and 4 is a ZMP constraint to avoid falling down by rotating around the foot edge (Figure 2). Equation 5 is a condition to prevent the foot from sliding around the Z-axis. The following two subsections describe the relationship between these constraints and each joint motion, in order to achieve stable motion.

Formulation of Center of Gravity Constraints

Let the joint angle vector of the robot be $\boldsymbol{\theta}$. $\boldsymbol{C}$, the center of gravity, is a function of $\boldsymbol{\theta}$ and it can be described as follows using a function $\boldsymbol{X}$ which is characterized by the individual robot.

$$\boldsymbol{C} = \boldsymbol{X}(\boldsymbol{\theta}) \tag{6}$$

The Jacobian of the center of gravity $\boldsymbol{J}$ is a partial derivative function matrix relating differential changes of $\boldsymbol{\theta}$ (i.e., the vector of rotation angles for all DOFs) to differential changes in $\boldsymbol{X}$. Then the velocity of the center of gravity $\dot{\boldsymbol{C}}$ and the angular velocity $\dot{\boldsymbol{\theta}}$ can be obtained as follows:

$$\dot{\boldsymbol{C}} = \boldsymbol{J}(\boldsymbol{\theta})\dot{\boldsymbol{\theta}}, \tag{7}$$

$$\boldsymbol{J}(\boldsymbol{\theta}) = \frac{\partial \boldsymbol{X}(\boldsymbol{\theta})}{\partial \boldsymbol{\theta}}. \tag{8}$$

Let the change in the joint angles at time t be $\delta\boldsymbol{\theta}(\boldsymbol{t})$, and suppose that each component is small enough. The differential changes of the center of gravity can be expressed as follows, using Equation 7.

$$\delta \boldsymbol{C} = \boldsymbol{J}(\boldsymbol{\theta})\delta\boldsymbol{\theta}(t) \tag{9}$$

Assuming that the center of gravity is on the perpendicular to the DPCGG, then the condition to restrict its derivatives on the perpendicular line of DPCGG can be represented as follows from Equation 9.

$$\delta C_x = \sum_{i=1}^{p} J_{x,i}\delta\theta_i(t) = 0, \tag{10}$$

$$\delta C_y = \sum_{i=1}^{p} J_{y,i}\delta\theta_i(t) = 0, \tag{11}$$

where i is the joint index, x, y are the motion directions, $\delta\theta_i$ is the differential angle change for joint i, $J_{x,i}, J_{y,i}$ are the components of $\boldsymbol{J}$, $\delta C_x, \delta C_y$ is the differential motion of the center of gravity, and p is the number of DOFs. Since $\boldsymbol{J}$ is a function of $\boldsymbol{\theta}$, it must be calculated at every control loop cycle. However, we can approximate $\boldsymbol{J}$ by setting the robot's current posture to $\boldsymbol{\theta}$ and perturbing each individual joint angle to measure differential changes in the relative position of $\boldsymbol{C}$. Since the calculation of the center of gravity is $O(p)$, the calculation of the approximate $\boldsymbol{J}$ is $O(p^2)$.

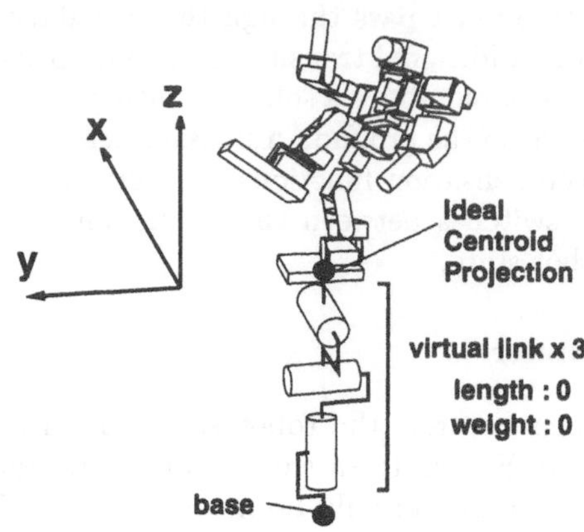

Figure 3: *Virtual Link for Calculating ZMP.*

Formulation of Moment of Inertia Constraints

In this section we introduce a system of three virtual links at DPCGG in order to calculate inertial moments. The axes of the virtual links are parallel to the coordinate system shown in Figure 2. The moments of inertia around DPCGG can be calculated as the joint torque of the virtual joints which keeps the virtual joint angle constant. In general, the motion equation can be represented as follows:

$$\boldsymbol{\tau} = \boldsymbol{H}(\boldsymbol{\theta})\ddot{\boldsymbol{\theta}} + \boldsymbol{b}(\boldsymbol{\theta}, \dot{\boldsymbol{\theta}}, \boldsymbol{g}), \tag{12}$$

where $\boldsymbol{\tau}$ is a vector of p joint torques, $\boldsymbol{H}$ is an inertia matrix of size $p \times p$, and $\boldsymbol{b}$ is a vector of size p which combines the centrifugal, coriolis, and gravity forces. However, we consider only the inertial components of $\boldsymbol{b}$, and ignore the gravitational component.

From Equation 12, the relationship between the moments of inertia around each axis M_x, M_y, M_z and the joint angle changes at each control cycle can be calculated. If the virtual joint does not move, $\delta\boldsymbol{\theta}$ can be obtained as follows:

$$M_x = -\left(\sum_{i=1}^{p} H_{x,i}\ddot{\theta}_i + b_x\right), \tag{13}$$

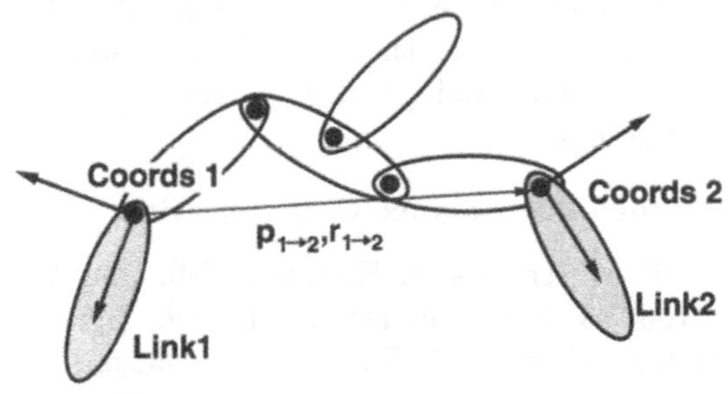

Figure 4: *Relative Position and Rotation between Link1 and Link2.*

$$M_y = -\left(\sum_{i=1}^{p} H_{y,i}\ddot{\theta}_i + b_y\right), \quad (14)$$

$$M_z = -\left(\sum_{i=1}^{p} H_{z,i}\ddot{\theta}_i + b_z\right). \quad (15)$$

Since $\dot{\theta}$ is described as a difference, $\ddot{\theta}$ is approximated as:

$$\dot{\theta}_i(t) \approx \frac{\theta_i(t) - \theta_i(t-\delta t)}{\delta t},$$
$$\ddot{\theta}_i(t) \approx \frac{\delta\theta_i(t) - \delta\theta_i(t-\delta t)}{\delta t^2}. \quad (16)$$

By plugging these expressions into Equations 3, 4, and 5, we see that the constraint equations for the moments of inertia are linear, just like the constraints on the position of the center of gravity. $\boldsymbol{H}, \boldsymbol{b}$ can be calculated using a fast algorithm for inverse dynamics based on the Newton-Euler method[8].

The inverse dynamics problem consists of computing an appropriate set of input torques $\boldsymbol{\tau}$ given a set of joint angles $\boldsymbol{\theta}$, joint angle velocities $\dot{\boldsymbol{\theta}}$, and joint angle accelerations $\ddot{\boldsymbol{\theta}}$. This algorithm solves the inverse dynamics problem for a serial-chain manipulator with rotational joints, but it can be applied to robots with a tree-like link structure (see Figure 4). Let the inverse dynamics algorithm be denoted by $INV()$. The solution torques can be computed as follows:

$$\boldsymbol{\tau} = INV(\boldsymbol{\theta}, \dot{\boldsymbol{\theta}}, \ddot{\boldsymbol{\theta}}). \quad (17)$$

Using this function, and setting some input components to zero [9], $\boldsymbol{H}$ and $\boldsymbol{b}$ can be solved as follows:

$$\boldsymbol{H}(\boldsymbol{\theta}) column\ i = INV(\boldsymbol{\theta}, \boldsymbol{O}, \boldsymbol{e}_i), \quad (18)$$
$$\boldsymbol{b}(\boldsymbol{\theta}, \dot{\boldsymbol{\theta}}, \boldsymbol{g}) = INV(\boldsymbol{\theta}, \dot{\boldsymbol{\theta}}, \boldsymbol{O}). \quad (19)$$

Where $\boldsymbol{e}_i$ is a vector whose ith component is 1 and all other components are 0. Equation 18 is calculated three times, once for each of the required components of the virtual linkage. Thus, $INV()$ is applied four times in calculating $\boldsymbol{H}$ and $\boldsymbol{b}$, and the computation is $O(p)$.

Balance Compensation as an Optimization Problem

Let $\delta\boldsymbol{\theta}_{\boldsymbol{cmd}}$ be the differential input joint angle vector, $\delta\boldsymbol{\theta}_{\boldsymbol{ret}}$ be the component to approach to the input joint angle vector, $\delta\boldsymbol{\theta}_{\boldsymbol{comp}}$ be the compensation component which satisfies the balance constraints. Then $\delta\boldsymbol{\theta}$ can be represented as follows:

$$\delta\boldsymbol{\theta} = \delta\boldsymbol{\theta}_{cmd} + \delta\boldsymbol{\theta}_{ret} + \delta\boldsymbol{\theta}_{comp}. \quad (20)$$

Given $\delta\boldsymbol{\theta}_{\boldsymbol{cmd}}$, we can calculate $\delta\boldsymbol{\theta}_{\boldsymbol{ret}}$. Using the condition of Equation 20, we can obtain constraint equations with $\delta\boldsymbol{\theta}_{\boldsymbol{comp}}$ as a variable. From the right side of Equation 20, it follows that $\delta\boldsymbol{\theta}$ becomes similar to the given input when $\delta\boldsymbol{\theta}_{\boldsymbol{comp}}$ is close to zero. Therefore, we can find a $\delta\boldsymbol{\theta}_{\boldsymbol{comp}}$ which minimizes an evaluation function as follows:

$$E(\delta\boldsymbol{\theta}_{\boldsymbol{comp}}) = \sum_{i=1}^{p} w_i(\delta\boldsymbol{\theta}_{\boldsymbol{comp},i})^2, \quad (21)$$

where w_i is a positive weight variable. Large values for w_i cause $\delta\boldsymbol{\theta}_{\boldsymbol{comp},i}$ to be closer to zero, which in turn causes the ith joint angle to more closely match the given input. The factor w_i essentially attempts to encode the importance of the ith joint angle.

The three inequalities of Equations 3, 4, and 5 form a second order programming problem. We can convert each inequality to an equality, and use Lagrange multipliers to solve this system.

$\delta\boldsymbol{\theta}$ can be obtained by solving Equation 21. Thus, balance compensation can be solved as a second-order programming problem in every control loop. The computational complexity of this problem is $O(3^n \times (p+c)^3)$ where n is the number of inequality constraints, p is the number of DOFs, and c is the total number of constraints. Since Equations 1–5 consist of exactly three inequalities, then n is a constant number.

Limiting the Momentum

We would like to impose limits on the moments of inertia by considering the accumulated angular momentum of the robot. Consider the x-axis direction. Let $M_{x,min}, M_{x,max}$ be the maximum and minimum moment of Equation 3. The limit values for the angular momentum of the robot $L_{x,min}, L_{x,max}$ can be defined as follows:

$$L_{x,min} = -M_{x,max} t_s, \tag{22}$$
$$L_{x,max} = -M_{x,min} t_s. \tag{23}$$

where t_s is the given time to stop the motion. The angular momentum L_x can be calculated as follows, by integrating both sides of Equation 12:

$$L_x = \sum_{i=1}^{p} H_{x,i} \dot{\theta}_i. \tag{24}$$

Therefore, the angular momentum can be limited by setting the moment limit values as follows, using δt:

$$\frac{L_{x,min} - L_x}{\delta t} \leq M_x \leq \frac{L_{x,max} - L_x}{\delta t} \tag{25}$$

Similarly, moment limit values can be defined for the y and z axes.

3.2 Single-leg Balance Compensation

Leg Interference

During single-leg motions, the free leg may collide with the supporting leg. To help prevent this, we limit the y-direction of the crotch joint, so that the angles are not changed from the given input:

$$\delta\theta_{left\_crotch\_y} + \delta\theta_{right\_crotch\_y} = 0. \tag{26}$$

However, this constraint is not sufficient to guarantee that no collision will occur, since all the cases that result in leg collisions are hard to formulate. In our current implementation, the user of the system must take care to detect and prevent possible interference between the legs.

Solving Single-leg Balance Compensation

Since all constraints in Equation 1-5, and 26 are represented as linear equations in $\delta\theta$, by simply substituting $\delta C_x, \delta C_y, M_x, M_y, M_z$ in Equation 20 and $\delta\theta_{left\_crotch\_y}, \delta\theta_{right\_crotch\_y}$ in Equation 26, the optimal $\delta\theta_{comp}$ can be obtained.

3.3 Dual-leg Balance Compensation

In the dual-leg state, both legs cannot move independently, so a geometrical joint constraint method is adopted to constrain the relative position and posture of both feet. We treat the dual-leg state as a special case of a single-leg state using the following assumptions: without loss of generality,

- one leg is designated as the supporting leg, while the other leg is free.
- the foot shape of the supporting leg is rectangular and included in the convex hull of both feet (we refer to this as the *virtual sole*, see Figure 5)
- both the relative position and posture are geometrically constrained between the two feet.
- the DPCGG is placed along the line segment connecting the DPCGG of each individual foot.

Formulation of Geometric Joint Constraints

When both legs are contacting the ground, a closed loop is formed and all the links are constrained geometrically. In this section, we describe a constraint method between two distant links. From the system of local coordinates $Coords_1$ which is attached to one link, the posture of the local coordinates $Coords_2$ of the other link can be represented by a translation vector $\boldsymbol{p}_{1\to2}$ and a rotation vector $\boldsymbol{r}_{1\to2}$ (Figure 4).

For every control cycle, the posture change between two links $\delta\boldsymbol{p}_{1\to2}, \delta\boldsymbol{r}_{1\to2}$ can be represented as a function of $\delta\boldsymbol{\theta}$ using the Jacobian $\boldsymbol{K}$ as follows:

$$\left\{ \begin{array}{c} \delta\boldsymbol{p}_{1\to2} \\ \delta\boldsymbol{r}_{1\to2} \end{array} \right\} = \boldsymbol{K}\delta\boldsymbol{\theta}. \tag{27}$$

Motion Name	Centroid Position	Supporting Polygon	Constraints	Complexity
OnLLeg OnRLeg	Fix above DPCGG on Supporting Foot	Sole of Supporting Leg	Equation.1–5 Equation 26	$O((p+6)^3)$
OnLegs	Fix above DPCGG on Virtual Sole	Virtual Sole	Equation.1–5 Equation 27	$O((p+11)^3)$
LLeg2LBoundary RLeg2RBoundary	Fix above DPCGG on Supporting Foot	Sole of Supporting Leg	Equation.1–5 Equation 26	$O((p+11)^3)$
LBoundary2LLeg RBoundary2RLeg	Fix above DPCGG on Supporting Foot	Sole of Supporting Leg	Equation.1–5 Equation 27	$O((p+11)^3)$
Legs2LBoundary Legs2RBoundary	Approach above DPCGG on One Leg	Virtual Sole	Equation.1–5 Equation 27	$O((p+11)^3)$
LBoundary2Legs RBoundary2Legs	Approach above DPCGG on Virtual Sole	Virtual Sole	Equation.1–5 Equation 27	$O((p+11)^3)$

Table 1: *Control Condition in Compensator.*

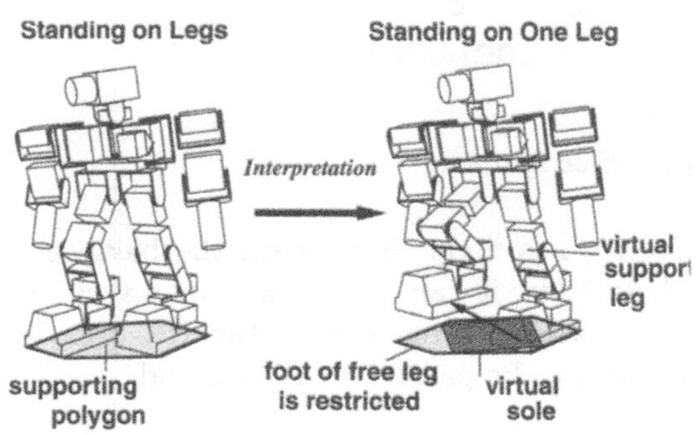

Figure 5: *Standing in the Dual-leg State.*

An approach similar to the computation of the balance constraints defined for the single-leg state is adopted. Equation 27 gives a linear equation of $\delta\boldsymbol{\theta}$. The Jacobian $\boldsymbol{K}$ is calculated at every control cycle using the 3D model of the robot.

Solving Dual-leg Balance Compensation

Since every constraint in Equations 1–5 and 27 is represented as a linear equation of $\delta\theta$, by assigning δC_x,δC_y,M_x,M_y,M_z to Equation 20 and $\delta\boldsymbol{p}_{1\to2}$,$\delta\boldsymbol{r}_{1\to2}$ to Equation 27, the optimal $\delta\theta_{comp}$ can be obtained.

3.4 Balance Compensation While Lowering the Foot of a Free Leg

In this section we describe the geometrical constraints used for the state transition from the single leg state to the boundary of the dual leg state. The constraints result in a motion that lowers the foot of the free leg to the ground. We name the foot landing position the *desired sole position* (DSP). The motion can be calculated as follows:

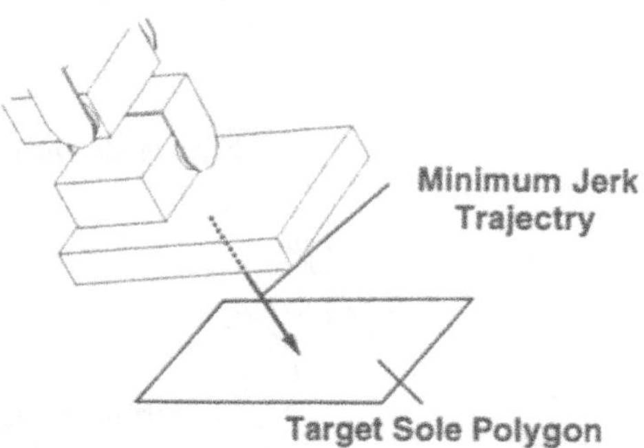

Figure 6: *Lowering a free leg to the ground.*

1. Plan a trajectory from the current sole position to the DSP.
2. Calculate the free leg position and posture $\delta\boldsymbol{p}_{1\to2}$ and $\delta\boldsymbol{r}_{1\to2}$ at every control cycle, and impose geometrical constraints, using Equation 27.

The trajectory of the free leg is calculated using a minimum jerk model[10] to generate a smooth motion (Figure 6).

3.5 Balance Compensation While Lifting a Foot

At the start of a lifting motion, the sole of the free leg is attached to the ground. Thus, the robot may cause

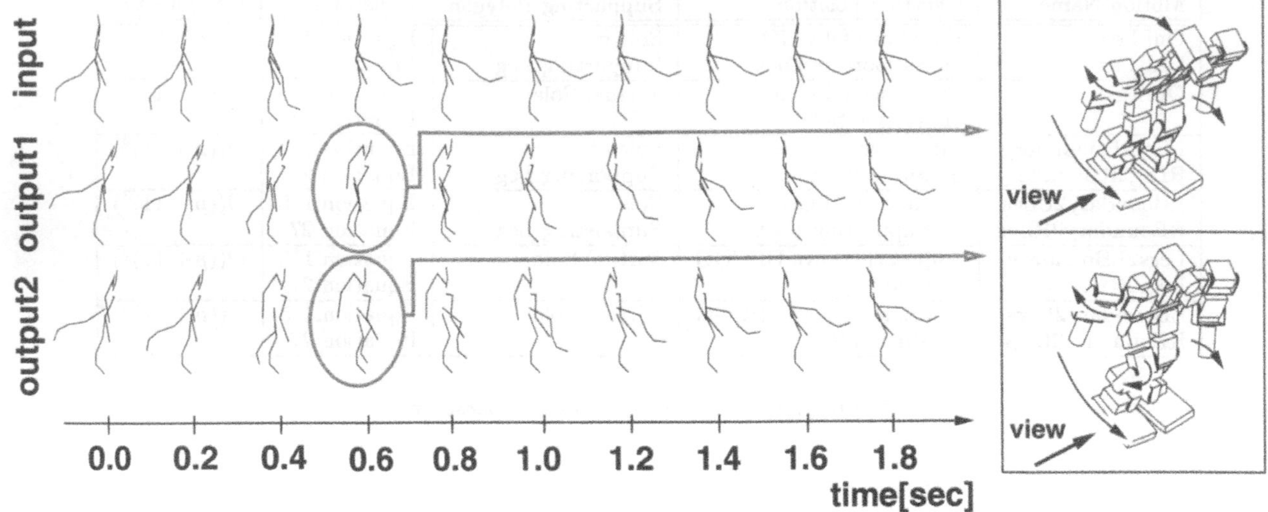

Figure 8: *Balance Compensation for Kick Motion.*

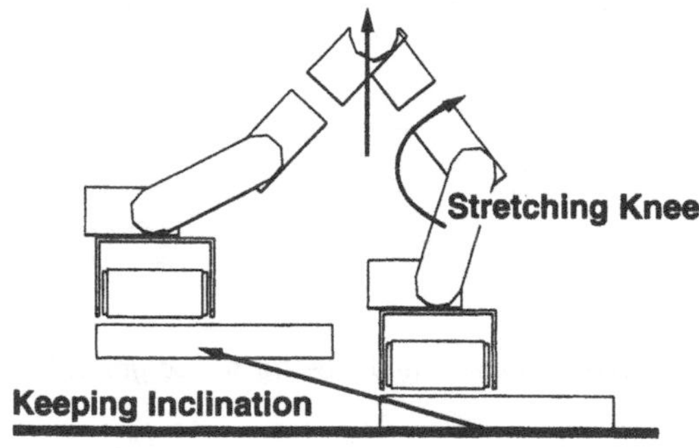

Figure 7: *Control in lifting a leg.*

itself to fall down by inadvertently kicking the ground. Therefore, we introduce the following constraints along with the single-leg state constraints, in order to not touch the ground during a lifting motion (see Figure 7).

a. ***Supporting Knee Joint***
The robot bends the knee of the supporting leg at the boundary of the standing state. By augmenting the variable w_i in Equation 21 corresponding to the weight of the knee joint temporarily, the knee can directly follow the given motion.

b. ***Free Sole Constraint***
Since the motion caused by the ankle joint of the free leg has little effect upon the balance of the whole body, it can follow the input very quickly. Thus, when lifting a foot, the free sole is constrained to be parallel to the ground.

After some pre-defined time, the system transitions to a single leg state.

3.6 Complexity of Balance Compensation

Table 1 shows the constraint equations for each state. In each state, the program solves 5 or 11 constraint equations including 3 inequalities (Equations 3, 4, and 5). Recall that the computational complexity for calculating J is $O(p^2)$, and the complexity for computing H, b are $O(p)$. Our program calculates an LU-decomposition with partial pivoting in order to solve the second order programming problem. The computational complexity of this process is $O(p + c)$.

4 Interactive Humanoid Experiments Using AutoBalancer

4.1 Humanoid Robot Model

A 3D robot model is required to calculate the dynamics. We treat the robot as a set of rigid bodies connected by rotational joints. For each link we have the mass, position of the center of gravity, inertia tensor matrix, and information about the adjacent

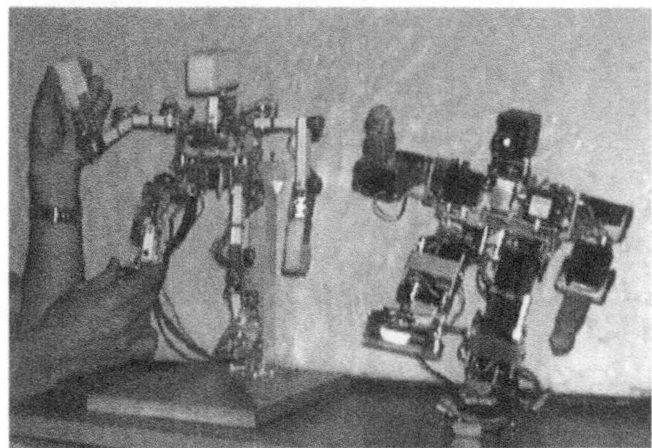

Figure 9: *Direct Operation Using a Puppet.*

links. This robot model is implemented in our robot model environment[11] based on Euslisp[12]. In order to achieve real-time motion, the robot model is compiled into C code, which is used to control the real robot.

4.2 Remote-Brained 16DOFs Humanoid "Akira" and Master-Slave Interface

We implemented our approach on the remote-brained 16DOF Humanoid robot "Akira." We controlled the robot using a puppet-based master-slave interface (see Figure 9). The robot is 320mm tall, and weighs 2.2kg. It has 4 DOFs in each leg, 3 DOFs in each arm, and 2 DOFs in the neck. The robot has 4 force sensors at the bottom of each sole which are used to measure the ZMP. The puppet has the same DOFs and link proportions as "Akira," and a human can control the robot intuitively. In order to designate the desired leg state, two touch switches are attached to the sole of the puppet.

Figure 10 shows the time series of inputs from the puppet, outputs to the robot, and the measured ZMP. It also describes the state transitions calculated by our Planner. There are long time delays from the given dual leg state designation input and output of AutoBalancer (about 2 seconds). Therefore, the human operator should work within these delays.

Figure 8 shows a kicking motion input and two types of balance-compensated motions. Every joint weight is equal in the middle row, and the crotch joint has 5 times the weight in the lower row. In both outputs, the x-axis moment is compensated by the upper body motion, the knee joint of the free leg is bent in the weighted version, and the z-axis moment is compensated by the arm motions.

4.3 Experiment using Humanoid "H5"

The remote-brained robot Akira has only 16DOFs. We also applied AutoBalancer to the 30DOF humanoid "H5". H5 stands 1300mm high, and weighs 33kg. It has 6 DOFs in each leg, 6 DOFs in each arm, 1 DOF for each gripper, and 4 DOFs in the head. It has 12 force sensors at the bottom of each sole which are used to measure the ZMP. It has a PentiumIII-333MHz PC/AT clone mounted on the body. It uses a 1msec motor servo loop, and 20msec Compensator. Figure 11 shows a high kick motion using H5.

5 Concluding Remarks

This paper describes "AutoBalancer" which reactively generates the stable motion of a standing humanoid robot on-line, from given motion patterns. The system consists of two parts: (1) a planner for transitions between states derived from the contacts between the legs and the ground, and (2) a dynamic balance compensator which maintains dynamic stability through solving a constrained second order nonlinear programming optimization problem. The dynamic balance compensator can compensate for the centroid position and the tri-axial moments of any standing motion, using all joints of the body in real-time. The complexity of AutoBalancer is $O((p+c)^3)$, where p is the number of DOFs and c is the number of constraint equations. Through experiments with real humanoid robots, we have shown that current hardware can treat such a problem in real-time. The algorithm can potentially be enhanced to control higher DOF humanoids, or other multiple-legged robots. The compensator can also be applied to stabilize dynamic walking trajectories for humanoid robots.

Currently, the planner component of AutoBalancer cannot generate state changes automatically. In addition, the time required to achieve the final configuration of the input motion may be increased by the necessity of having to pass through the boundary condition states. More research is needed to address these limitations. Another problem is that the current algorithm cannot handle rapid disturbances from the environment. Incorporating dynamic state changes in-

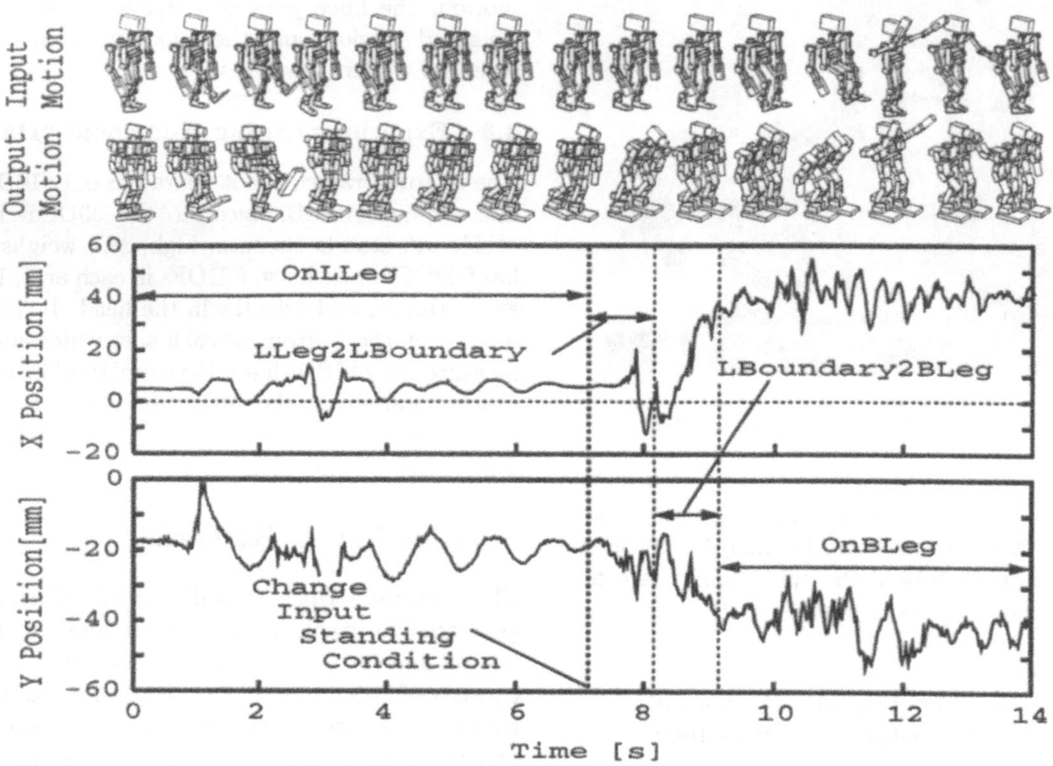

Figure 10: *Compensated Motion and ZMP Tracks.*

cluding reaction to sensor input would be a very useful improvement.

References

[1] M. H. Raibert, H. B. Brown, and M. Chepponis. Experiments in Balance with a 3D One-Legged Hopping Machine. *Robotics Research*, 3(2):75–92, 1984.

[2] Marc H. Raibert. Running with symmetry. *Robotics Research*, 5(4):3–19, 1986.

[3] Jessica K. Hodgins and Marc H. Raibert. Biped gymnastics. *Robotics Research*, 9(2):115–132, 1990.

[4] Vukobratović, A.A. Frank, and D.Juričić. On the Stability of Biped Locomotion. *IEEE Trans. on Biomedical Engineering*, 17(1):25–36, 1970.

[5] J. Yamaguchi, A. Takanishi, and I. Kato. Development of a Biped Walking Robot Compensating for Three-Axis Moment by Trunk Motion. *Journal of the Robotics Society of Japan*, 11(4):581–586, 1993.

[6] Kazuo HIRAI. Current and Future Perspective of Honda Humanoid Robot. In *Proc. of 1997 IEEE Intl. Conf. on Intelligent Robots and Systems (IROS'97)*, pages 500–508, 1997.

[7] K. Nagasaka, M. Inaba, and H. Inoue. Walking Pattern Generation for a Humanoid Robot based on Optimal Gradient Method. In *Proc. of 1999 IEEE Int. Conf. on Systems, Man, and Cybernetics No. VI*, 1999.

[8] J.Y.S.Luh, M.W.Walker, and R.P.C.Paul. On-Line Computational Scheme for Mechanical Manipulators. *ASME Journal of Dynamic Systems, Measurement, Control*, 102:69–76, 1980.

[9] M.W.Walker and D.E.Orin. Efficient Dynamic Computer Simulation of Robotic Mechanisms. *ASME Journal of Dynamic Systems,Measurement,Control*, 104:205–211, 1982.

[10] T.Flash and N.Hogan. The coordination of arm movements:an experimentally confirmed mathematical model. *The Journal of Neuroscience*, 5(7):1688–1703, 1985.

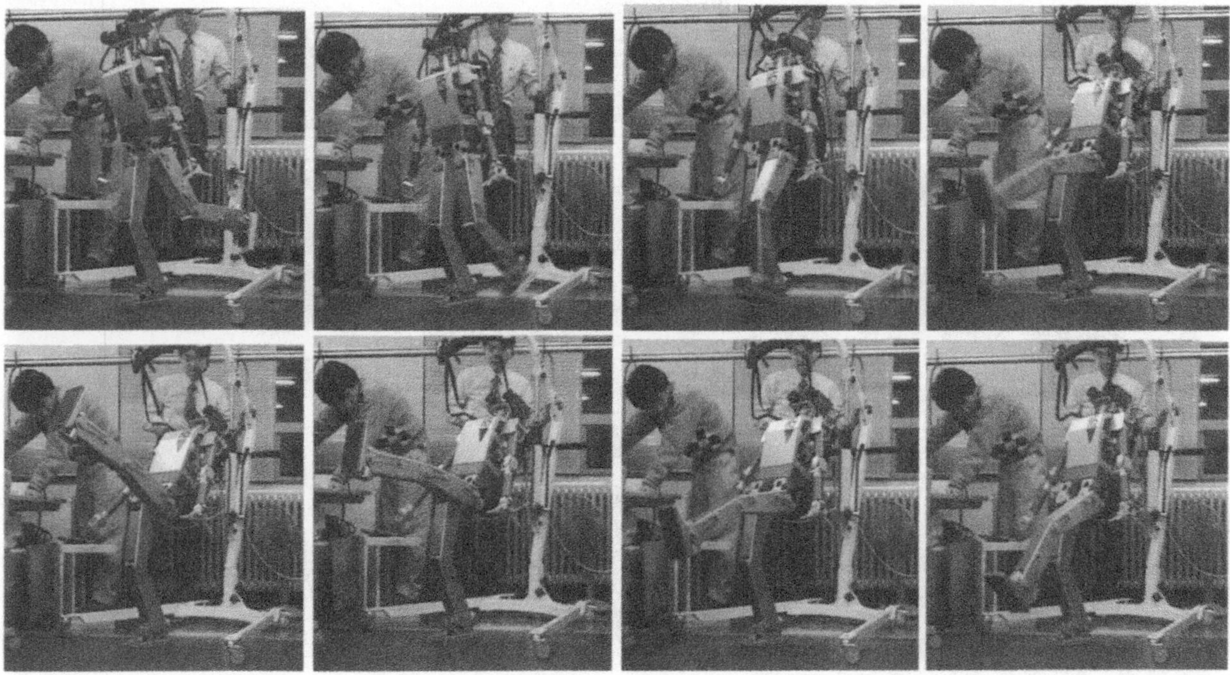

Figure 11: *High Kick Motion using H5*

[11] S. Kagami, M. Inaba, and H. Inoue. Design and Development of Software Platform using Remote-Brained Approach. *Journal of the Robotics Society of Japan*, 15(4):550–556, 1997.

[12] Toshihiro Matsui. Multithread Object-Oriented Language Euslisp for Parallel and Asynchronous Programming in Robotics. In *Workshop on Concurrent Object-based Systems, IEEE 6th Symposium on Parallel and Distributed Processing*, 1994.

Coupled Oscillators for Legged Robots

Matthew D. Berkemeier, *Utah State University, Logan, UT*

Biologists have uncovered some of the mechanisms at work in the control of legged animals. The animal CPG has received much attention and provides new inspiration for the control of legged robots. This paper presents the further development of a control method which involves coupled oscillators. In particular, the paper discusses a pair of coupled oscillators which operate in discrete time and which have been implemented on a pair of microcontrollers. Coupling between the oscillators only need occur twice each period (once for each oscillator), rather than continually. A simulation demonstrates phaselocking of the two oscillators. Experimental results are briefly described.

1 Introduction

Legged robots are thought to have great potential for locomotion over rough terrain. This belief depends heavily on observations of legged animals in rugged outdoor environments. Currently, there are no legged robots which can compare with humans and other animals in locomotion in unstructured environments. This is no surprise, as there are many other problems in which biological systems excel and engineered systems do not (and vice versa). This paper discusses continuing efforts to build a quadrupedal robot with the ability to run using several gaits. Simple biological principles (including central pattern generation) are being used to design and control the robot.

The current concept for a simple running quadruped model and mechanical design is shown in Figure 1. The discussion of this model will be divided into three parts.

Mechanical System This component is analogous to the animal musculoskeletal system. Rather than start with four legs, the model is restricted to a vertical plane and has the two front and two rear legs collapsed into one front and one rear leg. In addition, the legs can telescope only. There is no joint to move the leg forward and backward. Forward motions of the robot are not modeled yet. However, there is good reason to believe that the vertical motions of a running system are the most important. In our own analysis the vertical motions appear to be relatively unaffected by forward motions. However, the converse is not true: the forward motions rely greatly on the vertical dynamics. For additional evidence of this, see [8]. The obvious reason for these simplifications is to make analysis tractable. While it is easy to simulate a four-legged robot with many actuators and joints, it is usually not possible to derive analytical results with such a complicated system.

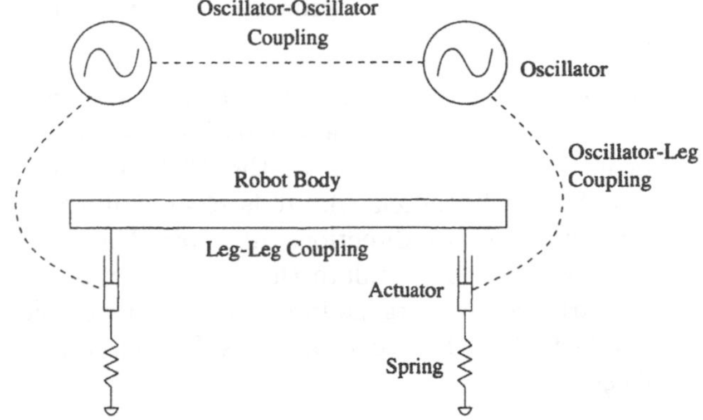

Figure 1: *Biomimetic Legged Robot Control.*

Coupled Oscillator Control Animals are thought to possess a Central Pattern Generator (CPG) made up of neurons in the spinal cord. The CPG produces patterns for the stimulation of muscles in rhythmic activities, such as locomotion. The CPG of (at least) simple animals can produce these patterns even when removed from the animal. Thus, it does not depend on feedback from sensory organs in the animal. However, when this feedback is present, the patterns are modulated to produce signals which are better adapted to the environment the animal is operating in. This is discussed in [6]. Applied mathematicians have modeled CPGs as coupled oscillators [5, 7]. This provides a convenient abstraction of the complex system of neurons present in a spinal cord. The model in Figure 1 contains two coupled oscillators. These oscillators are the primary focus of this paper.

Oscillator-to-Leg Coupling The connection between the animal CPG and muscles has not received as much attention as the CPG itself (at least by mathematicians). It is clear that there is coupling in both directions, however. The CPG must be able to stimulate the animal's muscles. Similarly, sensory organs in the legs must be able to affect the CPG to change its frequency and phase.

The approach taken in this paper is different from that typically found in engineering and control papers. Rather than attempting to construct an algorithm to produce specific leg motions and gaits (synthesis), the preliminary system in Figure 1 has been put together based on inspiration from biology and intuition. The intuition comes primarily from rigorous mathematical results which show that often, when two or more similar oscillators are coupled together in some manner, they will synchronize [11]. The legged robot model in Figure 1 consists of several oscillators which are coupled (note that the legs are mechanical oscillators). Intuition suggests that for some parameter choices, the oscillators in the system should synchronize. This behavior may correspond to gaits which animals use. While intuition was used to assemble the system in Figure 1, careful analysis of the model is being used to determine the resulting behavior of the system.

2 Background

In previous work, the first and third components of the model in Figure 1 were considered in detail. The mechanical system was analyzed in [1] and [2]. Two gaits were considered, the pronk and the bound. Using perturbation techniques, it was found that the stability of the bound depended solely on the dimensionless inertia of the robot body. [9] had earlier proved a similar result. On the other hand, the pronk's stability depended both on the inertia and the pronking height. Oscillator-to-leg coupling was next considered in [4] and [3]. It was shown that two-way coupling between an oscillator and a hopping leg provided better performance than one-way coupling from the oscillator to the leg.

[10] present a simple but elegant model for 2 coupled oscillators, in which only the phase of the oscillators is important. Let θ_i, $i = 1, 2$, be the phase of oscillator i. The model is:

$$\dot{\theta}_1 = \omega_1 + a_{12}\sin(\theta_2 - \theta_1) \quad (1)$$
$$\dot{\theta}_2 = \omega_2 + a_{21}\sin(\theta_1 - \theta_2). \quad (2)$$

Oscillator i has an intrinsic frequency ω_i. The parameters a_{ij} are constants which control the coupling of the oscillators to one another. If the constants are zero, then the oscillators' phase increases at the constant rate $\omega_i t$. For the case of non-zero coupling constants, let $\phi = \theta_1 - \theta_2$. The dynamics of ϕ (the phase difference) are then given by:

$$\dot{\phi} = \omega_1 - \omega_2 - (a_{12} + a_{21})\sin\phi.$$

The equilibria are given by:

$$\phi = \sin^{-1}\left(\frac{\omega_1 - \omega_2}{a_{12} + a_{21}}\right). \quad (3)$$

These are 1:1 phase-locked motions, in which the phase difference $\phi(t)$ is a constant over time. As an example, suppose $\omega_1 = \omega_2$. Then the two unique equilibria are $\phi = 0, \pi$. If $a_{12} + a_{21} > 0$, then in-phase oscillations ($\phi = 0$) are stable, but out-of-phase oscillations ($\phi = \pi$) are unstable. The reverse is true for $a_{12} + a_{21} < 0$. For a more detailed discussion see [10]. They also extend this model to a chain of N coupled oscillators with nearest neighbor coupling.

This model has some features, which are potentially useful in an engineered system (e.g., legged robot). These are:

- Each oscillator can function without the other. If the coupling mechanism fails for some reason, then an oscillator can still proceed at its intrinsic frequency. If one oscillator fails, then the other oscillator can still function at its intrinsic frequency.

- If the coupling constants are accessible to an operator (or higher level control system) then there is an easy way to specify whether in-phase or out-of-phase oscillations occur. As discussed above, if $a_{12} + a_{21} > 0$, then in-phase oscillations are the result, while if $a_{12} + a_{21} < 0$, then out-of-phase oscillations occur.
- The system is very simple. This one system has two patterns embedded in it. It is not necessary to have two distinct systems to generate in-phase and out-of-phase oscillations. The sign of the coupling constants simply determines this.
- The system is modular. Additional oscillators can be added to produce a chain [10] or a ring [5]. It is not necessary to modify the existing set of oscillators when new oscillators are added, except to possibly introduce additional coupling.

While the above model nicely demonstrates coupled oscillations, it is probably not the best for use in robotics. In order to implement it, one would either have to use a numerical integration routine on a microprocessor or build an analog circuit. Suppose microprocessors (or microcontrollers) were used. To obtain the benefits of fault tolerance, modularity, etc., it would make sense to use one microprocessor per oscillator. However, continuous coupling between oscillators is required. This would require a high-speed communications link between the two microprocessors. A better approach for robotics would be a modified set of equations which only require infrequent communication.

3 Coupled Oscillations from Difference Equations

Extending the model of [10] to difference equations is not difficult. The following is one possible approach.

Each oscillator will produce a signal with only 2 distinct values, say 0 and 1. We will only worry about the transition from 0 to 1. This will occur at time t_{fi}, where $i = 1, 2$ (the f is for "fire"). The transition from 1 to 0 is not important at this point, so long as it occurs. The firing time is given by a phase θ_i and a constant period T_i:

$$t_{f1}(j) = \theta_1(j) + T_1 j \quad (4)$$
$$t_{f2}(k) = \theta_2(k) + T_2 k. \quad (5)$$

It is significant that the intrinsic period, T_i, is fixed but the phase changes from one firing to the next. This is consistent with the model in equations (1, 2). Also, in previous work [3], the oscillator's period was adjusted in response to feedback from a leg sensor. Thus, in the full implementation (Figure 1), the plan is to adjust the period, but it will be in response to the leg's sensor, rather than in response to coupling between the oscillators.

The above oscillators would be easy to implement on a microcontroller. A timer on a microcontroller could be used to signal (interrupt) when the desired amount of time had elapsed (T_i). When the signal occurred, the microcontroller would change its state from 0 to 1.

Next, the form of coupling between the oscillators needs to be determined. There are many different couplings which could be used. For example,

- When an oscillator fired (made the transition from 0 to 1), it could check the state of the other oscillator. If the other oscillator was 1 (it had previously fired), then the phase should be reduced by some fixed amount, to try to get them to fire at the same time in the future. If the other oscillator was 0, the phase should be increased by some fixed amount.
- When an oscillator fired, it could immediately notify the other oscillator (i.e., interrupt the other microcontroller). The other oscillator would compare the firing time with its own. This could either be a previous firing, or a firing which was about to occur (whichever was closest). The difference would be considered an error quantity, and it could advance or retard the phase.

The first approach would be easier to implement, but the second has the advantage that there is an amplitude associated with the error. Performance might be better (particularly as in-phase operation was approached), if the error became smaller, the closer the firings came to being coincident.

An implementation of the second approach is as follows:

$$\theta_1(j+1) = \theta_1(j) + a_{12}[t_{f2}(k) - t_{f1}(j)] \quad (6)$$
$$\theta_2(k+1) = \theta_2(k) + a_{21}[t_{f1}(j) - t_{f2}(k)] \quad (7)$$

where a_{12} and a_{21} are the coupling constants. For a given j, the value of k is chosen to minimize:

$$|t_{f1}(k) - t_{f2}(j)| \quad (8)$$

In general, it is difficult to write down an explicit expression for k as a function of j which minimizes (8). On the other hand, the implementation of this form of coupling on a pair of microcontrollers would be relatively easy.

Between the previous and next firing of an oscillator (say, oscillator 1, with previous firing time corresponding to $j-1$ and next firing time corresponding to $j+1$), there is a window, during which the oscillator can receive notification that the other oscillator has fired. When the other oscillator fires, this time is compared to the current firing time (oscillator 1, firing time corresponding to j) and the difference produces a phase adjustment for the next firing time (oscillator 1, firing time corresponding to $j+1$). It is possible that no notification of firing is received during this window. For simplicity, this possibility is not considered in the present paper. Note that the above discussion might be easier to understand by referring to Figure 2.

Although each oscillator has an intrinsic period (T_1 and T_2), the actual period can be different if the phase is not constant. The effective period of oscillator 1 is given by:

$$T_{\text{eff}1}(j) = t_{f1}(j+1) - t_{f1}(j) = \theta_1(j+1) - \theta_1(j) + T_1.$$

Similarly, the effective period of oscillator 2 is given by:

$$T_{\text{eff}2}(k) = t_{f2}(k+1) - t_{f2}(k) = \theta_2(k+1) - \theta_2(k) + T_2.$$

In [10] the continuous-time oscillators could phase lock with a new frequency which was somewhere between the intrinsic frequencies of the 2 oscillators. This is also possible with the discrete-time system. Equating the two effective periods gives:

$$a_{12}[t_{f2}(k) - t_{f1}(j)] + T_1 = a_{21}[t_{f1}(j) - t_{f2}(k)] + T_2$$

or,

$$t_{f1}(j) - t_{f2}(k) = \frac{T_1 - T_2}{a_{12} + a_{21}} \tag{9}$$

Notice the similarity between this equation and Equation 3. If this equation held, then the phase changes in each oscillator would be constant and given by:

$$\begin{aligned} \theta_1(j+1) - \theta_1(j) &= -a_{12}\frac{T_1 - T_2}{a_{12} + a_{21}} \\ \theta_2(k+1) - \theta_2(k) &= a_{21}\frac{T_1 - T_2}{a_{12} + a_{21}}. \end{aligned}$$

Finally, the effective period of each oscillator would be:

$$T_{\text{eff}} = \frac{T_1 a_{21} + T_2 a_{12}}{a_{12} + a_{21}}. \tag{10}$$

Notice that this is just a weighted average of the two intrinsic periods.

While the above shows the possibility that phase locked patterns could exist with the discrete time system, it certainly doesn't demonstrate that they will exist. In order to give more confidence, we will assume that the value of k which minimizes Equation 8 for a given j is simply j. While this will not always be the case, in Section 4 a specific example is given where this holds.

Let $\tau(j) = t_{f1}(j) - t_{f2}(j)$ be the difference in firing times. Then,

$$\begin{aligned} \tau(j+1) &= t_{f1}(j+1) - t_{f2}(j+1) \\ &= \theta_1(j+1) + T_1(j+1) - \theta_2(j+1) \\ &\quad -T_2(j+1) \\ &= \theta_1(j) + T_1 j + a_{12}[t_{f2}(j) - t_{f1}(j)] + T_1 - \\ &\quad \theta_2(j) - T_2 j - a_{21}[t_{f1}(j) - t_{f2}(j)] - T_2 \\ &= t_{f1}(j) + a_{12}[t_{f2}(j) - t_{f1}(j)] + T_1 - \\ &\quad t_{f2}(j) - a_{21}[t_{f1}(j) - t_{f2}(j)] - T_2 \\ &= [1 - (a_{12} + a_{21})]\tau(j) + T_1 - T_2. \end{aligned} \tag{11}$$

Note that although Equation 11 is analogous to Equation 3, Equation 11 does not have the nonlinear sin() term that Equation 3 does. This makes the analysis simpler, although it also means that only one fixed point exists.

Setting $\tau(j+1) = \tau(j) = \tau$ gives the fixed point:

$$\tau = \frac{T_1 - T_2}{a_{12} + a_{21}}.$$

which is the same as Equation 9. This fixed point is stable provided that:

$$0 < a_{12} + a_{21} < 2. \tag{12}$$

Proposition 1 gives conditions under which equation 11 represents the dynamics of the discrete-time coupled oscillators (Equations 4–7, subject to k minimizing Equation 8). This result is somewhat preliminary, and efforts are still being made to refine it. Before discussing the proposition, note that certain self-consistency checks are possible. For example, we would expect that after the transient had died out, the firing time difference between the two oscillators would have to be less than half the effective period of each oscillator. Otherwise, Equation 8 would not be minimized. Using Equations 9 and 10 gives:

$$|T_1 - T_2| < \frac{|T_1 a_{21} + T_2 a_{12}|}{2}. \tag{13}$$

Equation 13 can be viewed as a necessary condition for Equation 11 to represent the dynamics of the coupled oscillators. Satisfying this equation doesn't mean that Equation 11 will actually represent the dynamics since it was applied by assuming steady-state conditions. Certainly, things could happen during the transient phase which would invalidate the assumption that the value of k which minimizes Equation 8 is j. Also, not satisfying Equation 13 doesn't mean that the oscillators will fail to phase lock. It just means that they won't phase lock according to the dynamics given in Equation 11.

Of course, Equation 11 can be solved exactly. The solution is

$$\tau(j) = \left[\tau(0) - \frac{T_1 - T_2}{a_{12} + a_{21}}\right][1-(a_{12}+a_{21})]^j + \frac{T_1 - T_2}{a_{12} + a_{21}}.$$

Using this result it is again possible look for self-consistency violations (i.e., cases where Equation 8 is not minimized). However, this result allows consideration of the transient portion of the dynamic evolution. If the inequality in Equation 12 is satisfied, a bound can be put on the firing time difference:

$$\left|\tau(j) - \frac{T_1 - T_2}{a_{12} + a_{21}}\right| \le \left|\tau(0) - \frac{T_1 - T_2}{a_{12} + a_{21}}\right| \ \forall j. \tag{14}$$

Next, we look at the main result.

Proposition 1 *Consider the system (4–7), subject to choosing k such that (8) is minimized for each j. Assume that (12) is satisfied. If:*

1. *The inequality*

$$|T_1 - T_2| < \frac{|T_1 a_{21} + T_2 a_{12}|}{2}$$

is satisfied,

2. *the quantity*

$$\left|t_{f1}(0) - t_{f2}(0) - \frac{T_1 - T_2}{a_{12} + a_{21}}\right|$$

is sufficiently small, and

3. *$|a_{12}|$ and $|a_{21}|$ are each sufficiently small,*

then $k = j$.

Proof: Consider the function:

$$\tilde{\tau}_j(k) = t_{f1}(j) - t_{f2}(k)$$

For a fixed j, this is a strictly decreasing function of k if the firing times $t_{f2}(k)$ are strictly increasing with k. Condition 3 is meant to ensure this, although the details are not included here. The function $|\tilde{\tau}_j(k)|$ will have a single minimum for each j, which will be given by one or possibly two values of k. The following two conditions are equivalent:

- $\bar{k}_j$ is the unique value of k which gives a minimum of $|\tilde{\tau}_j(k)|$.
- $|\tilde{\tau}_j(\bar{k}_j)| < |\tilde{\tau}_j(\bar{k}_j + 1)|$ and $|\tilde{\tau}_j(\bar{k}_j)| < |\tilde{\tau}_j(\bar{k}_j - 1)|$

Clearly, then, to show that $\bar{k}_j = j$, it is sufficient to show that:

$$|t_{f1}(j) - t_{f2}(j)| < |t_{f1}(j) - t_{f2}(j+1)| \ \forall j$$

and

$$|t_{f1}(j) - t_{f2}(j)| < |t_{f1}(j) - t_{f2}(j-1)| \ \forall j.$$

Previously, $\tau(j)$ was defined to be $t_{f1}(j) - t_{f2}(j)$. By making use of Equations 4–7 the above two inequalities become:

$$|\tau(j)| < |(1 - a_{21})\tau(j) - T_2| \ \forall j \tag{15}$$

and

$$|\tau(j)| < |(1 - a_{12})\tau(j) + T_1| \ \forall j. \tag{16}$$

From condition 1,

$$\left|\frac{T_1 a_{21} + T_2 a_{12}}{a_{12} + a_{21}}\right| > 2\left|\frac{T_1 - T_2}{a_{12} + a_{21}}\right|.$$

This implies both of the following:

$$\left|\frac{T_1 a_{21} + T_2 a_{12}}{a_{12} + a_{21}} + \frac{T_1 - T_2}{a_{12} + a_{21}}\right| > \left|\frac{T_1 - T_2}{a_{12} + a_{21}}\right| \tag{17}$$

and

$$\left|\frac{T_1 a_{21} + T_2 a_{12}}{a_{12} + a_{21}} - \frac{T_1 - T_2}{a_{12} + a_{21}}\right| > \left|\frac{T_1 - T_2}{a_{12} + a_{21}}\right|. \tag{18}$$

Now, consider the sequence given by $\tau(j)$ as defined in Equation 11. Since

$$|\tau(j)| = \left|\tau(j) - \frac{T_1 - T_2}{a_{12} + a_{21}} + \frac{T_1 - T_2}{a_{12} + a_{21}}\right|,$$

by using Equation 14 it is obvious that:

$$|\tau(j)| \leq \left|\tau(0) - \frac{T_1 - T_2}{a_{12} + a_{21}}\right| + \left|\frac{T_1 - T_2}{a_{12} + a_{21}}\right| \ \forall j. \quad (19)$$

Again using Equation 14 it is possible to show that:

$$\begin{aligned} &|(1 - a_{21})\tau(j) - T_2| \geq \\ &\left|\frac{T_1 - T_2}{a_{12} + a_{21}} - \frac{T_1 a_{21} + T_2 a_{12}}{a_{12} + a_{21}}\right| - \\ &|1 - a_{21}| \left|\tau(0) - \frac{T_1 - T_2}{a_{12} + a_{21}}\right| \ \forall j. \end{aligned} \quad (20)$$

By combining Equation 18, 19, and 20 and making use of condition 2 (define $\tau(0) = t_{f1}(0) - t_{f2}(0)$), the following result is obtained:

$$|\tau(j)| < |(1 - a_{21})\tau(j) - T_2| \ \forall j.$$

One can also show that:

$$|\tau(j)| < |(1 - a_{12})\tau(j) + T_1| \ \forall j.$$

This is done by first showing:

$$\begin{aligned} &|(1 - a_{12})\tau(j) + T_1| \geq \\ &\left|\frac{T_1 - T_2}{a_{12} + a_{21}} + \frac{T_1 a_{21} + T_2 a_{12}}{a_{12} + a_{21}}\right| - \\ &|1 - a_{12}| \left|\tau(0) - \frac{T_1 - T_2}{a_{12} + a_{21}}\right| \ \forall j. \end{aligned} \quad (21)$$

Then, combining Equation 17, 19, and 21 gives the result.

These two inequalities are the same as those in Equation 15 and 16, so this concludes the proof. ■

4 Simulation Results

In this section a specific example is considered. The parameter values are:

$$T_1 = 1, T_2 = 1.1, \theta_1(0) = \theta_2(0) = 0, a_{12} = a_{21} = 0.1.$$

The expected fixed point is given by Equation 9 as:

$$\tau = t_{f1} - t_{f2} = -0.5.$$

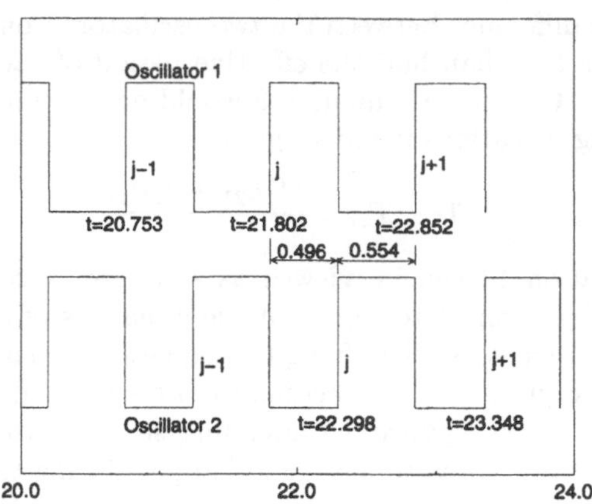

Figure 2: *Phase-Locked Oscillations. The difference between firing times at index j determine the firing times at index $j+1$. See text for further explanation.*

The effective period of each oscillator at steady state is:

$$T_{\text{eff}} = 1.05$$

from Equation 10. Note that the inequality in Equation 13 is just barely satisfied:

$$|T_1 - T_2| = 0.1 < 0.105 = \frac{|T_1 a_{21} + T_2 a_{12}|}{2}.$$

Also, notice that the parameters satisfy Equation 12 for the stability of Equation 11. The simulation was performed with Equations 4–7 (not with Equation 11). At each step it was verified that the k which minimized Equation 8 was indeed j.

Figures 2 and 3 show the results. In Figure 2, the oscillator outputs are shown after the transient behavior has essentially died out ($0.8^{20} = 0.01153$). The oscillators have phase-locked with a firing difference of -0.5, which means they are almost exactly out of phase (since the effective period of each oscillator is 1.05). Figure 2 also has labels for different firing times. Oscillator 1's firing time at index $j+1$ is determined by the difference in firing times at index j of both oscillators 1 and 2. The same is true for oscillator 2. Note that the firing of oscillator 1 at index $j+1$ is slightly farther away from the firing of oscillator 2 at index j ($22.852 - 22.298 = 0.554$) than the firing of oscillator 1 at index j ($22.298 - 21.802 = 0.496$). Thus, oscillator 2 adjusts its firing time at index $j+1$ based on oscillator 1's firing at index j, not $j+1$. The fact that the

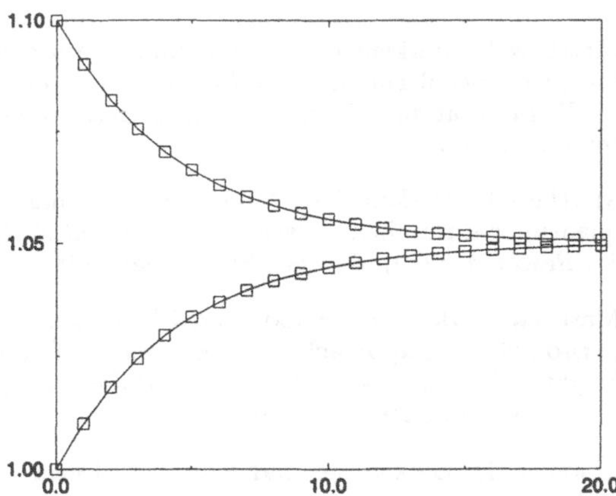

Figure 3: *Convergence of Periods to the Same Effective Period.*

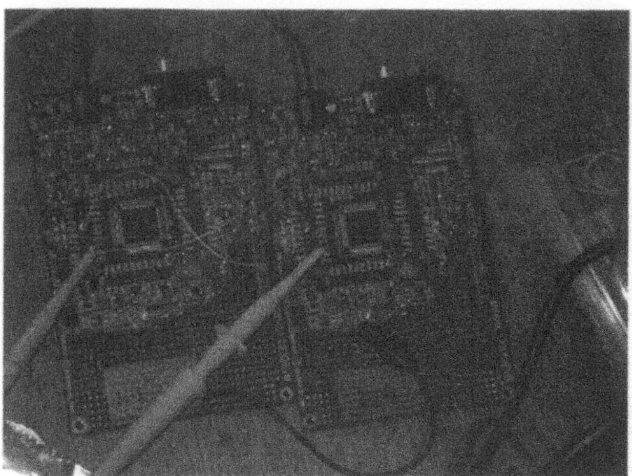

Figure 4: *A Pair of Coupled Oscillators Implemented on Microcontrollers.*

two differences in firing times are so close to each other (0.554 and 0.496) is because the choice of parameters just barely satisfied Inequality 13. For better parameter choices, the decision for oscillator 2 would not be dependent on such a small timing difference.

In Figure 3 The effective periods of each oscillator are seen to converge to the same value of 1.05. Note that the horizontal axis in this graph is the index j, not the time corresponding to the period.

5 Experiments

The algorithm was tested on a pair of Motorola 68HC12 evaluation boards. Figure 4 shows the setup. Note the pair of wires which provides bidirectional coupling. There were 3 main parts to the assembly language program:

1. Initialization code, followed by an infinite loop,
2. Timer interrupt service routine (ISR),
3. External capture ISR.

The evaluation boards used an internal clock speed of 8 MHz. One of the timers was used to interrupt the processor every 8000 counts, or every 1 ms. Then, in the timer ISR, a RAM variable was incremented until it reached the maximum count, which was nominally 500 for one board (giving 0.5 s elapsed time). At this point, the output state was changed. The external capture ISR compared previous firing times with current and saved the minimum firing time difference (to a resolution of 1 ms). The timer ISR used the stored difference to modify the count about the nominal 500.

One evaluation board had a nominal count of 500, which gave it a nominal frequency of 1 Hz. The other had a nominal frequency of 1.098 Hz. The expected frequency of the coupled pair was therefore 1.047 Hz. Figure 5 shows a frequency of 1.042 Hz for each microcontroller. It is likely that the difference is simply due to the limited resolution of the oscilloscope, although more careful experiments should be performed.

The implementation of the algorithm on the microcontrollers was very straightforward. This supports the earlier argument that the approach advocated here is more appropriate than, for example, attempting to numerically integrate ordinary differential equations. No multiplication or division was necessary. Scaling of the firing time difference (corresponding to the constants a_{12}, a_{21}) was done by right shifts. Certainly there would be plenty of processor time remaining to perform additional tasks, if desired. Alternatively, one could easily use a less expensive/less powerful microcontroller.

6 Conclusions

In this paper, inspiration was taken from the coupled oscillator model of [10] to develop a new discrete-time

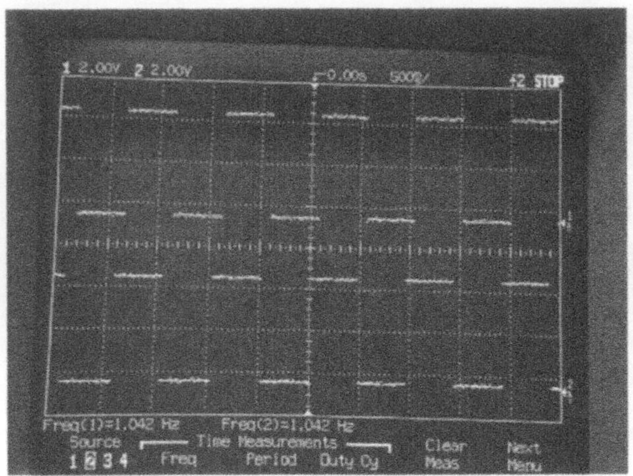

Figure 5: *Phase-Locked Oscillations Produced by Two Microcontrollers.*

coupled oscillator model. This new model was more appropriate for robotics use, as each oscillator could be easily programmed on a microcontroller. Moreover, communication between the oscillators was only required twice each period (once for each oscillator), rather than continually.

Preliminary analysis of the discrete model demonstrated that it had similarities to the model of [10]. Each oscillator had an intrinsic period, but under the right conditions, the coupled pair would phase lock and achieve an effective period which was somewhere between the intrinsic periods of the individual oscillators. A proposition provided precise conditions under which the oscillators phase-locked in the manner expected. A simulation demonstrated phase locking for one particular set of parameters. Lastly, experiments using a pair of microcontrollers demonstrated the expected operation of the algorithm.

The development of this model represents another step in achieving biologically-inspired control of a legged robot.

Acknowledgments

This work was supported by supported by NSF CAREER award nos. IIS-9996366, IIS-9733401.

References

[1] Matthew D. Berkemeier. Approximate return maps for quadrupedal running. In *Proceedings of the 1997 IEEE International Conference on Robotics and Automation*, 1997.

[2] Matthew D. Berkemeier. Modeling the dynamics of quadrupedal running. *International Journal of Robotics Research*, 17(9):971–985, September 1998.

[3] Matthew D. Berkemeier and Kamal V. Desai. A comparison of three approaches for the control of hopping height in legged robots. *International Journal of Robotics Research*, 1999. (accepted).

[4] Matthew D. Berkemeier and Kamal V. Desai. Control of hopping height in legged robots using a neural-mechanical approach. In *Proceedings of the 1999 IEEE International Conference on Robotics and Automation*, pages 1695–1701, 1999.

[5] J. J. Collins and I. N. Stewart. Coupled nonlinear oscillators and the symmetries of animal gaits. *Journal of Nonlinear Science*, 3:349–392, 1993.

[6] James Gordon. Spinal mechanisms of motor coordination. In Eric R. Kandel, James H. Schwartz, and Thomas M Jessel, editors, *Principles of Neural Science*, chapter 38. Elsevier, 3rd edition, 1991.

[7] N. Kopell and G. B. Ermentrout. Coupled oscillators and the design of central pattern generators. *Mathematical Biosciences*, 90:87–109, 1988.

[8] R. T. M'Closkey and J. W. Burdick. Periodic motions of a hopping robot with vertical and forward motion. *International Journal of Robotics Research*, 12(3):197–218, 1993.

[9] A. Neishtadt and Z. Li. Stability proof of Raibert's four-legged hopper in bounding gait. Technical Report 578, New York State University, 1991.

[10] Richard H. Rand, Avis H. Cohen, and Philip J. Holmes. Systems of coupled oscillators as models of central pattern generators. In Avis H. Cohen, Serge Rossignol, and Sten Grillner, editors, *Neural Control of Rhythmic Movements in Vertebrates*, chapter 9. Wiley, 1988.

[11] Steven H. Strogatz and Ian Stewart. Coupled oscillators and biological synchronization. *Scientific American*, 269(12):102–109, December 1993.

Reliable Mobile Robot Navigation From Unreliable Visual Cues

Amy J. Briggs, *Middlebury College, Middlebury, VT*
Daniel Scharstein, *Middlebury College, Middlebury, VT*
Stephen D. Abbott, *Middlebury College, Middlebury, VT*

Vision-based mobile robot navigation requires robust methods for planning and executing tasks due to the unreliability of visual information. In this paper we propose a new method for reliable vision-based navigation in an unmodeled dynamic environment. Artificial landmarks are used as visual cues for navigation. Our system builds a visibility graph among landmark locations during an exploration phase and then uses that graph for navigation. To deal with temporary occlusion of landmarks, long-term changes in the environment, and inherent uncertainties in the landmark detection process, we use a probabilistic model of landmark visibility. Based on the history of previous observations made, each visibility edge in the graph is annotated with an estimated probability of landmark detection. To solve a navigation task, our algorithm computes the expected *shortest paths between all landmarks and the specified goal, by solving a special instance of a Markov decision process. The paper presents both the probabilistic expected shortest path planner and the landmark design and detection algorithm, which finds landmark patterns under general affine transformations in real-time.*

1 Introduction

Vision-based mobile robot navigation is often planned using landmarks, either artificial or extracted from the environment. Such landmarks must be detected quickly and as reliably as possible. When landmark detection is unreliable due to factors such as temporary occlusion or varying lighting conditions, the planner should compute motion paths dynamically to always find the current best path. This paper makes two major contributions: (1) A navigation system that uses probability estimates of landmark visibility in order to compute expected shortest paths between landmarks; (2) a novel landmark pattern together with a real-time algorithm for landmark detection that can handle a wide range of affine transformations.

1.1 Navigation Using Expected Shortest Paths

Our system uses artificial landmarks as visual cues for navigation in an unknown environment. The robot first explores the environment to learn the relative locations of the landmarks and builds a graph of landmark locations that it subsequently uses for navigation. During navigation, the robot plans motion paths along edges of the landmark visibility graph. If it wants to navigate from landmark s to landmark g, it plans and executes a path starting with a landmark visible from s. As it moves through the environment, it continually updates the graph with any newly acquired data, such as measured distance between two landmarks, or changes to the landmark visibility information. As is discussed in Sections 2 and 3, the planner uses estimates of the probabilities of landmark visibility to find paths with *expected shortest length* by solving a Markov decision process.

1.2 Landmark-based Navigation

Many techniques have been employed for sensor-based navigation and localization. Industrial mobile robots have traditionally navigated by following painted lines on the floor or tracking buried wires or infrared beacons. The disadvantage of these approaches is that they require substantial engineering of the environment. Recently many researchers have employed landmarks — either artificial or extracted from the environment — to guide the motion of a mobile robot in indoor environments. The most commonly used approach with artificial landmarks is heuristic: Landmarks are designed and placed so that landmark detection under normal circumstances is straightforward. The problem with the heuristic approach is that it only works under certain restricted conditions that are enforced for the sake of speed: The patterns must be viewed from a narrow range of distances and angles and will not be recognized if partially occluded.

We propose a self-similar pattern specifically designed for the application of mobile robot navigation. The pattern is quickly recognizable under a variety of viewing conditions, even when partially occluded or mounted at an angle. In contrast to existing approaches that require two-dimensional analysis of an image, our method finds matches along individual scanlines, without any preprocessing, making it suitable for real-time applications.

1.3 Related Work

Traditionally, vision-based robot navigation has proceeded from three-dimensional maps of the environment, constructed, for example, using stereo vision techniques [3]. More recently, landmarks have been used to navigate without a full environment model. Techniques for mobile robot navigation based on landmarks include those that are primarily reactive [5], those planned within a geometric environment map enhanced with perceptual landmarks [9, 11], and those based on a topological description of landmark locations without a global map [8, 13, 17].

Our navigation system uses artificial landmarks placed throughout the environment as visual cues. A topological map of current landmark locations is first constructed during an exploratory phase and then used for navigation without requiring a global geometric map. To compensate for occlusion and unreliability of landmark detection, our navigation algorithm employs probabilistic techniques to construct reliable and efficient motion paths. Several different approaches to probabilistic path planning have been developed in related work. Blei and Kaelbling [2] describe Markov decision processes for finding shortest paths in stochastic graphs with partially unknown topologies. Their work differs from ours in that they assume that an edge is either passable or not, but that the state of each edge is only known with a certain probability. Kavraki and Latombe [7] propose a randomized method for configuration space preprocessing that generates a network of collision-free configurations in a known environment. Overmars and Švestka [12] describe a similar probabilistic learning approach that extends to a number of motion planning problems, including those for free flying planar robots, car-like robots, and robots with high degrees of freedom. Finally, a Markov model is used by Simmons and Koenig [18] to plan navigation strategies in partially observable environments.

Rather than relying on landmarks extracted from the environment [4, 5, 11, 13, 17], the approach taken in this paper and by a number of other research groups [1, 6, 10, 15, 19, 20] is to use artificial landmarks that can be easily and unobtrusively added to the environment. Becker et al. [1] use simple landmarks attached to the ceiling of the environment, and use a recognition algorithm that relies on a fixed distance of the pattern to the camera. Taylor and Kriegman [20] utilize the projective invariance of cross-ratios, but their approach cannot handle partial occlusion and requires specialized hardware for real-time performance. Lin and Tummala [10] propose three-dimensional landmarks consisting of two disks, which can be detected using Hough transforms from a restricted set of viewing angles. Unlike these approaches, our method uses simple 2D landmarks that can be recognized under a wide range of affine transformations in real-time without specialized hardware.

1.4 Outline of the Paper

Each of the two central themes of the paper — robust navigation and the design of artificial landmarks — is discussed in two sections. Section 2 presents our framework for planning based on unreliable sensor data and develops an algorithm for computing expected shortest paths. Section 3 discusses how a visibility graph can be annotated with estimates of the unreliability of observations. Section 4 then introduces self-similar functions for the design of an optimally recognizable intensity pattern. Finally, Section 5 presents an algorithm for finding such patterns in an image under general affine transformations in real time.

2 Robust Navigation Using Unreliable Sensors

The discussion in this section is based on the following scenario:

We assume an unknown environment augmented with visual landmarks $\{L_1, L_2, \ldots, L_N\}$ that can be detected by the robot, albeit unreliably. We will use lowercase letters $a, b, \ldots$ when referring to individual landmarks. The robot navigates the environment by traveling along the edges of the visibility graph defined by the landmarks. We assume an edge from landmark a to landmark b has associated probability $p_{ab} \in [0, 1]$ and length $l_{ab} > 0$. The probability p_{ab} represents the likelihood that landmark b can be detected from landmark a. The length l_{ab} can be, for example, the physical distance between a and b, or the time it takes the robot to travel from a to b. In this section we investigate the problem of robot navigation given such a visibility graph. In Section 3 we discuss how such a graph can be constructed, and how probability and length factors can be estimated.

Path planning in visibility graphs typically employs shortest-path algorithms. Given a directed graph whose edges have fixed lengths, the shortest path from a start node s to a goal node g can be computed easily, for example using Dijkstra's algorithm. In our scenario, landmark detection is unreliable, and thus the edges of the graph can only be traversed some of the time. Therefore, we must change the notion of a shortest path to that of a *path with shortest expected length*, or *expected shortest path*.

2.1 Navigation Using Expected Shortest Paths

Before explaining how these shortest expected lengths can be computed, let us see how the robot can use them to plan its path. Suppose that the robot at landmark s can currently see landmarks a and b, but not landmark c (see Figure 1). Let E_{ag} denote the expected length of the shortest path from a to goal g. The total expected length of the path through a will be $l_{sa} + E_{ag}$. Similarly, the total expected length of the path through b will be $l_{sb} + E_{bg}$. Thus, the smaller of those two sums will indicate a candidate shortest path. Note that these lengths are independent of the probabilities p_{sa} and p_{sb}, since *at the current moment*, both a and b are visible. The path with overall shortest expected length, however, may go through neither a nor b. It is possible that an expected shorter path to g goes through landmark c, which usually is visible from s, but at the current moment is not (for example, due to temporary occlusion). This would be reflected in a low expected length E_{sg}. In this case, it would be better to stay at s and wait for c to become visible, rather than going to either a or b. To prevent the robot from staying at a landmark indefinitely, we associate a non-zero cost with this option (for example, the time it takes to acquire a new image). That is, each node n in the graph has a self-edge $n \to n$ with cost $l_{nn} > 0$ and probability $p_{nn} = 1$ (staying is always an option).

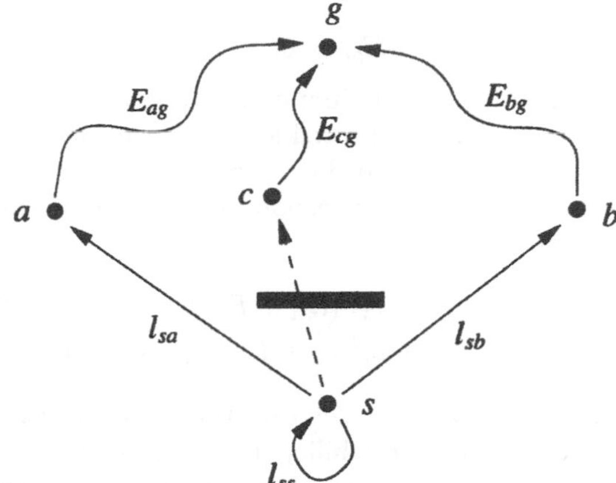

Figure 1: *Path planning example: The robot at landmark s can currently see landmarks a and b, but not c due to temporary occlusion.*

In summary, the robot will make its decision of whether to go to a, to go to b, or to stay at s based on which of the three sums $(l_{sa} + E_{ag})$, $(l_{sb} + E_{bg})$, and $(l_{ss}+E_{sg})$ is the smallest. If landmark c is permanently occluded, it may seem that the robot could "get stuck" at s. As will be discussed in Section 3, however, the current estimate of p_{sc} — the probability that c is visible from s — will decrease after a repeated failure to detect c. This will in turn increase the expected length of the path from s to g, until eventually the expected length of going through a or b will be shorter.

2.2 Deriving the Expected Lengths of the Shortest Paths

Given a designated goal g, we will now relate the unknown quantities E_{ng} (the expected lengths of the shortest paths from each node to the goal) in recursive equations. In the next section, we show that there is a unique solution to this system of equations if there is a path with non-zero probabilities from each node in the graph to goal g. It turns out that this problem is a special instance of a Markov decision process, which is discussed in section 2.4.

The following relations for the unknowns E_{ng} are motivated by the discussion in the previous section. To start, the expected length of the shortest path from the goal to itself is:

$$E_{gg} = 0. \qquad (1)$$

Next, let us consider a node n with only a single outgoing edge $n \to a$. The expected length E_{ng} of the shortest path from n to g can be expressed as a weighted sum of two terms that correspond to whether or not a is visible from n:

$$\begin{aligned} E_{ng} &= (1 - p_{na})\,(l_{nn} + E_{ng}) \\ &+ p_{na} \min(l_{na} + E_{ag},\ l_{nn} + E_{ng}). \end{aligned} \qquad (2)$$

The first term represents the case that a is not visible, which occurs with probability $1 - p_{na}$. In this case the only choice is to remain at n and acquire another image, which incurs cost l_{nn} and results in an expected length of $l_{nn} + E_{ng}$. The second term represents the case that a is visible, which occurs with probability p_{na}. In this case the expected length of the shortest path is the smaller of $l_{na} + E_{ag}$ and $l_{nn} + E_{ng}$, corresponding to the options of going to a or staying at n. Recall that we are assuming that the goal is reachable from any node, and thus $E_{ag} < \infty$. Given that the edge $n \to a$ is the *only* edge leaving n, we know that all paths from n to g have to go through a, and thus that $E_{ng} \geq l_{na} + E_{ag}$. This allows us to solve Equation 2 for E_{ng}, yielding:

$$E_{ng} = \frac{1 - p_{na}}{p_{na}}\, l_{nn} + l_{na} + E_{ag}. \qquad (3)$$

Now, let us consider a node n with two outgoing edges $n \to a$ and $n \to b$. The relation for the expected length of the shortest path from n to g can be expressed analogously, except that now there are four cases, depending on which of a and b are visible. Using the shorthand $\overline{p}$ to denote $(1 - p)$, we obtain the following relation for E_{ng}:

$$\begin{aligned} E_{ng} = &\ \overline{p_{na}}\,\overline{p_{nb}}\,(l_{nn} + E_{ng}) \\ +&\ p_{na}\,\overline{p_{nb}} \min(l_{na} + E_{ag},\ l_{nn} + E_{ng}) \\ +&\ \overline{p_{na}}\,p_{nb} \min(l_{nb} + E_{bg},\ l_{nn} + E_{ng}) \\ +&\ p_{na}\,p_{nb} \min(l_{na} + E_{ag},\ l_{nb} + E_{bg},\ l_{nn} + E_{ng}). \end{aligned} \qquad (4)$$

It is easy to see how the equation generalizes to nodes with more than two outgoing edges. A node n with k outgoing edges $n \to a,\ n \to b,\ \ldots,\ n \to z$ yields an equation with 2^k terms:

$$\begin{aligned} E_{ng} = &\ \overline{p_{na}}\,\overline{p_{nb}} \cdots \overline{p_{nz}}\,(l_{nn} + E_{ng}) \\ +&\ p_{na}\,\overline{p_{nb}} \cdots \overline{p_{nz}} \min(l_{na} + E_{ag},\ l_{nn} + E_{ng}) \\ +&\ \overline{p_{na}}\,p_{nb} \cdots \overline{p_{nz}} \min(l_{nb} + E_{bg},\ l_{nn} + E_{ng}) \\ +&\ p_{na}\,p_{nb} \cdots \overline{p_{nz}} \min(l_{na} + E_{ag},\ l_{nb} + E_{bg}, \\ &\qquad l_{nn} + E_{ng}) \\ +&\ \cdots \\ +&\ p_{na}\,p_{nb} \cdots p_{nz} \min(l_{na} + E_{ag},\ l_{nb} + E_{bg},\ \ldots, \\ &\qquad l_{nz} + E_{zg},\ l_{nn} + E_{ng}). \end{aligned} \qquad (5)$$

Unlike in the case for only a single edge, it is no longer possible to explicitly solve these equations for E_{ng} because of the minimum expressions, which recursively relate the unknown quantities E_{ig}. For example, to determine the value of the minimum expressions in Equation 4, we need to know the ordering among K_a, K_b, and K_n, where K_i denotes $l_{ni} + E_{ig}$. While one can still deduce that K_n cannot be the smallest of the three, each of the remaining 4 orderings is possible:

$$\begin{array}{ll} K_a < K_b < K_n & K_a < K_n < K_b \\ K_b < K_a < K_n & K_b < K_n < K_a \end{array} \qquad (6)$$

Note that if such orderings were known for all nodes, Equations 4 and 5 would simplify to linear equations involving the unknowns E_{ng}. The expected lengths of the shortest paths could then be computed simply by solving a system of $N-1$ linear equations, where N is the total number of landmarks (including the goal). The presence of the minima, however, prohibits this approach.

To summarize, the landmark visibility graph defines a system of $N-1$ equations — each of the form of Equation 5 — for the $N-1$ unknowns E_{ng}, $n \neq g$. We now turn to the question of how this system of equations can be solved.

2.3 An Algorithm for Computing the Expected Shortest Paths

For notational convenience, let us collect the $N-1$ unknowns E_{ng}, $n \neq g$, into a vector $\mathbf{X} = [x_1, x_2 \ldots, x_{N-1}] \in \mathbb{R}^{N-1}$, and rewrite Equation 5 in vector form:

$$x_n = f_n(\mathbf{X}). \qquad (7)$$

Collecting the individual functions $f_n : \mathbb{R}^{N-1} \to \mathbb{R}$ (each of which is a linear combination of minima of components of $\mathbf{X}$ plus constant offsets) into a function $\mathbf{F} : \mathbb{R}^{N-1} \to \mathbb{R}^{N-1}$,

$$\mathbf{F} = (f_1, f_2, \ldots, f_{N-1}), \tag{8}$$

we can then rewrite the entire system of equations concisely:

$$\mathbf{X} = \mathbf{F}(\mathbf{X}). \tag{9}$$

Thus, the desired solution $\mathbf{X}$ of this system of equations is a *fixed point* of function $\mathbf{F}$. The properties of $\mathbf{F}$ are summarized in the following theorem; a proof of the theorem can be found in Appendix A.

Theorem 1 *If there exists a path to the goal from every node in the graph, and if all edges in the graph have non-zero probabilities $p_{ij} \in (0, 1]$ and positive lengths $l_{ij} > 0$, then $\boldsymbol{F}$ has a unique fixed point $\boldsymbol{X}^*$ in $\mathbb{R}^{N-1}$. Moreover, the iterative process:*

$$\boldsymbol{X}^{(k+1)} = \boldsymbol{F}(\boldsymbol{X}^{(k)}) \tag{10}$$

converges to this fixed point given the initial value $\boldsymbol{X}^{(0)} = 0$:

$$\boldsymbol{X}^* = \lim_{k \to \infty} \boldsymbol{X}^{(k)}. \tag{11}$$

This theorem translates literally into an efficient algorithm for computing the expected shortest paths E_{ig}. Starting with an initial estimate $\mathbf{X}^{(0)}$, iterate Equation 10 until the value converges[1]. Convergence is geometric (i.e., the error decreases exponentially) once the current value is in the vicinity of the fixed point. In practice, usually no more than a few hundred iterations are necessary for the values to converge to within machine precision. (It is possible, however, to construct graphs for which convergence is arbitrarily slow.)

Instead of iterating Equation 10, it is also possible to repeatedly *solve* the linear system of Equations 9, using the previous solution to hypothesize which terms are the minima in each equation. Thus, instead of iterating over the *values* of the expected lengths, we can iterate over the *ordering* of the expected lengths of the outgoing paths at each node (as in Equation 6). The iteration terminates when the solution to the current system of equations yields the same ordering as the previous solution. It can be shown that this second algorithm (iterating over orderings) converges much faster than the first (iterating over values), but each iteration requires solving a system of linear equations, rather than just evaluating it.

[1]In fact, it can be shown that the process converges for any initial value $\mathbf{X}^{(0)} \in \mathbb{R}^{N-1}$.

2.4 Relation to Markov Decision Processes

The problem of expected shortest paths can be viewed as a special instance of a Markov decision process (MDP). Briefly, a MDP consists of a set of states S, and a set of allowable actions A_s associated with each state $s \in S$. Each action $a \in A_s$ taken in state s yields a reward $r(s, a)$, and results in a new (random) state s' according to a transition probability distribution $p(\cdot \mid s, a)$. The objective is to devise a *policy* or *decision rule* $d : S \to A_s$ that selects a certain (fixed) action in each state so as to optimize a function of the total reward. This brief discussion ignores many common variations of MDPs, including time-dependent or discounted rewards, and non-stationary policies. For a comprehensive introduction to Markov decision processes, see the book by Puterman [14].

Our problem of expected shortest paths translates into a non-discounted negative expected total-reward MDP. This means that each reward is interpreted as cost or penalty, and that the objective is to minimize the total expected cost. Upon reaching the goal g, no further cost is incurred. The key insight for relating our problem to a MDP is that the states in the MDP are not simply the nodes in the graph. Rather, each state encodes a node together with the set of outgoing edges currently passable. That is, a node with k outgoing edges contributes 2^k states. In each state, the allowable actions are to traverse any of the passable (visible) edges, including the self-edge. The cost associated with an action is the length of the edge traversed. The transition probabilities are the probabilities associated with the different states (visibility scenarios) of the *destination* node. For example, if the action is to go to a node x with 2 outgoing edges $x \to y$ and $x \to z$, the resulting state will be one of the 4 states encoding which of the 2 edges will be passable once x is reached; the corresponding probabilities are $\overline{p_{xy}}\,\overline{p_{xz}}$, $\overline{p_{xy}}\,p_{xz}$, $p_{xy}\,\overline{p_{xz}}$, and $p_{xy}\,p_{xz}$.

Note that while the corresponding MDP has many more states than there are nodes or edges, a policy can be specified by ordering the outgoing edges at each node (as in Equation 6). Then, for each subset of visible edges, the edge corresponding to the shortest path is chosen. Each such ordering can in turn be derived from a current estimate of the expected lengths of the shortest paths from each node to the goal. Thus, while the number of states $|S|$ of the MDP is:

$$|S| = \sum_{n=1}^{N} 2^{od(n)},$$

where $od(n)$ is the out-degree of node n, the entire MDP can be concisely described with the $N-1$ unknowns E_{ng}.

Our algorithms for computing the expected shortest paths presented in Section 2.3 are variants of two algorithms known as *value iteration* and *policy iteration* in the MDP community. An important difference is that in our case both algorithms require the iteration (or solution) of only $N-1$ equations, rather than of $|S|$ equations, as discussed in the previous paragraph.

3 Building the Visibility Graph

Now that we are armed with the ability to quickly compute the expected shortest paths from all nodes to a given goal node, we can apply the navigation strategy outlined in Section 2.1: At each node along the way, determine which landmarks are currently visible, and select among those the one that yields the overall expected shortest path (which includes the option of staying at the current position). Repeat the process at each subsequent node, until the goal is reached. If scanning for all visible landmarks is significantly more expensive than just looking for the next landmark on the predicted shortest path (for example, if scanning requires a 360 degree rotation of the robot), the strategy can be modified as follows: If possible, travel to each landmark in the sequence corresponding to the expected shortest path. Only if a landmark in the planned sequence cannot be detected, scan for all visible landmarks and replan.

3.1 Deriving Cost and Probability Estimates

The remaining problem is how the visibility graph can be constructed and how length (cost) and probability factors can be estimated and maintained. Let us first discuss how to estimate the lengths of edges. Clearly, once a visibility edge has been traversed by the robot, its length l_{ij} is known and can be stored. Note that "length" will typically refer to the *time* it takes to traverse the edge. When artificial landmarks with known size are used (such as the ones presented in Section 4), we can also estimate the lengths of edges that have not yet been traversed, based on the size of the landmark in the image. Such estimates are fairly accurate, and are immediately available with each new visibility edge; they can be replaced with the measured distance once the edge has been traversed.

We now address the less obvious problem of determining the probabilities p_{ij} (that landmark j is visible from landmark i). Since these probabilities are unknown, the best we can do is to compute estimates $\hat{p}_{ij}$ for the true p_{ij} as a function g of the history of observations $\mathbf{O}_{ij}$:

$$\hat{p}_{ij} = g(\mathbf{O}_{ij}). \tag{12}$$

The history of observations of landmark j from landmark i is:

$$\mathbf{O}_{ij} = [o_{ij}^{(1)}, o_{ij}^{(2)}, \ldots, o_{ij}^{(m_i)}], \tag{13}$$

where m_i is the total number of observations made from landmark i up to this point in time, and each $o_{ij}^{(k)} \in \{0, 1\}$ records whether landmark j was visible. Note that observation histories can have "leading zeros"; that is, even if j was not visible the first few times an observation was made, it is possible to reconstruct the complete observation history for j by keeping track of all observations ever made at landmark i.

Let us now turn to the function g: how can we derive probability estimates from observation histories? In the simplest scenario, assuming independent observations made with a fixed probability p_{ij}, the optimal estimate $\hat{p}_{ij}$ is given by a function g that returns the ratio of detections ("ones") to the total number of observations m_i. In reality, however, the observations will neither be independent, nor will the p_{ij} stay constant over time. While some failures to detect a landmark will have truly random causes (for example, occlusion by a person walking by), others will be caused by lighting changes throughout the day, or perhaps even by permanent changes to the environment (most extremely, the removal or addition of a landmark). Typically, observations closely spaced in time will be highly correlated. Therefore, in practice, a more sophisticated estimation function g should be used. It is also a good idea to record a time stamp with each observation, so that the temporal distribution of observations can be taken into consideration.

3.2 Exploration and Navigation

In our landmark-based navigation system, the robot operates in two modes: Exploration and navigation. In exploration mode, the robot explores the environment using a depth-first search among the unvisited landmarks. A visibility edge between two landmarks can be traversed by visual servoing, using the real-time recognition algorithm discussed in Section 5. At every newly-visited landmark, the robot scans for all landmarks visible from this position, records their relative angles and estimates of their distances, and starts an observation history for this landmark. As is discussed in Section 5.2, the landmarks all have unique barcodes, which are used as node labels in the graph. As mentioned above, in the process of exploring, the robot replaces distance estimates with more accurate odometry measurements. Also, as landmarks are revisited during the exploration phase, the observation histories are updated and the probability estimates are refined.

Once part of the environment has been explored, the robot can enter navigation mode, and accept navigation tasks from the user. For a given goal, the expected shortest paths are computed and used for path planning as described above. Navigation mode and exploration mode can be interleaved seamlessly; length and probability factors are continuously updated in both modes based on observations made and edges traversed. In summary, our navigation system is able to operate robustly in the presence of unreliable sensory input, and can cope both with the temporary occlusion of landmarks and with permanent changes to the environment, such as the removal and addition of new landmarks.

4 A Self-similar Landmark Pattern

Our vision-based navigation system relies on real-time detection of landmarks in the environment. We have designed a self-similar intensity pattern [16] that can be quickly and reliably detected in images taken by the robot. In this section we motivate our use of a self-similar pattern and describe the pattern we have developed.

4.1 Self-similar Functions

We say a function $f : \mathbb{R}^+ \rightarrow \mathbb{R}$ is *p-similar* for a scale factor p, $0<p<1$, if $f(x) = f(px)\ \forall x > 0$. The graph of a p-similar function is self-similar; that is, it is identical to itself scaled by p in the x-direction. Note that a p-similar function is also p^k-similar, for $k = 2, 3, \ldots$ Self-similar intensity patterns are attractive for recognition since the property of p-similarity is invariant to scale, and thus the distance of the pattern to the camera does not matter.

A p-similar pattern can be detected by comparing the observed intensity pattern to a version of itself scaled by p. We can accommodate patterns of limited spatial extent by restricting the comparison to a window of width w. Let $d_{p,w}(f)$ be the average absolute difference between the original and scaled functions over w:

$$d_{p,w}(f) = \frac{1}{w}\int_0^w |f(x) - f(px)|\, dx. \tag{14}$$

Then f is *locally* p-similar over w if and only if $d_{p,w}(f) = 0$. A simple method for detecting a locally p-similar pattern in a one-dimensional intensity function $I(x)$ would then be to minimize the above measure over translations $I_t(x) = I(x + t)$:

$$t_{\text{match}} = \arg\min_t d_{p,w}(I_t).$$

The minimal value of 0 is achieved only if I is p-similar over w at translation t. Unfortunately, all constant functions are p-similar for any p. Thus $d_{p,w}(I_t)$ would also be minimal in regions of constant intensity. To exclude locally constant functions, we must detect patterns that are self-similar *only* for scale p (and $p^2, p^3, \ldots$), and not for other scales. This can be achieved by *maximizing* the *mismatch* at scale $p^{1/2}$ (and $p^{3/2}, p^{5/2}, \ldots$). Let us assume without loss of generality that the range of observable intensities is $[0, 1]$. A maximal mismatch for scale $\sqrt{p}$ is then given if $|f(x) - f(\sqrt{p}x)| = 1$, or locally, if:

$$d_{\sqrt{p},w}(f) = \frac{1}{w}\int_0^w |f(x) - f(\sqrt{p}x)|\, dx = 1. \tag{15}$$

In this case we say that f is (locally) $\sqrt{p}$*-antisimilar.*

We can combine Equations 14 and 15, and revise our method for detecting self-similar patterns that are only p-similar for a given scale p. We will maximize the match function:

$$m(t) = d_{\sqrt{p},w}(I_t) - d_{p,w}(I_t) \tag{16}$$

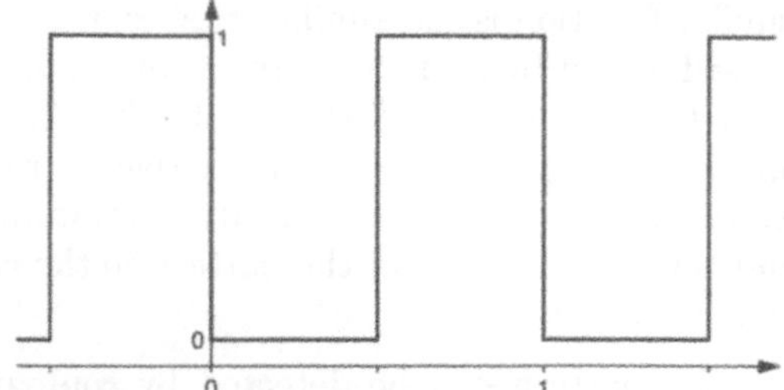

Figure 2: *The square wave function $S(x)$.*

over all translations t. Note that the range of $m(t)$ is $[-1, 1]$, since the range of both terms on the right-hand side of (16) is $[0, 1]$. A locally constant function will yield $m(t) = 0$. Similarly, a random intensity pattern that is neither p-similar nor $\sqrt{p}$-similar will yield a response close to zero. A significant positive response is only expected for intensity patterns that are p-similar but not $\sqrt{p}$-similar.

4.2 An Optimally Recognizable Ppattern

Given the match measure m defined in Equation 16, we now need to find an intensity function that yields the optimal response $m = 1$. That is, we seek a p-similar, $\sqrt{p}$-antisimilar function $s_p(x)$. To derive such a function, let us consider the periodic "square-wave" function S depicted in Figure 2:

$$S(x) = \begin{cases} 0, & x - \lfloor x \rfloor < \frac{1}{2} \\ 1, & x - \lfloor x \rfloor \geq \frac{1}{2} \end{cases} = \lfloor 2(x - \lfloor x \rfloor) \rfloor, \quad (17)$$

This function has the property that $S(x+1) = S(x)$ and $S(x+\frac{1}{2}) = 1 - S(x)$, i.e., it is a 1-periodic function that is similar under a *translation* of 1 and anti-similar under a translation of $\frac{1}{2}$. It is easy to show that we can transform S into a p-similar, $\sqrt{p}$-antisimilar function s_p by substituting $\log_p x$ for x:

$$s_p(x) = S(\log_p x) = \lfloor 2(\log_p x - \lfloor \log_p x \rfloor) \rfloor \quad (18)$$

with the desired properties of $d_{p,w}(s_p) = 0$ and $d_{\sqrt{p},w}(s_p) = 1$ for any w (see Figure 3).

4.3 Two-dimensional Landmarks

We now have all the components for landmark design and recognition. The key step for moving to two dimensions is to use a pattern that is p-similar in one direction (say, horizontally), and constant in the other direction. See Figure 4 for an illustration. If such a pattern is then sampled along any non-vertical line, the resulting intensity function is still p-similar because of the scale invariance of self-similarity. This allows us to detect two-dimensional p-similar patterns that have undergone an affine transformation by examining isolated scanlines.

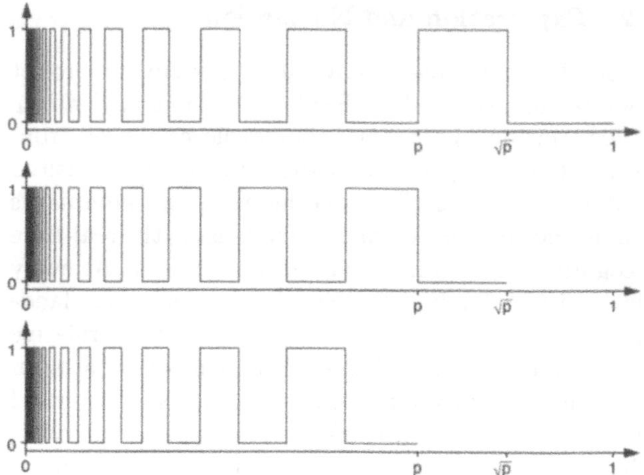

Figure 3: *Self-similar square wave $s_p(x)$ for $p = \frac{2}{3}$ (top); maximal mismatch at scale $p^{1/2}$ (middle); and match at scale p (bottom).*

Formally, let

$$L(x, y) = s_p(x) \quad (19)$$

be our two-dimensional landmark pattern, where $s_p(x)$ is the self-similar square wave function (Equation 18)

Figure 4: *Our self-similar landmark pattern with barcode.*

for a fixed p. An affine transformation yields:

$$A(L(x,y)) = L(ax+by+c, dx+ey+f) = s_p(ax+by+c).$$

Sampled at $y = y_0$, we get $s_p(ax + (by_0 + c)) = s_p(ax + t)$. Thus, the problem of finding an affine transformation of the two-dimensional pattern L has been reduced to finding a translation t of the one-dimensional pattern s_p.

5 The Landmark Recognition Algorithm

The idea underlying the recognition algorithm is to find locations (x_m, y_m) in the image at which a scanline is locally p-similar and $\sqrt{p}$-antisimilar. To do this, we adopt the matching function m from Equation 16 for scanlines in an image $I(x,y)$:

$$m_y(x) = \frac{1}{w}\int_0^w |I(x+\xi,y) - I(x+\sqrt{p}\,\xi,y)|\,d\xi - \frac{1}{w}\int_0^w |I(x+\xi,y) - I(x+p\xi,y)|\,d\xi. \quad (20)$$

The value of $m_y(x)$ depends on the intensities along the line (x,y) to $(x+w,y)$, and is constrained to the interval $[-1,1]$. If the pattern s_p is present at (x,y), then $m_y(x) = 1$. It is easy to see that the pattern cs_p with reduced contrast $c < 1$ (i.e., difference between maximal and minimal intensities) will only yield a response $m_y(x) = c$. Other (non p-similar) intensity patterns will yield responses close to or below zero. An algorithm for finding affine transformations of a landmark with intensities $L(x,y) = s_p(x)$ (for a known p) in an image can be formulated as follows:

for every k-th scanline y
  for all x
    compute $m_y(x)$
  mark all strong local maxima of m_y as matches.

This simple algorithm requires only $O(nw/k)$ operations for an n-pixel image. The computation of $m_y(x)$ can be adapted to discrete images by replacing the integrals in Equation 20 by summations, and determining inter-pixel intensities using linear interpolation. The two parameters, k, the spacing of scanlines to search, and w, the window size, only depend on the smallest expected size of the landmark pattern and can be fixed. Typical values for an image of size 640×480 are $k = 6$ and $w = 45$.

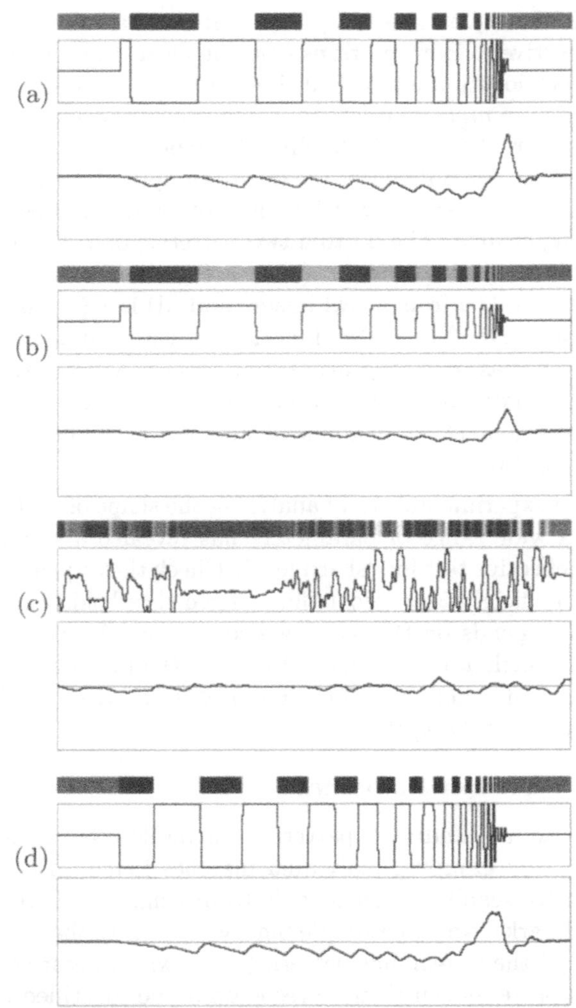

Figure 5: *Various intensity patterns $I(x)$ and match functions $m_y(x)$ for $p = \frac{2}{3}$ and $w = 50$. (See Section 5.1 for details.)*

5.1 Finding Matches Reliably

The interesting question is: What constitutes a "strong" local maximum? A simple answer would be to require that a local maximum be greater than a fixed threshold $c_{\min}$ (corresponding to a minimum contrast). More information, however, can be gained from observing the shape of $m_y(x)$ in the vicinity of a local maximum. Figure 5 shows several intensity profiles of scanlines that were synthesized from continuous functions using 10-fold oversampling. Below each intensity plot is a plot of $m_y(x)$ for $p = \frac{2}{3}$ and $w = 50$. The length of each scanline is 400. The top two patterns are locally

$\frac{2}{3}$-similar square waves s_p with full and half contrast, respectively. Both patterns result in clear peaks at the correct location $x_m = 350$; however, the value of the maximum $m_y(x_m)$ is less than expected. The observed values are 0.66 and 0.33, while the expected values are 1 and $\frac{1}{2}$. The differences are due to sampling and interpolation, in particular at locations of discontinuities or strong change. The bottom two patterns, on the other hand, are *not* $\frac{2}{3}$-similar: (c) is a "random" intensity pattern taken from a real image, and (d) is a $\frac{3}{4}$-similar square wave. The match function for the random pattern (c) has no strong peaks, but there is a distinctive local maximum in the match function for (d). It is, however, much more rounded and not as "sharp" as the top two peaks.

An experimental study analyzing the shape of $m_y(x)$ for a wide range of parameters has revealed a simple but effective test for "sharpness": Check that a peak at half its height is no wider than a fixed threshold, which only depends on the window size w. That is, given a local maximum $v = m_y(x_m)$ greater than a threshold $c_{\min}$, test whether $m_y(x_m \pm \delta_x) < v/2$, where δ_x is approximately $w/10$.

5.2 Grouping Matches

The actual landmark patterns can be detected in an image by grouping individual matches found on consecutive scanlines. To be able to distinguish different landmarks, we add a simple binary barcode to the right side of the self-similar pattern (as shown in Figure 4). The patterns can then be recognized and identified as follows: First, the position and orientation of all landmark patterns in the image is determined by fitting a straight line to each set of three or more individual matches detected on consecutive scanlines. The exact vertical extent of each pattern is then estimated from the intensity distribution to the left of this line. Once the locations of the patterns are known, their barcodes can be decoded easily. Figure 6 shows a Pioneer 2 mobile robot equipped with a camera, and two sample pictures taken by the robot. All landmarks are detected correctly, and all landmark numbers are decoded except for one that was too far away.

Figure 6: *The mobile robot, and two images taken with its camera. All landmarks have been found, and all but one have been identified by their barcodes.*

5.3 Achieving Real-time Performance

The recognition algorithm as described above is fairly fast: Typical running times for a straightforward implementation on a 450 MHz Pentium II machine are 0.33 seconds for 640×480 images and 0.08 seconds for 320×240 images (using parameters $w = 45$ and $k = 6$).

A dramatic speed-up can be achieved by further restricting the set of pixels for which $m_y(x)$ is computed. Recall that we are already considering only every k-th scanline. We can start by looking at only every $2k$-th scanline, and, only if a match is found, look at its neighboring scanlines as well. Since isolated matches are discarded anyway in the grouping process, no landmark will be missed. This technique can yield a speed-up of up to 2 if few or no landmark patterns are present in the image. Another opportunity for restricting the search is to make use of the fact that peaks corresponding to matches have a certain minimum width at a given height h (e.g., $h = c_{\min}/4$). Given a lower bound l for this width, it is possible to scan for peaks by looking only at every l-th pixel. A conservative bound for l is typically given by δ_x, i.e., half the allowable peak width used in the test for sharpness discussed in Section 5.1. Typical values for this number are 4 or 5. We have found, however, that even with much higher values for l (e.g., 10), only very few peaks are overlooked (typically those resulting from patterns with low contrast), so further speedup can be achieved with minimal impact on robustness.

These modifications result in new running times of 0.027 seconds for 640 × 480 images and 0.012 seconds for 320 × 240 images, corresponding to frame rates of 37 and 83 frames per second, respectively. Thus, the implementation runs at video frame rate even for full-size images, making it the first real-time method for landmark detection under affine transformations.

6 Conclusion

Artificial landmarks that can be unobtrusively added to an indoor environment can serve as practical and inexpensive aids for mobile robot navigation and localization. When detected quickly and reliably they can make task execution more robust by reducing uncertainty due to control and sensing errors. When detection is unreliable, the navigation strategies planned should be as robust as possible. This paper has addressed the robust planning problem by providing two major contributions. First, we have described a probabilistic path planner that computes paths of expected shortest length, given landmark visibility histories. Second, we have proposed a novel landmark pattern together with the first practical method employing full affine invariants for real-time detection of landmarks. The method operates on single scanlines without any preprocessing and runs at video frame rate without specialized hardware. An implementation of the landmark detection algorithm is publicly available at `http://www.middlebury.edu/~schar/landmark/`.

Acknowledgements

We are grateful to Darius Braziunas, Huan Ding, Cristian Dima, Cuong Nguyen, Sorin Talamba, and Peter Wall for their insights and their assistance in implementing the algorithms presented here. Many thanks to Tim Huang and Bill Peterson for helpful discussions and literature pointers, and to Leslie Pack Kaelbling and the anonymous reviewers for their insightful comments.

Support for this work was provided in part by the National Science Foundation under grant CCR-9902032, POWRE grant EIA-9806108, VT-EPSCoR grant OSR-9350540, by Middlebury College, and by a grant to Middlebury College from the Howard Hughes Medical Institute.

Appendix A: Proof of Theorem 1

Proof: The strategy of the proof is to show that each component of $\mathbf{X}^{(0)}$, $\mathbf{X}^{(1)}$, $\mathbf{X}^{(2)}$, ... forms a bounded increasing sequence, and thus has a limit. Setting $\mathbf{X}^{(0)} = [0, 0, \ldots, 0]$, it is straightforward to compute that the entries of the vector $\mathbf{X}^{(1)} = \mathbf{F}(\mathbf{X}^{(0)})$ are all strictly positive. Of particular importance here is the observation that every component of $\mathbf{X}^{(1)}$ is at least as big as (and in this case strictly greater than) the corresponding component of $\mathbf{X}^{(0)}$. With this base case verified, we can now argue by induction that each individual component entry of the sequence $\mathbf{X}^{(0)}$, $\mathbf{X}^{(1)}$, $\mathbf{X}^{(2)}$, ... forms a monotone increasing sequence. To see why, assume that the claim is true for $\mathbf{X}^{(0)}$, $\mathbf{X}^{(1)}$, ..., $\mathbf{X}^{(k)}$, and consider $\mathbf{X}^{(k+1)} = \mathbf{F}(\mathbf{X}^{(k)})$. The n-th component of $\mathbf{X}^{(k+1)}$ is given by $f_n(\mathbf{X}^{(k)})$ while the n-th component of $\mathbf{X}^{(k)}$ is given by $f_n(\mathbf{X}^{(k-1)})$. Using the induction hypothesis that every component of $\mathbf{X}^{(k)}$ is at least as big as that of $\mathbf{X}^{(k-1)}$, it is not hard to see from equation (5) that $f_n(\mathbf{X}^{(k)}) \geq f_n(\mathbf{X}^{(k-1)})$.

We must now make a case for boundedness of each component sequence. To begin, consider a node n in the graph containing an edge connected directly to the goal g. Looking at equation (5), the component function for this node would have the form:

$$f_n(\mathbf{X}) \quad = \overline{p_{na}} \ldots \overline{p_{ng}} \ldots \overline{p_{nz}}\,(l_{nn} + x_n) \qquad (21)$$

$$
\begin{aligned}
&+ p_{na} \cdots \overline{p_{ng}} \cdots \overline{p_{nz}} \min(l_{na} + x_a,\ l_{nn} + x_n) \\
&+ \cdots \\
&+ p_{na} \cdots p_{ng} \cdots p_{nz} \min(l_{na} + x_a, \ldots, \\
&\qquad\qquad l_{ng}, \ldots, l_{nz} + x_z,\ l_{nn} + x_n).
\end{aligned}
$$

The letters $a, \ldots, z$ represent nodes that are potentially visible from node n, so that x_n as well as $x_a, \ldots, x_z$ are just some subset of components from the vector $\mathbf{X} = [x_1, x_2, \ldots, x_{N-1}]$.

By choosing a node n that is adjacent to the goal g, we ensure that the length of the edge l_{ng} appears as a candidate in the final minimum of equation (21). The length l_{ng} also appears in several other minima of equation (21), but the special significance of this last minimum expression is that our hypothesis of non-zero probabilities guarantees that $p_{na} \cdots p_{ng} \cdots p_{nz} > 0$.

Now construct the function $u_n(\mathbf{X})$ from $f_n(\mathbf{X})$ in the following way. Consider each minimum expression appearing in f_n. If l_{ng} is among the options, replace the minimum with l_{ng} (regardless of whether or not it represents the minimum). If l_{ng} does not appear, then replace the minimum with the $l_{nn} + x_n$ option (present in every minimum expression). In terms of the underlying graph, this amounts to ignoring all outgoing edges from node n except for l_{ng} and l_{nn}. By replacing each minimum with a particular candidate, we have certainly made the value of the function larger, i.e., $u_n(\mathbf{X}) \geq f_n(\mathbf{X})\ \forall \mathbf{X}$. Moreover, combining terms, u_n reduces to a simpler equation for a node with only a single outgoing edge, similar to equation (2), and thus has the more accessible form:

$$u_n(\mathbf{X}) = p_{ng}\, l_{ng} + \overline{p_{ng}}\, (l_{nn} + x_n), \tag{22}$$

where $p_{ng} > 0$ and thus $\overline{p_{ng}} < 1$. For simplicity, we will write p instead of p_{ng} in the following discussion. The function u_n, although technically defined on $\mathbb{R}^{N-1}$, only depends on the n-th coordinate entry x_n. Also, since $\overline{p} < 1$, u_n is *geometrically contractive* in the sense that for any two points x_n and x_n' we have:

$$|u_n(x_n) - u_n(x_n')| = \overline{p}\, |x_n - x_n'|.$$

What this implies is that iterating a point x_n with u_n yields a sequence x_n, $u_n(x_n)$, $u_n^{(2)}(x_n)$, ... where the distance of the k-th iteration from the starting point x_n must satisfy:

$$
\begin{aligned}
|x_n - u_n^{(k)}(x_n)| \qquad\qquad\qquad\qquad\qquad\qquad (23)\\
\leq\ & |x_n - u_n(x_n)| + |u_n(x_n) - u_n^{(2)}(x_n)| + \ldots \\
& + |u_n^{(k-1)}(x_n) - u_n^{(k)}(x_n)| \\
=\ & |x_n - u_n(x_n)|\, (1 + \overline{p} + \overline{p}^2 + \ldots + \overline{p}^{\,k-1}) \\
<\ & |x_n - u_n(x_n)|\, \frac{1}{1 - \overline{p}}.
\end{aligned}
$$

In other words, the sequence is bounded. Now since for any $\mathbf{X} = [x_1, \ldots, x_n, \ldots, x_{N-1}]$ we have $u_n(x_n) \geq f_n(\mathbf{X})$, it follows that the monotone sequence in the n-th coordinate of $\mathbf{X}^{(0)}$, $\mathbf{X}^{(1)}$, $\mathbf{X}^{(2)}$, ... will also be bounded, and hence convergent.

To explicitly calculate the upper bound, set $x_n = 0$ in equation (23) and solve to get:

$$u_n^{(k)}(0) \ \leq\ (p\, l_{ng} + \overline{p}\, l_{nn}) \frac{1}{p} \ =\ l_{ng} + \frac{\overline{p}}{p}\, l_{nn}.$$

The observant reader will recognize this as the expected length of the shortest path from a node with a single outgoing edge to the goal (Equation 3).

The preceding argument shows that the sequence $\mathbf{X}^{(k)}$ converges in any component corresponding to a node that is adjacent to the goal. With this fact established, we can now repeat the proof for any component of $\mathbf{X}^{(k)}$ corresponding to a node adjacent to a node previously handled. More explicitly, assume node r has an edge to node n, and assume we have shown that the n-th component of $\mathbf{X}^{(k)}$ is bounded and hence converges. The final minimum in the expression for $f_r(\mathbf{X})$ will contain, among other options, the expression $l_{rn} + x_n$. Knowing that $l_{rn} + x_n$ is bounded, by say b_n, we construct $u_r(\mathbf{X}) = u_r(x_r)$ from $f_r(\mathbf{X})$ by replacing each minimum containing $l_{rn} + x_n$ with the upper bound b_n, and selecting $l_{rr} + x_r$ in all other cases. The remainder of the argument is the same.

Given our hypothesis that from every node there exists a path to the goal, this bootstrapping technique eventually leads to the conclusion that every component of our sequence $\mathbf{X}^{(0)}$, $\mathbf{X}^{(1)}$, $\mathbf{X}^{(2)}$, ... is an increasing bounded sequence of real numbers. By the Monotone Convergence Theorem, we may set $\mathbf{X}^* = (x_1^*, x_2^*, \ldots, x_{N-1}^*)$ where each x_n^* is the limit of the n-th components of $(\mathbf{X}^{(k)})$.

This can be summarized with the statement $\lim_{k\to\infty} \mathbf{X}^k = \mathbf{X}^*$. (Technically this is a coordinate-wise limit but we certainly get convergence using most any topology on $\mathbb{R}^{N-1}$.) Now the continuity of the component functions f_n which make up $\mathbf{F}$ allow us to conclude that:

$$\begin{aligned}\mathbf{F}(\mathbf{X}^*) &= \mathbf{F}(\lim_{k\to\infty}\mathbf{X}^{(k)}) = \lim_{k\to\infty}\mathbf{F}(\mathbf{X}^{(k)}) \\ &= \lim_{k\to\infty}\mathbf{X}^{(k+1)} = \mathbf{X}^*.\end{aligned}$$

Finally, to argue that $\mathbf{X}^*$ is the unique fixed point of $\mathbf{F}$, let $\mathbf{Y}^* \in \mathbb{R}^{N-1}$ also satisfy $\mathbf{F}(\mathbf{Y}^*) = \mathbf{Y}^*$. Consider a particular component function f_n of $\mathbf{F}$ (see Equation 21) and again pay special attention to the final minimum term where the variables of all potentially visible nodes are included. If $\mathbf{X}^*$ is fixed by f_n, i.e., $f_n(\mathbf{X}^*) = x_n^*$, then $l_{nn} + x_n^*$ *cannot* be the minimum here since otherwise it would be the minimum throughout and we would have $f_n(\mathbf{X}^*) = x_n^* + l_{nn}$. (This is were we need the hypothesis that all edge lengths are strictly positive.) The same observation of course holds for $l_{nn} + y_n^*$. But now, using the fact that the probability preceding this final minimum is strictly positive, we can show that:

$$|f_n(\mathbf{X}^*) - f_n(\mathbf{Y}^*)| < |x_n^* - y_n^*|.$$

Since $\mathbf{X}^*$ and $\mathbf{Y}^*$ are both assumed to be fixed by $\mathbf{F}$, this is only possible if $\mathbf{X}^* = \mathbf{Y}^*$. ■

References

[1] C. Becker, J. Salas, K. Tokusei, and J.-C. Latombe. Reliable navigation using landmarks. In *Proceedings of IEEE International Conference on Robotics and Automation*, pages 401–406, June 1995.

[2] D. M. Blei and L. P. Kaelbling. Shortest paths in a dynamic uncertain domain. In *Proceedings of the IJCAI Workshop on Adaptive Spatial Representations of Dynamic Environments*, 1999.

[3] O. Faugeras. *Three-Dimensional Computer Vision.* MIT Press, Cambridge, MA, 1993.

[4] C. Fennema, A. Hanson, E. Riseman, J. R. Beveride, and R. Kumar. Model-directed mobile robot navigation. *IEEE Transactions on Systems, Man, and Cybernetics*, 20(6):1352–1369, 1990.

[5] D. P. Huttenlocher, M. E. Leventon, and W. J. Rucklidge. Vision-guided navigation by comparing edge images. In Goldberg, Halperin, Latombe, and Wilson, editors, *1994 Workshop on the Algorithmic Foundations of Robotics, A. K. Peters*, pages 85–96, 1995.

[6] M. Kabuka and A. Arenas. Position verification of a mobile robot using standard pattern. *IEEE Journal of Robotics and Automation*, RA-3(6):505–516, December 1987.

[7] L. Kavraki and J.-C. Latombe. Randomized preprocessing of configuration space for fast path planning. In *Proceedings of IEEE International Conference on Robotics and Automation*, pages 2138–2145, May 1994.

[8] A. Kosaka and J. Pan. Purdue experiments in model-based vision for hallway navigation. In *Proceedings of the Workshop on Vision for Robots in IROS'95, Pittsburgh, PA*, pages 87–96, 1995.

[9] A. Lazanas and J.-C. Latombe. Landmark-based robot navigation. *Algorithmica*, 13(5):472–501, May 1995.

[10] C. Lin and R. Tummala. Mobile robot navigation using artificial landmarks. *Journal of Robotic Systems*, 14(2):93–106, 1997.

[11] B. Nickerson, P. Jasiobedzki, D. Wilkes, M. Jenkin, E. Milios, J. Tsotsos, A. Jepson, and O. N. Bains. The ARK project: Autonomous mobile robots for known industrial environments. *Robotics and Autonomous Systems*, 25:83–104, 1998.

[12] M. H. Overmars and P. Švestka. A probabilistic learning approach to motion planning. In Goldberg, Halperin, Latombe, and Wilson, editors, *1994 Workshop on the Algorithmic Foundations of Robotics, A. K. Peters*, pages 19–37, 1995.

[13] C. Owen and U. Nehmzow. Landmark-based navigation for a mobile robot. In *Proceedings of Simulation of Adaptive Behaviour.* MIT Press, 1998.

[14] M. Puterman. *Markov Decision Processes: Discrete Stochastic Dynamic Programming.* John Wiley & Sons, New York, NY, 1994.

[15] J. Salas, J. L. Gordillo, and C. Tomasi. Visual routines for mobile robots: Experimental results. *Expert Systems with Applications*, 14:187–197, 1998.

[16] D. Scharstein and A. Briggs. Fast recognition of self-similar landmarks. In *Workshop on Perception for Mobile Agents (in conjunction with IEEE CVPR'99)*, pages 74–81, June 1999.

[17] R. Sim and G. Dudek. Mobile robot localization from learned landmarks. In *Proceedings of IEEE/RSJ Conference on Intelligent Robots and Systems (IROS), Victoria, BC*, October 1998.

[18] R. Simmons and S. Koenig. Probabilistic navigation in partially observable environments. In *Proceedings of the International Joint Conference on Artificial Intelligence*, pages 1080–1087, 1995.

[19] K. Tashiro, J. Ota, Y. C. Lin, and T. Arai. Design of the optimal arrangement of artificial landmarks. In *Proceedings of IEEE International Conference on Robotics and Automation*, pages 407–413, June 1995.

[20] C. J. Taylor and D. J. Kriegman. Vision-based motion planning and exploration algorithms for mobile robots. In Goldberg, Halperin, Latombe, and Wilson, editors, *1994 Workshop on the Algorithmic Foundations of Robotics, A. K. Peters*, pages 69–83, 1995.

Toward Real-Time Path Planning in Changing Environments

Peter Leven, *University of Illinois, Urbana, IL*
Seth Hutchinson, *University of Illinois, Urbana, IL*

We present a new method for generating collision-free paths for robots operating in changing environments. Our approach is closely related to recent probabilistic roadmap approaches. These planners use preprocessing and query stages, and are aimed at planning many times in the same environment. In contrast, our preprocessing stage creates a representation of the configuration space that can be easily modified in real time to account for changes in the environment. As with previous approaches, we begin by constructing a graph that represents a roadmap in the configuration space, but we do not construct this graph for a specific workspace. Instead, we construct the graph for an obstacle-free workspace, and encode the mapping from workspace cells to nodes and arcs in the graph. When the environment changes, this mapping is used to make the appropriate modifications to the graph, and plans can be generated by searching the modified graph.

After presenting the approach, we address a number of performance issues via extensive simulation results for robots with as many as twenty degrees of freedom. We evaluate memory requirements, preprocessing time, and the time to dynamically modify the graph and replan, all as a function of the number of degrees of freedom of the robot.

1 Introduction

In this paper, we present a new method for generating collision-free paths for robots operating in changing environments. Our work builds on recent methods that use probabilistic roadmap planners [3, 8, 12, 22, 25]. The idea that the cost of planning will be amortized over many planning episodes provides a justification for spending extensive amounts of time during a preprocessing stage, provided the resulting representation can be used to generate plans very quickly during a query stage. Thus, these planners use a two-stage approach. During a preprocessing stage, the planner generates a set of nodes that correspond to random configurations in the configuration space (hereafter, C-space), connects these nodes using a (simple, local) path planner to form a roadmap, and, if necessary, uses a subsequent sampling stage to enhance the roadmap. During the second, on-line stage, planning is reduced to query processing, in which the initial and final configurations are connected to the roadmap, and the augmented roadmap is searched for a feasible path.

Our new approach is a descendant of the probabilistic roadmap methods. Our goal, like theirs, is a real-time planner that uses approximate representations such as those provided by computer vision or range sensors. However, unlike probabilistic roadmap methods, our method is intended for robots that will operate in changing environments, and therefore we cannot exploit the premise that planning will occur many times in the same environment.

Our method begins, as do the probabilistic roadmap planners, by constructing a roadmap that represents the C-space. Nodes are generated by a random sampling scheme, and connections between nodes are generated using a simple, straight-line planner. Unlike the probabilistic roadmap planners, we generate a roadmap that corresponds to an obstacle-free environment. Then, in a second phase of the preprocessing stage, we generate a representation that encodes the mapping from cells in the discretized workspace to nodes and arcs in the roadmap. These two phases are specific to the robot, but are independent of the environment in which the robot will operate. The fact that the preprocessing is completely independent of the robot's target environment removes constraints on preprocessing time. Indeed, with our approach, it is feasible that when a new robot is designed, an extended period of preprocessing could be performed, at the end of which, the robot would essentially be preprogrammed to construct path plans in any environment that it might encounter.

In the on-line planning phase, the planner first identifies the cells in the discretized workspace that correspond to obstacles (e.g., by using a range scanner or stereo vision system), and then uses the encoded mapping to delete the corresponding nodes and arcs from the roadmap. Planning is then reduced to connecting the initial and final configurations to the roadmap (again, as is also the case with the probabilistic roadmap planners), and then searching the roadmap for a path between these newly added nodes. Of course it is possible to add obstacles to the environment in such a way that the roadmap becomes disconnected. This is true for any of the probabilistic roadmap planners (once one knows how samples are selected, and how these samples are connected by local planners, it is fairly straightforward to construct environments that will thwart them), but, as we will describe in subsequent sections, there are a number of steps that can be taken to cope with this problem, both at runtime and during the preprocessing stages.

In the on-line planning stage, our method runs in real time; plans are generated in less than one second. Thus, it is feasible to use the planner even in the case when obstacles are moving in the environment with unknown trajectories, provided a sensing system can identify in real-time those regions of the workspace that are occupied by the obstacles.

2 Related Research

There has been much work on probabilistic roadmap planners, including planners for articulated robots [12, 22, 8], mobile robots in two dimensional environments [22], free-flying rigid objects in three dimensions [2], and flexible surfaces [7]. There are also versions of these planners that are geared toward single-shot motion planning [9, 15, 24]. Some modifications to the sampling methods have also been introduced: obstacle boundary [22, 3] and medial-axis [25].

The sample configurations generated during the construction of a probabilistic roadmap can be thought of as landmarks in the C-space. A family of approaches based on the Ariadne's Clew algorithm generate landmarks and search for paths between landmarks in a deterministic fashion by solving optimization problems. This method has been used to plan for manipulators [4, 18]. and for manipulation planning (i.e., constructing a sequence of transfer and transit paths) [1].

There have been several previous approaches to path planning in changing environments. In some cases, off-line planners have execution times that make it feasible to directly use them in some kinds of changing environments with no modifications. This is the case, e.g., for the Ariadne's Clew algorithm reported in [4, 18]. The Ariadne's Clew algorithm operates by generating landmarks (during an exploration phase) and then connecting them to the existing network (the search phase). Variations of this algorithm can be obtained by varying the search phase, and by using different optimization criteria to select candidate landmarks [1, 19]. The idea of incrementally expanding a network for single query planning has also been used in [9] and [15]. In both of these, networks are grown from both the initial and goal configurations until they can be connected. In [9] the notion of expansive C-spaces is used, while in [15], random trees are used. In [24] an adaptable approach that uses multiple local planners is described. At runtime, characteristics of the problem are used to determine which (combination of) local planners will be most effective.

We also note here that in two of these previous approaches ([8] and [19]), the idea of somehow representing the mapping from the workspace to the C-space was incorporated. In [8], during the off-line planning stage, the planner is aware of a set of obstacles that might be present in the environment (in their experiments, a single obstacle was used). The locations of these obstacles are specified a priori, and at runtime the robot sensor system determines which, if any, of the obstacles are present in the environment. During the off-line planning stage, the trajectories in the paths tree that cause collision with each of these obstacles is determined. When objects are detected at runtime, the corresponding arcs are deleted from the graph. In [19], the paths tree (created by the modified Ariadne's Clew algorithm) is augmented to generate a graph. Then, for each path in the graph, the corresponding workspace cells are identified. Thus, when a new obstacle is added to the environment, the set of paths that intersect that obstacle can be deleted from the graph. Because these authors are primarily interested in domains for which extensive preprocessing is not viable, they construct relatively small graphs (fewer than 100 landmarks, with fewer than 400 corresponding paths). Therefore, it is fairly easy to add obstacles that would disconnect the graph, even though these obstacles might not cause the free C-space to become disconnected. In their experimental evaluation of their planner, a single obstacle was added, and the addition of this obstacle resulted in disconnecting the graph into five components. In such cases, their algorithm resorts to the Ariadne's algorithm to reconnect the graph, which can take time long enough to prohibit the planner from being used in environments with moving obstacles.

We believe that our planner is the first planner that constructs a representation that can be used to replan in real time in changing environments. In the remainder of the paper we describe the specific details of the algorithm, give extensive evaluation of our methods via simulation results, and discuss the current trajectory of our research efforts.

3 Constructing the Roadmap

Our construction of a roadmap of the C-space is very similar to methods used in previous probabilistic roadmap planers [12]. Nodes are generated by generating sample configurations, and these nodes are then connected to form a graph. We will denote this graph by $\mathcal{G} = \langle \mathcal{G}_n, \mathcal{G}_a \rangle$, in which $\mathcal{G}_n$ is the set of nodes in the graph and $\mathcal{G}_a$ is the set of arcs in the graph.

Since the graph is constructed without the presence of obstacles in the workspace, local planning to connect the nodes is trivial. Generating samples is potentially more complex than for the traditional probabilistic roadmap planners, since we cannot exploit the geometry of the C-space obstacle region to guide the sampling. We now discuss sampling the C-space, and generating arcs to connect the resulting nodes.

3.1 Generating Sample Configurations

Since there are no obstacles to consider, it is fairly easy to generate samples in the C-space. The only hard constraint is that self-collision (i.e., collision between distinct links of the robot) is prohibited. At present, we generate configuration nodes by sampling from a uniform distribution on the C-space. This approach reflects a complete absence of prior knowledge about the environment in which the robot will ultimately operate. If prior knowledge, either about the environment or the set of tasks that the robot will perform, were available, an appropriate importance sampling scheme, or even a deterministic scheme (if the existence of certain obstacles were known in advance) could be used. More effective schemes for generating sample configurations are discussed in Section 6.

3.2 Connecting the Nodes

In order to create a graph, pairs of nodes are connected by arcs that correspond to trajectories of the arm. Like several of the previously mentioned approaches, we connect each node to its k-nearest neighbors. In our current implementation, the paths corresponding to arcs are not tested to ensure that no self-collisions take place along the path.

The distance metric used for determining the neighbors of a node plays a very important role in the probabilistic roadmap methods [2]. For these planners, good distance metrics provide a set of neighbors to which the local planner has a good likelihood of generating a collision free path. In addition, the distance metric must be fast to compute, as it will be called many times to evaluate pairs of nodes (this is particularly true for the single query planners).

For our application, the swept volume of a path in the workspace connecting two configurations would be an ideal metric. Unfortunately, as noted by others [2, 13], this metric is very expensive to compute, and therefore, in our current implementation, we have opted for approximations that are faster to compute. In [2], a number of metrics were evaluated for efficiency and effectiveness for the case of a rigid object translating and rotating in a three-dimensional workspace. In particular, as we will discuss below, we have performed extensive experimental evaluation with the following four distance metrics:

the 2-norm in C-space

$$\mathcal{D}_2^{\mathcal{C}}(\mathrm{q},\mathrm{q}') = \|\mathrm{q}' - \mathrm{q}\| = \left[\sum_{i=1}^{n} (q_i' - q_i)^2\right]^{\frac{1}{2}}$$

the ∞-norm in C-space

$$\mathcal{D}_\infty^{\mathcal{C}}(\mathrm{q},\mathrm{q}') = \max_n |q_i' - q_i|$$

the 2-norm in workspace

$$\mathcal{D}_2^{\mathcal{W}}(\mathrm{q},\mathrm{q}') = \left[\sum_{\mathrm{p}\in\mathcal{A}} \|\mathrm{p}(\mathrm{q}') - \mathrm{p}(\mathrm{q})\|^2\right]^{\frac{1}{2}}$$

the ∞-norm in workspace

$$\mathcal{D}_\infty^{\mathcal{W}}(\mathrm{q},\mathrm{q}') = \max_{\mathrm{p}\in\mathcal{A}} \|\mathrm{p}(\mathrm{q}') - \mathrm{p}(\mathrm{q})\|.$$

In each of these equations, the robot has n joints, q and q′ are the two configurations corresponding to different nodes in the graph, q_i refers to the configuration of the i-th joint, and p(q) refers to the workspace reference point p of the set of reference points of the robot $\mathcal{A}$ at configuration q. Versions of $\mathcal{D}_\infty^{\mathcal{W}}$ and $\mathcal{D}_2^{\mathcal{W}}$ were also used in [13].

Once a node's k-nearest neighbors have been identified, a local planner is used to connect the corresponding configurations. As a motion planner, this local planner in the preprocessing phase has very modest requirements. It should be reasonably fast, as that reduces the time needed to construct the data structures, but this is not a primary concern. It must be deterministic, i.e., it must always return the same path when given the same two nodes as input. This is required since our approach will use the volume swept by the robot when it traverses this path. If the path changed each time, we would be unable to guarantee that the path was not blocked during on-line planning. In addition, the local planner should consider self-collisions of the robot when determining whether two nodes can be connected (although we have not yet implemented this feature).

We also note here that, during the on-line planning phase, a second local planner will be used to connect the initial and final configurations to the roadmap. In contrast to the planner used to connect nodes in the graph, this planner does have to consider the obstacles around the robot. Since this problem is faced by all of the probabilistic roadmap planners, we plan to adapt existing methods for this problem (e.g., the technique presented in [24] may be used, which selects the local planner based on an evaluation of which planner is most likely to succeed).

4 Workspace to C-space Mapping

The ability to plan in real time in changing environments comes from our encoding of the relationship between the workspace and the C-space. To represent the workspace, we use a uniform, rectangular decomposition, which we denote by $\mathcal{W}$. We denote the C-space by $\mathcal{C}$, and define the mapping $\phi : \mathcal{W} \to \mathcal{C}$ as

$$\phi(w) = \{\mathrm{q} \mid \mathcal{A}(\mathrm{q}) \cap w \neq \emptyset\},$$

in which w is cell of the workspace and $\mathcal{A}(\mathrm{q})$ denotes that subset of $\mathcal{W}$ occupied by the robot at configuration q. We note that $\phi(w)$ is exactly the C-space obstacle region (often denoted by $\mathcal{CB}$) if w is considered as a polyhedral obstacle in the workspace. In our approach, we do not explicitly represent the C-space, but instead use the roadmap $\mathcal{G}$. Therefore, we define two additional mappings, one from the workspace to the nodes in the graph, and one from the workspace to the arcs in the graph:

$$\phi_n(w) = \{\mathrm{q} \in \mathcal{G}_n \mid \mathcal{A}(\mathrm{q}) \cap w \neq \emptyset\},$$
$$\phi_a(w) = \{\gamma \in \mathcal{G}_a \mid \mathcal{A}(\mathrm{q}) \cap w \neq \emptyset \text{ for some } \mathrm{q} \in \gamma\}.$$

4.1 Computation of ϕ_a and ϕ_n

In terms of the implementation, it is much easier to compute the inverse maps ϕ_n^{-1} and ϕ_a^{-1}. Therefore, we use the inverse maps for the construction of our representation of the mapping.

The construction of the representation for ϕ_n^{-1} is straightforward. For each $\mathrm{q} \in \mathcal{G}_n$ we note the mapping from each $w \in \phi_n^{-1}(\mathrm{q})$ to the corresponding q. The set of cells in $\phi_n^{-1}(\mathrm{q})$ is computed by expanding a "seed" cell in a set of shells surrounding the seed. The seed for fixed-base articulated robots is the origin of the robot, which is not stored as part of ϕ_n. The shell expands in each direction in the workspace until the collision tester determines that a cell is outside the robot. We use the collision-checking package V-Clip [20] to test for cell intersection with the robot.

The computation of ϕ_a^{-1} is more complex and time-consuming. It involves computing the swept-volume of the robot as it traverses a path computed by the local planner between two configurations. In our implementation, cells that are occupied by the robot at the endpoints of the arc γ (i.e., the configurations that correspond to the two nodes connected by the arc) are not included in $\phi_a^{-1}(\gamma)$. Computing the swept volume for a robot trajectory is not a trivial problem. A method for swept volume computation for three dimensional objects that can only translate is presented in [26]. An extension to this method was presented that could also accommodate rotation about a single axis, but no results were given. This approach could be used to simplify the computation of ϕ_a^{-1}.

We have developed a method for computing $\phi_a^{-1}(\gamma)$ as follows. First, ϕ_n^{-1} for the two endpoints of γ is computed. Then, the path corresponding to γ is sampled using a recursive bisection method, which proceeds as follows. First, the configuration corresponding to the midpoint of the segment connecting the two nodes is computed, and ϕ_n^{-1} for that configuration is computed. If this set contains any cells not already in ϕ_n^{-1} of the endpoints, these new cells are added to the swept volume, and the path is subdivided again on both sides of the midpoint. This process is repeated until no new cells of $\mathcal{W}$ are added.

4.2 Efficient Representations of ϕ

Even though computer memories are rapidly growing, it is beneficial to compress the representations of ϕ_n and ϕ_a, provided that this can be done without drastically increasing the computation required to compute plans online. Reducing the size of the representation will also enable us to consider larger graphs $\mathcal{G}$, which will increase the efficacy of the online planning.

From an information theoretic viewpoint, compression of a data set involves the reduction of redundancy in that data set. The amount of compression that can be performed is limited by the information content of the data set, which, in turn, is related to the degree of unexpectedness, or randomness, in the data set [5]. There are three sources of redundancy in the representation of ϕ_n and ϕ_a that can be exploited: 1) the spatial coherence of the set $\phi(w)$ in $\mathcal{C}$, 2) spatial coherence of $\phi(w)$ for neighboring w's in $\mathcal{W}$, and 3) the representation of the labels of the nodes and arcs in the graph. The third source provides only limited efficiency increases, as much as a factor of two for our experiments, depending on the size of the graph.

The spatial coherence of ϕ_a and ϕ_n is based on the following ideas. Since the robot is connected, and since w is connected and compact, then $\phi(w)$ is connected and compact. This follows from the application of well-known facts about the C-space obstacles [16], and that $\phi(w)$ is exactly the C-space obstacle region if w is considered as a polyhedral obstacle in the workspace.

The spatial coherence of $\mathcal{CB}$ has been exploited in previous collision checking approaches (e.g., [17, 20]). In our case, spatial coherence derives from the continuity of ϕ, namely, that small changes to w will cause only small changes to $\phi(w)$. Because of this, for some cell, say $w^* \in \mathcal{W}$, we expect that $\phi(w)$ will be very similar to $\phi(w^*)$ for $w \in \eta(w^*)$, with $\eta(w^*)$ some appropriate neighborhood of w^*. This spatial coherence presents a situation that is somewhat analogous to the situation confronted in video compression: in a stream of images, there will be only small variations between most adjacent images in the sequence. This is one of the premises for many modern video compression methods (e.g., MPEG [21]). Unfortunately, we cannot directly apply video compression techniques, since these techniques generally employ lossy compression (i.e., the original image sequence cannot be exactly reconstructed), and in our case this could lead to collisions.

The discussion above suggests the following approach: partition $\mathcal{W}$ into a set of neighborhoods, and, for each neighborhood (a) choose a representative w^*, (b) derive a compact representation of $\phi(w^*)$ and (c) for all $w \in \eta(w^*)$, express $\phi(w)$ in terms of $\phi(w^*)$. We postpone the discussion on step (b) to Section 6.3. In some cases, we may be able to improve upon this by selecting some other reference set in step (c), and we discuss this below.

Given the above, we can formulate the corresponding optimization problem. For specific choice of neighborhood system we have the cost functional:

$$\mathcal{L}(\mathcal{W}^*, \eta) = \sum_{w^* \in \mathcal{W}^*} \left\{ \text{cost}[\phi(w^*)] + \sum_{w \in \eta(w^*)} \text{cost}[\phi(w)] \right\}, \quad (1)$$

in which $\mathcal{W}^*$ is a set of representative cells in $\mathcal{W}$, $\eta(w^*)$ is the set of neighbor cells for w^*, and $\text{cost}[\phi(w)]$ denotes the cost of encoding the representation. This leads to the the optimization problem:

$$\left\{ \begin{array}{ll} \text{minimize} & \mathcal{L}(\mathcal{W}^*, \eta) \\ \text{subject to:} & \bigcup_{w^* \in \mathcal{W}^*} \eta(w^*) = \mathcal{W} \ \text{ and} \\ & \eta(w_i^*) \cap \eta(w_j^*) = \emptyset, i \neq j. \end{array} \right. \quad (2)$$

This particular formulation of the cost suggests an algorithm that first selects representatives in $\mathcal{W}$ and then builds the appropriate neighborhoods. The encoding of the neighborhoods is based on the idea discussed above that there will be only minor variations in ϕ over local neighborhoods of $\mathcal{W}$. Assuming for the moment that the neighborhoods have been determined, one way to encode ϕ over a neighborhood $\eta(w^*)$ is to first determine a representative $\phi^*(w^*)$ and to then specify $\phi(w)$ relative to $\phi^*(w^*)$ for each $w \in \eta(w^*)$. This is essentially a differential encoding, and can be specified as $\phi'(w) = \phi(w) \oplus \phi^*(w^*)$, where $\phi'(w)$ is the encoded representation of $\phi(w)$ and $\oplus$ is the set symmetric difference operator.

There are two methods for determining $\phi^*(w^*)$. The first of these is to use $\phi(w^*)$ itself, i.e., $\phi^*(w^*) = \phi(w^*)$. The second is to compute $\phi^*(w^*)$ as the set that minimizes:

$$\sum_{w \in \eta(w^*)} |\phi'(w)|,$$

where $|\cdot|$ denotes set cardinality. The first method may lead to a more efficient representation of $\phi^*(w^*)$ itself, while the second minimizes $|\phi'(w)|$ for each $w \in \eta(w^*)$. This is a tradeoff that will have to be evaluated when selecting which set to use for $\phi^*(w^*)$.

There is another possible differential encoding scheme that can also be used to take advantage of the spatial coherence of $\phi(w)$ over $\mathcal{W}$. This approach is similar to above, except that instead of choosing one $\phi^*(w^*)$ over $\eta(w^*)$, the ϕ^* is chosen individually for each $w \in \eta(w^*)$ from the set of cells adjacent to w that are closer to w^*. In this approach, $\phi(w^*)$ is encoded by itself, the cells adjacent to w^* are encoded using $\phi(w^*)$ as their reference, and for each of the other $w \in \eta(w^*)$, the neighbor w' to w that results in the smallest $|\phi'(w)|$ is chosen, with $\phi'(w) = \phi(w) \oplus \phi(w')$.

These two approaches may be combined in a hybrid approach, such that the neighborhood $\eta(w^*)$ is composed of two parts $\eta_1(w^*)$ encoded using the first scheme above and $\eta_2(w^*)$ encoded using the second scheme. The cell w^* is encoded as part of $\eta_1(w^*)$, and each $w \in \eta_2(w^*)$ is encoded with respect to the adjacent cell that is closest to a member of $\eta_1(w^*)$.

Still to be addressed is the problem of selecting the best set of representative cells in the workspace and the problem of choosing the best neighborhoods $\eta_1(w^*)$ and $\eta_2(w^*)$ for each $w^* \in \mathcal{W}^*$ that satisfies (2). This is a difficult combinatoric optimization problem, for which we have applied the following greedy expansion algorithm. The elements of $\mathcal{W}$ are placed into a priority queue, Ω in decreasing order of the size of the representation of $\phi(w)$ (i.e., $|\phi_n(w)| + |\phi_a(w)|$). Let w^* be the first element in Ω. Initialize $\eta(w^*) = \{w^*\}$. Then, until the cost of adding an additional neighbor to $\eta(w^*)$ exceeds the cost of directly encoding $\phi(w)$ or a size threshold is reached, expand the neighborhood, deleting the added w from Ω. In the hybrid approach, each cell w is placed in the neighborhood η_1 or η_2, depending on which minimizes the cost of its representation. This is repeated for the new head of Ω until the neighborhoods form a partition of the workspace.

Note that this operation is performed separately for both ϕ_n and ϕ_a.

5 Empirical Evaluation

For a real-time path planner, we are interested in three parameters: graph update time, planning time, and data structure size. We are also interested in the preprocessing time, though this is less important. In this section, we evaluate a preliminary version of the planner, studying the case of serial-link manipulators with revolute joints operating in a two-dimensional workspace. Note that our implementation is not limited to revolute joints; prismatic joints can be used as well. We limit the following discussion to revolute joints to simplify the analysis.

5.1 Experimental Set-up

The following experimental set-up was used to evaluate the approach. The robots tested are all serial-link manipulators with two to 20 revolute joints in a two-dimensional workspace. The workspace cells and the robot are given three-dimensional polyhedral descriptions to allow the use of standard collision checking packages for collision testing. Each cell of the workspace is modeled as a unit cube, and each link of the robot is modeled as a rectanguloid with a width of 2.1 units and a height of 1 unit. The total length of the robot was fixed at 70 units, with each link length being scaled appropriately. Each joint of the robot was given the full revolute range of motion, and the joints were allowed to wrap around. In addition, the robot was not tested for self-collision, except when the samples of C-space were generated. To evaluate how the data structures and computation times grow with the graphs, we tested graphs with 2048, 8192, and 16384 nodes.

5.2 Results for Serial-link Manipulators in a 2D Workspace

In this section we present some results from using the above setup. We start with an analysis of the data structures that are computed to implement our planning approach. Each of the four data structures—graph nodes, graph arcs, ϕ_n, and ϕ_a—was computed and stored separately to simplify analysis. We follow the analysis with some compression results for ϕ_n and ϕ_a, and we conclude with some planning examples.

5.2.1 C-space Sampling

The C-space for each of the robots was sampled uniformly at random, except for the first 129 samples. The first 129 samples were chosen such that the robot covers as much of the workspace as possible. For the robots tested, this means that 129 uniformly-spaced samples were taken of the range of the first joint, and the remaining joints were held fixed at a position that maximized the length of the robot. Of the remaining random samples, samples in which the robot collides with itself were rejected.

It is interesting to note that, when sampling the C-space of the robot uniformly, it became exponentially more difficult as the number of joints increased to find configurations in which the robot did not collide with itself. This is a consequence of a robot consisting of revolute joints operating in the plane: For each set of three links, there is a region of the C-space in which the third link is in collision with the first. For each additional joint beyond three, there is a combinatoric effect with combinations of links in collision, as well as an interaction among these links that tends to make the collision regions larger. For a non-planar robot operating in 3D, this effect is expected to be drastically reduced.

The exponential sampling effect can be seen in Figure 1, which shows the number of samples of C-space generated per node accepted. Notice that the y-axis scale of this graph is logarithmic. As expected, the number of samples generated per node accepted does not change as the desired number of nodes increases.

The size of the data structure for the configurations associated with the nodes in the graph is $8mn + 12$ bytes, where m is the number of nodes and n is the number of joints.

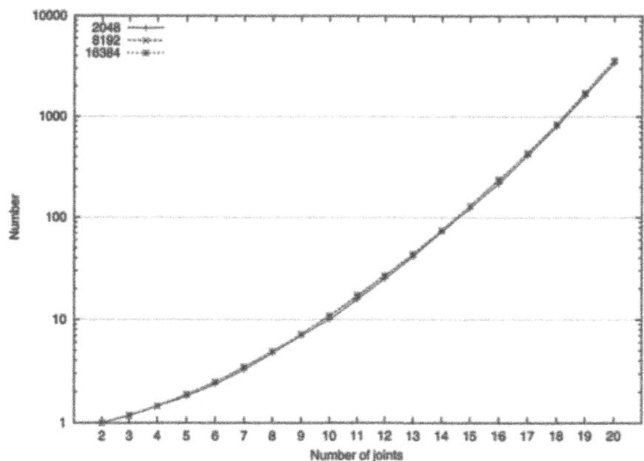

Figure 1: *Number of samples of C-space generated per node accepted.*

5.2.2 Graph Construction

As described above, arcs in the graph are constructed for the k-nearest neighbors of each node. For our evaluation, we tested the case of $k = 5$ nearest neighbors. We computed the nearest neighbors using the simple $O(n^2)$ method of calculating the distance between all pairs of nodes and keeping the closest five for each node. The paths connecting these neighbors are not checked for feasibility, i.e., whether the robot avoids self-collision while following the paths.

The computation times for computing the graph are shown in Figure 2. The increase in computation time with the number of joints reflects the increase in the cost of computing the metric. The sharp rise in the computation time after 16 joints seen in the graph for the $\mathfrak{D}^{\mathfrak{c}}_{\infty}$ and $\mathfrak{D}^{\mathfrak{c}}_{2}$ distance metrics is unexpected; this may be due to a memory effect in the implementation, where the amount of memory used for the calculation crosses some system threshold.

What also can be seen in Figure 2 is that there is not much difference in computation time between the 2-norm and the ∞-norm, except that the computations involving the ∞-norm tend to take somewhat more more time to perform (this is likely an artifact of the processor architecture, in which the cost of a branch is greater than the cost of a floating-point multiplication). A greater difference can be seen between the metric in C-space and the metric in the workspace, where the

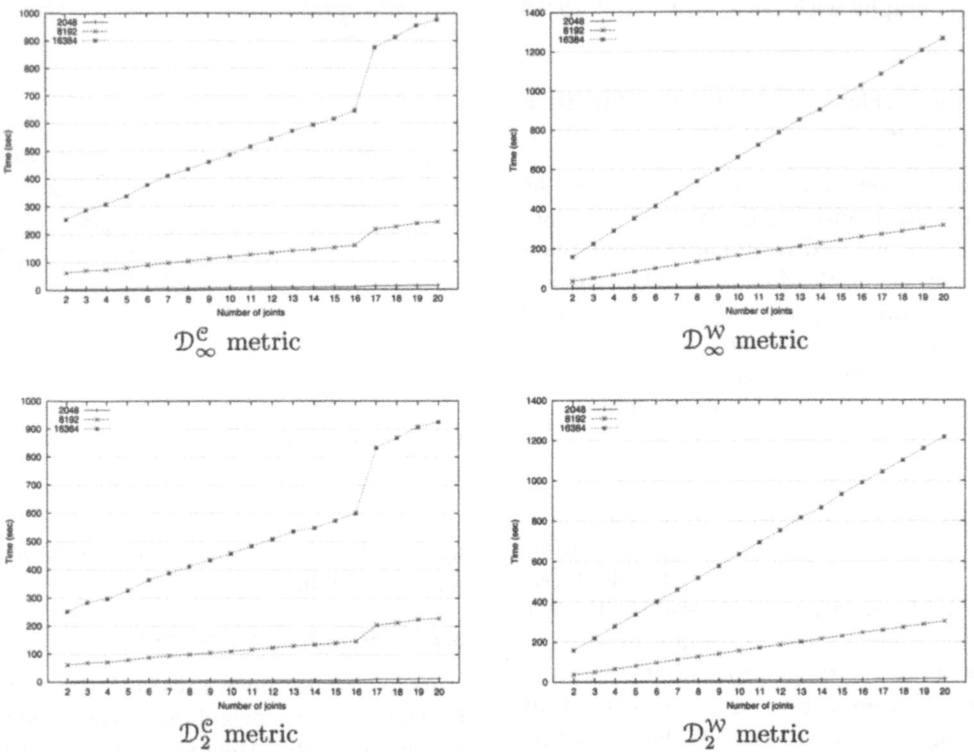

Figure 2: *Time in seconds to compute the k-nearest neighbors for the different distance metrics.*

latter appears to have a more linear increase in computation time as the number of joints increases. Overall, for robots with more than five joints, the workspace metric takes more time to compute.

The size of the graph varies linearly with the number of nodes, with an average size of 163kB for 2048 nodes, 654kB for 8192 nodes, and 1309kB for 16384 nodes. The variance in size from the average was less than 15%.

5.2.3 Computing ϕ_n

The overall size of ϕ_n is the product of the number of nodes and the average number of cells the robot covers in the workspace. For a robot with a fixed maximum length, this means that the size of ϕ_n is largely independent of the number of joints. For our experiments, the size of ϕ_n per node in $\mathcal{G}_n$ averages around 1000 bytes.

There is a sub-linear increase in the size of ϕ_n as the number of nodes increases that is a result of the overhead for maintaining the data structure being amortized over more nodes. The overhead for the data structure depends on the number of cells in the workspace that intersect with the robot at any node in the graph. This overhead does not depend on the number of nodes in the graph, as long as the graph contains nodes that span the reach of the robot in the workspace (i.e., the graph contains nodes that touch all the cells in the workspace that the robot can reach).

The time required per node to compute the node obstacle data structure increases linearly as the number of joints increases. This is expected because the size of the geometric description of the robot used for collision testing increases with the number of joints.

5.2.4 Computing ϕ_a

The size of ϕ_a for different numbers of nodes is shown in Figure 3. As can be seen in the figure, ϕ_a can become quite large. This is the reason that the graphs stop at 13 joints for the 8192 node case, and 8 joints for

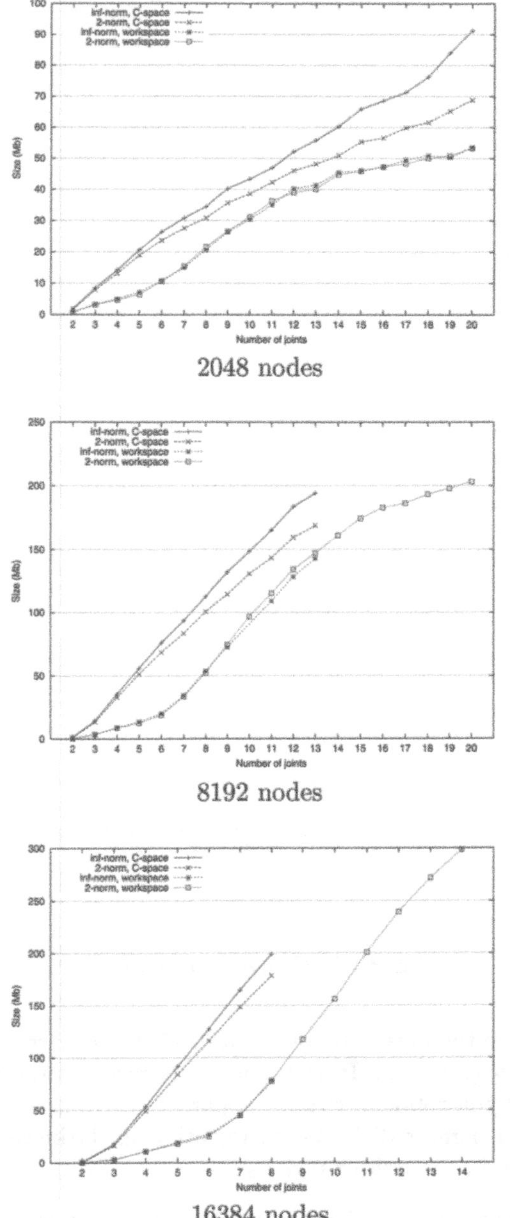

Figure 3: *Size of* ϕ_a.

the 16384 case. The limit in this case was the process size of the program computing ϕ_a; in order to keep the computation times comparable, the memory size of the program computing ϕ_a was limited to the physical memory available in the computer. More examples were computed for the $\mathcal{D}_2^{\mathcal{W}}$ metric for the 8192 and

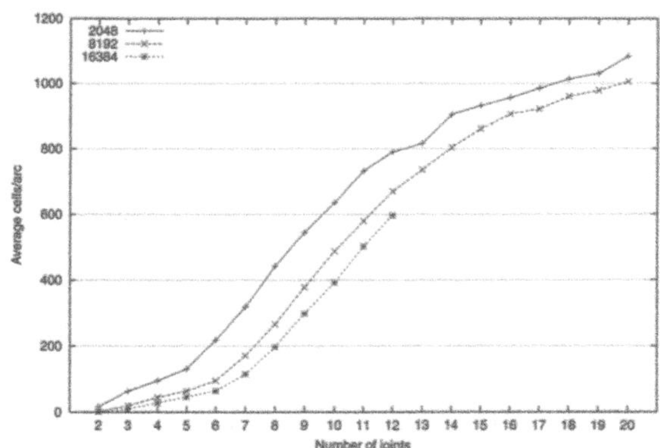

Figure 4: *Average of* $|\phi_a^{-1}(\gamma)|$ *per* γ *for* $\mathcal{G}_a$ *generated from the* $\mathcal{D}_2^{\mathcal{W}}$ *metric.*

16384 node cases in order to evaluate the compressibility of the resulting data files (see Section 5.2.5).

It is also apparent that the $\mathcal{D}_\infty^{\mathcal{C}}$ performs the worst in terms of the size of ϕ_a. The $\mathcal{D}_2^{\mathcal{C}}$ is the next worst, and the $\mathcal{D}_\infty^{\mathcal{W}}$ and $\mathcal{D}_2^{\mathcal{W}}$ produce the best results. It is expected that the two metrics defined on the workspace would produce better results than the C-space metrics, since the workspace metrics penalize the motion in the workspace more than do the C-space metrics. There is no apparent advantage to $\mathcal{D}_2^{\mathcal{W}}$ over $\mathcal{D}_\infty^{\mathcal{W}}$.

Another interesting point in Figure 3 is the apparent inflection point at six joints in the graphs for the size of ϕ_a when computed using the $\mathcal{D}_\infty^{\mathcal{W}}$ and $\mathcal{D}_2^{\mathcal{W}}$ metrics. This inflection point is more obvious in the 8192 and 16384 node graphs. An explanation for this is that, given the resolution of the workspace that we are using, the random sampling is better able to "fill" the C-space with samples for the robot with fewer joints. This results in the nodes being closer together, and, therefore, the paths between them do not cover as much of the workspace. This effect can also be seen in Figure 4, which shows the average size of $\phi_a^{-1}(\gamma)$ per $\gamma \in \mathcal{G}_a$ for the $\mathcal{D}_2^{\mathcal{W}}$ metric ($\mathcal{D}_\infty^{\mathcal{W}}$ produces similar results).

Shown in Figure 5 is the time it takes to compute ϕ_a for the $\mathcal{D}_2^{\mathcal{C}}$ and $\mathcal{D}_2^{\mathcal{W}}$ metrics (the $\mathcal{D}_\infty^{\mathcal{C}}$ and the $\mathcal{D}_\infty^{\mathcal{W}}$ produce similar results to the $\mathcal{D}_2^{\mathcal{C}}$ and $\mathcal{D}_2^{\mathcal{W}}$ metrics, respectively). These times correlate well with the increase in the size of ϕ_a: The larger ϕ_a is, the longer it takes to compute. Notice that the graph for the computation times for the workspace metric shows an inflection point like the graphs for the size of ϕ_a.

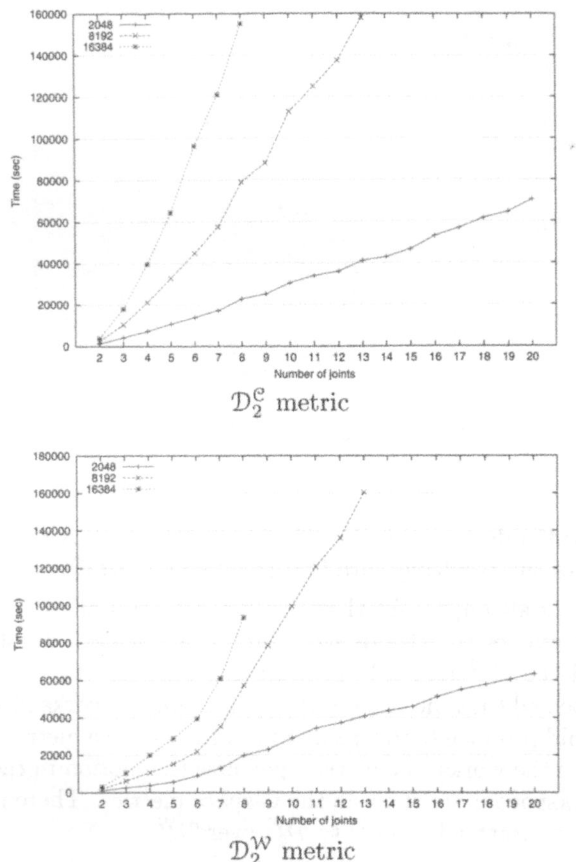

Figure 5: *Computation time for ϕ_a for two metrics.*

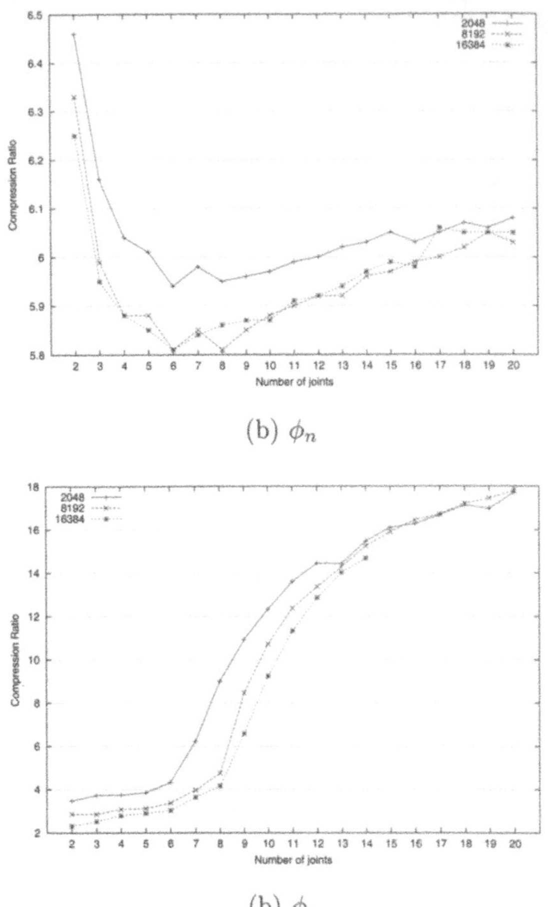

Figure 6: *Compression ratios.*

5.2.5 Some Compression Results

Shown in Figure 6 are the compression ratios for ϕ_n and ϕ_a that were achieved by differential encoding of the cells of the workspace and by using a more efficient encoding of the labels used to represent the nodes and arcs in the graph (roughly a factor of two of the compression ratio comes from this change in representation). The compression ratios are relatively flat for ϕ_n as the number of joints changes.

To gain some insight into the behavior of the compression ratios achieved for ϕ_n, a set of images showing the distribution of $|\phi_n(w)|$ over the workspace were generated. Figure 7 shows an example image, showing the two-link robot case with a graph of 2048 nodes. In the figure, the grid lines demark 11x11 collections of workspace cells, and the darkness of a cell is proportional to the logarithm of $|\phi_n(w)|$. Cells that are white have $|\phi_n(w)| = 0$. In the figure, the reach of the first link of the robot is readily apparent as a darker disk inside a lighter disk. Some variation in darkness can be seen in the lighter regions.

As the number of joints in the robot increases, the ring corresponding to the first link becomes smaller, and the links themselves have a greater tendency to cluster near the origin due to the uniform random sampling. This clustering also means that clearly-visible rings corresponding to links of the robot other than the first do not appear. The clustering also has an effect on the compressibility of the ϕ_n. At two joints, the distribution of $|\phi_n(w)|$ is relatively uniform, compared to robots with more joints. A greater uniformity of coverage of the workspace implies greater spatial co-

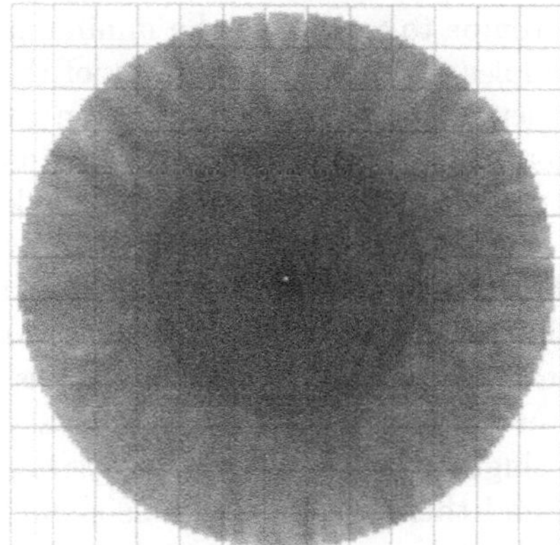

Figure 7: $|\phi_n(w)|$ *per cell for a two-link robot and a graph with 2048 nodes.*

herence between neighboring cells and that allows for greater compression. At the 20 joints, the distribution tends to be highly clustered near the origin, allowing for higher compression as the cells outside the cluster near the origin are grouped together in larger groups. In between the two extremes, the compression is lower.

On the other hand, the compression ratios for ϕ_a increase with the number of joints. This is expected because the average length of the paths corresponding to arcs in the graph tends to increase with the number of joints, and that in turn leads to an increase in the size of the volume of the workspace swept by the robot as it traverses the paths. As more cells are affected by the longer paths, the spatial coherence of the cells in the workspace increases. Notice that the inflection point seen around eight joints in this graph is similar to that seen in Figure 3. The same effect that minimizes the size of ϕ_a also reduces spatial coherence, and, therefore, the achievable compression ratio.

Shown in Figure 8 is the distribution of $|\phi_a(w)|$ per cell for the two link robot and the 2048-node and 16384-node graphs. These graphs have the same resolution as Figure 7. The sparseness of the distribution of the arcs in this case is the reason for the reduction in the compression ratio available using the differential encoding. Also shown in the figure is the effect of increasing the size of the graph from 2048 nodes to 16384 nodes. As noted earlier, the average distance between nodes decreases, decreasing the average swept volume of the paths corresponding to the arcs in the graph, and, in this case, dramatically increasing the sparseness of ϕ_a.

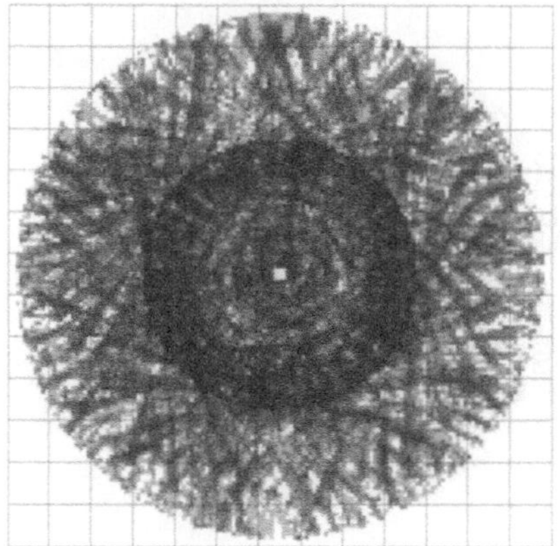

2048 node graph

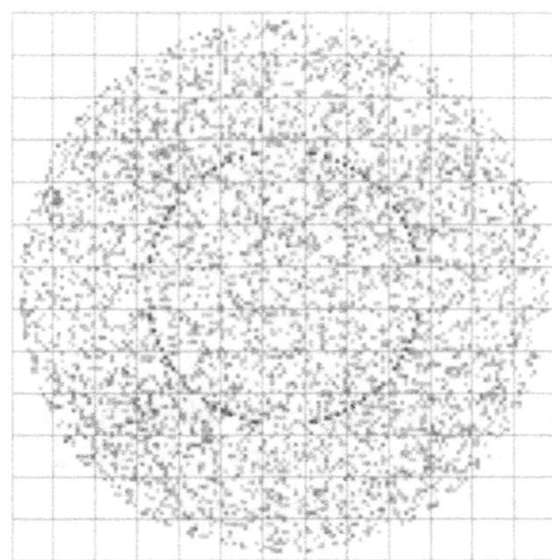

16384 node graph

Figure 8: $|\phi_a(w)|$ *per cell for a two-link robot.*

5.2.6 Some Planning Examples

The current implementation of our path planner tests the connectivity of the graph between two selected nodes in the modified graph; it does not connect arbitrary configurations to the graph. In all of the planning runs tested using the uncompressed representations of ϕ_n and ϕ_a, it takes less than one second to update the node and arc obstacle data structures and to find a path in the modified graph. Extensive tests using the compressed representations have not yet been performed.

Shown in Figure 9 is a 19-joint robot negotiating between two obstacles (black objects in the figure). The steps shown in the figure correspond to the nodes in the graph along the planned path. This plan was generated from the graph that used the $\mathcal{D}_2^{\mathcal{W}}$ distance metric with

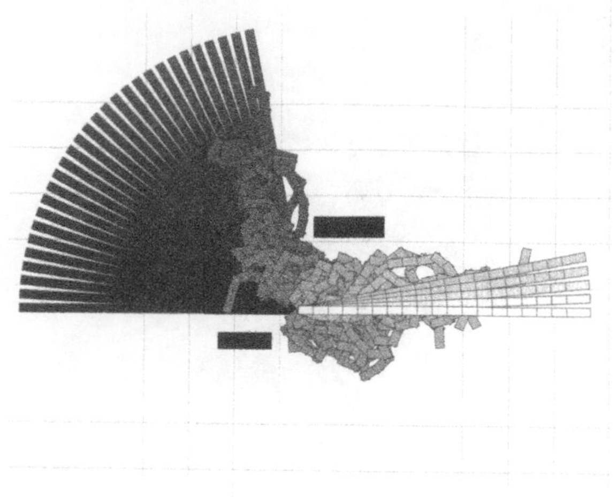

Figure 9: *A plan for a 19-joint robot passing through a (relatively) narrow corridor. The dark blocks are the obstacles.*

2048 nodes. The fan-like parts of the plan come from the set of nodes created to span the workspace of the robot, as described in Section 5.2.1.

6 Future Directions

Although we have an implemented version of our planner, there remain a number of issues that we are now beginning to investigate.

6.1 Generating Sample Configuration

The uniform sampling scheme described in Section 3 reflects complete ignorance about the environment. If prior knowledge, either about the environment or the set of tasks that the robot will perform, were available, an appropriate importance sampling scheme, or even a deterministic scheme (if the existence of certain obstacles were known in advance) could be used.

Even without prior knowledge about the environment, there are ways to improve the quality of the set of sample configurations. We could, for example, attempt to maximize the portion of the workspace that is covered by the sampled configurations (e.g., by generating configurations for a uniform distribution of end-effector positions in the workspace). In a similar spirit, we could attempt to maximize the minimum distance between samples as in [4]. The quality of the samples can be further improved by exploiting on the manipulability measure associated with the manipulator Jacobian matrix [27]. The basic idea is the following. In regions of the C-space where manipulability is high, the robot has great dexterity, and therefore relatively fewer samples should be required in these areas. Regions of the C-space where manipulability is low tend to be near (or to include) singular configurations of the arm. Near singularities, the range of possible motions is reduced, and therefore such regions should be sampled more densely.

These criteria can be posed as optimization problems with deterministic solutions; however, in each case, the complexity of the resulting problem would prohibit its direct solution. Therefore, we are currently investigating important sampling approaches that incorporate the criteria described above for sampling C-space.

6.2 Graph Enhancement

In previous probabilistic roadmap planners, if the initial roadmap was not singly connected, an enhancement stage was used to connect its various components. In our work, since there will be no obstacles in the environment, the roadmap will always have a single connected component, and therefore the traditional idea of enhancement does not apply. However, unlike previous approaches, we are concerned with maintaining the connectivity even if arcs or nodes are deleted from the roadmap. Therefore, we are investigating two methods to improve the robustness of roadmap connectivity to the addition of obstacles to the environment.

In the first method, the enhancement stage will attempt to increase the cardinality of the minimum cut set of $\mathcal{G}$. In particular, after an initial $\mathcal{G}$ is computed, minimal cut sets of both edges and vertices will be found, and then the C-space will be sampled in an attempt to add paths that would reconnect the graph if these cut edges or vertices were removed. This sampling would be driven by constructing the workspace volumes corresponding to the elements of the cut set (these are given by ϕ_n^{-1} and ϕ_a^{-1}). In order to disconnect the graph, an obstacle would need to touch each of these volumes. An algorithm such as the one proposed in [10] could be used to rapidly deduce a minimal set of cells in the workspace that satisfy this condition, and the mapping from workspace to C-space obstacles could then be used to seed an importance sampling algorithm that would add nodes to $\mathcal{G}$.

The second method relies on a quantitative evaluation of the quality of $\mathcal{G}$. We will investigate a property that we call ϵ-robustness (inspired by the notion of ϵ-goodness described in [14]). We say that $\mathcal{G}$ is ϵ-robust if no spherical obstacle in the workspace of radius ϵ (or less) can cause $\mathcal{G}$ to become disconnected. Using this concept, it will be possible to ensure after the enhancement stage that $\mathcal{G}$ can tolerate certain types of obstacles. In particular, for a specified value of ϵ, $\mathcal{G}$ can be tested for ϵ-goodness. This test can be performed in such a way that the specific workspace spheres that violate the ϵ-goodness criterion can be enumerated, and these can then be used, as above, to seed an importance sampling algorithm that will add nodes to $\mathcal{G}$.

6.3 Workspace to C-space Mapping

There are a number of possible improvements that can be made to reduce the computation time for the mapping and for more efficient representation of the mapping.

Improvement in computation time Using collision testing algorithms to compute ϕ_n and ϕ_a, while allowing for completely general geometric descriptions of the robot, is not the most efficient means available for computing $\phi_n^{-1}(q)$. An alternative is to consider 3D voxel scan-conversion techniques[11]. In our case, we do not need the full power of these algorithms as used to generate realistic images of three-dimensional scenes. We only need the components that compute the occupancy bitmaps for geometric data. This operation could also take advantage of any hardware acceleration available for this operation, as was done in [6] to accelerate the computation of Voronoi diagrams using polygon rasterizing hardware.

Efficient representations As described above, $\phi(w)$ is connected and compact. Exploiting the connectedness of regions for the purposes of compression has been a popular idea in the image processing and computer vision communities for many years [23]. Well known examples of this include using quadtrees and run length coding to compress images. In each case, the compression exploits an efficient representation of neighborhood structure, and then encodes homogeneous neighborhoods. For quadtrees, neighborhoods are defined by recursive partitions of the image, and in run length coding neighborhoods are defined in terms of individual horizontal rows in the image.

To compress the representations for $\phi_n(w^*)$, we will make use of the observation that often $\phi(w^*)$ is often fairly symmetric with respect to its center of mass. We plan to investigate using a prespecified graph traversal algorithm, and recording the depths at which the graph traversal algorithm exits or reenters $\phi(w^*)$. For most problems, breadth first traversal is likely to be the most effective. For example, if $\phi(w^*)$ were a hyper-sphere in an n-dimensional C-space, then this representation would require only the storage of a root node and a single integer to encode both $\phi_n(w^*)$ and $\phi_a(w^*)$. Breadth first traversal could then reconstruct the exact set of nodes and arcs in $\mathcal{G}$ that correspond to the cell w^*.

This approach is analogous to run length coding. In run length coding, the lengths of strings of ones are stored (for binary images). This approach essentially works by imposing an ordering on the pixels in the image (raster scan ordering), and then encoding when a region of ones is exited or reentered. In our proposed approach, we will use a breadth first tree traversal to impose an ordering on nodes and arcs in $\mathcal{G}$, and use the analogous idea of encoding tree depths at which $\phi(w^*)$ is exited or reentered. We note here that approaches analogous to 2^n−trees are not appropriate in our case, because these methods are very sensitive to small perturbations.

7 Conclusions

We have presented what we believe to be the first planner that is fully able to plan in real time in changing environments. We have presented extensive preliminary analysis of the planner via planning simulations. Although these preliminary results are quite promising, there are a number of open issues that remain, and we have outlined these along with our planned approaches.

References

[1] J. M. Ahuactzin, K. Gupta, and E. Mazer. Manipulation planning for redundant robots: A practical approach. *International Journal of Robotics Research*, 17(7):731–747, July 1998.

[2] N. M. Amato, O. B. Bayazit, L. K. Dale, C. Jones, and D. Vallejo. Choosing good distance metrics and local

planners for probabilistic roadmap methods. In *Proc. IEEE Conf. Robotics and Automation*, pages 630–637, 1998.

[3] N. M. Amato, O. B. Bayazit, L. K. Dale, C. Jones, and D. Vallejo. OBPRM: An obstacle-based PRM for 3D workspaces. In *Proceedings of the Workshop on Algorithmic Foundations of Robotics*, pages 155–168, 1998.

[4] P. Bessire, J.-M. Ahuactzin, E.-G. Talbi, and E. Mazer. The "Ariadne's Clew" algorithm: Global planning with local methods. In *Proceedings of the Workshop on Algorithmic Foundations of Robotics*, pages 39–47, 1994.

[5] D. Hankerson, G. A. Harris, and J. Peter D. Johnson. *Introduction to Information Theory and Data Compression.* Discrete Mathematics and its Applications. CRC Press, New York, 1998.

[6] K. E. Hoff, T. Culver, J. Keyser, M. Lin, and D. Manocha. Fast computation of generalized voronoi diagrams using graphics hardware. In *Proceedings of SIGGRAPH 1999 Conference on Computer Graphics*, pages 277–286, 1999.

[7] C. Holleman, L. E. Kavraki, and J. Warren. Planning paths for a flexible surface patch. In *Proc. IEEE Conf. Robotics and Automation*, pages 21–26, 1998.

[8] T. Horsch, F. Schwarz, and H. Tolle. Motion planning with many degrees of freedom — random reflections at c-space obstacles. In *Proc. IEEE Conf. Robotics and Automation*, pages 3318–3323, 1994.

[9] D. Hsu, J.-C. Latombe, and R. Motwani. Path planning in expansive configuration spaces. In *Proc. IEEE Conf. Robotics and Automation*, pages 2719–2726, 1997.

[10] Y.-B. Jia and M. Erdmann. Geometric sensing of known planar shapes. Technical Report CMU-RI-94-24, CMU, July 1994.

[11] A. Kaufman and E. Shimony. 3D scan-conversion algorithms for voxel-based graphics. In *Proceedings of the 1986 Workshop on Interactive 3D Graphics*, pages 45–75, New York, 1986. ACM.

[12] L. Kavraki and J.-C. Latombe. Randomized preprocessing of configuration space for fast path planning. In *Proc. IEEE Conf. Robotics and Automation*, volume 3, pages 2138–2145, 1994.

[13] L. Kavraki, P. Švestka, J.-C. Latombe, and M. Overmars. Probabilistic roadmaps for path planning in high-dimensional configuration spaces. *IEEE Transactions on Robotics and Automation*, 12(4):566–580, Aug. 1996.

[14] L. E. Kavraki, M. N. Kolountzakis, and J.-C. Latombe. Analysis of probabilistic roadmaps for path planning. In *Proc. IEEE Conf. Robotics and Automation*, volume 4, pages 3020–3025, 1996.

[15] J. J. Kuffner, Jr. and S. M. LaValle. RRT-connect: An efficient approach to single-query path planning. In *Proc. IEEE Conf. Robotics and Automation*, 2000.

[16] J.-C. Latombe. *Robot Motion Planning.* Kluwer Academic Publishers, Boston, 1991.

[17] M. Lin and D. Manocha. Efficient contact determination in dynamic environments. *International Journal of Computational Geometry and Applications*, 7(1):123–151, 1997.

[18] E. Mazer, J. M. Ahuactzin, and P. Bessire. The ariadne's clew algorithm. *Journal of Artificial Intelligence Research*, 9:295–316, 1998.

[19] A. McLean and I. Mazon. Incremental roadmaps and global path planning in evolving industrial environments. In *Proc. IEEE Conf. Robotics and Automation*, pages 101–107, 1996.

[20] B. Mirtich. V-Clip: Fast and robust polyhedral collision detection. Technical Report TR97-05, Mitsubishi Electric Research Laboratory, 201 Broadway, Cambridge, MA 02139, June 1997.

[21] J. L. Mitchell, W. B. Pennebaker, C. E. Fogg, and D. J. LeGall, editors. *MPEG Video Compression Standard.* Chapman and Hall, New York, 1997.

[22] M. H. Overmars and P. Švestka. A probabilistic learning approach to motion planning. In *Proceedings of the Workshop on Algorithmic Foundations of Robotics*, pages 19–37, 1994.

[23] A. Rosenfeld and A. Kak. *Digital Picture Processing.* Academic Press, New York, 1982.

[24] D. Vallejo, C. Jones, and N. M. Amato. An adaptive framework for 'single shot' motion planning. Technical Report 99-024, Department of Computer Science, Texas A&M University, College Station, TX, Oct. 1999.

[25] S. A. Wilmarth, N. M. Amato, and P. F. Stiller. Motion planning for a rigid body using random networks on the medial axis of the free space. In *Proceedings of ACM Symposium on Computational Geometry*, pages 173–180, 1999.

[26] P. G. Xavier. Fast swept-volume distance for robust collision detection. In *Proc. IEEE Conf. Robotics and Automation*, pages 1162–1169, 1997.

[27] T. Yoshikawa. Manipulability of robotic mechanisms. *International Journal of Robotics Research*, 4(2):3–9, 1985.

Geometric Construction of Time Optimal Trajectories for Differential Drive Robots

Devin J. Balkcom, *Carnegie Mellon University, Pittsburgh, PA*
Matthew T. Mason, *Carnegie Mellon University, Pittsburgh, PA*

We consider a differential drive mobile robot: Two unsteered coaxial wheels are independently actuated. Each wheel has bounded velocity, but no bound on torque or acceleration. Pontryagin's Maximum Principle gives an elegant description of the extremal trajectories, which are a superset of the time optimal trajectories. Further analysis gives an enumeration of the time optimal trajectories, and methods for identifying the time optimal trajectories between any two configurations. This paper recapitulates and refines the results of [1] and [2] and presents a simple graphical technique for constructing time optimal trajectories.

1 Introduction

This paper addresses the time optimal paths for differential drive mobile robots with bounded velocity. Differential drive (or diff drive) means there are two unsteered independently actuated coaxial wheels. Bounded velocity means that each wheel is independently bounded in velocity, but acceleration is not bounded. Even discontinuities in wheel velocity are permitted. The environment is planar and free of obstacles.

Under these assumptions, we will see that the time optimal paths are composed of straight lines alternating with turns "in place", i.e., turns about the center of the robot. Optimal paths contain at most three straights and two turns. There are a number of other restrictions, leading to a set of 40 different combinations arranged in 9 different symmetry classes. The simplest nontrivial motions are turn-straight-turn motions: turn to face the goal (or away from the goal); roll straight forward (or backward) to the goal; turn to the goal orientation. In some instances the optimal path passes through an intermediate "via" point. See Figure 1 for example motions from seven of the nine classes.

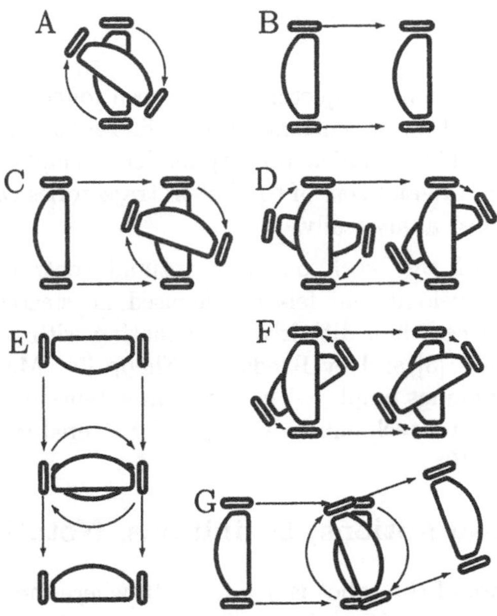

Figure 1: *The seven simplest optimal trajectory classes.*

To derive the optimal paths we will use Pontryagin's maximum principle to obtain a geometric program for the *extremal* trajectories, which are a superset of the optimal trajectories. We then derive some additional necessary conditions, leading to a complete enumeration of optimal trajectories and a planning algorithm. Finally we reformulate the analysis to give a more intuitive geometric procedure for constructing optimal trajectories.

Previous Work

This paper expands on the results presented in [1] and [2]. Other work on diff drive robots has assumed bounds on acceleration rather than on velocity; for example see papers by Reister and Pin [6] and Renaud and Fourquet [7]. For the bounded acceleration model,

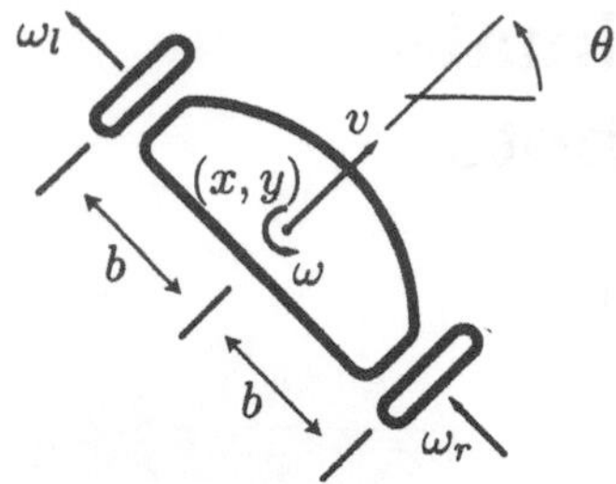

Figure 2: *Notation*

the time optimal trajectories have been found numerically, and there is current work to find a closed form solution. The bounded velocity model is simpler, and the structure and cost of the fastest trajectories can be determined analytically.

Most of the work on time optimal control with bounded velocity models has focused on steered vehicles rather than diff drives, originating with papers by Dubins [3] and by Reeds and Shepp [5]. Many of the techniques employed here are an extension of optimal control techniques developed for steered vehicles in [8, 9, 10].

2 Assumptions, Definitions, Notation

The state of the robot is $q = (x, y, \theta)$, where the robot reference point (x, y) is centered between the wheels, and the robot direction θ is 0 when the robot is facing along the x-axis (Figure 2). The robot's velocity in the forward direction is v and its angular velocity is ω. The robot's width is $2b$. The wheel angular velocities are ω_l and ω_r. With suitable choices of units we obtain:

$$v = \frac{1}{2}(\omega_l + \omega_r) \tag{1}$$

$$\omega = \frac{1}{2b}(\omega_r - \omega_l) \tag{2}$$

and

$$\omega_l = v - b\omega \tag{3}$$

$$\omega_r = v + b\omega. \tag{4}$$

The robot is a system with control input $w(t) = (\omega_l(t), \omega_r(t))$ and output $q(t)$. Admissible controls are bounded Lebesgue measurable functions from time interval $[0, T]$ to the closed box $W = [-1, 1] \times [-1, 1]$ (see Figure 3).

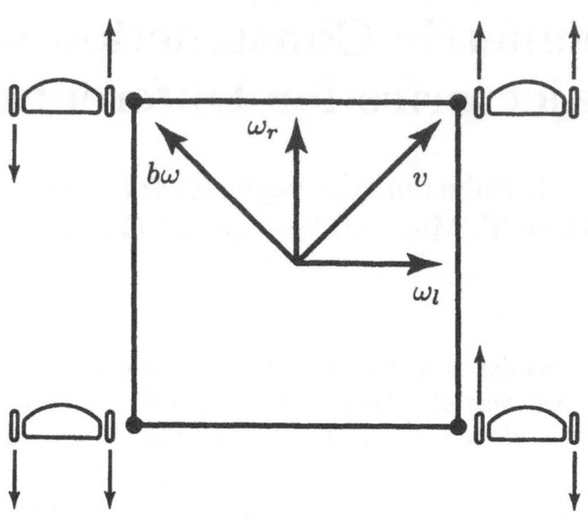

Figure 3: *Bounds on* (ω_l, ω_r)

It follows immediately that $v(t)$ and $\omega(t)$ are measurable functions defined on the same interval. Given initial conditions $q_s = (x_s, y_s, \theta_s)$ the path of the robot is given by:

$$x(t) = x_s + \int_0^t v\cos(\theta), \tag{5}$$

$$y(t) = y_s + \int_0^t v\sin(\theta), \tag{6}$$

$$\theta(t) = \theta_s + \int_0^t \omega. \tag{7}$$

We also define rectified path length in E^2, the plane of robot positions:

$$s(t) = \int_0^t |v| \tag{8}$$

and rectified arc length in S^1, the circle of robot orientations:

$$\sigma(t) = \int_0^t |\omega|. \tag{9}$$

It follows that θ, x, y, s, and σ are continuous, that their time derivatives exist almost everywhere, and that:

$$\dot{\theta} = \omega \quad \text{a.e.} \tag{10}$$

$$\dot{x} = v\cos(\theta) \quad \text{a.e.} \tag{11}$$

$$\dot{y} = v\sin(\theta) \quad \text{a.e.} \tag{12}$$

The admissible control region W also provides a convenient comparison with previously studied bounded

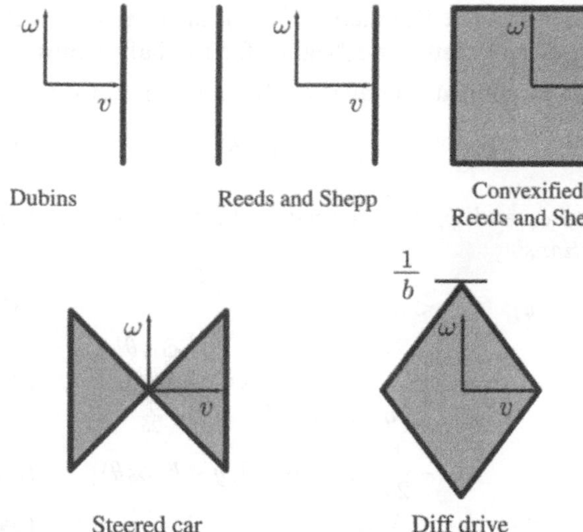

Figure 4: *Bounded velocity models of mobile robots*

velocity models. If we plot W in v-ω space, we obtain a diamond shape. Steered vehicles are typically modeled as having a bound on the steering ratio $\omega : v$, and on the velocity v (Figure 4).

We also need a notation for trajectories. Later sections show that extremal trajectories are composed of straight lines and turns about the robot's center. We will represent *forward* by ⇑, *backward* by ⇓, *left turn* by ↶, and *right turn* by ↷. Thus the trajectory ↶⇑↶ can be read "left forward left". When necessary, a subscript will indicate the distance or angle traveled. We will use t and s to represent turns and straights of indeterminate direction.

3 Pontryagin's Maximum Principle. Extremal Controls.

The existence of a time optimal trajectory for every pair of start and goal configurations is proven in [1]. This section uses Pontryagin's Maximum Principle [4] to derive necessary conditions for time optimal trajectories of the bounded velocity diff drive robot.

The robot system is described by:

$$\dot{q} = \begin{pmatrix} \dot{x} \\ \dot{y} \\ \dot{\theta} \end{pmatrix} = \begin{pmatrix} \frac{1}{2}(\omega_l + \omega_r)\cos(\theta) \\ \frac{1}{2}(\omega_l + \omega_r)\sin(\theta) \\ \frac{1}{2b}(\omega_r - \omega_l) \end{pmatrix} \qquad (13)$$

where our input is

$$w = \begin{pmatrix} \omega_l \\ \omega_r \end{pmatrix} \in W.$$

Equation 13 can be rewritten:

$$\dot{q} = \omega_l f_l + \omega_r f_r \qquad (14)$$

where f_l and f_r are the vector fields corresponding to the left and right wheels:

$$f_l = \begin{pmatrix} \frac{1}{2}\cos\theta \\ \frac{1}{2}\sin\theta \\ -\frac{1}{2b} \end{pmatrix} \qquad (15)$$

$$f_r = \begin{pmatrix} \frac{1}{2}\cos\theta \\ \frac{1}{2}\sin\theta \\ \frac{1}{2b} \end{pmatrix}. \qquad (16)$$

Vector field f_l corresponds to turning about a center located under the right wheel, and f_r corresponds to turning about a center located under the left wheel.

Define λ to be an R^3-valued function of time called the *adjoint vector*:

$$\lambda(t) = \begin{pmatrix} \lambda_1(t) \\ \lambda_2(t) \\ \lambda_3(t) \end{pmatrix}.$$

Let $H : R^3 \times SE^2 \times W \to R$ be the *Hamiltonian*:

$$H(\lambda, q, w) = < \lambda, \omega_l f_l + \omega_r f_r >$$

The maximum principle states that for a control $w(t)$ to be optimal, it is *necessary* that there exist a nontrivial (not identically zero) adjoint vector $\lambda(t)$ satisfying the *adjoint equation*:

$$\dot{\lambda} = -\frac{\partial}{\partial q}H \qquad (17)$$

while the control $w(t)$ minimizes the Hamiltonian at every t:

$$H(\lambda, q, w) = \min_{z \in W} H(\lambda, q, z) = \lambda_0. \qquad (18)$$

with $\lambda_0 \geq 0$. Equation 17 is called the *adjoint equation* and Equation 18 is called the *minimization equation*.

For the bounded velocity diff drive, the adjoint equation gives:

$$\dot{\lambda} = -\frac{\partial}{\partial q} < \lambda, \omega_l f_l + \omega_r f_r > \quad (19)$$

$$= \frac{\omega_l + \omega_r}{2} \begin{pmatrix} 0 \\ 0 \\ \lambda_1 \sin\theta - \lambda_2 \cos\theta \end{pmatrix}. \quad (20)$$

Fortunately these equations can be integrated to obtain an expression for the adjoint vector. First we observe that λ_1 and λ_2 are constant and define c_1 and c_2 accordingly:

$$\lambda_1(t) = c_1 \quad (21)$$

$$\lambda_2(t) = c_2. \quad (22)$$

For λ_3 we have the equation:

$$\dot{\lambda}_3 = \frac{\omega_l + \omega_r}{2}(\lambda_1 \sin\theta - \lambda_2 \cos\theta). \quad (23)$$

But we can substitute from Equations 1, 11, and 12 to obtain:

$$\dot{\lambda}_3 = c_1 \dot{y} - c_2 \dot{x}, \quad (24)$$

which is integrated to obtain the solution for λ_3:

$$\lambda_3 = c_1 y - c_2 x + c_3, \quad (25)$$

where c_3 is our third and final integration constant. It will be convenient in the rest of the paper to define a function η of x and y:

$$\eta(x, y) = c_1 y - c_2 x + c_3. \quad (26)$$

So then the adjoint equation is satisfied by:

$$\lambda = \begin{pmatrix} c_1 \\ c_2 \\ \eta(x, y) \end{pmatrix} \quad (27)$$

for any c_1, c_2, c_3 not all equal to zero.

Let the *η-line* be the line of points (x, y) satisfying $\eta(x, y) = 0$, and note that $\eta(x, y)$ gives a scaled directed distance of a point (x, y) from the η-line. Let the *right half plane* be the points satisfying:

$$\eta(x, y) > 0, \quad (28)$$

and let the *left half plane* be the points satisfying

$$\eta(x, y) < 0. \quad (29)$$

We also define a direction for the η-line consistent with the choice of "left" and "right" for the half planes.

The minimization equation 18 can be rewritten:

$$\omega_l \phi_l + \omega_r \phi_r = \min_{z_l, z_r} z_l \phi_l + z_r \phi_r, \quad (30)$$

where ϕ_l and ϕ_r are defined to be the two *switching functions*:

$$\phi_l = < \lambda, f_l > \quad (31)$$

$$= \begin{pmatrix} c_1 \\ c_2 \\ \eta(x, y, \theta) \end{pmatrix} \cdot \begin{pmatrix} \frac{1}{2}\cos\theta \\ \frac{1}{2}\sin\theta \\ -\frac{1}{2b} \end{pmatrix} \quad (32)$$

$$= -\frac{1}{2b}\eta(x + b\sin\theta, y - b\cos\theta) \quad (33)$$

$$\phi_r = < \lambda, f_r > \quad (34)$$

$$= \begin{pmatrix} c_1 \\ c_2 \\ \eta(x, y, \theta) \end{pmatrix} \cdot \begin{pmatrix} \frac{1}{2}\cos\theta \\ \frac{1}{2}\sin\theta \\ \frac{1}{2b} \end{pmatrix} \quad (35)$$

$$= \frac{1}{2b}\eta(x - b\sin\theta, y + b\cos\theta). \quad (36)$$

Note that the wheels' coordinates can be written:

$$\begin{pmatrix} x_l \\ y_l \end{pmatrix} = \begin{pmatrix} x - b\sin\theta \\ y + b\cos\theta \end{pmatrix} \quad (37)$$

$$\begin{pmatrix} x_r \\ y_r \end{pmatrix} = \begin{pmatrix} x + b\sin\theta \\ y - b\cos\theta \end{pmatrix}, \quad (38)$$

so the switching functions can be written:

$$\phi_l = -\frac{1}{2b}\eta(x_r, y_r) \quad (39)$$

$$\phi_r = \frac{1}{2b}\eta(x_l, y_l). \quad (40)$$

Now the minimization equation says that if the controls ω_l, ω_r are optimal then they minimize the Hamiltonian $H = \omega_l \phi_l + \omega_r \phi_r$. This implies the optimal controls can be expressed:

$$\omega_l \begin{cases} = 1 & \text{if right wheel} \in \text{right half plane} \\ \in [-1, 1] & \text{if right wheel} \in \eta\text{-line} \\ = -1 & \text{if right wheel} \in \text{left half plane} \end{cases} \quad (41)$$

$$\omega_r \begin{cases} = 1 & \text{if left wheel} \in \text{left half plane} \\ \in [-1, 1] & \text{if left wheel} \in \eta\text{-line} \\ = -1 & \text{if left wheel} \in \text{right half plane} \end{cases} \quad (42)$$

If $c_1 = c_2 = 0$, then the the entire plane is the left half plane or the right half plane, depending on the sign of

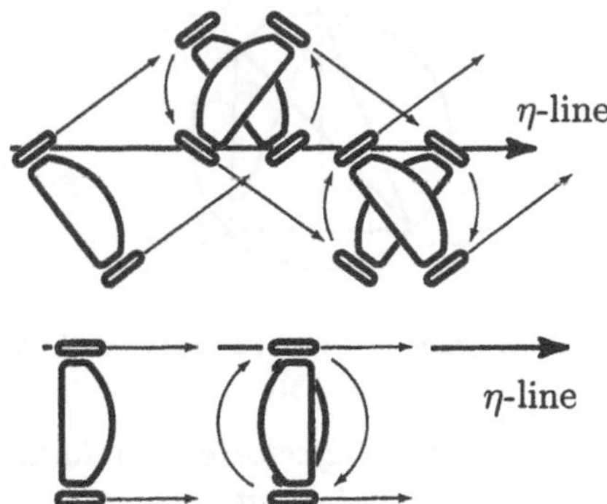

Figure 5: *Two extremals:* zigzag right *and* tangent CW. *Other extremal types are* zigzag left, tangent CCW, *and turning in place:* CW *and* CCW. *Straight lines are special cases of zigzags or tangents.*

c_3. (Recall that all three integration constants cannot be simultaneously zero.)

The location of the η-line depends on the apparently arbitrary integration constants. The maximum principle does not give the location of the line; it merely says that if we have an optimal control then the line exists and the optimal control must conform to the equations above. The question that naturally arises is how to locate the line properly, given the start and goal configurations of the robot. There seems to be no direct way of doing so. Rather, we must use other means to identify the extremal trajectory.

The robot switches only when a wheel touches the η-line, so the possibilities can be enumerated by constructing a circle whose diameter is the robot wheelbase, and considering the different possible relations of this circle to the η-line:

- `CCW` and `CW`: If the robot is in the left half plane and out of reach of the η-line, it turns in the counterclockwise direction (`CCW`). `CW` is similar.

- `TCCW` and `TCW` (Tangent CCW and Tangent CW): If the robot is in the left half plane, but close enough that a circumscribed circle is tangent to the η-line, then the robot may either roll straight along the line, or it may turn through any positive multiple of π. `TCW` is similar.

- `ZR` and `ZL`: If the circumscribed circle crosses the η-line, then a zigzag behavior occurs. The robot rolls straight in the η-line's direction until one wheel crosses. It then turns until the other wheel crosses, and then goes straight again. There are two non-degenerate patterns: ... ⇑↶⇓↷ ... called *zigzag right* (`ZR`), and ... ⇑↷⇓↶ ... called *zigzag left* (`ZL`). (Recall that *forward* and *backwards* actions are denoted by ⇑ and ⇓, and spins in place are denoted by ↶ and ↷.)

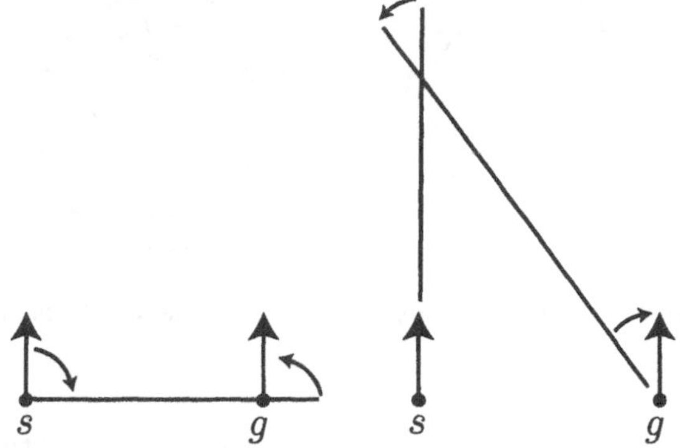

Figure 6: *Turn-straight-turn is not always optimal.*

Figure 5 shows examples of `ZR` and `TCW`.

Examining these classes, we see that

Theorem 1 *For an optimal trajectory,*

$$t = s(t) + b\sigma(t). \tag{43}$$

Proof: Extremal trajectories are composed only of turns and straight lines. ■

Note that in [2], we demonstrate that equation 43 actually holds for any trajectory such that $\max(|\omega_l|, |\omega_r|) = 1$ for almost all t; i.e., for trajectories in which one control is always saturated. This may provide some intuition for why turns and straights are faster than curves.

The fastest trajectories for a Reeds and Shepp car with zero turning radius are of the form `tst`. It might at first seem that Equation 43 implies the same is true for the diff drive. Trajectories of type `tst` certainly minimize translation time; however, they do not not necessarily minimize rotation time. Figure 3 shows an example: The robot is at the origin facing north, and the goal is some distance to the east, facing north. The

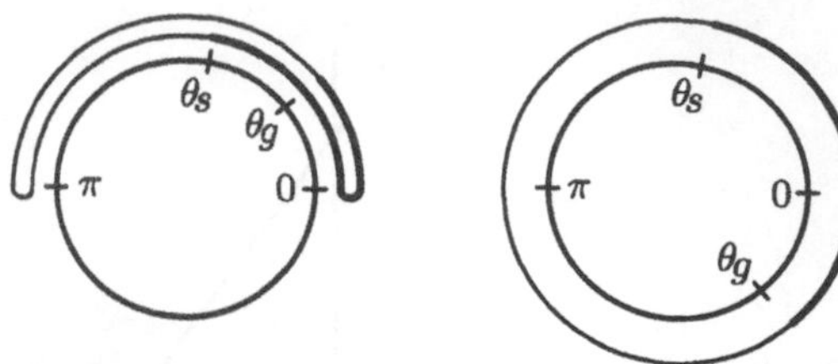

Figure 7: *There is always a roundtrip of length 2π visiting θ_s, θ_g, 0, and π.*

robot could follow a ↷⇑↶ trajectory, with a total turning angle of π. However, there is a ⇑↶⇓↷ trajectory that turns through a total angle of less than π. For a wide enough robot (b large enough), the second trajectory will be optimal.

4 Further Conditions for Optimality

The previous section showed that every time optimal path must be of type `CCW`, `CW`, `TCCW`, `TCW`, `ZR`, or `ZL`. However, the converse is definitely not true—not every such trajectory is optimal. For example, a robot turning in place for several revolutions is not time optimal. To keep the distinction clear, we refer to trajectories satisfying Pontryagin's Maximum Principle as *extremal*, and we note that the time-optimal trajectories are a subset of the extremal trajectories. In this section we find additional necessary conditions, ultimately finding that no time optimal path can have more than three straights and two turns.

Necessary conditions on `TCCW` and `TCW` trajectories

Theorem 2 *For every time-optimal trajectory $\sigma(T) \leq \pi$.*

Proof: Consider the fastest `tst` trajectory between a given start and goal. Obviously the straight action connects the start to the goal. Suppose we orient our coordinates so the angle from start to goal is 0. The robot's heading during the straight is either 0 or π. To plan the turns, we must consider different paths on the circle from θ_s to θ_g, passing through either 0 or π along the way. Note that in every instance there is a round trip from θ_s to 0, π, θ_g, and back to θ_s, of length 2π (Figure 7). Hence there is a one-way trip of length π or less. ∎

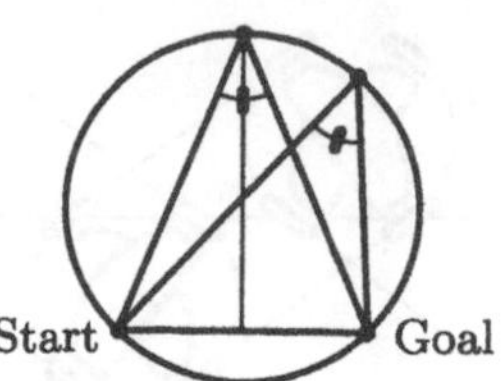

Figure 8: *Zigzags of three turns are not optimal*

Theorem 3 *Tangent trajectories containing more than three actions are not optimal.*

Proof: An extremal of type `TCW` or `TCCW` alternates turns and straights. Any full untruncated turn must be a multiple of π. If there are four actions, there is at least one untruncated turn of length at least π, and a second turn of nonzero length. The rectified arc length σ would be more than π, contradicting Theorem 2. ∎

Necessary conditions on `ZR` and `ZL` trajectories

Zigzag trajectories are composed of alternating turn or straight line actions. Successive turns or straights must be in opposite directions, but have the same magnitude if untruncated. Simple geometry also gives a relationship between ϕ, the angle of each turn, and d, the length of each straight. We have:

$$d = 2b\tan(\frac{\phi}{2}). \tag{44}$$

Theorem 4 *Zigzag subsections containing three turns are not optimal.*

Proof: Consider a zigzag subsection with three turns, and two straights. The straights are the same length, so the path comprises two legs of an isoceles triangle. Construct the circle containing the start, the goal, and the via point as in Figure 8. If we perturb the via point to a nearby point on the same circle the turning time is unchanged, and the translation is decreased. ∎

Zigzags can also be said to be periodic. Let τ be the smallest positive time such that:

$$\begin{aligned} \theta(t) &= \theta(t+\tau) \\ \eta(x(t), y(t)) &= \eta(x(t+\tau), y(t+\tau)). \end{aligned}$$

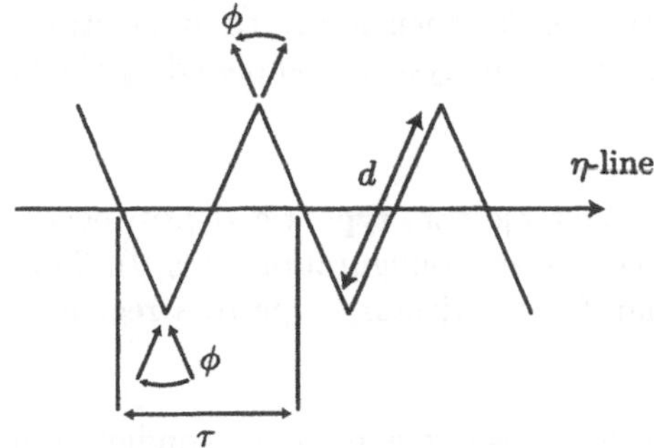

Figure 9: *Periodicity of a zigzag*

Theorem 5 *A zigzag trajectory of more than one period is not optimal.*

Proof: Consider a zigzag of more than one period, beginning at time 0 and ending at time $T > \tau$. By theorem 4, the zigzag is not optimal if $\sigma(T) > 2\phi$. If $s(T) > 2d$, then there are three straights. The first and last straights are parallel. If we reorder the actions to perform these consecutively, we get a straight with length greater than d. Thus we have a path which costs no more than the original but which is no longer a legitimate zigzag. Since it is not extremal, neither it nor the original path can be optimal. ■

Enumeration

Theorems 3, 4, and 5 allow a finite enumeration of the structure of optimal trajectories. The structure must be one of the following, or a subsection of one of the following:

Tangent	↷⇑↷	↷⇓↷	↶⇑↶	↶⇓↶
Tangent$_\pi$	⇑↷$_\pi$⇓	⇓↷$_\pi$⇑	⇑↶$_\pi$⇓	⇓↶$_\pi$⇑
Zigzag	⇑↶⇓↷⇑	⇓↶⇑↷⇓	⇑↷⇓↶⇑	⇓↷⇑↶⇓

5 Symmetries

Further analysis of the time optimal trajectories is difficult because of the large number of cases. This complexity is reduced using symmetries developed by Souères and Boissonnat [8] and Souères and Laumond [9] for steered cars.

The symmetries are summarized in Figure 10. Let "base" be any trajectory from $q = (x, y, \theta)$ to the origin. Then there are seven symmetric trajectories obtained by applying one or more of three transformations defined below. These transforms are isometries; if the base trajectory is optimal from the base configuration, the transformed trajectories are optimal from the transformed configurations.

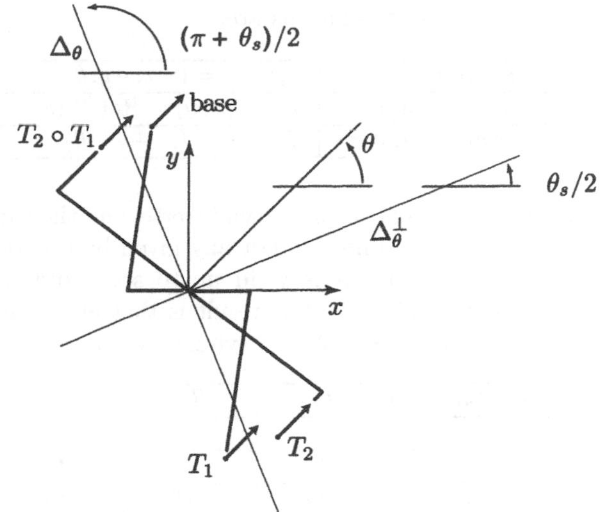

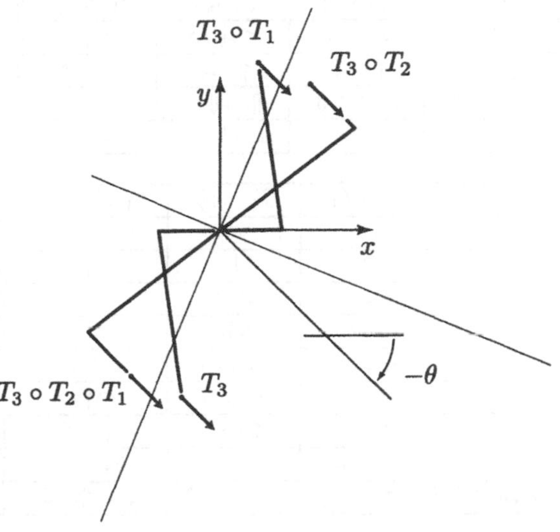

Figure 10: *Given an optimal trajectory from "base" with heading θ_s to the origin with heading $\theta_g = 0$, transformations T_1,T_2, and T_3 yield up to seven other optimal trajectories symmetric to the original.*

Geometrically, the transformations reflect the plane across the origin or across one of three other lines: the x-axis, a line Δ_θ at angle $(\pi + \theta_s)/2$, or the line $\Delta_\theta^\perp$ at angle $\theta_s/2$.

The three transformations are:

τ_1:	Swap ⇑ and ⇓	T_1:	$q = (-x, -y, \theta)$
τ_2:	Reverse order	T_2:	$(x, y) = \mathrm{Rot}(\theta)(x, -y)$
τ_3:	Swap ↷ and ↶	T_3	$q = (x, -y, -\theta)$

Each transformation is its own inverse, and the three transformations commute. For any given base trajectory, the transformations yield up to seven different symmetric trajectories. The result is that all optimal trajectories fall in one of nine symmetry classes.

	base	T_1	T_2	$T_2 \circ T_1$
A.	↷	↷	↷	↷
B.	⇓	⇑	⇓	⇑
C.	⇓↷	⇑↷	↷⇓	↷⇑
D.	↷⇓↷	↷⇑↷	↷⇓↷	↷⇑↷
E.	⇑↶$_{\pi}$⇓	⇓↶$_{\pi}$⇑	⇓↶$_{\pi}$⇑	⇑↶$_{\pi}$⇓
F.	↶⇓↷	↶⇑↷	↷⇓↶	↷⇑↶
G.	⇓↷⇑	⇑↷⇓	⇑↷⇓	⇓↷⇑
H.	↶⇓↷⇑	↶⇑↷⇓	⇑↷⇓↶	⇓↷⇑↶
I.	⇑↶⇓↷⇑	⇓↶⇑↷⇓	⇑↷⇓↶⇑	⇓↷⇑↶⇓

	T_3	$T_3 \circ T_1$	$T_3 \circ T_2$	$T_3 \circ T_2 \circ T_1$
A.	↶	↶	↶	↶
B.	⇓	⇑	⇓	⇑
C.	⇓↶	⇑↶	↶⇓	↶⇑
D.	↶⇓↶	↶⇑↶	↶⇓↶	↶⇑↶
E.	⇑↷$_{\pi}$⇓	⇓↷$_{\pi}$⇑	⇓↷$_{\pi}$⇑	⇑↷$_{\pi}$⇓
F.	↷⇓↶	↷⇑↶	↶⇓↷	↶⇑↷
G.	⇓↶⇑	⇑↶⇓	⇑↶⇓	⇓↶⇑
H.	↷⇓↶⇑	↷⇑↶⇓	⇑↶⇓↷	⇓↶⇑↷
I.	⇑↷⇓↶⇑	⇓↷⇑↶⇓	⇑↶⇓↷⇑	⇓↶⇑↷⇓

We can analyze all types of trajectories by analyzing just one type from each of the nine classes, and then applying the transformations T_1, T_2, T_3 to obtain the other members of the class.

6 Time Optimal Trajectories.

In this section we identify the time optimal trajectories between any given start and goal configuration. We introduce a "goal-centric" coordinate system, with the origin coincident with the goal position, and the x axis aligned with the goal heading.

The symmetries of the previous section greatly simplify our analysis. We only need to consider a "base" region; the results then apply to symmetric regions. In principle, the analysis is completed by the following steps:

1. For each trajectory type, we identify every feasible choice of start configuration (x, y, θ). This defines a map from trajectory type to a region of configuration space.

2. Now we consider a point in configuration space (x, y, θ). If it is in only one region, then the corresponding trajectory type is optimal from that point.

3. When regions overlap, we derive additional necessary conditions for optimality or calculate the actual times for each trajectory type to disambiguate.

To illustrate this procedure, we present the following example (Figure 11). The feasible regions for ⇓↷⇑↶ and ↶⇓↷⇑ overlap. For almost all q_s in the overlap, there are two possible extremals but only one true optimal path. The Δ_θ line is a decision boundary: For q_s to the right of Δ_θ the optimum is ↶⇓↷⇑, and to the left of Δ_θ the optimum is ⇓↷⇑↶. Figure 11 illustrates the proof. First we observe that the alternatives give equal time on the Δ_θ line, because that line is the axis of reflection for the $T_1 \circ T_2$ isometry. So *both* paths are optimal on Δ_θ.

Consider a ⇓↷⇑↶ path from the start pose shown. When the path crosses Δ_θ during the ⇓ action, the remaining cost is unchanged if it switches to ↶⇓↷⇑. But then the total path would have a structure of ⇓↶⇓↷⇑, and would not be a legitimate extremal. ■

Similar techniques can be applied to the other regions. The end result is a mapping that defines for each point in configuration space the set of optimal trajectories from that point to the origin. This mapping is illustrated by showing a slice at $\theta = \pi/4$ (Figure 12). The mapping from start configuration to optimal trajectory is usually, but not always, unique. At some boundaries in the figures there are two distinct trajectories that give the same time cost. More interesting is the case at $\theta = 0$ where a continuum of different trajectories of type A are all optimal, bounded by optimal trajectories of type B.

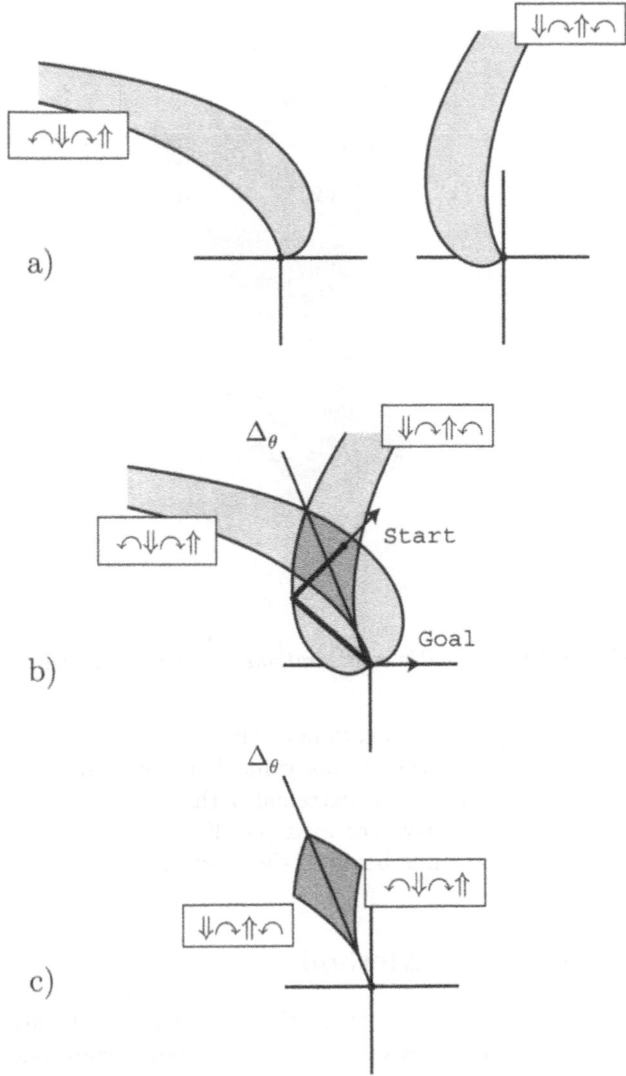

Figure 11: *An example of overlapping regions. The path shown in b) is extremal, but not optimal.*

7 Algorithm for Optimal Control and Value Function.

We now present an algorithm to determine the optimal paths between a given start and goal position, and the time cost of those paths. For each optimal path structure, the necessary conditions yield a region. (Twelve such regions are shown in Figure 12.) The determines which region(s) the start configuration (x, y, θ) falls in, and then calculates the value function for one of the optimal path structures. For example, the function

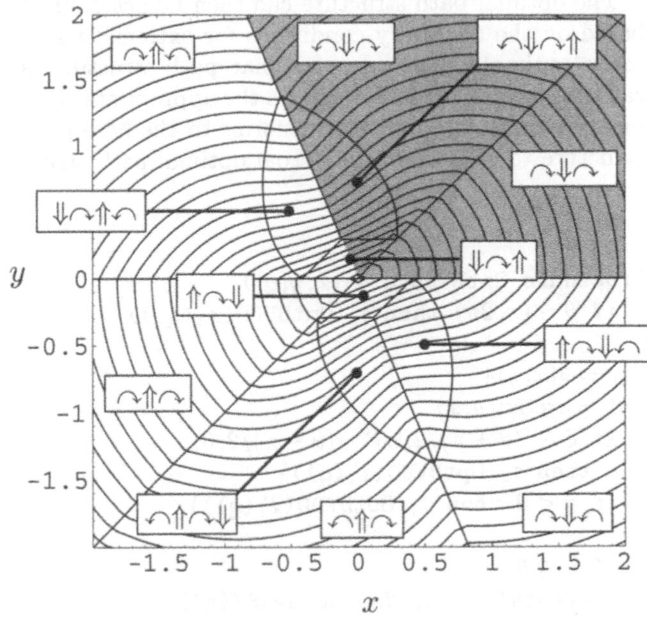

Figure 12: *Optimal control for start configuration $q_s = (x, y, \frac{\pi}{4})$ and goal configuration $q_g = (0, 0, 0)$, with isocost lines. Coordinates are measured in units of b.*

ValueBaseTSTS below calculates the cost of the fastest trajectory with a structure of ↶⇓↷⇑.

Procedure ValueBaseTSTS(q = (x, y, θ))
  $\arccos(1 - y) - \theta/2 - x + \sqrt{y(2 - y)}$
End ValueBaseTSTS

Procedure ValueBaseSTS(q = (x, y, θ))
  If $y = 0$ **then** $|x| + \theta/2$
  else $y(1 + \cos(\theta))/\sin(\theta) - x + \theta/2$
End ValueBaseSTS

Procedure ValueBaseTST(q = (x, y, θ))
  $r = \|(x, y)\|$
  $\zeta = \arctan(y, x)$
  $r + \min(|\zeta| + |\zeta - \theta|, 2\pi + |\zeta| - |\zeta + \theta|)$
End ValueBaseTST

We now can define OptBVDD (optimal bounded velocity diff drive). The function recursively applies symmetry transforms until the configuration is in the base region.

The optimal path structure can then be determined based on the necessary conditions for extremal paths to be optimal. The value for that path structure is calculated. The recursion applies the appropriate combination of τ_1, τ_2, and τ_3 transforms to the base path structure to determine the actual optimal path structure.

```
Procedure OptBVDD(q = (x, y, θ))
  if θ ∈ (π, 2π) then τ3(OptBVDD(T3(q)))

  r = ||(x, y)||
  ζ = arctan(y, x)
  if ζ ∈ ((θ + π)/2, π) ∪ ((θ − π)/2, 0)
    then τ2(OptBVDD(T2(q)))
  if y < 0 then τ1(OptBVDD(T1(q)))

  if ζ ≤ θ
    return(↷⇓↷, ValueBaseTST(q))
  else if y ≤ 1 − cos(θ)
    return(⇓↷⇑, ValueBaseSTS(q))
  else if r ≥ tan(ζ/2)
    return(↶⇓↷, ValueBaseTST(q))
  else
    return(↶⇓↷⇑, ValueBaseTSTS(q))
End OptBVDD
```

For the sake of brevity, certain special cases have been omitted from the pseudocode presented. Whenever two symmetric regions are adjacent, the fastest paths for both regions are optimal. For example, if the robot starts at $(0, 1, \pi)$, then both the paths ↷⇓↷ and ↶⇑↶ are optimal.

There are two other cases where multiple paths will be optimal. When $\theta_s = 0$, there may be a continuum of optimal five action paths, bounded by two different four-action paths. When $\theta_s = \pi$, there will be a continuum of optimal $\mathtt{t}_\pi\mathtt{s}$ paths (Class E), bounded by two-action paths of class C.

In all cases, the above algorithm will return a single optimal trajectory. Some additional bookkeeping would allow all of the optimal trajectories to be returned.

The level sets of the value function show the reachable configurations of the robot for some given amount of time. Figure 13 shows the shape of this region for

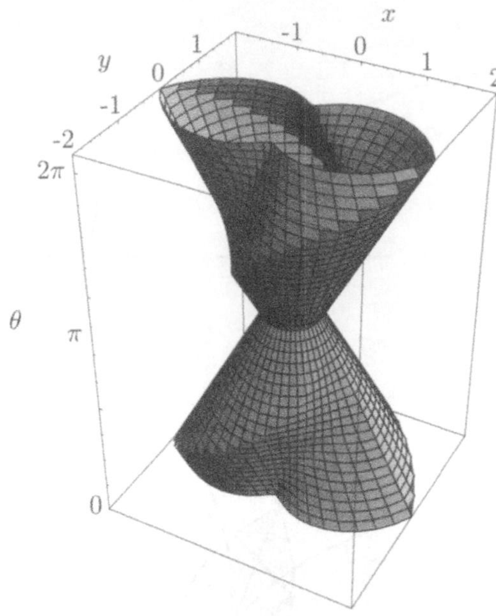

Figure 13: *Reachable configurations in normalized time 2.*

time 2. (x, y, and time are normalized by b, the width of the robot.) Slices of this value function allow the regions in which various extremal paths are optimal to be seen more clearly. For example, Figure 12 shows a slice where the angle between the start and goal robot is fixed at $\frac{\pi}{4}$.

8 Graphical Method

Graphical construction of the time optimal trajectories is usually straightforward. Figure 14 shows examples for each of the nine symmetry classes. The first five classes are really quite obvious. The last four classes are more challenging. In this section we will see a graphical way to identify which class gives the time-optimal trajectory, and to find the location of a via point if one exists.

The main idea is that the primary decision boundaries in the algorithm `OptBVDD` can be stated as conditions comparing the start heading θ_s, the goal heading θ_g, and the direction from start to goal. Based on this observation we can translate `OptBVDD` to a graphical procedure.

It is convenient to choose a different coordinate system, which we will call *bisector-centric* coordinates. We place the origin at the midpoint between the start and

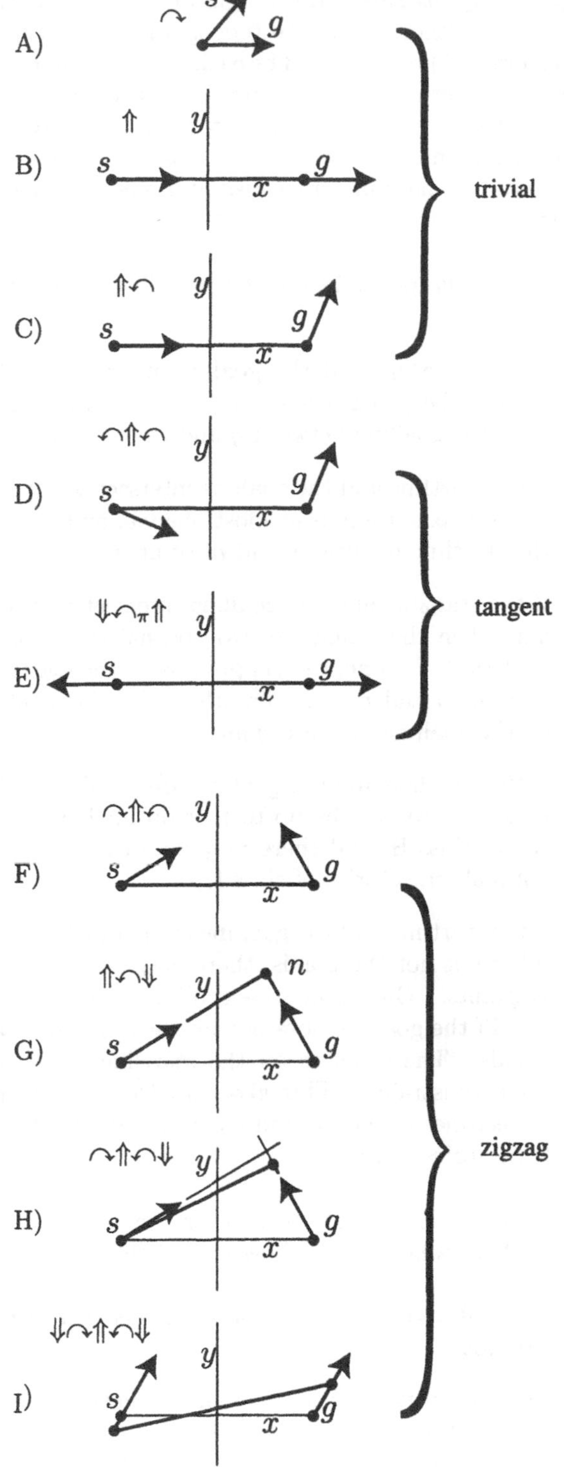

Figure 14: *Examples for each of the nine classes.*

goal, with the positive x axis directed toward the goal. The y axis is then the perpendicular bisector of the segment between x_s and x_g, oriented in the usual way. For convenience, we define l to be the distance between the start and goal; i.e, $l = 2x_g$. We define the range of θ_s and θ_g to be $(-\pi, \pi]$.

It is also convenient to define the *startline* and *goalline* to be the lines aligned with the robot heading at the start and goal, respectively. We define n to be the intersection of the startline and goalline, if they intersect.

For the rest of this section, we first walk through the cases from simplest to most complex. For the zigzags the most interesting part is finding the location of a via point, if any, and addressing some of the special cases.

Tangent trajectories. Although the time optimal trajectories for classes A, B, and C are obvious, we list them here for completeness:

A If s coincides with g, then turn in place.

B If $\theta_s = \theta_g = 0$ or $\theta_s = \theta_g = \pi$, then roll straight to the goal.

C If θ_s is 0 or π, then roll to the goal and turn to the goal heading. If θ_g is 0 or π then turn to the goal heading and then roll straight to the goal.

Next are the **tst** tangent and **sts** tangent$_\pi$ trajectories.

D If θ_s and θ_g are neither 0 nor π, and have different signs, a turn from θ_s to θ_g by the shortest arc will either pass through 0 or π. So turn to 0 (π), roll forward (backward) to the goal, then turn to θ_g. The cost of this trajectory is:

$$t_{\text{tangent}} = b \min(|\theta_s| + |\theta_g|, 2\pi - |\theta_s| - |\theta_g|) + l. \quad (45)$$

E If $(\theta_s, \theta_g) = (0, \pi)$ or $(\pi, 0)$, then roll partway toward the goal, turn through π or $-\pi$, then roll the rest of the way to the goal. This yields a continuum of optimal trajectories of class E, and four optimal trajectories of class C. The cost of trajectories in this region is also given by Equation 45.

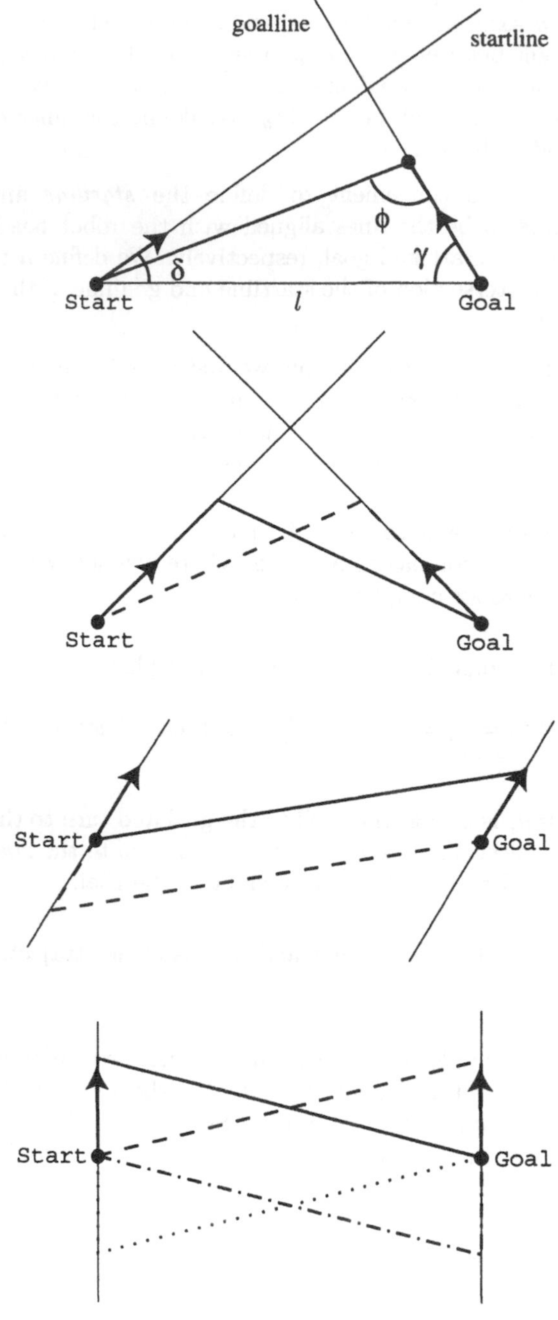

Figure 15: *Examples of optimal zigzags. If startline and goalline are symmetric across the y axis, there are at least two zigzags of equal cost. If startline and goalline are parallel, there are two zigzags of equal cost, and zigzags of class I may also be optimal.*

Zigzag trajectories. If the signs of θ_s and θ_g are the same, but neither is equal to 0 or π, then the optimal trajectory will be a zigzag. The easiest way to approach zigzags is in terms of their via points. If there is *no* via, then we have a simple `tst` trajectory. If there are one or two via points, then we have to look at special cases to determine the number of vias and where they can occur.

- Every via point is on the startline or on the goalline;
- If the startline and the goalline intersect to the right of the y axis, there is at most one via, and it is on the goalline between g and n, or at n.
- If the startline and the goalline intersect to the left of the y axis, there is at most one via, and it is on the startline between s and n, or at n.
- If the startline and the goalline intersect on the y axis, then there may be two optimal trajectories of class H. For one, the via point is on the startline between s and n. For the other, the via point is on the goalline between g and n.
- If the startline and the goalline are parallel to the y axis, there may be up to four optimal trajectories of class H, and there may be a continuum of optimal trajectories of class I.
- If the startline and the goalline are parallel to each other but not the y axis, there may be up to two viapoints. One is on the goalline, above the x axis iff the goalline does not intersect y above the x axis. The other is on the startline, with the same constraint. This gives up to two optimal trajectories of class H and a continuum of optimal trajectories of class I.

The above case analysis tells where to look for via points. Next we determine how to find the via points.

8.1 Graphical Construction of Class H Trajectories

For any trajectory with a via point we have a simple equation for the cost of the straight line actions. The above enumeration shows that there are four line segments where the via point may fall. Let γ be the magnitude of the internal angle between any line segment possibly containing a via point and the x axis, and let δ be the internal angle of the other line segment on the same side of the x axis. (See figure 15.) Recall that ϕ is the angle of the turns of the zigzag extremal; we wish to determine the optimal value for ϕ. Some simple geometry gives us:

$$t_{zz} = b(2\phi + \delta + \gamma - \pi) + \frac{l(\sin(\phi+\gamma) + \sin\gamma)}{\sin\phi}. \quad (46)$$

We take the derivative with respect to ϕ and set to zero. After some simplification, we get the following condition on ϕ:

$$\cos(\phi) = 1 - \frac{l\sin(\gamma)}{2b}. \quad (47)$$

ϕ must be no greater than the turning angle of the fastest **tst** trajectory, and cannot be less than the angle of the turn if the via point were at n. (Recall that the via does not fall past this intersection.)

$$\phi \leq \min(|\theta_s| + |\theta_g|, 2\pi - |\theta_s| - |\theta_g|) \quad (48)$$
$$\phi \geq \pi - \gamma - \delta \quad (49)$$

If Equation 47 has no solution, or requires that ϕ be larger than the turning angle of the fastest **tst** trajectory, the cost function is monotonically increasing with ϕ. In this case it is impossible to save turning time by using a four action trajectory; a trajectory of no more than three actions will be optimal. If Equation 47 requires that $\phi < (\pi - \gamma - \delta)$, the cost function is monotonically decreasing with ϕ and the optimal trajectory will have a via at n.

There is a graphical interpretation of Equation 47. For simplicity, we consider the case where n is above the x axis and to the left of the y axis, which means we should look for a via on the startline between s and g.

Put the robot at the origin, with angle θ_s. Roll it forward in a straight line. The right wheel rolls on a straight line, call it l_R. Now put the robot at the goal, with the right wheel on l_R. That is the configuration at which the robot should arrive at the goal. (See Figure 16.) We can determine the location of the via point by rolling the robot from this configuration in a straight line until it intersects the startline.

It may be impossible to place the right wheel on l_R, or the goal configuration of the robot may seem

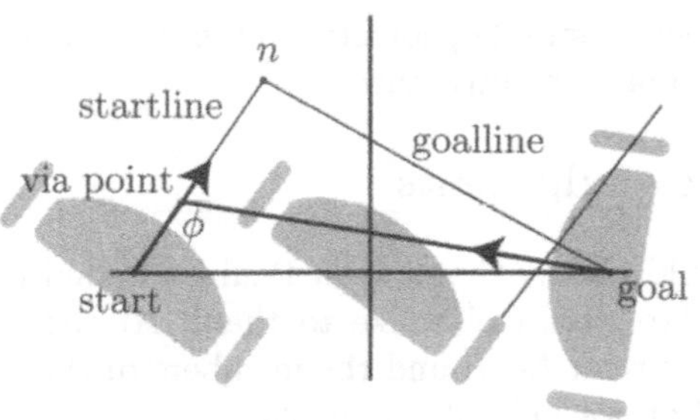

Figure 16: *Graphical construction of optimal path.*

to imply that the via point is not between s and n. This occurs when Equation 47 has no solution satisfying Equations 48 and 49, implying that either a trajectory of less than four actions is optimal, or the via point of the four action trajectory will be at n.

8.2 Construction of Class I Trajectories

Equation 47 allows the optimal value of ϕ to be calculated, and this determines the structure of the optimal zigzag extremal. The above graphical algorithm allows this extremal to be constructed. However, if the startline and the goalline are parallel, there is more than one subsection of this extremal that connects the start and goal. In this case, if trajectories of class H are optimal, there will therefore also be a continuum of optimal class I trajectories.

9 Summary and Conclusion.

The bounded velocity model of diff drive robots is simple enough that the set of time optimal trajectories between any two robot configurations may be found. Pontryagin's Maximum Principle provides an elegant geometric program describing extremal trajectories. These extremal trajectories are a superset of the optimal trajectories.

We derived conditions necessary for extremal trajectories to be optimal. These conditions require that optimal trajectories fall in one of 40 extremal trajectory classes. We then applied symmetries developed by Souères and Boissonnat, reducing the set of trajectory classes to be analysed to 9. We analysed each of the 9 classes to determine the start and goal configurations for which it was optimal. This yields a simple algorithm to determine the optimal trajectory structure and cost between any two configurations. The analysis also yields a simple geometric method of determining the optimal path structure.

Acknowledgments

We would like to thank Jean Paul Laumond for guidance. We would also like to thank Al Rizzi, Howie Choset, Ercan Acar, and the members of the Manipulation Lab for helpful comments.

References

[1] D. J. Balkcom and M. T. Mason. Extremal trajectories for bounded velocity differential drive robots. In *IEEE International Conference on Robotics and Automation*, 2000.

[2] D. J. Balkcom and M. T. Mason. Time optimal trajectories for bounded velocity differential drive robots. In *IEEE International Conference on Robotics and Automation*, 2000.

[3] L. E. Dubins. On curves of minimal length with a constraint on average curvature and with prescribed initial and terminal positions and tangents. *American Journal of Mathematics*, 79:497–516, 1957.

[4] L. S. Pontryagin, V. G. Boltyanskii, R. V. Gamkrelidze, and E. F. Mishchenko. *The Mathematical Theory of Optimal Processes*. John Wiley, 1962.

[5] J. A. Reeds and L. A. Shepp. Optimal paths for a car that goes both forwards and backwards. *Pacific Journal of Mathematics*, 145(2):367–393, 1990.

[6] D. B. Reister and F. G. Pin. Time-optimal trajectories for mobile robots with two independently driven wheels. *International Journal of Robotics Research*, 13(1):38–54, February 1994.

[7] M. Renaud and J.-Y. Fourquet. Minimum time motion of a mobile robot with two independent acceleration-driven wheels. In *Proceedings of the 1997 IEEE International Conference on Robotics and Automation*, pages 2608–2613, 1997.

[8] P. Souères and J.-D. Boissonnat. Optimal trajectories for nonholonomic mobile robots. In J.-P. Laumond, editor, *Robot Motion Planning and Control*, pages 93–170. Springer, 1998.

[9] P. Souères and J.-P. Laumond. Shortest paths synthesis for a car-like robot. *IEEE Transactions on Automatic Control*, 41(5):672–688, May 1996.

[10] H. Sussmann and G. Tang. Shortest paths for the reeds-shepp car: a worked out example of the use of geometric techniques in nonlinear optimal control. SYCON 91-10, Department of Mathematics, Rutgers University, New Brunswick, NJ 08903, 1991.

Printed and bound by CPI Group (UK) Ltd, Croydon, CR0 4YY

23/10/2024

01778248-0015